Chemical Reactors

H. Scott Fogler, Editor
The University of Michigan

Based on a symposium
sponsored by the Division of
Industrial and Engineering Chemistry
at the Second Chemical Congress
of the North American Continent
(180th ACS National Meeting),
Las Vegas, Nevada,
August 25–26, 1980.

ACS SYMPOSIUM SERIES 168

AMERICAN CHEMICAL SOCIETY
WASHINGTON, D. C. 1981

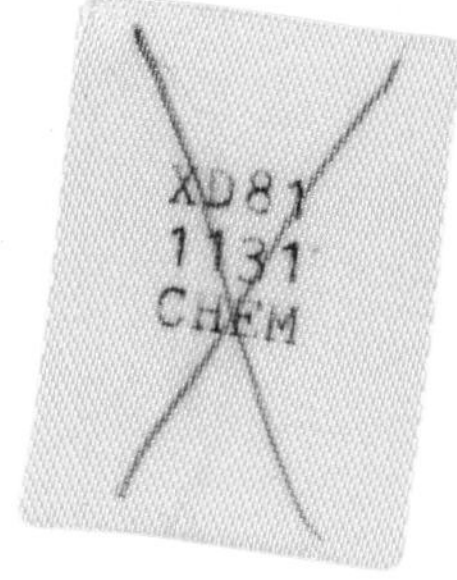

Library of Congress CIP Data

Chemical reactors.
(ACS symposium series, ISSN 0097-6156; 168)

"Based on a symposium sponsored by the Division of Industrial and Engineering Chemistry at the Second Chemical Congress of the North American Continent (180th ACS National Meeting), Las Vegas, Nevada, August 25-26, 1980."

Includes bibliographies and index.

1. Chemical reactors—Congresses.
I. Fogler, H. Scott. II. American Chemical Society. Division of Industrial and Engineering Chemistry. III. Chemical Congress of the North American Continent (2nd: 1980: Las Vegas, Nev.) IV. Series.

TP157.C423 660.2'8 81-12672
ISBN 0-8412-0658-9 AACR2 ACSMC8 168 1-396
1981

PRINTED IN THE UNITED STATES OF AMERICA

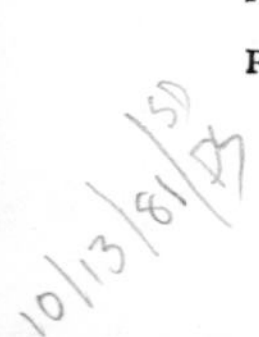

ACS Symposium Series

M. Joan Comstock, *Series Editor*

FOREWORD

The ACS SYMPOSIUM SERIES was founded in 1974 to provide a medium for publishing symposia quickly in book form. The format of the Series parallels that of the continuing ADVANCES IN CHEMISTRY SERIES except that in order to save time the papers are not typeset but are reproduced as they are submitted by the authors in camera-ready form. Papers are reviewed under the supervision of the Editors with the assistance of the Series Advisory Board and are selected to maintain the integrity of the symposia; however, verbatim reproductions of previously published papers are not accepted. Both reviews and reports of research are acceptable since symposia may embrace both types of presentation.

CONTENTS

PREFACE

The symposium upon which this volume is based focused on three areas in reaction engineering: fluidized bed reactors, bubble column reactors, and packed bed reactors. Each area comprises a section of this book. Professor J. R. Grace chaired and coordinated the fluidized bed sessions; Professors Y. T. Shah and A. Bishop, the bubble column reactor session; and Professor A. Varma, the packed bed reactor session. Each section in this book opens with a brief review chapter by the session chairman that includes an overview of the chapters in each session.

Fluidized bed reactors have received increased interest in recent years owing to their application in coal gasification. The section on fluidized beds discusses critical areas in fluid bed reactor modeling. Computer simulation of both solid-catalyzed gas phase reactions as well as gas–solid reactions are included.

In the section on bubble column reactors, the hydrodynamic parameters needed for scale-up are presented along with models for reaction and heat transfer. The mixing characteristics of columns are described as are the directions for future research work on bubble column reactors.

The packed bed reactors section of this volume presents topics of catalyst deactivation and radial flow reactors, along with numerical techniques for solving the differential mass and energy balances in packed bed reactors. The advantages and limitations of various models (e.g., pseudo-homogeneous vs. heterogeneous) used to describe packed bed reactors are also presented in this section.

H. Scott Fogler
The University of Michigan
Ann Arbor, MI 48109

June 1, 1981

FLUIDIZED BED REACTORS

1

Fluidized Bed Reactor Modeling

An Overview

J. R. GRACE

Department of Chemical Engineering, University of British Columbia,
Vancouver, Canada V6T 1W5

Critical areas in fluid bed reactor modeling are discussed in the light of papers in this symposium. There continues to be a wide diversity of assumptions underlying models. However, it is now clear that predictions are generally much more sensitive to some assumptions than to others. For example, proper modeling of interphase exchange is generally more critical than the assumptions adopted to describe axial gas dispersion in the dense or emulsion phase. For the 1980's advances are looked for in a number of areas, especially in more sophisticated computer models, unsteady state representations suitable for control purposes, models which describe high velocity regimes of fluidization, inclusion of grid and freeboard effects, and study of radial gradients.

This volume brings together a number of papers under the theme of fluidized bed reactor modeling. This field is of relatively recent origin. Table I gives the emphasis in research in successive decades beginning with the 1940's. It is seen that early research was devoted primarily to practical problems associated with the operation of fluidized bed reactors and to very simple models. With the passage of time models have been devised which are increasingly sophisticated. Reviews of the commercial development of fluidized beds as reactors have been prepared by Geldart (1,2). In the 1970's there were a number of reviews (3-7) which considered fluidized bed reactor modeling.

In order to be able to represent the behaviour of fluidized bed reactors with confidence, one must have a thorough understanding of the bed hydrodynamics and of the reaction kinetics. Almost all of the reactions carried out in fluidized beds are either solid-catalysed gas phase reactions or gas-solid reactions. (We will not consider here homogeneous gas phase reactions, reactions in liquid fluidized beds or reactions in three phase fluidized beds.) While the chemical kinetics can often be highly complex,

0097-6156/81/0168-0003$05.00/0

for example in the gasification or combustion of coal, the hydrodynamic aspects have given the greatest difficulty and have been subject to the greatest debate. While considerable progress has been made in achieving an understanding of many aspects of bed behaviour, there are many features which remain poorly understood. Some of these (e.g. regimes of bed behaviour, gas mixing patterns, and exchange of gas between phases) can affect profoundly the nature of the model adopted.

Table I: Focus of Research on Fluidized Bed Reactors

Decade	Emphasis
1940's	Practical design and operation problems. Single phase models only.
1950's	Simple two-phase models for gas-phase solid-catalysed reactions.
1960's	Incorporation of properties of single bubbles. Early models for gas-solid reactions.
1970's	Addition of end (grid and freeboard) effects. More sophisticated models for specific gas-solid reactions including energy balances. Consideration of complex kinetics.
1980's	? Probable emphasis on non-bubbling (turbulent and fast fluidization) regimes. Probable consideration of effects of aids to fluidization (e.g. centrifugal, magnetic and electrical fields, baffles). Increasing emphasis on more complex hydrodynamics and kinetics, with models requiring computers for solution.

The papers presented at the Las Vegas symposium, most of which are reproduced in this volume, both illustrate the diversity of modeling approach and show some new directions for reactor modeling in the 1980's. Before turning to these matters in detail, it is necessary to discuss briefly three of the papers which are fundamentally different in focus from the other eight.

The paper by Ramírez et al (8) considers the important question of particle-to-gas heat transfer in fluidized beds. In addition to the importance of this question in its own right, particle-to-gas heat transfer can be important for fluid bed reactors, for example in determining thermal gradients in the entry (grid) region, in establishing the surface temperature of particles undergoing reactions, and via the analogous case of gas-to-particle mass transfer. There has been considerable controversy over the fact that Nusselt and Sherwood numbers have been found to fall well below 2, the lower limit for a single sphere in a stagnant medium. Ramírez et al produce further evidence of Sh << 2 and Nu << 2 and consider these results in the light of transfer models in the literature.

The paper by Blake and Chen (9) represents an extension of

the novel approach adopted by the Systems, Science and Software group. In what must be the most ambitious and comprehensive fluidization modeling effort to date, this group has used modern computational techniques to solve a set of equations representing the physics and chemistry of fluidized bed coal gasifiers. Hydrodynamic fixtures are represented by a set of continuum equations and constitutive relationships, while chemical kinetics equations are written for key heterogeneous and homogeneous reactions based on studies reported in the literature. In previous papers, the authors have shown that the model gives a realistic simulation of a jet of gas issuing into a bed of solids. In the present paper they seek to duplicate results obtained in the IGT and Westinghouse pilot scale reactors. The results are of considerable interest, giving a good match with most of the experimental results.

A further paper by Gibbs (10) deals with design and modeling of centrifugal fluidized beds. In this case gas is fed radially inwards into a spinning bed. On account of the greatly augmented effective gravity force, greater through-puts of gas can be accommodated and entrainment is greatly lowered. This new technique has received attention in the late 1970's especially in connection with coal combustion. Some unique problems are encountered, e.g. the minimum fluidization velocity becomes a function of bed depth, while particles ejected into the "freeboard" by bubbles bursting at the bed surface travel initially nearly at right angles to the gas exit direction. This paper gives a preliminary scheme for dealing with some of these problems.

Classification of Reactor Models

There are many choices to be made in fluid bed reactor modeling and little unanimity among those who devise such models on the best choices. Table II lists some of the principal areas for decision and the corresponding choices of the other eight papers at this symposium (11-18).

Phases. Both two-phase and three-phase representations are widely used as shown schematically in Figure 1. In two-phase representations the dilute phase may represent bubbles alone, jets (in the grid region), or bubbles plus clouds. Three-phase representations generally use the scheme followed by Kunii and Levenspiel (19) whereby bubbles, clouds, and "emulsion" (i.e. that part of the non-bubble bed not included in the clouds) are each treated as separate regions. As shown in Table II, all of these possibilities are represented in the models adopted by the authors in this symposium. There appears, however, to be an increasing tendency to adopt three phase models, probably as a result of experimental results (20) which showed that the Kunii and Levenspiel model gave a better representation of measured concen-

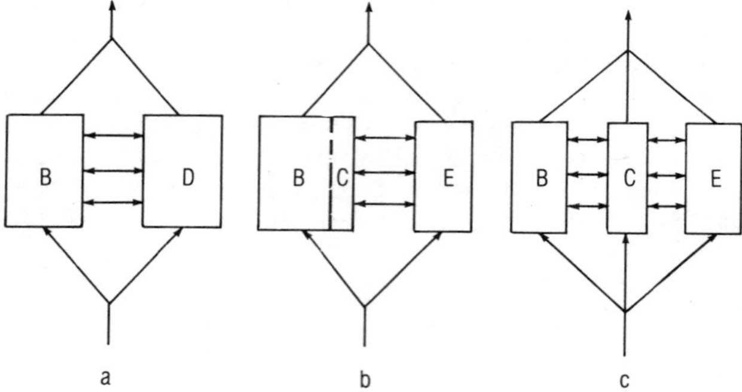

Figure 1. Schematic of two-phase and three-phase representations for fluidized beds operating in the bubble regime: B, bubble phase; C, cloud phase; D, dense phase; E, emulsion phase: Two-phase models, a and b; three-phase models, c

tration profiles for a particular particle size than other models tested. The bubbles themselves are usually treated as being completely devoid of particles, but it is important (21) that solids dispersed in the bubbles be included with the bubble phase for fast reactions, even though their concentration is small (typically < 1% by volume).

Gas Mixing in the Dense or Emulsion Phase. No other feature of fluidized bed reactor modeling has been subjected to so many alternative assumptions as axial mixing in the dense phase. At least eight possibilities have been tried as shown in Figure 2. These range from upward plug flow, through perfect mixing and stagnant gas, to downflow. Intermediate degrees of mixing have been represented by axial dispersion models and well-mixed stages in series. As shown in Tabe II many of these possibilities have been covered in the present symposium.

In view of the large number of disparate representations of dense phase axial mixing, one might easily conclude that this is one of the more important modeling features. In practice this is not the case, unless high conversions (e.g. 90% or greater in a single stage) are sought. For lower conversions, overall reactor performance tends to be insensitive to the pattern of axial mixing adopted (21). There are several illustrations of this point in this symposium. In the paper by Jayaraman et al (16), replacement of the downflow condition adopted by Fryer and Potter (22) by perfect mixing in the emulsion led to conversions which were barely distinguishable from those given by the earlier model. (At the same time solution became much simpler.) Jaffres et al (15) show that the two extreme cases of perfect mixing and plug flow in the Orcutt models (23) lead to similar results. (In their case, however, bubble properties were varied together with kinetic constants in their optimization so it is harder to distinguish the influence of the mixing assumptions alone.) Elnashaie and Elshishini (12) further show that the effect of axial dispersion is not only relatively slight in terms of overall conversion, but that dense phase mixing also plays a relatively minor role in determining selectivity for consecutive reactions and multiplicity of steady states.

In almost all previous modeling work, one-dimensional flow has been assumed in each phase, radial gradients being taken as negligible. There is some experimental evidence (24) that substantial radial gradients may exist, however. Radial gradients are especially important for fluid bed combustors with in-bed feeding of fresh coal via a series of nozzles. In this case the rapid devolatilization reactions will occur close to the distributed feed points, and radial dispersion of volatiles away from these points and oxygen towards them will be extremely important if the volatiles are to burn out within the bed. Fan and Chang (13) have considered this problem, coupling an assumption of perfect axial mixing with a diffusion-type mixing model in the ra-

Table II: Key features of models used in this symposium

	de Lasa *et al* (11)	Elnashaie & Elshishini (12)	Fan & Chang (13)	Fogler & Brown (14)
Phases	2: Bubble (or jet) & dense or 1: (CSTR)	2: Bub/Cloud & emulsion or 3: Bubble, cloud & emulsion	2: Bubble & dense	3: Bubble, cloud and emulsion
Gas mixing in dense or emulsion phase	perfect mixing	perfect mixing or plug flow upward	perfect axial mixing + radial dispersion	stagnant
Interphase transfer	jet/dense: Behie (43) bub/dense: (27) or (19)	(37) or (19)	K-L (19)	K-L (19)
Distribution of flow between phases	Two phase theory	2-ϕ theory + cloud or $U_{mf}A$ through emulsion	Two phase theory	Two phase theory
Bubble size	Basov equation (constant)	Kept as parameter (constant)	Kept as parameter (constant)	Mori & Wen (varies)
Heat balance?	Yes	Yes	Yes	No
Time variation	steady	steady	unsteady	steady
Reaction	gas-solid	catalytic	gas-solid	catalytic
Application	Catalyst regenerator	consecutive reactions	Coal combustion	general
Experimental data	large unit	none	none	none

Table II: Key features of models used in this symposium (Continued)

	Jaffres *et al* (15)	Jayaraman *et al* (16)	Peters *et al* (17)	Rehmat *et al* (18)
Phases	2: Bubble & dense	3: Bubble, cloud & emulsion	3: Bubble, cloud & emulsion	3: Gas, char & limestone
Gas mixing in dense or emulsion phase	plug flow or perfect mixing	perfect mixing	compartments in series	all gas in plug flow upward
Interphase transfer	D-H (27)	K-L (19)	K-L (19)	N.A.
Distribution of flow between phases	Two phase theory	as Fryer and Potter	New approach	N.A.
Bubble size	Mori & Wen & fitted values (constant)	specified value (constant)	Mori & Wen (varies)	N.A.
Heat balance?	Yes	No	No	Yes
Time variation	unsteady	steady	steady	steady
Reaction	catalytic	catalytic	pseudo-catalytic	gas-solid
Application	Maleic anhydride	general	Aerosol filtration	Coal combustion
Experimental data	cf. earlier data	none	cf. earlier data	none

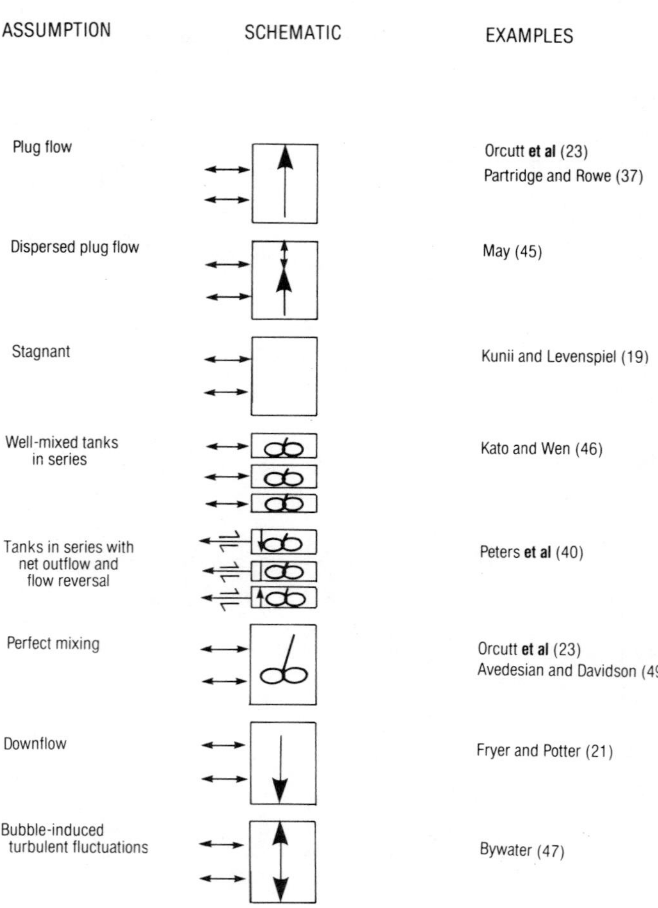

Figure 2. Alternative schemes used in reactor models to represent axial dispersion of gas in the dense or emulsion phase

dial direction. Although the underlying assumptions of diffusion-type models often seem inappropriate to axial mixing in fluidized beds (25), there is some evidence (e.g. 26) that lateral mixing can be described in this manner. Hence the paper of Fan and Chang (13) may represent a useful approach to the description of an important problem.

Interphase Gas Transfer. From the heavy reliance in this symposium (see Table II) on the mass transfer equations proposed by Davidson and Harrison (27) and by Kunii and Levenspiel (19), one might reasonably conclude that these approaches have been supported by at least the majority of experimental evidence. Nothing could be further from the truth.

The Davidson and Harrison approach concentrates solely on the resistance at the bubble/cloud boundary (or bubble/dense phase boundary for $\alpha < 1$). The transfer coefficient, referred to bubble surface area, is

$$k_{bc} = 0.75\ U_{mf} + 0.975[g\mathcal{D}^2/d_b]^{1/4} \quad (1)$$

Alternatively, on a bubble volume basis, this becomes

$$k_{bc}' = 4.5\ U_{mf}/d_b + 5.85\ [g\mathcal{D}^2/d_b{}^5]^{1/4} \quad (2)$$

The first term in each case arises from bulk flow of gas into the floor of an isolated bubble and out the roof, as required by the hydrodynamic model of Davidson and Harrison (27). The weight of experimental evidence, from studies of cloud size (28,29), from chemical reaction studies (e.g. 30), and from interphase transfer studies (e.g. 31,32), is that this term is better described by the theory proposed by Murray (33). The latter leads to a reduction in the first term by a factor of 3. Some enhancement of the bulk flow component occurs for interacting bubbles (34,35), but this enhancement for a freely bubbling bed is only of the order of 20-30% (35), not the 300% that would be required for the bulk flow term Equations (1) and (2) to be valid.

The second term in Equations (1) and (2) accounts for diffusional transfer across the bubble boundary. (A factor $\varepsilon_{mf}/(1+\varepsilon_{mf})$ is sometimes (e.g. 49) included in the bracket of Eq. 2 to account for the dense phase diffusional resistance.) There is some question (30) of the extent to which there is interference between the bulk flow and diffusion terms. Nevertheless, most experimental evidence suggests that the two terms are additive and that the diffusional term is described by the penetration theory. With these changes, and including a small enhancement factor for bubble interaction, Sit and Grace (35) have recommended the following equations as being in best accord with existing experimental data:

$$k_{bc} = U_{mf}/3 + [4\mathcal{D}\varepsilon_{mf}\bar{u}_b/\pi\bar{d}_b]^{1/2} \quad (3)$$

or $$k_{bc}' = 2U_{mf}/\bar{d}_b + 12\ [\mathcal{D}\varepsilon_{mf}\bar{u}_b/\pi\bar{d}_b{}^3]^{1/2} \qquad (4)$$

Kunii and Levelspiel (19) again use Equation (2) to describe bubble/cloud transfer. Based on the penetration theory, they propose the following expression for cloud/emulsion transfer:

$$k_{ce}' = 6.78\ [\mathcal{D}\varepsilon_{mf}u_b/d_b{}^3]^{1/2} \qquad (5)$$

Equation (5) considers gas diffusion to be the only mechanism of transfer across the outer cloud boundary. In practice there are at least three other important mechanisms not accounted for: (a) The cloud boundary is a streamline for gas elements but not for solid particles. Particles entering and leaving the cloud boundary will carry adsorbed species with them. (b) There is strong evidence of shedding of elements from the wakes. Photographs (28) indicate that these shed elements results in transfer of cloud gas to the emulsion. (c) The concept of the cloud is based on steady state analyses (27,33). While a mantel of gas appears to remain associated with bubbles as they coalesce, these "clouds", like the bubbles themselves, distort and undergo volume changes during bubble interaction and coalescence (28,36). This no doubt further enhances cloud/emulsion transfer.

For most practical conditions, a comparison of k_{bc}' and k_{ce}' from Equations (4) and (5) would suggest that the principal resistance to transfer resides at the outer cloud boundary. However, when (a), (b) and (c) are taken into account, this is no longer the case. In fact, experimental evidence (e.g. 30,31,32) indicates strongly that the principal resistance is at the bubble/cloud interface. With this in mind, it is probably more sensible to include the cloud with the dense phase (as in the Orcutt (23, 27) models) rather than with the bubbles (as in the Partridge and Rowe (37) model) if a two-phase representation is to be adopted (see Figure 1). If three-phase models are used, then Equations (2) and (5) appear to be a poor basis for prediction. Fortunately the errors go in opposite directions, Equation (2) overpredicting the bubble/cloud transfer coefficient, while Equation (5) underestimates the cloud/emulsion transfer coefficient. This probably accounts for the fact that the Kunii and Levenspiel model (19) can give reasonable predictions in specific instances (e.g.20).

Flow Distribution between Phases. One of the principal assumptions underlying many of the models of fluidized bed reactors is the "two-phase theory of fluidization". This theory, really no more than a postulate, holds that the flow beyond that required for minimum fluidization passes through the bed as translating void units. Although not included in what the originators of this postulate (38) appeared to have in mind, the two phase theory is often held to imply, in addition, that the dense phase voidage remains constant and equal to ε_{mf} for all $U > U_{mf}$.

Much has been written and said about the two phase theory

(e.g. see 39). For our purposes here it suffices to note that there is very little evidence indeed that the flow distribution really follows the theory. In fact, the weight of evidence (see 39) suggests that the theory seriously overestimates the flow accounted for by translation of bubbles, except in the limit as slug flow conditions are approached. Yet, despite all this evidence, the two phase theory continues as an underpinning for much of the serious modeling work, as is again evident from Table II. There are several probable reasons for the continuing popularity of the two phase theory in the face of contradicting evidence:

(i) There is a lack of alternative approaches.

(ii) There is confusion between "visible" and "invisible" (i.e. bulk flow or "throughflow") terms. Toomey and Johnstone (38) appeared to have in mind only the "visible" (i.e. flow due to void unit translation) term. As noted above, the theory then overestimates the bubble flow. However, if the bubble flow is taken to include the invisible throughflow, the theory may do better and may even underestimate the bubble flow. Many workers fail to distinguish clearly whether they are talking of visible or total bubble flow.

The paper by Peters *et al* (17) is welcome in that it attempts a new approach to the two phase flow distribution problem. Further details are given in another paper by the same authors (40). However, the authors fail to distinguish clearly between "visible" and invisible flow components in the bubble and cloud phases. At this time their approach must be regarded as a purely empirical method which appears to give a reasonable match with selected experimental data.

Bubble Size. A number of empirical and semi-empirical approaches are available for predicting mean bubble size as a function of height and other conditions in gas fluidized beds. Judging from Tabe II, the approach followed by Mori and Wen (41) appears to have become the favored method of predicting d_b. This equation is semi-empirical; predictions are bounded between an initial size produced at a distributor and a maximum size achieved only under slug flow conditions. Another recent mechanistically based equation due to Darton *et al* (42) is also receiving considerable attention, but has not been tested by any of the authors in this symposium. Both approaches seem to represent marked improvements over previous equations of a solely empirical nature in the literature. A method is still required for predicting bubble sizes in beds containing tubes as in the Type B combustor considered by Fan and Chang (13).

Five of the papers surveyed in Tabe II treat the bubble size as if it were independent of height. In two cases (14, 17) d_b is allowed to vary with height. While the latter assumption is certainly more realistic, assumption of a constant bubble size is defensible on the grounds of simplicity and limited sensitivity relative to some of the other assumptions discussed in this paper.

Heat Balance. For many years it has been customary to treat

fluid bed reactors as isothermal and to ignore energy balances in the modeling process. Recent emphasis on coal combustion and other gas-solid reactions with high heats of reaction has led to the inclusion of heat balances with more and more models. Heat balances are ignored in only three of the eight papers surveyed in Table II confirming this trend.

Steady versus Unsteady State Models. Until very recently, fluidized bed reactor models have dealt almost exclusively with steady state conditions. Steady state models are unsuitable for control purposes, for load following in fluid bed combustors, and for start-up and shutdown purposes. It is a welcome sign that two of the papers in this symposium (13,15) derive models which are potentially suitable for these purposes.

Type of Reaction and Application. An increased emphasis on gas-solid reactions has been evident for about a decade. Three of the papers in this symposium treat gas-solid reactions, two (13,18) dealing with coal combustion and the other (11) with catalyst regeneration. Of the four papers which consider solid-catalysed gas-phase reactions, one (15) deals with a specific application (production of maleic anhydride), and one (12) treats an unspecified consecutive reaction of the type $A \rightarrow B \rightarrow C$; the other two (14,16) are concerned with unspecified first order irreversible reactions. The final paper (17) considers a relatively recent application, fluidized bed aerosol filtration. Principles of fluid bed reactor modeling are directly applicable to such a case: Aerosol particles disappear by adsorption on the collector (fluidized) particles much as a gaseous component disappears by reaction in the case of a solid-catalysed reaction.

Experimental Data. While the emphasis in this session was on reactor modeling, models can only ultimately prove successful if they are compared to experimental data. This point may seem obvious, but it is worth making since modeling efforts too often seem to be intellectual exercises rather than efforts to represent reality. While there is a need to verify some of the models presented at this symposium, it is gratifying that three of the papers (11,15,17) have already been exposed to the test of experimental data.

Other Model Features

Some of the principal features common to the different models are discussed above. In this section some further features of reactor models are considered briefly with reference to individual papers in this symposium.

Only the paper by de Lasa *et al* (11) explicitly treats the entry or grid region as a non-bubbling region. This region is modeled in terms of discrete gas jets, an idea originated by Behie and Kehoe (43), but contested actively by Rowe *et al* (44). As indicated in the papers by Jaffres *et al* (15) and Rehmat *et al* (18), the grid region is clearly a zone of effective gas-solid contacting, but considerable work is required to achieve an understanding of the hydrodynamics and gas exchange processes therein.

None of the papers in this session explicitly considers the freeboard region although both de Lasa et al (11) and Jaffres et al (15) refer to previous work which has shown that the freeboard can play an important role in determining the overall reactor performance. None of the papers treats directly flow regimes other than the bubbling regime, although Rehmat et al (18) mention the turbulent flow regime (together with rapid interphase exchange in the grid region) as justification for using a model which treats the gas as a single phase in plug flow. As already suggested in Table I, efforts to model turbulent and fast fluidized beds are likely to be important features of the 1980's.

In modeling gas-solid reactions in fluid beds, provision must be made for dealing with particle size distributions and with solids mixing. Solids mixing is usually adequately described in terms of perfect mixing. To account for size distribution effects, population balances are generally required. These must take into account the size distribution of the feed, elutriation and losses of fines, attrition (if appreciable), and any changes in particle size due to chemical reaction. The paper by Rehmat et al (18) illustrates how these factors can be taken into account. Overall solid reaction rates must be determined by summing over all particle sizes, and conversion must be related to gas conversion via the stoichiometry of the reactions.

Concluding Remarks

It is clear that there are many unresolved questions in the field of fluidized bed reactor modeling. Only the bubble and slug flow regimes have received significant attention. While end effects (grid zone and freeboard region effects) are beginning to be treated, almost no efforts have been made to model high velocity fluidized beds operating in the turbulent and fast fluidization regimes. These regimes are of great importance industrially and for future applications. Even in the bubble flow regime, where there is a wealth of hydrodynamic and other data, some of the key aspects of behavior remain poorly understood.

It is clear from previous work and from the papers in this symposium that models are much more sensitive to assumptions in some areas than in others. For very slow reactions, rates become controlled by chemical kinetics and insensitive to whatever hydrodynamic assumptions are adopted (14,48). For intermediate reactions, interphase transfer generally becomes the key factor controlling the reactor performance, with the distribution of gas between phases also playing a significant role. As outlined above, advances have been made in understanding both areas, but models have generally been slow to adopt changes in the basic assumptions used in early bubble models. For fast reactions, the extent of axial mixing of gas in the dense or emulsion phase and the fraction of solids assigned to the dilute phase also become important. However, axial gas mixing is less important in general than might be indicated by the degree of attention devoted to this feature. On the other hand, radial mixing has received too little attention.

The papers presented in this symposium point to a number of advances that will be important in the 1980's. These include: (a) fundamentally new types of models using the power of modern computers to solve comprehensive governing equations (9); (b) continuing strong attention on gas-solid reactions as well as gas-phase solid-catalysed reactions; (c) unsteady state models suitable for control purposes (13,15); (d) attention to rate-limiting steps and to sensitivity analyses; (e) inclusion of grid and freeboard effects; (f) inclusion of energy balances; and (g) study of radial gradients and radial dispersion (13). Multiphase reactor models have chiefly been useful in the past as an educational tool in aiding understanding of fluid bed processes and, to a limited extent, for simulation of existing reactors and chemical processes. If these models are to become useful also for design, scale-up and control of new equipment and processes, advances in all of these areas may be very helpful.

Legend of Symbols

$\mathcal{D}$	molecular diffusivity
d_b	bubble diameter
$\bar{d}_b$	mean bubble diameter
g	acceleration of gravity
k_{bc}	bubble/cloud mass transfer coefficient based on bubble surface area
k_{bc}'	bubble/cloud mass transfer coefficient based on bubble volume
k_{ce}'	cloud/emulsion mass transfer coefficient based on bubble volume
Nu	Nusselt number
Sh	Sherwood number
U	superficial gas velocity
U_{mf}	superficial gas velocity at minimum fluidization
u_b	bubble rise velocity
$\bar{u}_b$	bubble rise velocity corresponding to $\bar{d}_b$
α	ratio of bubble velocity to remote interstitial velocity,= $u_b \varepsilon_{mf}/U_{mf}$
ε_{mf}	bed void fraction at minimum fluidization

Acknowledgement

Acknowledgement is made to the Donors of The Petroleum Research Fund, administered by the American Chemical Society, for the partial support of this research.

Literature Cited

1. Geldart, D. Chem. and Ind. 1967, 1474-1481.
2. Geldart, D. Chem. and Ind. 1968, 41-47.
3. Grace, J.R. AIChE Symp. Ser. 1971, 67, No. 116, 159.
4. Pyle, D.L. Adv. Chem. Ser. 1972, 109, 106.
5. Rowe, P.N. "Proc. 5th Europ./2nd Intern. Symp. on Chem. Reaction Engng.", Elsevier, Amsterdam, 1972, p.A9.
6. Yates, J.G. Chemical Engineer (London) Nov. 1975, 671.
7. Bukur, D.; Caram, H.S.; Amundson N.R. "Chemical Reactor Theory: a Review", ed. L. Lapidus and N.R. Amundson, Prentice-Hall, Englewood Cliffs, N.J., 1977, p.686.
8. Ramírez, J.; Ayora, M.; Vizcarra, M. This symposium volume.
9. Blake, T.R.; Chen, P.J. This symposium volume.
10. Gibbs, B.M. Paper presented at A.C.S. symposium, Las Vegas, 1980.
11. de Lasa, H.I.; Errazu, A.; Barreiro, E.; Solioz, S. Paper given at A.C.S. symposium, Las Vegas, 1980, and to be published in Can. J. Chem. Eng.
12. Elnashaie, S.S.E.H.; Elshishini, S.S. Paper presented at A.C.S. symposium, Las Vegas, 1980.
13. Fan, L.T.; Chang C.C. This symposium volume.
14. Fogler, H.S.; Brown, L.F. This symposium volume.
15. Jaffres, J.L.; Chavarie, C.; Patterson, W.I.; Laguerie, C. This symposium volume.
16. Jayaraman, V.K.; Kulkarni, B.D.; Doraiswamy, L.K. This symposium volume.
17. Peters, M.H.; Fan, L-S.; Sweeney, T.L. This symposium volume.
18. Rehmat, A.; Saxena, S.C.; Land, R. This symposium volume.
19. Kunii, D.; Levenspiel, O. "Fluidization Engineering", Wiley, New York, 1969.
20. Chavarie, C.; Grace, J.R. Ind. Eng. Chem. Fund. 1975, 14, 79.
21. Grace, J.R. Chapter 11 in "Gas Fluidization", ed. D. Geldart, Wiley, New York, 1982.
22. Fryer, C.; Potter, O.E. Ind. Engng. Chem. Fund., 1972, 11, 338.
23. Orcutt, J.C.; Davidson, J.F.; Pigford, R.L. Chem. Engng. Progr. Symp. Series 1962, 58, No. 38, 1.
24. Chavarie, C. Ph.D. thesis, McGill University, 1973.
25. Mireur, J.P.; Bischoff, K.B. A.I.Ch.E. Journal 1967, 13, 839.
26. Reay, D. "Proc. 1st International Symp. on Drying", ed. A.S. Mujumdar, Science Press, Princeton, 1978, p. 136.
27. Davidson, J.F.; Harrison, D. "Fluidized Particles", Cambridge University Press, Cambridge, England, 1963.
28. Rowe, P.N.; Partridge, B.A.; Lyall, E. Chem. Eng. Sci. 1964, 19, 973.

29. Anwer, J.; Pyle, D.L. "La Fluidisation et ses Applications", Soc. Chim. Ind., Paris, 1974, p.240.
30. Walker, B.V. Trans. Instn. Chem. Engrs., 1975, 53, 255.
31. Chavarie, C.; Grace, J.R. Chem. Eng. Sci. 1976, 31, 741.
32. Sit, S.P.; Grace, J.R. Chem. Eng. Sci. 1978, 33, 1115.
33. Murray, J.D. J. Fluid. Mech. 1965, 21, 465 & 22, 57.
34. Pereira, J.A.F. Ph.D. dissertation, University of Edinburgh, 1977.
35. Sit, S.P.; Grace, J.R. Effect of bubble interaction on interphase mass transfer in gas fluidized beds, Chem. Eng. Sci. 1981, 36, 327.
36. Clift, R.; Grace, J.R.; Cheung, L.; Do, T.H. J. Fluid Mech. 1972, 51, 197.
37. Partridge, B.A.; Rowe, P.N. Trans. Instn. Chem. Engrs. 1966, 44, 335.
38. Toomey, R.D.; Johnstone, H.P. Chem. Eng. Progr. 1952, 48,220.
39. Grace, J.R.; Clift, R. Chem. Eng. Sci. 1974, 29, 327.
40. Peters, M.H.; Fan, L-S.; Sweeney, T.L. A.I.Ch.E. National Meeting, Chicago, November, 1980.
41. Mori, S.; Wen, C.Y. A.I.Ch.E. Journal 1975, 21, 109.
42. Darton, R.C.; Lanauze, R.D.; Davidson, J.F.; Harrison, D. Trans. Instn. Chem. Engrs., 1977, 55, 274.
43. Behie, L.A.; Kehoe, P. A.I.Ch.E. Journal 1973, 19, 1070.
44. Rowe, P.N.; MacGillivray, H.J.; Cheesman, D.J. Trans Instn. Chem. Engrs. 1979, 57, 194.
45. May, W.G. Chem. Engng. Progr. 1959, 55, No. 12, 49.
46. Kato, K.; Wen, C.Y. Chem. Eng. Sci. 1969, 24, 1351.
47. Bywater, R.J. A.I.Ch.E. Symposium Series 1978, 74, No. 176, 126.
48. Grace, J.R. AIChE Symp. Series 1974, 70, No. 141, 21.
49. Avedesian, M.M.; Davidson, J.R. Trans. Instn. Chem. Engrs. 1973, 51, 121.

RECEIVED JUNE 25, 1981.

2

An Initial Value Approach to the Counter-Current Backmixing Model of the Fluid Bed

V. K. JAYARAMAN, B. D. KULKARNI, and L. K. DORAISWAMY

National Chemical Laboratory, Poona 411 008 India

The counter-current backmixing model of Fryer and Potter has been modified by assuming mixed flow in the emulsion phase. The terminal conversions obtained with the present model are compared with those of the original model and found to agree well except at very low values of bubble diameter. The assumption of complete mixing in the emulsion phase converts the original two-point boundary value problem into a simpler initial value problem, thereby considerably reducing the mathematical complexity.

The intensive gas mixing that occurs in a fluid bed due to the presence of bubbles and the associated circulatory movement of solids has been recognized for quite some time (1, 2). The rising bubbles carry wakes of solids along with them and release them subsequently on bursting at the surface (3, 4, 5). The released solids then move downwards for reasons of continuity and a simple circulatory pattern of movement of solids is set up. The studies on particle movement in deep fluidized beds (6) have indicated that solids move upwards in the center region of the bed and downwards at the periphery. The intensity of circulation of solids increases with increase in the fluidizing gas velocity, and at a critical velocity U_{cr} the velocity of down flowing solids exceeds the interstitial gas velocity, so that the interstitial gas is carried downwards as described by (7-10). A simple mechanism for gas mixing therefore seemed possible and several models - the so-called counter-current backmixing models that take into account this flow reversal - have been proposed (8, 11, 12).

It should, however, be noted that the solids movement pattern as mentioned above has been observed in beds with sufficiently large values of length to diameter ratio ($L_f/d_t \gg 1$). Industrial fluid beds normally operate with L_f/d_t values less than or close to unity and the solids flow pattern could be entirely different. More recent experimental studies such as

0097-6156/81/0168-0019$05.00/0

those of Okhi and Shirai (13) in shallow beds indicate a different flow pattern. Their experimental measurements have confirmed the fact that solids move downward in the central region of the bed. Earlier, Whitehead *et al* (14) had also made measurements of solids movement and demonstrated in some cases a strong down flow of solids in a small area at the center of the bed. Such solids circulatory pattern has also been reported by Werther (15) and Schmalfeld (16). Nguyen and Potter (9, 10) experimenting with a 30 cm diameter column have also observed that gas mixing is at its maximum in the center. Bubbles move in the area between the center and the wall, forcing the solids and the backmixed gas to move downwards in the central and near-to-wall region. The more recent experiments of Nguyen *et al* (17) in a large scale fluidized bed confirm this fact; however at very high velocities the stream becomes more unstable and flow is difficult to define.

It is clear from the foregoing discussion that a considerable extent of gas backmixing results due to the presence of bubble tracks and the associated solids movement. Besides this, the industrial units are normally operated with baffles and internals to remove the heat of reaction. The hinderances from these would lead to further enhancement of gas mixing in the emulsion phase. The common assumption of plug flow in the emulsion phase therefore seems incompatible with the situation prevailing in industrial reactors, and in the present work the original Fryer-Potter model (12) has been modified to take this reality into account. This has the additional advantage of converting the boundary value nature of the Fryer-Potter representation into an initial value problem, thus considerably simplifying the mathematical treatment.

Theoretical Development

Let us consider a simple reaction $A \longrightarrow R$ and make the following assumptions: the bubbles are uniform in size and free of particles. The emulsion phase voidage is constant, with the voidage of the bubbling bed equal to that at incipient fluidization. The voidage in the cloud is the same as in the emulsion. Plug flow prevails in the bubble and cloud phases, with the emulsion phase completely mixed. With these assumptions the material balance equations may be written as follows:

$$-U_{Gb}\frac{dC_b}{dz} + \epsilon_b K_{bc}(C_c - C_b) = 0 \qquad (1)$$

$$-U_{Gc}\frac{dC_e}{dz} + \epsilon_b K_{bc}(C_b - C_c) + \epsilon_b K_{ce}(C_e - C_c) - kf_w \epsilon_b C_c = 0 \qquad (2)$$

$$-U_{Ge}\,(C_c(1) - C_e) + K_{ce}\epsilon_b \int_0^{L_f} (C_c - C_e)\,dz$$

$$-k\,(1-\epsilon_b(1+f_w))\,L_f\,C_e = 0 \tag{3}$$

The appropriate initial conditions can be written as

$$C_b(0) = C_o \tag{4}$$

$$(U-U_{Gb})\,C_o + (-U_{Ge})\,C_e(0) = U_{Gc}\,C_c(0) \tag{5}$$

Equations 1–5 can be written in dimensionless forms as

$$-\frac{dC_1}{dl} + A_1\,(C_2 - C_1) = 0 \tag{6}$$

$$-\frac{dC_2}{dl} + A_2(C_1 - C_2) + A_3(C_3 - C_2) - A_4C_2 = 0 \tag{7}$$

$$-(C_2(1) - C_3) + A_5\int_0^1 (C_2 - C_3)\,dl - A_6C_3 = 0 \tag{8}$$

where the constants $A_1 - A_6$ are defined as follows :

$$A_1 = \frac{\epsilon_b K_{bc} L_f}{U_{Gb}} \qquad A_2 = \frac{\epsilon_b K_{bc} L_f}{U_{Gc}}$$

$$A_3 = \frac{\epsilon_b K_{ce} L_f}{U_{Gc}} \qquad A_4 = \frac{k f_w \epsilon_b L_f}{U_{Gc}}$$

$$A_5 = \frac{\epsilon_b K_{ce} L_f}{U_{Ge}} \qquad A_6 = \frac{k\,(1-\epsilon_b(1+f_w))L_f}{U_{Ge}} \tag{9}$$

The set of Equations 6–8 is accompanied by initial conditions

$$C_1 = 1 \text{ at } l = 0 \tag{10}$$

$$1 + B_1C_3 = B_2C_2 \quad \text{at } l = 0 \tag{11}$$

where

$$B_1 = \frac{-U_{Ge}}{(U-U_{Gb})} \quad \text{and} \quad B_2 = \frac{U_{Gc}}{(U-U_{Gb})} \tag{12}$$

The assumption of complete mixing in the emulsion phase renders the concentration C_3 constant in the bed, and Equations 6 and 7 can be rearranged as

$$D^2C_1 + (A_1+A_2+A_3+A_4)\ DC_1 + A_1(A_3+A_4)\ C_1 - A_1A_3C_3 = 0 \tag{13}$$

where the operator D refers to $d/d\ell$. The solution to this equation can be readily obtained as

$$C_1 = R_1e^{\lambda_1\ell} + R_2e^{\lambda_2\ell} + R_3 \tag{14}$$

where λ_1 and λ_2 are the roots of Equation 13 with the constant term $(A_1A_3C_3)$ removed, and R_3 is the particular solution of Equation 13 given by

$$R_3 = \frac{A_3}{A_3 + A_4}\ C_3 \tag{15}$$

Equation 14 can be substituted in Equation 6 to obtain

$$C_2 = R_1\alpha_1\ e^{\lambda_1\ell} + R_2\alpha_2\ e^{\lambda_2\ell} + R_3 \tag{16}$$

where the α's are defined as

$$\alpha_1 = \frac{\lambda_1 + A_1}{A_1} \quad \text{and} \quad \alpha_2 = \frac{\lambda_2 + A_1}{A_1} \tag{17}$$

It is interesting to note that Equations 14 and 16 require a knowledge of C_3 which can be obtained after some algebraic manipulations by substituting these equations in 8 as

$$C_3 = A_7A_8R_1 + A_7A_9R_2 \tag{18}$$

where A_7, A_8 and A_9 are constants defined as

$$A_7 = \left[1-A_6-A_5 + \frac{A_5A_3}{A_3+A_4} - \frac{A_3}{A_3+A_4}\right]^{-1}$$

$$A_8 = \alpha_1 e^{\lambda_1} + \frac{A_5 \alpha_1}{\lambda_1} (1 - e^{\lambda_1})$$

$$A_9 = \alpha_2 e^{\lambda_2} + \frac{A_5 \alpha_2}{\lambda_2} (1 - e^{\lambda_2}) \qquad (19)$$

Equations 14 and 16 along with 18 give the concentration profiles in the bubble and cloud phases. The constants R_1 and R_2 appearing in these equations can be evaluated subject to initial conditions given by Equations 9 and 10 and can be written in matrix form as

$$\begin{bmatrix} B_3 & B_4 \\ B_5 & B_6 \end{bmatrix} \begin{bmatrix} R_1 \\ R_2 \end{bmatrix} = \begin{bmatrix} 1 \\ 1 \end{bmatrix} \qquad (19a)$$

where parameters B_3–B_6 are defined as

$$B_3 = 1 + \frac{A_3 A_7 A_8}{A_3 + A_4} \qquad B_4 = 1 + \frac{A_3 A_7 A_9}{A_3 + A_4}$$

$$B_5 = B_2 \alpha_1 + A_8 \left(\frac{B_2 A_3 A_7}{A_3 + A_4} - B_1 A_7 \right)$$

$$B_6 = B_2 \alpha_2 + A_9 \left(\frac{B_2 A_3 A_7}{A_3 + A_4} - B_1 A_7 \right) \qquad (20)$$

The gas concentration at the bed exit is given by

$$C(1) = \frac{U_{Gb}}{U} C_1(1) + \left(1 - \frac{U_{Gb}}{U} \right) C_2(1) \qquad (21)$$

The concentration profiles in the bubbles, cloud and emulsion phases are plotted in Figure 1 for a set of parameter values. For the sake of comparison, the profiles for the same values of parameters obtained using the Fryer-Potter model are shown in Figure 2. Figures 3-6 show the influence of parameters such as bubble diameter, U/U_{mf}, H_o and rate constant on the extent of

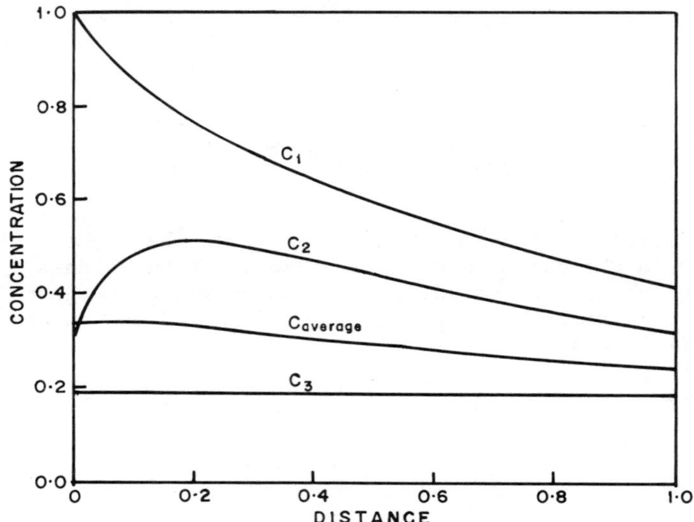

Figure 1. Concentration profiles using present model: $H_o = 50$ *cm,* $d_b = 5$ *cm,* $U = 10$ *cm/s,* $U_{mf} = 1$ *cm/s,* $k = 0.5$ s^{-1}, $\epsilon_{mf} = 0.5$, $D_e = 0.2$ cm^2/s

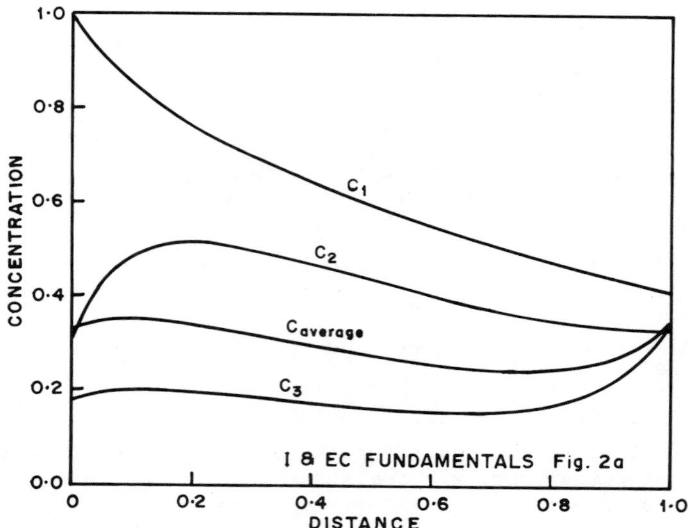

Figure 2. Concentration profiles using Fryer–Potter model (12): $H_o = 50$ *cm,* $d_b = 5$ *cm,* $U = 10$ *cm/s,* $U_{mf} = 1$ *cm/s,* $k = 0.5$ s^{-1}, $\epsilon_{mf} = 0.5$, $D_e = 0.2$ cm^2/s

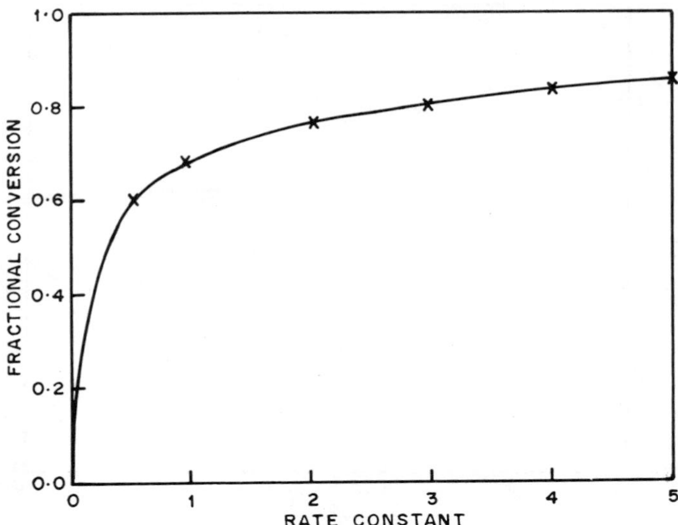

Figure 3. Comparison of the present and the FP model effect of rate constant: H_o = *50,* d_b = *5 cm,* U = *10 cm/s,* U_{mf} = *1 cm/s,* D_e = *0.2* cm^2/s, ϵ_{mf} = *0.5, (×) FP model, (——) present model*

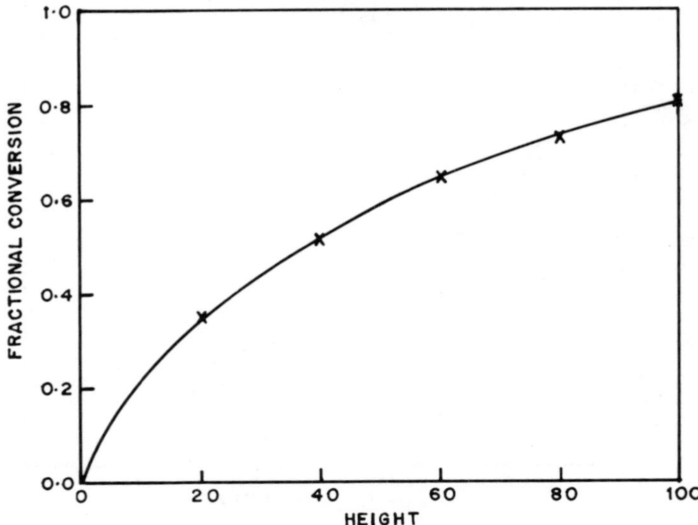

Figure 4. Comparison of the present and the FP model effect of the bed height: d_b = *5 cm,* U = *10 cm/s,* U_{mf} = *1 cm/s,* ϵ_{mf} = *0.5,* D_e = *0.2* cm^2/s, k = *0.5* s^{-1}, *(×) FP model, (——) present model*

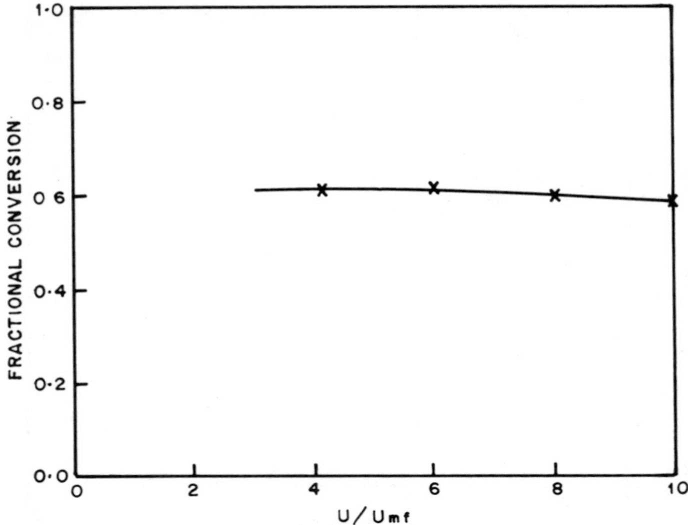

Figure 5. Comparison of the present and the FP model effect of U/U_{mf} *on conversion:* $H_o = 50$ *cm,* $d_b = 5$ *cm,* $U_{mf} = 1$ *cm/s,* $D_e = 0.2$ *cm²/s,* $k = 0.5$ s^{-1}, $\epsilon_{mf} = 0.5$*, (×) FP model, (——) present model*

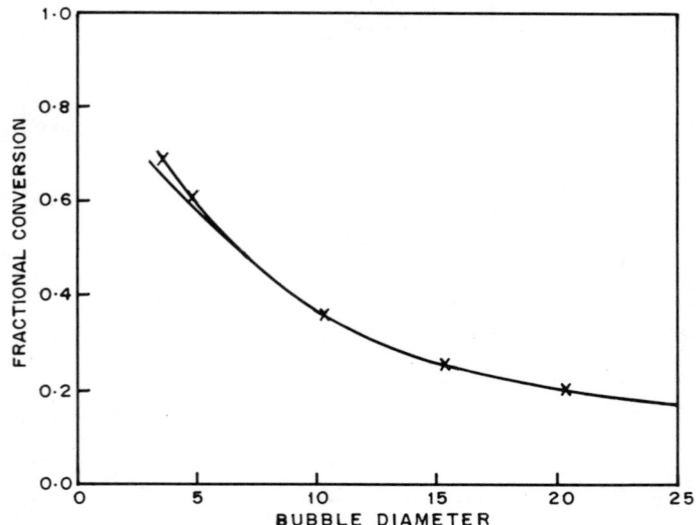

Figure 6. Effect of bubble diameter on conversion: $H_o = 50$ *cm,* $k_1 = 0.5$ s^{-1}, $D_e = 0.2$ *cm²/s,* $U = 10$ *cm/s,* $U_{mf} = 1$ *cm/s,* $\epsilon_{mf} = 0.5$*, (×) FP model, (——) present model*

conversion. Again, for the sake of comparison, the results obtained using the Fryer-Potter model are also presented.

It can be seen from these figures that the results obtained using the two models are almost indistinguishable from each other except at smaller values of d_b. The smaller bubble diameters are however unlikely in large industrial fluid beds, and therefore for all practical purposes the predictions of the two models are identical.

The large industrial fluid beds are normally operated with U/U_{mf} exceeding 10, so that a large portion of the gas bypasses the bed in the form of bubbles. Also the diameter of the bubbles is fairly large, so that interphase mass transport is small compared to the rate of reaction. Under these conditions the extent of mixing in the emulsion phase is rather an unimportant parameter as far as the prediction of conversion is concerned. It would, however, have significant influence when non first-order reactions are involved.

The formulation of the model as above has the advantage that mathematically it picturizes the bed as an initial value problem in contrast to the more complicated boundary value representation of the Fryer-Potter model. The implications of this reduced complexity become more evident (and considerably more important) when the reactions involved are nonlinear. While the initial value problem can be readily solved for such a case, the boundary value presentation leads to severe stability and convergence problems.

Conclusions

The behavioural features of the fluidized bed have been modeled based on a modified representation of the Fryer-Potter model. The restrictive assumption of plug flow of the emulsion gas has been removed, and model equations developed based on complete mixing of the emulsion gas. This simplification, in addition to bringing the model closer to reality, has led to the conversion of a boundary value problem (Fryer-Potter model) to a simpler initial value problem. Except at very low bubble diameters, the predictions of the two models (based on terminal conversion) agree closely with each other. On the other hand, agreement between the average concentration profiles in the bed predicted by the two models is less satisfactory. While therefore the modified model proposed in this work has the advantage of simplicity and is perhaps closer to reality, further experimental work on industrial size equipment is necessary for a firmer opinion on the latter (nature of gas flow in the emulsion phase).

Legend of Symbols

A_1 to A_6	constants defined by Equation (9)
A_7 to A_9	constants defined by Equation (19)
B_1, B_2	constants defined by Equation (12)
B_3 to B_6	parameters defined by Equation (20)
C_b, C_c, C_e	concentration in the bubble, cloud-wake and emulsion phase respectively
C_1, C_2, C_3	dimensionless concentration in the bubble, cloud and emulsion
$C_1(1)$, $C_2(1)$	dimensionless bubble phase and cloud-wake phase concentration at the bed exist
$C(1)$	dimensionless gas concentration at the exit
d_b	bubble diameter cm
d_t	diameter of the bed cm
f_w	ratio of wake volume to bubble volume
H_o	height of the bed at incipient fluidization
K_{bc}	volumetric rate of gas exchange between bubble and cloud-wake per unit bubble volume s^{-1}
K_{ce}	volumetric rate of gas exchange between cloud-wake and emulsion per unit volume s^{-1}
L_f	height of bubbling bed cm
l	dimensionless height above distributor
k	first order reaction rate constant, based on unit volume of dense phase, s^{-1}
R_1, R_2	parameters defined by Equation (19a)
R_3	parameter defined by Equation (15)
U	superficial gas velocity cm s^{-1}
U_{cr}	critical velocity cm s^{-1}
U_{Gb}	superficial velocity in bubble phase cm s^{-1}
U_{Gc}	superficial velocity in cloud-wake phase cm s^{-1}
U_{Ge}	superficial velocity in emulsion phase cm s^{-1}
Z	length parameter along the bed height

Greek Letters

α_1, α_2	constants defined by Equation (17)
ϵ_b	fraction of bed volume occupied by bubbles
λ_1, λ_2	roots of Equation (13)
ϵ_{mf}	void fraction in bed at minimum fluidization conditions

Acknowledgement

The financial support received from Indian Petrochemicals Corporation Limited, Baroda, is gratefully acknowledged.

Literature Cited

1. Gilliland, E.R.; Mason, E.A. Ind. Eng. Chem. 1949, 41, 1191.
2. Gilliland, E.R.; Mason, E.A. Ind. Eng. Chem. 1952, 44, 218.
3. Noble, P.J. "Mineral Dressing Research Symposium Annual Conference"; Australian Instn. Min. Met. 1962.
4. Rowe, P.N. Trans. Inst. Chem. Engrs. 1961, 39, 175.
5. Woollard, I. N. M.; Potter, O.E. AIChE J. 1968, 14, 388.
6. Marscheck, R.M.; Gomezplata, A. AIChE J. 1965, 11, 167.
7. Stephens, G.K.; Sinclair, R.J.; Potter, O.E. Powder Tech. 1967, 1, 157.
8. Latham, R.L.; Hamilton, C.J.; Potter, O.E. Brit. Chem. Eng. 1968, 13, 666.
9. Nguyen, H.V.; Potter, O.E. "Advances in Chemistry Series No.133"; American Chemical Society : Washington, DC. 1974; p 290.
10. Nguyen, H.V. Ph.D. thesis, Monash University, Australia, 1975.
11. Kunii, D.; Levenspiel, O. Ind. Eng. Chem. Fundam. 1968, 7, 446.
12. Fryer, C.; Potter, O.E. Ind. Eng. Chem. Fundam. 1972, 11, 338.
13. Ohki, K.; Shirai, T. "Fluidization Technology I"; Hemisphere Publishing Corporation : Washington DC, 1976; p 95.
14. Whitehead, A.B.; Gartside, G.; Dent, D.C. Chem. Eng. J. 1970, 1, 175.
15. Werther, J. Preprint, paper presented at GVC/AIChE Meeting, Munchen, 1974.
16. Schmalfeld, V.J. V.D.I-Z. 1976, 118, 65.
17. Nguyen, H.V.; Whitehead, A.B.; Potter, O.E. AIChE J. 1977, 23, 913.

RECEIVED June 3, 1981.

Predictions of Fluidized Bed Operation Under Two Limiting Conditions: Reaction Control and Transport Control

H. S. FOGLER

Department of Chemical Engineering, The University of Michigan, Ann Arbor, MI 48109

L. F. BROWN

Department of Chemical Engineering, The University of Colorado, Boulder, CO 80309

Some aspects of fluidized-bed reactor performance are examined using the Kunii-Levenspiel model of fluidized-bed reactor behavior. An ammonia-oxidation system is modeled, and the conversion predicted is shown to approximate that observed experimentally. The model is used to predict the changes in conversion with parameter variation under the limiting conditions of reaction control and transport control, and the ammonia-oxidation system is seen to be an example of reaction control. Finally, it is shown that significant differences in the averaging techniques occur for height to diameter ratios in the range of 2 to 20.

There has been increased interest in recent years in the science and engineering of fluidized-bed reactors. Part of this interest can be attributed to the projected extensive use of fluidized-bed coal gasifiers, but the development of magnetically-stabilized fluidized beds and centrifugal beds also has contributed significantly to rejuvenating fluidized-bed research and modeling. Some of the many recent reviews and evaluations of fluidized-bed modeling are those of Bukur (1974), Chavarie and Grace (1975), Yates (1975), Van Swaaif (1978), Weimer (1978), and Potter (1978). Of these, Yates gives an unusually good comparison of the theoretical similarities and differences among currently popular models, while Chavarie and Grace compare the predictions of various models with the experimentally-observed internal behavior of a fluidized-bed reacting system. These latter authors conclude that the Kunii-Levenspiel (K-L) model gives the most realistic estimate of behavior within a fluidized bed. Yates points out that the countercurrent-backmixing model of Fryer and Potter, not considered by Chavarie and Grace, is more rigorously founded than the K-L model. On the other hand, Potter shows that when the average bubble size is smaller than 8-10 cm, there is little difference between the countercurrent-

0097-6156/81/0168-0031$06.00/0

backmixing and the K-L models, both of which give good predictions of fluidized-bed performance.

The K-L model, because of its greater simplicity, thus seems to be the model of choice for systems with smaller bubbles. In this paper we shall show how the K-L model can be used to predict the experimental results obtained by Massimilla and Johnstone (1961) on the catalytic oxidation of ammonia. It will be seen that the performance of their system was largely controlled by reaction limitations within the bed's phases. The effects of various parameters on bed performance are examined for such a reaction-limited system, and then the effects of these parameters for a transport-limited system are also discussed. Finally, we consider the effect of using average values of the bubble diameter and transport coefficients on model predictions.

Applying the Kunii-Levenspiel Model

The Kunii-Levenspiel Model will be used in conjunction with the correlations of Broadhurst and Becker (1975) and Mori and Wen (1975) to analyze the ammonia oxidation of Massimilla and Johnstone (1961). The reaction

$$4NH_3 + 7O_2 \rightarrow 4NO_2 + 6H_2O$$

was carried out in an 11.4 cm diameter fluidized-bed reactor containing 4kg of catalyst particles. The particles had a diameter, d_p, of 105 μm, and a density, ρ_p, of 2.06 g/cm^3. The particle sphericity, ψ, was taken to be 0.6 as is typical of published values (Kunii and Levenspiel, 1969).

A mixture of 90% oxygen and 10% ammonia was fed to the reactor at a rate of 818 cm^3/s, a temperature of 523 K, and a pressure of 0.11 MPa (840 torr). The reaction is first order in ammonia. The reaction is apparently zero order in oxygen owing to the excess oxygen. Thus

$$-r_A = k_{cat} C_A \tag{1}$$

From fixed-bed studies, k_{cat}=0.0858 cm^{3}gas/[(cm^3 catalyst)(s)].

The catalyst weight, W, and corresponding expanded bed height, h, necessary to achieve a specified conversion, X, are

$$W = Ah(1-\varepsilon_{mf})(1-\delta)\rho_p \tag{2}$$

$$h = \frac{u_b}{k_{cat}K_R} \ln\left(\frac{1}{1-X}\right) \tag{3}$$

in which

A is the cross-sectional area
K_R is the overall dimensionless reaction rate constant
u_b is the velocity of bubble rise, cm/s
ε_{mf} is the bed porosity at minimum fluidization conditions
δ is the fraction of the column occupied by bubbles

Calculating the Fluidization Parameters. The porosity at minimum fluidization is obtained from the Broadhurst and Becker correlation (1975):

$$\varepsilon_{mf} = 0.586\psi^{-0.72}\left(\frac{\mu^2}{\rho_g \eta d_p^3}\right)^{0.029}\left(\frac{\rho_g}{\rho_p}\right)^{0.021} \quad (4)$$

resulting in ε_{mf}=0.657. At first sight, this value appears higher than void fractions of 0.35-0.45 normally encountered in packed beds (Drew et al., 1950). The catalyst used by Massimilla and Johnstone used an impregnated cracking catalyst, however, and a value of ε_{mf} of 0.657 is consistent with the numbers reported for materials of this type by Leva (1959) and by Zenz and Othmer (1960).

The corresponding minimum fluidization velocity is

$$u_{mf} = \frac{(\psi d_p)^2 \eta}{150\,\mu}\,\frac{\varepsilon_{mf}^3}{1-\varepsilon_{mf}} \qquad \text{Re} < 20 \text{ (Kunii and Levenspiel, 1969)} \quad (5)$$

which gives u_{mf} = 1.48 cm/s.

The entering volumetric flow rate of 818 cm^3/s corresponds to a superficial velocity of 8.01 cm/s. Therefore

$$\frac{u_o}{u_{mf}} = 5.4$$

In order to calculate the expanded bed height, h, for the given catalyst weight of 4 kg, one needs to calculate the fraction of bed occupied by bubbles, δ. From the K-L model

$$\delta = \frac{u_o - u_{mf}}{u_b - u_{mf}(1+\alpha)} \quad (6)$$

For 0.1 mm particles, Kunii and Levenspiel (1969) state that $\alpha = 0.4$ is a reasonable estimate.

At this point, however, there is a difficulty. To calculate the velocity of bubble rise, u_b, the bubble diameter at the midpoint in the column, d_b, is required:

$$u_b = u_o - u_{mf} + u_{br} \tag{7}$$

and

$$u_{br} = (0.71)(gd_b)^{1/2} \quad (u_{br} = \text{rise velocity of a single bubble})$$

In Mori and Wen's (1975) correlation, the bubble diameter is a function of distance up the column, L:

$$d_b = d_{bm} - (d_{bm} - d_{bo})[\exp(-0.3L/D)] \tag{8}$$

Thus, in order to obtain the bubble diameter at L=h/2, the height of the expanded bed, h, is needed. Equation (8) also contains a maximum bubble diameter, d_{bm},

$$d_{bm} = (0.652)[A(u_o - u_{mf})]^{0.4} \tag{9}$$

(A in cm^2, u's in cm/s, and d_{bm} in cm)

and a minimum diameter for a porous plate:

$$d_{bo} = (0.00376)(u_o - u_{mf})^2 \tag{10}$$

(u's in cm/s, and d_{bo} in cm)

Equations (6) and (2) are used to obtain h, and the bubble-rise velocity u_b, appears in Eq. (6). Consequently, we see a predicament has arisen, in that h is needed to calculate d_b, which is needed to calculate u_b, which in turn is needed to calculate h.

To overcome this difficulty, the sequence in Figure 1 is normally adopted. The unexpanded bed height was 39 cm, so the expanded bed height will probably be around 60 cm and the average bubble size will first be calculated at L = h/2 = 30 cm. Using Eqs. (9) and (10),

$$d_{bo} = 0.16 \text{ cm}$$

and

$$d_{bm} = 8.79 \text{ cm}$$

The bubble diameter at L = h/2 calculated from Eq. (8) is 4.87 cm. Using this value, one can now calculate u_b, δ, and h using Eqs. (2), (6) and (7). These values are given in Table 1.

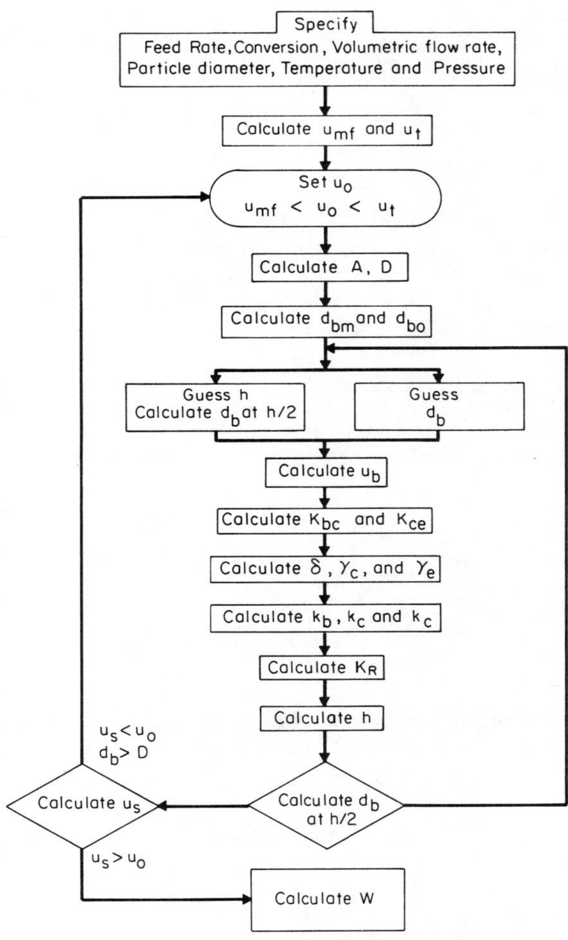

Figure 1. Computational algorithm for fluid bed reactor design

Table I
Fluidized-Bed Characteristics of Ammonia Oxidation Reactor

u_o	= 8.01 cm/s	d_{bo}	= 0.16 cm
ε_{mf}	= 0.657	d_{bm}	= 8.79 cm
u_{mf}	= 1.48 cm/s	d_b	= 4.87 cm
α	= 0.4	u_b	= 55.6 cm/s
δ	= 0.122	h	= 63.2 cm

Since the estimated bed height of 60 cm is sufficiently close to the calculated value of 63.2 cm, we can proceed in the calculations without making a new estimate of h. The only remaining parameter needed to calculate the conversion is the overall dimensionless reaction rate constant K_R.

<u>Calculating The Reaction Parameters</u>. The overall dimensionless rate constant is expressed in terms of exchange coefficients between the bubble, cloud, and emulsion, and in terms of the volumes of catalyst per volume of bubble in the bubble, cloud, and emulsion:

$$K_R = \gamma_b + \cfrac{1}{\cfrac{k_{cat}}{K_{bc}} + \cfrac{1}{\gamma_c + \cfrac{1}{\cfrac{1}{\gamma_e} + \cfrac{k_{cat}}{K_{ce}}}}} \tag{11}$$

The exchange coefficients between the bubble and the cloud, K_{bc}, and the cloud and the emulsion, K_{ce}, are respectively

$$K_{bc} = (4.5)\left(\frac{u_{mf}}{d_b}\right) + (5.85)\left(\frac{D_{AB}{}^2 g}{d_b{}^5}\right)^{1/4} \tag{12}$$

$$K_{ce} = (6.78)\left(\frac{\varepsilon_{mf} D_{AB} u_b}{d_b{}^3}\right)^{1/2} \tag{13}$$

Using these formulas, we obtain

$$K_{bc} = 4.92\ s^{-1}$$

$$K_{ce} = 3.00\ s^{-1}$$

The 0.618 cm^2/s used for the value of D_{AB}, the molecular diffusivity, in Eqs. (12) and (13) was calculated from the Fuller, Schettler, and Gidding correlation (Reid et al., 1977). The volumes of catalyst particles per volume of bubble in the bubble phase, γ_b, the cloud phase, γ_c, and the emulsion phase, γ_e, are given by the equations

$$\gamma_b = 0.01 \text{ cm}^3 \text{ of catalyst in bubbles/cm}^3 \text{ of bubble} \quad (14)$$

(this is a typical value which is frequently assumed)

$$\gamma_c = (1-\varepsilon_{mf})\left[\frac{\left(\dfrac{3u_{mf}}{\varepsilon_{mf}}\right)}{\left(u_{br} - \dfrac{u_{mf}}{\varepsilon_{mf}}\right)} + \alpha\right] \quad (15)$$

$$\gamma_e = (1-\varepsilon_{mf})\left(\frac{1-\delta}{\delta}\right) - \gamma_c \quad (16)$$

Substituting the indicated values into the equations yields

$$\gamma_c = 0.187 \text{ cm}^3 \text{ catalyst in clouds and wakes/cm}^3 \text{ of bubble}$$

and

$$\gamma_e = 2.28 \text{ cm}^3 \text{ catalyst in emulsion/cm}^3 \text{ of bubble}$$

When the values obtained from Eqs. (12) through (16) are substituted into Eq. (11),

$$K_R = 0.01 + \cfrac{1}{\cfrac{0.0858}{4.92} + \cfrac{1}{0.187 + \cfrac{1}{\cfrac{1}{2.28} + \cfrac{0.0858}{3.0}}}}$$

$$K_R = 0.01 + \cfrac{1}{0.0174 + \cfrac{1}{0.187 + \cfrac{1}{0.4386 + 0.0286}}}$$

Solving this equation gives the numerical value of the dimensionless reaction rate constant

$$K_R = 2.25$$

Equation (3) may be solved for the conversion, X:

$$X = 1 - \exp\left(\frac{k_{cat}K_R h}{u_b}\right) \tag{17}$$

Substituting the values we have determined into this equation gives

$$X = 0.197$$

The 20% conversion calculated using the Kunii-Levenspiel model compares quite well with the experimental value of 22% measured by Massimilla and Johnstone.

Limiting Situations. As engineers, it is important to deduce how a bed will operate if one were to change operating conditions such as gas flow rate, catalyst particle size, etc. To give some general guides as to how changes will affect bed behavior, we shall consider the two limiting circumstances of reaction control and transport control.

In the K-L model, reaction occurs within the bed's phases, and material is continuously transferred between the phases. Two limiting situations thus arise. In one, the interphase transport is relatively fast and transport equilibrium is maintained, causing the system performance to be controlled by the rate of reaction. In the other, the reaction rate is relatively fast and the performance is controlled by interphase transport. It will be shown that the ammonia oxidation example used above is essentially a reaction-limited system.

The overall reaction rate in the bed is proportional to K_R, so the reciprocal of K_R can be viewed as an overall resistance to the reaction. The different terms and groups on the RHS of Eq. (11) can be viewed as individual resistances which can be arranged in series or parallel to give the overall resistance.

$$R_o = \frac{1}{K_R} = \cfrac{1}{\cfrac{1}{\cfrac{1}{\gamma_b}} + \cfrac{1}{\cfrac{k_{cat}}{K_{bc}} + \cfrac{1}{\cfrac{1}{\cfrac{1}{\gamma_c}} + \cfrac{1}{\cfrac{1}{\cfrac{1}{\gamma_e}} + \cfrac{k_{cat}}{K_{ce}}}}}} \tag{18}$$

$$R_o = \cfrac{1}{\cfrac{1}{R_{rb}} + \cfrac{1}{R_{tbc} + \cfrac{1}{\cfrac{1}{R_{rc}} + \cfrac{1}{R_{re} + R_{tce}}}}} \tag{19}$$

in which:

$$R_{rb} = \frac{1}{\gamma_b} = \text{resistance to reaction in the bubble}$$

$$R_{tbc} = \frac{k_{cat}}{K_{bc}} = \text{resistance to transfer between bubble and cloud}$$

$$R_{rc} = \frac{1}{\gamma_e} = \text{resistance to reaction in cloud}$$

$$R_{re} = \frac{1}{\gamma_e} = \text{resistance to reaction in the emulsion}$$

$$R_{tce} = \frac{k_{cat}}{K_{ce}} = \text{resistance to transfer between cloud and emulsion}$$

The analog electrical resistance for the system is shown in Figure 2 along with the corresponding resistances for this reaction. As with its electrical analog, the reaction will pursue the path of least resistance, which in this case is along the right hand side branch of Figure 2. If the major resistance in this side, the resistance to reaction in the emulsion, R_{re}, could be reduced, a greater conversion could be achieved for a specified catalyst weight. To reduce R_{re}, one needs to look for ways of increasing γ_e

$$\gamma_e = (1-\varepsilon_{mf})\left[\frac{1-\delta}{\delta} - \frac{\dfrac{3u_{mf}}{\varepsilon_{mf}}}{(.71d_b g)^{1/2} - (u_{mf}/\varepsilon_{mf})} - \alpha\right] \tag{20}$$

Examination of equation (20) shows that decreasing the bubble size, d_b, and fraction, δ, while decreasing the minimum fluidization velocity would increase γ_e and hence the conversion. The minimum fluidization velocity could be decreased by decreasing the particles size. We now will investigate how the various parameters will affect the conversion for different limiting situations.

The Slow Reaction. In addition to the obvious way of increasing the temperature to increase the conversion, there are other ways the conversion may be increased when the reaction is slow. From equation (3) we know the conversion depends upon h, k_{cat}, u_b and K_R. We will first determine K_R under this situation. For a slow reaction, k_{cat} is small when compared to K_{bc} and K_{ce}, so that resistance to transport is essentially zero, i.e.

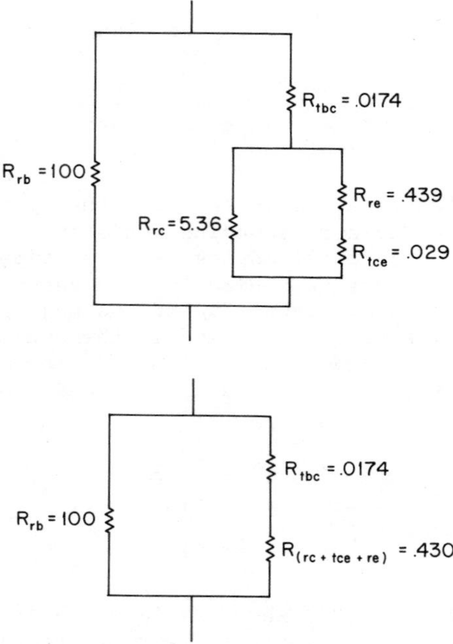

Figure 2. Electrical analog of transport and reaction resistances in the Kunni–Levenspiel model using the data of Massimillia and Johnstone (9)

$$\frac{k_{cat}}{K_{bc}} \simeq 0$$

and

$$\frac{k_{cat}}{K_{ce}} \simeq 0$$

then

$$K_R = \gamma_b + \cfrac{1}{0 + \cfrac{1}{\gamma_c + \cfrac{1}{\cfrac{1}{\gamma_e} + 0}}} = \gamma_b + \gamma_c + \gamma_e \tag{21}$$

Using Eq. (16) to substitute for γ_e we have

$$K_R = \gamma_b + (1-\varepsilon_{mf})\ (\frac{1-\delta}{\delta}) \tag{22}$$

neglecting γ_b w.r.t. the second term gives

$$K_R = (1-\varepsilon_{mf})\ (\frac{1-\delta}{\delta}) \tag{23}$$

Consequently we see that K_R can be increased by decreasing δ, the volume fraction of bubbles. For the ammonia oxidation example, this would give

$$K_R \simeq 2.47$$

or about 11% higher than the value obtained by the more elaborate calculations which included the transport. This would predict a conversion of 21.4%, very close to the 19.7% given by the method which includes the transport limitations. Thus the ammonia oxidation system of Massimilla and Johnstone is essentially a reaction-limited system.

The conversion and catalyst weight are related by

$$W = Ah\rho_p(1-\varepsilon_{mf})\ (1-\delta) = \frac{Au_b\rho_p(1-\varepsilon_{mf})\ (1-\delta)}{k_{cat}K_R} \ln\frac{1}{1-X} \tag{24}$$

Substituting for K_R

$$W = \frac{A\rho_p u_b \delta}{k_{cat}} \ln(\frac{1}{1-X}) \tag{25}$$

Recalling

$$\delta = \frac{u_o - u_{mf}}{u_b - u_{mf}(1+\alpha)} \tag{6}$$

In most all instances u_b is significantly greater than $u_{mf}(1+\alpha)$ so that Equation (6) is approximately

$$\delta = \frac{u_o - u_{mf}}{u_b} \tag{26}$$

combining Eq. (24) and (25)

$$W = \frac{A\rho_p(u_o - u_m)}{k_{cat}} \ln \frac{1}{1-X} \tag{27}$$

Therefore one observes that to reduce the catalyst weight for a specified conversion, u_o and u_{mf} should be as close as possible. One can now ask what ways may the catalyst weight be reduced for a specified conversion. The answer to this question is the same as to the question, "How may one increase the conversion for a fixed catalyst weight?"

For example, suppose you are operating at 5 times the minimum fluidization velocity, $u_o = 5u_{mf}$.

<u>Case 1</u>

$$W_1 = \frac{A\rho_p 4u_{mf1}}{k_{cat}} \ln \frac{1}{1-X_1} \tag{28}$$

What would be the effect of doubling the particle diameter on the catalyst weight for the same throughput and conversion?

<u>Case 2</u>

$$W_2 = \frac{\rho_p A(u_{o2} - u_{mf2})}{k_{cat2}} \ln \frac{1}{1-X_2} \tag{29}$$

Since the temperature $k_{cat1} = k_{cat2}$, the throughput ($u_{o1} = u_{o2}$), and conversion ($X_1 = X_2$) are the same for Cases 1 and 2, the ratio of equation (28) and (29) yield

$$\frac{W_2}{W_1} = \frac{u_{o1} - u_{mf2}}{4u_{mf1}} = \frac{5u_{mf1} - u_{mf2}}{4u_{mf1}} \tag{30}$$

Recalling Eq. (5)

$$u_{mf} = \frac{(\psi d_p)^2 \eta}{150\mu} \frac{\varepsilon_{mf}^3}{1-\varepsilon_{mf}} \tag{5}$$

and neglecting the dependence of ε_{mf} on d_p we see that the only parameters which vary between Case 1 (d_p) and Case 2 $(d_{p2}=2d_{p1})$ are u_{mf} and W.

$$\frac{u_{mf2}}{u_{mf1}} = \left(\frac{d_{p2}}{d_{p1}}\right)^2 = \left(\frac{2d_{p1}}{d_{p1}}\right)^2 = 4$$

and therefore

$$\frac{W_2}{W_1} = \frac{5u_{mf1} - 4u_{mf1}}{4u_{mf1}} = 0.25$$

Thus in the situation we have postulated, with a first-order reaction and reaction limiting the bed behavior, doubling the particle size will reduce the catalyst by approximately 75% and still maintain the same conversion.

The slow-reaction situation has been treated before (Grace, 1974), using a model of bed performance developed well before the K-L model (Orcutt et al., 1962). This earlier work concluded that when the reaction was very slow, the hydrodynamics and the way the hydrodynamics were modeled were unimportant. The analysis given above, using the more sophisticated K-L model, shows that the hydrodynamics can be very important indeed, even when the reaction is slow. In the situation cited, a reduction of 75% in catalyst requirement can be attained by expoitation of the bed hydrodynamics.

The Rapid Reaction. To analyze this limiting situation we shall assume the particles are sufficiently small so that the effectiveness factor is essentially one and that the rate of transfer from the bulk fluid to the individual catalyst particles is rapid in comparison with the rate of transfer between the fluidization phases. For the case of rapid reaction

$$\frac{k_{cat}}{K_{bc}} \text{ and } \frac{k_{cat}}{K_{ce}} >> 1$$

Using these approximations in the equation for K_R which is

$$K_R = \gamma_b + \cfrac{1}{\cfrac{k_{cat}}{K_{cb}} + \cfrac{1}{\gamma_c + \cfrac{1}{\cfrac{k_{cat}}{K_{ce}} + \gamma_e}}} \qquad (11)$$

one observes the first term to be neglected is

$$\frac{1}{\frac{k_{cat}}{K_{ce}} + \frac{1}{\gamma_e}} \quad \frac{1}{(\text{Large No.}) + \frac{1}{\gamma_e}} \simeq 0$$

Then neglecting the reciprocal of γ_c w.r.t. k_{cat}/K_R gives

$$K_R \cong \gamma_b + \frac{1}{\frac{k_{cat}}{K_{bc}} + \frac{1}{\gamma_c}} \simeq \gamma_b + \frac{K_{bc}}{k_{cat}} \qquad (31)$$

There are two situations one can analyze here

Situation 1: $\gamma_b << \frac{K_{bc}}{k_{cat}}$ Resistance to transport small w.r.t. resistance to reaction inside the bubble

Situation 2: $\gamma_b >> \frac{K_{bc}}{k_{cat}}$ Resistance to transport large w.r.t. resistance to reaction inside the bubble

Only situation 1 will be analyzed in the test and the analysis of situation 2 is left as an exercise for the interested reader.

Assuming very few particles are present in the bubble phase

$$K_R \cong \frac{K_{bc}}{k_{cat}} \qquad (32)$$

The catalyst weight is given by combining Eqs. (2) and (32)

$$W = \frac{Au_b\rho_p(1-\delta)\ (1-\varepsilon_{mf})\rho_p}{K_{bc}} \ln\left(\frac{1}{1-X}\right) \qquad (33)$$

Neglecting δ w.r.t. 1 in the numerator

$$W = \frac{A u_b \rho_p (1-\varepsilon_{mf})}{k_{cat} K_{bc}} \ln\left(\frac{1}{1-X}\right) \tag{34}$$

On observing the equation for K_{bc}, Eq. (12) is the sum of two terms A and B

$$K_{bc} = 4.5\frac{u_{mf}}{d_b} + 5.85\frac{D_{AB}^{1/2} g^{1/4}}{d_b^{5/4}} \tag{12}$$

$$K_{bc} = A_0 + B_0$$

One finds the problem can be further divided.

Case A: $A_0 >> B_0$

Case B: $B_0 >> A_0$

Only Case A will be considered here and Case B again will be left as an exercise for the interested reader.

For Case A

$$K_{bc} \cong 4.5\frac{u_{mf}}{d_b} \tag{35}$$

Then

$$W = \frac{u_b d_b}{4.5 u_{mf}} \rho_p A (1-\varepsilon_{mf}) \ln\left(\frac{1}{1-X}\right) \tag{36}$$

Recalling the equation for u_b and neglecting other terms in the equation w.r.t. the velocity of rise of a single bubble, i.e.,

$$u_b \simeq u_{br}$$

and

$$u_{br} = 0.71 g^{1/2} d_b^{1/2}$$

$$W = \frac{0.71 g^{1/2} d_b^{3/2}}{4.5 u_{mf}} A\rho_p (1-\varepsilon_{mf}) \ln\left(\frac{1}{1-X}\right)$$

$$W = 4.9\,\frac{d_b^{3/2}}{u_{mf}} A\rho_p (1-\varepsilon_{mf}) \ln\left(\frac{1}{1-X}\right) \tag{37}$$

The average bubble diameter is a function of the tower diameter (thus A), height, u_o, and u_{mf}. As a first approximation, we assume the average bubble diameter is some fraction, (say 0.75) of the maximum bubble diameter.

$$d_b = 0.75\, d_{bm}$$

Then, from Eq. (9),

$$d_b = (0.75)(0.652)[A(u_o - u_{mf})]^{0.4} \tag{38}$$

and

$$W = 1.69 \frac{A^{1.6}(u_o - u_{mf})^{0.6}}{u_{mf}} \rho_p (1-\varepsilon_{mf}) \ln(\frac{1}{1-X}) \tag{39}$$

We again consider the effect of doubling particle size while keeping all other variables the same.

$$\frac{W_2}{W_1} = \frac{(u_{o2} - u_{mf2})^{0.6}}{(u_{o1} - u_{mf1})^{0.6}} \frac{u_{mf1}}{u_{mf2}} \tag{40}$$

Recalling

$$u_{o2} = u_{o1} = 5u_{mf1}$$

$$u_{mf2} = 4u_{mf1}$$

then

$$\frac{W_2}{W_1} = \left(\frac{5u_{mf1} - 4u_{mf1}}{5u_{mf1} - u_{mf1}}\right)^{0.6} \frac{u_{mf1}}{4u_{mf1}} \tag{41}$$

or

$$\frac{W_2}{W_1} = 0.11 \tag{42}$$

In this case we see that doubling the particle diameter decreases the catalyst weight by 89% while still maintaining the same conversion. However, for a fast reaction, a significant decrease in effectiveness factor could offset this advantage.

It may be noted that the situation considered here, in which the bulk flow term >> diffusion term in Eq. (12), is a somewhat restricted one. For A_0 >> B_0 in small-particle systems, the binary diffusion coefficient must be on the order of 0.01 cm^2/s or less. Systems involving heavy hydrocarbons frequently have diffusion coefficients this low, but systems with lighter components do not. Systems using larger particles also have $A_0 >> B_0$, but then Re > 20, Eq. (5) cannot be used, and the example analysis given above is not directly applicable. Thus the example is limited to systems with small particles and low binary diffusion coefficients.

Evaluation of the Average Transport Coefficient and Bubble Size. A constant bubble size is used when evaluating the properties of the fluidized bed, and since bubbles in real beds vary in size, it is important to ask what bubble size should be used. Fryer and Potter (1972), using the model of Davidson and Harrison, reported that a bubble size found at about 0.4h could be used as the single bubble size in that model. Earlier in this paper, the bubble size found at 0.5h was used arbitrarily in calculating the conversion in an ammonia oxidation system using the K-L model.

The average bubble size $\bar{d}_b$ in a bed can be found using Eq. (8):

$$\bar{d}_b - d_{bo} = (d_{bm} - d_{bo}) \int_0^h (1-e^{-0.3L/D})dL \Big/ \int_0^h dL \quad (43)$$

Integrating:

$$(\bar{d}_b - d_{bo})/(d_{bm} - d_{bo}) = 1 - [(1-e^{-0.3h/D})/(0.3h/D)]$$

$$= 1 - (1-e^{-\beta})\beta \quad (44)$$

At midpoint, $(\hat{d}_b - d_{bo})/(d_{bm} - d_{bo}) = 1-e^{-\beta/2}$, and therefore

$$(\bar{d}_b - d_{bo})/(\hat{d}_b - d_{bo}) = [1 - (1-e^{-\beta})/\beta]/(1-e^{-\beta/2}) \quad (45)$$

A plot of the ratio of the mean bubble size to the bubble size evaluated at the midpoint in the column is shown in Fig. 3 as a function of h/D. The mean bubble size is at least 90% of the bubble size evaluated at h/2 for almost all the height-to-diameter ratios of practical interest.

Evaluation of the Transport Coefficient

We now wish to determine the difference between the average exchange coefficient evaluated at the midpoint in the columns. The dependence of the transport coefficient between the bubble and the cloud on the bubble diameter,

$$K_{bc} = \left[4.5u_{mf} + 5.85 \left(\frac{D_{AB}\ g}{d_b}\right)^{1/4}\right] \frac{1}{d_b} \tag{46}$$

will be approximated as

$$K_{bc} \cong \frac{A_1}{d_b} \tag{47}$$

owing to the weak dependence of K_{bc} and d_b in the second term in Eq. (46). From the Mori and Wen correlation

$$d_b = d_{bm} - (d_{bm}-d_{bo})e^{-\beta x} \tag{48}$$

where x = L/h. The local transport coefficient takes the form

$$K_{bc} = \frac{A_1}{d_{bm}-(d_{bm}-d_{bo})e^{-\beta x}} = \frac{A_2}{1-(1-\frac{d_{bo}}{d_{bm}})e^{-\beta x}} \tag{49}$$

At the midpoint in the column, L = h/2, x = 1/2

$$\hat{K}_{bc} = \frac{A_2}{1-(1-\frac{d_{bo}}{d_{bm}})e^{-\beta x}} \cong \frac{A_2}{1-(1-\frac{d_{bo}}{d_{bm}})e^{-\beta/2}} \tag{50}$$

The average transport coefficient

$$\overline{K}_{bc} = \frac{\int_0^1 K_{bc}dx}{\int_0^1 dx} \tag{51}$$

$$\overline{K}_{bc} = A_2 \int_0^1 \frac{dx}{1-(1-\frac{d_{bo}}{d_{bm}})e^{-\beta x}} \tag{52}$$

$$\overline{K}_{bc} = A_2\left(1 + 1/\beta \ln\left[\frac{d_{bm}}{d_{bo}} - \left(\frac{d_{bm}}{d_{bo}} - 1\right)e^{-\beta}\right]\right) \tag{53}$$

Taking the ration of Eq. (50) to Eq. (50) and letting r be the ration of maximum to minimum bubble diameter, $r = \frac{d_{bm}}{d_{bo}}$,

$$\frac{\overline{K}_{bc}}{\hat{K}_{bc}} = \left(1+1/\beta \ln[r-(r-1)e^{-\beta}]\right) \quad [1-(1-1/r)e^{-\beta/2}] \tag{54}$$

as $\beta \to \infty$

$$\frac{\overline{K}_{bc}}{\hat{K}_{bc}} \cong 1.0 \tag{55}$$

A plot of the ratio of the transport coefficients is shown in Figure 4 as a function of β for various values of the parameter r. For the ammonia oxication discussed earlier,

$$r = \frac{d_{bm}}{d_{bo}} = 55 \tag{56}$$

For large values of r

$$\frac{\overline{K}_{bc}}{\hat{K}_{bc}} \cong \left(1 + 1/\beta \ \ln[r(1-e^{-\beta})]\right) \ (1-e^{-\beta/2}) \tag{57}$$

and for larger values of β

$$\frac{\overline{K}_{bc}}{\hat{K}_{bc}} \cong 1 + 1/\beta \ \ln r \tag{58}$$

One notes the greatest disparity between the two transport coefficients for large ratios of the maximum to minimum bubble diameter and for columns with h/D ratios in the range of 2 to 20 ($.6<\beta<6$).

The exchange coefficient between the cloud and the emulsion is

$$K_{ce} = 6.78\left(\frac{\varepsilon_{mf}D_{AB}u_o}{d_b^3}\right)^{1/2}$$

We have shown that one can make the approximation

$$u_b \cong u_{br}$$

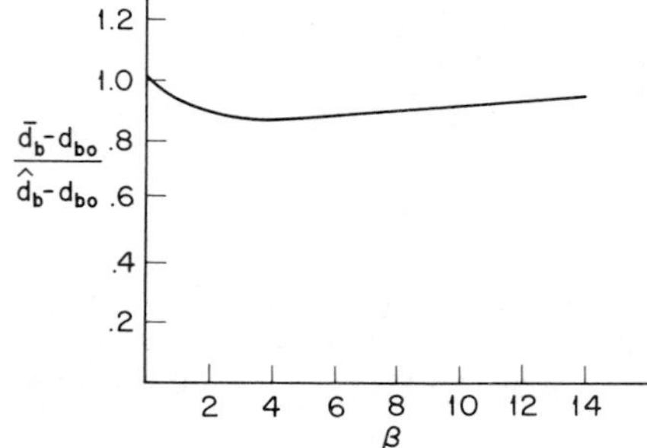

Figure 3. Ratio of the average bubble diameter to the bubble diameter evaluated at h/2 as a function of the aspect ratio, $\beta = .3h/D$

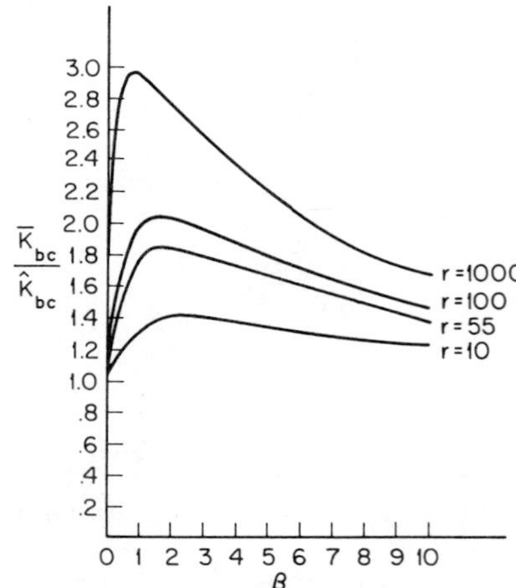

Figure 4. Ratio of the average bubble-cloud transport coefficient to the transport coefficient evaluated at h/2 as a function of the aspect ratio for various ratios the maximum to minimum bubble diameter, r

consequently

$$u_b \cong d_b^{1/2}$$

then

$$K_{ce} = A_3 \left(\frac{d_b^{1/2}}{d_b^{3}} \right)^{1/2} = A_3 \frac{1}{d_b^{5/4}} \tag{59}$$

$$K_{ce} = \frac{A_3}{d_b^{1/4}} \left(\frac{1}{d_b} \right) \tag{60}$$

Since the 1/4 power is a relatively weak functional dependence, we will consider the term

$$A_4 = \frac{A_3}{d_b^{1/4}} \tag{61}$$

to be a constant, in which case the averaging technique for

$$K_{ce} = \frac{A_4}{d_b} \tag{62}$$

gives the same result as obtained for K_{bc}, i.e.

$$\frac{\overline{K}_{ce}}{\hat{K}_{ce}} = \left(1+1/\beta \ln[r-(r-1)e^{-\beta}] \right) [1-(1-1/r)e^{-\beta/2}] \tag{63}$$

for the ammonia oxidation, r = 55 and

$$\beta = \frac{(.3)(63.2)}{(11.4)} = 1.66$$

then

$$\frac{\overline{K}_{ce}}{\hat{K}_{ce}} = 2.1$$

The average transport coefficients, K_{bc} and K_{ce} are twice the coefficient used in the L-K model when evaluated at h/2. The corresponding values of K_r and conversion are

$$K_r = 2.33$$

$$X = 0.204$$

Owing to the fact that the ammonia oxidation is mostly reaction limited, the correction factor for using the averaging technique is slight. However, in cases of rapid reaction, the two techniques $\hat{K}_{bc}$ will give significantly different results.

Summary

In this paper we have shown that the Kunii-Levenspiel model can be used to accurately predict the results of Massimilla and Johnstone. In addition, we have used the K-L model to predict the changes in conversion with parameter variation under the limiting conditions of reaction control and transport control. Finally, we have shown that significant differences in the averaging techniques occur for height to diameter ratios in the range of 2 to 20.

A	cross-sectional area of column, cm^2
	collection of terms in Eq. (12), cm/s
C	concentration, gmoles/cm^3
d	diameter, cm
D	diameter of column or bed, cm
	molecular diffusivity, cm^2/s
g	gravitational constant, cm/s^2
h	height of expanded bed, cm
k	reaction rate constant, s^{-1}
K	overall dimensionless reaction rate constant
	exchange coefficient between phases, s^{-1}
L	distance up the bed from distributor plate, cm
r	reaction rate, gmoles/(cm^3) (s)
Re	Reynolds number, dimensionless
u	axial velocity, cm/s
	superficial axial velocity, cm/s
W	mass of catalyst, g
x	dimensionless distance from distributor plate, L/h
X	fractional conversion, dimensionless

Greek

α	volume of wake per volume of bubble, dimensionless
β	dimensionless collection of terms in Eq. (44), 0.3h/D
γ	volume of catalyst in a particular phase per volume of bubble, dimensionless
δ	fraction of total bed in bubble phase (not including wakes), dimensionless
ε	void fraction of bed, dimensionless

η group of terms, $g(\rho_p - \rho_g)$, $g/(cm^2)(s^2)$
μ viscosity of gas, poise
ρ density, g/cm^3
ψ sphericity of particle, dimensionless

Subscripts

A of substance A
AB of substance A through substance B
b of bubbles or of bubble phase
bc between bubble phase and cloud phase
bm of bubbles at the maximum point
bo of bubbles at the distributor plate
br of a bubble in isolation from other bubbles
c of the cloud phase
cat of the catalyst
ce between cloud phase and emulsion phase
e of emulsion phase
g of gas
mf at minimum fluidization conditions
o at distributor plate
p of solid particle
R referring to reaction rate
s referring to slugging conditions

Superscripts

^ evaluated at midpoint in column
– average value

Literature Cited

1. Broadhurst, T. E.; Becker, H. A. AIChE J. 1975, 21, 238-247.
2. Bukur, D. B.; Wittman, C. V.; Amundson, N. R. Chem. Eng. Sci. 1974, 29, 1173-1192.
3. Chavarie, C.; Grace, J. R. Ind. Eng. Chem. Fundam. 1975, 14, 79-86.
4. Drew, T. B.; Dunkle, H. H.; Genereaux, R. P. in "Chemical Engineers' Handbook," 3rd ed. (J. H. Perry, Ed.), McGraw-Hill, New York, 1950, p. 394.
5. Fryer, C.; Potter, O. E. Powder Technol. 1972, 6, 317-322.
6. Grace, J. F. AIChE Symp. Ser. 1974, 70, No. 141, 21-6.
7. Kunii, D.; Levenspiel, O. "Fluidization Engineering," Wiley, New York, 1969.
8. Leva, M. "Fluidization," McGraw-Hill, New York, 1959, p. 21.
9. Massimilla, L.; Johnstone, H. F. Chem. Eng. Sci. 1961, 16, 105-112.
10. Mori, S.; Wen, C. Y. "Estimation of Bubble Diameter in Gaseous Fluidized Beds," AIChE J. 1975, 21, 190-115.
11. Orcutt, J. C.; Davidson, J. F.; Pigford, R. L. Chem. Eng. Progr. Symp. Ser. 1962, 58, No. 38, 1-15.
12. Potter, O. E. Catal. Rev.-Sci. Eng. 1978, 17, 155-202.

13. Reid, R. C.; Prausnitz, J. M.; Sherwood, T. K. "The Properties of Gases and Liquids," 3rd ed. McGraw-Hill, New York, 1977.
14. Weimer, A. W. M.S. Thesis, University of Colorado, Boulder, CO, 1978.
15. Yates, J. G. Chem. Eng. (London) 1975, 671-677.
16. Zenz, F. A.; Othmer, D. F. "Fluidization and Fluid-Particle Systems," Reinhold, New York, 1960, pp. 231, 234.
17. van Swaaij, W. P. M. in "Chemical Reaction Engineering Reviews-Houston" (D. Luss and V. W. Weekman, Eds.), ACS Symposium Series #72, American Chemical Society, Washington, DC, 1979, pp. 193-222.

RECEIVED June 3, 1981.

4

Simulation of a Fluidized Bed Reactor for the Production of Maleic Anhydride

J. L. JAFFRÈS, W. IAN PATTERSON, C. CHAVARIE, and C. LAGUÉRIE[1]

Ecole Polytechnique, Montréal, Canada

The simulation of a fluidized bed preheater-fluidized bed reactor system for the catalytic oxidation of benzene to maleic anhydride was attempted. The experimental apparatus and results of Kizer et al (7) together with the kinetics proposed by Quach et al (8) formed the basis for the simulation. It was determined that the rate constants and activation energies would not successfully describe the experimental results, and these parameters were estimated using a portion of the results. The rate constants and activation energies found in this manner were close to those reported by other workers for similar catalysts. The simulation using these estimated parameters gave reasonable agreement with the complete experimental results for conversion and selectivity as functions of temperature, air flow rate and bed height, except for selectivity versus bed height. An unsteady-state simulation agreed qualitatively with the limited data available.

The production of maleic anhydride by the catalytic oxidation of benzene is an established industrial process. While C_4 hydrocarbons are often suggested as a feedstock, it has been pointed out recently by De Maio (1) that they are an alternative but not necessarily a substitute. The benzene oxidation is done commercially in fixed bed reactors and, because of its exothermicity, is difficult to control in any optimal sense. The process is thus a natural candidate for a fluidized-bed reactor. The reaction has been studied in both fixed bed (2, 3) and fluidized bed (4-7) reactors. These studies, with the exception of that of Kizer et al (7) do not give sufficient information for simulation purposes. The availability of the reaction data of Kizer et al and the kinetic studies of Quach et al (8) using a similar catalyst suggested the possibility of simulating the process.

[1] Institut du génie chimique, Toulouse, France

0097-6156/81/0168-0055$05.00/0

Reaction Kinetics

The key to good reactor simulation is undoubtedly a knowledge of the reaction kinetics. The kinetics of the catalytic oxidation of benzene to maleic anhydride has been studied for different catalysts and conditions by many workers (8-13) however only Quach et al (8) examined a catalyst, FX203, of a type similiar to that employed by Kizer et al (FB203-S). Both catalysts are fabricated by Halcon Catalyst Industries, but are of different formulation.

Quach et al studied the catalyst (in the form of 0.4 cm granules) in a Carberry-type reactor. Reaction conditions were: a temperature range of 280°C to 430°C and a benzene to air feed ratio variation of 0.45 to 8.23 mol percent. Their results dictated a two-step oxidation of the form:

$$C_6H_6 + 4O_2 \rightarrow C_4H_2O_3 + CO + CO_2 + 2H_2O \qquad (1)$$

$$C_4H_2O_3 + 2O_2 \rightarrow 2CO + 2CO_2 + H_2O$$

Both reactions are exothermic and essentially irreversible. The maleic anhydride formation occurs only at the catalyst surface while its degradation takes place in the gas phase (8). It is therefore expected that the selectivity and the conversion will be equally important in the operation of fluidized bed reactor. Quach et al found that the benzene conversion rate was best described by the Langmuir-Hinshelwood relation:

$$r_B = \frac{k_B k_O p_B p_O^{1/2}}{k_O p_O^{1/2} + 4k_B p_B} \; ; \quad \begin{array}{l} k_B = 5.01 \exp(-24600/RT) \\ k_O = 3490 \exp(-64300/RT) \end{array} \qquad (3)$$

where: r_B = reaction rate in gmol · g^{-1} · h^{-1}

The form of equation (3) indicates that oxygen dissociation occurs before its adsorption on the catalyst. When the reaction has a large excess of air ($\frac{\text{benzene}}{\text{air}} \approx 1$ mol %) equation (3) can be rewritten as:

$$r_B = \frac{k_B p_B}{1 + 4\,\dfrac{k_B p_B}{k_O p_O^{1/2}}} = k_B' \, p_B \qquad (4)$$

and first order kinetic behaviour will be observed.

The gas phase degradation of the maleic anhydride is described by:

$$r_M = k_M p_M^{1/2} \; ; \; k_M = 90000 \exp(-33400/RT) \qquad (5)$$

where r_M = reaction rate in gmol · m^{-3} · h^{-1}

Pilot Reactor

The reactor used by Kizer and simulated in this work is illustrated in Figure 1. It consists of a fluidized bed preheater section feeding directly the fluidized bed reactor section. Each section was a 0.4 m high cylinder of 0.184 m diameter. The preheater contained sand and was heated by an external electrical element. The FB203S catalyst is a powder of 0.173 mm diameter particles (weight average) and has a minimum fluidization velocity, U_{mf}, of 0.021 $m \cdot s^{-1}$ at normal temperature and pressure. The reactor was cooled by ambient air blown through a jacket. The reactor distributor was made from a 50 mm thick fixed bed of 5 mm diameter pebbles supported on a perforated plate with the benzene introduced at its centre. Nickel particles (0.53 mm diameter) to a depth of 25 mm on top of a second perforated plate formed a second fixed bed and completed the distributor. The reactor was completely insulated with glass wool.

Experimental Results

The effects of the reaction temperature, T, the air flow rate F_a (reported at 20°C and 1 atm), the depth of the catalyst bed, H_{mf}, and the molar concentration of benzene, c, on the conversion, selectivity and production were reported by Kizer et al (14). The experiments were performed according to a factorial plan of 2^4 experiments within the following limits:

$430^oC \leqslant T \leqslant 490^oC$

$4 \leqslant F_a \leqslant 8\ m^3 \cdot h^{-1}$

$3 \leqslant H_{mf} \leqslant 7$ cm

$0.5 \leqslant c \leqslant 1.5$ mol percent, $\frac{C_6H_6}{air}$

The results for conversion, selectivity and production were expressed as:

$$Y_c = 74.79 + 0.29(T - 460) - 10.52(c - 1) - 3.91(F_a - 6) + 3.83(H_{mf} - 5) \quad (6)$$

$$Y_s = 51.34 - 0.22(T - 460) - 3.48(F_a - 6) - 3.76\ (H_{mf} - 5) \quad (7)$$

$$Y_p = 38.11 - 6.40(c - 1) \quad (8)$$

Reactor Model

The fluidized bed characteristics of high solids heat capacity, large interfacial heat transfer area, and good solids mixing allow the assumptions of thermal equilibrium between the solids and the gas, uniform bed temperature and negligible heat capacitance of the gas. An additional assumption required to use equation (9) is that the reactions do not change the gas volume.

The reactor and preheater each divide naturally into three types of thermal zone. These are:

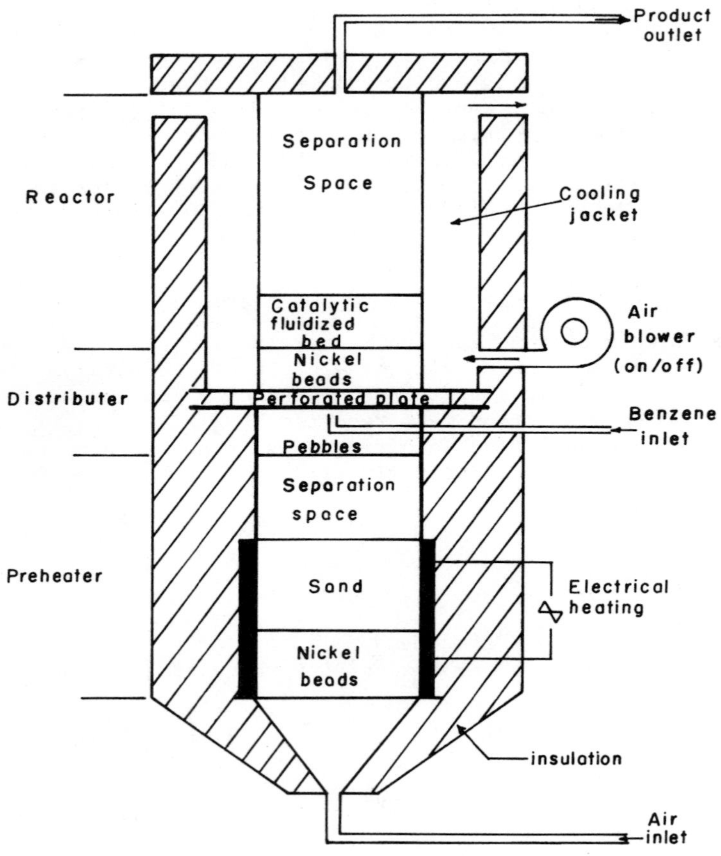

Figure 1. Experimental apparatus: preheater–reactor system

1) the fluidized beds of sand (preheater) or catalyst (reactor),
2) the fixed distributor beds and
3) the separation space above the fluidized beds.

Some of these zones have been divided into isothermal regions. This is shown in Figure 2 which shows that the preheater consists simply of the above three zones whereas the reactor distributor and separation space have been represented by three and five regions respectively. The reactor was cooled by forced air from a fan controlled in an on-off manner. Heat transfer to the cooling air was modelled as either forced or natural convection depending on whether the fan was on or off.

The reactor was simulated for both steady and transient behaviour. The steady-state model is straightforward and will not be discussed in detail. The unsteady-steady state simulation took advantage of the fact that the rate of reaction is much faster than the thermal response rate. The concentration transient response can thus be modelled as pseudo-steady state in the actual fluidized bed; this pseudo-steady state then follows the slowly changing temperature profile. A mass balance on the species, j, for each region (see Figure 2) is written as:

$$-\left(\frac{\partial c_i}{\partial t}\right) V_R = 0 = \sum_i V_R \nu_{ij} r_{ij} + Fc_{i,in} - Fc_i \qquad (9)$$

where: i refers to the reacting species
j refers to the product species.

Reaction Considerations

The reaction kinetics suggest the separation of the reactor into the fluidized-bed and separation space zones. The conversion of benzene to maleic anhydride and the degradation of the maleic anhydride both occur within the fluidized bed. Only the degradation reaction takes place in the space above the bed which has been divided into five regions, each of which is treated as a perfectly mixed, homogeneous gas-phase reactor.

It has been shown by Chavarie and Grace (15) that the decomposition of ozone in a fluidized-bed is best described by Kunii and Levenspiel's model (16) but that the Orcutt and Davidson models (17) gave the next best approximation for the overall behaviour and are easier to use and were chosen for the simulation. They suppose a uniform bubble size distribution with mass transfer accomplished by percolation and diffusion. The difference between the two models is the presumption of the type of gas flow in the emulsion phase: piston flow, PF, for one model and a perfectly mixed, PM, emulsion phase for the other model. The two models give the following expressions at the surface of the fluidized bed for first-order reaction mechanism:

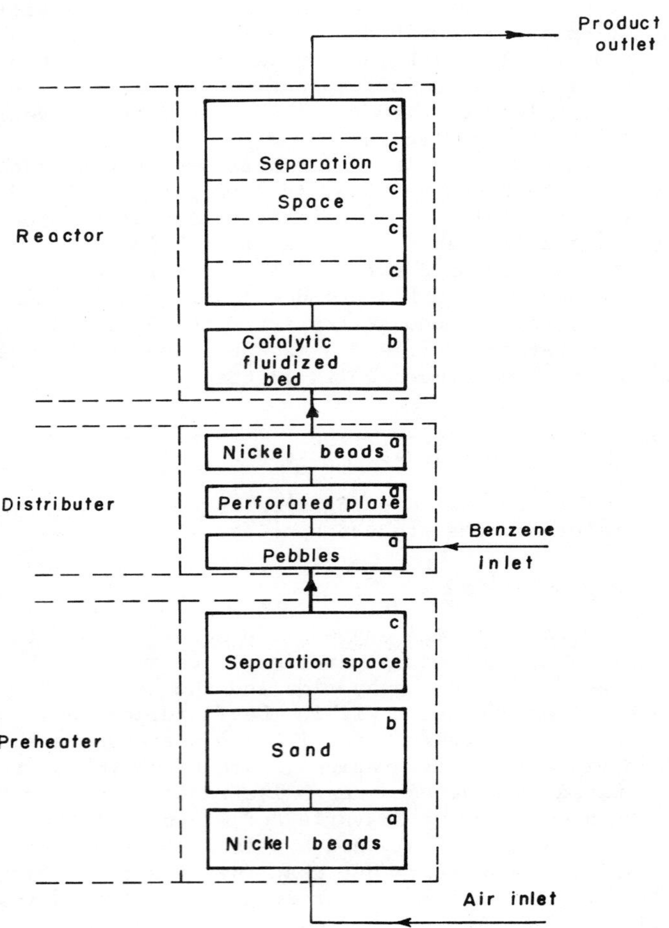

Figure 2. Physical model of the preheater–reactor system. Isothermal regions are indicated as: a, fixed beds; b, fluidized beds; c, gaseous regions.

$$\text{PM:}\quad Y_c = 1 - \left[\beta e^{-X} + \frac{(1 - \beta e^{-X})^2}{1 + K' - \beta e^{-X}} \right] \tag{10}$$

$$\text{PF:}\quad Y_c = 1 - \left[\frac{1}{m_2 - m_1} \left(m_2 e^{-m_1 H}\left(1 - \frac{m_1 HU_{mf}}{XU}\right) - m_1 e^{-m_2 H}\left(1 - \frac{m_2 HU_{mf}}{XU}\right)\right) \right] \tag{11}$$

where m_1 and m_2 are the roots of:

$$(1 - \beta)\, m^2 H^2 - (X + K')\, mH + XK' = 0 \tag{12}$$

Kinetics other than first-order require the numerical integration of the differential mass balances and the conversions cannot be expressed in simple equations.

The fluidized bed reactor model requires a description of the bubble diameter, D_b. The relationship of Mori and Wen (18) was chosen using the D_{bo} of a porous plate distributer:

$$\frac{D_{bm} - D_b}{D_{bm} - D_{bo}} = \exp\left(\frac{-0.3H}{D_R}\right) \tag{13}$$

Equation (13) was checked using the expression of Yacono (19) which was obtained from a distributor configuration similar to that employed by Kizer. Values from the two relationships were compared at bed mid-height, H/2, for typical reaction conditions and differed by 3%.

Reactor Simulation: Thermal Aspects

The energy balances on the different zones and regions of the preheater-reactor system yield the following types of terms:

I. heat introduced by convection from the zone ($\alpha - 1$) to the zone α, ΔQ_c;

$$\Delta Q_c = F_\alpha \rho_\alpha \left[\sum_i \tilde{H}_{i\alpha} c_{i\alpha} \right] - F_{(\alpha-1)} \rho_{(\alpha-1)} \left[\sum_i \tilde{H}_{i(\alpha-1)} c_{i(\alpha-1)} \right]$$

II. heat lost to the surroundings, ΔQ_ℓ;

$$\Delta Q_\ell = ha\Delta T$$

III. heat introduced by the chemical reactions of species i producing j, ΔQ_R;

$$\Delta Q_R = \sum_i \nu_{ij} r_{ij} \Delta H_{ij} V_R$$

IV. accumulation;

$$\frac{\partial Q}{\partial t} = \frac{\partial}{\partial t}\left[(W_s C_s + V_R \sum_i C_i c_i) T_\alpha \right]$$

which comprise the thermal balance:

$$- \frac{\partial Q}{\partial t} = \Delta Q_c + \Delta Q_\ell + \Delta Q_R \tag{14}$$

The simplifying assumption that the properties of the reaction mixture are those of air is justified by the maximum benzene concentration of 1.5 mol percent. It has also been assumed that the gas volume is unchanged by the reactions. Heat transfer to the walls from the distributer was evaluated by Froment's expression for fixed beds (20):

$$\frac{hD_R}{\lambda_f} = 0.813 \left[\frac{D_p G}{\mu_f}\right]^{0.9} \exp(-6D_p/D_R) \tag{15}$$

while that of Wen and Leva was used for the fluidized bed (21):

$$\frac{hD_p}{\lambda_f} = 0.16 \left[\frac{C_f \mu_f}{\lambda_f}\right]^{0.4} \left[\frac{D_p \rho_f U}{\mu_f}\right]^{0.76} \left[\frac{\rho_p C_p}{\rho_f C_f}\right]^{0.4} \left[\frac{U^2}{gD_p}\right]^{-0.2} \left[\frac{\eta}{\xi}\right]^{0.36} \tag{16}$$

The relationship of Pohlhausen was used for the heat transfer in the separation space (22):

$$Nu = \frac{RePrD_R}{4H_S} \ln \left[\frac{1}{1 - \dfrac{2.6}{Pr^{0.167}\left[\dfrac{RePrD_R}{H_S}\right]^{0.5}}} \right] \tag{17}$$

The relationship of Mac Adams was used to estimate the heat transfer due to natural convection (23):

$$h = 1.42 \left[\frac{\Delta T}{H_R}\right]^{0.25} \tag{18}$$

Activating the cooling blower causes air to enter the jacket tangentially to the wall of the reactor and is assumed to follow a helical path to the exit. The heat transfer coefficient was calculated from Perry (24):

$$h = h_{av} \left[1 + 3.5 \frac{D_{in}}{D_m} \right] \tag{19}$$

The thermal simulation was verified by choosing a benzene concentration of zero (no reaction) and natural convection cooling only. An ambient temperature of 20°C was assumed and, to minimise calculation time, the accumulation terms in the separation regions were neglected. For a 1.2 kW power input, the model predicted a steady-state catalyst temperature of 473°C which was reached about seven hours after heating was begun. A temperature loss of 42°C between the pebble benzene mixer and the catalyst was predicted while the difference between the catalyst and the fluidized bed preheater was 57°C. This loss was attributed to the increased heat transfer through the flanges used to attach the preheater to the reactor. The simulation results agreed to within 10% with the observed behaviour of the apparatus and are presented in Figure 3.

Reactor Simulation: Steady State

The results of the thermal simulation were sufficiently encouraging for us to proceed to the reactor simulation for a number of steady-state operating conditions, but neglecting the maleic anhydride degradation in the fluidized bed. Both the simplified kinetic expression (equation (4)) and the more exact equation (3) were used and the results are shown in Table I as Case 1 and Case 2 respectively.

TABLE I
PREDICTED CONVERSIONS: ORCUTT-DAVIDSON PM MODEL

T (^{o}C)	X	βe^{-X}	Conversion given by Kizer's model	Case 1		Case 2	
				K'	Y_c	K'	Y_c
430	2.10	0.113	66%	0.364	27%	0.327	25%
460	1.91	0.138	76%	0.432	30%	0.398	29%
490	1.73	0.165	83%	0.506	33%	0.475	32%

Operating conditions: $c = 1\%$, $H_{mf} = 5$ cm, $F_a = 6\ m^3h^{-1}$,
$D_{bo} = 0.9$ cm

It is obvious that the simulation predicts conversions for below those obtained by Kizer and this cannot be due solely to the neglect of the maleic anhydride degradation. There may be several possible causes for the low predicted values: the Orcutt-Davidson PM model may not be sufficiently accurate or the bubble size estimate may be incorrect. Alternatively, neither equation (3) nor (4) correctly describe the reactor kinetics. The number of possibilities may be reduced by considering Figure 4 which plots conversion versus the non-dimensional reaction rate constant, K', with βe^{-X} as a parameter. Two possible zones of operation are shown in the figure, zones A and B. Zone A is bounded by the Orcutt-Davidson PF model and the values of βe^{-X} from Table 1 together with Quach's kinetics allowing for a 10% error in the kinetic parameters. Zone B is delineated by the PF model, the maximum value of βe^{-X} from Table 1 and the values of conversion obtained by Kizer. Evidently an increase in K' is required to allow the two regions to overlap and furthermore, for the range of βe^{-X} reported in Table 1, both kinetics and bed hydrodynamics (bubble diameter) play a significant role in determining reactor conversion.

It was at this point that the simulation, per se, was

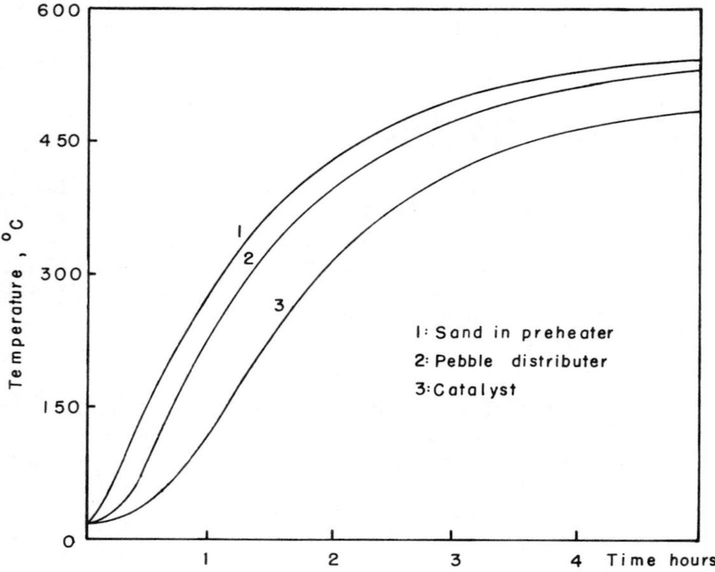

Figure 3. Thermal unsteady response of the apparatus during reactor start-up: 1, sand in preheater; 2, pebble distributer; 3, catalyst

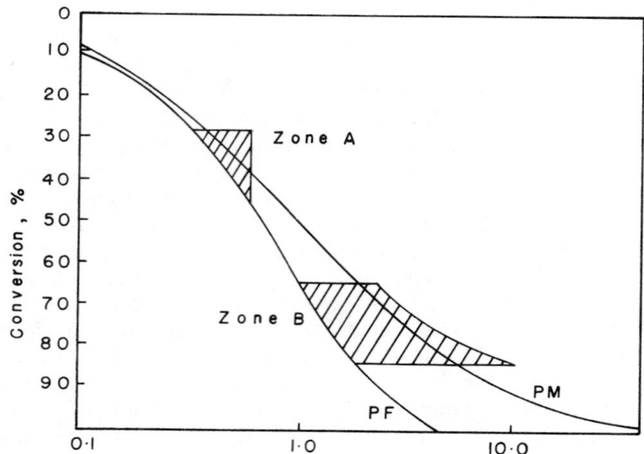

Figure 4. Conversion vs. nondimensional reaction rate constant, K′. *The two limiting cases of one phase (Orcutt–Davidson) PM and PF models are the solid lines. Zone A is the limit of operation allowing for a 10% error in the kinetic parameters of Quach et al. Zone B is the experimental limit of operation.*

abandoned. The previously cited results of other workers gave evidence that the form of the kinetic expression of Quach et al was probably adequate to simulate Kizer's reactor. Thus, we undertook to force the conversions obtained from the "simulation" to coincide with those obtained experimentally by Kizer for a number of operating conditions. The Marquardt (25) algorithm was chosen for this non-linear least-squares minimization problem in which the rate constants, activation energies and initial bubble diameter were the variables manipulated to obtain the minimal deviations for three combinations of reactor regime and kinetic expressions (cases A, B and C) as shown in Table II. It is seen that, for a given set of kinetic parameters the value of D_{bo} is almost independent of the gas flow rate and the assumption of first-order kinetics (equation (4)) gives a conversion that is independent of the benzene concentration in the feed. This latter feature significantly reduced the computation time and was retained for the selectivity and transient behaviour calculations. This was justified by the agreement of the conversions over the range of operating conditions.

The simulation could now be advanced to include the maleic anhydride degradation. This gas phase reaction takes place only in the bubble phase, the interstitial gas and the separation space. The interstitial gas and the bubbles account for about 15% of the total free volume of the reactor and therefore cannot be neglected. Moreover the degradation kinetics depend on a fractional power of the maleic anhydride concentration (equation (5)); hence the fluidized bed cannot be integrated analytically to yield a simple relationship. However, it has been shown by Grace (26) that for a fast reaction the major part of conversion occurs in the first few millimeters close to the distributer. The maleic anhydride concentration in the bed is thus very nearly constant and can be estimated from the conversion since the degradation reaction is relatively slow. This permits the fluidized bed to be modelled as a bipartite reactor as shown in Figure 5, and avoids the computer-time consuming subdivision of the bed into regions.

Despite this simplistic treatment the simulation has become quite complex and yielded selectivity and production values that differed significantly from those obtained by Kizer. Again, it was apparent that the kinetic parameters for equation (5) needed adjustment to reconcile the differences. This was done by a simple trial and error method.

Discussion and Conclusions: Kinetics

It was possible to determine a set of kinetic relations which gave the best possible simulation of the reported results. These relations are:

TABLE II
RESULTS OF OPTIMIZATION OF MODEL PARAMETERS

	Kizer's data		Y_c, %	75	66	83	83	67	67	82	80	69	
	Operating conditions		c(mol %) H_{mf} $F_a(m^3h^{-1})$ T(°C)	*1 5 6 460	1 5 6 430	1 5 6 490	1 5 4 460	1 5 8 460	1 3 6 460	1 7 6 460	0.5 5 6 460	1.5 5 6 460	Optimized parameters
Case A	Orcutt-Davidson PM	Eq. 4	Y_c	75	66	82	82	68	65	80	75	75	f_B =37500 E_B =66800 D_{bo}=0.043
			D_b(cm)	1.22	1.18	1.25	0.93	1.49	0.82	1.59	1.22	1.22	
Case B	Orcutt-Davidson PF	Eq. 4	Y_c	75	67	80	87	65	65	80	75	75	f_B =10260 E_B =60000 D_{bo}=0.420
			D_b(cm)	1.53	1.49	1.56	1.25	1.79	1.13	1.89	1.53	1.53	
Case C		Eq. 3	Y_c	75	66	80	87	65	62	81	77	72	f_B =11900 E_B =58100 f_O =2900 E_O =65900 D_{bo}=0.643
			D_b(cm)	1.71	1.68	1.75	1.44	1.97	1.32	2.08	1.71	1.71	

Kinetics: Eq. 3 $$r_B = \frac{k_B k_O p_B p_O^{1/2}}{k_O p_O^{1/2} + 4k_B p_B} \; ; \; k_B = f_B\, e^{-E_B/RT} \qquad k_O = f_O\, e^{-E_O/RT}$$

Eq. 4 $$r_B = k'_B p_B$$

* These conditions are the central point of the factorial plan and are counted four times for the purposes of optimization.

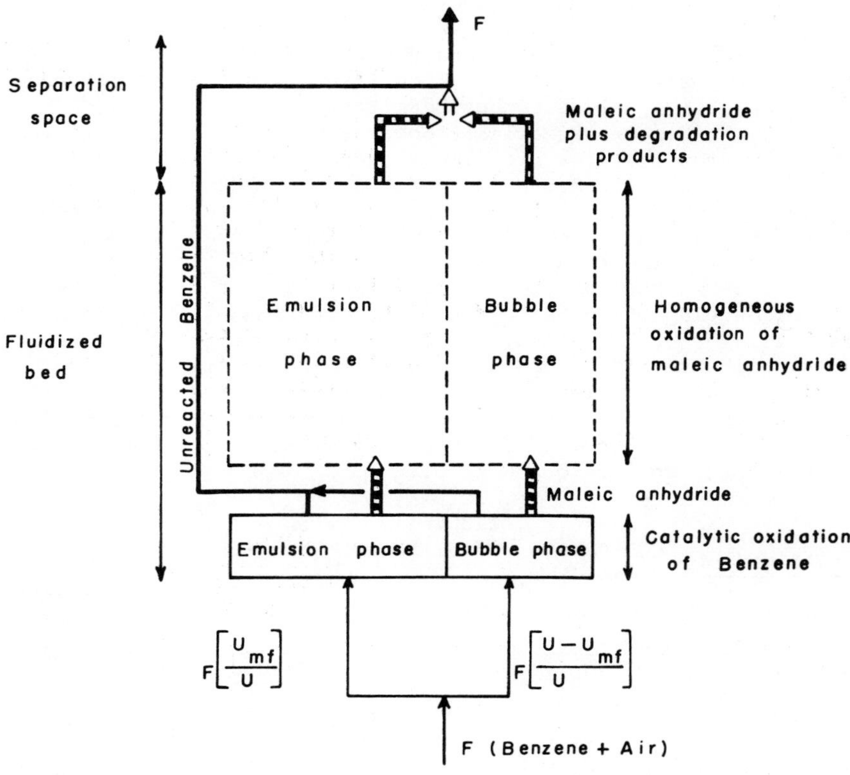

Figure 5. Model of the fluidized bed. Benzene flow is shown by the heavy solid line and maleic anhydride is represented by the heavy dashed line.

$$r_B = \frac{k_B k_O p_B p_O^{1/2}}{k_O p_O^{1/2} + 4k_B p_B}, \quad \text{where } k_B = 11900\ e^{-58100/RT},\ k_O = 2900\ e^{-65900/RT} \tag{20}$$

for the catalytic oxidation, and

$$r_M = k_M p_M^{1/2}, \text{ where } k_M = 237000\ e^{-30000/RT} \tag{21}$$

Implicit in these equations are the successive oxidations of benzene and maleic anhydride. The direct oxidation of benzene to water and carbon oxides is not permitted.

The optimization results reported in table II indicate that the activation energies are almost independent of the model chosen to represent the fluidized bed reactor. Furthermore, the activation energy obtained in this manner agree with those reported by Holsen, Steger and Germain et al while those given by Quach are much smaller. The data are summarized in table III below. Moreover, it is known from the catalyst fabricator that fixed bed reactors having an inlet benzene concentration of 1.5% and a residence time of 0.72 s. give conversions on the order of 93 to 95%. The kinetics required for this result coincide with the kinetics obtained from the numerical experimentation. Finally, we note that the energy of activation for the homogeneous decomposition of maleic anhydride obtained from the optimization is in good agreement with the work of Quach et al.

TABLE III
COMPARISON OF ACTIVATION ENERGIES

Worker	Catalyst	Temperature range ^oC	Activation energy kJ/mole
Holsen	V_2O_5/Al_2O_3	325-450	81-82
Steger	Ag_2O, V_2O_5, MoO_3, Al_2O_3/SiC	450-530	63
Germain et al	V_2O_5/MoO_3	380-500	92-42
Quach et al	V_2O_5/SiO_2	280-430	24
Our numerical optimization	V_2O_5/SiO_2	430-490	60-67

Discussion and Conclusion: Fluidized Bed Model

The optimized values given in table II include the values of the mean bubble diameter. These values are consistently smaller than those calculated from the Mori and Wen equation. For example, at the central point of the factorial plan, a value of D_b = 2.1 cm is predicted by Mori and Wen's equation while the "optimized" values for D_b vary between 1.22 and 1.71 cm depending on the simulation case.

This discrepancy is not entirely unexpected since the bubble diameters identified from the fluidized bed models are apparent or effective values intimately linked to the mass transfer mechanism of the model. The smaller bubble size values obtained by our procedure may simply mean that the actual mass transfer is larger than that suggested by the Orcutt-Davidson models. This is compatible with the fast reaction assumption that implies a disproportionately high conversion close to the distributer and a much higher mass transfer rate in this zone. Calculations of conversion and selectivity have good general agreement with Kizer's results as shown in Figure 6. An exception is the selectivity-bed height relationship. Our calculations show selectivity to be insensitive to bed height, but Kizer found a strong inverse relation between selectivity and H_{mf}. Kizer explains this by proposing a direct oxidation of benzene to water and carbon oxides which is in competition with the oxidation to maleic anhydride. We note that the heterogeneous depletion of maleic anhydride may also explain the above behaviour.

Kizer et al (14) claimed that the combined effects of bed height and flow rate could be replaced by the residence time. This implies that simple fixed bed models could be used to adequately describe this reactor. Table II and Figure 4 shows that this could be the case for the Orcutt-Davidson PM model, however the model demands the unrealistic value of D_{bo} = 0.043 cm (case A). The PF model requires operation away from the limiting conversions and is thus in conflict with Kizer's claim, although more realistic values of D_{bo} are estimated. It seems probable that the reactor operation is somewhere between that of a single-phase perfectly mixed reactor and plug flow in the same reactor. It is precisely in this region that both bed hydrodynamics and kinetics are important. Thus, it is not useful to further analyse our results without possessing independant knowledge of the hydrodynamic or kinetic parameters.

A number of points have become apparent as a result of our efforts to simulate the fluidized-bed reactor-preheater system studied by Kizer. Two of the most important are: it is imperative to have good kinetic data for the reaction(s) that occur. It has been demonstrated that the interpretation of the results is profoundly affected by relatively small changes in the kinetics. The second important point is the recognition that there are regions of operation where both the reaction kinetics and the bed hydrodynamics influence the overall performance of the reactor. The coupling of kinetic and hydrodynamic effects is strong such that both must be known to properly describe the reactor behaviour.

We note that this model is not suited to process control purposes. The computational resources and time required are simply too great to allow real-time control algorithms to use this

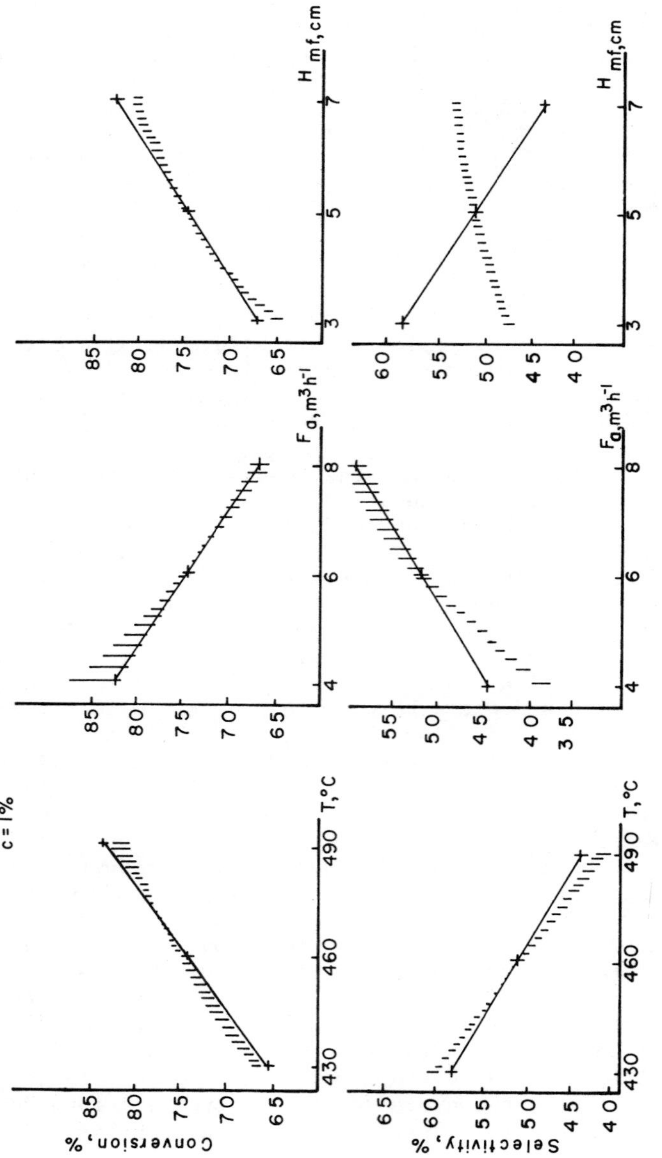

Figure 6. Conversion and selectivity vs. catalyst temperature, air flowrate, and bed height. The results of Kizer et al. are the solid lines and our calculations are shown as the hatched area. Operating conditions are: $F_a = 6 m^3h^{-1}$, $H_{mf} = 5$ *cm,* T = 460°C, *benzene concentration,* c = 1 *mol percent except when the variable appears on the abscissa.*

model in spite of the many simplifying assumptions made to reduce the computer load.

Legend of Symbols

A	- cross-sectional area of reactor, m^2
a	- heat-transfer area, m^2
C	- specific heat, cal · g^{-1}
c	- concentration, mol percent
D	- diameter, m
E	- activation energy, J · mol^{-1}
F, f	- volumetric flow rate, $m^3 \cdot h^{-1}$
G	- mass flow rate, g · h^{-1}
h	- heat transfer coefficient, cal · $m^{-2} \cdot s^{-1} \cdot {}^oC^{-1}$
$\tilde{H}$	- enthalpy, cal · g^{-1}
H	- height, m
ΔH	- heat of reaction, cal · mol^{-1}
K'	= $(k'_B RTW_s)/(AU)$, dimensionless reaction rate
k, k'	- reaction rate constant, sec^{-1}
Pr	- Prandtl number, dimensionless
p	- (partial) pressure, N · m^{-2}
Q	- heat, cal
Re	- Reynolds' number, dimensionless
R	- gas constant
r	- reaction rate, mol · h^{-1}
T	- temperature, ^oC
U	- velocity, m · sec^{-1}
V	- volume, m^3
w	- mass, g
X	= $(xH)/(U_b V_b)$, number of transfer units, equations (10) and (11)
x	- overall rate of exchange between bubble and dense phase
Y_c, Y_p, Y_s	- reactor conversion, production and selectivity
β	= 1 - (U_{mf}/U), equations (10) and (11), dimensionless
η	- parameter of equation (16)
λ	- thermal conductivity, cal · $sec^{-1} \cdot m^{-1} \cdot {}^oC^{-1}$
μ	- viscosity, Pa · sec^{-1}
ν	- stoichiometric coefficient, dimensionless
ξ	- parameter of equation (16)
ρ	- density, g · cm^{-3}

Subscripts

a	- air (at NTP; 20°C and 1 atm)
B	- benzene
b, bo, bm	- bubble, initial, mean
c	- convection
f	- fluid
in	- inlet

i, j	- summation indices
ℓ	- lost
M	- maleic anhydride
m	- mean
mf	- minimal fluidization
p	- particle
R	- reactor, reaction
S	- separation
s	- solid

Acknowledgements

The authors gratefully acknowledge the research grants provided by the Province of Quebec (FCAC), the Natural Sciences and Engineering Research Council of Canada and the Institut du génie chimique de Toulouse.

Literature Cited

1. De Maio, D.A., "Will Butane Replace Benzene as a Feedstock for Maleic Anhydride", Chem. Eng., 1980, May 19, p. 104.
2. Germain, J.E., Graschka, F., Mayeux, A., "Cinétique de l'oxydation catalytique du benzène sur oxydes de vanadium-molybdène", Bull. Soc. Chim. Fr., 1965, p. 1445.
3. Vaidyanathan, K., Doraiswamy, L.K., "Controlling Mechanism in Benzene Oxidation", Chem. Eng. Sci., 1968, 23, 537.
4. Badarinarayana, M.C., Ibrahim, S.M., Kuloor, N.R., "Single Step Catalytic Vapor Phase Oxidation of Benzene", Ind. J. Techn., 1967, 5, 314.
5. Kullavinajaya, P., "Statistical Study of the Benzene Oxidation Process in a Fluidized Bed Reaction", Ph.D. thesis, Ohio State University, Columbus, 1966.
6. Ahmad, S.I., Ibrahim, S.M., Kuloor, N.R., "Kinetic Studies on the Oxidation of Maleic Anhydride", Ind. J. Techn., 1971, 9, 251.
7. Kizer, O., Laguérie, C., Angelino, H., "Experimental Study of the Catalytic Oxidation of Benzene to Maleic Anhydride in a Fluidized Bed", Chem. Eng. Journal, 1977, 14, 205.
8. Quach, T.Q.P., Rouleau, D., Chavarie, C., Laguérie, C., "Catalytic Oxidation of Benzene to Maleic Anhydride in a Continuous Stirred Tank Reactor", Can. J. Chem. Eng., 1978, 56, 72.
9. Hammar, C.G.B., "Reaction Kinetics of the Catalytic Vapor-Phase Oxidation of Benzene to Maleic Anhydride", Svensk. Kem. Tid., 1952, 64, 165.
10. Holsen, J.N., "An Investigation of the Catalytic Vapor Phase Oxidation of Benzene", Ph.D. thesis, Washington University, St.Louis, Missouri, 1954.

11. Ioffe, I.I., Lyubarski, A.G., "Kinetics of Catalytic Oxidation of Benzene to Maleic Anhydride", Kin. i Kat., 1963, 3, 261.
12. Dmuchovsky, B., Freerks, M.C., Pierron, E.D., Munch, R.H., Zienty, F.B., "Catalytic Oxidation of Benzene to Maleic Anhydride", J. Catalysis, 1965, 4, 291.
13. Steger published in Catalysis, edited by P.H. Emmet, Reinhold, New York, 7, Chap. 3, 1960, pp. 186-194.
14. Kizer, O., Chavarie, C., Laguérie, C., Cassimatis, D., "Quadratic Model of the Behaviour of a Fluidized Bed Reactor: Catalytic Oxidation of Benzene to Maleic Anhydride", Can. J. Chem. Eng., 1978, 56, 716.
15. Chavarie, C., Grace, J.R., "Performance Analysis of a Fluidized Bed Reactor", IEC Fund., 1975, 14, 75, 79, 86.
16. Kunii, D., Levenspiel, O., "Bubbling Bed Model for Kinetic Processes in Fluidized Bed-Gas-Solid Mass and Heat Transfer and Catalytic Reactions", IEC Proc. Des. Dev., 1968, 7, 481.
17. Orcutt, J.C., Davidson, J.F., Pigford, R.L., "Reaction Time Distributions in Fluidized Catalytic Reactors", Chem. Eng. Progr. Symp. Ser., 1962, 58, 1.
18. Mori, S., Wen, C.Y., "Estimation of Bubble Diameter in Gaseous Fluidized-Beds", AIChE Journal, 1975, 11, 109.
19. Yacono, C., Angelino, H., "The Influence of Gas Distributor on Bubble Behaviour; Comparison between Ball Distributor and Porous Distributor", published in Fluidization, Cambridge University, Cambridge, 1978, p. 25.
20. Froment, G.F., "Fixed-Bed Catalytic Reactors - Current Design Status", Ind. Eng. Chem., 1967, 59, 18.
21. Wen, C.Y., Leva, M., "Fluidized-Bed Heat Transfer - a Generalized Dense Phase Correlation", AIChE Journal, 1956, 2, 482.
22. Pohlhaussen, K. published in Principles of Heat Transfer, 3rd ed. Intext Press, New York, 1976, p. 442.
23. Mac Adams, W.H., Heat Transmission 2nd ed. MacGraw Hill Book Company, New York, 1942, p. 241.
24. Perry, R.M., Chilton, C.H., Chemical Engineers Handbook, 5th ed., 1973, p. 10-15.
25. Marquardt, D.W., "An Algorithm for Least Squares Estimation of Non-Linear Parameters", Journ. Soc. Ind. Appl. Math., 1963, 11, 431.
26. Grace, J.R., De Lasa, H.I., "Reaction near the Grid in Fluidized Beds", AIChE Journal, 1978, 24, 364.

RECEIVED JUNE 30, 1981.

5

A Model for a Gas–Solid Fluidized Bed Filter

MICHAEL H. PETERS, THOMAS L. SWEENEY, and LIANG-SHIH FAN

Department of Chemical Engineering, The Ohio State University, Columbus, OH 43210

A general mathematical model for simulating particulate removal in gas-solid fluidized beds is presented. Model predictions of the fluidized bed filtration efficiencies, which include the possibility of electrical effects, are shown to compare well to the experimental results of various investigators. Because of the general formulation of the proposed model it is believed to be applicable in the design of both single and multistage fluidized bed filters.

Fluidized beds have been employed in many industrial processes such as coal combustion, gasification and liquefaction, solid residue pyrolysis, catalytic cracking and reforming, and polymer production. In addition, the possibility of using fluidized beds for fine particulate removal has recieved growing attention over recent years (1 - 12). Typically, the fluidized bed is of the gas-solid type and the particulates may be liquid or solid aerosols. Note that in this application the bed medium solids function as the collecting medium and particle removal is accomplished through particle-collector contacting.

Our approach to the problem of predicting the performance of fluidized bed filters involves logically coupling models that describe the flow behavior of the fluidized state with models that describe the mechanisms of particle collection. The collection mechanisms analysis leads to expressions for determining the collection efficiency of a single filter element. An example of a collection mechanism is inertial impaction by which a particle deviates from the gas stream lines, due to its mass, and strikes a collector. It should be noted that because particle collection mechanisms are functions of the fluid flow behavior in the vicinity of a collector, there exists an interdependency between fluidization mechanics and particle collection mechanisms.

In a previous paper, the importance of fluidization mechanics on the performance of fluidized bed filters was demonstrated (13). To accomplish this, classical methods were employed for evaluating

0097-6156/81/0168-0075$05.00/0

the single spherical collector efficiencies. In the present paper our analysis is extended by considering more realistic methods for estimating particle removal efficiencies for a single collector element.

Model Background

The model presented here for quantitatively describing the mechanics of the fluidization process is a simplified version of a more complex scheme recently proposed by Peters et al (14), and is largely based on bubble assemblage concepts (15). In brief, the bubble assemblage concept considers an aggregative fluidized bed to be divided axially into a number of compartments. Each compartment consists of a bubble, cloud, and emulsion phase. The size of each compartment, which varies throughout the fluidized bed, is based on the cloud diameter computed at a given bed height. The key features of the present analysis lie in the reduction in independencies among the relationships as well as elimination of major two phase theory assumptions (14).

Model

Figure 1 shows the present model representation of the gas-solid fluidized bed. Making a steady-state material balance on particulates over the nth compartment results in the equation

$$\overline{U}_{is} S(C_{i_{n-1}} - C_{i_n}) + F_{i(i+1)_n} V_{1_n} (C_{i+1_n} - C_{i_n}) + F_{(i-1)i_n} V_{1_n} (C_{i-1_n} - C_{i_n}) = \eta_{i_n} C_{i_n} \overline{U}_{is} \frac{3(1-\varepsilon_i) V_{i_n}}{2D_c} \tag{1}$$

Where, i = 1 for the bubble phase, i = 2 for the cloud phase, and i = 3 for the emulsion phase. Note from the term on the right-hand side of Eqn. (1) that a first order rate equation for particulate collection is assumed (10). The inlet gas corresponds to the $zero^{th}$ compartment, thus,

$$C_{1_o} = C_o$$

$$C_{2_o} = C_o \tag{2}$$

$$C_{3_o} = C_o$$

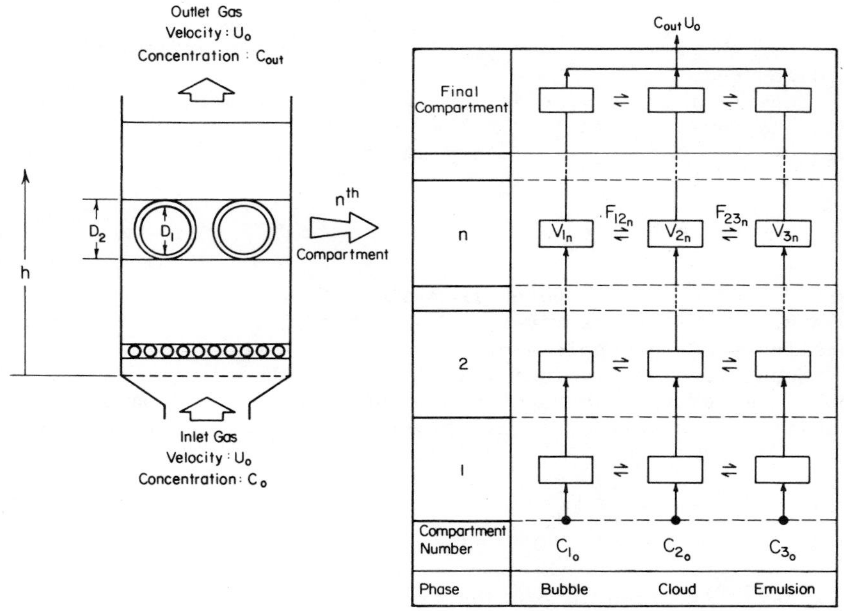

Figure 1. Schematic of the present model

Estimation of the Parameters of the Model

As presented below, the parameters in the model may be estimated in terms of a relatively small number of fundamental parameters that characterize either the bubbling phenomenon, mass conservation, or particulate collection mechanisms. For those parameters not based on average properties the subscript n has been omitted for clarity in many cases.

A. Superficial gas velocity, U_o. The superficial gas velocity can be expressed as

$$U_o = \overline{U}_{1s} + \overline{U}_{2s} + \overline{U}_{3s} \tag{3}$$

where $\overline{U}_{1s}$, $\overline{U}_{2s}$, and $\overline{U}_{3s}$ are based on average properties in the fluidized bed.

B. Superficial gas velocity in the bubble phase, $\overline{U}_{1s}$. The superficial gas velocity in the bubble phase is related to the average linear bubble phase gas velocity and the average bubble phase volume fraction by

$$\overline{U}_{1s} = \overline{U}_1 \overline{\delta}_1 \varepsilon_1 \tag{4}$$

where $\overline{U}_1$ and $\overline{\delta}_1$ are computed from the relationships given in sections E. and M., respectively. Note that Eqn. (4) represents the so-called visible bubble flow rate.

C. Superficial gas velocity in the cloud phase, $\overline{U}_{2s}$. Since a bubble and its associated cloud rise together at the same linear velocity, the superficial gas velocity in the cloud phase is given by

$$\overline{U}_{2s} = \frac{\overline{\delta}_2}{\overline{\delta}_1} \frac{\varepsilon_2}{\varepsilon_1} \overline{U}_{1s} \tag{5}$$

where $\overline{\delta}_2/\overline{\delta}_1$ is given in section F.

D. Superficial gas velocity in the emulsion phase, $\overline{U}_{3s}$. Substituting Eqns. (4) and (5) into Eqn. (3) gives the superficial gas velocity in the emulsion phase, as

$$\overline{U}_{3s} = U_o - \overline{U}_1 (\overline{\delta}_1 \varepsilon_1 + \overline{\delta}_2 \varepsilon_2) \tag{6}$$

subject to the stipulation that

$$U_o > \overline{U}_1 (\overline{\delta}_1 \varepsilon_1 + \overline{\delta}_2 \varepsilon_2) \tag{7}$$

E. Linear gas velocity in the bubble phase, U_1. The linear gas velocity in the bubble phase may be computed from the commonly accepted relationship proposed by Davidson and Harrison (16).

$$U_1 = (U_o - U_{mf}) + 0.71\sqrt{GD_1} \tag{8}$$

The average linear gas velocity in the bubble phase may be expressed as

$$\overline{U}_1 = (U_o - U_{mf}) + 0.71\sqrt{G\overline{D}_1} \tag{9}$$

F. Volume ratio of cloud to bubble phases, δ_2/δ_1. The volume ratio of the cloud phase to the bubble phase may be estimated from the model of Murray (17)

$$\frac{\delta_2}{\delta_1} = \frac{U_{mf}}{\varepsilon_{mf} U_1 - U_{mf}} \tag{10}$$

and the average volume ratio may be expressed as

$$\frac{\overline{\delta}_2}{\overline{\delta}_1} = \frac{U_{mf}}{\varepsilon_{mf} \overline{U}_1 - U_{mf}} \tag{11}$$

G. Bubble Diameter, D_1. A recent correlation by Mori and Wen (18), which considers the effects of bed diameters and distributor types, is utilized. This correlation, based on the bubble diameter data appearing in the literature prior to 1974 is

$$\frac{D_{1_m} - D_1}{D_{1_m} - D_{1_o}} = \exp\,(-0.3h/D_R) \tag{12}$$

where

$$D_{1_m} = 0.652\,[S\,(U_o - U_{mf})]^{2/5} \tag{13}$$

and

$$D_{1_o} = 0.347\left[\frac{S\,(U_o - U_{mf})}{N_D}\right]^{2/5} \tag{14}$$

(for perforated distributer plates)

$$D_{1_o} = 0.00376\,(U_o - U_{mf})^2 \tag{15}$$

(for porous distributor plates)

This correlation is valid over the following variable ranges:

$$0.5 < U_{mf} < 20 \quad , \text{cm/s}$$

$$0.006 < D_c < 0.045 \quad , \text{cm}$$

$$U_o - U_{mf} < 48 \quad , \text{cm/s}$$

$$D_R < 130 \quad , \text{cm}$$

H. Gas interchange coefficients. Gas interchange coefficients given here are based on the Murray model (12). The analysis parallels the two step transfer mechanism proposed by Kunii and Levenspiel (19) which is based on the Davidson model (16). Assuming an average bubble throughflow (20) and neglecting the film diffusional contribution between bubble and cloud phases, which is usually small compared to the bulk flow term, the gas interchange coefficients can be expressed as

$$F_{12} = 1.5 \left(\frac{U_{mf}}{D_1}\right) \tag{16}$$

and

$$F_{23} = 6.78 \left(\frac{D_G \, \varepsilon_{mf} \, U_1}{D_1^3}\right)^{1/2} \tag{17}$$

Note that these expressions have been previously given from an overall standpoint by Chavarie and Grace (21).

I. Expanded Bed Height. The height of bed expansion can be approximated as (14)

$$L = L_{mf} + \frac{Y L \,(U_o - U_{mf})}{U_o - U_{mf} + 0.71 \sqrt{G\overline{D}_1}} \tag{18}$$

where

$$\overline{D}_1 = D_{1_m} - (D_{1_m} - D_{1_o}) \exp(-0.15 \, L_{mf}/D_R), \tag{19}$$

and

$$Y = 0.76$$

J. Volume Fraction Gas in Each Phase. The volume fraction of gas in the cloud and emulsion phases is assumed to be equal to that at minimum fluidization throughout the entire bed:

$$\varepsilon_2 = \varepsilon_3 = \varepsilon_{mf} \tag{21}$$

The model assumes a value of 1.0 for the volume fraction of gas in the bubble phase.

K. Cloud diameter, D_2. The diameter of the cloud may be easily obtained by rearranging Eqn. (10) to give

$$\left(\frac{D_2}{D_1}\right)^3 = \frac{\varepsilon_{mf} U_1}{\varepsilon_{mf} U_1 - U_{mf}} \tag{22a}$$

as well as the average cloud diameter as

$$\left(\frac{\overline{D}_2}{\overline{D}_1}\right)^3 = \frac{\varepsilon_{mf} \overline{U}_1}{\varepsilon_{mf} \overline{U}_1 - U_{mf}} \tag{22b}$$

L. Number of Bubbles in a Compartment, N. With compartment height based on the diameter of the cloud, the number of bubbles can be computed from material balance considerations as well as some assumptions concerning the average solids volume fraction in the bed (14)

$$N = \frac{6\, S\, D_{2_n} (\varepsilon - \varepsilon_{mf})}{\pi\, D_{1_n}^{\;3} (1 - \varepsilon_{mf})} \tag{23}$$

where

$$1 - \varepsilon = \frac{L_{mf}}{L} (1 - \varepsilon_{mf}) \tag{24}$$

for $h \leq L_{mf}$

and

$$1 - \varepsilon = \frac{L_{mf}}{L} (1 - \varepsilon_{mf}) \left\{ \exp\left[- \left(\frac{h - L_{mf}}{L - L_{mf}}\right) \right] \right\} \tag{25}$$

for $L_{mf} \leq h$

M. Volume fraction of each phase, δ_i. The volume fraction of the bubble, cloud and emulsion phases may be computed as

$$\delta_{i_n} = V_{i_n} / SD_{2_n} \tag{26}$$

where i = 1 for the bubble phase, i = 2 for the cloud phase, i = 3 for the emulsion phase, and

$$V_{1_n} = N\,(1/6)\,\pi\,D_{1_n}^{\;3} \tag{27}$$

$$V_{2_n} = V_{1_n}\left[\frac{U_{mf}}{\varepsilon_{mf}\,U_1 - U_{mf}}\right] \tag{28}$$

$$V_{3_n} = SD_{2_n} - V_{2_n} - V_{1_n} \tag{29}$$

Note that,

$$\overline{\delta}_1 = \overline{V}_1/S\overline{D}_2 \tag{30}$$

and

$$\overline{V}_1 = \overline{N}\,(1/6)\,\pi\,\overline{D}_1^{\;3} \tag{31}$$

where $\overline{N}$ is evaluated at $h = L_{mf}/2$ and $\overline{D}_1$ is given by Eqn. (19).
N. Single Spherical Collector Efficiencies. Four collection mechanisms are considered in the present analysis: inertial impaction, interception, Brownian movement and Coulombic forces. Although in our previous analysis the electrical forces were considered to be of the induced nature (13), there is evidence that it is the Coulombic forces which dominate the electrical interactions between the particle and collector (7, 12, 22). Taking the net effect as the simple summation of each collection mechanism results in the single spherical collector efficiency equation,

$$\eta_i = \eta_{IMP} + \eta_{INT} + \eta_{BD} + \eta_E \tag{32}$$

where

$$\eta_{IMP} = -0.19133 + 1.7168\,Stk - 1.2665\,Stk^2 + 0.31860\,Stk^3 \tag{33}$$

for $\varepsilon_i \cong 0.4$ and $Stk_c \cong 0.12$

$$\eta_{INT} = 1.5 \left(\frac{1.31}{\varepsilon_i}\right)^3 N_R^2 \tag{34}$$

$$\eta_{BD} = 4 \left(\frac{1.31}{\varepsilon_i}\right) Pe^{-2/3} \tag{35}$$

$$\eta_E = 4.4\, K_c^{0.87} \tag{36}$$

for $\varepsilon_i \cong 0.4$

The particle diffusion coefficient is calculated from the Stokes-Einstein equation (24)

$$D_G = \frac{KT}{6\pi\mu r_p}\left\{1 + \frac{\lambda}{r_p}\left[1.257 + 0.4 \exp\left(-1.10\, r_p/\lambda\right)\right]\right\} \tag{37}$$

Equations (33) - (35) are taken from Tardos et al (23), and are based on a low Reynold's number analysis. Eqn. (33) is the result of a "best-fit" of the theoretically computed values taken from Figure 7 of that same work. Similarly, Eqn. (36) for the electrical deposition, is obtained from a "best-fit" of the theoretically computed values taken from Figure 3 of Tardos and Pfeffer (21). Note that if the particle and collector charges are of the same sign, the electrical deposition efficiency becomes the negative of Eqn. (36). Consistent with the flow field models used in the development of Eqns. (33) - (36), the velocity employed is an assembly averaged velocity for each phase. For the multi-phase situation that exists in the fluidized bed, this is given by the superficial or empty-tower velocity divided by the phase volume fraction,

$$\hat{U}_i = \frac{\overline{U}_{is}}{\delta_i} \tag{38}$$

Note from Eqn. (38) that since the volume fraction of each phase varies throughout the bed, so will the assembly average velocities and hence, the single collector efficiencies.

O. Volumetric average particulate concentration at the exit of the bed, C_{out}, and the overall collection efficiency, X.

The volumetric average particulate concentration at the exit of the bed is expressed by

$$C_{out} = C_1 \frac{\overline{U}_{1s}}{U_o} + C_2 \frac{\overline{U}_{2s}}{U_o} + C_3 \frac{\overline{U}_{3s}}{U_o}, \tag{39}$$

and the overall collection efficiency, in percent, is defined as

$$X = 100\left(\frac{C_o - C_{out}}{C_o}\right) \quad (40)$$

At ratios of superficial to minimum fluidization velocities greater than three to five, local flow reversal of gas in the emulsion phase can occur (14). In the present analysis the divisions of gas flow among the phases are based on average values, and thus are taken to be constant throughout the fluidized bed. Equation (7) states that only an average upward flow of gas in the emulsion phase is considered here. It is assumed that the equations describing the flow of gas in a fluidized bed are also applicable to the flow of particulates, and that the particulates contacting a collector adhere to it and are not re-entrained by the gas flow. Relative changes in particle velocities due to the motion of the collectors in the fluidized bed are neglected.

Method of Solution

Calculations of the overall collection efficiency for the fluidized bed filter begin with specification of the values of the superficial gas velocity, U_o, minimum fluidization velocity, U_{mf}, bed height at minimum fluidization, L_{mf}, void fraction at minimum fluidization, ε_{mf}, column diameter, D_R, gas viscosity, μ, collector diameter, D_c, density of particulate, ρ_p, and particulate diameter, D_p. There are no adjustable parameters in the present model. The charge acquired on both the particulates and collectors, Q_p and Q_{AC}, respectively, remain as experimentally determined input parameters in the present analysis.

Because bubble diameter is a function of the height from the distributor, and the height from the distributor is taken to the center of the bubble in question, an iterative procedure is used to determine D_1. The initial guess is taken to be the bubble diameter computed for the previous compartment. For each compartment there are three material balance equations with three unknowns, the concentrations in each phase (bubble, cloud and emulsion). The total number of equations then is three times the total number of compartments. These may be solved by a matrix reduction scheme or a trial and error procedure. The average superficial gas velocities in each phase are first determined from Eqns. (4) - (6). The computational sequence for the remaining parameters in Eqn. (1) is given in Table 1.

It is assumed here that the size of the last compartment is determined from the difference between the cummulative compartments size and the height of the expanded bed. However, for consistency, gas interchange coefficients and the linear bubble phase gas velocity are based on a hypothetical bubble diameter predicted from Eqn. (12). The computational scheme also takes into consideration the possibility of only two phases in any compartment. This can result from both cloudless and cloud overlap compartments,

Table I. Computational sequence for parametric evaluation at the nth compartment.

Sequence	Eqn. Number	Calculated Parameter
1	12	D_1
2	8	U_1
3	10	δ_2/δ_1
4	22a	D_2
5	24, 25	ε
6	23	N
7	27, 28, 29	V_1, V_2, V_3
8	26	δ_i
9	16, 17	F_{12}, F_{23}
10	38	$\hat{U}_2$, $\hat{U}_3$
11	32	η_i

typically occuring for larger minimum fluidization velocities. Figure 2 shows a typical situation that can occur along with the appropriate simplified equations. Gas interchange in a two phase compartment is taken to be solely Eqn. (16), based on the so-called invisible bubble flow rate. The unsteady-state diffusional contribution, Eqn. (17), is neglected.

Results and Discussion

The potency of the present model lies in predicting the performance of fluidized bed filters over a relatively wide range of operating conditions. Our previously reported sensitivity studies and comparisons with experimental results (13) are extended here.

Comparisons with the Experimental Results of Tardos et al (12).

Figure 3 shows a comparison of the model prediction of the overall collection efficiency as a function of superficial gas velocity versus the experimental data of Tardos et al (12). Since the charge acquired on the collectors was not reported, assumed values shown in Figure 3 were employed. It should be noted that this assumed functional dependency between Q_{AC} and U_o was not entirely arbitrary, but qualitatively suggested by experimental measurements of the electric potential in the fluidized bed (12). An important aspect of Figure 3 is both the model prediction and experimental

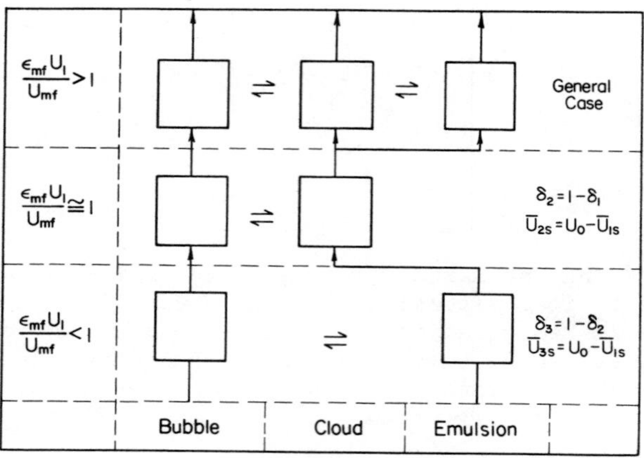

Figure 2. Compartments representation of cloudless and cloud overlap compartments

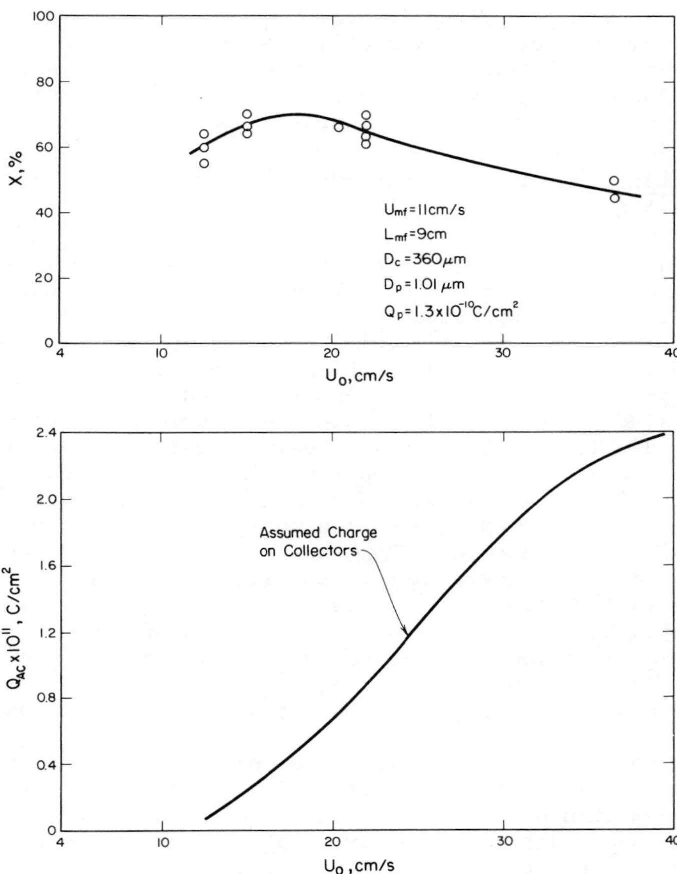

Figure 3. Comparison of (——) model prediction and (○) experimental data (12) for the overall collection efficiency as a function of superficial gas velocity: $D_p = 1.01\ \mu m$

observation of a maximum in the overall collection efficiency as a function of superficial gas velocity. Model analysis shows that this is due to the competing effects of electrostatic collection and gas by-passing. In Figure 4, the same charge distribution assumed in Figure 3 is employed for comparisons at a slightly larger particle diameter. Model analysis indicates that the higher efficiencies observed in Figure 4 over Figure 3 are due solely to the higher predicted interception efficiencies. Increases in the single collector efficiencies due to increases in the specific charge density outweigh gas by-passing effects up to a superficial gas velocity of about 18 cm/s in Figures 3 and 4.

Comparisons with the Experimental Results of Gutfinger and Tardos (11).

In addition to the effects of superficial gas velocity on the overall collection efficiency, the direct effects of particle diameter are also of importance. Figure 5 shows the present model predictions of the overall collection efficiency as a function of particle diameter compared to the experimental data of Gutfinger and Tardos (11). Since experimental care was taken to neutralize electrical effects for this system, these were not included in the model predictions. Thus, only three mechanisms were considered in Figure 5, namely, inertial impaction, interception and Brownian motion. In Figure 5 reasonable agreement is seen at small particle diameters ($< 0.3\ \mu m$) where Brownian motion is predominent, and at large particle diameters ($> 3\ \mu m$) where interception effects are controlling. In the vicinity of the minimum overall collection efficiency ($\sim 1\ \mu m$) the agreement is not as good. It is also in this region that the predicted results are very sensitive to the values of the single collector efficiencies. In Figure 5 the experimental data would indicate higher single collector efficiencies in the vicinity of the minimum than predicted by the equations employed here.

For completeness it should be noted that the minimum overall collection efficiencies in Figures 3 and 4 occur for particle diameters less than $0.5\ \mu m$. Thus, the particle diameters employed in Figures 3 and 4 are sufficiently displaced from the minimum so that the results are not considered fortuitous.

Comparisons with the Multistage Efficiencies of Patterson and Jackson (8).

For highly reactive systems in which the majority of particulate collection in the emulsion phase occurs in a relatively short distance from the distributor plate, multistage fluidized beds have been employed (8, 4). Because of the general formulation of the present model, it may be employed for determining multistage fluidized bed filtration efficiencies. This includes a variation in the characteristics of each stage such as bed depth and collector size.

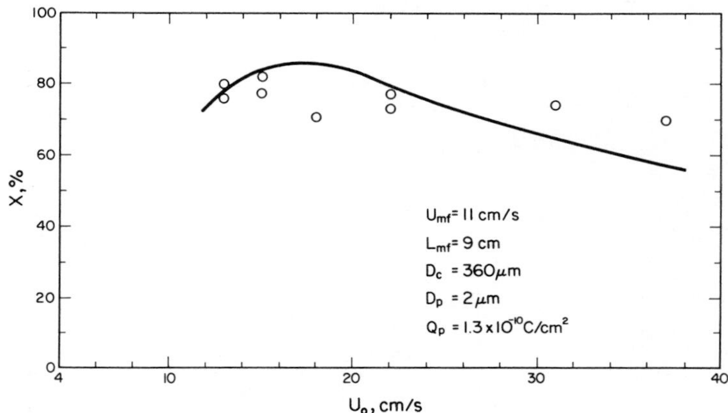

Figure 4. Comparison of (——) model prediction and (○) experimental data (12) for the overall collection efficiency as a function of superficial gas velocity: $D_p = 2$ *μm*

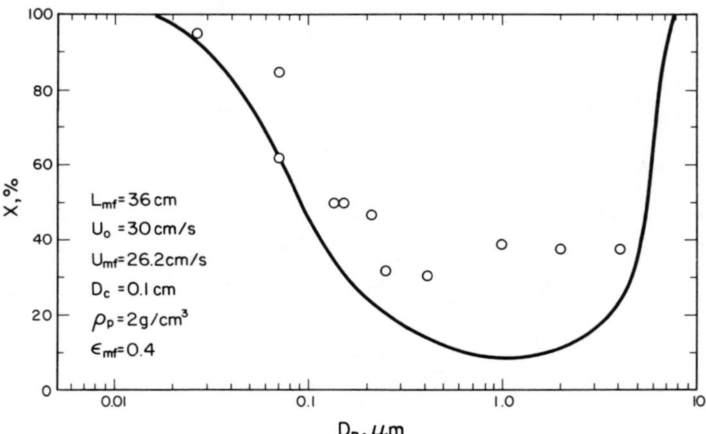

Figure 5. Comparison of (——) model prediction and (○) experimental data (11) for the overall collection efficiency as a function of particle diameter

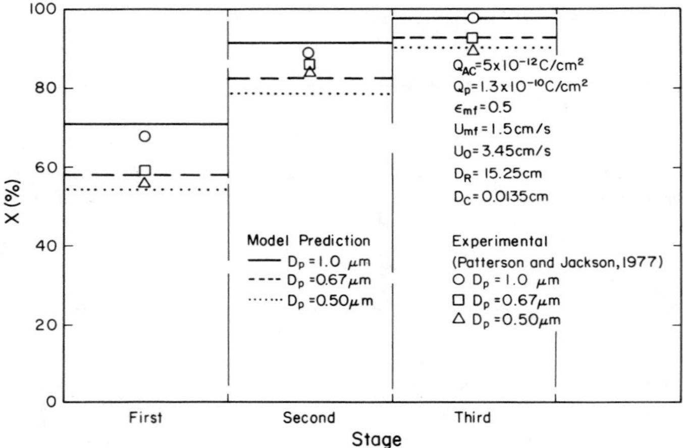

Figure 6. Comparison of the predicted multistage efficiencies and the experimental results (8). Model prediction: (——) D_p = 1.0 μm; (– – –) D_p = 0.67 μm; (· · ·) D_p = 0.50 μm. Experimental: (○) D_p = 1.0 μm; (□) D_p = 0.67 μm; (△) D_p = 0.50 μm.

Under the simplifications that each stage has identical characteristics and that the particulates are of a single size, model predictions of the single stage efficiencies may be directly used to calculate multistage efficiencies by (8)

$$X_M = 100 \left[1 - \left(1 - \frac{X_1}{100}\right)^M\right] \tag{41}$$

In Fig. 6 the present model predictions of the multistage efficiencies calculated from Eqn. (41) are shown to compare closely to the experimental data of Patterson and Jackson (8). Because of the importance of electrical effects noted for this system, (12) the Coulombic force term in Eqn. (32) was included. Values of Q_P and Q_{AC} were arbitrarily set as shown in Figure 5. It should be noted that along with pressure drop information the present model may be used for optimizing the depths of each stage in a multistage fluidized bed filter.

Conclusion

In the present paper our previous analysis of fluidized bed filtration efficiencies has been extended by considering more realistic methods for estimating the single collector efficiencies as well as more recently reported experimental results. In general the predicted values of the fluidized bed filtration efficiencies compare favorably to the experimental values. For electrically active fluidized beds, direct measurements of the particle and collector charges would be necessary to substantiate the results given here.

The present model appears to be useful in the design of fluidized bed filters. It does not address questions concerning the quality of fluidization, stickiness of the particles, solids regeneration rates and agglomeration effects. In order to optimize the fluidized bed filter these effects must be considered in conjunction with those aspects to the problem elucidated here.

Legend of Symbols

C_{i_n} = concentration of particles in n^{th} compartment in phase i, g/cm^3

C_o = inlet particle concentration, g/cm^3

C_{out} = outlet particle concentration, g/cm^3

D_c = collector diameter, cm

D_G = molecular diffusion coefficient of particulate, cm^2/s

D_P = particulate diameter, cm

D_R = fluid bed diameter, cm

D_1 = equivalent spherical bubble diameter having the same volume as that of a bubble, cm

D_2 = equivalent spherical cloud diameter, cm

$\overline{D}_1$ = average equivalent spherical bubble diameter, cm

$\overline{D}_2$ = average equivalent spherical cloud diameter, cm

D_{1_o} = initial bubble diameter, cm

D_{1_m} = maximum bubble diameter, cm

F_{12} = gas interchange coefficient between phase 1 and phase 2 per unit volume of phase 1, 1/s

F_{23} = gas interchange coefficient between phase 2 and phase 3 per unit volume of phase 1, 1/s

G = gravitational acceleration, cm/s^2

h = height from distributor plate, cm

K = Boltzman's Constant, 1.38×10^{-16} erg/molecule ^oK

K_C = dimensionless characteristic particle mobility for Coulombic force,

$$\frac{D_P\, Q_{AC}\, Q_P}{3\mu\, \hat{U}_i\, \varepsilon_f}$$

L = expanded bed height, cm

L_{mf} = bed height at U_{mf}, cm

N = number of bubbles in a compartment

$\overline{N}$ = average number of bubbles in a compartment

N_D = number of orifice openings on the distributor

Pe = Peclet number, $\hat{U}_i\, D_c/D_G$

N_R = dimensionless interception parameter, D_p/D_c

Q_{AC} = charge on collector, C/cm^2, } assumed of opposite signs throughout this work

Q_P = charge on particle, C/cm^2 }

r_p = particle radius, cm

S = cross sectional area of bed, cm^2

Stk = Stoke's number

$$\frac{1}{9}\, \frac{D_p^2\, \hat{U}_i\, \rho_p}{\mu\, D_c}$$

Stk_c = critical Stoke's number, below which there can be no collection by inertial impaction

$\hat{U}_i$ = assembly averaged velocity of gas in phase i, cm/s

$\overline{U}_1$ = average linear gas velocity in bubble phase, cm/s

$\overline{U}_{is}$ = average superficial velocity of gas in phase i, cm/s

U_{mf} = minimum fluidization velocity, cm/s

U_o = superficial gas velocity, cm/s

V_i^n = volume of phase i in n^{th} compartment, cm^3

$\overline{V}_1$ = average volume of bubble phase, cm^3

X = overall collection efficiency (%)

X_M = overall collection efficiency for M^{th} stage (%)

Greek Symbols

ε_i = void fraction of gas in phase i

ε_{mf} = void fraction in bed at U_{mf}

ε_f = permittivity of free space, 8.85×10^{-21} C^2/dyne - cm^2

δ_i = volume fraction of bed occupied by phase i

$\overline{\delta}_1$ = average bubble phase volume fraction

η_i = single spherical collector efficiency in phase i

η_{BD} = single spherical collector efficiency for Brownian motion

η_{IMP} = single spherical collector efficiency for impaction

η_{INT} = single spherical collector efficiency for interception

η_E = single spherical collector efficiency for Coulombic forces

ρ_p = particle density, g/cm^3

ρ = gas density, g/cm^3

μ = gas viscosity, g/cm-s

λ = mean free path of gas, ~ 6.5×10^{-6} cm for air at 20°C

Acknowledgements

The authors wish to acknowledge G. Tardos and R. Pfeffer for their helpful comments during the course of this work. L. - S.F. was financially supported in part by the Battelle Memorial Institute under the University Distribution Program.

Literature Cited

1. Knettig, P.; Beeckmans, J.M. Can. J. Chem. Eng. 1974, 52, 703.
2. Tardos, G.; Gutfinger, C.; Abuaf, N. Israel J. Tech. 1974, 12, 184.

3. McCarthy, D.; Yankel, A.J.; Patterson, R.G.; Jackson, M.L. Ind. Eng. Chem. Proc. Des. Dev. 1976, 15, 266.
4. Svrcek, W.Y.; Beeckmans, J.M. Tappi 1976, 59, 79.
5. Tardos, G.; Gutfinger, C.; Abuaf, N. AIChE J. 1976, 22, 1147.
6. Zahedi, K.; Melcher, J.R. J. Air Poll. Cont. Ass. 1976, 26, 345.
7. Ciborows, J.; Zakowski, L. Int. Chem. Eng. I. 1977, 17, 529.
8. Patterson, R.G.; Jackson, M.L. AIChE Symp. Ser. No. 161 1977.
9. Zahedi, K.; Melcher, J.R. Ind. Eng. Chem. Fund 1977, 16, 248.
10. Doganoglu, Y.; Jog, V.; Thambimuthu, D.V.; Clift, R. Trans. Int. Chem. Eng. 1978, 56, 239.
11. Gutfinger, C.; Tardos, G.I. Atm. Env. 1979, 13, 853.
12. Tardos, G.; Gutfinger, C.; Pfeffer, R. Ind. Eng. Chem. Fun. 1979, 18, 433.
13. Peters, M.H.; Fan, L.-S.; Sweeney, T.L. AIChE J. 1981, (in press).
14. Peters, M.H.; Fan, L.-S.; Sweeney, T.L. Reactant Dynamics in Catalytic Fluidized Bed Reactors with Flow Reversal of Gas in the Emulsion Phase, presented at 1980 AIChE Meeting, Chicago, Ill.
15. Kato, K.; Wen, C.Y. Chem. Eng. Sci. 1969, 24, 1351.
16. Davidson, J.F.; Harrison, D., "Fluidized Particles," 1963, Cambridge University Press.
17. Murray, J.D. J. Fluid Mech. 1965, 21, 465.
18. Mori, S.; Wen, C.Y. AIChE J. 1975, 21, 109.
19. Kunii, D.; Levenspiel, O. Ind. Eng. Chem. Fund. 1968, 7, 446.
20. Lockett, M.J.; Davidson, J.F.; Harrison, D. Chem. Eng. Sci. 1967, 22, 1059.
21. Chavarie, C.; Grace, J.R. Chem. Eng. Sci. 1976, 31, 741.
22. Tardos, G.I.; Pfeffer, R. Proc. 2nd World Filt. Cong. Sept. 18-20, 1979, London, U.K.
23. Tardos, G.I.; Yu, E.; Pfeffer, R.; Squires, M. J. Coll. Int. Sci. 1979, 71, 616.
24. Friedlander, S.K. J. Coll. Int. Sci. 1967, 23, 157.

RECEIVED June 3, 1981.

6

Modeling and Simulation of Dynamic and Steady-State Characteristics of Shallow Fluidized Bed Combustors

L. T. FAN and C. C. CHANG

Department of Chemical Engineering, Kansas State University, Manhattan, KS 66506

The dynamic and steady-state characteristics of a shallow fluidized bed combustor have been simulated by using a dynamic model in which the lateral solids and gas dispersion are taken into account. The model is based on the two phase theory of fluidization and takes into consideration the effects of the coal particle size distribution, resistance due to diffusion, and reaction. The results of the simulation indicate that concentration gradients exist in the bed; on the other hand, the temperature in the bed is quite uniform at any instant in all the cases studied. The results of the simulation also indicate that there exist a critical bubble size and carbon feed rate above which "concentration runaway" occurs, and the bed can never reach the steady state.

Fluidized bed combustion is believed to be one of the most promising methods for direct burning of coal in an environmentally acceptable and economically competitive manner. Many mathematical models have been proposed to predict the performance of fluidized bed combustors (see, e.g., 1-7). A review of the models has been presented by Carretto (8). Most of these models are steady-state ones, and, furthermore, assume that concentration and temperature variations do not exist in the lateral direction of the bed. However, it has been shown that there could be significant variation in the carbon concentration across a large fluidized bed reactor (9). Fan *et al.* (10) have proposed an isothermal dynamic model for estimating the lateral carbon concentration distribution in a shallow fluidized bed combustor. Simulation based on the model has indicated that an appreciable carbon concentration gradient can exist in the lateral direction in the bed. The objective of this work is to improve the model by eliminating the assumption of isothermal condition in the bed.

0097-6156/81/0168-0095$05.25/0

In the present work, the transient and steady-state characteristics of a fluidized bed combustor are studied by solving numerically a dynamic model in which lateral solids and gas dispersion, lateral temperature distribution and wide size distribution of coal feed are taken into account. The influences of bubble size, excess air rate, specific area of heat exchangers and coal feed rate on the performance of the fluidized combustor are examined by means of simulation with the model.

Mathematical Formulation

Let us consider a shallow fluidized bed combustor with multiple coal feeders which are used to reduce the lateral concentration gradient of coal (11). For simplicity, let us assume that the bed can be divided into N similar cylinders of radius R_B, each with a single feed point in the center. The assumption allows us to use the symmetrical properties of a cylindrical coordinate system and thus greatly reduce the difficulty of computation. The model proposed is based on the two phase theory of fluidization. Both diffusion and reaction resistances in combustion are considered, and the particle size distribution of coal is taken into account also. The assumptions of the model are: (a) The bed consists of two phases, namely, the bubble and emulsion phases. The voidage of emulsion phase remains constant and is equal to that at incipient fluidization, and the flow of gas through the bed in excess of minimum fluidization passes through the bed in the form of bubbles (12). (b) The emulsion phase is well mixed in the axial direction, and the solids mixing can be described by the diffusion equation in the lateral direction (9). The bed can be characterized by an effective bubble size, and the bubble flow is of the plug type (10). (c) No elutriation occurs. (d) The convective transport of coal particles in the lateral direction can be neglected. (e) Ash is continuously withdrawn from the bed at the same rate as that in the feed. (f) The only combustion reaction is

$$C + O \longrightarrow CO_2$$

(g) The solids and gas are at the same temperature. (h) The size distribution of coal particles in the bed is the same as that in the feed; the sizes of coal particles are widely distributed (1). These assumptions give rise to the following equations:

Material balances in the emulsion phase

$$\frac{\partial[CP_b(R)]}{\partial t} = \Psi_F P_o(R) + \frac{1}{r}\frac{\partial}{\partial r}\left(rD_s \frac{\partial[CP_b(R)]}{\partial r}\right) + \frac{\partial[CP_b(R)S(R)]}{\partial R} - \frac{3CP_b(R)}{R} S(R) \quad (1)$$

$$\frac{\partial C}{\partial t} = \Psi_F + \frac{1}{r}\frac{\partial}{\partial r}(rD_s \frac{\partial C}{\partial r}) - \int_0^{R_{max}} \frac{3S(R)}{R} CP_b(R)dR \qquad (2)$$

$$\varepsilon_{mf}\frac{\partial C_{ae}}{\partial t} = \frac{U_{mf}}{(1-\delta_b)L}(C_{ao} - C_{ae}) + \frac{1}{r}\frac{\partial}{\partial r}(rD_{ae}\frac{\partial C_{ae}}{\partial r})$$

$$+ \frac{\delta_b}{1-\delta_b}\frac{1}{L}\int_0^L K_{be}[C_{ab}(z) - C_{ae}]dz$$

$$- \int_0^{R_{max}} \frac{3S(R)}{RM_C} CP_b(R)dR \qquad (3)$$

Material balance in the bubble phase

$$\frac{\partial C_{ab}}{\partial t} = -U_b \frac{\partial C_{ab}}{\partial z} - K_{be}(C_{ab} - C_{ae}) \qquad (4)$$

Energy balance

$$\rho_m C_{pm}\frac{\partial T}{\partial t} = \frac{1}{r}\frac{\partial}{\partial r}(rk_e\frac{\partial T}{\partial r}) + \frac{U_g\rho_g C_{pg}}{L}(T_{go} - T) + \Psi_F(1-\delta_b)C_{ps}(T_{so}-T)$$

$$+ \{(1-\delta_b)\int_0^{R_{max}} \frac{3S(R)}{R} CP_b(R)\, dR\}(-\Delta H)$$

$$- ha(T - T_w) + \Phi(t) \qquad (5)$$

where

$$\rho_m C_{pm} = \rho_g C_{pg}\delta_g + \rho_s C_{ps}(1 - \delta_g)$$

The appropriate initial and boundary conditions are

$$t = 0; \quad C_{ab} = C_{ae} = C_{ao}, \quad T = T_o, \quad C = 0$$

$$t > 0; \quad C_{ab} = C_{ao} \quad \text{at } z = 0$$

$$\frac{\partial C}{\partial r} = \frac{\partial C_{ab}}{\partial r} = \frac{\partial C_{ae}}{\partial r} = \frac{\partial T}{\partial r} = 0 \quad \text{at } r = 0$$

$$\frac{\partial C}{\partial r} = \frac{\partial C_{ab}}{\partial r} = \frac{\partial C_{ae}}{\partial r} = \frac{\partial T}{\partial r} = 0 \quad \text{at } r = R_B$$

The feeding rate function, Ψ_F, is defined as:

$$\Psi_F \begin{cases} = \dfrac{F}{\pi r_f^2 (1 - \delta_b)L} = \dfrac{F}{\pi r_f^2 L_{mf}} & \text{at } 0 \leq r \leq r_f \\ = 0 & \text{at } r_f < r \leq R_B \end{cases}$$

and the size reduction function of coal particles is defined as:

$$S(R) = -\frac{dR}{dt}$$

The unreacted core shrinking model gives rise to the size reduction function of the following form (13);

$$S(R) = -\frac{dR}{dt} = \left(\frac{1}{\frac{1}{k_g} + \frac{1}{k_s}}\right) \frac{C_{ae} M_c}{\rho_c} \tag{6}$$

The heat source function, $\Phi(t)$, is defined as:

$$\Phi(t) = Q[U(t) - U(t - t_s)] \tag{7}$$

A fluidized bed combustor can be used as a process heater (the type A combustor) or a steam generator (the type B combustor) as shown in Fig. 1. The combustor usually has no built-in heat exchangers when it is used as a process heater, and it operates with a very high excess air flow rate.

Numerical Simulation

Numerical calculation has been carried out using a software interface which is based on the so-called "Method of lines" (14). Gear's backward difference formulas (15) are used for the time integration. A modified Newton's method with the internally generated Jacobian matrix is utilized to solve the nonlinear equations.

To simulate the start-up of the combustor, an external heat source with a constant strength of 30 cal/sec-cm^2, i.e., $Q = 30$ in equation (7), is applied to the system at the onset of operation. The coal particles are fed into the bed when the bed temperature reaches 1300°K, and the heat source is removed ten seconds later. Thus,

$$\Phi(t) = 30[U(t) - U(t - 10)]$$

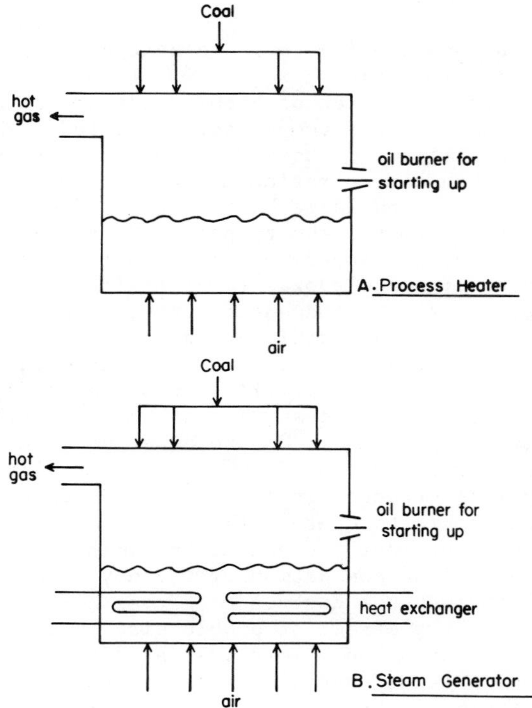

Figure 1. Two types of shallow fluidized bed combustor

It is assumed that circulation of the cooling steam through the heat exchangers is started fifty seconds after feeding of the coal particles into the bed is initiated. The functional relationships among variables and the nominal values of various parameters employed in numerical computation are listed in Table I and II, respectively. Numerical simulation has been carried out separately for the type A combustor and the type B combuster in order to emphasize the differences between them.

Results and Discussion

Type A combustor [see Figure 1; no heat exchangers in the bed, i.e., $a = 0$ in equation (5)].

Figure 2 shows the effect of bubble size on the steady-state carbon concentration and bed temperature profiles. The concentration profiles are similar to those obtained previously (10) in that an appreciable carbon concentration gradient is generated along the lateral direction, and large bubbles can reduce this concentration gradient. Furthermore, the temperature profiles are almost flat, indicating that the superior heat transfer property of a fluidized bed overcomes the poor lateral mixing effect and that an isothermal model is probably adequate for representing the steady-state characteristics of a fluidized bed combustor.

The effect of excess air rate on the steady-state carbon concentration and temperature profiles is shown in Figure 3. As can be seen, the effect of excess air rate on the carbon concentration profiles is not profound. On the other hand, even though the steady-state bed temperature profiles are essentially flat, the temperature level is strongly influenced by the excess air rate. It is usually desirable to operate the combustor at certain optimal temperature ranges in order to control the emission of NO_x and SO_2 or to obtain the maximum combustion efficiency. This can probably be accomplished through the control of excess air flow rate.

Figure 4 shows the effect of bubble size on the transient, average carbon concentration. Note that a critical bubble size exists, above which a concentration runaway occurs, i.e., the bed cannot reach a steady state. This is the result of an insufficient rate of oxygen transfer from the bubble phase to the emulsion phase. It can also be seen in the figure that the steady-state average carbon concentration is strongly influenced by the bubble size; it increases sharply when the bubble size exceeds a certain value, e.g., 5 cm in this case.

The effects of bubble size and excess air rate on the transient average bed temperature are illustrated in Figure 5. The effect of bubble size is almost negligible under stable operating conditions, while the effect of excess air has a strong influence on the temperature change. It can be seen in both Figures 4 and 5 that the bed reaches a steady state at about 2000 s after initiation of the operation. This value is very different from the value, 200 s, obtained based on an isothermal dynamic model (10).

Table I. Variable Relationships

$$U_{mf} = \left(\frac{\mu}{d_p \rho_g}\right) \{[33.7^2 + 0.0408 d_p^3 \rho_g(\rho_s - \rho_g) g/\mu^2]^{0.5} - 33.7\} \text{ cm/s} \quad (16)^*$$

$$\delta_b = \frac{U_0 - U_{mf}}{U_b} = \frac{L - L_{mf}}{L}$$

$$U_b = U_0 - U_{mf} + 0.711 (g d_B)^{0.5} \text{ cm/s}$$

$$K_{be} = \frac{K_1 K_2}{K_1 + K_2} \text{ 1/s}$$

$$\text{where } K_1 = 4.5 \frac{U_{mf}}{d_B} + 5.85 \left(\frac{D^{0.5} g^{0.25}}{d_B^{1.25}}\right)$$

$$K_2 = 6.78 \left(\frac{\varepsilon_{mf} D U_b}{d_B^3}\right)^{0.5} \quad (17)^*$$

$$D_s = 0.187\ \delta_b U_{mf} \frac{d_B}{(1-\delta_b)\varepsilon_{mf}} \text{ cm}^2\text{/s} \quad (18)^*$$

$$k_e = D_s(\rho_s C_{ps}) \frac{\text{cal}}{\text{cm}\cdot\text{s}\cdot{}^\circ\text{K}} \quad (17)^*$$

$$P_o(R),\ P_b(R) \sim N(\mu_c, \sigma_c)$$

$$\frac{\mu_1}{\mu_2} = \left(\frac{T_1}{T_2}\right)^{1/2}$$

$$D = \frac{10^{-3} T^{1.75} (1/M_{air} + 1/M_{O_2})^{0.5}}{P[(\Sigma v)_{air}^{1/3} + (\Sigma v)_{O_2}^{1/3}]} \text{ cm}^2\text{/s} \quad (19)^*$$

$$\frac{\rho_{g_1}}{\rho_{g_2}} = \frac{T_2}{T_1}$$

*References

Table II. Numerical Values of Parameters Used in Computation

C_{ao} = 2.38×10^{-6} mole/cm^3

C_{pg} = $0.238 + 1.753 \times 10^{-3}$ (T-273) cal/g°C

C_{ps} = 0.215 cal/g°C

d_B = 2, 5, 10 cm

d_p = 0.05 cm

Excess air = 100, 150, 200% (Type A combustor); 5, 10, 20% (Type B combustor)

$(-\Delta H)$ = 7831 cal/g

L_{mf} = 15 cm

R = 40 cm

T_o = 1300°K

T_{go} = T_{so} = 300°K

T_w = 600°K

ε_{mf} = 0.5

ρ_s = 1.4 g/cm^3

Ψ_F = 0.02, 0.03, 0.04, 0.05 (Type A combustor)
0.01, 0.02, 0.03, 0.04 (Type B combustor)

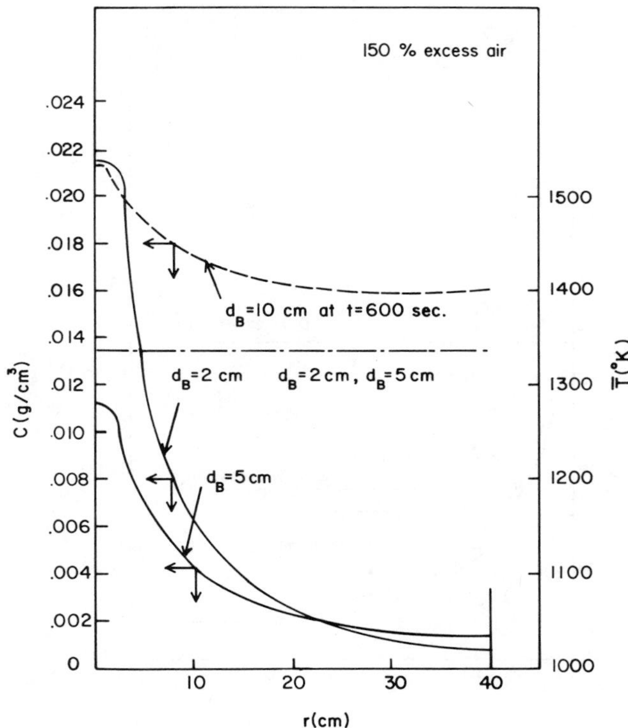

Figure 2. Effect of bubble size on the (——) steady-state carbon concentration and (– – –) bed temperature profiles in the type A combustor

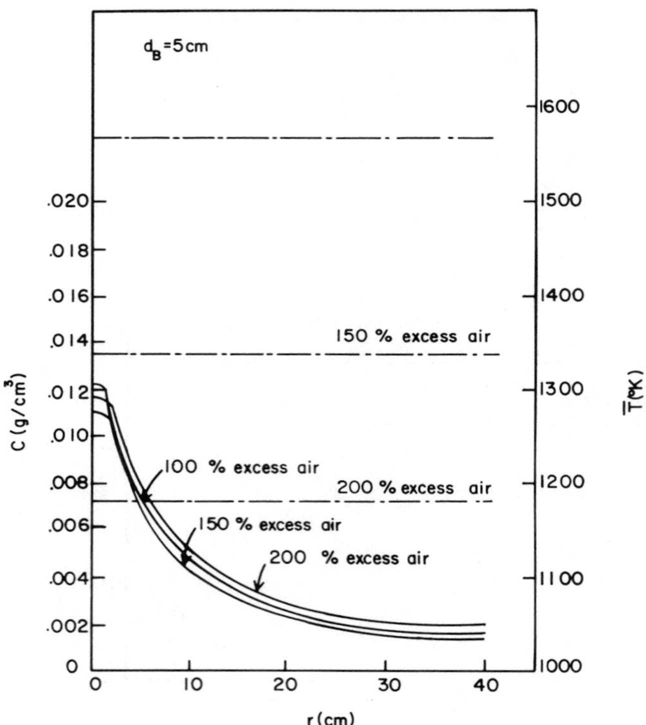

Figure 3. Effect of excess air rate on the (——) steady-state carbon concentration and (— – —) bed temperature profiles in a type A combustor

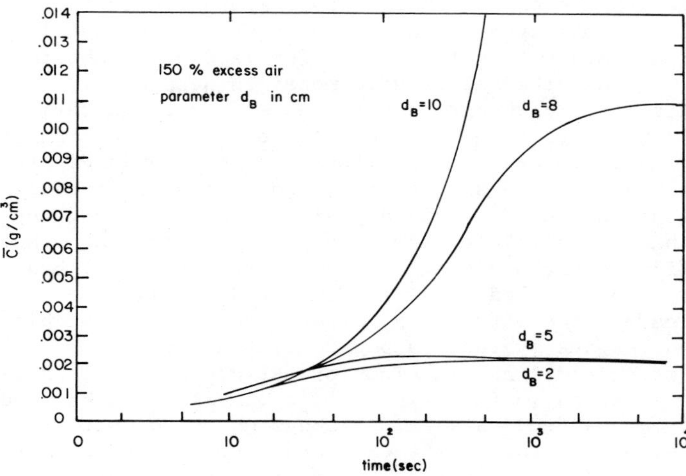

Figure 4. Effect of bubble size on the transient average carbon concentration in the type A combustor

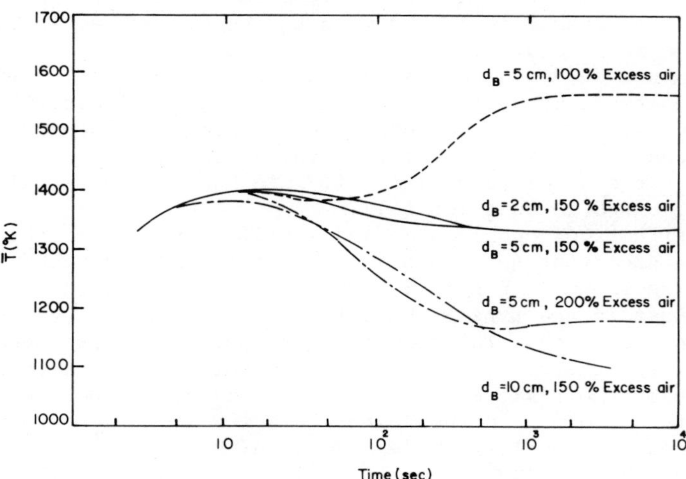

Figure 5. Effect of bubble size and excess air rate on the transient average bed temperature in the type A combustor

This implies that even though the bed temperature is fairly uniform, an isothermal dynamic model can not represent sufficiently well the temperature transient of a fluidized bed combustor, and therefore, a nonisothermal dynamic model is necessary.

The effect of carbon feed rate, as expressed in terms of the carbon feed rate function, Ψ_f, on the steady-state carbon concentration and bed temperature profiles is presented in Figure 6. It can be seen that the shapes of the concentration profiles remain almost unchanged. This appears to indicate that the carbon feed rate has a negligible effect on the concentration gradient; it only influences the average amount of carbon in the bed. This result was also observed in a previous study (11).

In Figure 7 the effects of carbon feed rate and bubble size on the steady-state average carbon concentration are shown. The existence of critical bubble size for a fixed carbon feed rate can clearly be observed in this figure. It can also be observed that a critical carbon feed rate exists above which concentration runaway occurs, and a stable or steady-state condition can not be reached for a given bubble size. The value of the critical feed rate increases with a decrease in the bubble size. Under the critical condition, the maximum attainable rate of oxygen transfer from the bubble phase to the emulsion phase is reached, and it becomes the rate determining step for combustion as explained previously. To increase the carbon feed rate to a fluidized bed combustor, either the oxygen concentration in the air (gas) stream or the rate of mass transfer between the bubble and emulsion phase needs to be increased.

It should be noted that a carbon concentration of 0.014 g/cm^3 corresponds to 1% by weight in the present study, indicating that the steady-state carbon concentrations in all cases studied are far less than 1% by weight for the type A combustor.

Type B combustor [see Figure 1: with heat exchangers in the bed, i.e., $a \neq 0$ in equation (5)].

The effects of bubble size and specific area of heat exchangers on the steady-state carbon concentration and bed temperature profiles are shown in Figure 8. Obviously, the carbon concentration gradient in the type B combustor is much greater than that in the type A combustor. The result might imply that more feeders are needed for the type B combustor than for the type A combustor, provided that their sizes are the same. The bed temperature is again quite uniform, and the temperature level in the type B combustor appears to be largely dependent on the characteristics of the heat exchangers instead of the excess air rate, as in the case of the type A combustor.

The effects of bubble size and specific areas of heat exchangers on the transient average carbon concentration and bed temperature are presented in Figure 9. It can be seen that the critical bubble size is about 5 cm, which is much smaller than that for the type A combustor. This is because of the relatively small excess air rate used and the large carbon concentration gradient

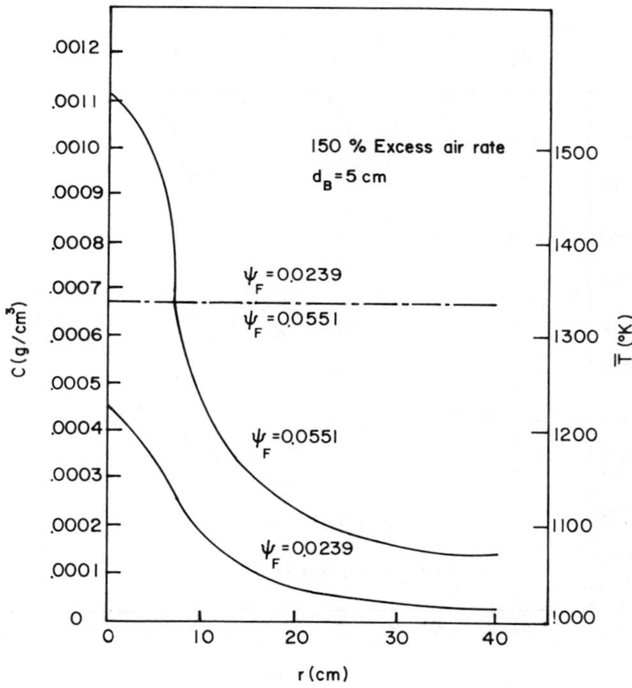

Figure 6. Effect of carbon feed rate on the (——) steady-state carbon concentration and (– – –) bed temperature profiles in the type A combustor

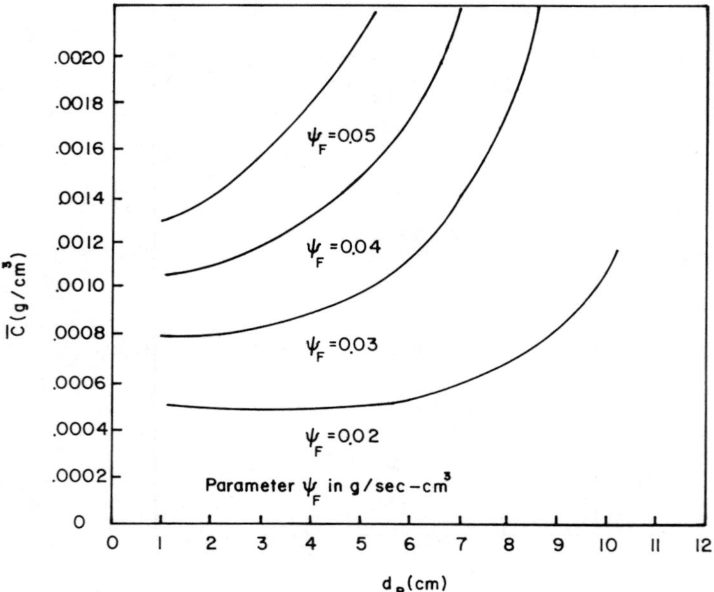

Figure 7. Effect of carbon feed rate and bubble size on the steady-state average carbon concentration in the type A combustor

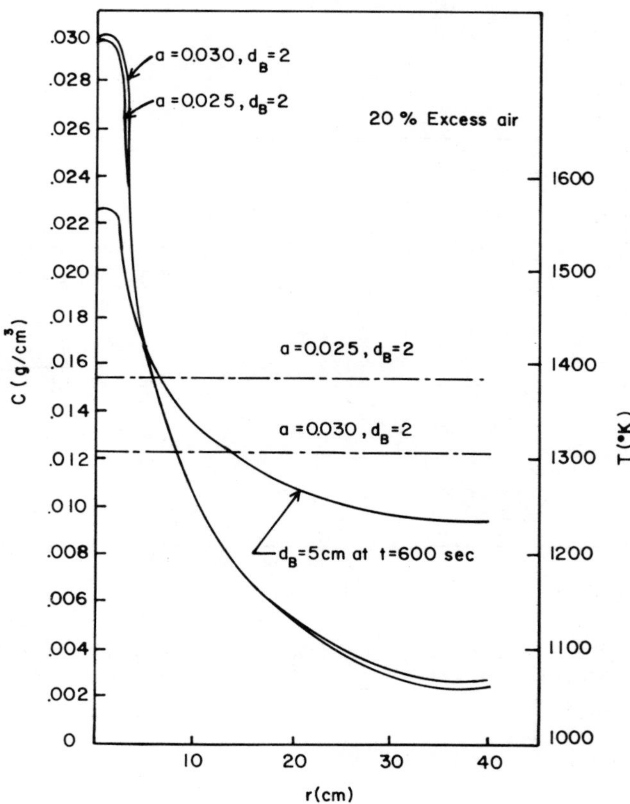

Figure 8. Effect of bubble size and specific area of heat exchangers on the (——) steady-state carbon concentration and (— — —) bed temperature profiles in the type B combustor

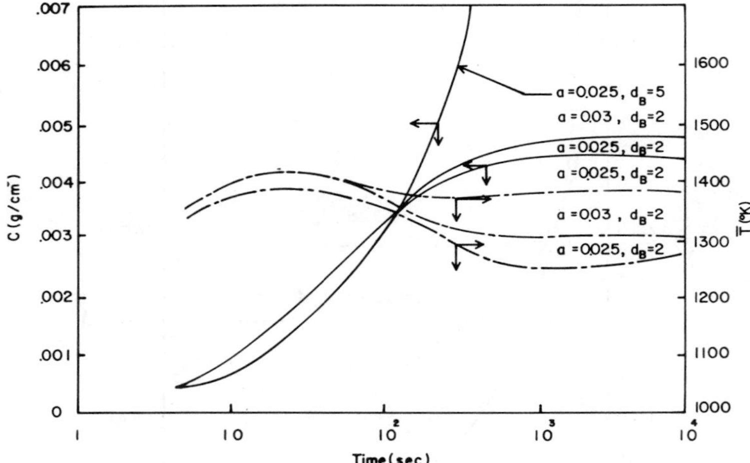

Figure 9. Effect of bubble size and specific area of heat exchangers on the (——) transient average carbon concentration and (— — —) bed temperature in the type B combustor

obtained. Since most of the heat generated in the type B combustor is removed through the heat exchangers, the excess air rate can usually be kept low to save the blowing costs and the elutriation loss. Therefore, the excess air rate should be controlled so that it is neither too large nor too small.

A parametric study on the carbon feed rate was also made for the type B combustor. The results obtained for the effect of carbon feed rate on the steady-state carbon concentration and bed temperature profiles are similar to those in the type A combustor and will not be presented here. The effects of carbon feed rate and bubble size on the steady-state average carbon concentration are shown in Figure 10. It shows that the average carbon concentration in the type B combustor is much higher than that in the type A combustor for the given values of the carbon feed rate function and bubble size. Apparently, both the critical bubble size and carbon feed rate in the type B combustor are smaller than those in the type A combustor under the same set of operating conditions except the excess air rate. This indicates the performance of a fluidized bed combustor with built-in heat exchangers is much more sensitive to variations in the operating conditions; greater effort is required to ensure its smooth operation.

The assumption (d) imposed in deriving the model appears to be valid for both types of combustors since the feed rate of coal is relatively small under normal operating conditions. The order of magnitude analysis shows that the convective flux, uc, is indeed much smaller than the dispersion flux, $-D \frac{\partial c}{\partial r}$, under the conditions simulated.

Conclusion

A non-isothermal dynamic model has been developed for a shallow fulidized bed combustor, which can be used to predict, at least qualitatively, the transient and steady-state characteristics of such systems. Parametric studies have been conducted to examine the effects of excess air flow rate, bubble size and carbon feed rate. It has been shown that an appreciable carbon concentration gradient does exist in the bed. This explains why it is necessary to use multiple feed points in large fluidized bed combustors. A surprising result obtained is that the temperature in the bed is essentially uniform under all conditions studied even though the carbon concentration is not uniform laterally.

For a combustor without heat exchangers, the bed temperature is strongly influenced by the excess air flow rate. On the other hand, for a combustor with heat exchangers, the bed temperature is mainly dependent on the characteristics of the heat exchangers.

It has been illustrated that the bubble size has strong influences on both the transient and steady-state carbon concentrations. The effects of the carbon feed rate, expressed as the carbon feed rate function, on the steady-state carbon concentration and bed temperature profiles are negligible under the conditions

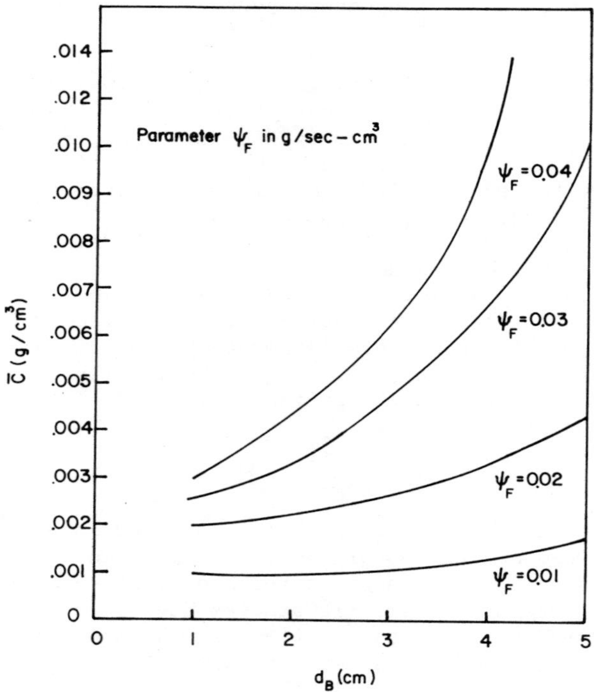

Figure 10. Effect of carbon rate and bubble size on the steady-state average carbon concentration in the type B combustor; 20% excess air, a = *0.025* cm^2/cm^3

simulated. It has also been illustrated that a critical bubble size and a critical carbon feed rate exist, above which a stable steady-state operating condition can never be attained. Such critical values for a combustor with heat exchangers (type B) have been found to be much smaller than those for the corresponding combustor without heat exchangers (type A).

Nomenclature

a = surface area of the heat exchanger per unit bed volume, cm^{-1}

C = carbon concentration in the emulsion phase, g/cm^3

C_{ab} = oxygen concentration in the bubble phase, mol/cm^3

C_{ae} = oxygen concentration in the emulsion phase, mol/cm^3

C_{ao} = initial oxygen concentration in the feed gas, mol/cm^3

C_{pg} = heat capacity of gas, cal/g°K

C_{ps} = heat capacity of the bed solid particles (limestone or dolomite), cal/g°K

D = gas diffusivity in the solid-gas boundary, cm^2/s

D_{ae} = effective dispersion coefficient of oxygen in the emulsion phase, cm^2/s

D_s = effective dispersion coefficient of solids, cm^2/s

d_B = bubble diameter, cm

F = feeding rate of coal particles, g/s

$(-\Delta H)$ = heat of reaction, cal/g

K = gas interchange coefficient, 1/s

k_e = effective thermal conductivity, cal/cm•s•°K

k_g = oxygen mass transfer coefficient in the solid-gas boundary, cm/s

L = bed height, cm

L_{mf} = bed height at minimum fluidization, cm

M_c = molecular weight of carbon, g/mole

$P_b(R)$ = coal size distribution function in the bed, -

$P_o(R)$ = coal size distribution function in the feed stream, -

Q = magnitude of the external heat source, $cal/s \cdot cm^2$

R = radius of the coal particles, cm

R_B = radius of equivalent bed radius per feed point, cm

r = coordinate in the radial direction, cm

r_f = radius of the feeder, cm

Sh = Sherwood number, -

S(R) = rate of particle shrinkage as defined in equation (6), cm/s

T = bed temperature, °K

T_{go} = temperature of the air in the feed stream, °K

T_o = initial bed temperature, °K

T_{so} = temperature of the coal particles in the feed stream, °K

t = time, s

t_s = time interval in which the external heat source is applied during start-up, s

U_b = bubble velocity, cm/s

U_g = superficial velocity of gas, cm/s

U_{mf} = incipient fluidization velocity, cm/s

U(t) = step function, -

z = coordinate in the axial direction, cm

δ_b = fraction of the bubble phase, -

δ_g = fraction of the gas phase in the bed, -

ε_{mf} = incipient void fraction of the emulsion phase, -

μ = gas viscosity, g/cm s

ρ_c = density of the coal particles, g/cm^3

ρ_s = density of the bed solid particles, g/cm^3

ρ_g = density of the gas mixture in the bed, g/cm^3

$\Phi(t)$ = external heat source function, $cal/s{\cdot}cm^2$

$\Psi_F(t)$ = carbon feed rate function, $g/s{\cdot}cm^3$

Acknowledgement

This work was conducted under the sponsorship of the Engineering Experiment Station (Kansas Energy Study Project) of Kansas State University.

Literature Cited

1. Gibbs, B.M., Paper A-5 Institute of Fuel Symposium Series, 1975.
2. Gordon A. L., Amundson, N. R., Chem. Eng. Sci., 1976, 31, 1163.
3. Beer, J. M., paper presented at 16th Symposium on Combustion, The Combustion Institute, 1976; p. 439.

4. Horio, M., Rengarajan, P., Krishnan, R., Wen, C. Y., "Fluidized Bed Combustor Modeling", Report to NASA Lewis Research Center: Cleveland, January, 1977.
5. Baron, R. E., Hodges, J. L., Sarofim, A. F., paper presented at AIChE 70th Annual Meeting: New York, November, 1977.
6. Borghi, G., Sarofim, A. F., Beer, J. M., paper presented at the AIChE 70th Annual Meeting: New York, November, 1977.
7. Horio, M., Wen, C. Y., AIChE Symposium Series, 1977 (161), 73, 9.
8. Caretto, L. S., paper presented at the 1977 Fall Meeting, Western States Section/The Combustion Institute: Stanford University, October, 1977.
9. Highley, J., Merrick, D., AIChE Symposium Series, 1971 (116), 69, 219.
10. Fan, L. T., Tojo, K., Chang, C. C., Ind. Eng. Chem. Process Des. Dev., 1979, 18, 333.
11. Tojo, K., Chang, C. C., Fan, L. T., paper submitted to Ind. Eng. Chem. Process Des. Dev., for publication, 1979.
12. Davidson, J. F., Harrison, D., "Fluidized Particles," Cambridge University Press: New York, NY, 1963.
13. Yagi, S., Kunii, D., 5th Symposium (International) on Combustion: Reinhold, New York, 1955; p. 231.
14. Sincovec, R. F., Madsen, N. K., ACM Trans. Math. Software, 1975, 1, 232.
15. Gear, C. W., "Numerical Initial Value Problems in Ordinary Differential Equations"; Chapter 9, Prentice-Hall: Englewood Cliffs, N.J., 1971.
16. Wen, C. Y., Yu, Y. H., AIChE J., 1966, 12, 610.
17. Kunii, D., Levenspiel, P., J. Chem. Eng. Jpn., 1969, 2, 122.
18. Kunii, D., "Kagaku Kikai Gijutsu"; Maruzen, 1966 (18), p. 161.
19. Fuller, E. N., Schettler, P. D., Giddings, J. G., Ind. Eng. Chem., 1966, 58, 18.

RECEIVED June 3, 1981.

7

Modeling of Fluidized Bed Combustion of Coal Char Containing Sulfur

A. REHMAT[1], S. C. SAXENA[2], and R. H. LAND

Argonne National Laboratory, 9700 South Cass Avenue, Argonne, IL 60439

A mathematical model is developed for coal char combustion with sulfur retention by limestone or dolomite in a gas fluidized bed employing noncatalytic single pellet gas-solid reactions. The shrinking core model is employed to describe the kinetics of chemical reactions taking place on a single pellet whose changes in size as the reaction proceeds are considered. The solids are assumed to be in back-mix condition whereas the gas flow is regarded to be in plug flow. The model is strictly valid for the turbulent regime where the gas flow is quite high and classical bubbles do not exist. Formulation of the model includes setting up heat and mass balance equations petaining to a single particle exposed to a varying reactant concentration along the height of the bed with accompanying changes in its size during the course of reaction. These equations are then solved numerically to account for particles of all sizes in the bed to obtain the overall carbon conversion efficiency and resultant sulfur retention. In particular, the influence of several fluid-bed variables such as oxygen concentration profile, additive particle size, reaction rate for sulfation reaction, sulfur absorption efficiency are examined on additive requirement.

[1]Current Address: Institute of Gas Technology, Chicago, IL

[2]Current Address: University of IL at Chicago Circle, Chicago, IL

0097-6156/81/0168-0117$09.75/0

Many models have been used to describe fluidized-bed operation (1-7). Several additional models have been proposed during the last three years and these will be referred to later in this report. It is commonly assumed that the bed is composed of two distinct phases, *viz.*, a dense phase (emulsion) consisting of solid particles and interstitial gas, and a bubble phase consisting of rising voids with almost no solids. The most advanced models (1, 2, 3) also consider additional phases, *viz.*, a cloud and wake associated with each bubble. Further variations appear in the characterization of gas flow within each phase, and mode of exchange among the phases, and the bubble shapes, velocities, and growth rates. It is generally assumed that in the two-phase theory of fluidization (8), the flow rate of bubble voids through the fluidized bed is equal to the excess gas flow rate above that required for minimum fluidization. Chemical reactions in the bed are assumed to occur entirely in the emulsion phase.

In the present analysis, we shall develop a basic model for fluidized bed operation by extending our earlier analysis (9, 10, 11) for a single pellet reaction to model the noncatalytic gas-solid reactions taking place in a fluidized bed. The earlier results have been derived with the assumption of constant gaseous reactant concentration surrounding the pellet. However, in a fluidized bed, the pellet encounters a considerable variation in the gaseous reactant concentration due to its movement. Also, the fluidized bed is composed of particles of different sizes, each of which will behave differently. The solid material in the bed is constantly being consumed due to chemical reactions and is being depleted by entrainment and overflow. This solid material should be replenished continuously by feeding fresh reactant particles. In order to develop a realistic model, the particle size distribution of the feed and the bed must be taken into account as also the fact that the fluidized bed operates in a continuous mode with solids addition to the bed by feed and removal by overflow and elutriation.

The model presented here takes into account the changes in the size of a particle as a result of chemical reactions in a fluidized bed. A number of modeling studies related to the noncatalytic reactions and to coal combustion in particular, taking place in a fluidized bed have been reported (2-25). A review of these studies indicate that the coal combustion process is primarily diffusion controlled. The amount of gaseous reactant diffusing through the gas film surrounding the particle, will depend on its size. In most of the models referred to above, the particle size is assumed to be constant throughout the reaction insofar as the mass transfer process is concerned. The shrinkage of particles in those cases where either no solid product is formed or ash is flaked off from the surface is used only in calculating the particle size distribution in the bed, carryover, and overflow streams.

To account for particle growth or shrinkage as the reaction progresses in the reactor, a parameter, Z, is introduced. The theory developed by Kunii and Levenspiel (1) is for a reacting system in which the particles maintain a constant size (Z = 1). It (1) deals with the derivation of relations giving the particle size distribution in the bed, overflow, and carryover streams and their respective weights. This theory will be extended to include the effects of particle growth or shrinkage ($Z>1$ or $Z<1$). For typical combustion of char containing sulfur followed by sulfur dioxide absorption by limestone, relations will be derived to determine the extent of sulfur retention. The reaction, carryover, and overflow rates will be evaluated with particular attention to their dependence on Z.

Description of Char Particle Combustion

The combustion of sulfur-rich char is accompanied by the production of an undesirable reaction product, *viz.*, sulfur dioxide. However, most of the sulfur dioxide should be removed from the combustion gases before they leave the combustor. This may be accomplished by the introduction into the combustor of suitable additives which can absorb sulfur dioxide. Limestone is such an additive. The limestone reacts with sulfur dioxide in the presence of oxygen to form calcium sulfate, which is a solid product and can be easily removed from the reactor. In this work, a model is proposed for the prediction of sulfur dioxide removal from the combustion gases, based on knowledge of gas-solid reactions taking place on a single pellet.

The kinetics of gas-solid reactions obtained from single-particle studies are utilized to calculate the generation and utilizaton of sulfur dioxide for many particles present in a fluidized-bed reactor. For simplicity, char (*i.e.*, coal with almost all volatiles removed) will be the basic feed to the reactor and it is assumed to contain carbon, ash, and sulfur. Carbon and sulfur react with oxygen to form their respective oxides. Sulfur dioxide subsequently reacts with limestone and excess oxygen to form calcium sulfate. Char and limestone particles undergo change in size as they react, and this will be included in obtaining average conversions. Ultimately, this model predicts the average concentration of sulfur dioxide in the combustion gas stream, solid flow rates, and the particle size distributions in the reactor and in the streams leaving the reactor.

The following assumptions are made in the mathematical formulation of the process:

1. The particles are completely mixed in the reactor.
2. Gases do not mix vertically, *i.e.*, the gas flow through the bed is in plug flow. Further, no gas concentration gradients exist transverse to the direction of flow.
3. The gas flow is statistically uniform over the bed cross section at a given bed height and is equal to a certain mean value.

4. The temperature is uniform throughout the bed.
5. The reactor is operated in the steady state mode.
6. Sulfur is uniformly distributed in the coal char particles.
7. Combustion and consequent generation and absorption of sulfur dioxide occur throughout the bed.
8. The reaction rates are independent of reaction product concentrations.
9. The gas-solid reaction follows a shrinking core model.

Char and limestone (calcium carbonate) particles are fed to the reactor continuously at rates, F_1 and F_1', respectively, and their particle size distributions in the feed are given by $P_1(\xi_s)$ and $P_1'(\xi_s')$, respectively. The mass of these solid components in the bed, the overflow and the carryover rates, and their respective size distributions are shown in Figure 1. The carry-over and the overflow particles are not recycled. The following reactions are considered to take place in the fluidized bed. The subscripts by which the reactants and products are referred to throughout in this work are given in parentheses.

$$\text{carbon (J)} + \text{oxygen (A)} \xrightarrow{k_1} \text{carbon dioxide (D)} \tag{1}$$

$$\text{sulfur (S)} + \text{oxygen (A)} \xrightarrow{k_2} \text{sulfur dioxide (B)} \tag{2}$$

$$\text{calcium carbonate (N)} + \text{sulfur dioxide (B)} + \text{oxygen (A)} \xrightarrow{k_3} \text{calcium sulfate (E)} + \text{carbon dioxide (D)} \tag{3}$$

When char reacts with oxygen, solid product ash is formed and it adheres to the particle. Similarly, in the case of limestone, the solid product calcium sulfate adheres to the limestone particle. The change in the overall size of the particle depends on the amount of solid formed and is related to the amount of solid reactant consumed in the following manner:

$$Z = \frac{\text{volume of solid product formed}}{\text{volume of solid reactant consumed}} \tag{4}$$

The average radius, $\bar{r}$, is defined as follows:

$$\bar{r} = \frac{1}{\Sigma_i(w_i'/r_i} \tag{5}$$

and is employed to normalize the distances from the center of the pellet.

In the following sections, the equations for a single pellet involving one and two independent reactions are presented. First, we shall derive equations pertaining to a single reaction and then extend the derivation to the two gas-solid reactions taking place with one gas and two solids. Gas-solid reactions given by Eqs. 1, 2, and 3 will be used.

Single Pellet: One Reaction. The sulfation reaction which is considered here for calcium carbonate is given by Eq. 3, and the temperature and concentration profiles of a typical growing limestone particle are shown in Figure 2. The rate of disappearance of sulfur dioxide is assumed to be the first order and is given by

$$-r_B = k_3(T_c') \, C_{NO} C_{BC} \tag{6}$$

The differential equation for the mass balance of gaseous reactant B reacting with a pellet of radius R_o' under the pseudo-steady-state assumption is

$$\frac{d}{d\xi'}\left[\xi'^2 \frac{d\omega_B}{d\xi'}\right] = 0 \tag{7}$$

with the following boundary conditions:

at $r' = R'$, i.e., at $\xi' = \xi_s'$,

$$\left.\frac{d\omega_B}{d\xi'}\right|_{\xi' = \xi_s'} = N'_{SH\bar{r}}(\omega_{BH} - \omega_{BS}) \tag{8}$$

at $r' = r_c'$, i.e., at $\xi' = \xi_c'$,

$$\left.\frac{d\omega_B}{d\xi'}\right|_{\xi' = \xi_s'} = \phi_{3\bar{r}} \frac{(\omega_{BC})}{U_c'} \exp\left\{\frac{E_3}{\bar{R}T_o}\left(1 - \frac{1}{U_c'}\right)\right\} \tag{9}$$

The solution of the above Equation 7 gives

$$\frac{\omega_{BH}}{\omega_{BC}} = \left[1 + \frac{\phi_{3\bar{r}}\xi_c'^2 \exp\left\{\frac{E_3}{\bar{R}T_o}\left(1 - \frac{1}{U_c'}\right)\right\}}{U_c' \, N_{SH\bar{r}} \, \xi_c'^2} + \frac{\phi_{3\bar{r}}\xi_c'\left(1 - \frac{\xi_c'}{\xi_s'}\right)}{U_c'} \cdot \exp \frac{E_3}{\bar{R}T_o}\left\{\left(1 - \frac{1}{U_c'}\right)\right\}\right] \tag{10}$$

The heat balance is

$$\frac{d}{d\xi'}\left[\xi'^2 \frac{dU'}{d\xi'}\right] = 0 \tag{11}$$

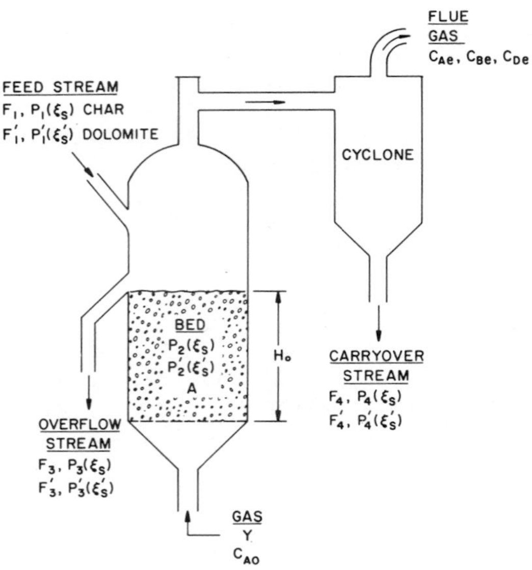

Figure 1. Feed and exit streams of the fluidized-bed combustor

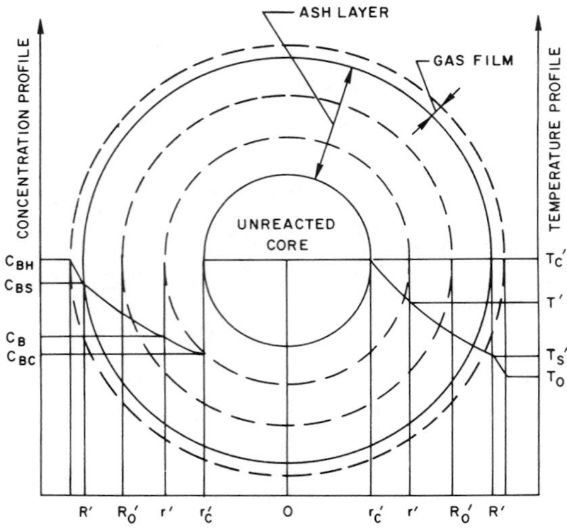

Figure 2. Gas–solid reaction of a growing limestone particle at height H *in the fluidized bed: concentration and temperature profiles*

with the following boundary conditions:

at $r' = R'$, i.e., $\xi' = \xi'_s$,

$$-\left.\frac{dU'}{d\xi'}\right|_{\xi' = \xi'_s} = N'_{Nur}\,(U'_s - 1) \tag{12}$$

and at $r' = r'_c$, i.e., $\xi' = \xi'_c$

$$-\left(\frac{\bar{R}T_o}{E_3}\right)\left.\left(\frac{dU'}{d\xi'}\right)\right|_{\xi' = \xi'_c} = \phi_{3\bar{r}}\beta_3\left(\frac{\omega_{BC}}{U'_c}\right)\exp\left\{\frac{E_3}{\bar{R}T_o}\left(1 - \frac{1}{U'_c}\right)\right\} \tag{13}$$

U'_c is then given by the following expression:

$$U'_c - 1 = \phi_{3\bar{r}}\beta_3\,\frac{\omega_{BC}}{U'_c}\,\frac{E_3}{\bar{R}T_o}\,\exp\left\{\frac{E_3}{\bar{R}T_o}\left(1 - \frac{1}{U'_c}\right)\right\}\cdot$$

$$\cdot\left[\frac{1}{N'_{Nur}}\,\frac{\xi'^2_c}{\xi'^2_s} + \xi'_c\left(1 - \frac{\xi'_c}{\xi'_s}\right)\right] \tag{14}$$

If we assume that the gas in the reactor is ideal and the gas pressure is constant, the following relation holds true throughout the reactor:

$$CT = \text{constant} \tag{15}$$

Thus, we can write the expression for the rate of conversion for the pellet in terms of its core radius as follows:

$$-\frac{d\xi'_c}{d\theta_3} = \frac{\omega_{BC}}{U'_c}\exp\left\{\frac{E_3}{\bar{R}T_o}\left(1 - \frac{1}{U'_c}\right)\right\} \tag{16}$$

where

$$\tau_3 = \frac{\rho_N\bar{r}}{M_N k_3(T_o)C_{NO}C_{AO}} \tag{17}$$

$$\text{and } \theta_3 = \frac{t}{\tau_3} \tag{18}$$

Since the behavior of the fluidized bed depends upon the overall particle size, it is necessary to derive an expression for $d\xi'_s/d\theta_3$. It is shown (9) that for spherical particles,

$$\frac{d\xi'_s}{d\theta_3} = (1 - Z')\,\frac{\xi'^2_c}{\xi'^2_s}\,\frac{d\xi'_c}{d\theta_3} \tag{19}$$

Substituting the expression for $d\xi_c/d\theta_3$ from Eq. 16 into Eq. 19, we get the required relation, viz.,

$$\frac{d\xi_s'}{d\theta_3} = (1 - Z')\frac{\xi_c'^2}{\xi_s'^2} \cdot \frac{\omega_{BC}}{U_c'}\left\{\exp\ \frac{E_3}{\bar{R}T_o}\left(1 - \frac{1}{U_c'}\right)\right\} \tag{20}$$

Single Pellet: Two Independent Reactions. The two independent reactions considered for char combustion are given by Eqs. 1 and 2, and their respective rates of reaction are

$$r_D = k_1 C_{JO} C_{AC}, \text{ and} \tag{21}$$

$$r_B = k_2 C_{SO} C_{AC} \tag{22}$$

Reactions 1 and 2 take place independently within a single pellet which contains both of the solid reactants, J and S.

The material balance for the gaseous reactant A in the ash layer of the pellet under the pseudo-steady-state assumption is represented by:

$$\frac{d^2\omega_A}{d\xi^2} + \frac{2}{\xi}\frac{d\omega_A}{d\xi} = 0 \tag{23}$$

The boundary condition at $\xi = \xi_s$ is

$$\left.\frac{d\omega_A}{d\xi}\right|_{\xi = \xi_s} = N_{Sh\bar{r}}(\omega_{AH} - \omega_{AS}) \tag{24}$$

and at $\xi = \xi_c$ is

$$\left.\frac{d\omega_A}{d\xi}\right|_{\xi = \xi_c} = \phi_{1\bar{r}}\left(\frac{\omega_{AC}}{U_c}\right)\exp\left\{\frac{E_1}{\bar{R}T_o}\left(1 - \frac{1}{U_c}\right)\right\} + \phi_{2\bar{r}}\left(\frac{\omega_{AC}}{U_c}\right)\exp\left\{\frac{E_2}{\bar{R}T_o}\left(1 - \frac{1}{U_c}\right)\right\} \tag{25}$$

Solution of the above equations gives the following result for ω_{AC}:

$$\frac{\omega_{AH}}{\omega_{AC}} = 1 + \frac{1}{U_c}\left[\phi_{1\bar{r}}\exp\left\{\frac{E_1}{\bar{R}T_o}\left(1 - \frac{1}{U_c}\right)\right\} + \phi_{2\bar{r}}\exp\left\{\frac{E_2}{\bar{R}T_o}\left(1 - \frac{1}{U_c}\right)\right\}\right]\left[\frac{1}{N_{Sh\bar{r}}}\frac{\xi_c^2}{\xi_s^2} + \xi_c\left(1 - \frac{\xi_c}{\xi_s}\right)\right] \tag{26}$$

The heat balance equation is

$$\frac{d^2U}{d\xi^2} + \frac{2}{\xi}\frac{dU}{d\xi} = 0 \tag{27}$$

with the following boundary conditions:
At $\xi = \xi_s$

$$-\left.\frac{dU}{d\xi}\right|_{\xi = \xi_s} = N_{Nu\bar{r}} (U_s - 1) \tag{28}$$

and at $\xi = \xi_c$

$$-\left(\frac{\bar{R}T_o}{E_1}\right)\left(\frac{dU}{d\xi}\right)\bigg|_{\xi = \xi_c} = \left(\frac{\omega_{AC}}{U_c}\right)\left[\phi_{1\bar{r}}{}^{\beta_1} \exp\left\{\frac{E_1}{\bar{R}T_o}\left(1 - \frac{1}{U_c}\right)\right\} + \phi_{2\bar{r}}{}^{\beta_2}\left(\frac{E_2}{E_1}\right) \exp\left\{\frac{E_2}{\bar{R}T_o}\left(1 - \frac{1}{U_c}\right)\right\}\right] \tag{29}$$

U_c is obtained from the following expression:

$$U_c - 1 = \frac{\omega_{AC}}{U_c}\left[\frac{1}{N_{Nu\bar{r}}}\frac{\xi_c^2}{\xi_s^2} + \xi_c\left(1 - \frac{\xi_c}{\xi_s}\right)\right] \phi_{1\bar{r}}{}^{\beta_1} \frac{E_1}{\bar{R}T_o} \exp\left\{\frac{1}{\bar{R}T_o}\left(1 - \frac{1}{U_c}\right)\right\} + \phi_{2\bar{r}}{}^{\beta_2}\left[\frac{E_2}{\bar{R}T_o} \exp\left\{\frac{E_2}{\bar{R}T_o}\left(1 - \frac{1}{U_c}\right)\right\}\right] \tag{30}$$

The conversion of the solid pellet expressed in terms of core radius is given by

$$-\frac{d\xi_c}{dt} = \frac{M_J}{\rho_{J\bar{r}}} k_1(T_o)C_{JO}C_{AO} \frac{\omega_{AC}}{U_c} \exp\left\{\frac{E_1}{\bar{R}T_o}\left(1 - \frac{1}{U_c}\right)\right\} + \frac{M_S}{\rho_{S\bar{r}}} k_2(T_o)C_{SO}C_{AO} \frac{\omega_{AC}}{U_c} \exp\left\{\frac{E_2}{\bar{R}T_o}\left(1 - \frac{1}{U_c}\right)\right\} \tag{31}$$

Let $$\tau_1 = \frac{\rho_{J\bar{r}}}{M_J k_1(T_o) C_{JO} C_{AO}}, \tag{32}$$

$$\tau_2 = \frac{\rho_{S\bar{r}}}{M_S k_2(T_o) C_{SO} C_{AO}}, \tag{33}$$

and

$$\theta_1 = t/\tau_1 \tag{34}$$

Substituting Eqs. 32, 33, and 34 into 31, we get:

$$-\frac{d\xi_c}{d\theta_1} = \frac{\omega_{AC}}{U_c}\left[\exp\left\{\frac{E_1}{\bar{R}T_o}\left(1 - \frac{1}{U_c}\right)\right\} + \frac{\tau_1}{\tau_2} \exp\left\{\frac{E_2}{\bar{R}T_o}\left(1 - \frac{1}{U_c}\right)\right\}\right] \tag{35}$$

On the basis of Eqs. 19 and 35, the rate of change of overall particle size is given by

$$-\frac{d\xi_s}{d\theta_1} = (1 - Z)\frac{\xi_c^2}{\xi_s^2}\frac{\omega_{AC}}{U_c}\left[\exp\left\{\frac{E_1}{\bar{R}T_o}\left(1 - \frac{1}{U_c}\right)\right\} + \frac{\tau_1}{\tau_2} \exp\left\{\frac{E_2}{\bar{R}T_o}\left(1 - \frac{1}{U_c}\right)\right\}\right] \tag{36}$$

The above equations will be employed in the mathematical modeling of fluidized bed presented below. When two solid reactants (char and limestone) are present, we shall use primed (limestone) and unprimed (char) symbols to distinguish between them.

Mathematical Model for Fluidized-Bed Combustion Process

The development of mathematical models to describe the thermochemical process occurring in a fluidized bed involves setting up the material and energy balance equations. The total process is represented in terms of a set of independent equations which are solved simultaneously to obtain such quantities as combustion efficiency, sulfur retention, oxygen utilization, oxygen and sulfur dioxide concentration profiles in the bed, etc.

The relationship between various streams, flow rates and particle sizes will be derived following the method of Kunii and Levenspiel (1). First, the relations are derived for char, whose particles generally shrink as they react with oxygen. The overall mass balance for char particles of the system is given by:

$$F_1 - F_3 - F_4 = \left\{ \begin{array}{c} \text{mass of carbon and sulfur} \\ \text{consumed by chemical reaction} \end{array} \right\} \tag{37}$$

In order to evaluate the right side of Eq. 37, we will calculate the mass loss for a single pellet due to reaction and sum up such losses for all particles present in the fluidized bed. Upon combustion, char leaves behind a layer of ash having a different density than that of coke. Thus, the mass of a single char particle, w_i, of size ξ_s in the bed is given by:

$$w_i = \frac{4}{3} \pi \bar{r}^3 \rho_Q \left[\frac{\rho_a}{\rho_Q} \left(\xi_s^3 - \xi_c^3 \right) + \xi_c^3 \right] \tag{38}$$

Therefore, the rate of change of mass of a single particle size ξ_s is,

$$\frac{dw_1}{dt} = 4 \pi \bar{r}^3 \rho_Q \left[\alpha_1 \xi_s^2 \frac{d\xi_s}{dt} + (1 - \alpha_1) \xi_c^2 \frac{d\xi_c}{dt} \right] \tag{39}$$

where

$$\alpha_1 = \frac{\rho_a}{\rho_Q} \tag{40}$$

The volume, dV, of the fluidized bed of cross-sectional area A_o and elemental height dH is $A_o dH$. Let f_Q be the fraction of char particles in the bed voidage ε. The volume of char particles of size ξ_s in the elemental volume dV is $f_Q(1 - \varepsilon)A_o P_2(\xi_s)d\xi_s dH$. The number of particles of size ξ_s in this elemental volume is

$$dN^* = \frac{f_Q(1 - \varepsilon)A_o P_2(\xi_s)d\xi_s dH}{(4/3)\, \bar{r}^3\, \xi_s^3} \tag{41}$$

Therefore, the mass depletion rate of char particles of size ξ_s in dV due to reaction is $(dw_1/dt)dN^*$. Substituting from Eqs. 41 and 39 into this relation we get

$$\frac{dw_1}{dt} dN^* = 3f_Q(1-\varepsilon)A_oP_2(\xi_s)\rho_Q \left\{\frac{\alpha_1}{\xi_s}\frac{d\xi_s}{dt} + \frac{(1-\alpha_1)}{\xi_s^3}\xi_c^2\frac{d\xi_c}{dt}\right\} d\xi_s dH \quad (42)$$

Thus, the total mass of carbon and sulfur consumed by chemical reaction, needed to evaluate the right side of Eq. 37, can be obtained by integrating Eq. 42 for all ξ_s, ξ_o, and for the entire height of the bed, H_o. The overall mass balance of Eq. 37 can, therefore, be written as

$$F_1 - F_3 - F_4 = \int_o^{H_o}\int_{\xi_{s,min}}^{\xi_{s,max}} 3f_Q(1-\varepsilon)A_oP_2(\xi_s)\rho_Q \left\{\frac{\alpha_1}{\xi_s}\frac{d\xi_s}{dt} + \frac{(1-\alpha_1)}{\xi_s^3}\xi_c^2\frac{d\xi_c}{dt}\right\} d\xi_s dH \quad (43)$$

The feed rate of char F_1, the overflow rate F_3, and the carryover rate F_4 are given in terms of their corresponding volumetric flow rates as follows.

$$F_1 = V_1(1-\varepsilon_1)\rho_Q \quad (44)$$

$$F_3 = \int_{\xi_{s,min}}^{\xi_{s,max}} V_3(1-\varepsilon_3)P_3(\xi_s)\,\rho_Q\left\{\alpha_1 + (1-\alpha_1)\frac{\xi_c^3}{\xi_s^3}\right\}d\xi_s \quad (45)$$

$$F_4 = \int_{\xi_{s,min}}^{\xi_{s,max}} V_4(1-\varepsilon_4)P_4(\xi_s)\,\rho_Q\left\{\alpha_1 + (1-\alpha_1)\frac{\xi_c^3}{\xi_s^3}\right\}d\xi_s \quad (46)$$

The strategy of the calculations involves the manipulation of Eqs. 45 and 46 so that F_3 and F_4 could be determined in terms of the only unknown f_Q. Equation 43 is then employed to establish the value of f_Q. In the following we develop the mathematical relations expressing F_3 and F_4 in terms of various quantities and f_Q.

Next we consider the mass balance at the steady state for char particles of size between ξ_s and $\xi_s + \Delta\xi_s$ for the entire system at a particular instant as shown in Figure 3. The corresponding relation for a system of shrinking particles is:

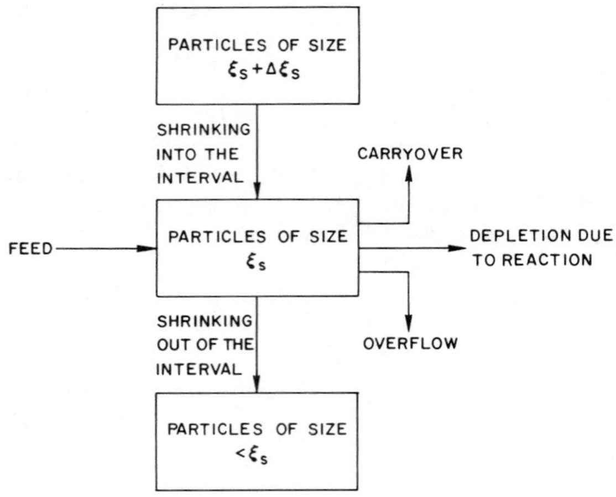

Figure 3. Mass balance for shrinking char particles of size ξ_s

$$F_1P_1(\xi_s)\Delta\xi_s - F_3P_3(\xi_s)\Delta\xi_s - F_4P_4(\xi_s)\Delta\xi_s + W_QP_2(\xi_s)\left.\frac{d\xi_s}{dt}\right|_{\xi_s+\Delta\xi_s} - W_QP_2(\xi_s)\left.\frac{d\xi_s}{dt}\right|_{\xi_s}$$
$$+\left[\int_o^{H_o} 3f_Q(1-\varepsilon)A_o\rho_QP_2(\xi_s)\left\{\frac{\alpha_1}{\xi_s}\frac{d\xi_s}{dt} + \frac{(1-\alpha_1)}{\xi_s^3}\xi_c^2\frac{d\xi_c}{dt}\right\}dH\right]\Delta\xi_s = 0 \tag{47}$$

Similarly for a system consisting of growing particles

$$F_1P_1(\xi_s)\Delta\xi_s - F_3P_3(\xi_s)\Delta\xi_s - F_4P_4(\xi_s)\Delta\xi_s + W_QP_2(\xi_s)\left.\frac{d\xi_s}{dt}\right|_{\xi_s}$$
$$- W_QP_2(\xi_s)\left.\frac{d\xi_s}{dt}\right|_{\xi_s+\Delta\xi_s} + \left[\int_o^{H_o} 3f_Q(1-\varepsilon)A_o\rho_QP_2(\xi_s)\right.$$
$$\left.\left\{\frac{\alpha_1}{\xi_s}\frac{d\xi_s}{dt} + \frac{(1-\alpha_1)}{\xi_s^3}\xi_c^2\frac{d\xi_c}{dt}\right\}dH\right]\Delta\xi_s = 0 \tag{48}$$

For simplicity we define

$$X_1 = \alpha_1 + (1-\alpha_1)(\xi_c^3/\xi_s^3) \tag{49}$$

Dividing Eq. 47 by $\Delta\xi_s$, taking limits as $\Delta\xi_s \to 0$ and substituting for F_1, F_3, F_4, and W_Q we get:

$$V_1(1-\varepsilon_1)\rho_QP_1(\xi_s) - V_3(1-\varepsilon_3)\rho_QX_1P_3(\xi_s) - V_4(1-\varepsilon_4)$$
$$\rho_QX_1P_4(\xi_s) + \int_o^{H_o} f_Q(1-\varepsilon)A_o\rho_Q\frac{d}{d\xi_s}\left(X_1P_2(\xi_s)\frac{d\xi_s}{dt}\right)dH$$
$$+\int_o^{H_o} 3f_Q(1-\varepsilon)A_o\rho_QP_2(\xi_s)\left\{\frac{\alpha_1}{\xi_s}\frac{d\xi_s}{dt} + \frac{(1-\alpha_1)}{\xi_s^3}\xi_c^2\frac{d\xi_c}{dt}\right\}dH = 0 \tag{50}$$

Similarly on dividing Eq. 48 by $\Delta\xi_s$, taking limits as $\Delta\xi_s \to 0$ and substituting for F_1, F_3, F_4, and W_Q we get:

$$V_1(1-\varepsilon_1)\rho_QP_1(\xi_s) - V_3(1-\varepsilon_3)\rho_QX_1P_3(\xi_s) - V_4(1-\varepsilon_4)$$
$$\rho_QX_1P_4(\xi_s) - \int_o^{H_o} f_Q(1-\varepsilon)A_o\rho_Q\frac{d}{d\xi_s}\left(X_1P_2(\xi_s)\frac{d\xi_s}{dt}\right)dH$$
$$+\int_o^{H_o} 3f_Q(1-\varepsilon)A_o\rho_QP_2(\xi_s)\left\{\frac{\alpha_1}{\xi_s}\frac{d\xi_s}{dt} + \frac{(1-\alpha_1)}{\xi_s^3}\xi_c^2\frac{d\xi_c}{dt}\right\}dH = 0 \tag{51}$$

It may be noted that for growing particles $(d\xi_s/dt) > 0$ and for shrinking particles $(d\xi_s/dt) < 0$. Comparing Eqs. 50 and 51, which represent the mass balances for particles of size ξ_s for shrinking and growing particles respectively, it is seen that they both differ from each other only in the sign for the fourth term. In order to represent both of them by a single equation, we may substitute the absolute value of $d\xi_s/dt$ in Eq. 51, so that

$$V_1(1 - \varepsilon_1)\rho_Q P_1(\xi_s) - V_3(1 - \varepsilon_3)\rho_Q X_1 P_3(\xi_s) - V_4(1 - \varepsilon_4)\rho_Q X_1 P_4(\xi_s) - \int_o^{H_o} f_Q(1 - \varepsilon)A_o\rho_Q \frac{d}{d\xi_s}\left(X_1 P_2(\xi_s)\left|\frac{d\xi_s}{dt}\right|\right) dH$$
$$+ \int_o^{H_o} 3f_Q(1 - \varepsilon)A_o\rho_Q P_2(\xi_s)\left\{\frac{\alpha_1}{\xi_s}\frac{d\xi_s}{dt} + \frac{(1 - \alpha_1)}{\xi_s^3}\xi_c^2\frac{d\xi_c}{dt}\right\} dH = 0 \tag{52}$$

Simplification of Eq. 52 gives

$$V_1(1-\varepsilon)P_1(\xi_s) - V_3(1-\varepsilon_3)X_1P_3(\xi_s) - V_4(1-\varepsilon_4)X_1P_4(\xi_s) - \int_o^{H_o} f_Q(1-\varepsilon)A_o \left[X_1P_2(\xi_s)\frac{d}{d\xi_s}\left|\frac{d\xi_s}{dt}\right| + X_1\left|\frac{d\xi_s}{dt}\right|\frac{dP_2(\xi_s)}{d\xi_s} - P_2(\xi_s)\cdot\left|\frac{d\xi_s}{dt}\right|\left(\frac{3(\alpha_1 - 1)\xi_o^3 Z}{(1 - Z)\xi_s^4}\right)\right] dH + \int_o^{H_o} 3f_Q(1 - \varepsilon)A_oP_2(\xi_s)\left\{\frac{\alpha_1}{\xi_s}\frac{d\xi_s}{dt} + \frac{(1 - \alpha_1)\xi_c^2}{\xi_s^3}\frac{d\xi_c}{dt}\right\}dH = 0 \tag{53}$$

Equation 53 consists of three unknown quantities, <u>viz.</u>, $P_2(\xi_s)$, $P_3(\xi_s)$, and $P_4(\xi_s)$. Fortunately, for fluidized bed operations these quantities are inter-related and this simplifies the calculation procedure. The relationships between these quantities are discussed below.

The solids in the bed are assumed to be backmixed and, therefore,

$$P_2(\xi_s) = P_3(\xi_s) \tag{54}$$

The elutriation constant for char particles is:

$$\kappa_Q(\xi_s) = \frac{V_4(1 - \varepsilon_4)P_4(\xi_s)}{f_Q(1 - \varepsilon)A_oH_oP_2(\xi_s)} \tag{55}$$

Therefore,

$$P_4(\xi_s) = \frac{f_Q(1-\varepsilon)A_oH_o\kappa_Q(\xi_s)P_2(\xi_s)}{V_4(1-\varepsilon_4)} \tag{56}$$

Substituting Eqs. 54 and 56 in Eq. 53 we get on simplification and rearrangement:

$$-f_Q(1-\varepsilon)A_oX_1\frac{dP_3(\xi_s)}{d\xi_s}\int_o^{H_o}\left|\frac{d\xi_s}{dt}\right|dH = -V_1(1-\varepsilon_1)P_1(\xi_s) + \left\{V_3(1-\varepsilon_3)\right.$$

$$\left.+f_Q(1-\varepsilon)A_oH_o\kappa_Q(\xi_s)\right\}X_1P_3(\xi_s) + f_Q(1-\varepsilon)A_oX_1P_3(\xi_s)\int_o^{H_o}\frac{d}{d\xi_s}\left|\frac{d\xi_s}{dt}\right|dH$$

$$-f_Q(1-\varepsilon)A_o\frac{3(\alpha_1-1)\xi_o^3Z}{(1-Z)\xi_s^4}P_3(\xi_s)\int_o^{H_o}\left|\frac{d\xi_s}{dt}\right|dH - 3f_Q(1-\varepsilon)A_oP_3(\xi_s)\int_o^{H_o}$$

$$\left\{\frac{\alpha_1}{\xi_s}\frac{d\xi_s}{dt} + \frac{(1-\alpha_1)\xi_c^2}{\xi_s^3}\frac{d\xi_c}{dt}\right\}dH \tag{57}$$

This analysis for particles of size between ξ_s and $\xi_s + d\xi_s$ can be easily extended to the feed with wide size distribution. The feed with wide size distribution can be looked upon as the sum of narrow cuts of solids and it will be reasonable to expect the outflow stream to be the sum of outflow streams from these narrow cuts. To achieve this we shall first examine the system using a single size feed.

Consider a feed of size $R_o(\xi_o)$. The particles change size as the reaction proceeds. Thus, for the entire system all particles will be less than or equal to ξ_o for shrinking particles, and equal to or greater than ξ_o for the growing particles. The above analysis will be performed for this feed for particles in the size range ξ_s and $\xi_s + \Delta\xi_s$ (not including ξ_o). Equation 57 can be applied directly as the feed size is constant and not included in the above size range we have,

$$F_1P_1(\xi_s)\Delta\xi_s = 0 \tag{58}$$

Therefore, from Eq. 57 we can write as follows:

$$\frac{dP_3^*(\xi_s)}{P_3^*(\xi_s)} = \frac{-\left\{\frac{V_3(1-\varepsilon_3)}{f_Q(1-\varepsilon)A_o} + H_o\kappa_Q(\xi_s)\right\}}{\int_o^{H_o}\left|\frac{d\xi_s}{dt}\right| dH} d\xi_s - \frac{\int_o^{H_o}\frac{d}{d\xi_s}\left|\frac{d\xi_s}{dt}\right| dH}{\int_o^{H_{max}}\left|\frac{d\xi_s}{dt}\right| dH} d\xi_s$$

$$+ \frac{1}{X_1}\left(\frac{3(\alpha_1-1)\xi_o^3 Z}{(1-Z)\xi_s^4}\right) d\xi_s + \frac{3\alpha_1 d\xi_s}{X_1\xi_s^3} + \frac{3(1-\alpha_1)\xi_c^2}{X_1\xi_s^3}\left(\int_o^{H_o}\frac{d\xi_c}{dt}\, dH\right)\Bigg/$$

$$\left(\int_o^{H_o}\left|\frac{d\xi_s}{dt}\right| dH\right) d\xi_s \qquad (59)$$

$P_3^*(\xi_s)$ is the size distribution of the particles in the bed and overflow stream for a feed of fixed particle size ξ_o. The relation connecting ξ_s, ξ_c, ξ_o and Z is,

$$\xi_s^3 = Z\xi_o^3 + (1-Z)\xi_c^3 \qquad (60)$$

Therefore

$$\left(\frac{d\xi_c}{dt}\right)\Bigg/\left(\frac{d\xi_s}{dt}\right) = \frac{\xi_s^2}{(1-Z)\xi_c^2} \qquad (61)$$

Substituting Eqs. 49, 60, and 61 into Eq. 59 we finally get,

$$\frac{dP_3^*(\xi_s)}{P_3^*(\xi_s)} = \frac{-\left\{\frac{V_3(1-\varepsilon_3)}{f_Q(1-\varepsilon)A_o} + H_o\kappa_Q(\xi_s)\right\}}{\int_o^{H_o}\left|\frac{d\xi_s}{dt}\right| dH} d\xi_s - \frac{\int_o^{H_o}\frac{d}{d\xi_s}\left|\frac{d\xi_s}{dt}\right| dH}{\int_o^{H_o}\left|\frac{d\xi_s}{dt}\right| dH} d\xi_s$$

$$+ \frac{3(\alpha_1-1)\xi_o^3 Z}{X_1(1-Z)\xi_s^4} d\xi_s + \frac{3(1-\alpha_1 Z)}{X_1(1-Z)\xi_s} \qquad (62)$$

where $X_1 = \alpha_1 + (1-\alpha_1)(\xi_s^3 - Z\xi_o^3)/(1-Z)\xi_s^3$ (63)

Integration with respect to ξ_s leads to the following:

$$\ln \frac{P_3^*(\xi_s)}{P_3^*(\xi_o)} = - \int_{\xi_o}^{\xi_s} \frac{\left\{\frac{V_3(1-\varepsilon_3)}{f_Q(1-\varepsilon)A_o} + H_o\kappa_Q(\xi_s)\right\}}{\int_o^{H_o} \left|\frac{d\xi_s}{dt}\right| dH} d\xi_s - \ln \frac{\int_o^{H_o} \left|\frac{d\xi_s}{dt}\right| dH}{\int_o^{H_o} \left|\frac{d\xi_s}{dt}\right| dH \Big|_{\xi_s=\xi_o}} + \ln \frac{\xi_s^3}{\xi_o^3} \tag{64}$$

The above expression gives $P_3^*(\xi_s)$ if we could evaluate $P_3^*(\xi_o)$ which comes out to be

$$P_3^*(\xi_o) = \frac{V_1(1-\varepsilon_1)}{f_Q(1-\varepsilon)A_o \int_o^{H_o} \left|\frac{d\xi_s}{dt}\right| dH} \tag{65}$$

Substituting Eq. 65 into Eq. 64 we get,

$$P_3^*(\xi_s) = \frac{V_1(1-\varepsilon)\xi_s^3 I(\xi_s,\xi_o)}{f_Q(1-\varepsilon)A_o\xi_o^3 \int_o^{H_o} \left|\frac{d\xi_s}{dt}\right| dH} \tag{66}$$

where

$$I(\xi_s,\xi_o) = \exp\left[- \int_{\xi_o}^{\xi_s} \frac{\left\{\frac{V_3(1-\varepsilon_3)}{f_Q(1-\varepsilon)A_o} + H_o\kappa_Q(\xi_s)\right\}}{\int_o^{H} \left|\frac{d\xi_s}{dt}\right| dH} d\xi_s \right] \tag{67}$$

The above results apply to a single size feed. For a wide distribution

$$P_3(\xi_s)\Delta\xi_s = \sum_{\xi_{o,min}}^{\xi_{o,max}} P_3^*(\xi_s)\Delta\xi_s P_1(\xi_o)\Delta\xi_o \tag{68}$$

where $\xi_{o,max}$ and $\xi_{o,min}$ are the largest and the smallest size particles in the feed.

The output distribution function $P_3^*(\xi_s)$ for constant input size ξ_o is already derived and is given by Eq. 66, dividing Eq. 68 by $\Delta\xi_s$ and taking limits as $\Delta\xi_s \rightarrow 0$, we get

$$P_3(\xi_s) = \int_{\xi_{o,min}}^{\xi_{o,max}} P_3(\xi_s)\ P_1(\xi_o)d\xi_o \tag{69}$$

Substituting Eq. 66 into Eq. 69 we get

$$P_3(\xi_s) = \frac{V_1(1 - \varepsilon_1)}{f_Q(1-\varepsilon)A_o} \xi_s^3 \int_{\xi_{o,min}}^{\xi_{o,max}} \frac{P_1(\xi_o)I(\xi_s,\xi_o)d\xi_o}{\xi_o^3 \int_o^H \left|\frac{d\xi_s}{dt}\right| dH} \tag{70}$$

Equation 72 defines $P_3(\xi_s)$ in which there are two unknowns viz., V_3 and f_Q. One of them can be eliminated by utilizing the normalization condition for $P_3(\xi_s)$. Integration of Eq. 70 for all sizes of particles in the bed ($\xi_{s,min}$ to $\xi_{s,max}$) yields

$$\frac{f_Q(1-\varepsilon)A_o}{V_1(1-\varepsilon_1)} = \int_{\xi_{s,min}}^{\xi_{s,max}} \int_{\xi_{o,min}}^{\xi_{o,max}} \frac{\xi_s^3 P_1(\xi_o)I(\xi_s,\xi_o)}{\xi_o^3 \int_o^H \left|\frac{d\xi_s}{dt}\right| dH} d\xi_o d\xi_s \tag{71}$$

It may be noted that $\xi_{s,max} = \xi_{o,max}$ for shrinking particles. If H is nondimensionalized such that: $\eta = (H/H_o)$. The resulting equations are then utilized to compute F_3 from Eq. 45 as a function of f_Q. Next we proceed to establish a mathematical framework in a somewhat analogous fashion for calculating F_4 as a function of f_Q.

Combining Eqs. 54 and 56 and substituting for $P_3(\xi_s)$ from Eq. 70 we get,

$$P_4(\xi_s) = \kappa_Q(\xi_s) \frac{V_1(1-\varepsilon_1)}{V_4(1-\varepsilon_4)} \xi_s^3 \int_{\xi_{o,min}}^{\xi_{o,max}} \frac{P_1(\xi_o)I(\xi_s,\xi_o)}{\xi_o^3 \int_o^1 \left|\frac{d\xi_s}{dt}\right| d\eta} d\xi_o \tag{72}$$

The normalization of the size distribution function $P_4(\xi_s)$ finally yields

$$\frac{V_4(1-\varepsilon_4)}{V_1(1-\varepsilon_1)} = \int_{\xi_{s,min}}^{\xi_{s,max}} \int_{\xi_{o,min}}^{\xi_{o,max}} \frac{\xi_s^3 \kappa_Q(\xi_s) P_1(\xi_o) I(\xi_s,\xi_o)}{\xi_o^3 \int_o^1 \left|\frac{d\xi_s}{dt}\right| d\eta} \tag{73}$$

Equations 72, 73 and 71 can be solved simultaneously to obtain $P_4(\xi_s)$ and V_4 as a function of f_Q. These relations are then utilized in Eq. 46 to obtain F_4 as a function of f_Q only.

The mathematical development presented so far enables us to employ the mass balance of Eq. 43 to determine uniquely the fraction of char present in the bed for a given set of feed and operating conditions.

A similar set of equations can be derived for the various streams of limestone. For the general case when the size of the particle changes as it reacts with oxygen and sulfur dioxide, the following equations apply. The overall mass balance is given by the following relation which is analogous to Eq. 43 developed above for char particles:

$$F_1' - F_3' - F_4' = \int_{\xi'_{s,min}}^{\xi'_{s,max}} \int_o^1 3f_N(1-\varepsilon)A_oH_oP_3'(\xi_s')\rho_N \left\{\frac{\alpha_2}{\xi_s'}\frac{d\xi_s'}{dt} + \frac{(1-\alpha_2)\xi_c'^2}{\xi_s'^3}\frac{d\xi_c'}{dt}\right\} d\eta \, d\xi_s' \tag{74}$$

where the feed rate of limestone F_1', the overflow rate F_3', and the carryover rate F_4', may be expressed in terms of the corresponding volumetric flow rates.

Adopting the approach developed above for the char particles combustion, the size distribution function of limestone particles as a result of sulfation reaction in the overflow stream which is the same as in the bed is given by,

$$P_3'(\xi_s') = \frac{V_1'(1-\varepsilon_1')}{f_N(1-\varepsilon)A_oH_o}\xi_s'^3 \int_{\xi'_{o,min}}^{\xi'_{o,max}} \frac{P_1'(\xi_o')I'(\xi_s',\xi_o')d\xi_o'}{\xi_o'^3 \int_o^1 \left|\frac{d\xi_s'}{dt}\right| d\eta} \tag{75}$$

The normalization property of $P_3'(\xi_s')$ leeds to the following relationship between V_1' and V_3':

$$\frac{f_N(1-\varepsilon)A_oH_o}{V_1'(1-\varepsilon_1')} = \int_{\xi'_{s,min}}^{\xi'_{s,max}} \int_{\xi'_{o,min}}^{\xi'_{o,max}} \frac{\xi_s'^3 P_1'(\xi_o')I'(\xi_s',\xi_o')d\xi_o' d\xi_s'}{\xi_o'^3 \int_o^1 \left|\frac{d\xi_s'}{dt}\right| d\eta} \tag{76}$$

Here

$$I(\xi_s', \xi_o') = \exp - \left[\frac{\int_{\xi_o'}^{\xi_s'} \left\{ \frac{V_3'(1-\varepsilon_3)}{f_N(1-\varepsilon) A_o H_o} + \kappa_N(\xi_s') \right\} d\xi_s'}{\int_o^1 \left| \frac{d\xi_s'}{dt} \right| d\eta} \right] \tag{77}$$

and

$$\kappa_N(\xi_s') = \frac{V_4'(1-\varepsilon_4) P_4'(\xi_s')}{A_o H_o (1-\varepsilon) f_N P_2'(\xi_s')} \tag{78}$$

The size distribution of limestone in the elutriated stream, $P_4'(\xi_s')$, is obtained from Eq. 78 in which $P_2'(\xi_s')$ is replaced with $P_3'(\xi_s')$ (backmix approximation) as defined by Eqs. 75 and 77.

In the present formulation the size of the fluidized bed is kept constant by the presence of an overflow pipe, Figure 1, and consequently A_o and H_o are constant. The fraction of dolomite in the bed, f_N, is related to the fraction of char in the bed such that

$$f_N = 1 - f_Q \tag{79}$$

f_Q has already been determined and hence it may be assumed that f_N is known.

In case the size and density of the limestone particles remain the same as a result of chemical reaction, the above relations are simplified.

We next develop the mass balance equations for the gaseous reactant (oxygen) and the product (sulfur dioxide). The gas flow in the reactor is assumed to be in plug flow and hence the concentration of these gases will depend only on the height H, in the bed above the distributor plate. The rate of consumption of oxygen by reactions 1, 2 and 3 can be obtained from Eqs. 43 and 80 and the stoichiometry of these reactions. We will first examine Eq. 43 which may be rewritten as follows after appropriate substitutions.

$$\text{RHS of Eq. 43} = \frac{3V_1(1-\varepsilon_1)\rho_Q(1-\alpha_1)Z}{\tau_1} \int_{\xi_{s,min}}^{\xi_{s,max}} \int_{\xi_{o,min}}^{\xi_{o,max}} \int_o^1 \frac{\omega_{AC}}{U_c}$$

$$\frac{I(\xi_s,\xi_o) P_1(\xi_o) \exp\left\{ \frac{E_1}{RT_o} \left(1 - \frac{1}{U_c}\right) \right\}}{\int_o^1 \left| \frac{d\xi_s}{dt} \right| d\eta} \, d\eta \, d\xi_o \, d\xi_s$$

$$+ \frac{3V_1(1-\varepsilon_1)\rho_Q(1-\alpha_1 Z)}{\tau_2} \int_{\xi_{s,min}}^{\xi_{s,max}} \int_{\xi_{o,min}}^{\xi_{s,max}} \int_{o}^{1} \frac{\omega_{AC}}{U_c} \frac{I(\xi_s,\xi_o)P_1(\xi_o) \exp\left\{\frac{E_2}{\bar{R}T_o}\left(1-\frac{1}{U_c}\right)\right\}}{\int_o^1 \left|\frac{d\xi_s}{dt}\right| d\eta} d\eta d\xi_o d\xi_s \tag{80}$$

The first term of Eq. 80 represents the mass loss of char particles due to carbon combustion and the second term represents the mass loss of char particles due to sulfur reaction with oxygen. Using the stoichiometry of reactions 1 and 2, we can obtain the moles of oxygen used up in these respective reactions for any arbitrary height η as,

$$\begin{bmatrix}\text{Moles of oxygen} \\ \text{used upto bed} \\ \text{height } \eta \text{ due to} \\ \text{reaction 1}\end{bmatrix} = \frac{3V_1(1-\varepsilon_1)\rho_Q(1-\alpha_1 Z)}{M_J\ \tau_1} \int_{\xi_{s,min}}^{\xi_{s,max}} \int_{\xi_{o,min}}^{\xi_{o,max}} \int_{o}^{\eta} \frac{\omega_{AC}}{U_c} \frac{I(\xi_s,\xi_o)P_1(\xi_o) \exp\left\{\frac{E_1}{\bar{R}T_o}\left(1-\frac{1}{U_c}\right)\right\}}{\int_o^1 \left|\frac{d\xi_s}{dt}\right| d\eta} \tag{81}$$

$$\begin{bmatrix}\text{Moles of oxygen} \\ \text{used upto bed} \\ \text{height } \eta \text{ due to} \\ \text{reaction 2}\end{bmatrix} = \frac{3V_1(1-\varepsilon_1)\rho_Q(1-\alpha_1 Z)}{M_S\ \tau_2} \int_{\xi_{s,min}}^{\xi_{s,max}} \int_{\xi_{o,min}}^{\xi_{o,max}} \int_{o}^{\eta} \frac{\omega_{AC}}{U_c} \frac{I(\xi_s,\xi_o)P_1(\xi_o) \exp\left\{\frac{E_2}{\bar{R}T_o}\left(1-\frac{1}{U_c}\right)\right\}}{\int_o^1 \left|\frac{d\xi_s}{dt}\right| d\eta} d\eta d\xi_o d\xi_s \tag{82}$$

$$\begin{bmatrix}\text{Moles of oxygen}\\ \text{consumed upto bed}\\ \text{height } \eta \text{ due to}\\ \text{reaction 3}\end{bmatrix} = \frac{3V_1'(1-\epsilon_1')\rho_N(1-\alpha_2 Z')}{2M_S\,\tau_3} \int_{\xi_s',min}^{\xi_s',max} \int_{\xi_o',min}^{\xi_o',max} \int_o^{\eta} \frac{\omega_{BC}}{U_c'}$$

$$\frac{I'(\xi_s',\xi_o')P_1'(\xi_o)\exp\left\{\frac{E_3}{\bar{R}To}\left(1-\frac{1}{U_c'}\right)\right\}}{\int_o^1 \left|\frac{d\xi_s}{dt}\right| d\eta} d\eta\, d\xi_o'\, d\xi_s' \tag{83}$$

Since there is only a small change in the total moles of gases in the fluidized bed combustor as a result of chemical reactions, we can assume that the gas flow remains unchanged. Let this flow be Y. Also the bulk temperature of the fluidized bed remains constant, total gas concentration remains constant throughout the reactor and hence

$$\omega_{AH} = C_{AH}/C_{AO} \tag{84}$$

The rate of change of oxygen concentration with height in the fluidized bed is given by the following relation.

$$YC_{AO}\frac{d\omega_{AH}}{d\eta} = -\frac{3V_1(1-\epsilon_1)\rho_Q(1-\alpha_1 Z)}{M_J\,\tau_1} \int_{\xi_{s,min}}^{\xi_{s,max}} \int_{\xi_{o,min}}^{\xi_{o,max}} \frac{\omega_{AC}}{U_c}$$

$$\frac{I(\xi_s,\xi_o)P_1(\xi_o)\exp\left\{\frac{E_1}{\bar{R}To}\left(1-\frac{1}{U_c}\right)\right\}}{\int_o^1 \left|\frac{d\xi_s}{dt}\right| d\eta} d\xi_o d\xi_s$$

$$\frac{-3V_1(1-\epsilon_1)\rho_Q(1-\alpha_1 Z)}{\tau_2\, M_S} \int_{\xi_{s,min}}^{\xi_{s,max}} \int_{\xi_{o,min}}^{\xi_{o,max}} \frac{\omega_{AC}}{U_c}$$

$$\frac{I(\xi_s,\xi_o)P_1(\xi_o)\exp\left\{\frac{E_2}{\bar{R}T_o}\left(1-\frac{1}{U_c}\right)\right\}}{\int_o^1 \left|\frac{d\xi_s}{dt}\right| d\eta} d\xi_o d\xi_s$$

$$- \frac{3V_1'(1-\varepsilon_1')\rho_N(1-\alpha_2 Z')}{2M_S\ \tau_3} \int_{\xi'_{s,min}}^{\xi'_{s,max}} \int_{\xi'_{o,min}}^{\xi'_{o,max}} \frac{\omega_{BC}}{U_c'} \frac{I'(\xi_s',\xi_o')P_1'(\xi_o') \exp\left\{\frac{E_3}{\bar{R}T_o}\left(1 - \frac{1}{U_c'}\right)\right\}}{\int_o^1 \left|\frac{d\xi_s'}{dt}\right| d\eta} d\xi_o' d\xi_s' \quad (85)$$

The boundary condition for Eq. 85 is

$$\omega_{AH} = 1 \text{ at } \eta = 0 \quad (86)$$

The solution of Eqs. 85 with 86 will yield the oxygen concentration profile along the fluidized bed combustor.

Similarly the mass balance equation for sulfur dioxide is

$$YC_{AO} \frac{d\omega_{BH}}{d\eta} = \frac{3V_1(1-\varepsilon_1)\rho_Q(1-\alpha_1 Z)}{\tau_2\ M_S} \int_{\xi_{s,min}}^{\xi_{s,max}} \int_{\xi_{o,min}}^{\xi_{o,max}} \frac{\omega_{AC}}{U_c} \frac{I(\xi_s,\xi_o)P_1(\xi_o) \exp\left\{\frac{E_2}{\bar{R}T_o}\left(1 - \frac{1}{U_c}\right)\right\}}{\int_o^1 \left|\frac{d\xi_s}{dt}\right| d\eta} d\xi_o d\xi_s$$

$$- \frac{3V_1'(1-\varepsilon_1')\rho_N(1-\alpha_2 Z')}{\tau_3\ M_S} \int_{\xi'_{o,max}}^{\xi'_{s,max}} \int_{\xi'_{o,min}}^{\xi'_{o,max}} \frac{\omega_{BC}}{U_c'} \frac{I'(\xi_s',\xi_o')P_1'(\xi_o') \exp\left\{\frac{E_3}{\bar{R}T_o}\left(1 - \frac{1}{U_c'}\right)\right\}}{\int_o^1 \frac{d\xi_s'}{dt} d\eta} d\xi_o' d\xi_s' \quad (87)$$

The boundary condition for Eq. 87 is given by

$$\omega_{BH} = 0 \text{ at } \eta = 0 \tag{88}$$

The solution of Eq. 87 with Eq. 88 will establish the sulfur dioxide concentration profile along the fluidized bed combustor.

There are two more quantities that must be defined to complete the description of the fluidized bed combustor viz., the carbon combustion efficiency, η_{CCE}, and sulfur absorption efficiency, η_{SAE}. These are:

$$F_1 f_c \eta_{CCE} = \frac{3V_1(1-\varepsilon_1)\rho_Q(1-\alpha_1 Z)}{\tau_1} \int_{\xi_{s,min}}^{\xi_{s,max}} \int_{\xi_{o,min}}^{\xi_{o,max}} \int_o^1$$

$$\frac{\omega_{AC}}{U_c} \frac{I(\xi_s,\xi_o)P_1(\xi_o) \exp\left\{\frac{E_1}{\bar{R}T_o}\left(1 - \frac{1}{U_c}\right)\right\}}{\int_o^1 \left|\frac{d\xi_s}{dt}\right| d\eta} d\eta\, d\xi_o\, d\xi_s \tag{89}$$

and

$$\eta_{SAE} = 1 - \frac{C_{Ao} M_S \omega_{BH}(\eta=1)}{F_1 f_s \eta_{CCE}} \tag{90}$$

The sulfur adsorption efficiency, η_{SAE}, is defined as the ratio of moles of sulfur dioxide consumed by the sulfation reaction to the moles of sulfur dioxide produced due to char combustion. The above equations will be utilized to analyze the parametric sensitivity of the fluidized-bed combustion operation.

Numerical Calculations: Parametric Investigations

The mathematical model for char combustion described in the previous two sections is applicable to a bed of constant volume, i.e., to a fluidized bed of fixed height, H_o, and having a constant cross-sectional area, A_o. The constant bed height is maintained by an overflow pipe. For this type of combustor operating for a given feed rate of char and limestone particles of known size distributions, the model presented here can predict the following:

(1) the fraction of char particles in the bed, f_Q;
(2) the fraction of limestone particles in the bed, f_N;
(3) the size distribution of char particles in the bed or in the overflow, $P_3(\xi_s)$;

(4) the size distribution of char particles in the bed or in the overflow, $P_3'(\xi_s')$;
(5) the overflow rate of char particles, F_3;
(6) the overflow rate of limestone particles, F_3';
(7) the flow rate of char particles in the carryover stream, F_4;
(8) the particle size distribution in the carryover stream, $P_4(\xi_s)$;
(9) the flow rate of limestone particles in the carryover stream, F_4';
(10) the particle size distribution of limestone particles in the carryover stream, $P_4'(\xi_s')$;
(11) the concentration profile of oxygen along the fluidized bed, ω_{AH};
(12) the concentration profile of sulfur dioxide along the fluidized bed, ω_{BH};
(13) the carbon conversion efficiency, η_{CCE}; and
(14) the sulfur absorption efficiency, η_{SAE}.

The above calculation is quite tedious and gets complicated by the fact that the properties which ultimately control the magnitude of these fourteen unknown quantities further depend on the physical and chemical parameters of the system such as reaction rate constants, initial size distribution of the feed, bed temperature, elutriation constants, heat and mass transfer coefficients, particle growth factors for char and limestone particles, flow rates of solid and gaseous reactants. In a complete analysis of a fluidized bed combustor with sulfur absorption by limestone, the influence of all the above parameters must be evaluated to enable us to optimize the system. In the present report we have limited the scope of our calculations by considering only the initial size of the limestone particles and the reaction rate constant for the sulfation reaction.

Further, it is not necessary to carry out excessive calculations to investigate the parametric sensitivity of the combustor operation. The same goal can be accomplished by assuming some of the fourteen unknowns and determining the remaining by the solution of the above mentioned equations. This procedure is adopted here. We assume a form for the oxygen profile, values of the carbon combustion and sulfur absorption efficiencies, char feed rate, and the various constants of the system and then the framework of mathematical model is employed to evaluate the amount of dolomite, F_1', needed to obtain such an operation. In general, the functional form for oxygen profile in a combustor is as follows:

$$\omega_{AH} = a^* + (1 - a^*)e^{-b^*\eta} \quad (91)$$

Here the constants a* and b* are to be specified. The constant a* is generally related to the excess oxygen in the flue gas whereas b* establishes the slope of the profile. A larger value of b* signifies that a major degree of combustion reactions take place near the bottom of the reactor.

The fluidized-bed coal combustion calculations described below according to the mathematical model developed here will employ the parameter values given in Table 1. The analysis is confined to a fixed temperature of 1225 K for the bed. Figure 4 gives the normalized particle size distribution for char and dolomite feeds for a particular case when $\bar{r} = \bar{r}' = 0.04$ cm. The particle size distribution for char in the feed is held constant, whereas for dolomite it is changed such that $\bar{r}'$ varies from 0.02 to 0.08 cm. The reaction rate between sulfur dioxide and dolomite is changed by varying the reaction rate constant, $k_3(T_o)$, from 380 to 960 $cm^4/mol \cdot s$. The changes in the dolomite size and reaction rate constant reflect in τ_3 and $\phi_{3\bar{r}}$ as seen from Table 1. The variation in the heat and mass transfer coefficients due to change in particle size within the range considered here is found to be negligible (±5%) and, therefore, is ignored in the present calculations. In one of our test runs the sulfur absorption efficiency is kept constant at 0.99 while the carbon combustion efficiency is varied from 0.7 to 0.995. In all of the remaining runs, the carbon combustion efficiency is held constant at 0.995 while the sulfur absorption efficiency is varied from 0.7 to 0.99. It may be pointed out that the dolomite feed rate is directly proportional to the rate of change of size of the dolomite particle, Eq. 73, and consequently the dolomite requirement is strongly dependent upon the factors influencing the rate of change of the dolomite particle. Let us examine the effect of changing oxygen profile on the dolomite requirement.

If the values of η_{CCE} and η_{SAE} are fixed, the parameter a* of Eq. 91 has a definite value. The oxygen concentration profiles are then changed by altering the value of the parameter b*. By assuming b* as 2.5, 4.5 and 6.5, the computed oxygen distribution in the bed is obtained as shown in Figure 5. The effect of these profiles is examined for two values of $k_3(T_o)$, 480 and 960 $cm^4/mol \cdot s$, and two values of $\bar{r}'$ (0.04 and 0.08 cm) on limestone requirement. The results are shown in Figures 6 and 7 for two values of η_{SAE}, <u>viz.</u>, 0.8 and 0.99. It is seen that for the change in oxygen profile considered here, the variation in limestone requirements, F_1', is within 10%. The trend in the change of the dolomite requirement with the change in the value of b* is always the same, regardless of the value of $k_3(T_o)$, $\bar{r}'$ and η_{SAE}. It is, therefore, appropriate to fix a value of b* while investigating the parametric sensitivity of the proposed mathematical model.

The concentration profile of oxygen in the bed is fixed by establishing apriori a value for b* as 4.5 and that of a* as obtained from the assumed values of carbon conversion and sulfur absorption efficiencies. For a given oxygen profile the reaction rate constant, $k_3(T_o)$, and the size of the dolomite feed are varied. The changes in both of these parameters affect the value

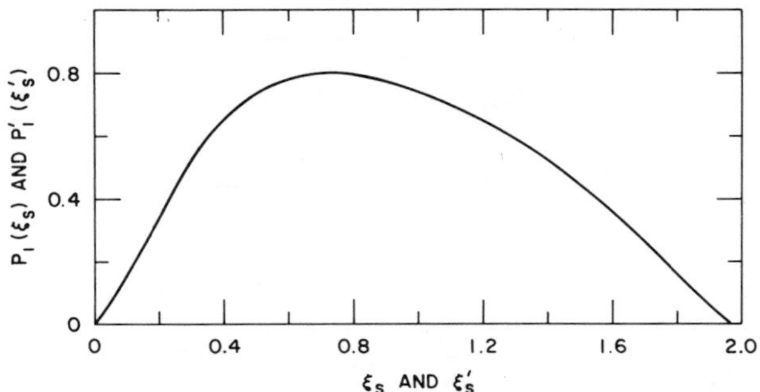

Figure 4. Particle size distribution of char, $P_1(\xi_s)$, and limestone, $P_1'(\xi_s')$, feeds: $\bar{r} = \bar{r}' = 0.04$ cm

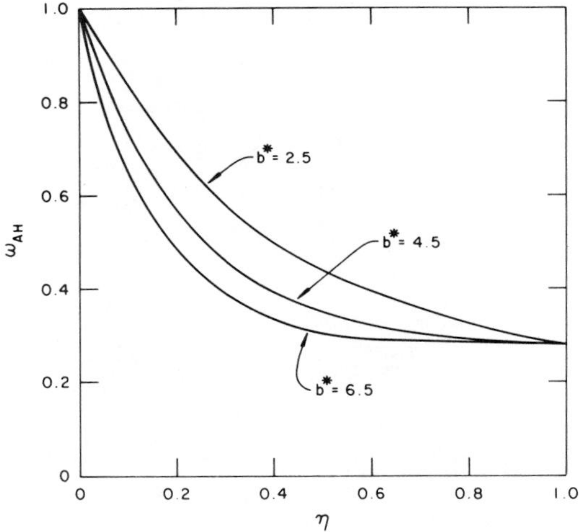

Figure 5. Oxygen profiles in the fluidized-bed combustor corresponding to $\eta_{CCE} = 0.995$ and $\eta_{SAE} = 0.99$

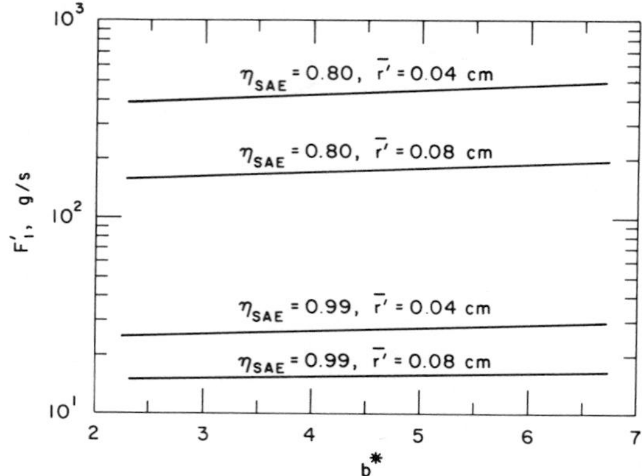

Figure 6. Effect of oxygen concentration profile on limestone requirement for different assumed η_{SAE} and $\bar{r}'$

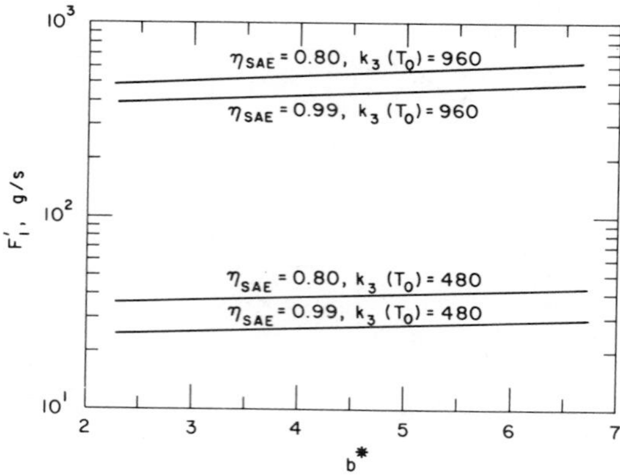

Figure 7. Effect of oxygen concentration profile on limestone requirement for different assumed η_{SAE} and $k_3(T_o)$

Table 1. Constants Used in Coal Combustion Model Calculations:

$Z = 0.05$, $Z' = 1.25$, $K_1 = 1.0$, $K_3 = 0.4$, $K_4 = 0.2$, $\tau_1 = 7.6$ s, $\tau_2 = 7.6$ s, $E_1/\bar{R}T_o = 13.8$, $E_2/\bar{R}T_o = 12.3$, $E_3/\bar{R}T_o = 5.9$, $\beta_1 = 0.0057$, $\beta_2 = 0.0068$, $\beta_3 = 0.0086$, $\phi_{1\bar{r}} = 4.9$, $\phi_{2\bar{r}} = 0.21$, $F_1 = 52$ g/s, $\varepsilon = 0.50$, $\varepsilon_1 = \varepsilon_1' = \varepsilon_3 = \varepsilon_4 = 0.45$, $u = 76.2$ cm/s, $H_o = 140$ cm, $\alpha_1 = 2.1$, $\alpha_2 = 1.36$, $\rho_Q = 1.0$ g/cm^3, $\rho_N = 2.45$ g/cm^3, $\rho_S = 0.035$ g/cm^3, $M_J = 12$ g/gmole, $M_N = 100$ g/gmole, $M_S = 32$ g/gmole, $Y = 4.25 \times 10^5$ cm^3/s, $C_{AO} = 1.2 \times 10^{-4}$ mole/cm^3, $\kappa_Q(\xi_s) = 1.6 \times 10^{-5}\ u^4/\xi_s^3$, $\kappa_N(\xi_s') = 1.6 \times 10^{-5}\ u^4/\xi_s'^3$, $f_c = 0.82$, and $\bar{r} = 0.04$ cm.

$\bar{r}'$ cm	$k_3(T_o)$ $cm^4/s \cdot mol$	$\phi_{3\bar{r}}$ (-)	τ_3 s	η_{CCE} (-)	η_{SAE} (-)
0.04	380	1.29	8.6	0.995	0.7-0.99
0.04	480	1.63	6.8	0.995	0.7-0.99
0.04	580	1.97	5.8	0.995	0.7-0.99
0.04	960	3.26	3.4	0.995	0.7-0.99
0.02	480	0.79	3.3	0.995	0.7-0.99
0.06	480	2.36	9.9	0.995	0.7-0.99
0.08	480	3.14	13.2	0.995	0.7-0.99
0.04	480	1.63	6.8	0.7-0.995	0.99

of $d\xi_s'/dt$ according to Eq. 20. The value of $d\xi_s'/dt$ is directly proportional to $k_3(T_o)$ and also to the dolomite feed size, $\bar{r}'$. Furthermore, the dolomite requirement is directly proportional to $d\xi_s'/dt$. Consequently, F_1' is altered in magnitude which is directly proportional to the changes made in $k_3(T_o)$ and $\bar{r}'$. The changes in the limestone requirement are also related to the residence time of the limestone in the bed. If the rate of reaction $k_3(T_o)$ is increased, less reaction time is needed to achieve the same degree of sulfur retention. Shorter residence times are obtained by increasing the limestone feed rate for the same bed volume. Thus, F_1' will increase with an increase and $d\xi_s'/dt$. Alternately, if for the same volumetric feed rate $d\xi_s'/dt$ is increased, an improved sulfur retention will result.

Figure 8 represents the variation in limestone requirement as a function of sulfur absorption efficiency for various values of $k_3(T_o)$. The results emphasize that for a given value of $k_3(T_o)$, if the lmestone feed rate is increased which for a bed of fixed size implies a reduction in residence time, the sulfur absorption efficiency is correspondingly decreased. The implication of this result for an actual operating plant is important. It is implicit in these plots that if the limestone feed rate is held constant, η_{SAE}, increases with an increase in $k_3(T_o)$.

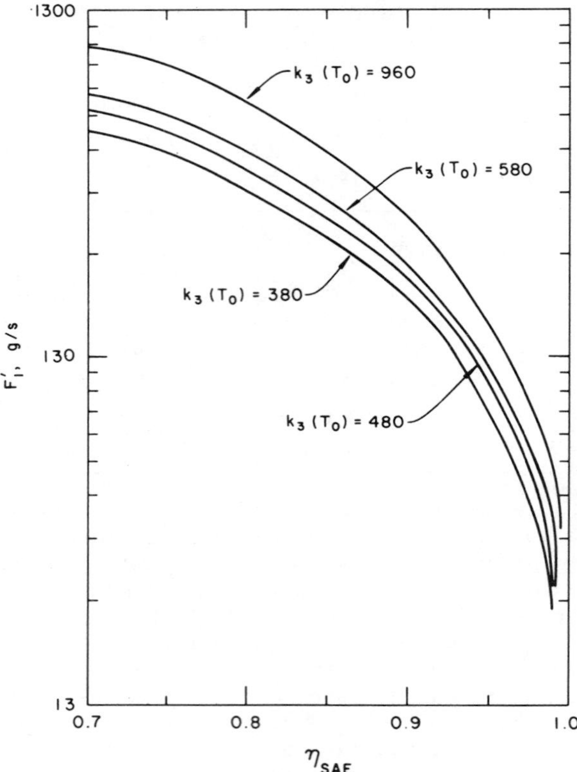

Figure 8. Limestone requirement, F_1', as a function of η_{SAE} and $k_3(T_o)$ in $cm^4/mol \cdot s$ corresponding to $\bar{r} = \bar{r}' = 0.04$ cm and $\eta_{CCE} = 0.995$

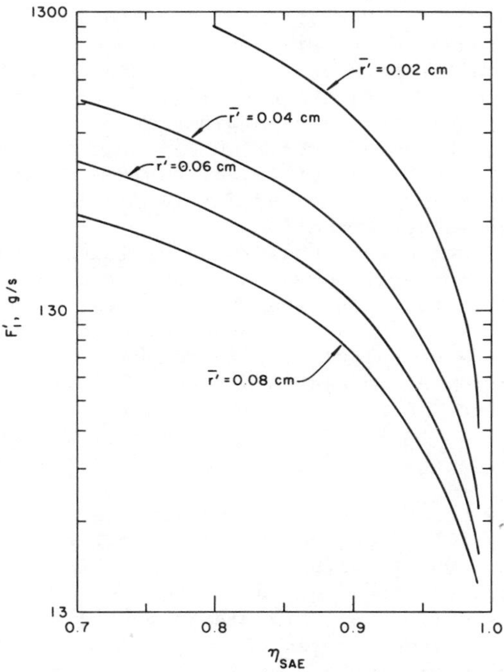

Figure 9. Limestone requirement, F_1', as a function of η_{SAE} and $\overline{r'}$ corresponding to $\bar{r} = 0.04$ cm, $\eta_{CCE} = 0.995$ and $k_3(T_o) = 480$ cm⁴/mol · s

Figure 9 illustrates the effect of changing limestone average size, $\bar{r}'$, in the feed stream on the dependence of limestone feed rate, F_1', and on sulfur absorption efficiency, η_{SAE}. All the plots refer to a constant value of $k_3(T_o)$. These results suggest that if the feed size of limestone is kept fixed, an increase in the limestone feed rate will result in the reduction of sulfur absorption efficiency. These results also emphasize that if the same sulfur retention is to be obtained when the size of the limestone particles is decreased the feed rate must be increased. However, for the same feed rate of limestone, a decrease in the size of limestone particles results in an increased sulfur retention. This may be explained on the basis of an increase in the overall surface area per unit volume of the bed when the average diameter of the particles decreases. It may be noted from Figure 9 that regardless of the limestone particle size, if sufficient residence time is allowed for limestone particles in the bed, it is possible to obtain sufficiently high sulfur retention.

The influence of carbon conversion efficiency on the requirement of limestone for a fixed value of sulfur absorption efficiency is also computed. The generation of sulfur dioxide is found to be directly related to the amount of carbon combusted. The generation rate of sulfur dioxide reduces with the decrease in carbon conversion efficiency and hence the limestone requirement also decreases. A reduction in the carbon conversion efficiency from 99.5 to 70.0% causes a reduction in dolomite requirement from 27.5 to 18.9 g/s for a 99% sulfur absorption efficiency.

In Figure 10 the dependence of oxygen profile, ω_{AH}, and sulfur dioxide profile, ω_{BH}, on the dimensionless bed height, η, for the case of 99.5% sulfur absorption efficiency is presented. The particle size distributions in the bed and in the elutriated stream for char and limestone are also computed but for the sake of brevity, these are not presented here. A more detailed discussion of these results are available elsewhere (26).

Acknowledgments

This work is partly supported by the U.S. Department of Energy and the U.S. Environmental Protection Agency. The authors are grateful to Drs. Irving Johnson and K. M. Myles for their continued interest and many helpful discussions. They are also thankful to A. A. Jonke, D. S. Webster, and L. Burris for thier interest and encouragement.

Nomenclature

A_o = cross sectional area of reactor, m^2

a^* = constant in oxygen concentration profile, eq. 91, dimensionless

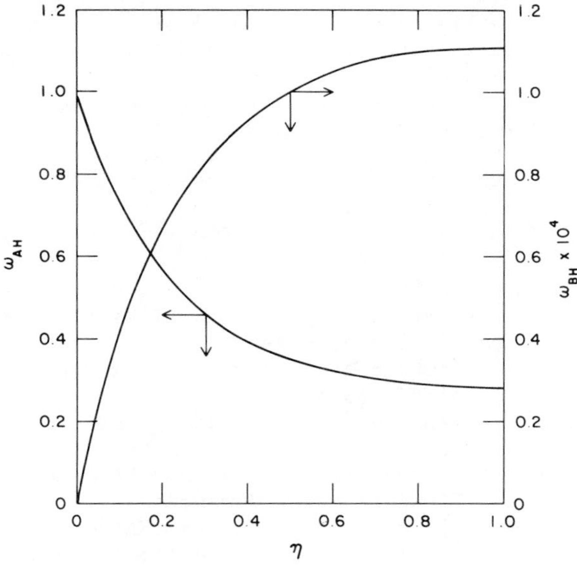

Figure 10. Concentration profiles of oxygen, ω_{AH}, and sulfur dioxide, ω_{BH}, in the bed corresponding to $\eta_{CCE} = 0.995$, $\eta_{SAE} = 0.99$, $\overline{r} = \overline{r'} = 0.04$ cm and $k_3(T_o) = 480\ cm^4/mol \cdot s$

b^* = constant in oxygen concentration profile, eq. 91, dimensionless

C = total concentration of gases, mol/m^3

C_A, C_B = concentrations of the gas components A and B, mol/m^3

C_{AO} = initial concentration of gas component A, mol/m^3

C_{AC}, C_{AS} = concentration of A at the core and surface of char particles, mol/m^3

C_{Ae}, C_{Be} = concentrations of A and B in the off-gas, mol/m^3

C_{AH}, C_{BH} = concentrations of A and B in the bulk gas at height H in the fluidized bed, mol/m^3

C_{BC}, C_{BS} = concentration of B at the core and surface of limestone particle, mol/m^3

C_{JO}, C_{SO}, C_{NO} = initial concentration of solid reactant J and S in the char particle, and N in the limestone particle

D_A, D_B = molecular diffusiviy of the components A and B in the bulk gas phase, m^2/s

D_{eA}, D_{eB} = effective diffusivity of the component A in the ash layer, and of the component B in the layer of solid product E, m^2/s

E_1, E_2, E_3 = activation energy of reactions given by Eqs. 1, 2, and 3, J/mol

F_1, F_1' = feed rate of char and limestone, kg/s

F_3, F_3' = overflow rate of char and limestone from the fluidized bed, kg/s

F_4, F_4' = carryover rate of char and limestone from the fluidized bed, kg/s

f_c, f^* = fraction and weight fraction of carbon in char particles in the fluidized bed, dimensionless

f_Q, f_N = volume fraction of char and limestone particles in the fluidized bed, dimensionless

h, h' = overall convective and radiative heat transfer coefficient for char and limestone particles W/m^2K

H = distance along the bed, m

H_o = height of the fluidized bed, m

$\Delta H_1, \Delta H_2, \Delta H_3$ = heat of reaction per mole of reactant for reaction 1, 2, and 3, J/mol

k = thermal conductivity of the bulk gas, W/mK

k_e, k_e' = effective thermal conductivity of ash layer and solid product, W/mK

k_{mA}, k_{mB} = mass transfer coefficient for the components A and B across the gas film, m/s

k_1, k_2, k_3 = reaction rate constants for Eqs. 1, 2, and 3, $m^4/mol \cdot s$

M_J, M_N, M_S = molecular weight of the solid reactants, J, N, and S

N^* = number of particles of size ξ_s in the fluidized bed

N_{Nu}, N'_{Nu}	= Nusselt Number for char and limestone particles, 2Rh/k, 2R'h'/k, dimensionless
$N_{Nu\bar{r}}, N'_{Nu\bar{r}}$	= $N_{Nu}(\bar{r}/2R)(k/k_e)$, $N'_{Nu}(\bar{r}/2R')(k/k_e')$, dimensionless
N_{Sh}, N'_{Sh}	= Sherwood number for char and limestone particles, $2Rk_{mA}/D_A$, $2R'k_{mB}/D_B$, dimensionless
$N_{Sh\bar{r}}, N'_{Sh\bar{r}}$	= $N_{Sh}(\bar{r}/2R)(D_A/D_{eA})$, $N'_{Sh}(\bar{r}/2R')(D_e/D_{eB})$, dimensionless
$P_1(\xi_s), P'_1(\xi'_s)$	= size frequency distribution of char and limestone feed, 1/m
$P_2(\xi_s), P'_2(\xi'_s)$	= size frequency distribution of char and limestone in the fluidized-bed, 1/m
$P_3(\xi_s), P'_3(\xi'_s)$	= size frequency distribution of char and limestone in the overflow, 1/m
$P^*_3(\xi_s)$	= value of $P_3(\xi_s)$ corresponding to particle of size ξ_o in the feed, 1/m
$P_4(\xi_s), P'_4(\xi'_s)$	= size frequency distribution of char and limestone in the carryover, 1/m
Q	= char,--
r	= radial position in the limestone particle, m
r'	= radial position in the limestone particle, m
$\bar{r}, \bar{r}'$	= average radius general, limestone defined by Eq. 5, m
r_i	= radius of the ith size fraction of feed, m
r_B	= rate of formation of gas product B, $mol/m^2 \cdot s$ mol of solid reactant
r_c, r'_c	= radius of the unreated core of the char and limestone particles, m
r_D	= rate of formation of gas product D, $mol/m^2 \cdot s$ mol of solid reactant
R, R'	= instantaneous radius of char and limestone particles, m
$R_{o,max}(R_{o,min})$	= radius of the largest (smallest) char particle in the feed, m
$R_{s,max}(R_{s,min})$	= radius of the largest (smallest) char particle after complete reaction, m
$R'_{o,max}(R'_{o,min})$	= radius of the largest (smallest) limestone particle in the feed, m
$R'_{s,max}(R'_{s,min})$	= radius of the largest (smallest) limestone particle after complete reaction, m
$\bar{R}$	= gas constant, J/mol K
R_o, R'_o	= initial radius of the char and limestone particles in the feed, m
t	= time, s
T, T'	= temperature of the char particle at radius r, limestone particle at r', K
T_o	= temperature of the fluidized bed, K
T_c, T'_c	= temperature of the unreacted core surface of char and limestone, K
T_s, T'_s	= temperature of the outer surface of the char and limestone particles, K

U, U'	= reduced temperature of the char and limestone particles, T/T_o, dimensionless
U_c	= reduced core temperature of the char and limestone particles, T_c/T_o, dimensionless
U_s	= reduced surface temperature of the char and limestone particles, T_s/T_o, T_s'/T_o, dimensionless
u	= superficial velocity, m/s
V	= volume of the fluidized bed, m^3
V_1, V_3, V_4	= volumetric feed, overflow and carryover rates of char, m^3/s
V_1', V_3', V_4'	= volumetric feed, overflow and carryover rates of limestone, m^3/s
w_i'	= weight fraction of particles in the feed of radius, r_i, dimensionless
W_Q, W_N	= weight of char and limestone particles in the fluidized-bed, kg
x_A, x_B	= mole fraction of components A and B, dimensionless
x_{AC}, x_{AO}, x_{AS}	= value of x_A at the unreacted core surface, bottom of the bed and outer surface of the char particle, dimensionless
x_{AH}	= value of x_A at a height H in the fluidized bed, dimensionless
x_{BC}, x_{BS}, x_{BH}	= value of x_B at the unreacted core surface, outer surface of limestone particle and at height H in the bed, dimensionless
Y	= average rate of gas flow through the reactor bed, m^3/s
Z, Z'	= parameter defining particle growth or shrinkage of char and limestone defined by Eq. 4, dimensionless

Greek Letters

$\varepsilon, \varepsilon_1, \varepsilon_3, \varepsilon_4$	= average void fraction of the bed, char feed, overflow and carryover, dimensionless
ε_1'	= average void fraction of limestone feed, dimensionless
ξ, ξ'	= any reduced distance for char and limestone particles, $r_o/\bar{r}$ and r_o'/r, dimensionless
ξ_c, ξ_c'	= reduced unreacted core radius of char and limestone particles, $r/\bar{r}$ and $r'/\bar{r}$, dimensionless
ξ_s, ξ_s'	= reduced radius of the char and limestone particles, $R/\bar{r}$ and $R'/\bar{r}$, dimensionless
$\xi_{o,max} (\xi_{o,min})$	= reduced radius of the largest (smallest) particle in the char feed, dimensionless
$\xi_{s,max} (\xi_{s,min})$	= reduced radius of the largest (smallest) char particle after complete reaction, dimensionless
$\xi_{o,max}' (\xi_{o,min}')$	= reduced radius of the largest (smallest) limestone particle in the feed, dimensionless

$\xi'_{s,max}(\xi'_{s,min})$ = reduced radius of the largest (smallest) limestone particle after complete reaction, dimensionless

ξ_o, ξ'_o = reduced value of R_o and R'_o, $R_o/\bar{r}$ and $R'_o/\bar{r}$, dimensionless

ρ_a, ρ_E = density of the solid product ash and calcium sulfate, kg/m^3

ρ_J, ρ_N = density of the solid reactant J, and limestone N, kg/m^3

ρ_Q, ρ_S = density of char and solid reactant S, kg/m^3

$\phi_{1\bar{r}}$ = a parameter to characterize the rates of intraparticle diffusion resistance to the reaction residence for reaction 1 = $\bar{r}k_1(T_o)C_{JO}/D_{eA}(T_o)$, dimensionless

$\phi_{2\bar{r}}$ = a parameter to characterize the rates of intraparticle diffusion resistance to the reaction residence for reaction 2 = $\bar{r}k_2(T_o)C_{SO}/D_{eA}(T_o)$, dimensionless

$\phi_{3\bar{r}}$ = a parameter to characterize the rates of intraparticle diffusion resistance to the reaction residence for reaction 3 = $\bar{r}k_3(T_o)C_{NO}/D_{eA}(T_o)$, dimensionless

$\omega_A, \omega_{AC}, \omega_{AH}$ = reduced values of x_A, x_{AS}, x_{AH}, x_A/x_{AO}, x_{AC}/x_{AO}, x_{AH}/x_{AO}, dimensionless

ω_{AS}, ω_B = reduced value of x_{AS} and x_B, x_{AS}/x_{AO}, x_{BC}/x_{AO}, dimensionless

$\omega_{BC}, \omega_{BH}, \omega_{BS}$ = reduced value of x_{BC}, x_{BH}, and x_{BS}, x_{BC}/x_{AO}, x_{BH}/x_{AO}, x_{BS}/x_{AO}, dimensionless

$\kappa_Q(\xi_s)$ = elutriation constant for char particles of size ξ_s, 1/s

$\kappa_N(\xi'_s)$ = elutriation constant for limestone particles of size ξ'_s, 1/s

η = reduced bed height, H/H_o, dimensionless

η_{CCE} = carbon conversion efficiency defined by Eq. 89, dimensionless

η_{SAE} = sulfur absorption efficiency defined by Eq. 90, dimensionless

Literature Cited

1. Kunii, D.; Levenspiel, O. "Fluidization Engineering"; John Wiley: New York, 1969.
2. Chen, T. P. ; Saxena, S. C. AIChE Sym. Series 1968, No. 176, 74, 149-161.
3. Chen, T. P.; Saxena, S. C. Fuel, 1977, 56, 401-413.

4. Saxena, S. C.; Chen, T. P.; Jonke, A. A. "A Slug Flow Model for Coal Combustion with Sulfur Emission Control by Limestone or Dolomite"; Presented at the 70th Annual AIChE Meeting, New York City, 1977, Paper No. 104d.
5. Rengarajan, P.; Krishnan, R; Wen, C. Y. "Simulation of Fluidized Bed Coal Combustors"; Report No. NASA CR-159529, February 1979, 199 pp.
6. Becker, H. A.; Beer, J. M.; Gibbs, B. M. "A Model for Fluidized-Bed Combustion of Coal"; Inst. of Fuel Symposium Series No. 1, Proc. Fluidized Combustion Conference 1, Paper No. AI, 1975, A1·1-A1·10.
7. Horio, M; Mori, S.; Muchi, I. "A Model Study for the Development of Low NO_x Fluidized-Bed Coal Combustors"; Proc. 5th Intern. Conf. on Fluidized-Bed Combustion, Vol. II, 1977, 605-624.
8. Davidson, J. F.; Harrison, D. "Fluidized Particles"; Cambridge University Press: New York, 1963.
9. Rehmat, A.; Saxena, S. C. "Single Nonisothermal Noncatalytic Gas-Solid Reaction. Effect of Changing Particle Size"; Ind. Eng. Chem. Process Des. Dev. 1975, 16, 343-350.
10. Rehmat, A.; Saxena, S. C. "Multiple Nonisothermal Noncatalytic Gas-Solid Reaction. Effect of Changing Particle Size"; Ind. Eng. Chem. Process Des. Dev. 1977, 16, 502-510.
11. Rehmat, A.; Saxena, S. C.; Land, R. H.; Jonke, A. A. "Noncatalytic Gas-Solid Reaction with Changing Particle Size: Unsteady State Heat Transfer"; Canadian J. Chem. Eng. 1978, 56, 316-322.
12. Yagi, S.; Kunii, D. "Studies on Combuston of Carbon Particles in Flames and Fulidized Beds"; 5th Symposium (Int'l) on Combustion, 1955, 231-244.
13. Burovoi, I. A.; Eliashberg, V. M.; D'yachko, A. G.; Bryukvin, V. A. "Mathematical Models for Thermochemical Processes Occurring in Fluidized Beds"; Int. Chem. Eng. 1962, 2, 262-258.
14. Strel'tsov, V. V. "Approximate Relationships for Calculating the Kinetics of Reactions of a Solid Phase in a Fluidized Bed"; Int. Chem. Eng. 1969, 9, 511-513.
15. Bethell, F. B.; Gill, D. W.; Morgan, B. B. "Mathematical Modelling of the Limestone-Sulfur Dioxide Reaction in a Fluidized-Bed Combustor"; Fuel 1973, 52, 121-127.
16. Koppel, L. "A Model for Predicting the Extent of Reaction of Limestone and Sulfur Dioxide During Fluidized-Bed Combustion of Coal"; Appendix C, pp. 60-77, in Jonke, A. A. "Reduction of Atmospheric Pollution by the Application of Fluidized Bed Combustion"; Argonne National Laboratory Annual Report, ANL/ES-CEN-1002, July 1969-June 1970.
17. Avedesian, M. M.; Davidson, J. F. "Combustion of Carbon Particles in a Fluidized Bed"; Trans. Inst. Chem. Engrs. 1973, 51, 121-131.

18. Evans, J. W.; Song, S. "Application of a Porous Pellet Model to Fixed, Moving, and Fluidized Bed Gas-Solid Reactors"; Ind. Eng. Chem. Process Des. Dev. 1974, 13, 146-152.
19. Campbell, E. K.; Davidson, J. F. "The Combustion of Coal in Fluidized Beds"; Inst. of Fuel Sym. Series No. 1, Proc. Fluidized Combustion Conference 1, Paper No. A2, 1975, A2·1-A2·9.
20. Gibbs, B. M.; "A Mechanistic Model for Predicting the Performance of a Fluidized Bed Coal Combustor"; Inst. of Fuel Sym. Series No. 1, Proc. Fluidized Combustion Conference 1, Paper No. A5, 1975, A5·1-A5·10.
21. Beer, J. M.; "The Fluidized Combustion of Coal"; XVIth Sym. (Int'l) on Combustion, 1976, 439-460.
22. Hovmand, S.; Davidson, J. F. Chapter 5 in "Fluidization"; Editors J. F. Davidson and D. Harrison; Academic Press: London, 1971.
23. Horio, M.; Wen, C. Y. "Simulation of Fluidized Bed Combustors: Part I. Combustion Efficiency and Temperature Profile"; AIChE Symp. Series 1978, No. 176, Vol. 74, 101-111.
24. Mori, S.; Wen, C. Y. "Estimation of Bubble Diameter in Gaseous Fluidized Beds"; AIChE J 1975, 21, 109-115.
25. Gibbs, B. M.; Pereira, F. J.; Beer, J. M. "Coal Combustion and NO Formation in an Experimental Fluidized Bed"; Institute of Fuel Symp. Series No. 1, Proc. Fluidized Combustion Conference 1, Paper No. D6, 1975, D6·1-D6·13.
26. Rehmat, A.; Saxena, S. C.; Land, R. H. "Application of Noncatalytic Gas-Solid Reactoins for a Single Pellet of Changing Size to the Modeling of Fluidized-Bed Combustion of Coal Char Containing Sulfur"; Argonne National Laboratory Report, ANL/CEN/FE-80-13, September 1980, 86 pp.

RECEIVED July 15, 1981.

8

Computer Modeling of Fluidized Bed Coal Gasification Reactors

T. R. BLAKE[1] and P. J. CHEN

Fossil Energy Program, Systems, Science and Software, P.O. Box 1620, La Jolla, CA 92038

The application of large scale computer simulations in modeling fluidized bed coal gasifiers is discussed. In particular, we examine a model wherein multidimensional predictions of the internal gas dynamics, solid particle motion and chemical rate processes are possible.

A computer model has been developed to provide numerical simulations of fluidized bed coal gasification reactors and to yield detailed descriptions, in space and time, of the coupled chemistry, particle dynamics and gas flows within the reactor vessels. Time histories and spatial distributions of the important process variables are explicitly described by the model. With this simulation one is able to predict the formation and rise of gas bubbles, the transient and quasi-steady temperature and gas composition, and the conversion of carbon throughout the reactor.

The effects of gas and coal/char feeds and reactor geometries upon these internal processes and, hence, upon the performance of the reactor, can be simulated with this numerical model. The model incorporates representations of particle-particle and particle-gas interactions which account for finite rate heterogeneous and homogeneous chemistry as well as the hydrodynamical processes associated with particle collisions and drag between the particles and the gas flow. The important influences of multicomponent gas phase properties as well as solid particle properties, such as shape and size, are included in the representations.

[1]Current address: Consultant, also Professor, Mechanical Engineering Department, University of Massachusetts, Amherst.

0097-6156/81/0168-0157$06.75/0

Nature of The Finite Difference Model

It is useful and appropriate to compare some of the specific capabilities of the present model with other representations of reactors. We note that existing models of fluidized beds are typified by the two-phase models of fluidization (e.g., 1-4) which are of great utility, but which do not predict the gas dynamics and solid particle transport in the reactor. Rather, these models require input prescriptions for such transport and provide, in general, one-dimensional axial representations of temperature, carbon consumption and gas composition. The present fluidized bed computer model provides a transient multi-dimensional field description of these process variables and also provides predictions of the gas dynamics and solid particle motion. Within this context, we note that there have been analogous multi-dimensional field descriptions and numerical models for the study of related processes, such as those associated with nuclear reactor safety (e.g., 5, 6, 7) and entrained flow combustion and gasification (e.g., 8, 9, 10). However, physical and chemical mechanisms in such models are rather different from fluidized bed coal gasification reactors and consequently the capabilities of the respective numerical representations are also different. For example, the particle-particle and particle-gas forces dominate the fluidized bed flows. The numerical representations of these physical phenomena, together with the particle-gas mass and energy exchange produced by the heterogeneous combustion and gasification reactions, require the Eulerian-Lagrangian and finite difference implicit capabilities which have been specifically developed for and incorporated into the present fluidized bed coal gasifier computer model.

The computer model is based upon a continuum description of fluidization in coal gasification reactors. In general, fluidized flows are dominated by specific physicochemical processes and, hence, require particular theoretical representations. For example, in the heavily loaded gas-particle regime appropriate to fluidization, the solid particles dominate the transport of momentum and energy. This aspect of fluidization is reflected in the mathematical descriptions which have been used in the fluidized bed model.

These mathematical representations are complex and it is necessary to use numerical techniques for the solution of the initial-boundary value problems associated with the descriptions of fluidized bed gasification. The numerical model is based on finite difference techniques. A detailed description of this model is presented in (11-14). With this model there is a degree of flexibility in the representation of geometric surfaces and hence the code can be used to model rather arbitrary reactor geometries appropriate to the systems of interest. [The model includes both two-dimensional planar and axisymmetric geometries.] That is, the use of finite difference computational zones permits the numerical "construction" of a wide variety of reactor geometries because the walls, orifices, etc., are resolved incrementally into computational zones. Consequently, the code is not restricted to a specific reactor; rather, it is designed to model classes of reactor flows through the specific nature of the gas dynamic, solid particle and thermochemistry representations which are incorporated into the numerical analogs of the differential equations. This specificity of the conservation equations is based upon the theoretical considerations for fluidized bed flows and upon the incorporation of physicochemical data into the models to define constitutive equations and interaction functions in the theoretical representations.

Within the context of these code applications, simulations of both local flow regimes and flows on the scale of the entire reactor are possible. It is to be noted that these computer codes are designed to provide a resolution of the gas dynamics, solid particle motion and the major coupling of the chemistry and the flow field on time scales which measure the gas residence time in the reactor, but are not, at the present, envisioned to provide a detailed inventory of process variables and gas composition within the gasifier on time scales of hours.

To illustrate this point, one may consider the detailed mass and energy balance calculations that are an integral part of fluidization gasification modeling. For example, Weil and his co-workers, at the Institute of Gas Technology (15), are interpreting PDU and pilot plant data related to high pressure fluidized bed steam-oxygen, steam-air and hydrogasification. An important parameter, in their semi-empirical interpretation of the mass and energy balance in these fluidized beds, is the bubble size. The present fluidized beds model can be used to predict the bubble size in the high pressure and high temperature regime of the experiments and such bubble size predictions can be used in and thereby complement the detailed kinetics studies of Weil.

In this paper we will discuss some numerical calculations related to the IGT bench scale data and also examine some numerical calculations of the Westinghouse Agglomerating Combustor/Gasifier. The emphasis in both of these studies is upon the hydrodynamic mixing processes and the coupling of that hydrodynamic mixing to the chemical reactions. For example, flow visualization experiments performed at Westinghouse are used to verify some of the model predictions.

It is to such applications that the numerical model studies are directed. That is, the predictions of the model are compared with bench scale and pilot plant data and with flow visualization experiments. Such comparisons serve to validate the model and to provide guidance for the experimental trials. It is likely that modifications of the physicochemical aspects of the model will be required to predict specific reactor environments. When the model has been verified and, if necessary, modified through such comparative studies, then it can be used to predict reactor performance and, in particular, examine questions such as reactor scaleup.

Fluidized Bed Gasifier Theory and Numerical Model

We now discuss the theory and some of the numerical aspects of the model. Again, this model is based upon the theoretical formulation presented in (12, 13, 14). The theoretical formulation, including differential equations and appropriate initial and boundary conditions, defines a complicated initial value problem which, in general, must be solved with numerical methods. A finite difference computer model has been developed to provide such a solution. The mathematical character of the system of equations is of the mixed hyperbolic-parabolic type; consequently, we have used a numerical technique based upon an iterative, implicit, finite difference scheme. While there is an extensive literature related to such techniques and further documentation exists in text books (e.g., 16), the development of an iterative, implicit method for this coupled solids-gas system of equations is unique to the present investigation.

Differential Equations for Fluidized Bed Gasifier Model. In a hydrodynamical sense, the processes in fluidized bed gasifiers involve the interaction of a system of particles with flowing gas. The motion of these particles and gas is, at least in principle, completely described by the Navier-Stokes equations for the gas and by the Newtonian equations of motion for the particles. Solution of these equations together with appropriate boundary and initial conditions would determine the mechanics of the fluidized bed gasifier. However, such fluidized bed gasifiers contain a large number of closely spaced particles; consequently, such systems are far too complex to permit direct solution. For practical purposes, it is therefore necessary to simplify the governing equations so that the gas-particle system is described by a smaller number of differential equations.

Such a simplification is possible through the introduction of a continuum mathematical description of the gas-solid flow processes where this continuum description is based upon spatial averaging techniques. With this methodology, point variables, describing thermohydrodynamic processes on the scale of the particle size, are replaced by averaged variables which describe these processes on a scale large compared to the particle size but small compared to the size of the reactor. There is an extensive literature of such derivations of continuum equations for multiphase systems (17, 18, 19). In the present study, we have developed (12, 13, 14) a system of equations for compressible gas flow in a fluidized bed, based upon the method of Anderson and Jackson (18) and we have used laboratory data to define interaction functions and constitutive equations.

These equations have further been coupled to the kinetics and transport relationships associated with the heterogeneous and homogeneous reactions of coal gasification. This coupled system of equations provides the theoretical basis of our computer model of coal gasification reactors.

The derivation of these equations (cf., 14) involves the important assumptions that the gas phase inertia is negligible compared with that of the solid, the temperatures of the solid and gas phases have the same local values and the kinetic energy of the system is small compared with the thermal energy.

Closure of such differential equations requires the definitions of both constitutive relations for hydrodynamical functions and also kinetic relations for the chemistry. These functions are specified by recourse both to theoretical considerations and to rheological measurements of fluidization. We introduce the ideal gas approximation to specify the gas phase pressure and a caloric equation-of-state to relate the gas phase internal energy to both the temperature and the gas phase composition. It is assumed that the gas and solid phases are in local thermodynamic equilibrium so that they have the same local temperature.

A solid phase internal energy is related, again through a caloric equation-of-state, to the temperature. The solid phase pressure is defined as a function of the solid volume fraction where the functional relationship (cf., 14) is based upon the fluidized bed stability measurements of Rietma and his coworkers (20).

The gas phase viscosity is defined by the temperature and the gas composition through a semi-empirical function. For the solid phase shear viscosity, we (cf., 14) use semi-empirical relations based upon the viscometric measurements of Schugerl (21). The solid phase bulk viscosity is, at present, inaccessible to measurement; consequently, we define it to be a multiple of the shear viscosity.

One of the most important functions in the description of fluidization is the drag function which measures the ratio of pressure gradient to gas volume flux. The definition of this drag function is discussed in (14) through recourse to the correlation of Richardson (22).

A thermal conductivity of the gas-solid particle mixture is determined (cf., 14) by the correlation of Gelperin and Einstein (23). We use a law of mixtures to define a radiation diffusion coefficient and, for the present, we consider only the limits of (1) opaque gas and opaque particles and (2) transparent gas and opaque particles.

Chemistry of Combustion and Gasification. The present model is designed to examine the combustion and gasification regions associated with steam-oxygen gasification in the fluidized bed. Consequently, we consider that the solid phase is composed of char particles. That is, any devolatilization is assumed to occur upon feeding of the coal into the bed and this devolatilization is instantaneous, relative to the hydrodynamic time scales of interest in the present model. The overall mass balance, associated with the evolution of coal to char through devolatilization, can be easily accounted for by a stoichiometric analysis similar to that suggested in (11), viz., the coal feed is simulated by a feed of char and gas with the gas composition representing the volatile yield.

For the present, the following reactions are assumed to occur between the carbon in the char and the gaseous reactants:

$$\gamma C(s) + O_2(g) \rightarrow (2-\gamma)\ CO_2(g) + 2(\gamma-1)\ CO(g) \tag{R1}$$

$$C(s) + H_2O(g) \rightarrow CO(g) + H_2(g) \tag{R2}$$

$$C(s) + CO_2(g) \rightarrow 2CO(g) \tag{R3}$$

$$C(s) + 2H_2(g) \rightarrow CH_4(g)\quad . \tag{R4}$$

The first of these reactions is the combustion reaction, where γ is a parameter determining the distribution of carbon dioxide and carbon monoxide in the overall combustion process (24). To some extent, it reflects an uncertainty in defining the respective roles of heterogeneous and homogeneous oxidation processes. Naturally, we expect that the heterogeneous reaction involves the production of carbon monoxide at the temperatures of interest (25, 26, 27). However, the extent of mixing and oxidation of this carbon monoxide in the gas phase can certainly occur anywhere from the particle surface (γ=1) to the ambient gas flow (γ=2). We do not include any additional homogeneous reaction of carbon monoxide and oxygen in the gas phase, so this latter limit corresponds to no oxidation of the CO. The above statement is, thereby, a parametric expression which measures the extent of carbon monoxide oxidation in the gas phase at the particle surface.

The remaining three heterogeneous reactions involve gasification of carbon by steam, carbon dioxide and hydrogen (28-35).

The homogeneous reaction is that of water gas shift:

$$CO + H_2O \rightleftarrows CO_2 + H_2\quad . \tag{R5}$$

Again, the homogeneous reaction of carbon monoxide and oxygen is implicitly included in the parameter, γ, of Equation (R1). Further, we neglect the gas phase oxidation of hydrogen and methane; briefly, with the possible exception of the Westinghouse gasifier, we do not expect significant quantities of these species to be present in that region of the reactor where there is significant oxygen (35).

The water gas shift reaction is considered to be in equilibrium. However, the heterogeneous reactions are influenced by both chemical kinetics and diffusive transport of reactants. Further, in the case of the carbon-steam reaction, the inhibition by both carbon monoxide and hydrogen is also included.

The nature of carbon mass loss is, in general, a complicated function of intra- and extra-particle reactant transport and chemical kinetics (26). While the numerical structure incorporated into the fluidized bed gasifier computer model can include a broad range of particle mass loss configurations. We, for the calculations herein, used an equivalent unreacted shrinking core particle configuration for both combustion and gassification reactions. Thus, the heterogeneous reaction rate, where $\dot{R}_i$ is the rate of carbn mass loss per unit surface area due to reaction i, is

$$\dot{R}_i = \frac{f_i}{\frac{1}{k_i} + \frac{1}{k_0} + \frac{1}{k_{0E}}}$$

where k_i, k_0 and k_{0E} represent the velocity coefficients associated with kinetics, extra-particle diffusion and intra-particle ash layer diffusion (36), and f_i is a function of the partial pressures of the reactants in the gas phase.

With the equation for R_i we obtain the source terms representing heterogeneous chemistry equations for the solid and gas phase species. Those source terms are discussed briefly in (13).

Numerical Formulation. Let us now consider the general character of the numerical solution of the equations. The model involves a combined Eulerian-Lagrangian formulation which permits calculation of large displacements while at the same time maintaining sharp interfaces. As shown in Figure 1, the x-y plane is divided into a number of rectangular zones of size Δx by Δy. This Eulerian grid is fixed in space. Upon this grid, a large collection of Lagrangian marker particles is superimposed. These marker particles, each of which actually describes the average behavior of a large number of physical particles, move through the Eulerian grid with the local instantaneous solid velocity as the calculation proceeds. Each such particle has assigned to it an amount of mass, horizontal momentum, vertical momentum, and energy, all of which change

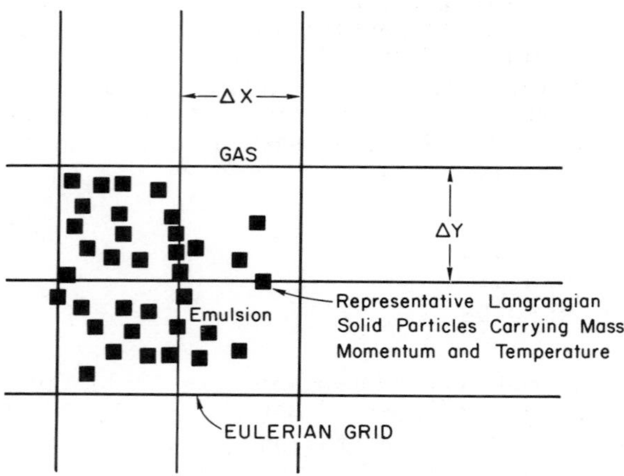

Figure 1. Eulerian/Langrangian formulation of solid–gas motion

with time. Thus, the motion of these particles automatically takes into account all solid advection effects.

For a particular time step, the positions of the solid particles are first changed by an amount ($\vec{u} \cdot \Delta t$), and the field variables assigned to the Eulerian grid are altered to reflect the new particle distribution. Then, the additional terms in the field equations for mass, momentum and energy conservation (viscous stresses, pressure forces, heat conduction, etc.) are taken into account using the Eulerian grid. Finally, the field variable quantities assigned to the representative particles are changed to reflect the effect of these latter terms. This general procedure (or "time cycle") may be repeated as many times as desired, with each such repetition advancing the solution further in time. The use of a superposed Lagrangian representative particles to treat advective effects avoids the computational "smearing" of field variables which often occurs in purely Eulerian computations.

Applications of Fluidized Bed Computer Model

A major factor in fluidized bed behavior is the interaction between the gas flow from individual orifices and the particle and gas mixture within the bed. The jet penetration and the subsequent bubble formation have an important influence upon solids and gas mixing and, ultimately, upon the usefulness of the bed for reactor purposes. While flow visualization data are available at ambient pressures and temperatures, the natures of jet penetration and bubble development at high pressures and temperatures are not easily measured. Typical data on bubble size and bubble velocity at ambient conditions are shown, represented by the small size symbols, in Figure 2. It is well known that bubble volume can be correlated as a function of gas volumetric flow rate (37) and that bubble velocity is related to the size of the bubble radius (38). Such semi-empirical correlations are indicated as solid lines in that figure.

A quantitative comparison between the numerical model and experimental data can be made using those measurements of bubble volume and bubble rise velocity. The calculated values of bubble volume and bubble rise velocity for both ambient and high pressure conditions and also for some complex geometries such as the Westinghouse Cold Flow 30 cm diameter semi-circular model, are shown as large symbols in Figure 2. The specific geometries and flow conditions for the calculations are listed in Table I.

We find that the numerical fluidized bed model predicts bubble size and velocity, at ambient pressure and temperature, which agrees with the data. Further in that figure we show the results of three calculations at high pressure (40 atm) and room temperature (293°K) which, when correlated in the same fashion, yield predictions of bubble size and velocity which also agree with the ambient pressure data. This agreement between high

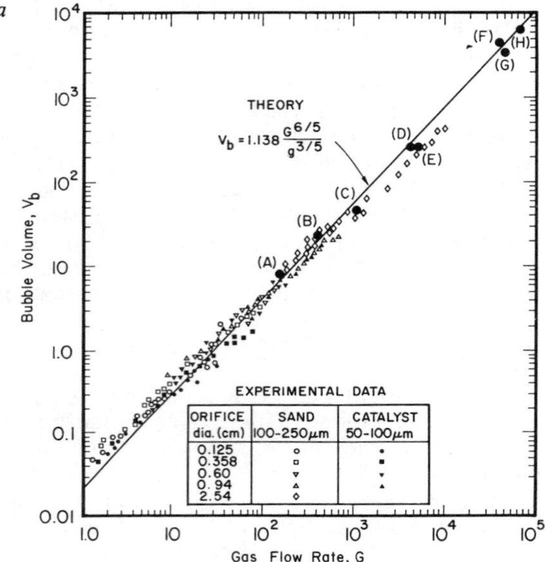

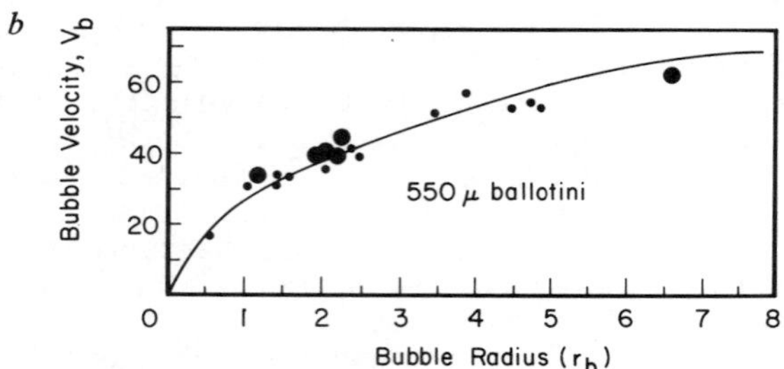

Figure 2. a, Bubble volume—comparison between numerical model and experiment (37): (●) numerical calculations; b, Bubble velocity as function of equivalent spherical radius—comparison between numerical model and experiment (38): (●) numerical calculations

TABLE I
FLUIDIZATION PARAMETERS FOR CALCULATIONS WITH FLUIDIZED BED COAL GASIFIER MODEL OF BUBBLE VOLUME AND BUBBLE RISE VELOCITY IN FIGURE 2

CALCULATION	A	B	C	D	E	F	G	H
DESCRIPTION	Single Jet in Cylindrical Annulus					Westinghouse Geometry		
Pressure (Atm)	40	40	40	40	32	1	1	1
Gasification Chemistry	No	No	No	Yes	Yes	No	No	No
Particle Diameter (?m)	250	250	750	750	400	2800	2800	2800
Particle Density (gm/cm^3)	1.5	1.5	1.5	1.5	1.5	0.2	0.2	0.2
Maximum Temperature (K)	293	293	293	1200	1200	293	293	293
Mass Flow-Jet (gm/sec)	1.21	3.25	8.26	15.33	35.4	18.3	8.88	30.05
Mass Flow-Annulus (gm/sec)	1.60	1.60	7.07	-	-	12.3	9.66	9.66
Mass Flow-Cone (gm/sec)	-	-	-	-	-	0.22	16.22	16.22

pressure calculations and low pressure data can be explained: first, the influences of pressure and temperature upon the relationship between bubble size and gas flow are implicitly normalized when one correlates the bubble volume with the gas volumetric flow rate through the orifice. Second, in the comparison between high pressure predictions of bubble velocity and the low pressure data there is good agreement because bubble velocity is relatively insensitive to gas density.

The use of the numerical model at high pressures and the comparison of the high pressure calculations with the data suggests scaling relationships between bubble size and velocity, which should be useful in applying such a low pressure data to the reactor environment. Finally, the agreement between bubble size predictions for the complex geometry of the Westinghouse Cold Flow 30 cm Diameter Semi-Circular Model and the data suggests the application of the data to a wide range of conditions. Of course, there are limitations such as suggested in (37), but the broad agreement between calculation and data tends to verify the model.

Simulation of Institute of Gas Technology Six-Inch Diameter Bench Scale Reactor. This model has also been used in preliminary calculations of the Institute of Gas Technology (IGT) six-inch diameter bench scale experiments (39) on steam-oxygen gasification of char. In this reactor the steam, oxygen and nitrogen are injected at the base of the column through a six-cone feed gas distributor and, in the case of nitrogen, also around the distributor to maintain the state of fluidization.

In the numerical calculation of that experiment we use the axisymmetric version of the fluidized bed computer model to reproduce the cylindrical geometry of the reactor. The gas feed is simulated by a fully mixed stream of oxygen, steam and nitrogen which is injected at the base of the reactor within a radius of two-inches, corresponding to the radius through the centers of the injection cones in the actual six-cone feed gas distributor.

A time sequence of bubbling from such a calculation of the IGT six-inch EGO-33 run is shown in Figure 3. The sequence is that of a stationary or quasi-steady pattern of bubble evolution subsequent to the start-up transient. In each individual "frame" of the time sequence a reactor section, bounded by the centerline axis on the left and the reactor radius on the right, is shown. The representative particles are indicated by the black dots while the bubbles and the voids are white. The time corresponding to discrete frames is indicated on the base of the figure and the rise of the bubbles, together with the solids mixing, can be discerned in that sequence of frames. This calculation suggests that a relatively small number of large

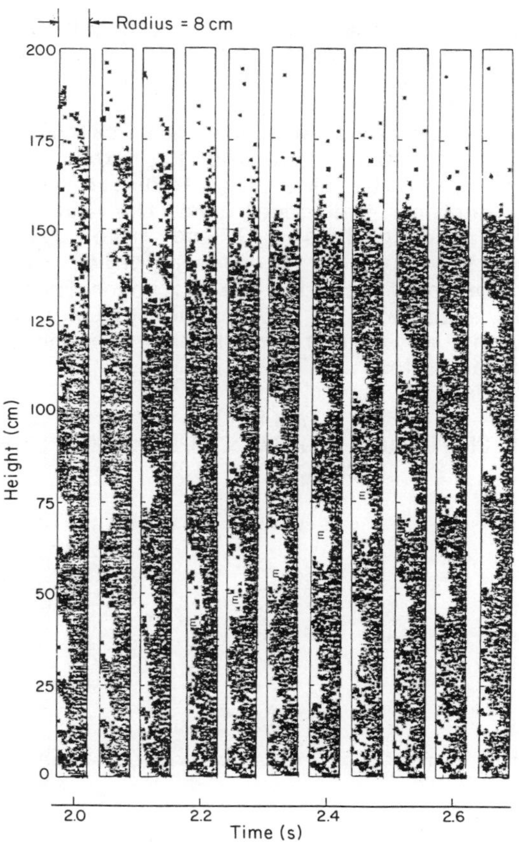

Figure 3. Time sequence of bubble evolution during steam oxygen gasification in IGT 6-in. diameter bench scale reactor

bubbles, in a sequential train, rise through the bed and produce solids and gas mixing. The bubbles are large because of bubble coalescence at the base of the reactor.

We find that while the gas composition in the bubble is quite different from that in the mix of gas and solid particles surrounding the bubbles; near the base of the reactor these respective gas compositions are almost identical by the time that bubbles approach the top of the bed. There is significant exchange of gas between the bubbles and the surrounding emulsion which causes a good mixing of the reactants and products of combustion and gasification. Some particles are entrained into the bubble through the wake.

Preliminary comparisons between the calculated exit gas composition and that measured in the IGT experiments have been made. For example, in the case of IGT Run EGO-33, we find very good agreement between the model and the data for CO, CH_4, H_2 and N_2 but predict less H_2O and more CO_2 in the exit gas than is indicated by our interpretation of the experimental measurements. For example, in the nitrogen-free product gas, including steam, a comparison between our evaluation of the data and a one-dimensional simulation of the reactor process gives the composition mass flows in Table II.

TABLE II
COMPARISON OF PRODUCT AND GAS COMPOSITION IN EXPERIMENT AND CALCULATION

Species	Mass Flow (lb/hr)	
	IGT	Calculation
CO	38.9	38.5
CO_2	62.1	83.0
H_2	3.9	4.7
CH_4	3.2	3.7
H_2O	93.1	79.0

where the approach factors expressed in N_2-free mol percent X are:

$$\frac{X_{CO_2}\, X_{H_2}}{X_{CO}\, X_{H_2O}} = 0.377,\ 0.730$$

for the IGT and present cases respectively. There is approximately a 20 percent discrepancy between the IGT data and the present simulation with regard to CO_2 and H_2O but there is excellent agreement on the other species. We also note that the total oxides of carbon are within 20 percent agreement. The reason for the difference in steam consumption between calculation and experiment is not clear. We note, however that the approach factor in the IGT experiments (specifically our interpretation of that data), if indicative of equilibrium, would suggest a temperature of approximately 2230°F which is about 300°F higher than the largest temperature measured.

Simulation of Westinghouse Agglomerating Combustor/Gasifier. The design of the Westinghouse Agglomerating Combustor/Gasifier includes a nozzle introducing a jet of oxidant to create locally high temperatures for ash sintering and agglomeration. The agglomerates are subsequently removed by falling countercurrently through a cylindrical annulus surrounding the oxidizing jet.

There is relatively little data obtained in the hot reactor environment. Typical measurements include wall temperature records and bed densities. However, there is a continuing cold flow visualization test program at Westinghouse (40). Data from that test program can be used to validate the hydrodynamic aspects of the present model. Then, with that validation the limited hot flow measurements can often provide sufficient information about the validity of hot flow hydrodynamic predictions. Indeed, in the work at Westinghouse some limited hot flow data (40) was used to verify a jet penetration correlation which had been developed from cold flow visualization experiments. Such engineering judgements also apply to the numerical "experiments" which comprise our simulation studies. The advantage of these numerical "experiments" is that they are more cost effective and provide more detailed information than the actual hot flow hydrodynamic experiments.

In the following paragraphs we shall examine some cold and hot flow calculations related to the hydrodynamics of the Westinghouse Agglomerating Combustor/Gasifier.

In Figure 4 we show the cross-section of the Westinghouse cold flow geometry and the corresponding geometry of the computational grid in the numerical simulation. These are both axisymmetric configurations where the axes of symmetry are to the left of the respective figures. Typical cold flow reactor conditions for an experiment and a calculation are shown in Table III.

TABLE III
EXPERIMENTAL AND NUMERICAL COLD FLOW IN WESTINGHOUSE REACTOR

Particle Diameter (μm)	2800
Particle Density (gm/cm^3)	0.21
Mass Flow, Air Tube (gm/sec)	18.3
Mass Flow, Annulus (gm/sec)	12.3
Mass Flow, Cone (gm/sec)	0.22
Reactor Pressure (atm)	1.0
Reactor Temperature (°K)	293

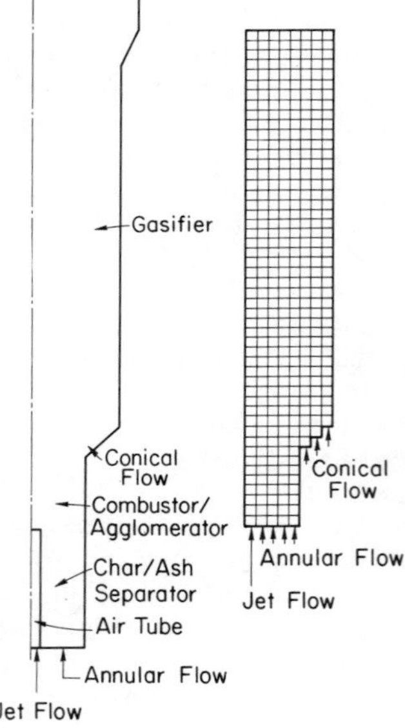

Figure 4. Schematic of Westinghouse cold flow 30-cm diameter semicircular model and finite difference grid for numerical calculations

The numerical simulation of the cold flow experiment is in the form of an initial value problem. The computational grid is initially filled with particles and air, at atmospheric pressure and ambient temperature. A flow field is established by gas injections through the air tube, annulus and conical distributors. Subsequently in the calculation a time varying jet is established at the air tube, and the interaction of this jet with the mass flows through the annulus and conical distributors produces particle mixing within the reactor. We find that good agreement exists between this numerical simulation and the Westinghouse cold flow studies (40).

A time sequence of the calculated particle motion is shown in Figure 5 where the centerline of the reactor is the lefthand side and the outer radius of the reactor is the righthand side of each frame in the sequence. The jet penetration is observed qualitatively from the evolution of that flow region where there are no particles. This jet is quite diffused in character and is clearly a transient phenomenon. We define the jet length as the voidage from the jet inlet. Thus, the bubble, before it detaches from the jet, will be considered as part of the jet. The calculation and experiment are in agreement with regard to the bounds of jet penetration into the mix of particles and gas surrounding the air tube. Further, as shown in Figure 6 the mean jet penetrations in the calculation and in the experimental observations are in good agreement.

Associated with the time dependent particle displacement and transient jet is a gas velocity field. In Figure 7 the time average of the gas velocity profiles, normalized by the inlet jet velocity, as a function of the radial distance, normalized by the air tube radius, in the calculation are shown as the solid symbols centered on the axis of the air tube. The calculated jet initially decays in magnitude and disperses radially; but, at larger normalized axial distance $y/r_0 > 10$, this decay decreases. To some extent this behavior reflects a transition in the jet from a flow dominated by gas flow alone for $y/r_0 < 10$ to a flow dominated by solid particle motion for $y/r_0 > 10$. The particles tend to drag the gas with them and, since these particles are large, they maintain their momentum and prevent the jet from decaying. In addition, this jet is confined by the annulus in which the air tube is positioned (c.f., Figure 4). This confinement also limits the dispersion of the jet. There are not comparable measurements of gas velocity in the experiments reported in (40). However, some recent jet gas velocity measurements have been obtained in (41). In these latter experiments the air tube was positioned in the wide part of the cold flow rig, above the conical grid. The velocities were larger than in the present calculation, by a factor of 3 or more, and the bed was composed of denser particles (0.9 gm/cm^3). The measured velocity profiles, normalized and shown in Figure 7 as open symbols, reflect the

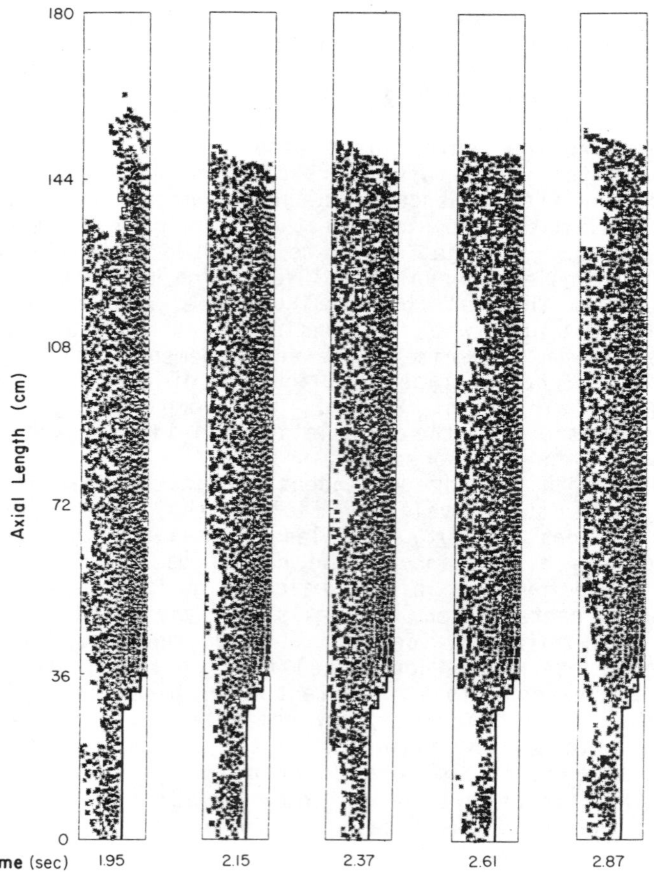

Figure 5. Time sequence in numerical prediction of cold flow jet penetration and particle mixing in Westinghouse reactor

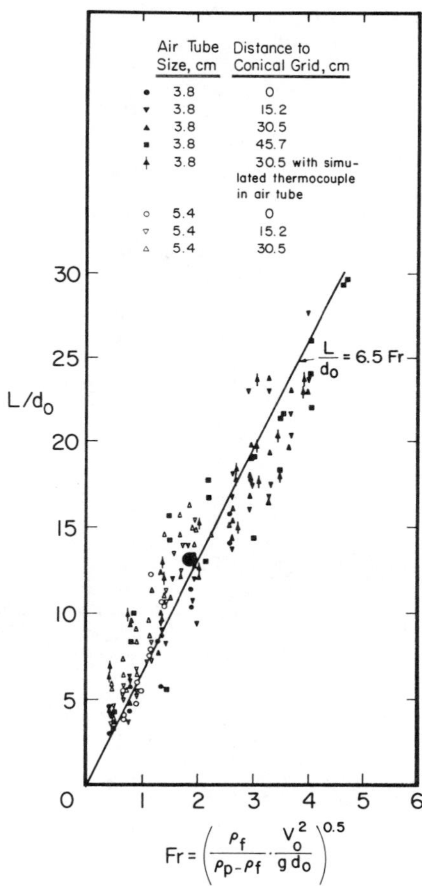

Figure 6. Comparison of the Westinghouse experimental data with (●) calculation, L/d_o *= mean jet penetration/air tube diameter,* ρ_f *= gas density,* ρ_p *= particle density*

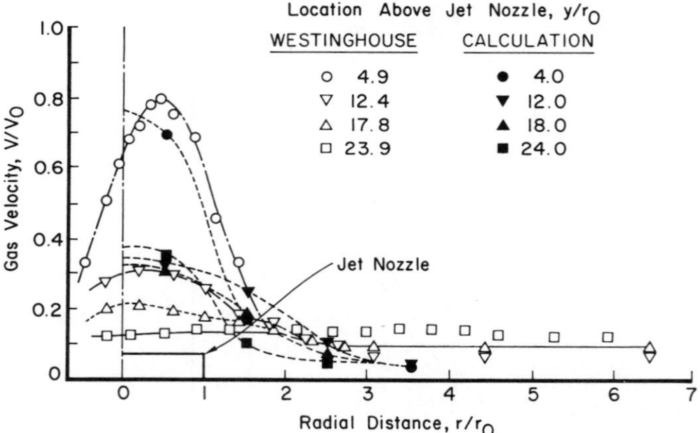

Figure 7. Comparison between Westinghouse experimental data and numerical calculation of gas phase velocity in jet

similarities and the differences between the latter experiment and the numerical calculation of the jet. The dispersion of the experimental jet is enhanced by the positioning of the jet in the wider section of the reactor. However, despite such differences the normalized calculational and experimental velocity profiles in Figure 7 do show remarkable similarities with regard to both the initially rapid dispersion of the jet between $y/r_0 = 4$ and $y/r_0 = 12$ and to the radial extent of the jet velocity field at most axial (y/r_0) locations above the air tube. Such comparisons tend to validate the hydrodynamic mixing predictions of the numerical model.

This numerical model has also been used to predict the hydrodynamic mixing in a hot reactive environment. The objective of the calculation is to demonstrate the influence of coupled chemistry and hydrodynamics upon the hydrodynamic mixing processes.

In this latter calculation, we use the same geometry as in the previous cold flow calculation (c.f., Figure 4) but change the reactor operating and feed conditions. While there are geometrical differences between the Westinghouse cold and hot flow rigs we, for this present hot flow calculation, keep the geometries the same. The conditions for the hot flow calculation are shown in Table IV where because of the higher nominal gas density in the reactor we increase, relative to the cold flow conditions, the mass flow through the air tube, annulus and conical distributors, thereby maintaining the same nominal superficial velocity in the reactor. Further, we load the reactor with char particles having a particle density of 1 gm/cm^3 (in the present calculation the injection of coal was not included in the simulation).

TABLE IV
REACTOR AND FEED CONDITIONS FOR NUMERICAL SIMULATION OF HOT, REACTING FLOW IN WESTINGHOUSE REACTOR

Particle Diameter (μm)	2800
Particle Density (gm/cm^3)	1.0
Mass Flow, Air Tube (air, gm/sec)	68.5
Mass Flow, Annulus (H_2O, gm/sec)	6.5
(CO_2, gm/sec)	46.2
Mass Flow, Cone (H_2O, gm/sec)	2.13
Reactor Pressure (atm)	15.0
Reactor Temperature (°K)	1290.0

Thus the basic concept of scaling between cold and hot flows which is used herein and implied by the choice of mass flows and gas to particle density ratios in Tables III and IV is that related to the gas density ratio between the hot and cold flows. This scaling criterion, previously used by Westinghouse (40) is, according to our present numerical calculations, a good one. This can be seen in Figure 8, where we show the numerical prediction of the jet penetration. The mean jet penetrations in the hot and cold flow cases are in agreement and both agree with the Westinghouse data on jet penetration. Further, in that Figure 8, we indicate a single hot flow measurement by Westinghouse (40) which coincides with their cold flow data and by implication suggests that the hot flow numerical calculation is valid.

In addition, the numerical calculation can provide descriptions of the temperature, gas composition, and particle conversion within the reactor. In Figure 9 we show a time sequence of temperature distribution in the gasifier. The oxygen is injected through the air tube and is rapidly consumed by reaction with the char. This oxygen enters the reactor at a temperature of 840°K and, through the combustion which occurs where oxygen and carbon meet, the products of reaction are heated to provide the initial high temperature in the vicinity of the air tube. Subsequent mixing and the endothermic gasification reactions moderate the temperature to provide a fairly uniform temperature distribution in the gasifier.

Concluding Remarks

The fluidized bed coal gasifier computer model is designed to provide a description of the hydrodynamic mixing and coupled chemistry within the reactor. This model should permit a designer to predict:

bubble size,
gas composition and temperature,
stagnant particle regimes,
elutriation,
jet penetration,
and solids mixing.

It is basically a hydrodynamic model, including particle scale effects, which can, therefore, be used to study scale-up and optimization of fluidized bed gasifiers. The hydrodynamic component of the model has been validated through comparison with cold flow visualization data and limited hot flow measurements.

The chemistry component of the model is, in most aspects, identical to the chemistry of the classical models of fluidized bed gasification. A major difference between the classical reactor models and the present fluidized bed coal gasifier computer model is that the classical models require specification of the bed hydrodynamics, such as bubble size. The present model can predict bubble size and the associated solids mixing. Again it is expected that the two types of models are complimentary. The present model can be used to define the hydrodynamics in the hot reactive environment and these hydrodynamics (e.g., bubble size) can then be used as

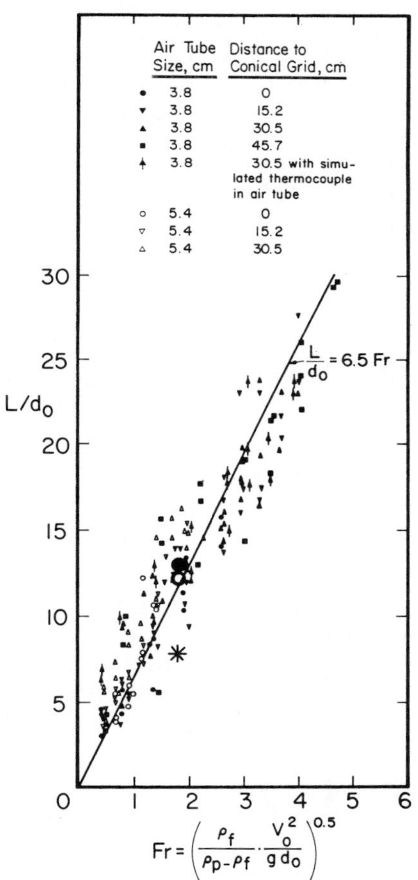

*Figure 8. Comparison of the Westinghouse experimental data with (●) cold and (○) hot calculations: Westinghouse cold flow data are indicated by small symbols; Westinghouse hot flow data are indicated by **

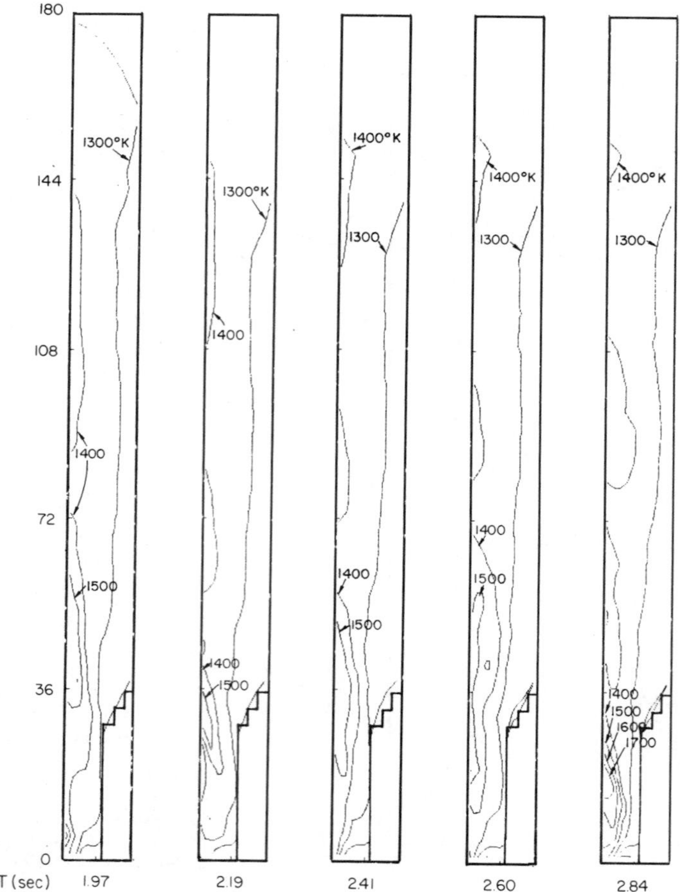

Figure 9. Numerical calculation of temperature distribution in hot, high pressure, Westinghouse reactor

input to the classical models for a detailed engineering calculation of, say, mass and energy balances for the reactor.

Legend Of Symbols

d_o	=	air tube diameter (cm)
f_i	=	function of partial pressures of reactants in reaction i.
F_r	=	Froude number
G	=	gas volume flow rate (ml/sec)
g	=	acceleration of gravity
k_i	=	kinetic rate coefficient for heterogeneous reaction i (gm/dyne sec)
k_o	=	intra-particle diffusion coefficient (gm/dyne sec)
k_{oE}	=	extra-particle diffusion coefficient, i.e., ash layer of particle (gm/dyne sec)
L	=	mean jet penetration length - the mean of the maximum and minimum jet lengths
$\dot{R}_i$	=	rate of carbon mass loss per unit surface area in i^{th} reaction
r	=	radial distance (cm)
r_o	=	air tube radius (cm)
r_b	=	bubble radius (cm)
V_o	=	air tube gas velocity (cm/sec)
$\vec{U}$	=	velocity vector of a solid particle (cm/sec)
V	=	gas velocity
ν_b	=	bubble volume (ml)
V_b	=	bubble velocity (cm/sec)
X	=	mol percent (N_2 free)
x	=	spatial variable in horizontal or radial direction (cm)
y	=	spatial variable in vertical direction (cm)
γ	=	mole ratio of carbn to oxygen involved in reaction
Δt	=	time increment
$\Delta x, \Delta y$	=	spatial increment in x and y, respectively
ρ_f, ρ_p	=	mass densities of gas and solid phases (gm/cm^3)

Acknowledgements

This research has been supported by the U.S. Department of Energy through contract EX-76-C-01-1770.

Literature Cited

1. Toomey, R. D.; Johnston, H. F.; Chem. Eng. Progr., 1952, 48, p. 220

2. Davidson, J. F.; Harrison, D.; "Fluidized Particles," Cambridge University Press, 1963.
3. Kunii, P.; Levenspiel, O.; Ind. Eng. Chem. Fund., 1968, 7, p. 446.
4. Kato, K.; Wen, C. Y.; Chem. Eng. Sci., 1969, 24, p. 1351.
5. Harlow, F. H.; Amsden, A. A.; J. Comp. Phys., 1974, 61, p.1.
6. Gidaspow, D.; Solbrig, C. W.; Presented at AIChE 81st National Meeting, Kansas City, April 11-14, 1976.
7. Sha, W. T.; Soo, S. L.; "Multidomain Multiphase Mechanics," Argonne National Laboratory Technical Memo, ANL-CT-77-3, 1977.
8. Blake, T. R.; Brownell, D. H., Jr.; Schneyer, G. P.; "A Numerical Simulation Model for Entrained Flow Coal Gasification, I. The Hydrodynamical Model," Proceedings, Miami International Conference on Alternative Energy Sources, 1978.
9. Smith, P. J.; Fletcher, T. H.; Smoot, L. D.; Eighteenth Symposium on Combustion, the Combustion Institute, Pittsburgh, 1980.
10. Chan, R. K.-C.,; Dietrich, D. E.; Goldman, S. R.; Levine, H. B.; Meister, C. A.; Scharff, M. F.; Ubhayakar, S. K.; "A Computer Model for the BI-GAS Gasifier," Jaycor Report J510-80-008A-2183, 1980.
11. Blake, T. R.; Garg, S. K.; Levine, H. B.; Pritchett, J. W.; "Computer Modeling of Coal Gasification Reactors," Annual Report June 1975 - June 1976, U.S. Energy Research and Development Administration Report FE-1770-15, 1976.
12. Blake, T. R.; Brownell, D. H., Jr.; Garg, S. K.; Henline, W. D.; Pritchett, J. W.; Schneyer, G. P.; "Computer Modeling of Coal Gasification Reactors - Year 2," 1977, Department of Energy Report FE-1770-32.
13. Blake, T. R.; Brownell, D. H., Jr.,; Chen, P. J.; Cook. J. L.; Garg, S. K.; Henline, W. D.; Pritchett, J. W.; Schneyer, G. P.; "Computer Modeling of Coal Gasification Reactors, Year 3,: Vol. I," 1979, Department of Energy Report FE-1770-49.
14. Pritchett, J. W.; Blake, T. R.; and Garg, S. K.; AIChE Progress Symposium Series, 1978, 176, (4), p. 134.
15. Weil, S., Private communication, 1977.
16. Richtmyer, R. D.; Morton, K. W.; "Difference Methods for Initial Value Problems," 2nd ed., Interscience, New York, 1967.
17. Murray, J. D.; J. Fluid Mech., 1965, 21, Part 2, pp465-493.
18. Anderson, T. B.; Jackson, R.; Ind. Eng. Chem. Fund., 1967, 6. p. 527.
19. Garg, S. K.; Pritchett, J. W.; J. of Appl. Physics, 1975, 46, p. 4493.
20. Rietema, K.; Chem. Eng. Sci., 1973, 28, pp. 1493-1497.
21. Schugerl, K.; in "Fluidization," eds. J. F. Davidson and D. Harrison, Academic Press, London, 1971, pp. 261-292.

22. Richardson, J. F.; in "Fluidization," eds., J. F. Davidson and D. Harrison, Academic Press, London and New York, 1971, pp. 26-64.
23. Gelperin, N. I.; Einstein, V. G.; in "Fluidization,: eds. J. F. Davidson and D. Harrison, Academic Press, London, 1971, pp. 471-540.
24. Yoon, H.; Wei, J.; Denn, M. M.; AIChE Journal, 1978, 24, p. 885.
25. Field, M. A.; Gill, D. W.; Morgan, B. B.; Hawksley, P. B.; "Combustion of Pulverized Coal," BCURA, Leatherhead, 1967.
26. Gray, D.; Cogoli, J. G.; Essenhigh, R. H.; Adv. in Chemistry, 131, ACS., 1974, p. 72-91.
27. Libby, P. A.; Blake, T. R.; Comb. and Flame, 1979, 36, p. 139.
28. Batchelder, H. R.; Busche, R. M.; Armstrong, W. P.; I EC, 1953, 45, No. 9, p. 1856.
29. Blackwood, J. D.; McGrory, R.; Australia Journal of Chem., 1958, 11, p. 16.
30. Blackwood, J. D.; Ingeme, A. H.; Australia Journal of Chem., 1960, 13, p. 194.
31. von Fredersdorf, C. F.; Elliott, M. A.; "Chemistry of Coal Utilization," Supplement Volume, H. H. Lowry, Editors, John Wiley, New York, 1963, pp. 892-1022.
32. Johnson, J. L.; Advances in Chemistry, 1974, 131, p. 145.
33. Dobner, S.; "Modelling of Entrained Bed Gasification: The Issues," EPRI Internal Report, 1976.
34. Wen, C. Y.; Dutta, S.; "Solid-Gas Reactions in Coal Conversion Processes," unpublished report, Department of Chemical Engineering, University of West Virginia, 1976.
35. Wen, C. Y.; Chaung, T. Z.; "Entrained-Bed Coal Gasification Modeling," 71st Annual AIChE Meeting, Miami, 1978.
36. Essenhigh, R. M.; Froberg, R.; Howard, J. B.; Ind. Eng. Chem., 1965, 57 p. 33.
37. Davidson, J.F.; Harrison, D.; Guedes de Carvalho, J.R.F.; in Annual Review of Fluid Mechanics, 1977, 9, p. 55-86.
38. Rowe, P. N.; in "Fluidization," eds. Davidson and Harrison, Academic Press, New York, 1971, p. 121a.
39. Anonymous; "HYGAS; 1964-1972, Pipeline Gas from Coal-Hydrogenation (IGT Hydrogasification Process)" Final Report OCR Contract 14-01-0001-381, 1975, 1-4.
40. Salvador, L. A.; Keairns, D. L.; "Advanced Coal Gasification System for Electric Power Generation Research and Development" Quarterly Progress Report: October-December, 1976; April-June 1977, (ERDA Reports 1514-61; 1514-69), 1977.
41. Yang, W. C.; Keairns, D. L.; "Momentum Dissipation of and Gas Entrainment into a Gas Jet in a Fluidized Bed," 72nd AIChE Annual Meeting, Nov., 1979, San Francisco, 1979.

RECEIVED June 3, 1981.

Study of the Behavior of Heat and Mass Transfer Coefficients in Gas–Solid Fluidized Bed Systems at Low Reynolds Numbers

J. RAMÍREZ, M. AYORA, and M. VIZCARRA

Departamento de Ingeniería Química, División de Estudios de Posgrado, Facultad de Química, Universidad Nacional Autónoma de México

Correlations to estimate heat and mass transfer coefficients in gas-solid fluidized beds operating in the controversial low Reynolds numbers zone are proposed. The correlations incorporate the influence of particle diameter to bed length and particle diameter to bed diameter ratios and gas flowrate. Also, the experimental data are used to analyse the models proposed by Kato and Wen , and Nelson and Galloway in order to explain the behaviour of fluid bed systems operating at low Reynolds numbers.

In spite of the amount of research effort directed towards the determination of the fluid to particle heat and mass transfer coefficients in fluidized beds of fine particles, there is a wide spread in the correlations proposed to estimate them.

A close look at the available experimental data on heat and mass transfer coefficients (1), shows that in the low Reynolds numbers zone exists the peculiar fact that both, the heat and mass transfer coefficients fall well below the value predicted by Ranz (2) for a single sphere submerged in a fluid in laminar flow (Sh=2). In this zone, the numerical results from the different studies also show major disagreement. In general, this is not the case in the high Reynolds numbers zone.

Literature correlations to estimate heat and mass transfer coefficients are generally of the form: $Sh=a\ Re^m$ (3). In general, they do not take into account the scale factors dp/D and dp/L which should be important, especially in the case of fluidized beds, given the complex hydrodynamics of these systems.

From studies on the behaviour of fluidized beds it is already known that bubbles are of great importance if one seeks to describe these systems . Mori and Wen (4) have shown an influence of the ratio dp/D on the growth of bubbles, and it is well known that bubbles grow when they rise through the bed. Clearly, it

0097-6156/81/0168-0185$05.00/0

should be important to include the factors dp/D and dp/L in the correlations to estimate fluid to particle mass and heat transfer coefficients in fluidized beds.

Kato and Wen (5) found, for the case of packed beds, that there was a dependency of the Sherwood and Nusselt numbers with the ratio dp/L. They proposed that the fall of the heat and mass transfer coefficients at low Reynolds numbers is due to an overlapping of the boundary layers surrounding the particles which produces a reduction of the available effective area for transfer of mass and heat. Nelson and Galloway (6) proposed a new model in terms of the Frossling number, to explain the fall of the heat and mass transfer coefficients in the zone of low Reynolds numbers. The model was developed to show that if the proper boundary conditions are used, one should not expect at low Reynolds numbers that the Nusselt and Sherwood numbers approach the limiting value of two, which is valid for a sphere in an infinite static medium. Since the particles are members of an assemblage, they assume in their model that there is a concentric spherical shell of radius $R > r_o$, on which the radial derivative of temperature or concentration is zero. This change in the boundary condition has a profound effect on the character of the transfer process, especially at low Reynolds numbers. In this case, the dependence of Sherwood number on Reynolds number becomes linear at low enough flowrates, and the limiting zero flow value of the Sherwood number is zero.

The model proposed by Nelson and Galloway can be stated as follows:

$$\mathcal{D} \frac{1}{r^2} \frac{\partial}{\partial r} \left(r^2 \frac{\partial c}{\partial r} \right) = \frac{\partial c}{\partial t}$$

$$c(r,0) = C_o$$

$$c(r_o,t) = C_o'$$

$$\frac{\partial c(R,t)}{\partial r} = 0$$

which for the limit of interest when Re → 0 gives:

$$\lim_{Re \to 0} Sh = \frac{1}{(1-\varepsilon)^{1/3}} \left[\frac{1}{(1-\varepsilon)^{1/3}} - 1 \right] \frac{\alpha^2}{2} Re\ Sc^{2/3}$$

Nelson and Galloway (6) propose a value of $\alpha = 0.6$ as an aproximation to compare experimental data. Clearly, the proposals of Kato and Wen (5) and Nelson and Galloway (6) are the most interesting ones.

In the present work, fluidized bed studies of simultaneous

heat and mass transfer are performed and analyzed in an effort to add some light into the zone of low Reynolds numbers.

To this end, experimental heat and mass transfer coefficients were determined in a fluidized bed. Nusselt and Sherwood numbers were obtained in terms of Reynolds number and aspect ratios dp/L and dp/D. The results are also analyzed in terms of the Kato and Wen(5) and Nelson and Galloway(6) models.

Experimental Work

Figure 1 depicts the experimental apparatus used in the determination of heat and mass transfer coefficients.A compressor (A) feeds the air to a tank (B), to minimize pulse fluctuations in flowrate. The air is dried as it passes trough a bed of silica gel (C). Air flowrate is measured with a rotameter and in addition with a calibrated capillary meter(E). The inlet air moisture content is measured by means of a dry and wet bulb thermometer system (D) prior to its entrance to a coil submerged in a constant temperature bath (F). From here, the air enters the bottom of the fluidized bed (G) where its temperature is measured.The fluidized bed consisted of an insulated QVF glass tube 2 inches in diameter and 12 inches in length. A system for the collection of fines(I) was installed after the bed to evaluate entrainment, although at all experimental conditions used in this work entrainment was absent. A thermometer placed on top of the bed of solids was used to measure the temperature of the bed exit.Air moisture content was also determined at the outlet of the fluidized bed by means of a hygrometer and a wet and dry bulb temperature system(J).

At time intervals, a sample of solid was taken out of the bed for moisture content analysis. This was determined by weight using an analytical scale.

The data used for the calculation of heat and mass transfer coefficients were taken only from the constant rate drying period. It was assumed that during this time the solid surface was well saturated with moisture.

The solid used in this work was silica gel of two different mean particle diameters. The fine silica gel ranged between 0.0058 to 0.0304 cm., and the coarser between 0.020 to 0.050 cm. in particle diameters. The mean particle diameters were 0.0125 cm. and 0.035 cm. respectively. Minimum fluidization velocities were 0.4113 cm/sec. and 1.67 cm/sec. respectively for the small and big particle diameters.Solid particle density was 1.25 g/cm^3.

In order to obtain uniform moisture content in the solid, this was humidified in the same fluidized bed. The time interval for data recording was normally 15 min.

The fluid bed distributor was an aluminium perforated disc whose holes were 0.5 mm. in diameter arranged in one centimeter square pitch. A stainless steel 325 mesh screen was fixed at the entrance side of the perforated disc. The thermometers used for temperature measuring had 0.1 degree centigrade divisions.In order

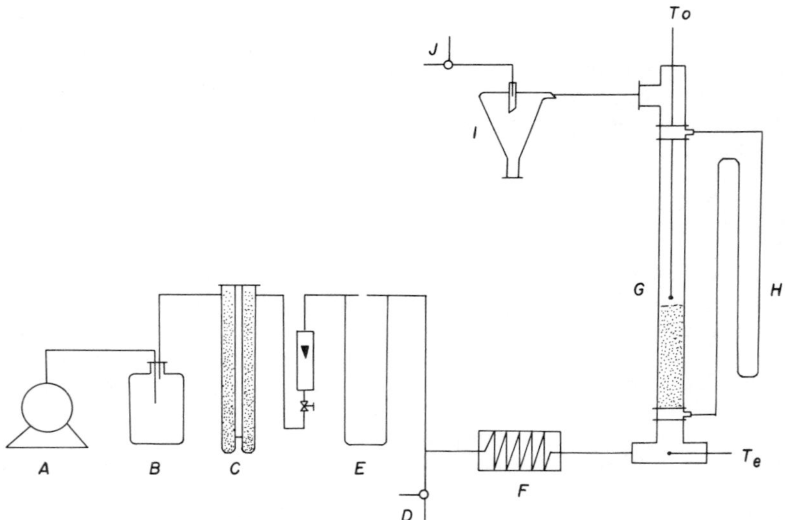

Figure 1. Experimental apparatus: A, air compressor; B, tank; C, silica gel trap; D, dry and wet bulb thermometer system; E, capillary meter; F, constant temperature bath; G, fluidized bed; I, fines collector; J, dry and wet bulb thermometer system; T_e, entrance temperature thermometer; H, U tube manometer; T_o, outlet temperature thermometer

to avoid condensation problems and to maintain uniform operating temperatures, all lines after the constant temperatures, all lines after the constant temperature bath as well as the fluidized bed were well insulated with glass fibre and asbestos tape.

At time intervals, air and solid moisture content was determined in order to construct the drying curves.A typical set of drying curves is presented in Figure 2.

By carrying out mass and heat balances over the constant rate drying period, mass and heat transfer coefficientes can be obtainded:

$$-\rho_s \frac{dH}{d\Theta} = km\ A\ (X_{bh} - X)_{LM}\ \rho_a \qquad 1$$

from this equation; the mass transfer coefficient can be readily obtained if one knows the slope of the graph of solid moisture content versus time and the humidities of the air:

$$km = \frac{\rho_s \left(-\frac{dH}{d\Theta}\right)}{A\ (X_{bh} - X)_{LM}\ \rho_a} \qquad 2$$

Similarly, the heat balance equation is:

$$-\rho_s \left(\frac{dH}{d\Theta}\right) = \frac{hA}{\lambda}\ (T - T_{bh})_{LM}$$

accordingly the heat transfer coefficient will be expressed as:

$$h = \frac{\rho_s \left(-\frac{dH}{d\Theta}\right) \lambda}{A\ (T - T_{bh})_{LM}} \qquad 3$$

Experimental Results and Conclusiones

During the experimental runs, the air flowrate was varied over the range 5.3 to 20.29 l/min. (Re=0.28 to Re=3.0), the bed height was also varied between 2 and 8 cm. Also, two different particle diameters were used. This gave a variation of the dp/L ratio from 0.0016 to 0.0152 and the dp/D ratio from 0.025 to 0.007 that is an order of magnitud in both ratios.

From the experimental data, taken at time intervals and equations 2 and 3, experimental heat and mass transfer coefficients were obtained. Tables I and II show the values of the experimental heat and mass transfer coefficients.

Figures 3 to 5 show the effect of introducing the scale factors dp/L, dp/D into the correlations to predict heat and mass transfer coefficients.

It comes up clearly from these figures that one way not to correlate heat and mass transfer coefficients, at least in fluidized bed systems, at low Reynolds is a correlation of the

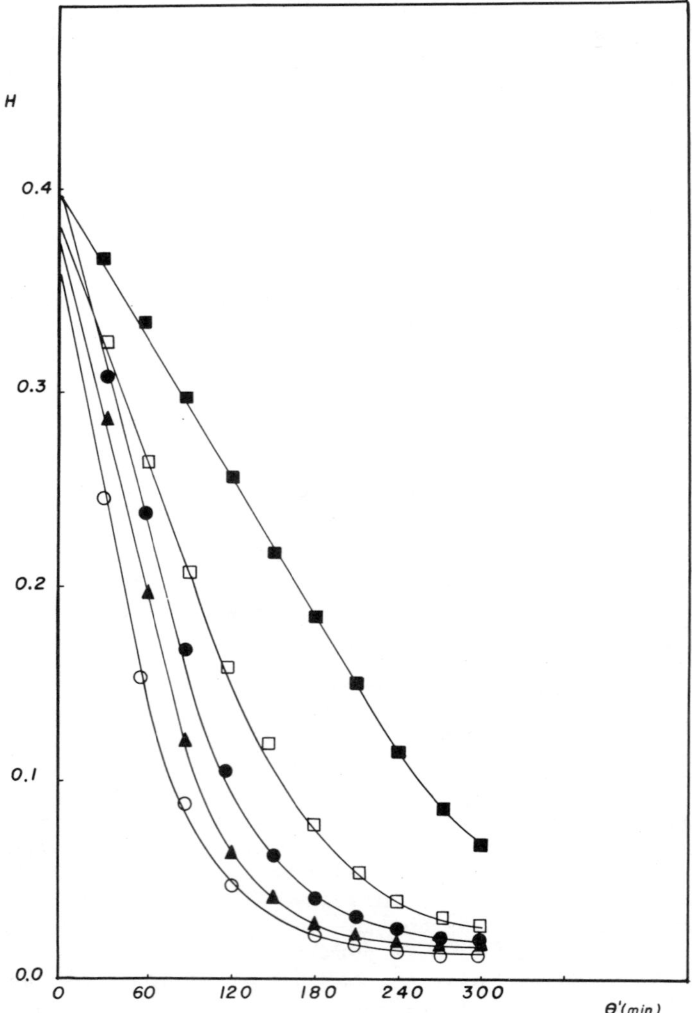

Figure 2. Typical drying curves; L = *4;* D = *5.08;* dp = *0.0125: (■) Re = 0.285; (□) Re = 0.462; (●) Re = 0.663; (▲) Re = 0.85; (○) Re = 0.94*

Table I. Calculated values of the mass transfer coefficient at the different operating conditions.

	dp=0.035 cm. km'(m/hr)				dp=0.0125 cm. km'(m/hr)		
Re	L/D=0.39	L/D=0.76	L/D=1.49	Re	L/D=0.39	L/D=0.76	L/D=1.49
3.05	35.55	19.62	19.14	1.06	48.06		
2.48		18.64		0.94	37.52	31.34	10.42
2.20	25.36		16.90	0.85		17.31	7.33
1.74	17.18	11.25	7.93	0.66	23.75	16.19	6.64
1.40	13.08	7.76		0.46	15.79	11.47	5.87
1.22			4.58	0.28	13.34	6.58	4.48

Table II. Calculated values of the heat transfer coefficient at the different operating conditions.

	dp=0.035 cm. km'(m/hr)				dp=0.0125 cm. km'(m/hr)		
Re	L/D=0.39	L/D=0.76	L/D=1.49	Re	L/D=0.39	L/D=0.76	L/D=1.49
3.05	8.09	4.92	5.37	1.060	11.97		
2.48		3.92		0.941	8.56	7.72	2.39
2.20	6.10		3.64	0.850		4.48	1.89
1.74	3.99	2.52	1.68	0.663	5.12	3.84	1.68
1.40	3.09	1.82		0.462	3.36	2.90	1.38
1.22			1.09	0.285	3.08	1.60	1.08
0.826		1.03					

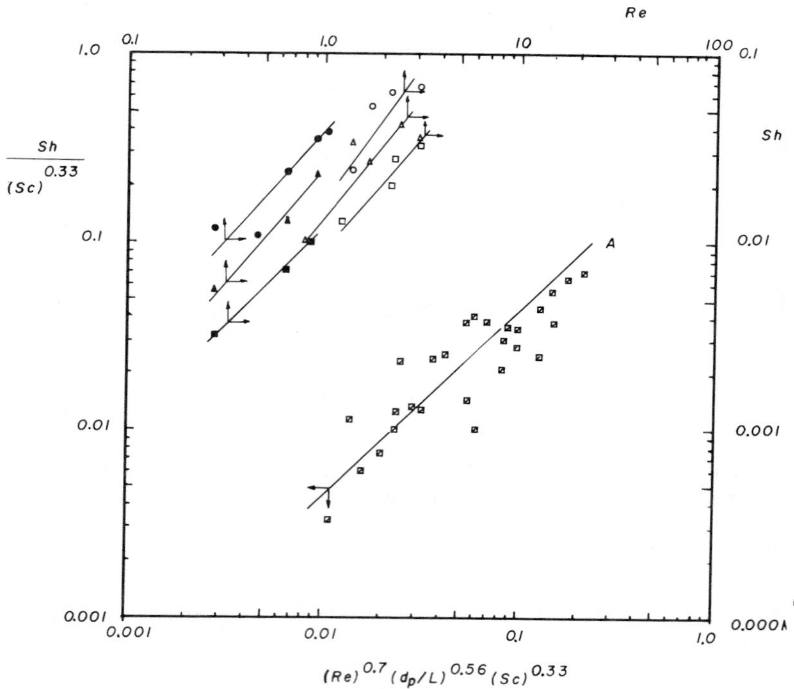

Figure 3. Effect of the dp/L *ratio on the estimation of Sherwood numbers: (●)* L/D = *0.39,* dp = *0.0125; (▲)* L/D = *0.76,* dp = *0.125; (■)* L/D = *1.49,* dp = *0.0125; (○)* L/D = *0.39,* dp = *0.035; (△)* L/D = *0.76,* dp = *0.035; (□)* L/D = *1.49,* dp = *0.035. The crossed squares (◪) encompass all the experimental data and line A shows the fit to the correlation:* $Sh = 0.4329\ (Re)^{0.7}\ (dp/L)^{0.56}\ (Sc)^{0.33}$

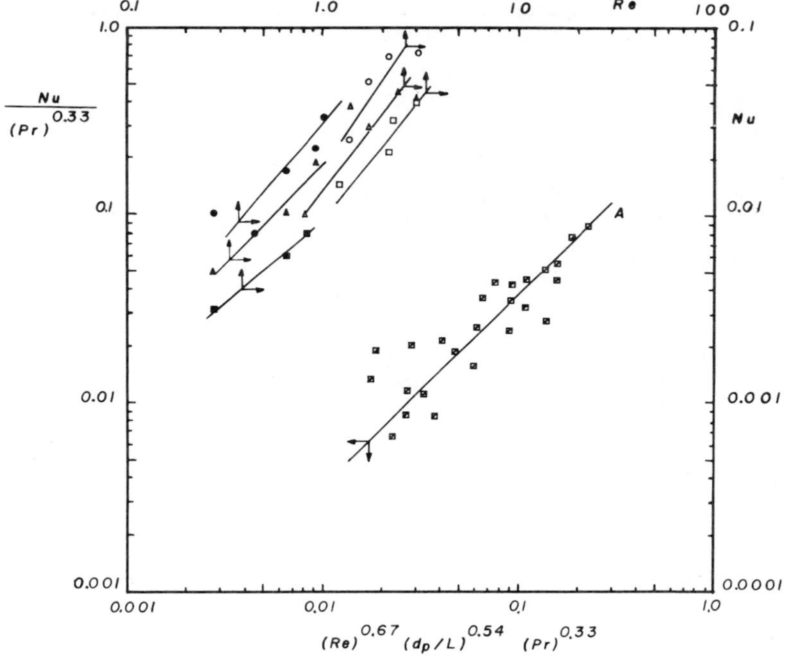

Figure 4. Effect of the dp/L *ratio on the estimation of Nusselt numbers:* (●) L/D = *0.39,* dp = *0.0125;* (▲) L/D = *0.76,* dp = *0.0125;* (■) L/D = *1.49,* dp = *0.0125;* (○) L/D = *0.39,* dp = *0.035;* (△) L/D = *0.76,* dp = *0.035;* (□) L/D = *1.49,* dp = *0.035. The crossed squares* (⊠) *encompass all the experimental data and line A shows the fit to the correlation:* $Nu = 0.3726\ (Re)^{0.67}\ (dp/L)^{0.54}\ (Pr)^{0.33}$

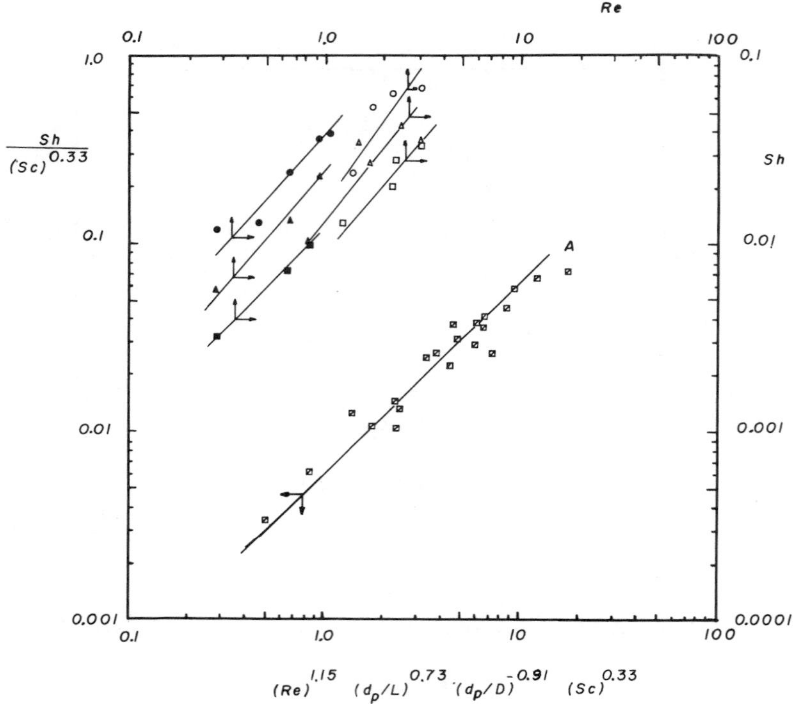

Figure 5. Effect of introducing both aspect ratios dp/L *and* dp/D *on the correlation to estimate the Sherwood number: (●)* L/D = *0.39,* dp = *0.0125; (▲)* L/D = *0.76,* dp = *0.0125; (■)* L/D = *1.49,* dp = *0.0125; (○)* L/D = *0.39,* dp = *0.036; (△)* L/D = *0.76,* dp = *0.035; (□)* L/D = *1.49,* dp = *0.035. The crossed squares (⊠) encompass all the experimental data and line A shows the fit to the correlation:* $Sh = 0.00632\,(Re)^{1.15}\,(dp/L)^{0.73}\,(dp/D)^{-0.91}\,(Sc)^{0.33}$

form Sh=a Re^n or Nu=a Re^n. Unfortunately,this is the way in which the data are presented in the literature.

Exclusion of the dp/L and dp/D ratios from the correlations lead to gross errors up to 200% in the prediction of heat and mass transfer coefficients. Correlations of this form will be too particular to be useful for design.

The final correlations obtained which encompass all the experimental data in this work are:

$$Sh= 0.00632\ (Re)^{1.15}(dp/L)^{0.73}(dp/D)^{-0.91}(Sc)^{0.33}$$

$$Nu=0.004948\ (Re)^{1.14}(dp/L)^{0.71}(dp/D)^{-0.94}(Pr)^{0.33}$$

One can see from the values of the exponents in the aspect ratios dp/D and dp/L, that both parameters are important and their influence cannot be neglected.

As for the Nelson and Galloway model, Figures 7 and 8 show that the proposed model predictions get closer to the experimental results as long as one uses a different value of the parameter α. The value of α=0.6 proposed by Nelson and Galloway predicts too high values of the mass transfer coefficients. If one uses a value of α=0.3, the model predictions get closer to the experimental results of this work.

In other works(7,8) , Galloway and Sage state that the Frossling number (α) varies in the turbulent regimes($900 \leq Re \leq 75000$) from 0.5 to 1.6.Given the low Reynolds numbers in this work it is quite possible that α should have a value lower than 0.5.This is in agreement to the value of 0.3 which makes the model predict in the range of our experimental values.

Finally, Kato and Wen(5) have proposed that the drastic fall observed for heat and mass transfer coefficients in the zone of low Reynolds numbers is due to an overlapping of the boundary layers surrounding the solid.

This overlapping will in fact reduce the available area for heat and mass transfer. During the present work , some boundary layer thicknesses were estimated for the experimental conditions of this work. As a result , the boundary layers only overlap for Reynolds numbers below 0.826. For the case of Reynolds numbers of 1.74 and 3.05 using the particle diameter of 0.035 cm. , the boundary layers do not overlap.Table III shows some of the values obtained.Clearly, this effect cannot explain completely the low heat and mass transfer coefficients at low Reynolds numbers.

Summing up, for design purposes one should continue using empirical correlations and from this, the ones which include the effect of the scale factors dp/L, dp/D, as the ones proposed in this work , especially at low Reynolds numbers.

As for the theoretical explanation of lower heat and mass transfer coefficients:

The Kato and Wen proposal does not seems to explain all the experimental findings,and the Nelson and Galloway model seems

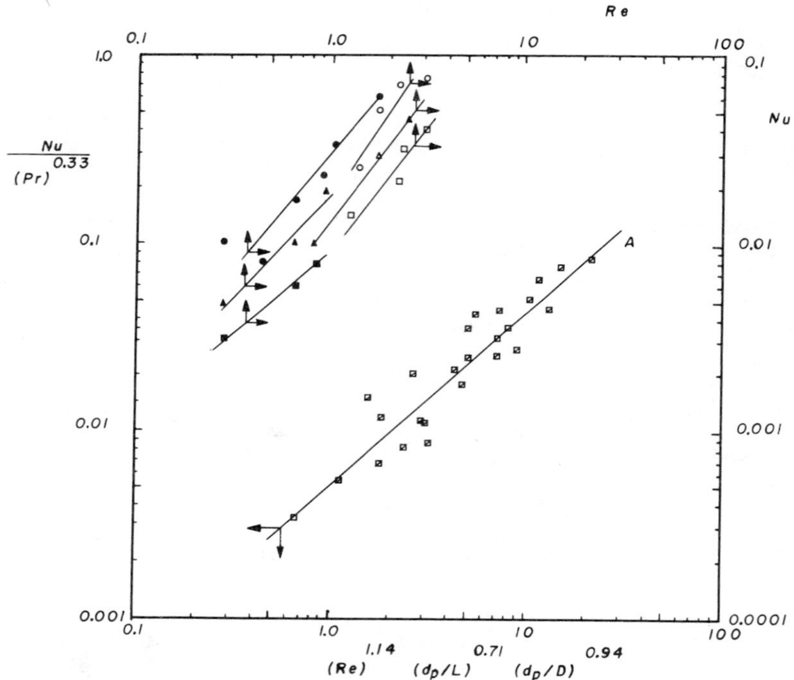

Figure 6. Effect of introducing both aspect ratios dp/L *and* dp/D *on the correlation to estimate the Nusselt number: (●)* L/D = *0.39,* dp = *0.0125; (▲)* L/D = *0.76,* dp = *0.0125; (■)* L/D = *1.49,* dp = *0.0125; (○)* L/D = *0.39,* dp = *0.035; (△)* L/D = *0.76,* dp = *0.035; (□)* L/D = *1.49,* dp = *0.035. The crossed squares (⊠) encompass all the experimental data and line A shows the fit to the correlation:* $Nu = 0.004948\,(Re)^{1.14}\,(dp/L)^{0.71}\,(dp/D)^{-0.94}\,(Pr)^{0.33}$

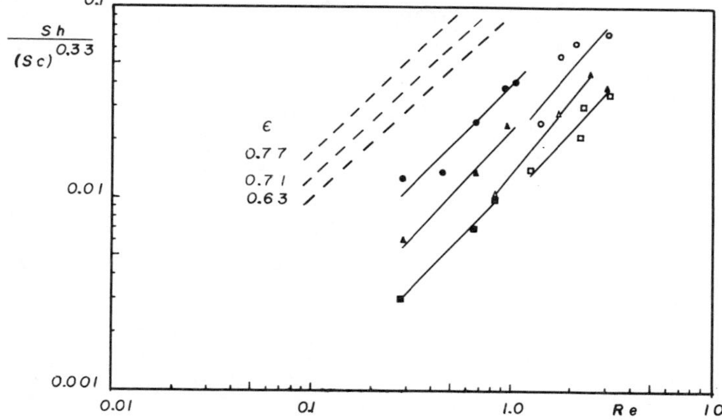

Figure 7. Experimental and predicted values of the Sherwood number using Nelson and Galloway model with $\alpha = 0.6$: (●) L/D = 0.39, dp = 0.0125, $\epsilon = 0.77$; (▲) L/D = 0.76, dp = 0.0125, $\epsilon = 0.74$; (■) L/D = 1.49, dp = 0.0125, $\epsilon = 0.71$; (○) L/D = 0.39, dp = 0.035, $\epsilon = 0.69$; (△) L/D = 0.76, dp = 0.035, $\epsilon = 0.66$; (□) L/D = 1.49, dp = 0.035, $\epsilon = 0.63$. The broken lines show the model predictions.

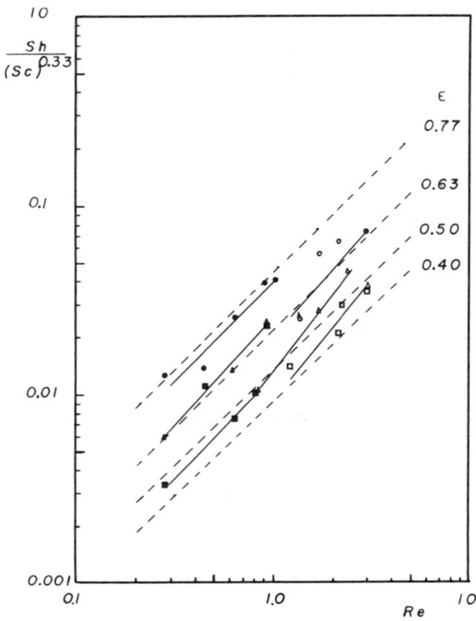

Figure 8. Experimental and predicted values of the Sherwood number using Nelson and Galloway model with $\alpha = 0.3$: (●) L/D = 0.39, dp = 0.0125, $\epsilon = 0.77$; (▲) L/D = 0.76, dp = 0.0125, $\epsilon = 0.74$; (■) L/D = 1.49, dp = 0.0125, $\epsilon = 0.71$; (○) L/D = 0.39, dp = 0.035, $\epsilon = 0.69$; (△) L/D = 0.76, dp = 0.035, $\epsilon = 0.66$; (□) L/D = 1.49, dp = 0.035, $\epsilon = 0.63$. The broken lines show the model predictions.

Table III. Momentum (S) and mass (Sm) boundary layer thinckness calculated at the different operating conditions.

dp=0.035 cm			dp=0.0125 cm.		
Re	S(cm)	Sm(cm)	Re	S(cm)	Sm(cm)
0.2845	0.0147	0.0154	0.100	0.1600	0.1800
0.2845	0.0144	0.0151	0.826	0.0287	0.0301
0.2845	0.0138	0.0145	0.826	0.0274	0.0287
0.4616	0.0096	0.0101	0.826	0.0262	0.0274
0.4616	0.0094	0.0098	1.74	0.0136	0.0142
0.4616	0.0090	0.0094	1.74	0.0130	0.0136
1.06	0.0076	0.0080	1.74	0.0124	0.0130
1.06	0.0075	0.0079	3.05	0.0078	0.0082
1.06	0.0072	0.0075	3.05	0.0074	0.0078
3.5	0.0038	0.0040	3.05	0.0071	0.0074
3.5	0.0037	0.0039			
3.5	0.0037	0.0038			

to point in the right direction. More study about the value of the Frossling numbre (α) with Reynolds number will be necessary before one takes up this model.

The dependence of heat and mass transfer coefficientes on the scale factors dp/L and dp/D can also be rationalized in terms of gas bypassing through the bed in the form of bubbles. Since bubbles coalesce and grow as the rise from the distributor, a longer bed, big L/D values, will operate with larger bubbles in its upper part. This will lead to smaller values of the Sherwood and Nusselt numbers since the interchange coefficient between bubble and emulsion phase varies inversely with bubble diameter.

It seems clear that a two phase model will be able to predict low values of the heat and mass transfer coefficients as Kunii has done (1). The trouble with this approach will be an accurate estimate of the equivalent bubble bed diameter. Thus, improved empirical correlations are still useful for design purposes, when one looks for estimates of gas-solid heat and mass transfer coefficients.

Notation

A area per unir volume, cm^2/cm^3
c concentration, gm/cm^3
C_o initial concentration, gm/cm^3
D bed diameter, cm.
dp particle diameter, cm.
h heat transfer coefficient $cal/sec\text{-}cm^2\text{-}°C$
h' heat transfer coefficient $Kcal/hr\text{-}m^2 - °C$
H solid moisture content, gm H_2O/gm. dry solid
km mass transfer coefficient cm/sec.
km' mass transfer coefficient m/hr,
L bed length, cm.
r radial coordinate, cm.
r_o particle radius, cm.
S momentum boundary layer thickness, cm.
Sm mass boundary layer thickness, cm.
T fluid temperature °C
T_{bh} wet bulb temperature °C
X air moisture content gm. H_2O/gm. dry solid
y logarithmic mean fraction of inert nondiffusing component
P_r Prandtl number $c_p\, \mu / k$
R_e Reynolds number $dp\, \rho_a\, \nu/\mu$
Sc Schmidt number $\mu/\rho_a \mathcal{D}$
Sh Sherwood number km dp $y/\mathcal{D}$
α Frossling number, $(Sh-2)/Re^{0.5}\, Sc^{0.33}$
$\mathcal{D}$ Diffusion coefficient, cm^2/sec.
ε bed porosity
θ time, sec
ρ_a air density, gm/cm^3
ρ_s solid density, gm/cm^3
λ heat of vaporization, cal/gm

Literature Cited

1. Kunii, D.; Levenspiel, O., "Fludization Engineering",Wiley, New York 1969, 199,215.
2. Ranz W.E. Chem.Eng.Progr. 1952, 48, 247.
3. Ayora, M.; "Determinación Experimental de parámetros de diseño para secadores de lecho fluidizado sólido-gas" Tesis de Maestría, Fac.de Química, UNAM, México 1978.
4. Mori, S.; Wen, C.Y. AIChE J. 1975, 21, 190.
5. Kato, K.; Wen, C.Y. Chem. Eng. Prog. Symp. Ser.1970,66,100.
6. Nelson, P.; Galloway, T. Chem. Eng.Sci. 1975,30, 1-6.
7. Galloway, T.; Sage,B.H. Chem. Eng. Sci. 1970,25 ,495.
8. Galloway,T.; Sage,B.H. AIChE, J. 1967, 13 , 563.

RECEIVED June 3, 1981.

BUBBLE COLUMN REACTORS

10

Bubble Column

An Overview

YATISH T. SHAH

Chemical and Petroleum Engineering Department, University of Pittsburgh, Pittsburgh, PA 15261

Bubble columns have been widely used in chemical and petroleum industries. This paper presents a brief overview on the present state of art of vertically sparged bubble columns. Hydrodynamics, mixing and transport characteristics of the bubble column are briefly evaluated. Recommendations for the future experimental work are also made.

Bubble column is a term used to describe a vertical column wherein gas is bubbled through either a moving or stagnant pool of liquid or liquid-solid slurry. The name is rather loosely used because the gas is not retained in the column in the form of bubbles at high gas velocity. The column need not be vertical either; it can be horizontal or even coil shaped. This brief overview examines the present state of art of the vertically sparged bubble columns. The discussion is restricted to gas-liquid systems only.

The important variables that affect the bubble dynamics and flow regime in a bubble column are gas velocity, fluid properties (e.g. viscosity, surface tension etc.), nature of the gas distributor, and column diameter. Generally, at low superficial gas velocities (approximately less than 5 cm/sec) bubbles will be small and uniform though their nature will depend on the properties of the liquid. The size and uniformity of bubbles also depends on the nature of the gas distributor and the column diameter. Bubble coalescence rate along the column is small, so that if the gas is distributed uniformly at the column inlet, a homogeneous bubble column will be obtained.

At high superficial gas velocities (greater than approximately 5 cm/sec), the bubble coalescence rate increases significantly, the gas-liquid flow becomes non-homogeneous (see Figure I) and the bubble column contains a mixture of large and small bubbles. The fraction of gas occupied by large bubbles increases with the gas velocity. The size of the large bubbles depends on the nature of the gas distributor, column diameter and physical

0097-6156/81/0168-0203$05.00/0

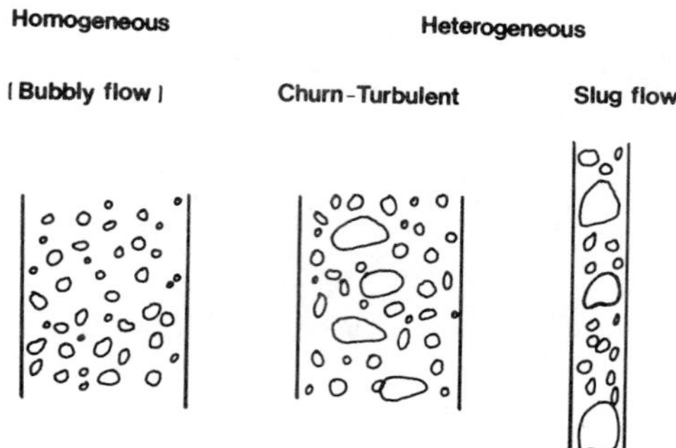

Figure 1. Different flow regimes in a bubble column

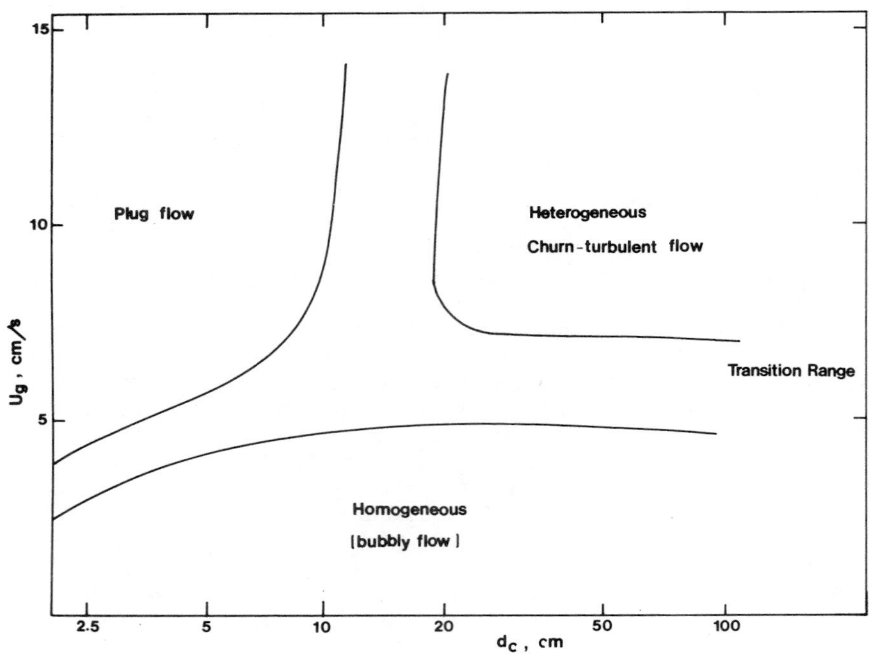

Figure 2. Flow regime plot for the bubble column

properties of the liquid. The hydrodynamic, mixing and transport properties of a non-homogeneous bubble column should be considerably different from that of a homogeneous column.

A bubble column can be divided into three regions. In the first region near the bottom (i.e. entrance region), the behavior and the properties of the bubbles are determined by the sparger design and the gas flow rate. The second region, above the first, occupies a large fraction of column volume. In this region, bubble properties are determined by the liquid flow pattern and the liquid properties. In the third region, close to the top, bubble coalescence occurs. The position of the regions are affected by superficial gas velocity. As the velocity increases, the coalescence occurs at earlier stage.

A plot of the flow regime prevailing in the major part (i.e., second region in the above classification) of the column in terms of gas velocity and column diameter is described in Figure II. Future work should clearly illustrate the effects of fluid properties and the nature of the gas distributor on the flow regime. From the practical point of view, both homogeneous and non-homogeneous flow regimes are important. The present state of art on hydrodynamics, mixing and transport characteristics of a bubble column in these flow regimes are separately outlined below.

Homogeneous Flow

The hydrodynamic, transport and mixing characteristics in this flow regime are reasonably well understood.

For small liquid velocities, gas holdup is essentially independent of liquid velocity. Provided the ratio of the column to bubble diameter is large, say > 40, the column diameter does not significantly affect the holdup. Usually a column diameter of 10 cm is sufficient to yield holdup values which are close to the ones obtained in larger diameter columns under the same conditions. Dependence of gas holdup on the gas velocity is generally of the form

$$\varepsilon_G \propto u_g^{\,n} \tag{1}$$

where in values from 0.7 to 1.2 are reported (1-11). The drift flux theory (12) would give n = 1. The holdup also depends on the fluids employed and trace impurities. The effects of fluid properties such as density, viscosity and surface tension on the gas holdup have been empirically correlated (1-11,13,14). Liquids that have been examined are organic liquids (mostly alcohols and halogenated hydrocarbons), aqueous solutions of glycol, glycerol, ethanol, NaCl, Na_2SO_3, synthetic fermentation media and a variety of electrolytes. The holdup may depend on the liquid phase composition.

Future work should concentrate on large diameter (both short and tall) bubble columns. The applicability of the existing correlation should be examined for these columns for a variety of organic liquids and electrolytes and different types of gas distributors. Very little is known about the bubble dynamics and the gas holdup in a bubble column containing highly viscous polymeric solution and water soluble polymer solutions. This area should be explored.

The gas-liquid interfacial area has been measured by both physical and chemical methods. The accuracy of these measurements is generally very poor (15). The interfacial area has been related to the energy input per unit volume and the gas holdup by the expression (6-19)

$$a \propto \left(\frac{E}{V_R}\right)^{0.4} (1 - \varepsilon_G)^n \qquad (2)$$

If physico-chemical properties of the system are changed, the above relation may not be completely valid (13). Future efforts should be concentrated in examining the validity of the existing correlations for large diameter (tall and short) bubble columns operated with a variety of gas spargers.

The dependence of the volumetric gas-liquid mass transfer coefficient on gas velocity can be expressed as (26)

$$K_L a = b u_g^{\,n} \qquad (3)$$

For water and electrolyte solutions, n = 0.82. The constant b largely depends on the nature of distributor and the liquid media (26). For the coalescing media and if strong absorption takes place, $K_L a$ would be the largest near the gas distributor and would decrease along the length. In industry, porous plates are usually not used even though they provide high mass transfer rates. The gas is either sparged by two-phase nozzles of injector, ejector and slot types (1,16,18-22) or by single and multi-orifice distributors. Two phase nozzles with liquid recirculation (jet and slot injectors) provide intimate mixing and high interfacial areas, and hence high mass transfer rates (1,18,19,20). The performance charcteristics and design principles of various gas spargers are treated in the literature (16,18,20-25). For single and multi-orifice spargers, the correlation of Akita and Yoshida (6) is recommended for oxygen in water and aqueous solutions. $K_L a$ data for aqueous solutions of alcohols and glucose, with and without the presence of inorganic electrolytes, are given by Schugerl et al. (1) and for CMC solutions by Buchholz et al. (27), Deckwer et al. (28), Schumpe and Deckwer (29) and Nakanoh and Yoshida (30). $K_L a$ data for large diameter bubble columns, for non-aqueous organic media like alcohols, ketones, esters and hydrocarbons and for gases such as hydrogen, chlorine etc., are needed.

In the homogeneous bubble flow regime, the gas phase is generally assumed to move in plug flow and the liquid phase axial mixing is characterized by the axial dispersion coefficient. The axial dispersion coefficient is dependent upon gas velocity and column diameter according to (26,41,42)

$$E_L \alpha u_g^{1/3} d_c^{4/3} \tag{4}$$

The dispersion coefficient is essentially independent of liquid velocity. The effect of liquid properties on the dispersion coefficient has been found to be very mild and correlated by Cova (31) and Hikita and Kikukawa (32) as

$$E_L \alpha \rho_L^{0.07} (1/\mu_L)^{0.12} \tag{5}$$

The dispersion coefficient was found to be essentially independent of liquid surface tension. The effects of liquid properties on the dispersion coefficient were also examined by Kato and Nishiwaki (33), Akita (34), and Ulbrecht and Sema Baykara (35). A number of hydrodynamics and mixing models have been reported in the literature and are reviewed by Joshi and Shah (36). Future work should include the testing of these models for large columns.

The heat transfer in a bubble column is very high. Numerous data reported in the literature are reviewed and analyzed by Joshi et al. (37). Further work should be carried out in large diameter columns only.

Non-Homogeneous Flow

As mentioned earlier, for superficial gas velocities larger than approximately 5 cm/sec, flow in a bubble column generally becomes non-homogeneous. This flow regime prevails in most industrial bubble columns. The flow becomes unstable and coalescence occurs. Large bubbles with high rise velocities coexist in the presence of small bubbles. The large bubbles are no longer spherical but take the form of spherical caps of varying form with a very mobile and flexible interface. These large bubbles can grow up to diameters of about 10 cm. In small diameter columns, large bubbles can be stabilized by the column wall which leads to the formation of slugs. In tall columns slugs can be observed for column diameters as large as 15 cm. The rising bubbles cause liquid to flow downwards resulting in a circulation of liquid within the column. The non-homogeneous flow is, therefore, sometimes called recirculating flow (38).

In non-homogeneous flow, the knowledge of gas holdup is not very meaningful unless it is divided into two parts: (a) holdup occupied by large bubbles and (b) holdup occupied by small bubbles. The large bubbles will collapse much more rapidly than small bubbles as the gas flow is turned off. Figure IIIa shows the experimental data of Beinhauer (39,40) in a bubble column of 10 cm diameter using water as the liquid phase. At large gas velocities, a large fraction of the column is occupied by large bubbles. As shown in Figure IIIb, large bubbles move faster than smaller bubbles. A proper hydrodynamic characterization of the bubble column under non-homogeneous flow conditions require the holdups occupied by both size bubbles. The literature shows a very low value of n in Equation (1) in non-homogeneous flow regime.

Just as with the gas holdup, gas-liquid interfacial area should also be divided into two parts. The literature, however, gives a unified correlation. The same is true for volumetric gas-liquid mass transfer coefficients and mixing parameters for both gas and liquid phases. The fundamental mechanism for inter-phase mass transfer and mixing for large bubbles is expected to be different from the one for small bubbles. Future work should develop a two phase model for the bubble column analogous to the two phase model for fluidized beds.

General Remarks

The literature on bubble columns is abundant (42,43,44). Future experimental work should be concentrated on large diameter, tall bubble columns. Organic and non-newtonian fluids should be examined. High pressure, high temperature operations need to be emphasized. Theoretical work needs to be concentrated on non-homogeneous flow regime. Solutions to various scale-up problems would be facilitated with a better fundamental understanding of the non-homogeneous flow regime.

Nomenclature

a gas-liquid interfacial area
b a constant in Equation (3)
d_c column diameter
E/V_R energy per unit volume
E_L liquid phase axial dispersion coefficient
K_La volumetric gas-liquid mass transfer coefficient
n coefficients in Equations (1) and (3)
u_g superficial gas velocity
$u_g{}^*$ bubble rise velocity
ε_G gas holdup
ρ_L liquid density
μ_L liquid viscosity

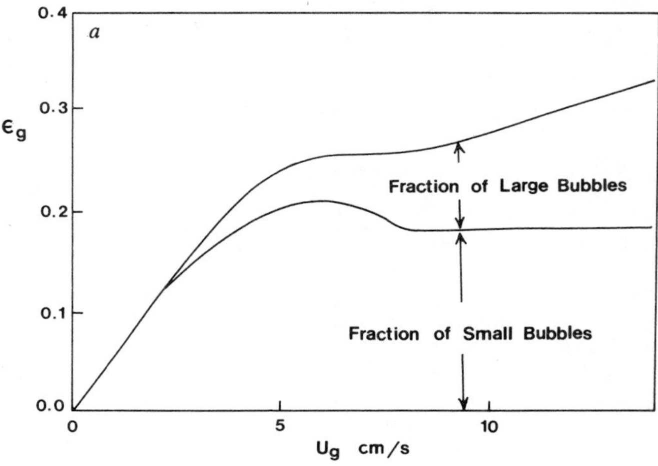

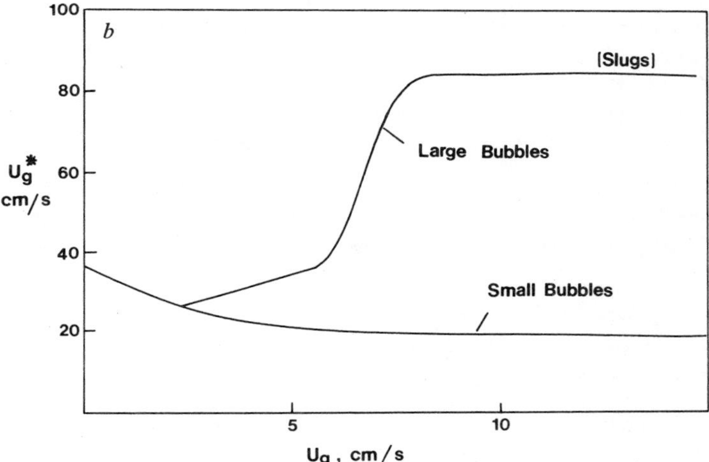

Figure 3. a, Fractional gas holdup of large and small bubbles (39); b, rise velocity of large and small bubbles (39)

References

1. Schugerl, K.; Lucke, J,; Oels, U. Adv. Biochem. Engng. Ed. by Ghose, T.K.; Fiechter, A.; Balkeborogh, N. 1977, 7, 1.
2. Hammer, H.; Rahse, W. Chem.-Ing.-Tech. 1973, 45, 968.
3. Zlokarnik, M. Chem.-Ing. Tech. 1971, 43, 329.
4. Deckwer, W. D.; Louisi, Y.; Zaidi, A.; Ralek, M. I&EC Process Des. Dev. 1980, 19, 699.
5. Reith, T.; Renken, S.; Israel, B. A. Chem. Eng. Sci. 1968, 23, 619.
6. Akita, K.; Yoshida, F. I&EC Proc. Des. Dev. 1973, 12, 76.
7. Botton, R.; Cosserat, D.; Charpentier, J. C. Chem. Eng. J. 1978, 16, 107.
8. Mersmann, A. Ger. Chem. Eng. 1978, 1, 1.
9. Bach, H. F., Dr.-Ing. Thesis, TU Munich 1977.
10. Bach, H. F.; Pilhofer, T. Ger. Chem. Eng. 1978, 1, 270.
11. Sharma, M. M.; Mashelkar, R. A. Instn Chem. Engrs. Sympos. Ser. 1968, 28, 10.
12. Wallis, G. "One-Dimensonal Two-Phase Flow"; McGraw-Hill: New York, NY, 1969.
13. Oels, U.; Lucke, J.; Buchholz, R.; Schugerl, K. Germ. Chem. Eng. 1978, 1, 115.
14. Hikita, H.; Asai, S.; Tanigawa, K.; Segawa, K. paper presented at CHISA'78, Prague 1978.
15. Schumpe, A.; Deckwer, W.-D. Chem. Eng. Sci. 1980, 35, 1275.
16. Nagel, O.; Kurten, H.; Hegner, B. Chem.-Ing.-Tech. 1973, 45, 913.
17. Nagel, O.; Kurten, H. Chem.-Ing.-Tech. 1976, 48, 513.
18. Nagel, O.; Hegner, B.; Kurten, H. Chem.-Ing.-Tech. 1978, 50, 934.
19. Nagel, O.; Kurten, H.; Hegner, B. in "Two-Phase Momentum, Heat, and Mass Transfer in Chemical Process, and Engineering Systems," ed. by Durst, F.; Tsiklauri, G. V.; Afgan, N. H. Hemisphere Publ. Corp.: Washington, DC 1979, 2, 835.
20. Zlokarnik, M. Chem. Eng. Sci. 1979, 34, 1265.
21. Zlokarnik, M., Verfahrenstechn. (Mainz) 1979, 13, 601.
22. Witte, J.H., Brit. Chem. Eng. 1965, 10, 602.
23. Fair, J. R. Chem. Eng. 1967, 67.
24. Litz, W. J. Chem. Eng. 1972, 162.
25. Bhavaraju, S. M.; Russel, T. W. F.; Blanch, H. W. AIChE J. 1978, 24, 454.
26. Deckwer, W.-D.; Burckhart, R.; Zoll, G. Chem. Eng. Sci. 1974 29, 2177.
27. Buchholz, H.; Buchholz, R.; Lucke, J.; Schugerl, K. Chem. Eng. Sci. 1978, 33, 1061.
28. Deckwer, W.-D.; Schumpe, A.; Nguyen-Tien, K. Biotechn. Bioeng. to be published 1981.
29. Schumpe, A.; Deckwer W.-D. paper presented at ACS Meeting, Atlanta, March 30-April 3, 1981.

30. Nakanoh, M.; Yoshida, F. I&EC Process Des. & Develop. 1980, 19, 190.
31. Cova, D.R. I&EC Process Des. & Develop. 1974, 13, 392.
32. Hikita, H.; Kikukawa, H. Chem. Eng. J. 1974, 8, 191.
33. Kato, Y.; Nishiwaki, A. Int. Chem. Eng. 1972, 12, 182.
34. Akita, K., Dr.-Ing. Dissertation, Kyoto University, zitiert in /26/, 1973.
35. Ulbrecht, J. J.; Baykara, Z. Sema "Liquid Phase Mixing in Bubble Columns with Dilute Polymer Solutions," paper presented at ACS Las Vegas Meeting, August 1980.
36. Joshi, J. B.; Shah, Y. T. "Hydrodynamic and Mixing Models for Bubble Column Reactors," paper presented at ACS Las Vegas Meeting, August 1980.
37. Joshi, J. B.; Shah, Y. T.; Sharma, M. M. "A New Model for Heat Transfer in Bubble Column," paper presented at ACS Las Vegas Meeting, August 1980.
38. Ueyama, K.; Miyauchi, T. AIChE J. 1979, 25, 258.
39. Beinhauer, R., Dr.-Ing. Thesis, TU Berlin 1971.
40. Kolbel, H.; Beinhauer, R.; Langemann, H. Chem.-Ing.-Tech. 1972, 44, 697.
41. Baird, M. H. I.; Rice, R. G. Chem. Eng. J. 1975, 9, 17.
42. Shah, Y. T.; Stiegel, G. J.; Sharma, M. M. AIChE J. 1978, 24, 369.
43. Shah, Y. T.; Deckwer, W.-D. a chapter on "Fluid-Fluid Reactors" in "Scaleup in the Chemical Process Industries" ED. Bisio, A; Kabel, R. J. Wiley & Sons: New York, NY to be published.
44. Shah, Y. T.; Kelkar, B. G.; Godbole, S.; Deckwer, W.-D. "Bubble column Reactors" submitted to AIChE J. 1981.

RECEIVED June 16, 1981.

11

Access of Hydrodynamic Parameters Required in the Design and Scale-Up of Bubble Column Reactors

WOLF-DIETER DECKWER

Institut für Technische Chemie, Universität Hannover, Callinstrasse 3, D-3000 Hannover, Federal Republic of Germany

The general difficulties in design and scale-up of bubble column reactors concern reaction specific data, such as solubilities and kinetic parameters as well as hydrodynamic properties. The paper critically reviews correlations and new results which are applicable in estimation of hydrodynamic parameters of two-phase and three-phase (slurry) bubble column reactors.

Bubble column reactors (BCR) are widely used in chemical process industries to carry out gas-liquid and gas--liquid-solid reactions, the solid suspended in the liquid phase being most frequently a finely divided catalyst (slurry reactor). The main advantages of BCR are their simple construction, the absence of any moving parts, ease of maintenance, good mass transfer and excellent heat transfer properties. These favorable properties have lead to their application in various fields: production of various chemical intermediates, petroleum engineering, Fischer-Tropsch synthesis, fermentations and waste water treatment.

Owing mainly to their simple construction and the absence of moving parts and installments (which usually provide for a somewhat settled flow pattern) the hydrodynamic behavior of BCR is rather complex and changes considerably with variations in physico-chemical properties and operational conditions. This causes difficulties in the design and scale-up of BCR and leads to errors and unreliabilities which, in turn, often result in overdimensioning of the reactors.

0097-6156/81/0168-0213$07.25/0

Flow Regimes

If a gas is distributed in a liquid by means of a certain sparger the bubbles are uniform in size and uniformly distributed provided the gas velocity is low, say less than 5 cm/s. There is no or little interaction between the bubbles, and this regime is called bubbly or homogeneous flow, see Fig. 1. The bubble size distribution is narrow, and the rise velocities of the bubbles in the swarm lie between 20 and 30 cm/s.

At higher gas velocities with an increasing number of bubbles, the pseudohomogeneous bubble-in-liquid dispersion can no longer be maintained. The flow becomes unstable and coalescence sets in leading to larger diameter bubbles. This flow regime where large bubbles with high rise velocities coexist with small bubbles is called heterogeneous or, owing to the movement of the large bubbles, churn-turbulent flow regime. The large bubbles are nonspherical and of varying form with very mobile and flexible surfaces, for instance, spherical caps. These large bubbles can have diameters of about 8 to 10 cm. Though small in number the large bubbles carry the bulk of the gas through the columns as their rise velocity is large ($\geq$1 m/s).

A peculiar situation occurs in small diameter columns. At high gas flow rates, the larger bubbles are stabilized by the column wall leading to the formation of bubble slugs. In tall columns bubble slugs can be observed for column diameters as large as 20 cm.

A rough indication of the prevailing flow regime for a known column diameter and gas velocity, which are thought to be the major variables, can be taken from Fig. 2. However, other parameters, such as the sparger design, physico-chemical properties and the liquid velocity can effect the transition from one flow regime to the other. For instance, it is well known that in highly viscous solutions, with Newtonian as well as non-Newtonian behavior, large bubbles occur at gas velocities considerably less than 5 cm/s. The type of gas sparger also decisively influences the flow regimes and the transition range between them. Porous spargers with mean pore sizes less than 15o μm commonly lead to bubbly flow up to gas velocities of about 5 to 8 cm/s. On the other hand, if perforated plates, single or multinozzle distributors with orifice diameters larger than 1 mm are used homogeneous flow can only be realized at very low gas velocities. At larger orifice diameters

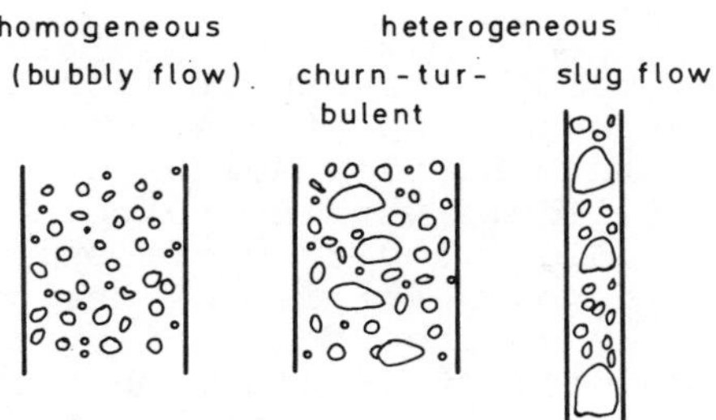

Figure 1. Flow regimes in BCR

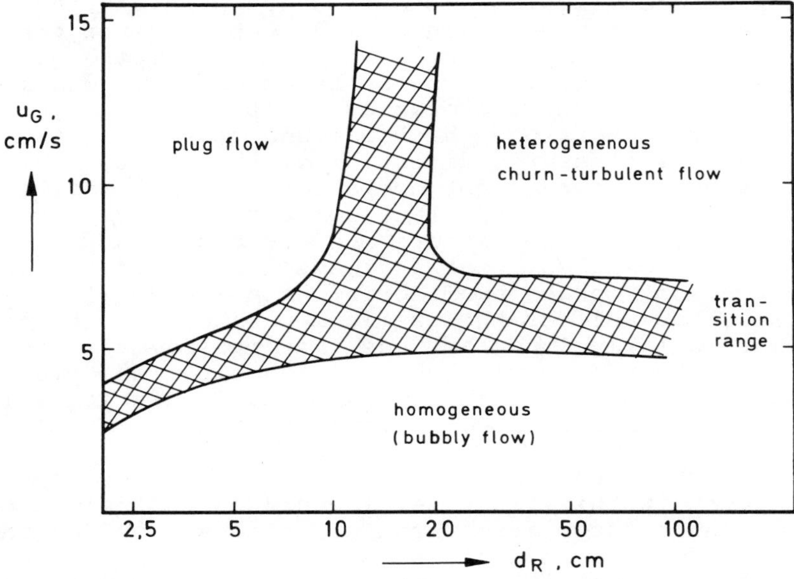

Figure 2. Dependence of flow regime on gas velocity and column diameter (qualitatively for low viscosity liquids)

bubbly flow may not occur at all if pure liquids are aerated. For mixtures the situation may change again. Therefore, the boundaries between the different flow regimes are not fixed.

The rising bubbles transport liquid on their rear from the bottom to the top of the column. Therefore, for continuity, liquid must flow downwards. This results in a circulation of liquid within the column which leads to pronounced liquid velocity profiles. The liquid circulation causes a nonuniformity of the gas holdup at any cross-section. The voidage is usually very large in the center of the column. It is clear that the circulation velocity increases with increasing gas throughput. Therefore, the heterogeneous flow is also called recirculating flow regime.

Parameters Involved in BCR Design

In general, the procedure for designing a bubble column reactor (BCR) (1) should start with an exact definition of the requirements, i.e. the required production level, the yields and selectivities. These quantities and the special type of reaction under consideration permits a first choice of the so-called adjustable operational conditions which include phase velocities, temperature, pressure, direction of the flows, i.e. cocurrent or countercurrent operation, etc. In addition, process data are needed. They comprise physical properties of the reaction mixture and its components (densities, viscosities, heat and mass diffusivities, surface tension), phase equilibrium data (above all solubilities) as well as the chemical parameters. The latter are particularly important, as they include all the kinetic and thermodynamic (heat of reaction) information. It is understood that these first level quantities (see Fig. 3) are interrelated in various ways.

For the case of single phase reactors the information given in this first level would be sufficient to design the reactor from first principles provided additional knowledge on the residence time distribution is available. In multiphase reactor design hydrodynamic properties constitute another group of important parameters. These are more or less "nonadjustable" or self-adjusting" quantities dependent on the chosen reactor geometry, the adjustable operational conditions as well as the process data. Under this notion we summarize the phase holdups, the interfacial areas, the heat and mass transfer properties and the dispersion coefficients. All these quantities, i.e. the geometry, the process data, the adjustable and nonadjustable parameters, are then introduced in the reactor model equations derived on the basis of the physical and chemical phenomena which are suspected to take place within the reactor. Usually the model equations have to be solved numerically as they contain strong nonlinearities (temperature dependency of reaction rates and solubilities, phase flow variation). It is understood that the general scheme outlined in Fig. 3 must be run through several times since the desired optimal reactor design cannot be obtained explicitly and is commonly subjected to various economic choices. The complete design model consisting of the model equations and the outlined scheme is usually embedded in effective optimization procedures. Such optimization techniques can lead to different choices depending on the specific objectives and the conditions to which the chemical process may be subjected.

The use of models and model simulations are extremely useful in all design and scale-up considerations. Mathematical methods to solve model equations of any degree of complexity are available now, and fast numerical techniques have been developed. In addition, almost everywhere abundant computer facilities are at hand. Therefore, a reliable design and scale-up should use mathematical models formulated on the basis of first principles, even if these models are very sophisticated. Such models and simulations based on them present the most efficient and probably the cheapest way in today's design works.

The use of models and particularly those of a sophisticated nature is, however, seriously restricted by the limitations of the parameters involved in the model equations. It is actually the determination of certain parameters which becomes the crucial point in designing. The major uncertainties originate from two sources. Firstly, the process data, i.e. estimation of phase equilibria (solubilities), diffusivities and especially kinetic rate data,involves inaccuracies. The second major source of large uncertainties is the reliability of the nonadjustable hydrodynamic quantities.

Correlations for Fluiddynamic Properties

In the following sections, some correlations for the important hydrodynamic parameters involved in BCR design are presented and critically discussed. The brief review includes recent data of the author as well as various other investigators.

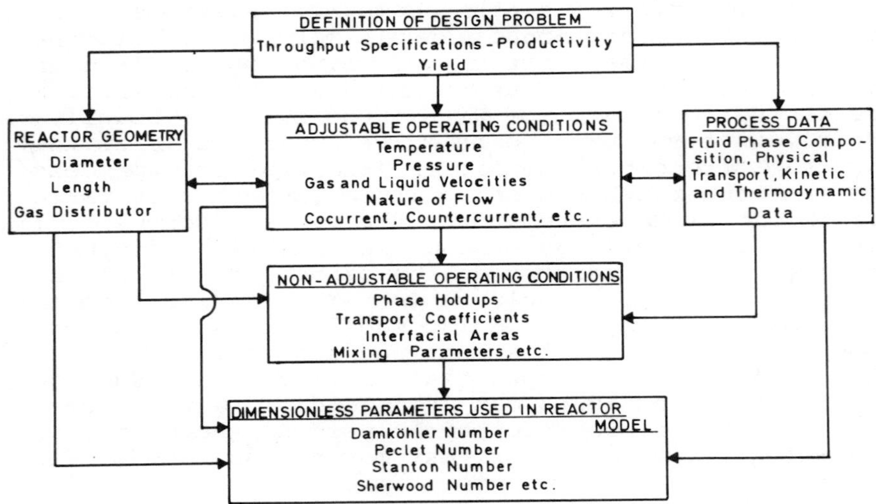

Figure 3. General procedure to design and scale-up BCR

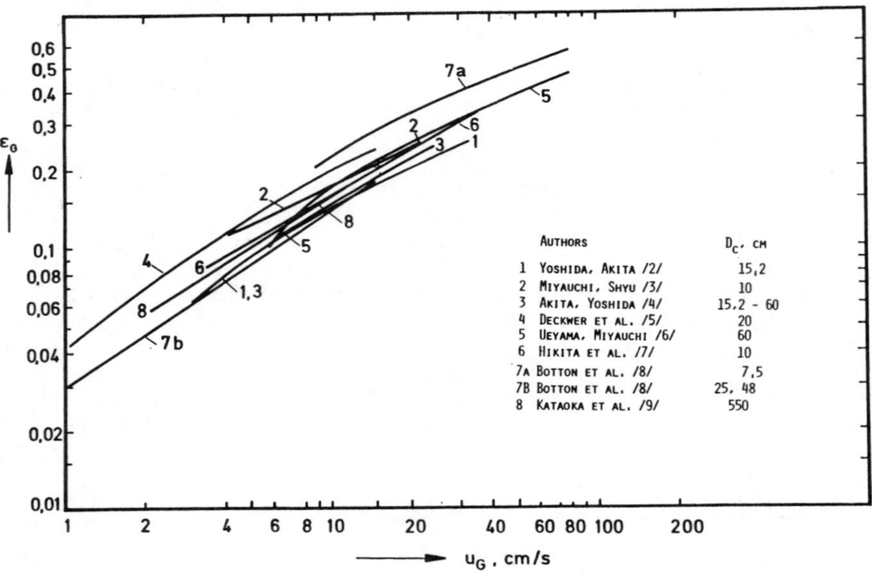

Figure 4. Measured gas holdup in water for single and multinozzle spargers

In general, little arbitrary and independent variation of the hydrodynamic parameters is possible for a certain reaction system. These properties are determined by complex relationships and interactions of all the quantities given in the first level of parameters in Fig. 3. For estimating the nonadjustable hydrodynamic parameters numerous empirical correlations have been proposed in the literature. Only those correlations are reported here which, at least under certain conditions, seem to be of broader applicability and yield rather reliable results. Particular emphasis is placed on those correlations which are based on sound theoretical concepts.

Gas Holdup

An important parameter to characterize gas-in-liquid dispersions is the gas holdup, ε_G. It depends mainly on the gas throughput and, to a small extent, on the sparger and physico-chemical properties. The column diameter has no influence on ε_G provided the ratio of column to bubble diameter is large, say ≥ 40. This condition is usually fulfilled for column diameters larger than 10 cm.

The dependency of ε_G on the gas velocity u_G can be expressed by

$$\varepsilon_G \propto u_G^{\,n} , \tag{1}$$

the value of the exponent, n, depends on the flow regime. In homogeneous bubbly flow n varies from 0.7 to 1.2, while for churn-turbulent flow n is in the range from 0.4 to 0.7. Fig. 4 presents some data for single and multinozzle spargers ($d_n \geq 1$ mm), for water and aqueous systems. The flow is heterogeneous for these conditions and the exponent in Eqn. (1) is about 0.6.

From the numerous correlations for ε_G reported in the literature only two will be given here. Akita and Yoshida (4) correlated their data for water and aqueous solution of glycol, glycerol, methanol, NaCl and Na_2SO_3 by

$$\frac{\varepsilon_G}{(1-\varepsilon_G)^4} = 0.2 \left(\frac{g d_c^2 \rho_L}{\sigma}\right)^{1/8} \left(\frac{g d_c^3}{\nu_L^2}\right)^{1/12} \left(\frac{u_G}{\sqrt{g d_c}}\right) \tag{2}$$

The column diameter d_c in Eqn. (2) is only included to make the various terms dimensionless. Generally, the correlation of Akita and Yoshida gives a conservative estimate.

Bach and Pilhofer /10/ proposed the following relation

$$\frac{\varepsilon_G}{1-\varepsilon_G} = 0.115 \left(\frac{u_G^3}{\nu_L g \Delta\rho/\rho_L}\right)^{0.23} = 0.115\ (\mathrm{ReFr})^{0.23} \tag{3}$$

which describes holdup data for water and various pure organic liquids.

In many practical applications the above correlations as well as others often fail (11-13). This is particularly the case with liquid mixtures and porous spargers. As the holdup can easily be measured it is recommendable to carry out some measurements in labscale equipment with a column diameter of 10 cm or larger.

Interfacial Area

Like the gas holdup the gas-liquid interfacial area, a, represents an important quantity. If the reaction takes place in the fast reaction regime of diffusion-reaction theory, the interfacial area is the main design criterion. Gas holdup and interfacial area are related by

$$a = \frac{6\,\varepsilon_G}{d_s} \tag{4}$$

where d_s is the Sauter (volume-to-surface mean) bubble diameter defined by

$$d_s = \frac{\sum N_i d_{bi}^3}{\sum N_i d_{bi}^2} \tag{5}$$

Values of d_s usually vary between 2 and 6 mm. For water and electrolyte solutions Deckwer et al. (14) recommend a value of d_s = 2.9 mm which was obtained from a lot of data in various bubble columns. For jet spargers in aqueous solutions of organic substances lower values of d_s (1 mm or even less) are also found. Correlations for d_s and the interfacial area are reported by Akita and Yoshida (15).

Various physical and chemical methods can be applied to determine interfacial areas. Unfortunately, the different methods yield largely differing results. The methods used most often are photography and sulfite oxidation. Schumpe and Deckwer (16) recently showed that in the bubble flow regime, i.e. at $u_G \leqslant 5$ cm/s, the photographic area and that obtained from sulfite oxidation are related by

$$a_{photo} = 1.35\ a_{chem} \tag{6}$$

though the measurements were carried out in equal columns. In churn-turbulent flow, the difference between the two methods is about 100 % and maybe even larger, in agreement with the findings of other authors (17,18).

Different chemical methods do not lead to equal interfacial areas either. This is demonstrated in Fig. 5 where areas found with sulfite oxidation and CO_2 absorption in alkali in equal or similar equipment are plotted against u_G. The data considered in Fig. 5 and an explanation of these discrepancies is given in detail elsewhere (19). The main reasons are nonlinear dependency of conversion on a and nonuniformity of bubble sizes and their rise velocities. The results presented in Fig. 5 and also comparison of various physical methods lead to the conclusion that the interfacial area in gas-in-liquid dispersions has serious errors and only rough estimates are available. However, Schumpe and Deckwer (19) presented some reasonable guidelines for efficient experimentation and recommended the use of sulfite oxidation. The sulfite oxidation method may be particularly useful to compare the efficiency of different chemical reactors. Indeed, sulfite oxidation was extensively used by Nagel and coworkers (20-24) to determine interfacial areas in various gas-liquid contactors. On the basis of Kolmogoroff's theory of isotropic turbulence Nagel et al. (22, 24) derived an expression for a as a function of the energy dissipation rate per unit volume of the reactor (E/V_R). Fig. 6 presents their results as a plot of a vs. (E/V_R). In various articles (21-24) Nagel et al. demonstrated the usefulness of their findings for design and scale-up considerations of gas-liquid reactors. It should, however, be pointed out that Fig. 6 refers solely to interfacial areas obtained from the sulfite oxidation system. The picture may be changed considerably if other liquid systems are taken into consideration (25).

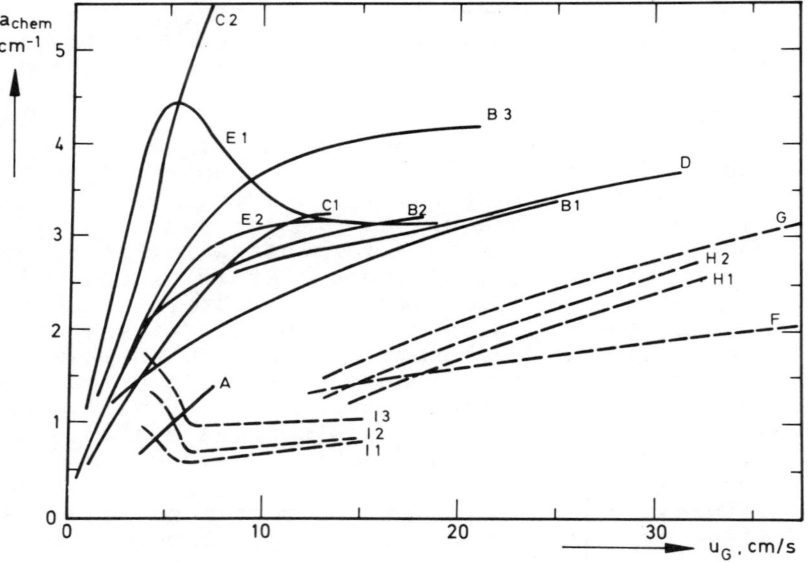

Figure 5. Interfacial areas from (——) sulfite oxidation and (– – –) CO_2 absorption in alkali

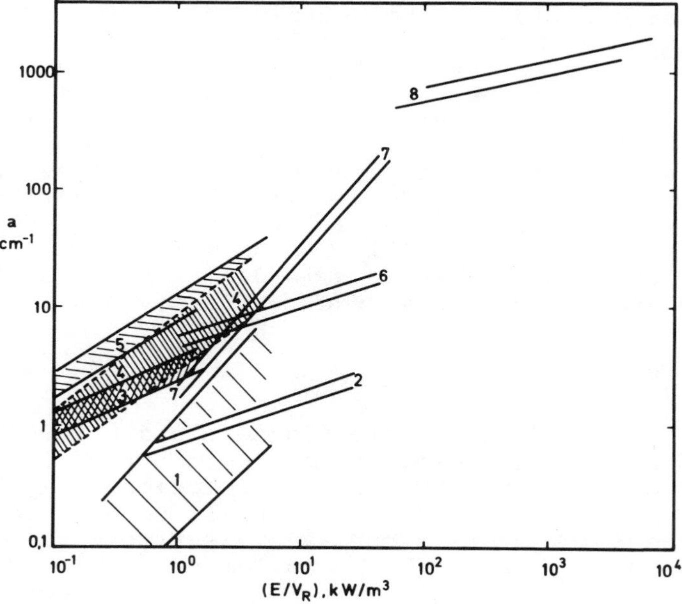

Figure 6. Interfacial area vs. energy dissipation rate (23): 1, dual-flow column; 2, pipe flow; 3, bubble column; 4, stirred tank; 5, bubble column with 2-phase nozzle; 6, co-current packed bed; 7, jet tube washer (2-phase nozzle); 8, tube reactor with 2-phase nozzle

Volumetric Mass Transfer Coefficients

The determination of volumetric mass transfer coefficients, k_La, usually requires additional knowledge on the residence time distribution of the phases. Only in large diameter columns the assumption is justified that both phases are completely mixed. In tall and smaller diameter bubble columns the determination of k_La should be based on concentration profiles measured along the length of the column and evaluated with the axial dispersed plug flow model (5, 12).

Since gas flow strongly influences a it also mainly affects k_La. Fig. 7 shows a log-log plot of k_La vs. the mean gas velocity, u_G, for oxygen mass transfer in water. The data refers to different column sizes, gas spargers and operational conditions. The k_La values are not influenced by column geometry, direction of flow, i.e. cocurrent or countercurrent flow, and mass transfer, i.e. absorption or desorption. But the nature of the gas sparger obviously exerts a major influence though the columns are tall. The data for sintered plate spargers and nozzles can be well described by straight lines in Fig. 7. The least square fit gives

$$k_La = b\ u_G^{0.82} \tag{7}$$

where b is 0.0107 for nozzle spargers and 0.0296 for sintered plates. The exponent of u_G in Eqn. (7) is in full agreement with a theoretical relation derived by Kastanek (26) on the basis of Higbie's penetration theory and Kolmogoroff's theory of isotropic turbulence. The value of b in Eqn. (7) is influenced not only by the sparger but also by the physico-chemical properties of the fluid phases. The k_La values are independent of the liquid velocity except at unusually high liquid flow rates (27).

For the case of the less effective single and multinozzle spargers Akita and Yoshida (4) proposed the following correlation

$$k_La\ \frac{d_c^{\ 2}}{D_L} = 0.6\ \varepsilon_G^{\ 1.1}\left(\frac{\nu_L}{D_L}\right)^{0.5}\left(\frac{gd_c^2\ \rho_L}{\sigma}\right)^{0.62}\left(\frac{gd_c^3}{\nu_L^{\ 2}}\right)^{0.31} \tag{8}$$

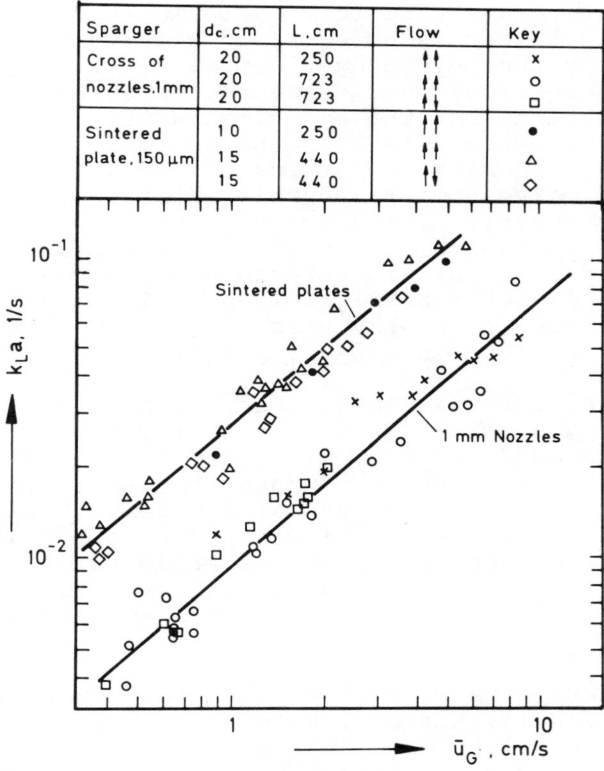

Figure 7. Volumetric mass transfer coefficients vs. gas velocity in various BCR

Fig. 8 demonstrates that this empirical correlation describes fairly well the oxygen mass transfer data of various authors and for large-scale equipment. Therefore the equation of Akita and Yoshida can be recommended for low mass transfer rates, i.e. the mass transfer of slightly soluble gases. It usually gives a conservative estimate of k_La.

k_La data for oxygen mass transfer in aqueous solutions of alcohols and glucose in the presence and absence of inorganic electrolytes (simulated fermentation media) in bubble columns equipped with various spargers are presented by Schügerl et al. (12,25,28). The somewhat surprising results cannot be correlated by simple equations but the experimental findings can be explained qualitatively by means of coalescence promoting and hindering properties of the liquid media. The studies give some reasonable guidelines for estimating the relative influence of various substances on k_La (12,25,28) which are in general agreement with the findings of Zlokarnik in aerated stirred vessels (29,30) and larger diameter bubble columns (31). k_La data for oxygen transfer in highly viscous and non-Newtonian media can be found in refs. (32-34).

Special effects can be observed at high mass transfer rates, i.e. with gases of high solubilities, e.g. CO_2 in water and aqueous solutions. In such cases the evaluation of mass transfer data from measured profiles requires the use of sophisticated models (14,15). Due to variations in gas velocity and gas holdup along the column the volumetric mass transfer coefficient is not constant (14,35). The situation becomes even more complex if mass transfer of two gases takes place simultaneously (36).

Volumetric mass transfer data for nonaqueous media such as alcohols, ketones, esters and hydrocarbons (which are widely used solvents in the chemical process industry) are scarce. Recently, k_La values for CO mass transfer in molten higher paraffins were reported. This data refers to conditions prevailing in the Fischer--Tropsch slurry process (37).

Liquid Side Mass Transfer Coefficient

For mass transfer processes accompanied by slow chemical reaction it is not required to separate the volumetric mass transfer coefficient into its individual quantities, i.e. k_L and a. If the reaction in the

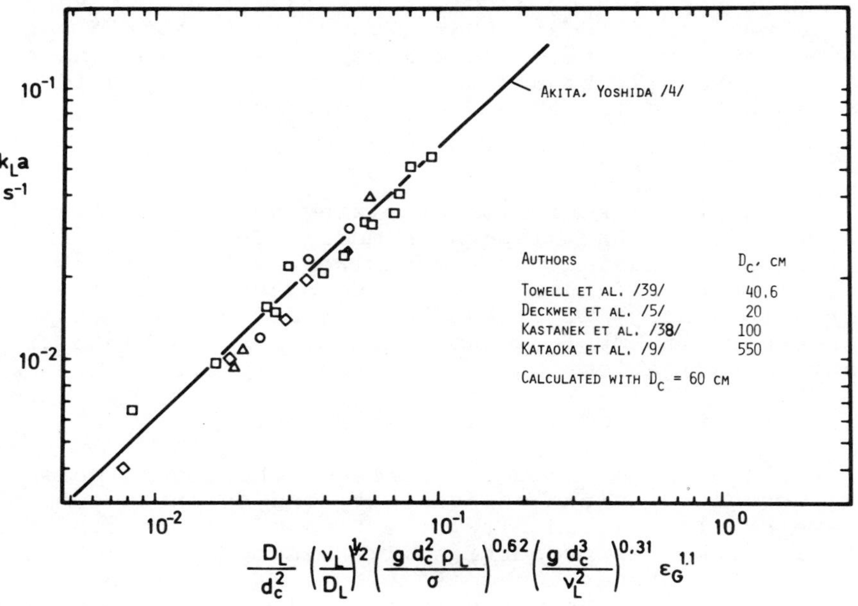

Figure 8. Check of correlation of Akita and Yoshida (water–air system)

liquid is fast and instantaneous the locale of the reaction is within the liquid film and knowledge of k_L is required to calculate the enhancement factors.

A number of empirical and theoretical equations for evaluating k_L, is available in the literature (12, 15,40-44). However, the predictions of the various correlations scatter considerably. The correlation proposed by Akita and Yoshida (15) seems to give rather low values. The empirical equations developed by Hughmark (42) and by Calderbank and Moo-Young (41) are widely used and their application can be recommended. For instance, the data reported by Zaidi et al. (37) for CO transfer in molten paraffin at 260 to 290 °C is in sufficient agreement with both correlations. As a rule of thumb it can generally be assumed that for gas-in-liquid dispersions k_L varies from 0.01 to 0.03 cm/s.

Mixing

A useful description of mixing in bubble columns is provided by the dispersion model. The global mixing effects are generally characterized by the dispersion coefficients E_L and E_G of the two phases which are defined in analogy to Fick's law for diffusive transport. Dispersion in bubble columns has been the subject of many investigations which have recently been reviewed by Shah et al. (45). Particularly, plenty of data are available for liquid-phase dispersion.

The dispersion coefficients of the liquid phase are dependent on the gas velocity and on the column diameter. The liquid flow, the type of gas sparger and physico-chemical properties like viscosity do not influence the dispersion coefficient, E_L, or,at best, these parameters are of very minor importnace. For instance, Hikita and Kikukawa (46) found only a slight viscosity influence, i.e. $E_L \propto u^{-0.12}$.

Baird and Rice (47) derived a theoretical equation for E_L on the basis of Kolmogoroff's theory of isotropic turbulence. Their expression is

$$E_L = A\, d_c^{4/3}\, (u_G g)^{1/3} \qquad (9)$$

where A is a constant. The exponents of d_c and u_G are in striking agreement with the experimental results (5). By introducing dimensionless groups

$$Pe_L = \frac{u_G d_c}{E_L} \tag{10}$$

and

$$Fr = \frac{u_G^2}{g d_c} \tag{11}$$

eqn. (9) can be written as

$$Pe_L = c\ Fr^{1/3} \tag{12}$$

A plot of Pe_L vs Fr based on the data of various authors (5,46,48-54) is shown in Fig. 9. A least square fit of all the data gives

$$Pe_L = 2.83\ Fr^{0.34} \tag{13}$$

which is indeed in excellent agreement with the theoretical prediction. Empirical equations in terms of Pe_L and Fr were also proposed by Kato and Nishiwaki (51) and by Akita (55), see Fig. 9. Another theoretical approach to calculate the liquid dispersion coefficient is based on the circulation velocity in bubble columns which can be obtained from an overall energy balance (56,57).

Data on gas phase dispersion are rather scarce, and in general they reveal considerable scatter. Towell and Ackerman (52) proposed the following empirical equation

$$E_G = 0.2\ d_c^2 u_G \tag{14}$$

which includes the data provided by Kölbel et al. (58) and Carleton et al. (59). On the basis of comprehensive experiments with various liquids (water, propanol, glycerol) Mangartz and Pilhofer (60) concluded that the bubble rise velocity in the swarm ($u_G = u_G/\varepsilon_G$) is a characteristic variable which mainly influences the gas dispersion. Mangartz and Pilhofer correlated their results by

$$E_G = 5 \times 10^{-4}\ u_G^3\ d_c^{1.5}\ . \tag{15}$$

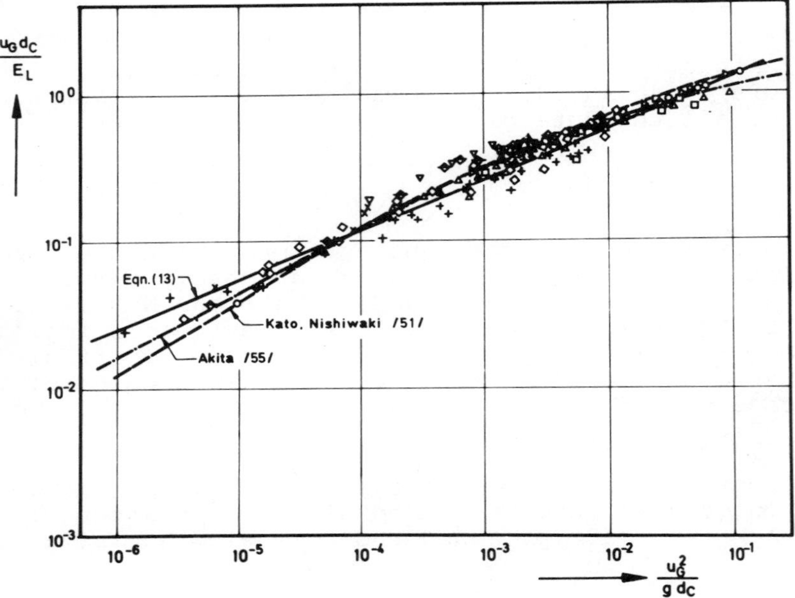

Figure 9. Correlation of liquid phase dispersion coefficients

Though the gas phase dispersion coefficients are large and often larger than those of the liquid phase the influence of the gas phase dispersion on conversion should not be overestimated. One has to consider that it is not the dispersion coefficient itself but the Peclet number which is the governing parameter in the model equations. The Peclet number has to be formulated under consideration of the fractional gas holdup

$$Pe_G = \frac{u_G L}{\varepsilon_G E_G} \ . \tag{16}$$

As ε_G is usually small the detrimental effect of gas phase dispersion on the performance of bubble columns can be neglected in columns less than 20 cm in diameter (61). For illustrating the influence of gas phase dispersion some computed conversions are presented in Fig. 10 (1). The simulations refer to CO_2 absorption in carbonate buffer in a column 5 m in length. E_G was calculated from eqn. (15). The liquid phase dispersion does not affect the conversion in the present case as the process takes place in the diffusional regime of mass transfer theory. As shown in Fig. 10, the decrease in conversion due to gas phase dispersion increases with increasing diameter and gas velocity. However, in the favorable bubbly flow regime and in small diameter columns the effect is less pronounced.

Heat transfer

BCR are particularly useful to carry out high exothermic gas-liquid reactions like oxygenations, hydrogenations, and chlorinations. One reason for applying BCR certainly are their favorable heat transfer properties, i.e. the large heat capacity of the liquid phase and high wall-to-dispersion heat transfer coefficients which are commonly larger by about one order of magnitude than for single phase flow (62). Deckwer (63) proposed a theoretical heat transfer model by combining Higbie's penetration model with Kolmogoroff's theory of isotropic turbulence. The final result can be expressed in dimensionless form by

$$St = 0.1\ (ReFrPr^2)^{-1/4} \tag{17}$$

where

$$St = \frac{h}{\rho c_p u_G} \tag{18}$$

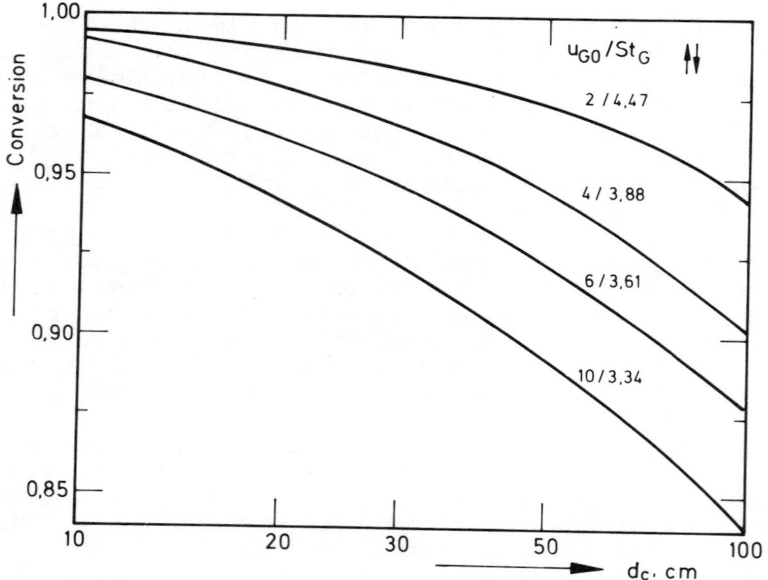

Figure 10. Effect of gas phase dispersion on conversion by increasing BCR diameter and gas flow rate

$$\mathrm{ReFr} = \frac{u_G^{\,3}}{\nu g} \tag{19}$$

$$\mathrm{Pr} = \frac{\nu \rho c_p}{k} \tag{20}$$

Fig. 11 shows that eqn. (17) describes measured data of various authors (62,64-66) with surprising agreement for a large range of Prandtl numbers.

Effect of Solids

Bubble columns are convenient for catalytic slurry reactions also (67). It is therefore important to know how the hydrodynamic properties of the gas-in-liquid dispersion is influenced by the presence of suspended solid particles. In the slurry reactor absorption enhancement due to chemical reaction cannot be expected. However, if particle sizes are very small, say less than 5 μm, and if, in addition, the catalytic reaction rate is high a small absorption enhancement can occur (68). Usually the reaction is in the slow reaction regime of mass transfer theory. Hence, it is sufficient to know the volumetric mass transfer coefficient, k_La, and there is no need to separate k_La into the individual values.

In typical slurry reactions like hydrogenations and oxidations the particle sizes are usually smaller than 200 μm and their concentration is less than 10 wt. percent. Under such conditions, the variations in k_La due to the presence of solids reported (31,69-72) do not commonly exceed 10 to 20 %. If the particles are small ($\leq$ 50 μm) the suspended solid and the liquid behave as a pseudohomogeneous phase. This can be concluded from a study on the CO conversion reaction on a catalyst suspended in molten paraffin where no significant effect on ε_G and k_La could be observed (13,37).

The above result is correct as far as slurries of non-porous particles are concerned. If particles of small size and high porosity are suspended a significant increase in k_La over that for the gas-liquid system can be observed (68,73,74). Kars et al. (74) and Alper et al (68) studied physical absorption in the presence of suspended activated carbon and found increased k_L values. Alper et al. (68) explained their findings with the absorption capacity of charcoal which leads to increased absorption if the particles enter the boundary layer at the gas-liquid interface.

Heat transfer in bubble column slurry reactors was studied by Kölbel and coworkers (75-77) and Deckwer et al. (13). The addition of solids increases the wall-to--suspension heat transfer coefficient. However, this increase is only due to changes in the physico-chemical properties and represents no independent contribution of the particles. Therefore, the heat transfer model, i.e. eqn. (17), developed by Deckwer (63) for two-phase BCR also applies to slurry reactors as was proved for particle sizes up to 120 μm. This confirms that solids and liquid in the slurry can be regarded as a pseudo-homogeneous phase provided the gas velocity is large enough to provide for complete fluidization of the particles.

Due to density differences the particles have the tendency to settle. Thus, solid concentration profiles result which can be described on the basis of the sedimentation-dispersion model (78,79,80). This model involves two parameters, namely, the solids dispersion coefficient, E_S, and the mean settling velocity, u_S, of the particles in the swarm. Among others Kato et al. (81) determined E_S and u_S in bubble columns for glass beads 75 and 163 μm in diameter. The authors propose correlations for both parameters, E_S and u_S. The equation for E_S almost completely agrees with the correlation of Kato and Nishiwaki (51) for the liquid phase dispersion coefficient.

In catalytic slurry reactors the locale of the reaction is the catalyst surface. Hence, in addition to the mass transfer resistance at the gas-liquid interface a further transport resistance may occur at the boundary layer around the catalyst particle. This is characterized by the solid-liquid mass transfer coefficient, k_S, which has been the subject of many theoretical and experimental studies. Brief reviews are given by Shah (82). In general, the liquid-solid mass transfer coefficient is correlated by expressions like

$$\frac{k_s d_p}{D} = 2 + \alpha \left(\frac{\nu}{D} \right)^n \left(\frac{u_s d_p}{\nu} \right)^m \tag{21}$$

or

$$Sh = 2 + \alpha \, Sc^n Re_p^{\,m} \; . \tag{22}$$

As the slip velocity, u_S, of the particles is difficult to estimate it is now common to compute the Reynolds number on the basis of Kolmogoroff's theory which leads to

$$Re_p = C \left(\frac{\epsilon d_p^4}{\nu^3}\right)^p \tag{23}$$

where the value of exponent p depends on the ratio of the particle size to the scale of the microeddies. ϵ is the energy dissipation rate per unit mass which in the case of BCR can simply be calculated from

$$\epsilon = u_G g \tag{24}$$

While many studies on k_S in the two-phase liquid--solid system have been carried out only few have been reported for three-phase bubble columns (83,84). Most recently, Deckwer and Sänger (85) investigated liquid--solid mass transfer on suspended ionic resin beads in a bubble column, the range of the Schmidt number was varied from 137 to 5 x 10^4 by using aqueous solutions of polyethylene glycol. The findings were correlated by

$$Sh = 2 + 0.545\ Sc^{1/3} \left(\frac{\epsilon d_p}{\nu^3}\right)^{0.264} \tag{25}$$

Fig. 12 shows that the description of the measured data is fairly good. This is one more example which demonstrates the usefulness of Kolmogoroff's turbulence theory to correlate hydrodynamic parameters in BCR. Eqn. (25) is in reasonable agreement with the results of Sano et al. (84), and compares also well with correlations established for two-phase systems, i.e. stirred liquid-solid suspensions (1).

Summary

The hydrodynamic parameters involved in BCR design and scale-up are mainly dependent on adjustable operational conditions, physico-chemical properties and geometrical sizes. In general, little arbitrary variations are possible with respect to changes in chemical processes. Though a large amount of data has been reported the parameters which characterize the gas-liquid mass transfer properties are still subject to considerable error and unreliabilities. Only for aqueous systems

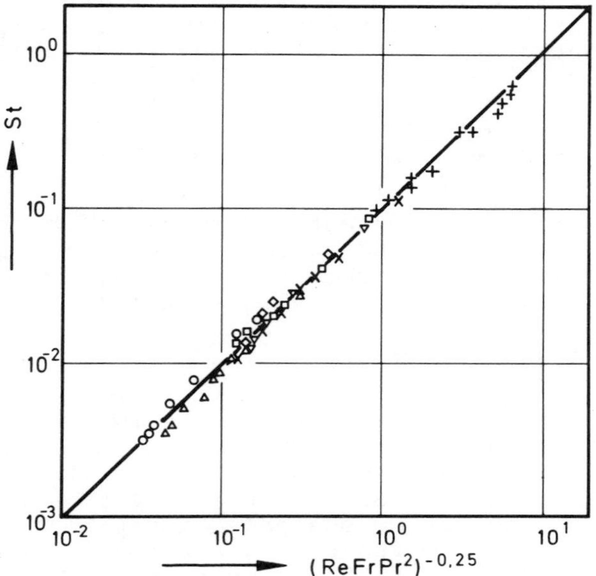

Figure 11. Correlation of heat transfer data by Equation 17

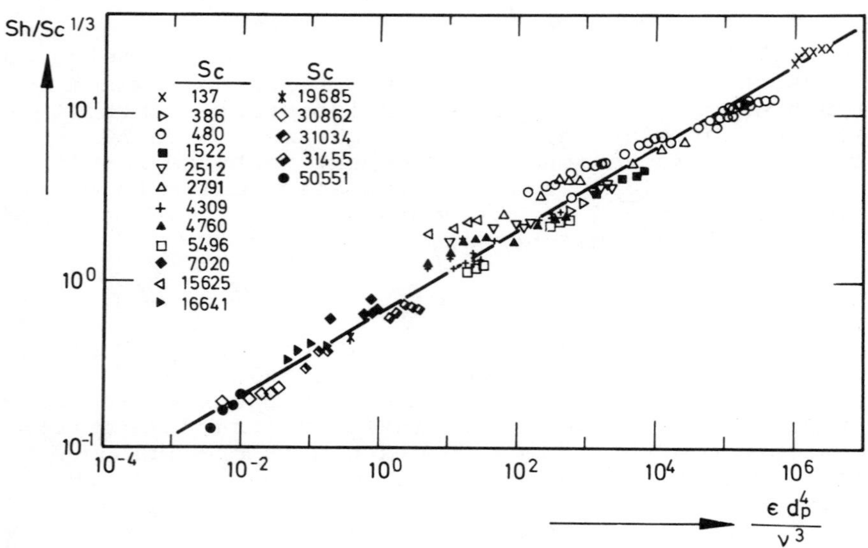

Figure 12. Correlation of liquid–solid mass transfer coefficients, Equation 25

do the correlations give rather accurate estimates. Contrary to gas-liquid mass transfer parameters the dispersion coefficients of both phases and the wall-to-dispersion heat transfer coefficients can be predicted reliably from theoretical models which have been validated experimentally. Also the influence of solids on the hydrodynamic parameters in three-phase systems is predictable provided the particle sizes are less than about 200 µm and the solids concentration is lower than 10 wt.%. Special effects can only be observed in powdered microporous suspensions.

Legend of Symbols

a	specific gas-liquid interfacial area
c_p	heat capacity
d_c	column diameter
d_{bi}	bubble diameter (of class i)
d_p	particle diameter
d_s	Sauter bubble diameter
D	diffusivity
E	dispersion coefficient
E/V_R	energy dissipation rate per volume
Fr	Froude number, eqn. (11)
g	gravitational acceleration
h	heat transfer coefficient
k	thermal conductivity
k_L	liquid side mass transfer coefficient
k_s	liquid-solid mass transfer coefficient
L	column length
N_i	number of bubbles of diameter class i
Pe_G	gas phase Peclet number, defined by eqn. (16)
Pe_L	Peclet number, defined by eqn. (10)
Pr	Prandtl number, defined by eqn. (20)
Re	Reynolds number, $u_G d/$
Re_p	particle Reynolds number, see eqs. (21) to (23)
Sc	Schmidt number, ν/D
Sh	Sherwood number, $k_s d_p/D$

St	Stanton number, defined by eqn. (18)
St_G	gas phase Stanton number, k_La $(L/u_{Go})(RT/H)$
u_G	linear gas velocity
u_S	slip velocity
u_G	rise velocity of bubbles in swarm, u_G/ε_G
ε	fractional holdup
ϵ	energy dissipation rate per unit mass, see eq.(24)
ν	kinematic viscosity
ρ	density
σ	surface tension

Indices

o	refers to inlet
G	refers to gas phase
L	refers to liquid phase

Acknowledgment

The investigations on BCR were supported by grants from Deutsche Forschungsgemeinschaft and Stiftung Volkswagenwerk. They are gratefully acknowledged.

Literature Cited

1. Shah, Y.T.; Deckwer, W.-D. In: "Scale-up in the Chemical Process Industries". Ed. by R. Kabel and A. Bisio, John Wiley & Sons, New York, to be published
2. Yoshida, F.; Akita, K. AIChE-J. 1965, 11, 9
3. Miyauchi, T.,; Shyu, C.N. Kagaku Kogaku 1970, 34, 958
4. Akita, K.; Yoshida, F. Ind. Eng. Chem. Proc. Des. Dev. 1973, 12, 76
5. Deckwer, W.-D.; Burckhart, R.; Zoll, G. Chem. Eng. Sci. 1974, 29, 2177
6. Ueyama, K.; Miyauchi, T. Kagaku Kogaku Ronbunshu, 1977, 3, 19
7. Hikita, H.; Asai, S.; Tanigawa, K.; Segawa, K. Paper presented at CHISA '78, Prague 1978
8. Botton, R.; Cosserat, D.; Charpentier, J.C. Chem. Eng. J., 1978, 16, 107

9. Kataoka, H.; Takuchi, H.; Nakao, K.; Yagi, H.; Tadaki, T.; Otake, T.; Miyauchi, T.; Washimi, K.; Watanabe, K.; Yoshida, F. J. Chem. Eng. Japan 1979, 12, 105
10. Bach, H.F.; Pilhofer, T. Ger. Chem. Eng. 1978, 1, 270
11. Hammer, H.; Rähse, W. Chem.-Ing.-Tech. 1973, 45, 968
12. Schügerl, K.; Lücke, J.; Oels, U. Adv. Biochem. Engng. 1977, 7, 1
13. Deckwer, W.-D.; Louisi, Y.; Zaidi, A.; Ralek, M. Ind. Eng. Chem. Proc. Des. Dev. 1980, 10, 699
14. Deckwer, W.-D.; Adler, I.; Zaidi, A. Can. J. Chem. Eng. 1978, 56, 43
15. Akita, K.; Yoshida, F. Ind. Eng. Chem. Process Des. Dev. 1974, 13 84
16. Schumpe, A.; Deckwer, W.-D. Chem.-Ing.-Tech. 1980, 52, 468
17. Voyer, R.D.; Miller, A.I. Can. J. Chem. Eng. 1968, 46, 335
18. Weisweiler, W.; Rösch, S. Ger. Chem. Eng. 1978, 1, 212
19. Schumpe, A.; Deckwer, W.-D. Chem. Eng. Sci. 1980, 35, 2221
20. Nagel, O.; Kürten, H.; Sinn, R. Chem.-Ing.-Tech. 1972, 44, 367,399
21. Nagel, O.; Kürten, H.; Hegner, B. Chem.-Ing.-Tech. 1973, 45 913
22. Nagel, O.; Kürten, H. Chem.-Ing.-Tech. 1976, 48, 513
23. Nagel, O.; Hegner, B.; Kürten, H. Chem.-Ing.-Tech. 1978, 50, 934
24. Nagel, O.; Kürten, H.; Hegner, B. In: "Two-Phase Momentum, Heat, and Mass Transfer in Chemical Process and Engineering Systems". Ed. by F. Durst, G.V. Tsiklauri and H.H. Afgan, Vol. 2, p. 835, Hemisphere Publ. Corp., Washington DC, 1979
25. Oels, U.; Lücke, J.; Buchholz, R.; Schügerl, K. Ger. Chem. Eng. 1978, 1, 115
26. Kastanek, F. Coll. Czechoslov. Chem. Commun. 1977, 42, 2491
27. Alvarez-Cuenca, M.; Baker, C.G.J.; Bergeougnou, M. A. Chem. Eng. Sci. 1980, 35, 1121
28. Voigt, J.; Schügerl, K. Chem. Eng. Sci. 1979, 34, 1221
29. Zlokarnik, M. Chem.-Ing.-Techn. 1975, 47, 281
30. Zlokarnik, M. Adv. Biochem. Engng. 1978, 8, 133
31. Zlokarnik, M. Chem. Eng. Sci. 1979, 34, 1265
32. Buchholz, H.; Buchholz, R.; Lücke, J.; Schügerl, K. Chem. Eng. Sci. 1978, 33, 1061

33. Buchholz, H.; Buchholz, R.; Niebelschütz, H.; Schügerl, K. Eur. J. Appl. Microbiol. Biotechnol. 1978, 6, 115
34. Nakanoh, M.; Yoshida, F. Ind. Eng. Chem. Proc. Des. Dev. 1980, 19, 190
35. Deckwer, W.-D.; Hallensleben, J.; Popovic, M. Can. J. Chem. Eng. 1980, 58, 190
36. Hallensleben, J. Dr. thesis, Universität Hannover, 1980
37. Zaidi, A.; Louisi, Y.; Ralek, M.; Deckwer, W.-D. Ger. Chem. Eng. 1979, 2, 94
38. Kastanek, F.; Kratochvil, J.; Rylek, M. Coll. Czechoslov. Chem. Commun. 1977, 42, 3549
39. Towell, G.D.; Strand, C.P.; Ackerman, G.H. Preprints AIChE-Instn. Chem. Engrs. Joint Meeting, London, June 1965, Chem. Eng. Symp. Series 10 (1965), 97
40. Ranz, W.E.; Marshall, W.F. Chem. Eng. Progr. 1952, 48, 141
41. Calderbank, P.H.; Moo-Young, M.B. Chem. Eng. Sci. 1961, 16, 39
42. Hughmark, G.A. Ind. Eng. Chem. Proc. Des. Dev. 1967, 6, 218
43. Reuß, M. Dr.-Ing. thesis, TU Berlin, 1970
44. Gestrich, W.; Esenwein, H.; Krauss, W. Chem.-Ing.-Tech. 1976, 48, 399
45. Shah, Y.T.; Stiegel, G.J.; Sharma, M.M. AIChE-J. 1978, 24, 369
46. Hikita, H.; Kikukawa, H. Chem. Eng. J. 1974, 8, 191
47. Baird, M.H.I.; Rice, R.G. Chem. Eng. J. 1975, 9, 17
48. Reith, T.; Renken, S.; Israel, B.A. Chem. Eng. Sci. 1968, 23, 619
49. Aoyama, Y.; Ogushi, K.; Koide, K.; Kubota, H. J. Chem. Eng. Japan 1968, 1, 158
50. Ohki, Y.; Inoue, H. Chem. Eng. Sci. 1970, 25, 1
51. Kato, Y.; Nishiwaki, A. Int. Chem. Eng. 1972, 12, 182
52. Towell, G.D.; Ackerman, G.H. Proc. 2nd Int. Symp. Chem. React. Engng. B 3-1, Amsterdam, 1982
53. Deckwer, W.-D.; Graeser, U.; Langemann, H.; Serpemen, Y. Chem. Eng. Sci. 1973, 28, 1223
54. Badura, R.; Deckwer, W.-D.; Warnecke, H.J.; Langemann, H. Chem.-Ing.-Tech. 1974, 46, 399
55. Akita, K. Dr. thesis, Kyoto University, 1973
56. Joshi, J.B.; Sharma, M.M. Trans. Instn. Chem. Engrs. 1979, 57, 244
57. Joshi, J.B. Trans. Instn. Chem. Engrs., to be published
58. Kölbel, H.; Langemann, H.; Platz, J. Dechema-Monogr. 1962, 41, 225
59. Carleton, A.J.; Flain, R.J.; Rennie, J.; Valentin, F.H.H. Chem. Eng. Sci. 1967, 22, 1839

60. Mangartz, K.-H.; Pilhofer, Th. Verfahrenstechn. (Mainz) 1980, 14, 40
61. Deckwer, W.-D. Chem. Eng. Sci. 1978, 31, 309
62. Kast, W. Int. J. Heat Mass Transf. 1962, 5 329
63. Deckwer, W.-D. Chem. Eng. Sci. 1980, 35, 1341
64. Perner, D. Diplom-Arbeit, TU Berlin, 1960
65. Müller, D. Dr.-Ing. thesis, TU Berlin, 1962
66. Burkel, W. Dr.-Ing. thesis, TU Munich, 1974
67. Deckwer, W.-D.; Alper, E. Chem.-Ing.-Tech. 1980, 52, 219
68. Alper, E.; Wichtendahl, B.; Deckwer, W.-D. Chem. Eng. Sci. 1980, 35, 217
69. Sharma, M.M.; Mashelkar, R.A. Proc. Sym. Mass Transfer with Chemical Reaction 1968, No. 28, 10
70. Kato, Y.; Nishiwaki, A.; Kago, T.; Fikuda, T.; Tanaka, S. Int. Chem. Eng. 1973, 13, 582
71. Sittig, W. Verfahrenstechn. (Mainz) 1977, 11, 730
72. Joosten, G.E.H.; Schilder, J.G.M.; Janssen, J.J. Chem. Eng. Sci. 1977, 32, 563
73. Chandrasekaran, K.; Sharma, M.M. Chem. Eng. Sci. 1974, 29, 2130, 1977, 32, 669
74. Kars, R.L.; Best, R.J.; Drinkenburg, A.A. Chem. Eng. J. 1979, 17, 201
75. Kölbel, H.; Siemes, W.; Maas, R.; Müller, K. Chem.-Ing.-Tech. 1958, 30, 400
76. Kölbel, H.; Borchers, E.; Müller, K. Chem.-Ing.-Tech. 1958, 30, 792
77. Kölbel, H.; Borchers, E.; Martins, J. Chem.-Ing.-Tech. 1960, 32, 84
78. Cova, D.R. Ind. Eng. Chem. Proc. Des. Dev. 1966, 5, 21
79. Suganuma, T.; Yamanishi, T. Kakagu Kogaku 1966, 30, 1136
80. Imafuku, K.; Wang, T.-Y.; Koide, K.; Kubota, H. J. Chem. Eng. Japan 1968, 1, 153
81. Kato, Y.; Nishiwaki, A.; Fukuda, T.; Tanaka, S. J. Chem. Eng. Japan 1972, 5, 112
82. Shah, Y.T. Gas-Liquid-Solid Reactor Design. McGraw-Hill, New York, 1979
83. Kamawura, K.; Sasona, T. Kagaku Kokagu 1965, 29, 693
84. Sano, Y.; Yamaguchi, N: Adachi, T. J. Chem. Eng. Japan 1974, 7, 255
85. Sänger, P.; Deckwer, W.-D. to be published in Chem. Eng. J. 1981

RECEIVED June 3, 1981.

12

A New Model for Heat Transfer Coefficients in Bubble Columns

YATISH T. SHAH, J. B. JOSHI[1], and M. M. SHARMA[1]

Chemical and Petroleum Engineering Department, University of Pittsburgh, Pittsburgh, PA 15261

In many multiphase (gas-liquid, gas-solid, liquid-liquid and gas-liquid-solid) contactors, a large degree of circulation of both discrete and continuous phases occurs. This circulation causes a good degree of mixing and enhances heat and mass transfer between fluid and walls. The degree of circulation depends on a number of parameters such as the size of equipment, the nature of the phases involved, velocities of various phases, nature of the internals within the equipment and many others. The importance of circulation in bubble columns, gas-solid fluidized beds and agitated contactors has been extensively examined in the literature. In this paper we present a generalized procedure for the calculation of bed-wall heat transfer coefficient in bubble columns on the basis of their hydrodynamic behavior. It has been shown that the high values of heat transfer coefficient obtained in bubble columns, as compared to the single phase pipe flow, can be explained on the basis of the enhanced local liquid velocities in the presence of gas phase. A comparison between the predicted and experimental values of heat transfer coefficient is presented over a wide range of design and operating variables.

There are several industrially important multiphase reactors in which the chemical reaction is accompanied by large heat effects (Table I). Heat is either supplied or removed depending upon whether the reaction is endothermic or exothermic. For the heat to be transferred an area is provided in the form of either coils or vertical or horizontal bundle of tubes or the column

[1] Current address: Department of Chemical Technology, University of Bombay, Bombay, India.

0097-6156/81/0168-0243$05.00/0

TABLE I

THE REACTIONS OF INDUSTRIAL IMPORTANCE WHICH ARE CARRIED OUT IN BUBBLE COLUMNS AND SLURRY REACTORS AND ARE ACCOMPANIED BY LARGE HEAT EFFECTS

(i) Oxidation of organic compounds such as toluene, cumene, o-xylene, ethylene, acetaldehyde, butane, sec-butyl benzene.
(ii) Chlorination of benzene, toluene, phenol, ethylene, ethanol, acetic acid, and paraffin wax.
(iii) Hydrogenation of benzene, nitrobenzene, acetone, adiponitrile, butynediol and "oxo" aldehydes.
(iv) Hydration of olefins to alcohols.
(v) Fischer-Tropsch synthesis.
(vi) Manufacture of organic chemicals by alkylation such as cumeme and sec-butyl benzene.

wall. In the last two decades, several investigators have reported the experimental values of heat transfer coefficients in the multiphase or bubble columns and proposed correlations to explain their data. Table II gives a summary of the heat transfer studies made in bubble columns. The experimental observations may be summarized as follows:

(i) The heat transfer coefficient at the wall (h_W) is independent of the column diameter.

(ii) The values of h_W are 20 to 100 times larger than the single phase pipe flow and are comparable to those obtained from mechanically agitated contactors.

(iii) The values of h_W are practically independent of the sparger design.

(iv) The heat transfer coefficient does not increase indefinitely with the superficial gas velocity (V_G). The value of h_W levels off at some value of V_G depending upon the column diameter and the other physical properties of the gas-liquid system.

In the past, there have been two major approaches to analyze the problem of heat and mass transfer across the liquid-solid interface. The first approach can be broadly classified as "Analogies." This method essentially consists of (von Karman (1) and Wasan and Wilke (2)): (i) development of velocity profile near the interface, (ii) suitable assumption for the variation of eddy diffusivity with respect to the distance from the interface, and (iii) solution of the Reynolds modification of the Navier-Stokes equation for the turbulent flow.

For the case of single phase pipe flow, von Karman (1) selected the universal velocity profile. The value of eddy viscosity was obtained from the slope of the velocity profile. Further, it was assumed that the numerical values of eddy viscosity and the eddy diffusivity are the same. The following equation was obtained:

TABLE II

EXPERIMENTAL DETAILS OF HEAT TRANSFER STUDIES IN BUBBLE COLUMNS

Symbols in Figures 1 and 4	Tank Dia. T, m	$\frac{c_p\mu}{k}$	System	Investigator
○	0.46	7.6	air-water	Fair et al. (10)
⦶	1.06			
△	0.099	3.0	-do-	Hart (11)
▽	0.099	36.5	air-ethylene glycol	
□	---	6.0	air-water	Permer (12)
■	---	8.0	air-ethylene -alcohol	
⊖	0.19	6.0	air-water	Burkel (13)
-○-	0.09 0.19 0.29	6.0	air-water	Muller (14)
+	0.1	4.0	air-xylene	Louisi (15)
⊕	0.1	6.1	air-kogasene	
×	0.1	10.2	air-decalin	

$$\frac{1}{St} = \frac{2}{f} + \frac{2}{f} \cdot f\,(Pr) \qquad (1)$$

where f is the fanning fraction factor and its value is obtained from a suitable correlation. There are several other analogies available in the published literature. The equations widely used for the calculation of heat transfer coefficient are the dimensionless correlations which find their basis in analogies. For instance, Sieder and Tate (3) have proposed the following correlations:

$$\frac{h_w T}{k} = 0.027 \left\{\frac{TV_a\rho}{\mu}\right\}^{0.8} \left(\frac{c_p\mu}{k}\right)^{1/3} \left(\frac{\mu}{\mu_w}\right)^{0.14} \qquad (2)$$

A theoretical approach to analyze the problem of heat transfer is to develop some form of surface renewal model. The rate of surface renewal is found from the knowledge of energy input per unit mass. Recently Deckwer (4) has analyzed the problem of heat transfer in bubble columns on the basis of surface renewal model. The present study uses the earlier approach.

Mathematical Model

The enhancement in the bed-wall heat transfer coefficients in bubble columns as well as the two- and three-phase contactors as compared to the single phase pipe flow is likely because of the strong circulation flow pattern in the continuous phase. Joshi (5) has shown that the average continuous phase circulation velocities in multiphase contactors are 1 to 2 orders of magnitude larger than the net superficial continuous velocities. Joshi (5) has proposed the following equations:

The average liquid circulation velocity is given by the following equation.

$$V_c = 1.31' \{gT\ (V_G - \varepsilon V_{b\infty})\}^{1/3} \tag{3}$$

The average axial and radial components of the liquid velocity are given by the following equations:

$$V_a = 1.18' \{gT\ (V_G - \varepsilon V_{b\infty})\}^{1/3} \tag{4}$$

$$V_r = 0.42' \{gT\ (V_G - \varepsilon V_{b\infty})\}^{1/3} \tag{5}$$

The values of V_a predicted by Equation (4) as well as the predicted velocity profile agree with the experimental values reported by several investigators within 15 percent.

Since the velocity profile in bubble columns is known, the procedure for the calculation of heat transfer coefficient can be developed on a more rational basis. Substitution of Equation (4) in (1) gives:

$$\frac{h_w T}{k} = 0.031' \left\{ \frac{T^{1.33} g^{1/3} (V_G - \varepsilon V_{b\infty})^{1/3} \rho}{\mu} \right\}^{0.8} \left(\frac{c_p \mu}{k}\right)^{1/3} \left(\frac{\mu}{\mu_w}\right)^{0.14} \tag{6}$$

From Equation (6) it can be seen that h_w is practically independent of the column diameter ($T^{0.066}$) which agrees with the experimental observation. However, the values of h_w calculated from Equation (6) are about 25 to 35 percent of the experimental values. This may be because of the presence of radial component of the liquid velocity in bubble columns as against its absence in the pipe flow.

In the case of helical coils it is known that the enhancement in h_w occurs because of the presence of radial flow. The enhancement factor is given by the following equation (Perry and Chilton (6)):

$$E_c = 1 + 3.5\ d/D_H \tag{7}$$

where,

$$d/D_H = Re_c/De \tag{8}$$

$$Re_c = \frac{dV_a\rho}{\mu} \tag{9}$$

$$De = \frac{dV_r\rho}{\mu} \tag{10}$$

and

$$\frac{d}{D_H} = \frac{V_r^2}{V_a^2} \tag{11}$$

The above analysis can be applied for the case of bubble columns. From Figure I it can be seen that, near the wall the axial component of the liquid velocity is downwards, whereas the radial component of the liquid velocity is towards the wall in the top half of the circulation cell and away from the wall in the lower half of the circulation cell. As a result, for one circulation cell the enhancement factor is given by the following equation:

$$E_c = 1 + \left(\frac{2V_r}{V_a}\right)^2 \tag{12}$$

Substitution of Equations (4) and (5) in (12) gives:

$$E_c = 2.8 \tag{13}$$

Equation (12), therefore, takes the following form:

$$\frac{h_w T}{k} = 0.087' \left\{\frac{T^{1.33} g^{1/3} (V_G - \varepsilon_G V_{b\infty})^{1/3} \rho}{\mu}\right\}^{0.8} \left(\frac{c_p \mu}{k}\right)^{1/3} \left(\frac{\mu}{\mu_w}\right)^{0.14} \tag{14}$$

From Figure II it can be seen that the predicted and experimental values are within 30 percent.

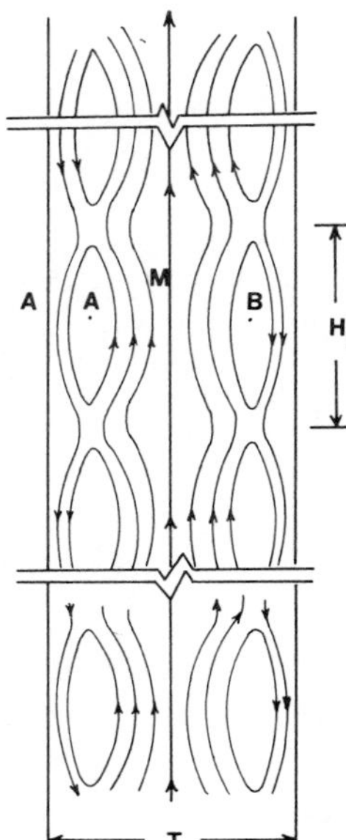

Figure 1. *Liquid flow pattern in a bubble column*

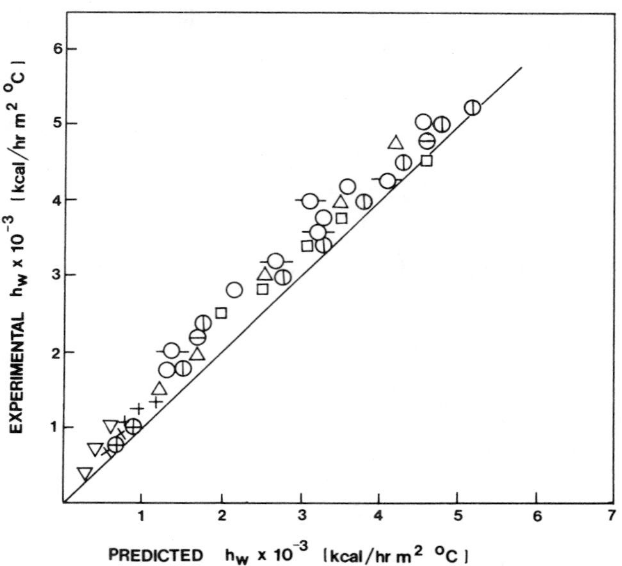

Figure 2. Comparison between experimental and predicted (Equation 24) heat transfer coefficients: bubble column (analogy with pipe flow); for symbol key, see *Table II*

Comparison with Mechanically Agitated Contactors. The heat transfer coefficient at the wall of mechanically agitated contactors is given by the following equation (Perry and Chilton (6)):

$$\frac{h_w T}{k} = C \left(\frac{ND^2 \rho}{\mu}\right)^{2/3} \left(\frac{c_p \mu}{k}\right)^{1/3} \left(\frac{\mu}{\mu_w}\right)^{0.14} \tag{15}$$

for

$$40 < \frac{ND^2 \rho}{\mu} < 3 \times 10^5$$

where N and D are the impeller speed and the impeller diameter, respectively. The value of C in Equation (15) is 0.6 for the case of six-bladed disk (Rushton) turbine.

The values of h_w for bubble columns can probably be calculated from the above correlation if the comparison between bubble columns and mechanically agitated contactors is based on the same value of average liquid circulation velocity, V_c.

The value of average liquid circulation velocity in the case of mechanically agitated contactors can be calculated from the data on mixing time. The flow pattern generated by a Rushton disk turbine is shown in Figure III. It can be seen from Figure III that four circulation cells (in a plane, two donut cells in three dimensions) are developed due to the impeller action. The average length of the path for a fluid element leaving the impeller and returning back to the impeller, for any circulation cell, depends on the vessel diameter, the liquid height and the impeller speed. The following value is selected for the central location of the impeller (H = 2h'), where h' is the distance between the impeller and the liquid surface:

$$L = T + H \tag{16}$$

The circulation time, therefore, is given by the following equation:

$$t_c = \frac{L}{V_c} \tag{17}$$

Holmes et al. (7) and Norwood and Metzner (8) have shown that the mixing time is about four to five times the circulation time or (say):

$$\theta_{mix} = 4t_c \tag{18}$$

The value of V_c can be calculated from correlations reported for the mixing time. Thus for instance, Holmes et al. (7) have reported the following equation:

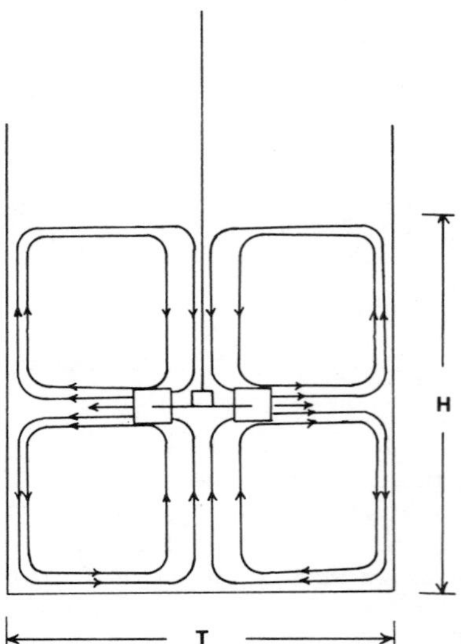

Figure 3. Liquid flow pattern generated by disk turbine

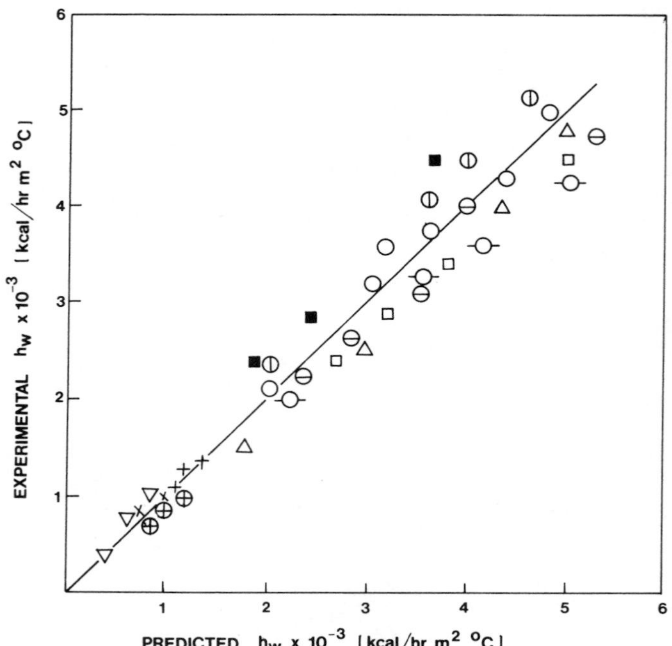

Figure 4. Comparison between experimental and predicted heat transfer coefficients in bubble columns (Equation 24) (analogy with mechanically agitated contactors); for symbol key, see *Table II*

$$\theta_{mix} = \frac{4.3(T/D)^2}{N} \quad (19)$$

Substitution of Equations (16), (17) and (18) in (19) gives:

$$V_c = \frac{(H+T)N}{1.075(T/D)^2} \quad (20)$$

$$= \frac{1.86ND^2}{T} \text{ for } T = H \text{ and } H = 2h' \quad (21)$$

Since, θ_{mix} α L, and for T $\neq$ 2h'

$$V_c = 3.72\ ND^2/(1+2h'/T)T \quad (22)$$

Substitution of Equation (21) in (15) gives:

$$\frac{h_w T}{k} = 0.4 \left(\frac{TV_c\rho}{\mu}\right)^{2/3} \left(\frac{c_p\mu}{k}\right)^{1/3} \left(\frac{\mu}{\mu_w}\right)^{0.14} \quad (23)$$

Equation (23) will be applicable for bubble columns if an appropriate value of V_c is incorporated. Substitution of Equation (3) in (23) gives:

$$\frac{h_w T}{k} = 0.48' \left\{\frac{T^{1.33} g^{0.33} (V_G - \varepsilon V_{b\infty})^{1/3} \rho^{0.66}}{\mu}\right\} \left(\frac{c_p\mu}{k}\right)^{1/3} \left(\frac{\mu}{\mu_w}\right)^{0.14} \quad (24)$$

Figure IV shows a comparison between the experimental (Table II) values of h_w and those predicted by Equation (24). It can be seen that the agreement is within 15 percent. Further, from Equation (24) it can be seen that h_w varies as $T^{-0.11}$. Because of low power on T, the value of h_w are found to be practically independent of T (the variation of column diameter from 1 m to 0.1 m gives only 22 percent reduction). Some more details on this comparison and the application of the present model to other multiphase contactors are recently outlined by Joshi et al. (9).

Nomenclature

c_p	specific heat of the liquid, kcal/kg^{o}C
D	impeller diameter, m
De	Dean's number, $dV_r\rho/\mu$

D_H helix diameter of a coil, m
d tube diameter, m
d_B average bubble diameter, m
E_b enhancement factor for heat transfer coefficient due to the presence of radial component of the liquid velocity: bubble column
E_c enhancement factor for heat transfer coefficient due to the presence of radial component of the liquid velocity: helical coil
Fr Froude number, V_G^2/gd_B
f fanning friction factor
g acceleration due to gravity, m/s^2
H liquid height in mechanically agitated contactors
h heat transfer coefficient, kcal/hr m^2 ^oC
h_w bed-wall heat transfer coefficient, kcal/hr m^2 ^oC
k thermal conductivity of the continuous phase, kcal/hr m ^oC
k_1, k_2 constants in Equations (30) and (32), respectively
L length of the path in a circulation cell, m
N impeller speed, revolutions/s
Pr Prandtl number, $c_p\mu/k$
Re Reynold's number for bubble (Equation (3))
Re_c Reynold's number for coil (Equation (15))
St Stanton number, $h_w/\rho C_p V_a$
T column diameter or vessel diameter, m
V_a average axial component of the continuous phase velocity, m/s
$V_{b\infty}$ terminal rise velocity of bubbles, m/s
V_c average continuous phase circulation velocity, m/s
V_G superficial gas velocity, m/s
V_{mf} minimum velocity for fluidication, m/s
V_0 average rise velocity of bubbles, m/s
V_r radial component of the continuous phase velocity
t_c circulation time, s
ρ density of the liquid, g/cc
ε fractional gas holdup, dimensionless
μ viscosity of the continuous phase, cp
θ_{mix} mixing time, s

Literature Cited

1. von Karman, Th. Trans. ASME 1939, 61, 705.
2. Wasan, D. T.; Wilke, C. R. Int. J. Heat and Mass Transfer 1964, 7, 87.
3. Sieder, E. N.; Tate, G. E. Ind. Eng. Chem. 1936, 28, 1429
4. Deckwer, W. D. Chem. Eng. Sci. 1980, 35, 1341.
5. Joshi, J. B. Trans. Instn. Chem. Engr. 1980, 58, 155.
6. Perry, R. H.; Chilton, C. H. "Chemical Engineers Handbook," 5th ed.; McGraw-Hill: New York, NY, 1973.
7. Holmes, D. B.; Voncken, R. M.; Dekker, J. A. Chem. Eng. Sci 1964, 19, 201.

8. Norwood, K. W.; Metzner, A. B. AIChE J. 1960, 6, 432.
9. Joshi, J. B.; Sharma, M. M.; Shah, Y. T.; Singh, C. P. P.; Ally, M.; Klinzing, G. E. "Heat Transfer in Multiphase Contactors," Chem. Eng. Communication 1981 (in press).
10. Fair, J. R.; Lambright, A. J.; Anderson, J. M. I&EC Proc. Des. and Develop. 1962, 1, 33.
11. Hart, W. F. I&EC Proc. Des. and Develop. 1976, 15, 109.
12. Permer, D. Dipl. Arbeit, TU Berlin, 1960.
13. Burkel, W. Dr. Ing. Thesis, TU Munchen, 1974.
14. Muller, D. Dr. Ing. Thesis, TU Berlin, 1962.
15. Louisi, Y. Dr. Ing. Thesis, TU Berlin, 1979.

RECEIVED June 16, 1981.

13

Dispersion and Hold-Up in Bubble Columns

RICHARD G. RICE

Department of Chemical Engineering, Louisiana State University, Baton Rouge, LA 70803

JORMA M. I. TUPPERAINEN and ROBYN M. HEDGE

Department of Chemical Engineering, University of Queensland, St. Lucia, Queensland, Australia

The performance of a new type rubber-sheet sparger is compared with the rigid perforated plate gas distributor Impulse response tests were analyzed using the weighted-moments method to determine voidage and dispersion coefficients in a countercurrent bubble column of 9.5 cm diameter. The flexible rubber sheet sparger produced more uniform emulsions, smaller bubbles and larger voidages than perforated plates, while dispersion coefficients were reduced for a range of superficial gas velocities. The mixing results are contrasted with predictions based on Taylor, entrainment and energy dissipation models.

Bubble columns are a cheap, simple gas-liquid contractor in which gas is bubbled co-currently or counter-currently through a liquid. Bubble columns have been used in the process industries for over twenty years for carrying out gas-liquid reactions. They are likely to find increasing use in the future as tower fermenters in the production of liquid fuels from biological materials. Furthermore, it has been shown that bubble columns are prime candidates for fine-particle flotation. Flotation columns can increase the number of effective stages relative to the same volume in a cell device, and moreover, gangue carry-over is reduced because tailings and froth are widely separated. One can foresee a significant increase in the use of bubble-column contactors in the future, but the fundamentals of the technology are only thinly developed.

In the present research, we study two fundamental properties of bubble columns: liquid hold-up and mixing. Both of these properties depend on the flow rates of the gas and liquid phases. These two properties may be considered response variables in the sense that their values depend on the way bubbles are formed. We present results for two types of bubble generating devices (or, for short, spargers) i.e. perforated rigid plates and perforated rubber sheets. An advantage of the rubber-sheet sparger is the self-cleaning feature. This is

0097-6156/81/0168-0255$05.50/0

because during operation the rubber sheet oscillates and deforms, thus preventing mud build-up on the spargers in practical applications such as mineral flotation. Moreover, the rubber sparger is self regulating in the sense the holes expand to accomodate high flow rates.

Axial mixing refers to the mechanism by which a phase can move or disperse against the direction of its main flow. In counter-current bubble columns, which are the most common physical configuration, liquid phase axial mixing is induced mainly by the rising gas bubbles dragging the liquid against its net downward flow. The bulk circulation pattern thus set up leads to a "spreading out" of solute or reactant and thus enhances mixing in the longitudinal direction. To a lesser extent, the trajectory, coalescence rate and deformations of a bubble also affect the mixing process. The overall effect of increasing the axial mixing in a contacting process is to reduce the number of theoretical stages available in the column as compared with simple plug flow.

One can visualize in the design of bubble columns that there must exist an optimization problem at hand. Thus, efforts to reduce bubble size should increase interfacial mass transfer rates. This of course means shorter columns. At the same time, smaller bubbles may tend to increase axial mixing, and this effect serves to increase column design length. In addition, column diameter seems to have an even more important effect on the magnitude of axial mixing as shown by Baird and Rice (1). It is well known that the general liquid flow pattern is that of upflow at the center and downflow near the walls. This circulation pattern induces the large axial mixing effect, and in fact it has been shown (2) that center-counted baffles can significantly reduce dispersion coefficient.

Flow Regimes

In characterizing upward movement of bubble swarms, Wallis (3) suggests there exists three separate flow regimes. These regimes occur in order of increasing gas rates as follows:

(i) Bubbly Flow is characterized by a constant bubble size, bubble velocity and gas hold-up throughout the column.

(ii) Churn-Turbulent Flow exhibits an unsteady flow pattern with channeling due to bubbles following in each other's wakes. The bubble sizes and velocities vary throughout the column. One might tend to model the system dynamics as an equilibrium between rate of bubble coalescence and rate of breakage.

(iii) Slug Flow is characterized by a series of individual large bubbles which almost fill the column cross-section.

Of the above regimes, Bach and Pilhofer (4) suggest the churn-turbulent "heterogeneous" regime is the one commonly encountered in industrial scale bubble columns. Transition effects have been reported by Anderson and Quinn (5) in the presence of minute amounts of trace contaminants in tap water. It is now well known that small amounts of surface active agents have a very strong effect on the flow regime and the bubble size distrubution. For example, Lockett and Kirkpatrick (6), under special conditions, have demonstrated ideal bubbly flow at gas hold-ups up to 66%.

Gas-liquid systems are classified as coalescing or non-coalescing, depending on the behavor exhibited by the bubbles. Water is classed as coalescing by Konig (7). Examples of systems exhibiting lower coalescence rates are 0.5% propanol solutions, dilute solutions of surfactants and many of the media used in fermentation reactions.

The degree of coalscence plays an important part in determining the bubble size distribution. Otake et al. (8) showed that the interaction between the wake of a leading bubble and a trailing bubble could lead to either coalescence of the two bubbles, or alternately to the breakup of the trailing bubble. The net balance between rates of coalescence and rates of breakage must determine the ultimate bubble size distribution, and this in turn determines the properties such as liquid hold-up, interfacial area, swarm velocity and finally, axial mixing. On a small scale, bubble columns are inherently non-linear and unsteady in behavior. This suggests a non-linear analysis to uncover the multiple steady-states would be a productive avenue to follow.

Characterization of Liquid Holdup or Voidage

In this section, we give a brief overview and the important developments relating to the characterization of liquid hold-up (or gas voidage) in bubble columns.

Wallis (3) introduced the concept of "drift-flux" analysis as a means to relate phase flow rates, voidage and certain physical properties. The slip velocity for counter-current flow is defined as:

$$(1)\quad v_s = \frac{u_{og}}{\varepsilon} + \frac{u_{ol}}{1-\varepsilon}$$

where ε is the gas voidage. Wallis defines the drift-flux as simply the product of slip velocity times the respective hold-ups, so that

$$(2)\quad j_D = v_s(1-\varepsilon)\varepsilon = u_{og}(1-\varepsilon) + u_{ol}\varepsilon$$

To complete the drift-flux model, the phenomenological theory of Lapidus and Elgin (9) is used, wherein it was suggested for dispersed flow systems that the slip velocity depends directly on terminal bubble rise velocity, U_∞, so that

(3) $v_s = U_\infty \phi(\varepsilon)$.

Combining this with equation (2), the drift-flux becomes

(4) $j_D = U_\infty \varepsilon (1-\varepsilon) \phi(\varepsilon)$

and we see that

(5) $U_\infty \varepsilon (1-\varepsilon) \phi(\varepsilon) = u_{og}(1-\varepsilon) + u_{ol}(\varepsilon)$.

The usefulness of the drift-flux structure resides in the simplicity afforded in obtaining a graphical solution. Thus, one sees that the RHS of equation (5) is linear in ε, while the LHS is (usually) non-linear. Given the structure of $\phi(\varepsilon)$, along with the phase flow rates, one determines ε by finding the intersection of the linear RHS with the non-linear LHS. An example of this construction is shown in Figure 3.

Forms for $\phi(\varepsilon)$ proposed in the literature are summarized by Lockett and Kirkpatrick (6). These relationships vary significantly, but with voidages less than 0.2 they are quite close to one another. Lockett and Kirkpatrick (6) give the following list of slip functions:

Turner	$\phi(\varepsilon) = 1$
Davidson and Harrison	$\phi(\varepsilon) = 1/(1-\varepsilon)$
Wallis	$\phi(\varepsilon) = (1-\varepsilon)^{n-1}$
	where: n = 2 for small bubbles n = 0 for larger bubbles
Richardson and Zaki	$\phi(\varepsilon) = (1-\varepsilon)^{1.39}$
Marrucci	$\phi(\varepsilon) = (1-\varepsilon)/(1-\varepsilon^{5/3})$

In the presence of a surfactant (Terpineol), Rice et al. (2) obtained a very good fit between theory and experiment using $\phi(\varepsilon) = (1-\varepsilon)$ which is the Wallis model with n = 2. In the work cited, average bubble size was around 1 mm diameter. This particular structure shows, according to equation (3), that the slip velocity approaches terminal rise velocity as voidage becomes small, as one expects.

A different approach, first proposed by Towell et al. (10) and later confirmed by Reith et al. (11), suggests that slip velocity is apparently independent of voidage:

(6) $v_s = U_\infty + 2\ u_{og}$.

This equation has been tested at superficial gas velocities up to 45 cm/sec and superficial liquid velocities up to 2 cm/sec. If one compares equation (6) with equation (3), we see that

(7) $\phi = 1 + 2\ (\frac{u_{og}}{U_\infty})$.

Bubbles moving along straight trajectories in an otherwise quiescent liquid obey the mass balance:

(8) $\frac{u_{og}}{U_\infty} \cong \varepsilon$

which shows that, approximately, Towell's (<u>10</u>) slip velocity expression requires:

(9) $\phi \cong 1 + 2\ \varepsilon$

which is quite different from the structure of the researchers listed above. However, all the proposed functions $\phi(\varepsilon)$ have the property:

(10) $\lim_{\varepsilon \to 0} \phi(\varepsilon) \to 1.0$

which imples that slip velocity approaches the terminal rise velocity of a single bubble at low voidages.

<u>Axial Mixing</u>

When a steady stream of liquid enters a bubble column, the effect of axial mixing is to cause elements of the liquid, which enter the column together, to leave at times which differ from the mean residence time.

Two main types of models are in common use for describing axial mixing in bubble columns. The most commonly used model is the Dispersion Model. Here, a diffusion-like process is superimposed on piston or plug flow. The stirred tanks-in-series model has also been used to describe flow of liquids in bubble columns. Levenspiel (<u>12</u>) presents a number of models incorporating various combinations of mixed tanks to model stagnant regions and backflow.

In the present research, we use a dispersion-type model. Techniques for estimating dispersion coefficients are classed as non-steady or steady state. Steady-state methods usually incorporate the continuous addition of a fixed tracer concentration into the liquid stream followed by the measurement of the steady-state concentration profile along the column length.

This method has been used by Ohki and Inoue (13), Smith et al. (14) and Reith et al. (11)

Unsteady-state methods are experimentally easier to do, whereby a shot of tracer is added followed by the measurement downstream of the time variation of tracer concentration. This technique was used in bubble columns by Seher and Schumacher (15) and Rice et al. (2). Some researchers carried out studies using both methods as a double-check on parameters, cf. Konig et al. (7), Deckwer et al. (16) and Rice et al. (2)

Taking the liquid velocity profile to be uniform, the dispersion equation is represented by

$$(11)\quad \frac{\partial C}{\partial t} + v\,\frac{\partial C}{\partial z} = E\,\frac{\partial^2 C}{\partial z^2}$$

where C is the liquid composition, v is the liquid interstitial velocity, and E is the liquid interstitial dispersion coefficient. Special note should be taken of the fact that E is based on available liquid area, not on total column cross-sectional area. Wen and Fan (17) suggest the application of a dispersion-type model to bubble columns is not always satisfactory. However, it has been our experience (2) that the dispersion model can give a very satisfactory fit of response data, especially when bubble size is maintained small by the addition of surfactant. The fit is not as good when coalescence occurs, such as operation in the churn-turbulent regime. A poor fit to the dispersion equation also results for very low gas rates such that chain bubbling occurs, and the bubble swarm is not uniformly distributed across the column cross-section. Under such conditions, dead-water or unmixed pockets may occur. The pockets exchange mass with the bulk flow mainly by the slow process of molecular diffusion. Dead-water pockets are always present, even under uniform bubbly flow conditions. For example, the liquid entrapped in the region following the wake of bubbles causes the appearance of spreading of tracer in impluse response tests. One sees there are good reasons for treating the liquid phase as comprised of two parts: bulk flowing liquid and dead-water.

Suppose we denote the liquid in dead-water regions as having compositon C_D, and liquid within the main bulk flow to have composition C_B. The liquid hold-up in the two regions would in general be different, but the sum of the two would be taken to be equal to the hydrostatically measured holdup. Thus, a two-region dispersion model can be represented by, firstly an overall mass balance:

$$(12)\quad \varepsilon_D\,\frac{\partial C_D}{\partial t} + \varepsilon_B\,\frac{\partial C_B}{\partial t} + \varepsilon_B v\,\frac{\partial C_B}{\partial z} = \varepsilon_B E\,\frac{\partial^2 C_B}{\partial z^2}$$

and secondly a mass balance on the dead-water phase, assuming the deadpockets are fixed in space:

$$(13)\quad \varepsilon_D \frac{\partial C_D}{\partial t} = q_E\,(C_B - C_D)$$

The liquid holdups, ε_D and ε_B, are for deadwater and bulk liquid, respectively. Here, one would surmise that the exchange coefficient, q_E, is proportional to molecular diffusion. This model may have advantages in characterizing bubble column behavior, and may explain the long-tail response which often occurs in experimental response curves. However, the parameter estimation problem is unwieldy, reqiring the determination of four parameters (ε_D, ε_B, E, q_E) instead of the simple plug flow dispersion model which has only two parameters (ε, E).

In the present work, we use the weighted-moments method to estimate parameters by fitting the theoretical Laplace domain moments to the experimentally generated moments for the impulse response. Thus, for the model without deadwater (equation 11), the Laplace transform for the response variable at the exit is (2):

$$(14)\quad \bar{C}(s,L) = Q \exp\left[\tfrac{1}{2}\,Pe\left(1 - \sqrt{1 + \frac{4s\tau}{Pe}}\right)\right]$$

where

Q	=	strength of the impulse stimulus
Pe	=	Peclet number, vL/E
τ	=	liquid phase residence time, L/v

The deadwater model comprised of equation (12) and (13) is similar in many respects to the equations describing adsorption in packed columns. When equations (14) and (13) are expressed with dimensionless independent variables, we have

$$(15)\quad R\frac{\partial C_D}{\partial \theta} + \frac{\partial C_B}{\partial \theta} + \frac{\partial C_B}{\partial \zeta} = \frac{1}{Pe}\frac{\partial^2 C_B}{\partial \zeta^2}$$

$$(16)\quad \frac{\partial C_D}{\partial \theta} = \beta\,(C_B - C_D)$$

where

R	=	$\varepsilon_D/\varepsilon_B$
θ	=	t v/L, dimensionless time
ζ	=	z/L, dimensionless distance
Pe	=	vL/E, Peclet number
β	=	$q_E\ L/v\varepsilon_D$, dimensionless exchange coefficient

The Laplace solution for exit composition following an impulse at the inlet, written in terms of a "dimensional s" (as in equation 14), can easily be seen to be:

(17) $\bar{C}_B(s) = Q \exp [\frac{1}{2}Pe\ (1 - \sqrt{1 + \frac{4}{Pe}\ (\frac{R\beta}{s\tau+\beta} + s\tau)})]$

which of course reproduces the simple dispersion model when R or $\beta \to 0$. The unknown parameters usually would be taken to be τ, Pe, R, and β. In addition, an optimum weighting factor "s" must be found (2). One can use independently determined static hold-up measurements to check consistency,

(18) $\varepsilon_{static} = \varepsilon_D + \varepsilon_B$

and to apportion the relative hold-ups. However, this does not reduce the complexity of the moments analysis, since four unknown parameters still exist. This analysis requires at least four experimental moments.

Another model worth considering is to assume all the deadwater resides in the stagnant pockets in bubble wakes. Here, a moving coordinate system would be used, taking the bubble swarm velocity to be U_∞. For this model, equation (13) is replaced by the distributed parameter equation:

(19) $\varepsilon_D \frac{\partial C_D}{\partial t} + \varepsilon_D U_\infty \frac{\partial C_D}{\partial z} = q_E\ (C_B - C_D)$

The parameter estimation problem remains essentially the same as before, since U_∞ could be determined separately from hold-up measurements.

Correlations and Theories for Axial Mixing in Bubble Columns

Workers in the field have reached several generalizations regarding the behavior of axial dispersion coefficients. Firstly, there are indications that E depends strongly on superficial gas velocity. Secondly, an even stronger dependence of E on column diameter has been observed. The scale-up problem is thus quite sensitive to column diameter.

There is much discrepancy in the reported literature regarding the dependence of E on the two key variables, d, (column diameter) and u_{og} (superficial gas velocity). Thus, Reith et al. (11) observed, following a large number of experiments, an approximately constant value for a Peclet number:

(20) $\frac{v_s d}{E} \sim 3$

A short time later, Ohki and Inoue (13) produced the following dimensional correlation based on a large number of experiments

(21) $E = 0.3\ d^2\ u_{og}^{1.2} + 170\ \delta$

where

δ = sparger hole diameter (cm)
d = column diameter (cm)
u_{og} = superficial gas velocity (cm/sec)
E = dispersion coefficient (cm^2/sec)

Since sparger hole diameter in some sense relates to bubble diameter, this correlation suggests mixing increases with increased bubble diameter. These workers also suggest that dispersion coefficient for the coalescent bubble-slug flow conditions follows:

(22) $E = 14d/(1-\varepsilon)^2$

More recently, Deckwer et al. (16) have tested a correlation originally proposed by Towell et al. to give the dimensional correlation

(23) $E = 2.7d^{1.4}u_{og}^{0.3}$ (units are cm, sec)

This relationship shows quite different exponents compared to that of Ohki and Inoue, especially the exponent on u_{og}.

Using the isotropic turbulence theory of Kolmogoroff, Baird and Rice (1) deduced the following expression with a single arbitrary constant to describe mixing in large diameter bubble columns

(24) $E = K\ d^{4/3}(u_{og}g)^{1/3}$

Analysis of a broad range of published data produced a value of K, which is dimensionless, equal to 0.35. Later, Smith et al. (14) found their experiments exceeded by a factor of two the predictions of the isotropic turbulence model.

The exponents in the isotropic turbulence model (equation 24) are very close to those found in the comprehensive work by Deckwer et al. (equation 23). The isotropic turbulence model is also attractive owing to dimensional consistency, since any set of consistent units can be used to correlate dispersion data. While the diameter dependence of the varous researchers is not too wide afield (ranging from d to d^2), the dependence on gas velocity varies widely (ranging from $u_{og}^{.3}$ to $u_{og}^{1.2}$). While a comprehensive theory is not yet available, these authors suspect mixing follows the flow regime according to

(i) chain bubbling $E \propto u_{og}^{2}$

(ii) bubbly flow $E \propto u_{og}^{1/2}$ to u_{og}^{1}

(iii) churn-turbulence $E \propto u_{og}^{1/3}$ to $u_{og}^{1/2}$

(iv) slug flow $E \propto u_{og}^{o}$

This leads one to conclude that maxima in E with respect to u_{og} should exist, and this phenomenon has been reported in the recent literature (Smith et al. (14)). In experiments conducted at the University of Queensland, we observed a sharp maxima to exist when surfactant (terpineol) was present; presently we show an apparent maxima also exists for "clean" systems.

The behavior of E depends of course on the method of determining this parameter. None of the methods used to date have included the possibility of deadwater. The growth and decay of deadwater pockets clearly must affect the value deduced for E. Thus one has the expectation that deadwater is large at small gas rates, smaller at intermediate gas rates, but then becomes larger again at high (slugging) gas rates. The presence of deadwater manifests itself as a spreading of the Impulse Response Curve, hence this dispersive force gives an appearance of being caused by axial mixing. If deadwater is significant, one suspects there should be a difference between values of E obtained by unsteady-state and steady-state methods. Deckwer et al. (16) used both methods and there seemed to be no significant difference in mixing coefficients. However, the experiments may have been conducted in the flow region of small deadwater. It is not possible to quantify this any further at this time. It should be possible to measure deadwater effects in the various flow regimes and thereby deduce if a dispersion coefficient maxima is real or simply a consequence of the force-fitting of an incorrect model. Recently, Tuppurainen (18) found that a tanks-in-series model including backmix could be fitted to response curves with a smaller IAE than the dispersion model. However, three parameters (rather than two in the dispersion model) were required to be fitted. It would seem the loss of physical reality in using a tanks-in-series model is too high a price to pay.

Some Preliminary Experiments

Dispersion coefficients were determined using salt tracer response curves along with the weighted moments method of analysis applied to equation (14) (details in Appendix I). The circular column was 9.5 cm in diameter and 182.2 cm long.

Liquid hold-up was measured by simultaneously closing the liquid inlet and outlet valves and measuring the height of the collapsed column. As a check, manometer measurements along the column length were taken to deduce hold-up. Superficial liquid velocity was held relatively constant (.36 cm/sec ± 10%) for all experiments.

Three different spargers were used in the experiments. Details are given in Table I. The experimental arrangement and sparger details are shown in Figures 1 and 2, respectively.

Table I

Details of Spargers

Material & Designation	Thickness (mm)	Hole Number	Hole Diameter(mm)	Arrangement	Holes /cm^2
Polyethylene (PE)	12.5	68	2	6mm,sq.pitch	2.8
Rubber (R1)	1.4	300	2*	random	7.5
Rubber (R2)	1.4	120	2.8*	random	3.0

*This hole diameter refers to the size of nail used to pierce the vulcanized rubber sheet.

The drift flux curves for the three spargers are shown in Figure 3 for the 9.5 cm diameter column. The solid curves represents j_D versus ε, with U_∞ as a fitted parameter. Note the function $\phi(\varepsilon)$ used here was orginally suggested by Wallis (taking n = 2). The experimental points are generated, according to equation (5), as follows. Taking the left hand ordinate scale as $j_D = u_{og}$ (where $\varepsilon = 0$) and the right hand ordinate scale as $j_D = u_{ol}$ (where $\varepsilon = 1$), one simply draws a straight line between experimental values of u_{og} and u_{ol}, marking along this line the measured voidage (abcissa). The drift flux, $j_D(\varepsilon)$, is fitted to the first four points, which also gives a fitted value for U_∞.

When the data abruptly departs from the drift flux curve, a change in the flow regime is assumed to occur; flooding, according to Wallis (3). It was visually observed that increased coalescence and slugging occurred when the experimental data diverged from the drift flux curve.

We take note of the fact that the flexible rubber spargers produced uniform bubbly-flow up to a gas voidage of around 0.2. The perforated plate sparger (PE), however, sustained bubbly flow only up to a voidage 0.1. The points of departure from the drift flux curve highlights this transition. The gas velocity at the transition points were 2.8 cm/sec and 3.7 cm/sec for the perforated plate and flexible rubber spargers, respectively. The rubber spargers therefore sustain significantly higher gas rates and voidages before flooding occurs. Bubble size from the rubber sparger appeared to be about half that resulting from the perforated plate.

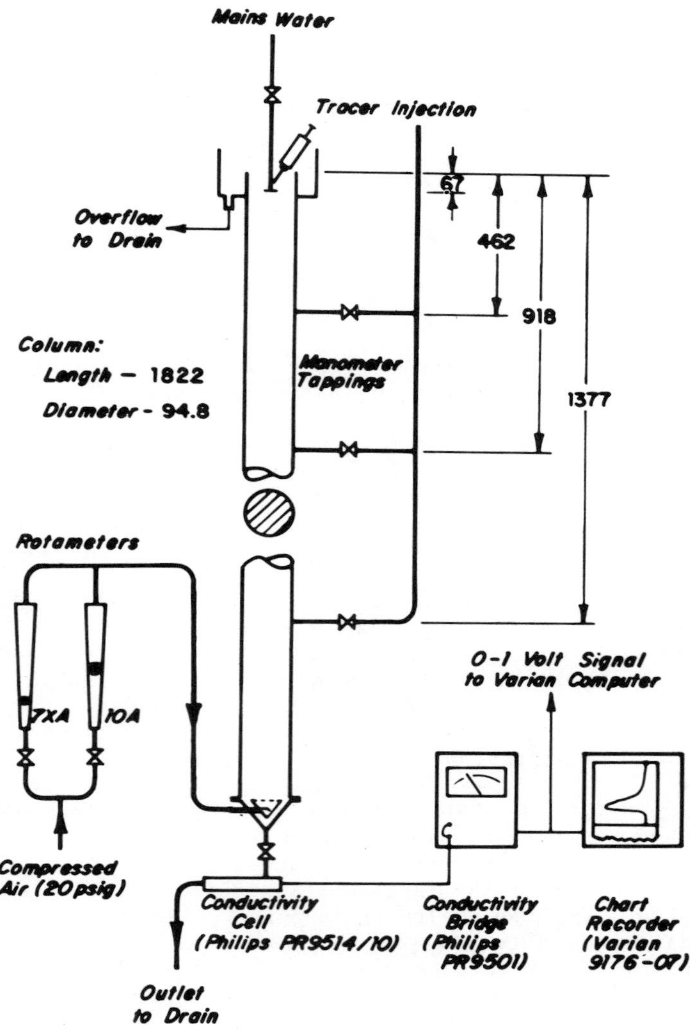

Figure 1. Schematic of typical equipment arrangement; all lengths given in millimeters

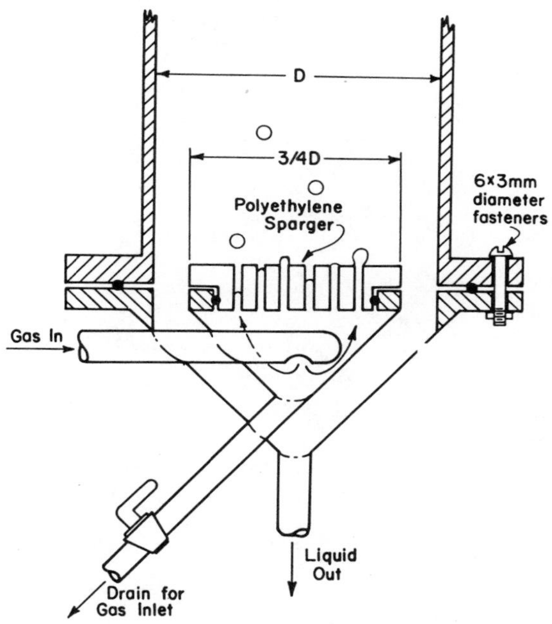

Figure 2. Sparger design

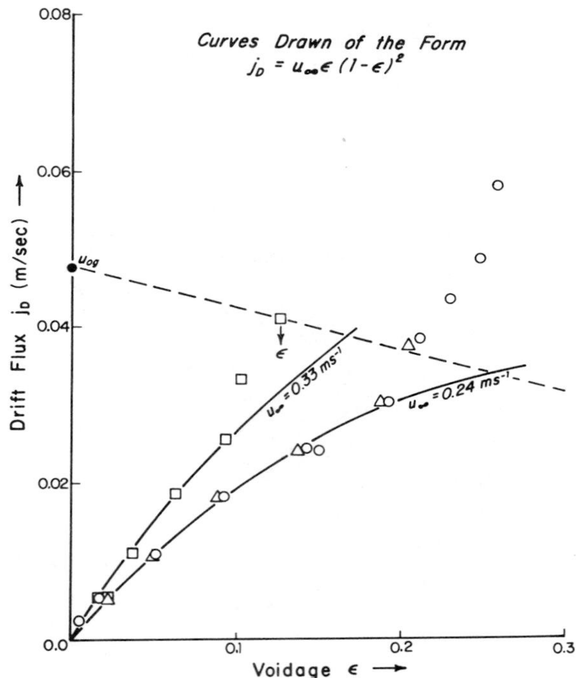

Figure 3. Drift–flux curves for 9.5-cm diameter column: (□) PE spargers, (○) R1 sparger, (△) R2 sparger; (– – –) placement of experimental point (u_{og} = LH ordinate; u_{ol} = RH ordinate; ϵ = abscissa)

Figure 4 shows how the hold-up data (d = 9.5 cm) compares with the limits of literature values (4). In Figure 5, we treat the data from the two types of spargers according to the slip velocity model of Towell et al (10). The slip velocity for the rubber spargers behaves rather curiously, sustaining a minimum when superficial gas velocity if ~ 3 cm/sec.

A detailed comparison of axial mixing coefficents in the 9.5 cm column for the two types of spargers is given in Figure 6. One would expect the smaller bubbles from the rubber sparger to produce increased axial mixing. This did no appear to happen; mixing coefficients from the perforated plate sparger were always larger than the flexible rubber sparger. None of the dispersion data seemed to follow the trends suggested by literature correlation. The data suggests that a maxima in E may occur around $u_{og} \sim 1$ cm/sec.

Discussion of Results

The dispersion coefficients are surprisingly lower than reported elsewhere (see Figure 6), especially those obtained using the flexible rubber sheet sparger. Apparently the energy dissipated in mixing is significantly less than the input energy. Thus, the Baird-Rice (1) model suggests

$$(25)\quad E = K\, \ell^{4/3}\, (P_m)^{1/3}$$

where

ℓ = turbulent mixing length (taken as column diameter by Baird and Rice)

P_m = specific energy dissipation rate per unit mass of liquid

The specific energy dissipation rate for mixing was equated to the input energy rate so that Baird and Rice (1) used

$$(26)\quad P_m = \frac{(u_{og}\ A)(\rho_L g\ L(1-\varepsilon))}{\rho_L\ (A\ L)(1-\varepsilon)} = u_{og}\ g$$

which is simply the power input per unit mass liquid. Field and Davidson (19) argue that the actual rate of energy dissipation which is available for mixing under true bubbly flow conditions (no radial variation in voidage) is given by

$$(27)\quad P_m = [\ g\ (\ u_{og} - \varepsilon\ v_s \pm \frac{\varepsilon}{1-\varepsilon} u_{ol})]$$

where the ± is positive for counter-current, negative for co-current flow. This contrasts with the result of Joshi and

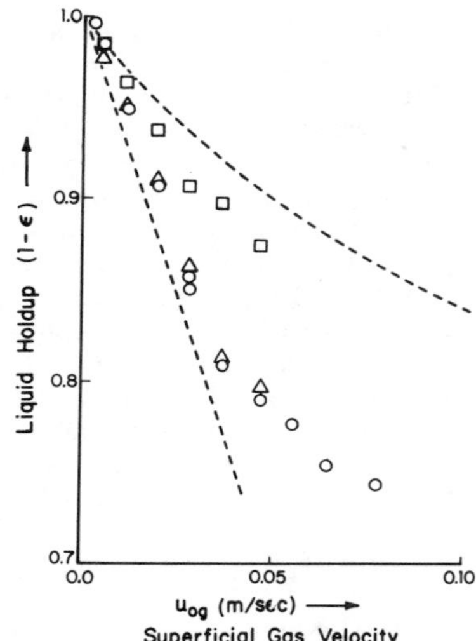

Figure 4. Comparison of liquid hold-up with literature (9.5-cm diameter column): (□) PE sparger, (○) R1 sparger, (△) R2 sparger; (- - -) reported limits (4)

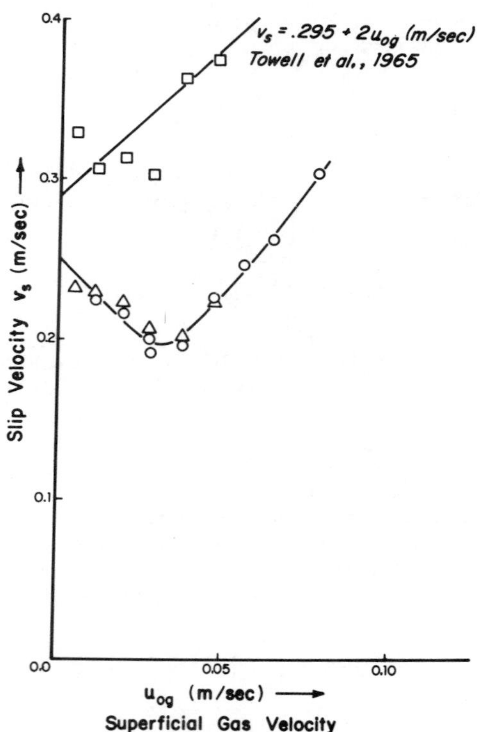

Figure 5. Comparative slip velocities for 9.5-cm diameter column: (□) PE sparger, (○) R1 sparger, (△) R2 sparger

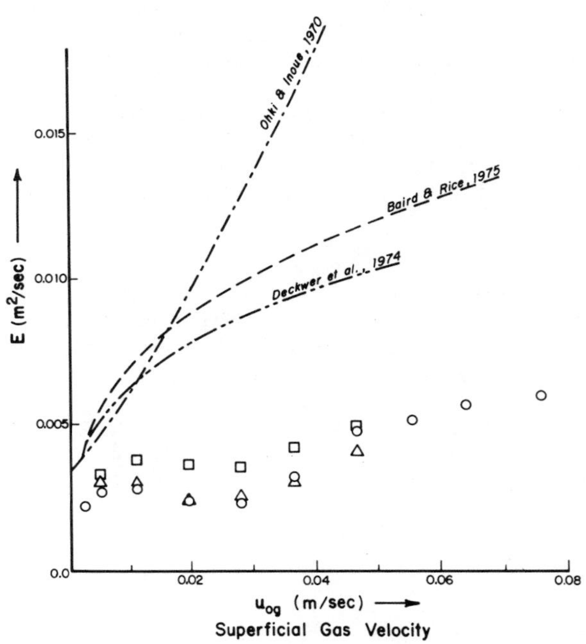

Figure 6. Comparative dispersion coefficients for 9.5-cm diameter column: (□) PE sparger, (○) R1 sparger, (△) R2 sparger

Sharma (20) who deduct the energy dissipated in bubble wakes to obtain

$$(28)\ P_m = [g\ (u_{og} - \varepsilon\ v_s)]$$

Using the definition of slip velocity given by equation (1), it is seen that the Field-Davidson model requires P_m to be identically zero, while the Joshi-Sharma modification suggests the unrealistic result that P_m is negative if the slip velocity definition is obeyed.

Presumably, when very uniform bubbly flow is maintained, the Field-Davidson (19) modification suggests that mixing owing to energy dissipation is expected to be quite small, and this was indeed the observation in the present work. Bubble emulsions were quite uniform, especially when using the rubber sheet sparger. Thus, for very uniform emulsions, mixing occurs mainly by entrainment in wakes (deadwater transfer) and by Taylor dispersion arising from the liquid velocity profile. If the Taylor mechanism is controlling, then one expects

$$(28)\ E \sim d\ v\ 3.57\ \sqrt{f}$$

where Taylor has taken the radial eddy diffusion equal to the eddy viscosity and f is the Fanning friction factor. For the present case, the average interstitial velocity varies but little and can be taken as $v \sim .4$ cm/sec, $d \sim 9.5$ cm and $f \sim .04$, from which the effective dispersion coefficient is estimated to be $E \sim 2.8$ cm²/sec. This is much lower than the observed values of order 25 cm²/sec. Joshi (21) has suggested a method to calculate the liquid circulation arising from wake entrainment to give:

$$(29)\ V_c = \frac{\varepsilon}{1-\varepsilon}\ \alpha\ U_\infty$$

where α is the ratio of wake to bubble volume and takes a value around 11/16. For the present work U_∞ varies between 24 and 33 cm/sec (see Figure 6) so that $v_c \sim 2.1$. Taking eddy dispersion to be the product of mixing length and turbulence velocity, and assuming the turbulence velocity is approximated by the circulation arising from wake entrainment (v_c), one obtains

$$(30)\ E \sim V_c \cdot \ell$$

If we take $\ell \sim d$ and $v_c \sim 2.1$ cm/sec as before, then equation (30) gives $E \sim 20$ cm²/sec, which is very close to the experimental observations for small superficial gas velocities. We note in passing that the entrainment model suggests $E \propto u_{og}$ since $U_\infty \sim u_{og}/\varepsilon$.

Conclusions

Dispersions coefficients were found to lie somewhere in between predictions based, on the low side, the entrainment model, and on the high side the isotropic turbulence model (energy dissipation). From the previous discussion, one is tempted to propose a two-regime model comprised of entrainment and energy dissipation mechanisms:

Entrainment: $E = d\ (\frac{\varepsilon}{1-\varepsilon}\ \alpha\ U_\infty)$

Energy dissipation: $E = K\ d^{4/3}\ P_m^{\ 1/3}$

The transition between the two mdoels is not yet clear, however it would seem departure from the drift-flux curve may indicate this transition. It remains to be determined the precise form that P_m should take, but for engineering estimates, P_m can be taken $\sim u_{og}g$. Combining the two models gives:

$$(31)\quad E = K_1\ (u_{og}d) + K_2\ (u_{og} \cdot g)^{1/3}\ d^{4/3}$$

where K_1, K_2 are dimensionless constants.

Finally, we conclude that the type of bubble generating device has a very significant effect on the flow regime, especially for a low superficial gas velocity less than 5 cm/sec. Apparently, to minimize mixing without using intervals such as baffles, one must use a sparger which produces uniform emulsions at the source. In this respect, the rubber sheet sparger is quite suitable, and moreover, this device is somewhat self-regulating since sparger holes expand as flow increases. Apparently, this work constitutes the first effort to study a thin, flexible perforated rubber sheet as a gas sparger.

Nomenclature

A	=	column cross-sectional area
C	=	tracer composition
d	=	column diameter
E	=	eddy dispersion coefficient
f	=	Fanning friction factor
j_D	=	drift-flux (eqn 2)
$K_k(s)$	=	kth weighted cumulant
L	=	column length
ℓ	=	liquid mixing length
$M_k(s)$	=	kth weighted moment
N	=	number of bubbles
Pe	=	liquid Peclet number (vL/E)
P_m	=	rate of energy dissipation per unit mass of liquid
s	=	Laplace operator, or weighting variable.

t = time
U_∞ = terminal rise velocity
u_{og} = superficial gas velocity
u_{ol} = superficial liquid velocity
v = interstitial liquid velocity
v_c = interstitial liquid circulation velocity
v_s = slip or relative velocity
z = axial position from injection point

Greek

α = ratio wake to bubble volume
ε = gas voidage
ζ = dimensionless axial coordinate (z/L)
τ = liquid residence time (L/v)
$\phi(\varepsilon)$ = slip function (eqn 3)
ρ_L = liquid density

Literature Cited

(1) Baird, M.H.I. and R.G. Rice, Chem. Eng. Journ. 1975, 9, 171-174.
(2) Rice, R.G., Oliver, A.D., Newman, J.P. and R.J. Wiles, Powder Tech. 1974, 10, 201-210.
(3) Wallis, G.B. "One Dimensional Two-Phase Flow" McGraw-Hill, New York (1969), Chap. 9.
(4) Bach, H.F. and T. Pilhofer, Ger. Chem. Eng. 1978, 1, 270-275.
(5) Anderson, J.L. and J.A. Quinn, Chem. Eng. Sci. 1970, 25, 373-380.
(6) Lockett, M.J. and R.D. Kirkpatrick, Trans. Inst. Chem. Eng. 1975, 53, 267-273.
(7) König, B., Bucholz, R., Lücke, J. and K. Schügerl, Ger. Chem. Eng. 1978, 1, 199-205.
(8) Otake, T., Tone, S., Nakao, K. and Y. Mutsuhashi, Chem. Eng. Sci. 1977, 32, 377-383.
(9) Lapidus, L. and J.C. Elgin, AIChE Journ. 1957, 3, 63-68.
(10) Towell, G.D., Strand, C.P. and G.H. Ackerman, AIChE - Inst. Chem. Engrs. Joint Mtg. Series No. 20, p. 97-105, 1968.
(11) Reith, T., Renken, S. and B.A. Israel, Chem. Eng. Sci. 1968, 23, 619-629.
(12) Levenspiel, O., "Chemical Reaction Engineering" 2nd ed., Wiley and Sons, New York, 1972.
(13) Ohki, Y. and H. Inoue, Chem. Eng. Sci. 1970, 25, 1-16.
(14) Smith, E.L., Fidgett, M. and J.S. Salek, BHRA Fluid Engineering, 2nd European Conf. on Mixing, Mar. 30-Apr. 1, paper G2, 1977.
(15) Seher, A. and V. Schumacher, Ger. Chem. Eng. 1979, 2, 117-122.

(16) Deckwer, W.D., Burckhart, R. and G. Zoll, Chem. Eng. Sci. 1974, 29, 2177-2188.
(17) Wen, C.Y. and L.T. Fan, "Models for Flow Systems and Chemical Reactors," Marcel Dekker, Inc., N.Y., 1975.
(18) Tuppurainen, J.M.I., "Liquid Holdup and Axial Dispersion in an Unbaffled bubble Column," B. Eng. Thesis, University of Queensland (Australia), 1979.
(19) Field, R.W. and J.F. Davidson, Trans. I. Chem. E. 1980, 58, 228.
(20) Joshi, J.B. and M.M. Sharma, Trans. I. Chem. E. 1979, 57, 244.
(21) Joshi, J.B., Trans. I. Chem. E., (in press, 1981).

APPENDIX I

Building on the Laplace solution given by equation (14), the moments method proceeds as follows; the kth weighted moment is defined:

$$\text{(A1)} \quad M_k(s) = \int_o^\infty e^{-st} t^k \, C(t,L)dt$$

We note that the $zero^{th}$ weighted moment is exactly the Laplace solution (equation (17)):

$$\text{(A2)} \quad M_o(s) = \int_o^\infty e^{-st} C(t,L)dt = \bar{C}(s,L)$$

By successive differentiation, it is easy to prove the recurrence relation

$$\text{(A3)} \quad \frac{dM_k(s)}{ds} = -M_{k+1}(s)$$

Applying (A3) to (A2), using equation (17), gives

$$\text{(A4)} \quad \frac{M_1(s)}{M_o(s)} = \frac{\tau}{\sqrt{1+4s\tau/Pe}}$$

This does not allow the separation of the two parameters v and E. However, this separation can be made by defining weighted cumulants as follows; define the $zero^{th}$ cumulant as $K_o(s) = \ln M_o(s)$ and noting as before the recurrence relation, there results:

$$\text{(A5)} \quad Pe = \frac{2}{\tau} K_1{}^3/K_2$$

$$\text{(A6)} \quad \tau = K_1/\sqrt{1-2s\,K_2/K_1}$$

where the cumulants are determined experimentally from impulse response curves using (A1) and (A2) along with the definitions:

(A7) $K_1 = M_1/M_o$

(A8) $K_2 = (\frac{M_2}{M_1} - M_1^2/M_o^2)$.

RECEIVED June 3, 1981.

PACKED BED REACTORS

14

Packed Bed Reactors

An Overview

ARVIND VARMA

Department of Chemical Engineering, University of Notre Dame, Notre Dame, IN 46556

Packed-bed reactors are discussed qualitatively, particularly with respect to their models. Features of the two basic types of models, the pseudohomogeneous and the heterogeneous models, are outlined. Additional issues -- such as catalyst deactivation; steady state multiplicity, stability, and complex transients; and parametric sensitivity -- which assume importance in specific reaction systems are also briefly discussed.

Packed-bed reactors are commonly used in industrial practice for conducting solid-catalyzed reactions. Most often, they physically consist of tube-bundles, which are packed with pellets on which the active catalyst is deposited. The reactants enter at one end of the tubes, and the reaction products are withdrawn from the other end. The reaction(s) proceed over the length of the tube, and so the species concentrations, as well as the fluid and solid temperatures, vary as a function of position within the tube. The tube bundles are stacked in a shell, and because most industrial reactions are exothermic, cooling medium flows in the shell to maintain a desired temperature distribution over the tube length.

At a fixed concentration of reacting species and temperature, the rate of solid-catalyzed reactions is directly proportional to the active catalyst surface area. The pellets are normally a means to support the catalytically active metal or metal oxide, and maintain it in dispersed form -- thus with a high surface area. Catalyst preparation is frequently described as an art, with doses of serendipity; there is, of course, more to it than that - as Satterfield (1) has recently described.

Some prominent industrial examples of packed-bed reactors are in ammonia, methanol or vinyl acetate synthesis, and in ethylene, methanol, naphthalene, xylene or SO_2 oxidation. In recent years (since the 1975 model year), an important application of packed-bed reactors has been as catalytic converters for pollution control from automotive exhausts.

0097-6156/81/0168-0279$05.00/0

Transport Processes

A variety of gradients in species concentrations and temperature exist within a packed-bed reactor. Since reaction(s) proceed along the tube length, there are obvious gradients in concentration, and fluid and solid temperatures in the axial direction. Because of heat transfer at the tube wall between the reacting mixture and the cooling medium, radial gradients in temperature and species concentrations also exist. At any location within the tube, there are concentration and temperature gradients between the fluid and solid phases. Finally, there are species concentration (but negligible temperature) gradients within each of the individual catalyst pellets, if the active catalyst is distributed throughout the pellet.

A variety of transport processes therefore occur in a packed-bed reactor, simultaneously with chemical reaction(s). Accurate modeling of these processes is essential to predict reactor performance.

Packed-Bed Reactor Models

A relatively large number of models can be written down for a packed-bed reactor, depending on what is accounted for in the model. These models, however, basically fall into two categories: pseudohomogeneous models and heterogeneous models. The various models are described in standard reaction engineering texts -- such as those of Carberry (2), Froment and Bischoff (3), and Smith (4), to cite just a few -- and in review articles (cf., 5-8), and so details of their equations will not be reported here. We will, instead, only make some qualitative remarks about the models.

Pseudohomogeneous Models. The basic assumption that is made in a pseudohomogeneous model is that the reactor can be described as an entity consisting only of a single phase. Since, in reality, two phases are present, the properties used in describing the reactor are so-called "effective" properties which respect the presence of two phases. A comprehensive review of estimating these effective properties has recently been published (9).

The simplest pseudohomogeneous model is the "plug-flow" model, in which the fluid is taken to move as a plug through the reactor tube, and the reaction rate - which depends on local species concentration and temperature - is described as rate of species generation or consumption per unit reactor volume. In the steady state, the model equations are a set of coupled first-order ordinary differential equations - one each for every independent reaction, and one for temperature - with prescribed initial conditions describing the fluid composition and temperature at the reactor inlet. These equations are, in general, nonlinear but can be readily and efficiently integrated numerically with modern-day digital computers to provide concentration and temperature profiles as a function of axial distance from the reactor inlet.

The plug-flow model can be augmented by including axial and/or radial dispersions, for both mass and heat transport. These dispersions are characterized by so-called Peclet numbers. It is generally agreed that axial dispersion of mass is not significant if the tube length/pellet diameter ratio is $\geq$ 50, while that for heat is also negligible if the same ratio is $\geq$ 300.

Radial dispersion, on the other hand, is generally more important than axial dispersion, since the ratio of tube/pellet diameters is frequently quite modest -- as compared with the tube length/pellet diameter ratio. The radial Peclet number for mass transport (ud_p/D_{mr}) is approximately 10, while for heat transport (ud_p/D_{hr}) it lies between 5-10 (2). Radial dispersion becomes negligible if the reactor is adiabatic, because there is then no driving force for long-range gradients to exist in the radial direction.

For non-adiabatic reactors, along with radial dispersion, heat transfer coefficient at the wall between the reaction mixture and the cooling medium needs to be specified. Correlations for these are available (cf. 9, 10); however, it is possible to modify the effective radial thermal conductivity (k_r), by making it a function of radial position, so that heat transfer at the wall is accounted for by a smaller k_r value near the tube-wall than at the tube center (11).

Inclusion of axial dispersion in the plug-flow model makes the model equations a boundary-value problem, so that conditions at both the reactor inlet and outlet need to be specified. The commonly used boundary conditions are the so-called Danckwerts type (12), although their origin goes back to Langmuir (13). When radial dispersion is included, even the steady state equations are partial differential equations -- in the axial and radial space variables. The dispersion model equations can be numerically solved by finite-difference schemes, or more efficiently, by orthogonal collocation methods (14, 15).

The basic plug-flow model, with or without dispersions, is a "continuous" model because the concentrations and temperature are described by differential equations. An alternative representation is by a discrete model - the so-called "cell" model (16, 17), in which it is assumed that the reactor can be broken down into several connected cells. It had long been assumed that the continuous and discrete models are equivalent ways of representing a reactor; however, this assumption has recently been questioned in two different contexts (18, 19).

Heterogeneous Models. The two-phase character of a packed-bed is preserved in a heterogeneous model. Thus mass and energy conservation equations are written separately for the fluid and solid phases. These equations are linked together by mass and heat transport between the phases.

The simplest heterogeneous model is one with plug-flow in the fluid phase, mass and heat transfer between the fluid and solid phases, and surface catalytic reaction on the solid -- if the catalyst is indeed deposited near the pellet external surface. More complex fluid phase behavior can be accommodated by axial and radial dispersion features, among which radial dispersion ones are again the more important -- and those only for a non-adiabatic reactor.

If the catalyst is dispersed throughout the pellet, then internal diffusion of the species within the pores of the pellet, along with simultaneous reaction(s) must be accounted for if the prevailing Thiele modulus $\gtrsim 1$. This aspect gives rise to the "effectiveness factor" problem, to which a significant amount of effort, summarized by Aris (20), has been devoted in the literature. It is important to realize that if the catalyst pellet effectiveness factor is different from unity, then the packed-bed reactor model must be a heterogeneous model; it cannot be a pseudohomogeneous model.

There are theoretically sound correlations available for estimating effective diffusion coefficients in porous catalyst pellets (cf., 21, 22). It has been shown that for most gas-solid catalytic reactions, the pellets are virtually isothermal, so that temperature gradients within them can safely be ignored (23, 24).

There are correlations available for estimating heat and mass transfer coefficients between the phases (2, 3, 4); they are generally cast in form of j-factors, as functions of the fluid Reynolds number. Caution must, however, be exercised in using these since most of the correlations were developed for non-reactive systems -- although successful attempts have been made for relatively simple reactive cases (25). In a specific experimental study in a packed-bed reactor (26), it was recently shown that because of increased convection between the catalyst pellet and the bulk gas, caused by relatively large temperature differences between the two phases when a highly exothermic reaction occurs, the transport coefficients increase considerably -- although their power dependence on Reynolds number, which arises from boundary layer arguments, remains the same as in cases without reaction.

Along with wall heat transfer coefficient in non-adiabatic reactors, another effect frequently added in models is that of thermal conduction in the solid phase (27, 28, 29). One should be particularly careful here, since most of the correlations available in the literature (9, 30, 31, 32) are for effective transport parameters to be used with pseudohomogeneous models, and not for the solid phase alone.

Intrinsic Reaction Kinetics

Either with pseudohomogeneous or with heterogeneous models,

the reaction rate term must always be included in the reactor model. This takes the form of specifying the rate of reaction, as a function of species concentration and catalyst temperature; this information is always obtained experimentally for each reaction system in kinetics experiments. It is crucial that when reaction kinetics are measured, that there be no transport effects present; otherwise the kinetic data would be influenced by such effects. For gas-solid reactions, the most commonly used reactors are the spinning-basket and the recycle reactors (2). Weekman (33), and Doraiswamy and Tajbl (34) have provided recent reviews summarizing advantages and limitations of various reactors used in the laboratory for procurement of intrinsic kinetic data.

Some Other Issues

Under this heading, some issues which assume importance in specific reaction systems, are briefly outlined.

Catalyst Deactivation. Most catalysts suffer from decay in their activity with time, which arises as a consequence, in general, of one among three causes. In "thermal sintering", purely as a result of high temperature, nature of the reactive atmosphere and of the support, smaller crystallites of the active catalyst grow into larger ones with time via various agglomeration processes (35). Thus the active surface area decreases, resulting in a loss of catalytic activity per unit weight of the catalyst. The second cause is "chemical poisoning", normally the result of chemisorption of reactants, reaction products, or impurities in the feedstream, whereby such species permanently occupy sites otherwise available for catalysis. Finally, "fouling" is a term commonly used for physical adsorption of a species upon the catalytic surface, thereby covering or blocking it from future catalytic action -- such as in overcracking of hydrocarbons to produce coke ("coking"), or in lead poisoning of noble metal catalysts in catalytic converters for automotive exhausts. A thorough review of catalyst deactivation is available (36).

With deactivation, the reactor model must immediately become a transient one, to account for change in catalyst activity with time. Among others, two successful instances of packed-bed reactor modeling, in the presence of catalyst deactivation and including comparisons with experiments, are found in the works of Weekman (37, 38) and Butt (39, 40).

Steady State Multiplicity, Stability, and Complex Transients. This subject is too large to do any real justice here. Ever since the pioneering works of Liljenroth (41), van Heerden (42), and Amundson (43) with continuous-flow stirred tank reactors, showing that multiple steady states -- among them, some stable to perturbations, while others unstable -- can arise, this topic has become one of the major ones in reaction engineering. Over the years, virtually all types of reactors have shown these features, in both experimental and modeling studies. These features arise either as a consequence of interactions between reaction and transport processes, or purely as a consequence of complex reaction kinetics; examples in the former category presently far outnumber those in the latter. An authoritative survey of the area was given in 1974 by Schmitz (44); more recent reviews are also available (45, 46, 47). These aspects in the diffusion-reaction context were treated comprehensively by Aris (20), and by Luss (48). For stirred tanks and empty tubular reactors, Varma and Aris (49) may be consulted.

Jensen and Ray (50) have recently tabulated some 25 experimental studies which have demonstrated steady state multiplicity and instabilities in fixed-bed reactors; many of these (cf., 29, 51, 52) have noted the importance of using a heterogeneous model in matching experimental results with theoretical predictions. Using a pseudohomogeneous model, Jensen and Ray (50) also present a detailed classification of steady state and dynamic behavior (including bifurcation to periodic solutions) that is possible in tubular reactors.

A feature related to steady state multiplicity and stability is that of "pattern formation", which has its origins in the biological literature. Considering an assemblage of cells containing one catalyst pellet each, Schmitz (47, 53) has shown how non-uniform steady states - giving rise to a pattern - can arise, if communication between the pellets is sufficiently small. This possibility has obvious implications to packed-bed reactors.

Parametric Sensitivity. One last feature of packed-bed reactors that is perhaps worth mentioning is the so-called "parametric sensitivity" problem. For exothermic gas-solid reactions occurring in non-adiabatic packed-bed reactors, the temperature profile in some cases exhibits extreme sensitivity to the operational conditions. For example, a relatively small increase in the feed temperature, reactant concentration in the feed, or the coolant temperature can cause the hot-spot temperature to increase enormously (cf. 54). This sensitivity is a type of instability, which is important to understand for reactor design and operation. The problem was first studied by Bilous and Amundson (55). Various authors (cf. 56, 57) have attempted to provide estimates of the heat of reaction and heat transfer parameters defining the parametrically sensitive region; for the plug-flow pseudohomogeneous model, critical values of these parameters can now be obtained for any reaction order rather easily (58).

A related phenomenon is the "wrong-way behavior" of packed-bed reactors, where a sudden reduction in the feed temperature leads to a transient temperature rise. This has been observed (52, 59) and satisfactorily analyzed using a plug-flow pseudohomogeneous model (60).

Acknowledgements

We are grateful for financial support by the National Science Foundation under Grant No. INT-7920843, and by the Nalco Foundation.

Literature Cited

1. Satterfield, C. N., "Heterogeneous Catalysis in Practice," McGraw-Hill, New York, 1980.
2. Carberry, J. J., "Chemical and Catalytic Reaction Engineering," McGraw-Hill, New York, 1976.
3. Froment, G. F.; Bischoff, K. B., "Chemical Reactor Analysis and Design," John Wiley, New York, 1979.
4. Smith, J. M., "Chemical Engineering Kinetics," Third Edition, McGraw-Hill, New York, 1980.
5. Amundson, N. R. Ber. Bunsen-Gesellschaft für Phys. Chemie 1970, 74, 90.
6. Froment, G. F. Adv. Chem. 1972, 109, 1.
7. Karanth, N. G.; Hughes, R. Catal. Rev.-Sci. Eng. 1974, 9, 169.
8. Hlaváček, V.; Votruba, J., Chapter 6 in "Chemical Reactor Theory - A Review," L. Lapidus and N. R. Amundson (Editors), Prentice-Hall, Englewood Cliffs, New Jersey, 1977.
9. Kulkarni, B. D.; Doraiswamy, L. K. Catal. Rev.-Sci. Eng. 1980, 22, 325.
10. Li, C.-H.; Finlayson, B. A. Chem. Eng. Sci. 1977, 32, 1055.
11. Ahmed, M.; Fahien, R. W. Chem. Eng. Sci. 1980, 35, 889.
12. Danckwerts, P. V. Chem. Eng. Sci. 1953, 2, 1.
13. Langmuir, I. J. Amer. Chem. Soc. 1908, 30, 1742.
14. Finlayson, B. A., "The Method of Weighted Residuals and Variational Principles," Academic Press, New York, 1972.
15. Villadsen, J.; Michelsen, M. L., "Solution of Differential Equation Models by Polynomial Approximation," Prentice-Hall, Englewood Cliffs, New Jersey, 1978.
16. Deans, H. A.; Lapidus, L. AIChE Jl. 1960, 6, 656.
17. Coste, J.; Rudd, D.; Amundson, N. R. Canad. Jl. Chem. Eng. 1961, 39, 149.
18. Sundaresan, S.; Amundson, N. R.; Aris, R. AIChE Jl. 1980, 26, 529.
19. Varma, A. Ind. Eng. Chem. Fundls. 1980, 19, 316.
20. Aris, R., "The Mathematical Theory of Diffusion and Reaction in Permeable Catalysts," Volumes I and II, Clarendon Press, Oxford, England (1975).
21. Feng, C.; Stewart, W. E. Ind. Eng. Chem. Fundls. 1973, 12, 143.
22. Luss, D., Survey paper on "Interactions between Transport Phenomena and Chemical Rate Processes," at ISCRE4, Heidelberg, Germany; DECHEMA, Frankfurt (1976), 487.
23. Carberry, J. J. Ind. Eng. Chem. Fundls. 1975, 14, 129.
24. Pereira, C. J.; Wang, J. B.; Varma, A. AIChE Jl. 1979, 25, 1036.

25. Sørensen, J. P.; Stewart, W. E. Chem. Eng. Sci. 1974, 29, 833.
26. Paspek, S. C.; Varma, A. Chem. Eng. Sci. 1980, 35, 33.
27. Eigenberger, G. Chem. Eng. Sci. 1972, 27, 1909.
28. Rhee, H.-K.; Foley, D.; Amundson, N.R. Chem. Eng. Sci. 1973, 28, 607.
29. Paspek, S. C.; Varma, A. AIChE Jl. (in press).
30. Yagi, S.; Kunii, D. AIChE Jl. 1957, 3, 373.
31. Kunii, D; Smith, J. M. AIChE Jl. 1960, 6, 71.
32. DeWasch, A.P.; Froment, G. F. Chem. Eng. Sci. 1971, 26, 629.
33. Weekman, V. W., Jr. AIChE Jl. 1974, 20, 833.
34. Doraiswamy, L. K.; Tajbl, D. G. Catal. Rev.-Sci. Eng. 1974, 10, 177.
35. Wanke, S. E; Flynn, P. C. Catal. Rev.-Sci. Eng. 1975, 12, 93.
36. Butt, J. B., Adv. Chem. 1972, 109, 259.
37. Weekman, V. W., Jr. Ind. Eng. Chem. Proc. Des. Dev. 1968, 7, 90.
38. Weekman, V. W., Jr.; Nace, D. M. AIChE Jl. 1970, 16, 397.
39. Weng, H. S.; Eigenberger, G.; Butt, J. B. Chem. Eng. Sci. 1975, 30, 1341.
40. Price, T. H.; Butt, J. B. Chem. Eng. Sci. 1977, 32, 393.
41. Liljenroth, F. G. Chem. Met. Eng. 1918, 19, 287.
42. van Heerden, C. Ind. Eng. Chem. 1953, 45, 1242.
43. Bilous, O.; Amundson, N. R. AIChE Jl. 1955, 1, 513.
44. Schmitz, R. A. Adv. Chem. 1975, 148, 156.
45. Gilles, E. D., Survey paper on "Reactor Models" at ISCRE4, Heidelberg, Germany; DECHEMA, Frankfurt (1976), 459.
46. Ray, W. H., in "Applications of Bifurcation Theory," Academic Press, 1977, 285.
47. Schmitz, R. A. Proc. JACC 1978, Vol. II, 21.
48. Luss, D., Chapter 4 in "Chemical Reactor Theory - A Review," L. Lapidus and N. R. Amundson (Editors), Prentice-Hall, Englewood Cliffs, New Jersey, 1977.
49. Varma, A; Aris, R., Chapter 2 in "Chemical Reactor Theory - A Review," L. Lapidus and N. R. Amundson (Editors), Prentice-Hall, Englewood Cliffs, New Jersey, 1977.
50. Jensen, K. F.; Ray, W. H., Paper presented at the AIChE Annual Meeting, Chicago, November 1980.
51. Hegedus, L. L.; Oh, S. H.; Baron, K. AIChE Jl. 1977, 23, 632.
52. Sharma, C. S.; Hughes, R. Chem. Eng. Sci. 1979, 34, 625.
53. Schmitz, R. A.; Tsotsis, T. T., Paper presented at the AIChE Annual Meeting, San Francisco, November 1979.
54. Emig, G.; Hofmann, H.; Hoffman, U.; Fiand, U. Chem. Eng. Sci. 1980, 35, 249.
55. Bilous, O.; Amundson, N. R. AIChE Jl. 1956, 2, 117.
56. Barkelew, C. H. Chem. Eng. Prog. Symp. Ser. 1959, 25 (55), 37.
57. Van Welsenaere, R. J.; Froment, G. F. Chem. Eng. Sci. 1970, 25, 1503.
58. Morbidelli, M.; Varma, A. AIChE Jl. (in press).
59. Van Doesburg, H.; DeJong, W. A. Chem. Eng. Sci. 1976, 31, 45.
60. Mehta, P. S.; Sams, W. N.; Luss, D. AIChE Jl. 1981, 27, 234.

RECEIVED June 3, 1981.

15

Solution of Packed Bed Heat-Exchanger Models by Orthogonal Collocation Using Piecewise Cubic Hermite Functions

A. G. DIXON[1]

Mathematics Research Center, 610 Walnut Street, Madison, WI 53706

An orthogonal collocation method for elliptic partial differential equations is presented and used to solve the equations resulting from a two-phase two-dimensional description of a packed bed. Comparisons are made between the computational results and experimental results obtained from earlier work. Some qualitative discrimination between rival correlations for the two-phase model parameters is possible on the basis of these comparisons. The validity of the numerical method is shown by applying it to a one-phase packed-bed model for which an analytical solution is available; problems arising from a discontinuity in the wall boundary condition and from the semi-infinite domain of the differential operator are discussed.

The choice of a model to describe heat transfer in packed beds is one which has often been dictated by the requirement that the resulting model equations should be relatively easy to solve for the bed temperature profile. This consideration has led to the widespread use of the pseudo-homogeneous two-dimensional model, in which the tubular bed is modelled as though it consisted of one phase only. This phase is assumed to move in plug-flow, with superimposed axial and radial effective thermal conductivities, which are usually taken to be independent of the axial and radial spatial coordinates. In non-adiabatic beds, heat transfer from the wall is governed by an apparent wall heat transfer coefficient.

The earliest heat-transfer studies neglected the effective axial conduction term as this was expected to be negligible by comparison with the bulk-flow term in the long beds typically used in industry. Axial dispersion was also neglected in mixing studies, and experiments by Hiby (1) confirmed the absence of axial

[1]Current address: Department of Chemical Engineering, Worcester Polytechnic Institute, Worcester, MA 01609

0097-6156/81/0168-0287$05.00/0

back-mixing. More recently it has been shown (2-4) that measurements of temperature profiles in non-reacting systems in laboratory packed bed heat exchangers can yield statistically meaningful heat transfer parameter estimates only if the measurements are made at relatively short bed depths, where significant axial and radial temperature gradients are present. The omission of axial conduction at such bed depths leads to systematic errors in the predicted temperature profiles, which cause the model to be statistically rejected when it is fitted to data taken at several bed depths. If the model is fitted depth-by-depth, the parameter estimates are found to have a depth-dependence, as noticed by De Wasch and Froment (5). In this case, they must be regarded as length-averaged values rather than point values. Li and Finlayson (6) argue that constant asymptotic values should be used, as obtained from data taken at long bed depths, although this would give badly-determined estimates.

When a chemical reaction is present, implying larger temperature gradients, Young and Finlayson (7) have shown that an effective axial dispersion term should be included, and Mears (8) has given criteria for the neglect of axial dispersion which show that increasing fluid velocity reduces axial effects. This is to be expected, since conduction through the solid, a static effect, is believed to be the major contributor to axial effects.

The disadvantage of including axial dispersion is that an exit boundary condition must be specified, and in cases where an analytical solution is not available, a numerical boundary-value problem must be solved in the axial direction, rather than an initial-value problem.

For steady-state heat transfer an elliptic partial differential equation is the result of using the one-phase model. Previous studies (7,9) have used the orthogonal collocation method due to Villadsen and Stewart (10) to determine the coefficients of trial-function expansions in both spatial coordinates. This method works well when the temperature gradients are moderate and few collocation points are required. For steep profiles, however, such as may be encountered at a "hot-spot" in the reactor, many collocation points may be required, especially as the generation of these points as roots of polynomials does not allow them to be placed in the region of interest. Such a collocation scheme is a global one, resulting in a collocation matrix which is large and not usefully sparse, so that the solution of the resulting algebraic equations may become costly.

The answer to this difficulty lies in the use of piecewise approximants, such as cubic splines, which are in general use in the mathematics literature (11). Carey and Finlayson (12) have introduced a finite-element collocation method along these lines, which uses polynomial approximants on sub-intervals of the domain, and apply continuity conditions at the break-points to smooth the solution. It would seem more straight-forward, however, to use piecewise polynomials which do not require explicit continuity equations; in this paper the use of piecewise cubic Hermite functions is considered, as described by Prenter and Russell (13).

The advantages of piecewise polynomials are firstly that the subintervals may be clustered in regions of interest, so that an improved approximation may be obtained where gradients are steep, and secondly that the collocation matrix is banded, allowing advantage to be taken of this special structure, both in work required for decomposition and in computer storage used. This second advantage was clearly demonstrated in a preliminary study on the one-phase model (14), where the cubic Hermite function method was shown to give an order-of-magnitude improvement in exit temperature profile over the polynomial collocation method even when the subintervals were chosen to be of equal length. A more extensive investigation using the one-phase model as a test case is described in the present work.

The use of two-phase homogeneous continuum models in packed bed modelling has often been avoided due to the computational difficulties. Recently, Paspek and Varma (15) have found a two-phase model to be necessary to describe an adiabatic fixed-bed reactor, while Dixon and Cresswell (16) have shown that the effective parameters of the one-phase model may be interpreted in terms of the more fundamental parameters of a two-phase model, thus demonstrating more clearly their qualitative dependencies on the operating and design characteristics of the bed. When two phases and several species are involved, the computational advantages of the cubic Hermite method may be anticipated to be high.

In this paper the coupled elliptic partial differential equations arising from a two-phase homogeneous continuum model of heat transfer in a packed bed are solved, and some attempt is made to discriminate between rival correlations for those parameters not yet well-established, by means of a comparison with experimental results from a previous study (3,4).

Collocation using piecewise bicubic Hermite functions

The use of piecewise bicubic Hermite functions in collocation schemes for the solution of elliptic partial differential equations has been described by Prenter (13,17); a short outline is presented here.

Consider partitioning the interval [a,b] into subintervals by $a = \xi_1 < \xi_2 < \ldots < \xi_p < \xi_{p+1} = b$. Then the piecewise cubic Hermite functions are defined for $1 \le i \le p+1$ by

$$\phi_i(\xi) = \begin{cases} -2((\xi-\xi_{i-1})/h_{i-1})^3 + 3((\xi-\xi_{i-1})/h_{i-1})^2 & (\xi_{i-1} \le \xi \le \xi_i) \\ 1 + 2((\xi-\xi_i)/h_i)^3 - 3((\xi-\xi_i)/h_i)^2 & (\xi_i \le \xi \le \xi_{i+1}) \\ 0 & \text{otherwise} \end{cases}$$

$$\psi_i(\xi) = \begin{cases} (\xi-\xi_i)((\xi-\xi_{i-1})/h_{i-1})^2 & (\xi_{i-1}\leq\xi\leq\xi_i) \\ (\xi-\xi_i)((\xi_{i+1}-\xi)/h_i)^2 & (\xi_i\leq\xi\leq\xi_{i+1}) \\ 0 & \text{otherwise} \end{cases}$$

The length of subinterval $[\xi_i,\xi_{i+1}]$ is denoted by h_i. The functions ϕ_1,ψ_1 are restricted to the interval $[\xi_1,\xi_2]$ and ϕ_{p+1}, ψ_{p+1} are restricted to the interval $[\xi_p,\xi_{p+1}]$.

If the domain $[a,b] \times [c,d]$ of an elliptic partial differential operator is partitioned into subrectangles by

$$a = y_1 < y_2 < \ldots < y_n < y_{n+1} = b$$

and

$$c = x_1 < x_2 < \ldots < x_m < x_{m+1} = d ,$$

then the piecewise bicubic Hermite interpolation polynomial to a function $T(y,x)$ is

$$\begin{aligned} T(y,x) &\doteq s_{n,m}(y,x) \\ &= \sum_{i=1}^{n+1} \sum_{j=1}^{m+1} \{T(y_i,x_j)\phi_i(y)\phi_j(x) + \frac{\partial T}{\partial y}(y_i,x_j)\psi_i(y)\phi_j(x) \\ &\quad + \frac{\partial T}{\partial x}(y_i,x_j)\phi_i(y)\psi_j(x) + \frac{\partial^2 T}{\partial y \partial x}(y_i,x_j)\psi_i(y)\psi_j(x)\} . \end{aligned} \quad (1)$$

This expression is used as a trial-function expansion for T in much the same way as the Lagrange interpolation polynomial is in the polynomial collocation method of Villadsen and Stewart (10,18). There are four unknown constants associated with each node, giving a total of $4(n+1)(m+1)$ unknowns in the expansion.

The Gaussian points of subinterval $[\xi_i,\xi_{i+1}]$ are

$$\xi_{i1} = \xi_i + (\sqrt{3}-1)h_i/2\sqrt{3} , \quad \xi_{i2} = \xi_{i+1} - (\sqrt{3}-1)h_i/2\sqrt{3} ;$$

combining the two points for each of $[y_i,y_{i+1}]$ and $[x_j,x_{j+1}]$ gives four Gaussian points for the subrectangle $[y_i,y_{i+1}] \times [x_j,x_{j+1}]$. Collocation at these points for each subrectangle yields a total of 4 nm equations. It should be noted that at any collocation point $(y_{ik},x_{j\ell})$ only sixteen bicubic product functions are non-zero, hence each collocation equation involves only sixteen unknowns.

The remaining $4n + 4m + 4$ equations required to determine the expansion coefficients are supplied by the boundary conditions as follows:

(i) on the lines $y = 0$ and $y = 1$, the boundary conditions given are differentiated with respect to x. Together with the original equations this yields two equations on each boundary, which may be applied at each of the m - 1 internal boundary nodes to obtain a total of 4m - 4 conditions. For example, given $\frac{\partial T}{\partial y} = 0$ at y = 0, then $\frac{\partial^2 T}{\partial y \partial x} =$ 0 at y = 0 also, and hence

$$\left.\begin{aligned} \frac{\partial T}{\partial y}(0,x_j) &= 0 \\ \frac{\partial^2 T}{\partial y \partial x}(0,x_j) &= 0 \end{aligned}\right\} \quad j = 2,3....m$$

(ii) a similar procedure may be followed on the lines x = 0, x = 1, to obtain 4n - 4 conditions.

(iii) at each corner both (i) and (ii) above may be applied, to give four conditions. However only three of these will be independent, and one must be eliminated, except when there is a corner discontinuity, when an arbitrary decision must be made.

The 4n + 4m + 4 conditions derived above may be used either to eliminate unknowns and thus reduce the size of the system of equations to be solved, as was done in the original paper (13), or to generate extra equations. The latter procedure is easier to apply when mixed boundary conditions are used, the resulting increase in computer time used being offset by the saving in programming effort.

A proper numbering of the equations and unknowns ensures that the matrix representing the linear part of the differential operator will be a band matrix with bandwidth proportional to the lesser of m and n. Standard methods for decomposing such matrices exist (19), which allow savings in both storage and time.

The application of the above method is facilitated by the definition of suitable notation and the use of some simple subroutines to produce the cubic Hermite functions and their derivatives. These are described in detail elsewhere (20), the approach used imitates that of Villadsen and Stewart (10).

One-phase continuum model

The packed bed heat exchanger considered here is that used in recent experimental studies (2,4) and shown schematically in Figure 1. The long unheated calming section, (a), and the heated test section, (b), are each considered to be semi-infinite and packed with similar solid particles. There is a step change in wall temperature at the plane z = 0, which is represented by the Heaviside step function in equation (6).

The model equations are well-known, and are, in dimensionless form,

$$\frac{\partial\theta}{\partial x} = \frac{1}{Pe_A}\frac{\partial^2\theta}{\partial x^2} + \frac{1}{Pe_R}\left(\frac{\partial^2\theta}{\partial y^2} + \frac{1}{y}\frac{\partial\theta}{\partial y}\right), \tag{2}$$

$$\frac{\partial\theta}{\partial y} = 0 \quad \text{at} \quad y = 0 \tag{3}$$

$$\theta \to 0 \quad \text{as} \quad x \to -\infty \tag{4}$$

$$\theta \to 1 \quad \text{as} \quad x \to \infty \tag{5}$$

$$\frac{\partial\theta}{\partial y} + (Bi)\theta = (Bi)H(x) \quad \text{at} \quad y = 1 \tag{6}$$

The downstream axial boundary condition, equation 5, was found to be consistent with experimental data in a previous study (4).

The equations (2) - (6) have an easily-determined analytic series solution in terms of Bessel functions (4); this is therefore a good test case upon which to try out the numerical method.

There are two points of interest associated with the numerical solution of equations (2) - (6): (i) the x-domain is infinite and (ii) there is a step-function in the wall boundary condition.

Transformation of the infinite domain. There are several ways of dealing with an infinite domain. Guertin et al. (21) chose perturbation solutions of the model as basis functions. This approach may be difficult to extend to more complicated equations than the non-linear initial-value problems which they considered; even so, the modeller must do a considerable amount of analytical work with this method.

Birnbaum and Lapidus (22) suggest the use of polynomials which are orthogonal over the infinite domain, obtaining these either by using a weighting function such as e^{-x^2}, which gives Hermite polynomials orthogonal over $(-\infty,\infty)$, or by transforming the infinite domain onto a finite domain, and using conventional polynomials such as shifted Legendre polynomials. The drawback to these methods is that there is no control of the placement of the collocation points, some of which are always included in regions where the profile is essentially flat, and are thus wasted.

The use of standard piecewise polynomials, together with an appropriate transformation of the infinite domain, overcomes these difficulties, provided some care is taken in the transformation.

Verhoff and Fisher (23) used the form $t = \frac{1}{\pi}\tan^{-1}(\frac{x}{\alpha})$ in their solution of the Graetz problem with axial conduction. Somewhat neater equations result from taking $t = \tanh(x/\alpha)$ giving

$$\left(\frac{1-t^2}{\alpha}\right)\left(1 + \frac{2t}{\alpha Pe_A}\right)\frac{\partial\theta}{\partial t} = \frac{1}{Pe_R}\left(\frac{\partial^2\theta}{\partial y^2} + \frac{1}{y}\frac{\partial\theta}{\partial y}\right) + \frac{(1-t^2)^2}{\alpha^2 Pe_A}\frac{\partial^2\theta}{\partial t^2} \tag{7}$$

$$\frac{\partial\theta}{\partial y} = 0 \quad \text{at} \quad y = 0 \tag{8}$$

$$\theta = 0 \quad \text{at} \quad t = -1 \tag{9}$$

$$\theta = 1 \quad \text{at} \quad t = 1 \tag{10}$$

$$\frac{\partial\theta}{\partial y} + (Bi)\theta = Bi\ H(t) \quad \text{at} \quad y = 0 \tag{11}$$

Initially it may appear simpler to let $\alpha = 1$, and use $t = \tanh x$. However this results in the region $x > 3$ in the bed being mapped onto a small interval [0.995,1]. This is undesirable for two reasons:

1) If subintervals are required in the bed downstream of $x = 3$, it becomes difficult to place the breakpoints in the t-domain
2) If axial gradients are present downstream of $x = 3$, they will become very sharp in the t-domain, since $\frac{\partial\theta}{\partial t} = \frac{\alpha}{(1-t^2)}\frac{\partial\theta}{\partial x}$ so $\frac{\partial\theta}{\partial t} \to \infty$ as $t \to 1$.

It can be seen that for a fixed choice of $\{t_i\}$, different choices of α will locate the induced partition $\{x_i\}$ in different physical parts of the bed. Thus α must be chosen so that the collocation points in the t-domain are placed in a way that makes physical sense in the x-domain. The appropriate value is found by trial and error; an empirical rule suggested by experience is to take $\alpha \approx Pe_R$.

The effect of an inappropriate value for α is shown in Figure 2. It should be noted that although Verhoff and Fisher used $\alpha = 1$ throughout, their computations were made for sufficiently short beds to ensure good results; larger values of α would be required for temperature profiles further downstream, which would be of interest in the presence of reaction.

Step-change in wall boundary condition. It was anticipated that the discontinuity in wall temperature at $x = 0$, and the resulting steep local gradients, would lead to a locally poor approximation which might have adverse effects further downstream. It was soon found that mesh refinement in the axial direction improved the results considerably over the use of an equally-spaced mesh, whereas mesh refinement in the radial direction had little effect, and a fairly coarse uniform radial mesh was always found to be adequate.

The mesh refinement was carried out by heuristically search-

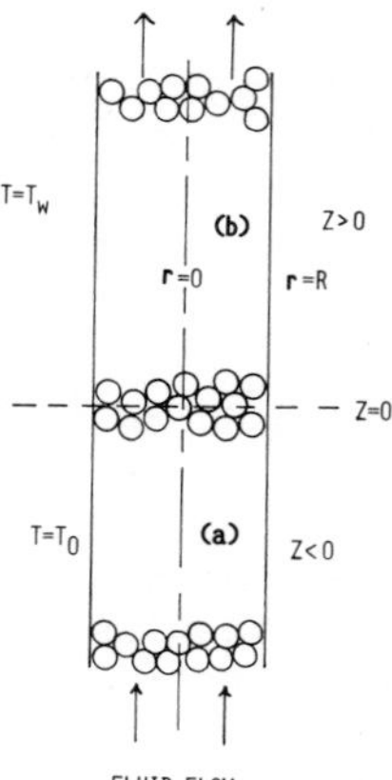

Figure 1. Schematic of packed bed

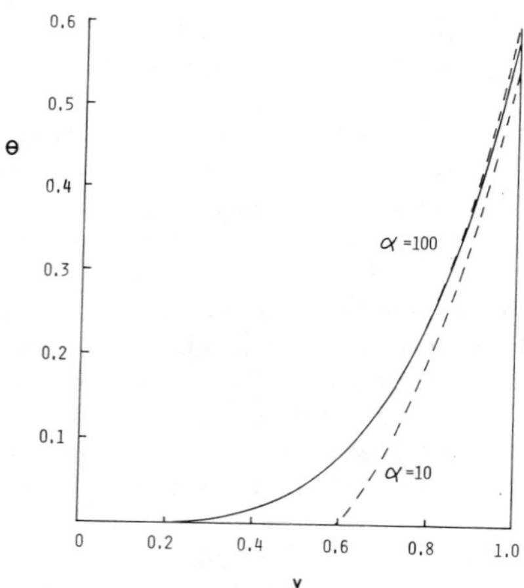

Figure 2. Effect of choice of scaling factor α: x = 4.0, $Pe_A = 4.0$, $Pe_R = 120.0$, $Bi = 5.0$; (——) analytic solution, (– – –) numerical solution

ing for good breakpoint distributions, the final choice (corresponding to m = 10) being $\{t_j\} = \{-0.1, -0.05, -0.025, 0, 0.025, 0.05, 0.1, 0.2, 0.6\}$. Any further refinement led to improved values only at extremely short bed depths.

The effect of the wall temperature discontinuity may also be mitigated by an alternative implementation of the wall boundary condition to that described in the previous section. The two equations at each interior boundary node $(1,t_j)$, $j = 2,3,\ldots,m$, are dropped, together with one equation at each of (1,-1) and (1,1). These 2m equations are replaced by application of the boundary conditions at two points within each subinterval on the line y = 1. The Gaussian points are not necessarily the optimal choice, but were used in the absence of any other guideline.

For the case of a uniform mesh, the Gaussian point implementation was an order-of-magnitude improvement over the breakpoint implementation. When the refined mesh was used, the two methods gave essentially the same results, except at low bed depths near the wall, where the Gaussian point method was slightly better. Consequently it was the method used in the rest of the work.

The reason for the above differences is not clear but appears to lie in the wall temperature specification. No difference between the methods was found when applied to the centre-line condition, and reduction of Bi to lessen effects of the discontinuity also greatly reduced the advantage of the Gaussian point method.

When considering the degree of accuracy to be required from a numerical method, it is necessary to take into account the potential uses of the model equations being solved. It would be inappropriate to require high accuracy in the present study, as bed temperature profiles are seldom accurately measured. Consequently errors of 1°C in comparing numerical and analytical results were considered reasonable.

The comparisons were made at y = 0.1,0.2...1.0 for each of 5 bed depths: x = 1.33,2.67,4.00,5.33,6.67. The parameter ranges covered were $0.25 \leq Pe_A \leq 20.0$, $2.5 \leq Pe_R \leq 120.0$ and $0.5 \leq Bi \leq 8.0$ which are based on a tube-to-particle diameter ratio in the range $5 \leq d_t/d_p \leq 20$, for flow rates corresponding to Reynolds number $Re \geq 50$.

The refined mesh given above was used in the axial direction; the radial mesh used n = 3, $\{y_i\} = \{0.25,0.5,0.75\}$. It was found possible to use the same mesh throughout in the t-domain, due to the freedom to vary the scaling factor α. This avoided heuristic searching for a new mesh for each new set of parameters. Automatic mesh generation was not felt to be worthwhile, in view of the extra costs involved and the relatively underdeveloped state of the art (24).

The 1°C criterion was met in all cases except for $Pe_A = 20$ using this method. For that case the centre-line discrepancy rose to approximately 3°C at lower bed depths. Presumably this error could be eliminated by taking higher order approximations. A typical computation time to produce one set of solutions (i.e.,

5 bed depths) was approximately 1.5 sec on a UNIVAC 1110 computer, a reasonable compromise between cost and accuracy.

Two-phase continuum model

Two-phase continuum models, in which the solid particles and their associated stagnant fillets of fluid are regarded as a continuous pseudo-solid phase, are to be preferred to the more traditional cell model of a particle bathed in fluid, which does not allow conduction from particle to particle. In previous studies, such models have been simplified by considering a one-dimensional model only (25). This study considers the full equations, which are, in dimensionless form:

$$\frac{1}{Pe_{RF}}\left(\frac{\partial^2 T_f}{\partial y^2} + \frac{1}{y}\frac{\partial T_f}{\partial y}\right) + \frac{1}{Pe_{AF}}\frac{\partial^2 T_f}{\partial x^2} - \frac{N_f}{Pe_{RF}}(T_f - T_s) = \frac{\partial T_f}{\partial x} \tag{12}$$

$$\left(\frac{\partial^2 T_s}{\partial y^2} + \frac{1}{y}\frac{\partial T_s}{\partial y}\right) + \frac{\partial^2 T_s}{\partial x^2} + N_s(T_f - T_s) = 0 \tag{13}$$

$$\frac{\partial T_f}{\partial y} = \frac{\partial T_s}{\partial y} = 0 \quad \text{at} \quad y = 0 \tag{14}$$

$$\left.\begin{aligned} \frac{\partial T_f}{\partial y} + (Bi_f)T_f &= (Bi_f)H(x) \\ \frac{\partial T_s}{\partial y} + (Bi_s)T_s &= (Bi_s)H(x) \end{aligned}\right\} \quad \text{at} \quad y = 1 \tag{15}$$

$$T_f, T_s \to 0 \quad \text{as} \quad x \to -\infty \tag{16}$$

$$T_f, T_s \to 1 \quad \text{as} \quad x \to \infty \ . \tag{17}$$

A semi-analytical solution to these equations was derived by Dixon and Cresswell (16), who then matched the fluid phase temperature profile to the one-phase model profile to obtain explicit relations between the parameters of the two models.

The numerical solution to the system of equations (12) - (17) parallels that of the one-phase model almost exactly, with longer computation times due to the increased size of the collocation matrix and its bandwidth. Typical computation times to produce fluid and solid temperature profiles at each of five bed-depths were 3 - 4 seconds.

A discussion of the correlations available for prediction of the parameters required in the two-phase model is presented in (16); those chosen in this study were

$$\frac{1}{Pe_{RF}} = \frac{2d_p}{d_t}\left(\frac{1}{Pe_{rf}(\infty)} + \frac{0.67\varepsilon}{(Re)(Pr)}\right) \tag{18}$$

$$\frac{1}{Pe_{AF}} = \frac{2d_p}{d_t}\left(\frac{1/Pe_{af}(\infty)}{1.0 + 9.7\varepsilon/(RePr)} + \frac{0.73\varepsilon}{(RePr)}\right) \tag{19}$$

$$Bi_f = 0.12\ Pe_{rf}(\infty)\ \frac{d_t}{d_p}\ (Pr)^{-0.67}(Re)^{-0.25} \tag{20}$$

$$N_s = \frac{1.5(1-\varepsilon)(d_t/d_p)^2}{(k_{rs}/k_g)(1/Nu_{fs} + 0.1k_g/k_p)} \tag{21}$$

$$N_f = \frac{2d_p}{d_t}\left(\frac{k_{rs}}{k_g}\right)\frac{N_s Pe_{RF}}{(RePr)} \tag{22}$$

$$Nu_{fs} = \frac{0.255}{\varepsilon}\ (Pr)^{0.33}(Re)^{0.665} \tag{23}$$

As the fluid was air, Pr = 0.72; the bed voidage ε was taken as 0.4. The ratios k_{rs}/k_g and k_p/k_g were related using the formulas of Zehner and Schlünder (27). Preliminary sensitivity tests showed that the parameter Nu_{fs} had very little effect on the profiles; in the presence of reaction this parameter would be more important. The correlations for the fluid-phase Peclet numbers, equations (18) and (19), were obtained by analogy from reported mass transfer work.

It is clear that by adjusting the parameters of this model by nonlinear regression, an excellent fit to the experimental data could be obtained. The value of such a procedure is rather dubious, however, and it is more useful to use the model to obtain qualitative information about the quantities $Pe_{rf}(\infty)$, $Pe_{af}(\infty)$ and Bi_s, which are poorly determined in the literature.

The value for the radial fluid-mixing Peclet number, $Pe_{rf}(\infty)$, is often given in the range 8.0-12.0, as determined by Fahien and Smith (28) and later workers. The radially-averaged Peclet numbers of such workers include mass transfer resistance all the way to the wall, however, whereas the extra resistance to fluid phase transport near the wall is covered in the present model by the use of the parameter Bi_f. Thus the appropriate value for present purposes would be nearer the bed-centre value of $Pe_{rf}(\infty) = 8.0$, also reported by Fahien and Smith (28).

Recent work of Hsiang and Haynes (29) shows the commonly-taken value of $Pe_{af}(\infty) = 2.0$ to be questionable in the (d_t/d_p) range considered here. It was decided to use the relationship $Pe_{af} \approx Pe_a$, which results from the model matching of (16) when Re $\geq$ 20-30. The values of Pe_a were poorly determined from measurements made at bed exit alone (4), so estimates from data which included profiles at z = 0 were used (3). These values led to great improvements in the slopes of the predicted profiles.

Values of Bi_s may lie between the theoretical lower bound derived by Olbrich (26):

$$Bi_s \geq (2.12)(\frac{R}{d_p})$$

and $Bi_s = \infty$, corresponding to no thermal resistance between solid and wall.

If equation (20) is used to predict Bi_f, then it is necessary to take $Bi_s = 1000$ for a good fit, as shown in Figure 3 for a typical case. However, it should be noted that equation (20) underestimates the values of Bi_f(= Bi) found in (4). This is probably due to the unreliable correlation used for Nu_{wf}, as pointed out in (3). If the experimental estimates of Bi are used instead of equation (20), then values of Bi_s in the range 10 - 20 are needed. Some of these results are shown in Figures 4 - 7.

This result indicates that only precise determination of Bi_f will allow any conclusions to be drawn on Bi_s, since these parameters may be mutually varied over fairly large ranges and similar results obtained.

The computed temperature profiles in Figures 4-7 show good general agreement with experiment; some deviation is apparent in the centre of the bed for Re = 73 and 224. Agreement is improved if lower values of $Pe_{rf}(\infty)$ are used, but there is no justification for this in the literature.

Conclusions

The orthogonal collocation method using piecewise cubic Hermite polynomials has been shown to give reasonably accurate solutions at low computing cost to the elliptic partial differential equations resulting from the inclusion of axial conduction in models of heat transfer in packed beds. The method promises to be effective in solving the nonlinear equations arising when chemical reactions are considered, because it allows collocation points to be concentrated where they are most effective.

The fluid-phase temperatures predicted from a two-phase pseudo-homogeneous model were shown to give reasonable agreement with experimental measurements, without explicitly adjusting the model parameters. It was demonstrated that more refined experimental measurements will be needed to determine the parameters of the model; in particular, the solid and fluid phase wall Biot numbers were mutually adjustable.

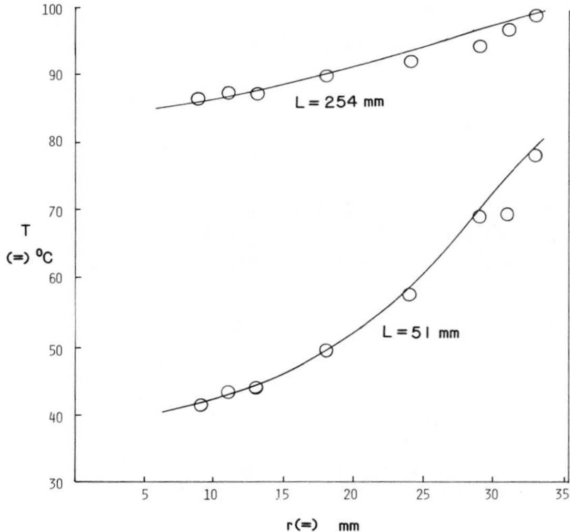

Figure 3. Temperature profiles for 9.5-mm ceramic beads: Re = 120, $Pe_{rf}(\infty)$ = 8.0, $Pe_{af}(\infty)$ = 0.49, Bi_f = 3.35, Bi_s = 1000, (○) experimental points, (——) calculated results

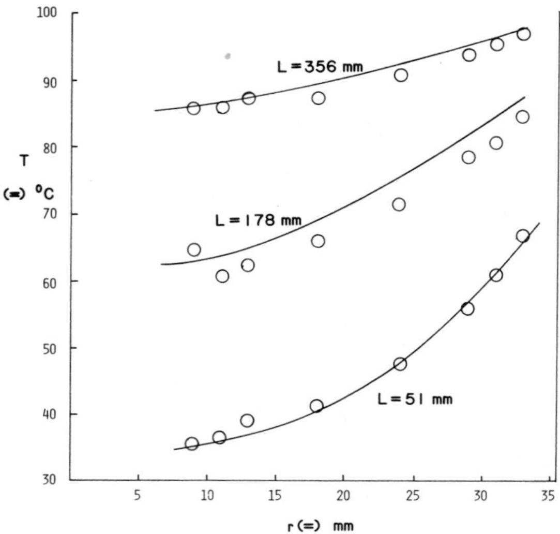

Figure 4. Temperature profiles for 12.7-mm ceramic beads: Re = 430, $Pe_{rf}(\infty)$ = 8.0, $Pe_{af}(\infty)$ = 0.45 Bi_f = 2.85, Bi_s = 10.0; (○) experimental points, (——) calculated results

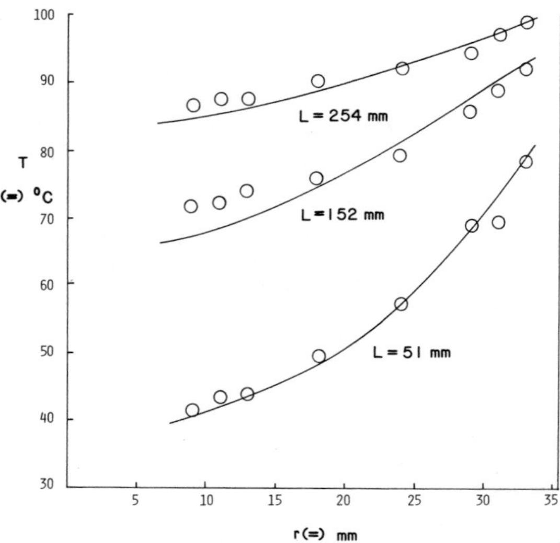

Figure 5. Temperature profiles for 9.5-mm ceramic beads: Re = 120, $Pe_{rf}(\infty)$ = 8.0, $Pe_{af}(\infty)$ = 0.49, Bi_f = 4.57, Bi_s = 20.0; (○) experimental points, (——) calculated results

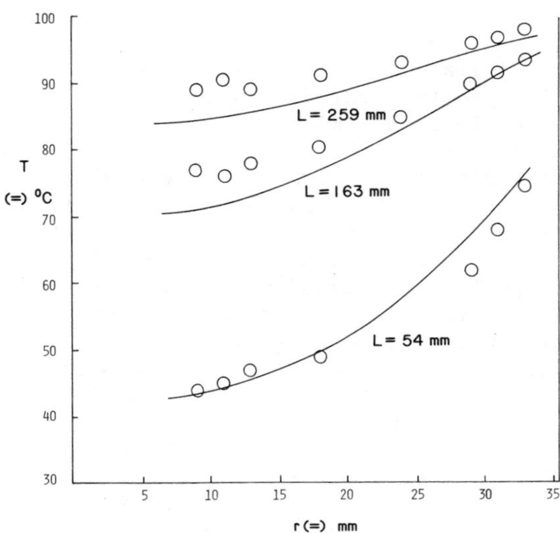

Figure 6. Temperature profiles for 9.5-mm steel balls: Re = 224, $Pe_{rf}(\infty) = 8.0$, $Pe_{af}(\infty) = 0.31$, $Bi_f = 3.90$, $Bi_s = 20.0$; (○) experimental points, (——) calculated results

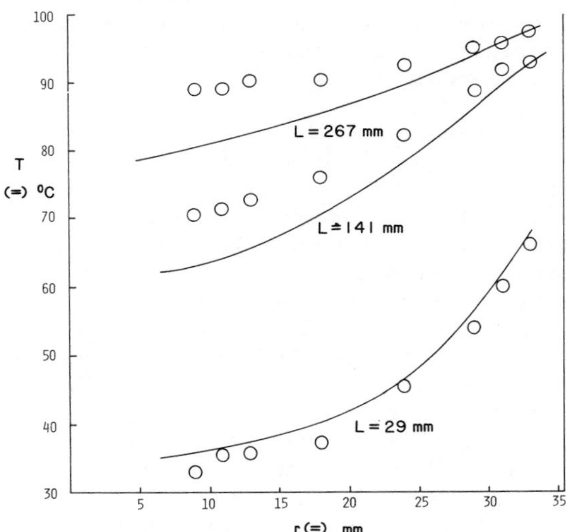

Figure 7. Temperature profiles for 6.4-mm ceramic beads: Re = 73, $Pe_{rf}(\infty) = 8.0$, $Pe_{af}(\infty) = 0.14$, $Bi_f = 5.85$, $Bi_s = 20.0$; (○) experimental points, (——) calculated results

Nomenclature

a	specific interfacial surface area (m^{-1})
c_p	fluid specific heat (kJ/kg°C)
d_p	pellet diameter (m)
d_t	tube diameter (m)
G	superficial mass flow rate (kg/m^2s)
h	apparent interphase heat transfer coefficient $(w/m^2{}°C)$
h_w	apparent wall heat transfer coefficient $(w/m^2{}°C)$
h_{wf}	wall-fluid heat transfer coefficient $(w/m^2{}°C)$
h_{ws}	wall-solid heat transfer coefficient $(w/m^2{}°C)$
h_{fs}	fluid-solid heat transfer coefficient $(w/m^2{}°C)$
k_a	axial effective conductivity (w/m°C)
k_r	radial effective conductivity (w/m°C)
k_{af}	axial conductivity of fluid phase (w/m°C)
k_{as}	axial conductivity of solid phase (w/m°C)
k_{rf}	radial conductivity of fluid phase (w/m°C)
k_{rs}	radial conductivity of solid phase (w/m°C)
k_g	molecular conductivity of fluid (w/m°C)
k_p	pellet conductivity (w/m°C)
L	length of packed test section (m)
R	tube radius (m)
r	radial coordinate (m)
T_b	bed temperature, one-phase model (°C)
T_{bf}	fluid phase temperature (°C)
T_{bs}	solid phase temperature (°C)
T_O	temperature of calming section wall (°C)
T_w	temperature of test section wall (°C)
u	superficial fluid velocity (m/s)
z	axial co-ordinate (m)

Dimensionless parameters

Bi	apparent wall Biot number, h_wR/k_r
Bi_f	fluid-wall Biot number, $h_{wf}R/k_{rf}$
Bi_s	solid-wall Biot number, $h_{ws}R/k_{rs}$
h_i	subinterval length
m	number of axial subintervals
n	number of radial subintervals
N_F	interphase heat transfer group, aR^2h/k_{rf}
N_S	interphase heat transfer group, aR^2h/k_{rs}
Nu_w	apparent wall Nusselt number, h_wd_p/k_g
Nu_{fs}	fluid-solid Nusselt number, $h_{fs}d_p/k_g$
Nu_{wf}	fluid-wall Nusselt number, $h_{wf}d_p/k_g$
Pe_a	effective axial Peclet number, Gc_pd_p/k_a
Pe_A	effective axial Peclet number (based on R), Gc_pR/k_a
Pe_r	effective radial Peclet number, Gc_pd_p/k_r
Pe_R	effective radial Peclet number (based on R), Gc_pR/k_r
Pe_{af}	axial fluid Peclet number, Gc_pd_p/k_{af}

Pe_{AF} axial fluid Peclet number (based on R), Gc_pR/k_{af}
Pe_{rf} radial fluid Peclet number, Gc_pd_p/k_{rf}
Pe_{RF} radial fluid Peclet number (based on R), Gc_pR/k_{rf}
$Pe_{rf}(\infty)$ asymptotic value of Pe_{rf} as $Re \rightarrow \infty$
$Pe_{af}(\infty)$ asymptotic value of Pe_{af} as $Re \rightarrow \infty$
Pr Prandtl number, $\mu c_p/k_g$
Re Reynolds number, Gd_p/μ
y normalized radial co-ordinate (r/R)
t transformed axial co-ordinate
T_f dimensionless fluid temperature $(T_{bf}-T_0)/(T_w-T_0)$
T_s dimensionless solid temperature $(T_{bs}-T_0)/(T_w-T_0)$
x normalized axial co-ordinate (z/R)

Greek symbols

α axial scaling factor
ε bed voidage
ϕ_i cubic Hermite basis function
ψ_i cubic Hermite basis function
θ dimensionless bed temperature, $(T_b-T_0)/(T_w-T_0)$
μ viscosity of fluid (kg/ms)
ρ density of fluid (kg/m^3)
ξ a general independent variable

Acknowledgments

This work was sponsored by the United States Army under Contract No. DAAG29-80-C-0041, and also by the National Science Foundation under Grant ENG76-24368.

The author would also like to record his thanks to Professor Warren E. Stewart, for many helpful and illuminating discussions on collocation methods.

Literature Cited

1. Hiby, J.W. Inst. Chem. Eng. Symp. Series 1962, 9, 312.
2. Gunn, D.J.; Khalid, M. Chem. Eng. Sci. 1975, 30, 261.
3. Dixon, A.G. Ph.D. Thesis, University of Edinburgh 1978.
4. Dixon, A.G.; Cresswell, D.L.; Paterson, W.R. ACS Symp. Series, 1978, 65, 238.
5. DeWasch, A.P.; Froment, G.F. Chem. Eng. Sci. 1972, 27, 5667.
6. Li, Chi-Hsuing; Finlayson, B.A. Chem. Eng. Sci. 1977, 32, 1055.
7. Young, L.C.; Finlayson, B.A. I.E.C. Fund. 1973, 12, 412.
8. Mears, D.E. I.E.C. Fund. 1976, 15, 20.
9. Stewart, W.E.; Sørensen, J.P. Proceedings 2nd ISCRE, Amsterdam, 1972: (B8) 75-88.
10. Villadsen, J.V.; Stewart, W.E. Chem. Eng. Sci. 1967, 22, 1483; 1968, 23, 1515.

11. de Boor, C. "A Practical Guide to Splines," Springer-Verlag, Berlin, 1978.
12. Carey, G.F.; Finlayson, B.A. Chem. Eng. Sci. 1975, 30, 587.
13. Prenter, P.M.; Russell, R.D. SIAM J. Number. Anal., 1976, 13, 923.
14. Paterson, W.R.; Dixon, A.G.; Cresswell, D.L. paper 3 in "Computers in Chemical Engineering. Recent Developments in Education and Practice." Meeting of Inst. of Chem. Eng. (Scottish Branch), Edinburgh 1977.
15. Paspek, S.C.; Varma, A. Chem. Eng. Sci. 1980, 35, 33.
16. Dixon, A.G.; Cresswell, D.L. AIChE J. 1979, 25, 663
17. Prenter, P.M. "Splines and Variational Methods," Wiley Interscience, New York, 1975.
18. Villadsen, J.V.; Michelsen, M.L. "Solution of Differential Equation Models by Polynomial Approximation," Prentice-Hall, Englewood Cliffs, 1978.
19. Martin, R.S.; Wilkinson, J.H. Numer. Math. 1967, 9, 279.
20. Dixon, A.G. MRC Technical Summary Report #2116, Mathematics Research Center, University of Wisconsin-Madison, 1980.
21. Guertin, E.W.; Sørensen, J.P.; Stewart, W.E. Comp. Chem. Eng., 1977, 1, 197.
22. Birnbaum, J.; Lapidus, L. Chem. Eng. Sci. 1978, 33, 455.
23. Verhoff, F.H.; Fisher, D.P. Trans. ASME, J. Heat Transfer, 1973, 95, 132.
24. Russell, R.D.; Christiansen, J. SIAM J. Numer. Anal., 1978, 15, 59.
25. Littmann, H.; Barile, R.G.; Pulsifer, A.H. I.E.C. Fund., 1968, 1, 554.
26. Olbrich, W.E. Proc. "CHEMECA" '70 Conf., Melbourne and Sydney, August 19-26, 1970, Butterworth, London, 1971, p.101.
27. Zehner, P.; Schlünder, E.U. Chemie-Ing.-Techn. 1970, 42, 933.
28. Fahien, R.W.; Smith, J.M. AIChE J. 1955, 1, 28
29. Hsiang, T.C.; Haynes, H.W. Jr. Chem. Eng. Sci. 1977, 32, 678.

RECEIVED June 3, 1981.

16

An Analysis of Radial Flow Packed Bed Reactors

How Are They Different?

HSUEH-CHIA CHANG

Department of Chemical and Nuclear Engineering, University of California, Santa Barbara, CA 93106

JOSEPH M. CALO

Department of Chemical Engineering, Princeton University, Princeton, NJ 08544

An analysis of radial flow, fixed bed reactor (RFBR) is carried out to determine the effects of radial flow maldistribution and flow direction. Analytical criteria for optimum operation is established via a singular perturbation approach. It is shown that at high conversion an ideal flow profile always results in a higher yield irrespective of the reaction mechanism while dependence of conversion on flow direction is second order. The analysis then concentrates on the improvement of radial profile. Asymptotic solutions are obtained for the flow equations. They offer an optimum design method well suited for industrial application. Finally, all asymptotic results are verified by a numerical experience in a more sophisticated heterogeneous, two-dimensional cell model.

The radial flow, fixed bed reactor (RFBR) was originally developed to handle the large gas flow rates in the catalytic synthesis of ammonia. Since then, RFBR's have been used, or considered for catalytic reforming, desulfurization, nitric oxide conversion, catalytic mufflers, and other processes in which fluids must be contacted with solid particles at high space velocity. The four basic types of flow configurations are shown in Figure 1. In all these configurations fluid enters parallel to the reactor axis either through the center pipe or the peripheral annulus, and then flows radially (perpendicular to the reactor axis) into the catalyst bed contained in the annular basket. The relatively large flow area offered by the inner or outer surface of the catalyst basket decreases fluid velocity through the bed, thereby permitting the use of a relatively short bed, which significantly reduces pressure drop.

Work on the fluid mechanics of radial flow reactors can be traced back to the calculations of radial velocity profiles by Soviet investigators (1,2). This analysis was later repeated by

0097-6156/81/0168-0305$06.25/0

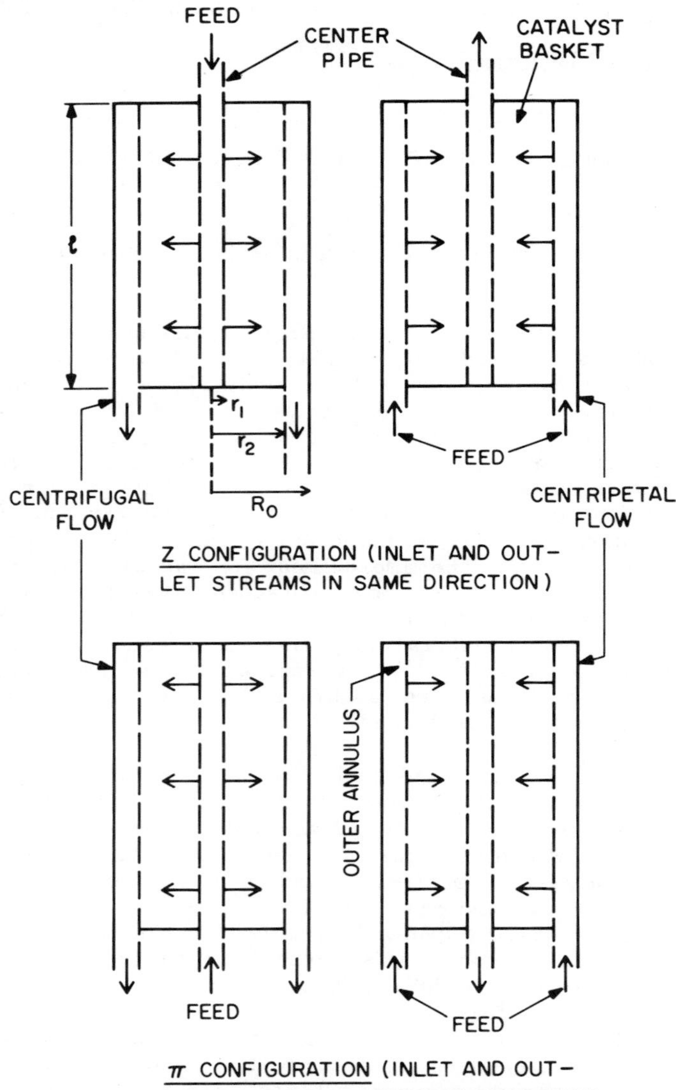

Figure 1. Schematic of radial flow, fixed bed reactor operation

Kaye (3) who also obtained experimental data on a larger scale radial reactor sector model.

Perhaps the first published analysis of an RFBR was by Raskin et al. (4), who developed a quasicontinuum model for a radial ammonia synthesis reactor and later applied it to carbon monoxide conversion (5).

Hlavacek et al. (6) and Dudukovic et al. (7) have shown that the direction of radial flow has little effect on conversion for isothermal and first order exothermic reactions if one assumes a perfect radial flow profile. No extension to more general kinetics was attempted by these workers. The effect of imperfect radial flow profiles on conversion was more recently investigated by Ponzi and Kaye (8). These authors found, numerically, using assumed flow profiles, that a perfect radial flow profile always yields higher conversion than an imperfect one in the high conversion regime, for the cases they considered. Calo (9) has used a cell model to investigate the stability and steady state multiplicity of the centrifugal (CFRF) and centripetal flow directions (CPRF). However, this study was limited to first order irreversible, exothermic reactions. In a previous paper (10) we extended Calo's approach to a two-dimensional model using flow profiles calculated using the earlier Soviet results (1,2).

The present work generalizes, unifies, and extends that done previously. Limiting analytical results are obtained, allowing generalizations and conclusions which are presently not available from the numerical work of previous investigators. These results can serve as design guidelines as well as tools for analyzing the effects of the physical parameters on total conversion. First, it is shown that the dependence of conversion on total flow direction for perfect distribution and near plug flow conditions is second order. Nevertheless, analytical criteria for the superiority (in the sense of conversion) of the two flow directions is established. Realizing that for perfect distribution, all operating modes are approximately equivalent, the analysis then concentrates on the effects of nonideal radial flow profiles, or fluid maldistribution. It is proven that at high conversion an ideal flow profile always results in a higher yield irrespective of the reaction mechanism. Moreover, the nonuniformity of radial velocity decreases conversion at an infinite rate near maximum conversion for some representative kinetics, thereby rigorously confirming the conjecture that flow maldistribution is the single most important variable in attaining optimal radial flow operation.

Armed with this knowledge, we then set out to examine the fluid mechanics of the reactor. Limiting asymptotic solutions are obtained for some of the flow equations which allow the determination of the optimum flow configuration for actual conditions which approximate those assumed. Outside the domain of validity of the asymptotic solutions, numerical integration must be applied, of course. These results substantiate conclusions arrived at using the asymptotic solutions, even in parametric regions where the asymptotic solutions are not valid. Finally, numerical experiments with a two-dimensional cell model are carried out to assess the effects of dispersion and interphase heat and mass transport, which are not considered in the analytical approach. Although the cell model represents a more detailed analysis of the problem, the results were found to be in agreement with the simpler models.

Flow Direction Dependence Of Perfect Radial Flow Distribution At Near-Plug Flow Conditions

In order to investigate the dependence of conversion on flow directions (CPRF vs. CFRF), the radial flow profile is assumed to be uniform and a one-dimensional quasicontinuum model is chosen. At high Reynolds numbers, as assumed here, experimental evidence indicates that D/ud_p, $k/\rho c_p ud_p$ is constant with respect to Reynolds number (4,5,7). This means that the heat and mass dispersion coefficients are directly proportional to the velocity, u. However, for a perfect radial profile, the continuity equation yields $ur = u_1 r_1 = u_2 r_2$ = constant. Thus for high Reynolds numbers D and $k/\rho c_p$ are inversely proportional to r. Introducing the following dimensionless variables:

$$c = \frac{c'}{c_o} \quad : \quad \theta = \frac{T}{T_o}; Pe_M = \frac{r_2 u_2}{D}; \; Pe_H = \frac{\rho c_p r_2 u_2}{k}$$

$$Le = \frac{Pe_H}{Pe_M} \quad ; \quad Da = \frac{k_o r_2 c_o'^{\,n-1}}{u_2} \quad ; \quad B = \frac{(-\Delta H)}{\rho \, C_p}; \; \bar{r} = \frac{r}{r_2}; \; f = \frac{R}{k_o c_o'^{\,n}}$$

The governing equation and boundary conditions can be unified for both flow directions, viz.,

$$Le \; \varepsilon \frac{d^2 c}{dy^2} - \left[\frac{dc}{dy} + Da \; h(y) \;\; f\,(c,\, \theta)\right] = 0 \tag{1}$$

$$\varepsilon \frac{d^2 \theta}{dy^2} - \left[\frac{d\theta}{dy} - Da \; B \; h(y) \; f\,(c,\theta)\right] = 0 \tag{2}$$

$$y = 0 \quad : \quad Le \; \varepsilon \frac{dc}{dy} = c - 1 \tag{3}$$

$$\varepsilon \frac{d\theta}{dy} = \theta - 1 \tag{4}$$

$$y = 1 - \rho \quad : \quad \frac{dc}{dy} = \frac{d\theta}{dy} = 0 \tag{5}$$

where

$$h(y) = \begin{cases} y + \rho \quad ; \quad \text{CFRF} & (6) \\ 1 - y \quad ; \quad \text{CPRF} & (7) \end{cases}$$

and $\varepsilon = 1/Pe_H$ becomes the perturbation parameter when $Pe_H >> 1$ (i.e., $\varepsilon << 1$).

Perturbation Solution for Le = 1. The simplest case of $Le = Pe_H/Pe_M = 1$ is analyzed here. This assumption yields the adiabatic invariant, $B(1 - c) = \theta - 1$ which can be used to further simplify Eqns. (1)-(5) to

$$\varepsilon \frac{d^2c}{dy^2} - \left[\frac{dc}{dy} + Dah(y)\ f(c)\right] = 0 \tag{8}$$

$$y = 0 : \varepsilon \frac{dc}{dy} = c - 1 \tag{9}$$

$$y = (1 - \rho): \frac{dc}{dy} = 0 \tag{10}$$

When $\varepsilon << 1$, this becomes a singular perturbation problem.

Outer Solution. Let the outer solution, $\bar{c}$, be expressed as the following power series in ε,

$$c = \sum_{n=o}^{\infty} \varepsilon^n \bar{c}_n \tag{11}$$

Consequently,

$$f = f_o + (\bar{c} - \bar{c}_o)\ f_{\bar{c}_o} + - - - - = f_o + \varepsilon\bar{c}_1\ f_{\bar{c}_o} + - - - - \tag{12}$$

$$\text{where} \quad f_o = f(\bar{c}_o) \quad \text{and} \quad f_{\bar{c}_o} = \left(\frac{\partial f(\bar{c})}{\partial \bar{c}}\right)_{\bar{c}_o} \tag{13}$$

ε^o Term. Substituting Eqns. (11), (12) into Eqns. (8)-(10), one obtains, for the zeroth order, the plug-flow equation

$$\frac{d\bar{c}_o}{dy} + Da\ h(y)\ f_o = 0 \tag{14}$$

$$\text{B.C.:} \quad y = o : \quad \bar{c}_o = 1 \tag{15}$$

Eqns. (14) and (15) can be integrated,

$$\int_1^{\bar{c}_{o_e}} \frac{d\bar{c}_o}{f_o} = - \text{Da} \int_o^{1-\rho} h(y)\, dy = - \frac{\text{Da}(1-\rho^2)}{2} \quad (16)$$

where $\bar{c}_{oe}$ is the value at y = 1 - ρ. Since Eqn. (16) is independent of flow direction, to the zeroth order (i.e., plug flow), there is no difference in conversion between the two flow directions, and any differences that do arise can only be of higher order. This is an important conclusion of this section.

<u>ε^1 Term.</u> $$\frac{d^2\bar{c}_o}{dy^2} - \frac{d\bar{c}_1}{dy} - \bar{c}_1 \text{ Da } h(y)\, f_{\bar{c}_o} = 0 \quad (17)$$

B.C.: $y = o \; ; \quad \frac{d\bar{c}_o}{dy} = \bar{c}_1$

Using the integral form of the solution to Eqn. (17),

$$\bar{c}_1 = - \text{Da } e^{-g(y)} \int_o^y e^{g(y')} \frac{dhf_o}{dy'} dy' - \text{Da } h(o) f_o(o) e^{-g(y)} \quad (18)$$

where $$g(y) = \ln \frac{f_o(y=o)}{f_o} \quad (19)$$

Substituting Eqn. (19) into Eqn. (18), one obtains

$$\bar{c}_{1_e} = - \text{Da } f_o(e) \left[h_e + \int_o^e h d\ln f_o \right] \quad (20)$$

where (e) denotes the value at the exit, y = 1 - ρ (e.g. $f_o(e) = f_o(\bar{c}_{oe})$). Higher order terms are more difficult to obtain, but do not contribute significantly.

<u>Inner Solution</u>. Rescaling the spatial coordinates at the boundary layer located at y = 1 - ρ, by z = [y-(1-ρ)]/ε, one obtains

$$\frac{d^2\hat{c}}{dz^2} - \left[\frac{d\hat{c}}{dz} + \varepsilon \text{ Da } h\, f(\hat{c})\right] = 0 \quad (21)$$

$$\text{B.C.:} \quad z = 0, \quad \frac{d\hat{c}}{dz} = 0 \tag{22}$$

where

$$h = \begin{cases} \varepsilon z + 1 & ; \quad \text{CFRF} \\ \rho - \varepsilon z & ; \quad \text{CPRF} \end{cases} \tag{23}$$

ε° Term. The zeroth order equation is simply $\hat{c}_o = c_1$, a constant to be determined by matching.

ε^1 Terms.

$$\frac{d^2\hat{c}_1}{dz^2} - \frac{d\hat{c}_1}{dz} - \text{Da } h(z)\ f_o = 0 \tag{24}$$

$$\text{B.C:} \quad z = 0, \quad \frac{d\hat{c}_1}{dz} = 0$$

Integrating Eqn. (24) and applying the appropriate boundary conditions, one obtains,

$$\hat{c}_1 = \text{Da } f_o\ (\hat{c}_o) h_e\ \left[e^z - z - 1\right] + \hat{c}_{1e} \tag{25}$$

Matching. The constants $\hat{c}_o$ and $\hat{c}_{1e}$ must be evaluated by matching with the outer solution:

$$\lim_{z\to\infty} \hat{c} = \hat{c}_o - \varepsilon \text{ Da } f_o(\hat{c}_o)\ h_e\ (z+1) + \varepsilon\ \hat{c}_{1e} + 0\ (\varepsilon^2) \tag{26}$$

$$\lim_{y\to-\rho} \bar{c} = \bar{c}_{oe} - \varepsilon \text{ Da } f_o\ (\bar{c}_{oe})\ h_e\ z + \varepsilon\bar{c}_{1e} + 0\ (\varepsilon^2) \tag{27}$$

The composite solution is thus

$$c = \bar{c} + \hat{c} - \lim_{\varepsilon\to o} \bar{c}\ (\varepsilon z + 1 - \rho)$$

$$= \bar{c}_o + \bar{\varepsilon} c_1 + \varepsilon \text{ Da } f_o(e)\ h_e\ e^{\frac{y-(1-\rho)}{\varepsilon}} - \frac{y-(1-\rho)}{\varepsilon} \tag{28}$$

and the difference in exit concentration for the two flow directions is

$$\Delta c_e = c_{e_{CF}} - c_{e_{CP}} = \varepsilon \left[\bar{\Delta} c_{1e} + \text{Da } f_o(e)\ (1-\rho)\right] + 0\ (\varepsilon^2) \tag{29}$$

But from Eqn. (20),

$$\Delta c_e = \varepsilon\left[- Da\ f_o(e)\ (1-\rho)\ -\ Da\ f_o \int_o^e (h_{CF}-h_{CP}) d\ln f_o\ +\ Da\ f_o(e)(1-\rho)\right]$$

$$= \varepsilon\ Da\ f_o(e) \int_o^e (h_{CP} - h_{CF})\ d\ln f_o \qquad (30)$$

Eqn. (30) requires a specific relationship between h(y) and f_o in order to be evaluated. Recalling the indefinite version of Eqn. (16), i.e.,

$$\int_1^{\bar{c}_o} \frac{d\bar{c}_o}{f_o} = -\ Da \int_o^y h(y)\ dy \qquad (31)$$

let

$$-\int_1^{\bar{c}_o} \frac{d\bar{c}_o}{f_o} = \Gamma\ (\bar{c}_o) \qquad (32)$$

since f_o, the reaction expression, is yet to be specified. Then

$$\frac{\Gamma}{Da} = \begin{cases} \frac{y^2}{2} + \rho y & ; \quad CFRF \\ y - \frac{y^2}{2} & ; \quad CPRF \end{cases} \qquad (33)$$

and

$$\Delta c_e = \varepsilon\ Da\ f_o(e) \int_1^{\bar{c}_{o_e}} \left[\sqrt{1 - \frac{2\Gamma}{Da}} - \sqrt{\rho^2 + \frac{2\Gamma}{Da}}\right] \frac{d\ln f_o}{d\bar{c}_o}\ d\bar{c}_o \qquad (34)$$

Eqn. (34) determines the difference in exit concentrations for the two flow directions, given the rate expression $f_o(\bar{c}_o)$. The relative effect of flow direction can be determined from the sign of Δc_e; i.e., $\Delta c_e > 0 \rightarrow$ CPRF has higher conversion, $\Delta c_e < 0 \rightarrow$ CFRF has higher conversion.

An important conclusion can be derived from Eqn. (34). Note that Δc_e is proportional to $f_o(e)$, the rate of reaction at the exit. For irreversible endothermic and isothermal positive order reactions, the rate at the exit is the lowest in the reactor and Δc_e is small. Consequently, if the flow direction is a factor for conversion, the effects should be most pronounced for negative order or highly exothermic reactions far removed from plug-flow conditions.

Under certain conditions the Lewis number may be close to unity. For these cases one can extend the analysis in the previous section to include the effect of Lewis number by perturbing about Le = 1. The results indicate that the effects of non-unity Le appear only in $0(\delta\varepsilon)$ and higher terms. Consequently, the solution in the previous section represented by Eqn. (34) is also valid here to order $(\delta\varepsilon)$, and since Eqn. (34) is only exact to order ε, it is also appropriate as the solution for non-unity Lewis number.

Isothermal nth Order Irreversible Reaction. Here $f = c^n$ and the zeroth order equation representing the outer solution is

$$\frac{d\bar{c}_o}{dy} = - Da\, h(y)\, \bar{c}_o^{\,n} \tag{35}$$

n = 0. For this case, f is independent of c and it is obvious, without further analysis, from Eqns. (1)-(5) that $\Delta c_e = 0$.

n = 1. $\Gamma = - \ln \bar{c}_o$ and $\Gamma_e = \frac{Da\,(1-\rho^2)}{2}$

Thus,
$$\Delta c_e = \varepsilon Da\, f_o(e) \int_o^{\Gamma_e} \sqrt{1-\frac{2\Gamma}{Da}} - \sqrt{\rho^2 + \frac{2\Gamma}{Da}}\; d\Gamma = 0 \tag{36}$$

n ≠ 1. The zeroth order outer solution is

$$\bar{c}_o = \left[1 - (1-n)\, Da \int_o^y h\, dy\right]^{\frac{1}{1-n}} \tag{37}$$

Note that for n < 1, the argument on the left hand side of Eqn. (37) can be negative which may make the solution meaningless. Thus, we define the restriction

$$2 \int_o^y h\, dy = \rho^2 - 1 > b \tag{38}$$

for orders of reaction less than unity, where b = 2/Da(n-1). Thus, from Eqn. (34),

$$\Delta c_e = - \varepsilon\, Da\, f_o(e) n \int_o^{\Gamma_e} \left[\frac{\sqrt{1- \frac{2\Gamma}{Da}} - \sqrt{\rho^2 + \frac{2\Gamma}{Da}}}{(n-1)\,\Gamma + 1}\right] d\Gamma \tag{39}$$

The integral in Eqn. (39) cannot be expressed in closed form. However, the sign of the integral can be obtained by making the substitution, $u = \Gamma/\Gamma_e - 1/2$ which yields

$$\Delta c_e = \frac{2n\ f_o(e)}{n-1}\ \varepsilon\ \chi\ (b,\ \rho) \tag{40}$$

where

$$\chi\ (b,\ \rho) = \int_{-\frac{1}{2}}^{\frac{1}{2}} \frac{g\ (u)}{q\ (u)}\ du \qquad q(u) = b + (1-\rho^2)\ (u + \tfrac{1}{2})$$

and $$g(u) = [u + \tfrac{1}{2} - \rho^2\ u + \rho^2/2]^{\frac{1}{2}} - [-u + \tfrac{1}{2} + \rho^2 u + \rho^2/2]^{\frac{1}{2}}$$

Note that g is asymmetric with respect to u, i.e. g(u) = -g(u). It can then be shown that

$$\chi = \int_0^{\frac{1}{2}} g(u)\ \rho(u)\ du \tag{41}$$

where $$\rho(u) = \frac{-2\ (1-\rho^2)u}{b + \frac{1}{2}\ (1-\rho^2) - (1-\rho^2)^2 u^2}$$

Note that also g(0) = 0 and the derivative of g with respect to u is positive for u between 0 and 1/2. Thus, g(u) > 0 for u in that range. The signs of Δc_e are obtained in the following theorems from the preceding information.

<u>Theorem 1</u>. $n > 1 \rightarrow \Delta c_e < 0$

Proof: $n > 1 \rightarrow b > 0 \rightarrow [b + \frac{1}{2}\ (1-\rho^2)]^2 - (1-\rho^2)^2\ u^2 > o$

for $u\ \varepsilon\ (o,\ \frac{1}{2})$

$$\rightarrow 2\ (b + \frac{1-\rho^2}{2}) \qquad b + \frac{1}{2}\ (1-\rho^2)^2 - (1-\rho^2)^2\ u \quad > 0$$

for $u\ \varepsilon\ (o,\ \frac{1}{2})$

$$\rightarrow \rho(u) < 0 \rightarrow \chi < o \rightarrow \frac{2n}{n-1}\ f_o(e)\ \varepsilon\ \chi < 0 \rightarrow \Delta c_e < 0 \qquad \text{Q.E.D.}$$

<u>Theorem 2</u>. $n < 1$ and $b < \rho^2 - 1 \rightarrow \Delta c_e > 0$

Proof: $b < \rho^2 - 1 \rightarrow b + \frac{1}{2}\ (1-\rho^2) < \frac{\rho^2-1}{2}$

$$\rightarrow\ b + \frac{1}{2}(1-\rho^2)^2 > \frac{\rho^2-1}{2}^2 \qquad \text{Since } \frac{\rho^2-1}{2} < 0$$

$$\rightarrow\ p(u) < 0 \qquad \rightarrow\ \chi < 0$$

$$\rightarrow\ \frac{2n}{n-1} f_o \varepsilon \chi > 0 \qquad \text{since } n-1<0 \qquad \rightarrow \Delta c_e > 0 \qquad \text{Q.E.D.}$$

Thus, in general, CFRF yields higher conversion for reaction orders greater than unity, while CPRF is the preferred direction for $0 < n < 1$. For zeroth and first order reactions, there is no difference between conversions for the two flow directions to the ε order.

Nonisothermal Conditions. For nonisothermal conditions,

$$f_o = \bar{c}_o^{\,n} e^{-\Upsilon/\theta_o} = \bar{c}_o^{\,n} e^{-\Upsilon/(1+B[1-c_o])} \quad \text{and } \Gamma = -\int_1^{\bar{c}_o} \frac{d\bar{c}_o}{f_o} \qquad (42)$$

Eqn. (42) cannot be integrated in closed form. Thus Eqn. (34) will be evaluated numerically. In Figure 2, the locus of $\Delta c_e = 0$ is plotted in the B-$\bar{c}_{oe}$ domain for different reaction orders. Note that at high conversions ($\bar{c}_{oe} \rightarrow 0$) CPRF yields a higher conversion for exothermic reactions with reaction order greater than unity. This is the same conclusion arrived at by Calo (9) using a simple cell model for $n = 1$.

It is interesting to note that the underlying cause of the difference in conversion for different flow directions is dispersion or mixing. It is well known that, at isothermal conditions, dispersion reduces conversion for reactions of order greater than unity and increases conversion for negative order reactions, while the first order reaction is completely unaffected (11). In the present context one expects CFRF to exhibit a lower degree of mixing relative to CPRF, since CFRF yields higher conversions for $n > 1$ and less for $n < 1$. However, it would be more appropriate to say that the major amount of dispersion in CPRF is near the exit of the bed. Apparently, the total amount of dispersion is the same for both configurations and it is the distribution of dispersion in the radial direction which is the actual cause of the difference in conversion. For nonisothermal reactors, such generalizations are not possible a priori, as shown by Ray et al. (12), and one must consult Figure 2 for any specific case. In any event, the large flow rates in industrial reactors tend to minimize the effects of dispersion and one should be able to model radial flow reactors adequately with using the plug flow assumption. On the other hand, for laboratory or small scale reactors care should be exercised.

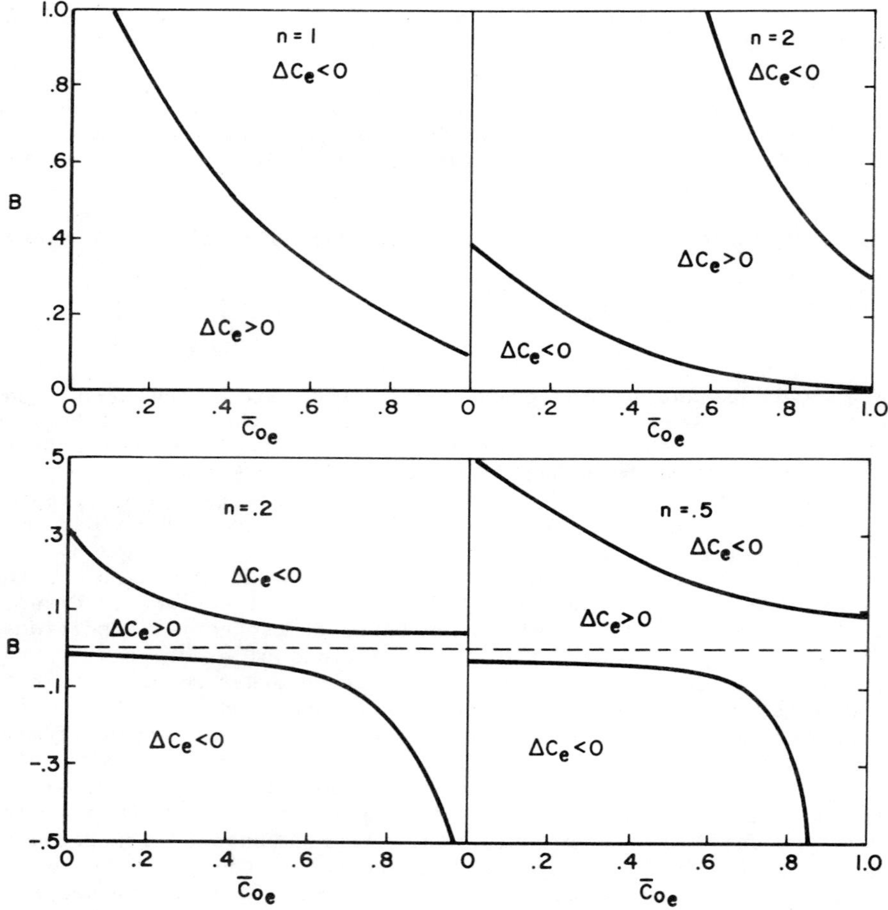

Figure 2. The $\Delta c = 0$ *locus in* $B - \bar{c}_{oe}$ *space for nth order, irreversible, exothermic reaction,* $\gamma = 10$, $\rho = 0.2$

Effects Of Maldistribution On Conversion

In the previous section, it was recognized that for perfect fluid distribution flow direction has only a second order effect on conversion. In this section, the effects of maldistribution are investigated. In order to eliminate the influence of flow direction, a pseudo-homogeneous plug flow model is used with purely radial flow through the catalyst basket. The governing equation is thus Eqn. (14), which in more convenient form is

$$\int_{o}^{x(\bar{v})} \frac{dx}{R(x)} = \frac{\Omega}{\bar{v}} \quad , \tag{43}$$

where $\Omega = \frac{2\ \tau\ 1}{C_o r_1}$ $\bar{w}_1 = w/w_o$; $\bar{v} = 2v1/r_1$; $z = z'/1$

For perfect fluid distribution, $\bar{v}(z) = 1$ and the conversion x_p, is defined by

$$\int_{o}^{x_p} \frac{dx}{R} = \Omega\ (x_p) \tag{44}$$

The deviation of the total conversion for maldistributed flow from that at perfect distribution is defined by:

$$\Delta = \frac{\int_o^1 x\bar{v}dz}{x_p} \tag{45}$$

Equating total radial flow to inlet axial flow, one obtains the following constraint on $\bar{v}$,

$$\int_{o}^{1} \bar{v}\ dz = 1 \tag{46}$$

The term Δ is expressible in terms of x_p if $\bar{v}(z)$ is known from the flow solutions delineated in Section 4. However, the effects of maldistribution on conversion can be assessed without any information on $\bar{v}(z)$ in the manner outlined below. We shall investigate the behavior of Δ near high conversion and near perfect distribution, which are both limits of practical interest.

Expanding x in a Taylor series of $\bar{v}$ about $\bar{v} = 1$,

$$x(\bar{v}) = x_p + \left.\frac{\partial x}{\partial v}\right|_{\bar{v}=1} (\bar{v} - 1) + - - - - \tag{47}$$

where from Eqn. (43),

$$\left(\frac{\partial x}{\partial \bar{v}}\right)_{v=1} = - \left(\frac{\Omega\ R(x)}{\bar{v}^{\,2}}\right)_{v=1} = - \Omega\ (x_p)\ R(x_p) \qquad (48)$$

Substituting Eqns. (47) and (48) into Eqn. (45) and invoking constraint (49) one obtains

$$\Delta \sim 1 - \frac{\Omega(x_p)\ R(x_p)}{xp} \left[\int_o^1 v^{-2} dz - 1\right] \qquad (49)$$

However, from Eqn. (44),

$$\frac{\partial \Omega}{\partial x_p} = \frac{1}{R(x_p)}$$

$$\text{Thus,} \quad \frac{\partial \Delta}{\partial x_p} \sim \frac{[\int_o^1 \bar{v}^{2} dz - 1]}{x_p^{\,2}} \left[x_p \Omega\ \frac{\partial R}{\partial x} + x_p - R\Omega\right] \qquad (50)$$

At high conversions, $x_p \rightarrow x_{eq}$ where x_{eq} is the equilibrium conversion, and

$$\lim_{x_p \rightarrow x_{eq}} \frac{\partial \Delta}{\partial x_p} \sim$$

$$- \frac{\int_o^1 v^{-2} dz \quad -1}{x_{eq}} \left[\Omega(x_p)\ \frac{\partial R}{\partial x_p}\Bigg|_{x_p = x_{eq}} + 1 - R\Omega\Bigg|\right]_{x_p = x_{eq}} \qquad (51)$$

From Eqns. (49) and (51) it can be seen that the behavior at Δ and $\partial\Delta/\partial x_p$ in the neighborhood of $c_p = 1$ is determined by the following two limits,

$$\lim_{x_p \rightarrow x_{eq}} R\Omega \quad \text{and} \quad \lim_{x_p \rightarrow x_{eq}} \Omega\ \frac{\partial R}{\partial x_p}$$

For positive order reactions, $R(x_{eq}) = 0$ and $\lim_{x_p \rightarrow x_{eq}} \Omega = \infty$. Thus, the evaluation of the limits (51) must be carried out with care. Expanding $R(x)$ about $x = x_e$:

$$R(x) = \left(\frac{d^m R}{d x^m}\right)_{x=x_{eq}} \frac{(x - x_{eq})^m}{m !} + - - - \tag{52}$$

where m is the first nonvanishing derivative of R at x_e, since it is possible that the first few derivatives might vanish at x_e. Then

$$\Omega(x) = \int_o^x \frac{dx}{R} = \frac{- m !}{\left.\frac{\partial^m R}{\partial x^m}\right|_{x_{eq}}} \left[\frac{1}{(m-1)(x-x_{eq})^{m-1}} - \frac{(-x_{eq})^{m-1}}{m-1}\right] + --- \tag{53}$$

The limits can be evaluated by using Eqns. (52) and (53)

$$\lim_{x \to x_{eq}} R\Omega = o \quad \text{for } m \geq 1$$

$$\lim_{x \to x_{eq}} \Omega \frac{\partial R}{\partial x} = \left.\begin{matrix} \frac{-m}{m-1} & \text{for } m > 1 \\ -\infty & \text{for } m = 1 \end{matrix}\right\}$$

Thus, $\lim_{x_p \to x_{eq}} \Delta = 1$

and $$\lim_{x_p \to x_{eq}} \frac{\partial \Delta}{\partial x_p} = + \infty \quad \text{for } m = 1$$

$$= \frac{1}{m-1} [\int_o^1 \bar{v}^2 dz - 1] > o \quad \text{for } m > 1$$

The expression is positive since from constraint (46),

$$\int_o^1 \bar{v}^2 dz \geq \int_o^1 \bar{v} dz \int_o^1 v dz = 1$$

$$\text{Thus,} \quad \lim_{x_p \to x_{eq}} \frac{\partial \Delta}{\partial x_p} > 0 \tag{54}$$

and Δ decreases with x_p near x_{eq}, sometimes infinitely fast (for the m = 1 case). Thus, the nonuniformity of the radial flow profile can cause drastic differences in conversion in the high conversion regime. Results for typical specific reaction kinetics indicate that for a first order reaction, regardless of whether it is isothermal and irreversible, or nonisothermal and irreversible, or isothermal and reversible; conversion would decrease at an infinitely fast rate upon the slightest deviation from perfect distribution in the high conversion region. (For detailed calculations, see (16).) This effect can also be seen in the results of Ponzi and Kaye (8), reproduced in Figure 3. By assuming that the nonuniform radial profile is the form $\bar{v} = sy^{s-1}$, the figure shows the predicted infinite slope as x_p approaches x_{eq}. Near x_{eq} (i.e. in the high conversion region), reduction by as much as 50% occurs for a first order exothermic irreversible reaction.

Radial Flow Profile Determination

In the previous section, the importance of the uniformity of the radial flow profile was established. In the present section, the fluid mechanical equations for all four flow configurations in Figure 1 are derived and solved for comparison. The development of equations closely follows the approach of Genkin et al. (1,2). Here we extend their work to include both radial and axial flow in the catalyst bed. Following our derivation in reference (16), the dimensionless equations for the axial velocity in the center-pipe for all four configurations are (the primes denote derivative with respect to the dimensionless axial coordinate),

$$a_1 \, \bar{w}_1' \, \bar{w}_1'' + a_2 \, \bar{w}_1 \, \bar{w}_1' + a_3 \, \bar{w}_1^{\,2} = 0 \tag{55}$$

B.C.: $\bar{w}_1(0) = 0 \qquad \bar{w}(1) = 1 \qquad$ (CPRF)

$\bar{w}_1(0) = 1 \qquad \bar{w}(1) = 0 \qquad$ (CFRF)

$$a_1 \bar{w}_1'' + a_4 \, \bar{w}_1 \bar{w}_1' + a_5 \bar{w}_1' + a_6 \, (1-\bar{w}_1)^2 + a_7 \, \bar{w}_1^{\,2} = 0 \tag{56}$$

B.C.: $\bar{w}_1(0) = 0 \qquad \bar{w}_1(1) = 1 \qquad$ (CPRF)

$\bar{w}(0) = 1 \qquad \bar{w}(1) = 0 \qquad$ (CFRF)

where

$$a_1 = \frac{1}{\ell^2}\left\{\frac{r_1^{\,4}}{4r_2^2\phi_2^2} - \frac{r_1^{\,2}}{4\phi_1^2} - \left[\frac{b^*\phi_2^2 r_2^2}{\chi^3}\left(\frac{r_2 - r_1}{r_1 r_2}\right) + \frac{1}{2}\left(1 - \left[\frac{\phi_2 r_2}{\phi_1 r_1}\right]^2\right)\right]2\left(\frac{r_1^{\,2}}{2\phi_2 r_2}\right)^2\right\}$$

$$a_2 = a_4 = \frac{3}{2}(\xi^2 - 1); \quad \xi = \frac{r_1^{\,2}}{R_o^2 - r_2^2}; \quad a_3 = -\frac{R_o}{R_o^2 - r_2^2}\,\xi^2 + \frac{1}{r_1}\,f\ell; \quad a_5 = -\frac{3}{5}\,\xi^2$$

$$a_6 = \frac{R_o}{R_o^2 - r_2^2} f\ell\, \xi^2 \; ; \quad a_7 = -\frac{f\ell}{r_1}$$

The radial velocities are related to the axial velocities by

$$\pm \frac{r_1}{2\phi_1} \bar{w}_1' = \bar{v}_1 \qquad \begin{cases} + & \text{CPRF} \\ - & \text{CFRF} \end{cases} \tag{57}$$

$$\pm \frac{R_o^2 - r_2^2}{2r_2\phi_2} \bar{w}_2' = \bar{v}_2 \qquad \begin{cases} + & \text{CPRF} - \Pi \\ & \text{CFRF} - \text{Z} \\ - & \text{DPRF} - \text{Z} \\ & \text{CFRF} - \Pi \end{cases} \tag{58}$$

Even before carrying out the actual numerical integration, certain qualitative features of the solutions are evident. It should be noted that if $\xi^2 = 1$ and $a_3 = 0$, the flow equation yields the solution for perfect distribution; i.e.,

$$\bar{w}_{1_{\text{ideal}}} = \begin{cases} 1 - z & \text{CF} \\ z & \text{CP} \end{cases} \tag{59}$$

upon inspection. This does not apply to the Z-flow equation which yields nonideal distribution at the same conditions. These conditions correspond to equal center-pipe and outer annulus cross-sectional area, and negligible feed and exit channel flow resistances. Consequently, one expects, a priori, that π-flow reactors will yield better profiles than Z-flow reactors.

The flow distribution superiority of the π-flow configuration has a simple physical interpretation. When frictional losses in the feed channel are low, pressure increases in the flow direction due to momentum loss. In the exit channel, the opposite is true. Thus in Z-flow a large radial pressure drop occurs at the exit end of the reactor, while a much smaller one occurs at the entrance end. This relatively large gradient in radial pressure drop is essentially what causes the flow maldistribution. In π-flow, however, due to the opposite flow of feed and product streams, pressure increases in the same direction in both channels and results in a more uniform pressure drop, and consequently, fluid distribution.

The effects of flow direction can be diagnosed likewise. Since the superiority of π-flow has already been established, we will only investigate the effects of flow direction for π-flow reactors. At near ideal distribution $a_2/a_1 \sim a_3/a_1 \rightarrow 0$. Thus, we can rewrite Eqn. (55) as

$$\bar{w}_1' \bar{w}_1'' + \varepsilon \bar{w}_1 \bar{w}_1' \pm \delta \bar{w}_1^2 = 0 \tag{60}$$

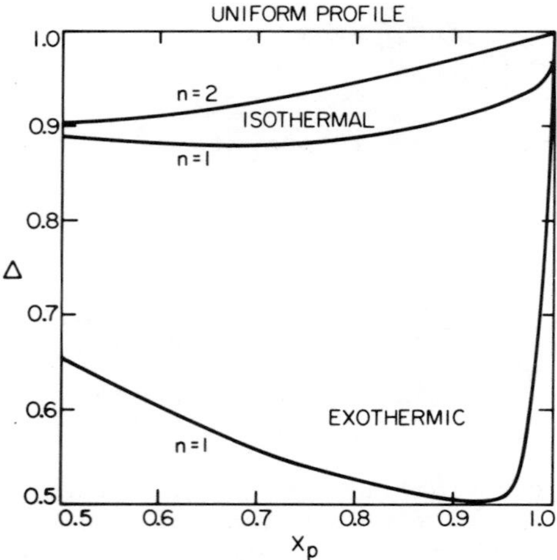

Figure 3. Results of Ponzi and Kaye (8) for Δ vs. x_p assuming the radial profile is in the form of $\overline{v} = sy^{s-1}$ for first- and second-order isothermal reactions and first-order exothermic reaction ($s = 2, \gamma = 2.4, \beta = 1$)

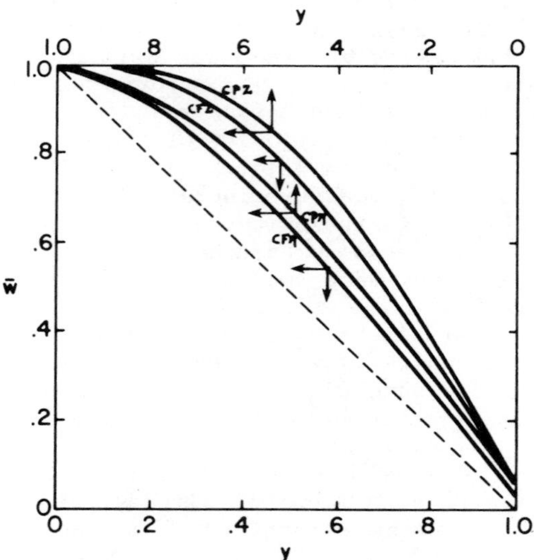

Figure 4. Dimensionless velocity distribution for all four configurations at "standard" conditions

where $\varepsilon = a_2/a_1$ and $\delta = a_3/a_1$. In deriving Eqn. (60), we have unified the CRFR and CFRF equations by defining $z = 1 - z_{CF}$ for CFRF. Consequently, the plus and minus signs in front of δ in Eqn. (60) indicate CPRF and CFRF configurations, respectively; and the unified boundary conditions become, $\bar{w}_1(o) = 0$; $\bar{w}_1(1) = 1$. Expanding $\bar{w}_1$ in a series in ε and δ and substitute into Eqn. (60), one obtains

$$\bar{w} \sim z - \frac{z}{6}(z^2-1)(\varepsilon \pm \delta) \tag{61}$$

where the positive and negative signs in front of δ correspond to CPRF and CFRF, respectively. For Eqn. (61), a perfectly uniform radial flow profile results in simply $\bar{w}_1 = z$. Thus the second term in Eqn. (61) actually represents the first order approximation of the deviation from uniformity. If both ε and δ are positive, one expects CFRF to yield a smaller deviation from uniformity than CPRF. Here $\delta = a_3/a_1$ is positive for all cases. However, ε is positive for $\xi < 1$. In summary, for $\varepsilon < 1$, i.e., viz. when the cross sectional area of the center pipe is smaller than that of the outer annulus, one expects CF-π to be the best flow configuration. For $\xi < 1$, on the other hand, CP-π is the superior configuration. Another interesting case occurs for $\xi > 1$. Since ε is negative in this case, the second term of Eqn. (61) can also be negative, which actually represents a negative deviation from perfectly uniform flow; i.e., a concave relationship between $\bar{w}_1$ and y, rather than convex.

The preceding analysis shows that certain general conclusions concerning flow distribution can be made using simple approximations. Eqn. (61) is especially useful as an analytical solution to the flow equations. However, perturbation techniques are only valid in regions where ε and δ are small. Outside this region, the conclusion stated above can be verified only with numerical techniques. A fourth-order (3,8) Runge-Kutta shooting technique was used to solve Eqns. (55) and (56). The "standard" set of parameters, derived in part from Genkin et al. (1,2), was chosen as a basis (see ref. (16)). Some of the results for these parameters and systematic variations thereof are summarized in Figure 4. Significant deviations from the ideal profile (represented by the dashed 45° line) are evident. In all the cases examined, the π-flow distribution was always closer to ideal than the corresponding Z-flow distribution (e.g., as shown in Figure 4). In addition, CFRF is superior to CPRF for $\xi < 1$ as predicted. The interesting concave behavior of the π-flow configuration for $\xi > 1.0$ is also witnessed in the numerical study. In conclusion, all the predictions derived from the perturbation analysis are substantiated by the results from numerical integration.

Cell Model Simulation

A two-dimensional cell model from an earlier publication (10) is utilized here to verify some of the results obtained from the analyses in the previous sections. This approach includes mixing in the fluid phase and interphase resistance to heat and mass transport. Consequently, it is a more sophisticated model than the models used in the preceding analyses. However, the equations must be solved numerically. The numerical techniques and the parameters used can be found in our previous works (10,16). Some typical two-dimensional particle temperature profiles are presented in Figures 5-7. The effect of flow maldistribution is evident in Figures 6 and 7. Both these cases were calculated for the same conditions except for the outer reactor radius, which was changed from the "standard" value of $R_o = 0.12$ m (Fig. 6) to $R_o = 0.11$ m (Fig. 7). This causes the distribution to become concave but closer to a perfect distribution. Thus for a very modest change in outer reactor radius, a significant improvement in exit temperature uniformity is achieved, in addition to increasing the overall conversion by 6.1% (see Table I). Of an even more dramatic nature is the improvement in fluid distribution and overall conversion caused by changing from CF-Z to CF-π flow as shown in Figure 5 and 6, respectively. For the same "standard" conditions, the change to π-flow increased overall conversion by 49.9% (See Table I). Total conversion results for various flow distributions are presented in Table I.

From the preceding analysis, the importance of radial flow distribution predicted in the previous section is substantiated. From the analysis for a perfect radial flow profile CPRF yields a higher conversion than CFRF for a highly exothermic, first order, irreversible reaction (see Fig. 2). This result is also evident in Case B of Table I. The isothermal cases, G and H, exhibit little dependence on flow direction, again confirming the results of the analysis. In general, flow maldistribution predominates and determines total conversion. However, for practically identical radial flow profiles for all flow configurations, the effects of flow direction become more evident.

The effect of axial flow in the catalyst bed is indicated in Cases I and J in Table I, where axial flow is shown to decrease conversion in all four reactor configurations. In fact, axial flow decreases conversion because it mixes fluid of different ages throughout the bed, introducing a backmixing effect. The magnitude of the axial velocity component is generally small relative to the radial component.

Conclusion

Generally, π-flow produces the most uniform radial profile at near perfect distribution or if the ratio of the area of the center pipe to outer annulus, ξ, is unity. If, however, $\xi \neq 1$, then

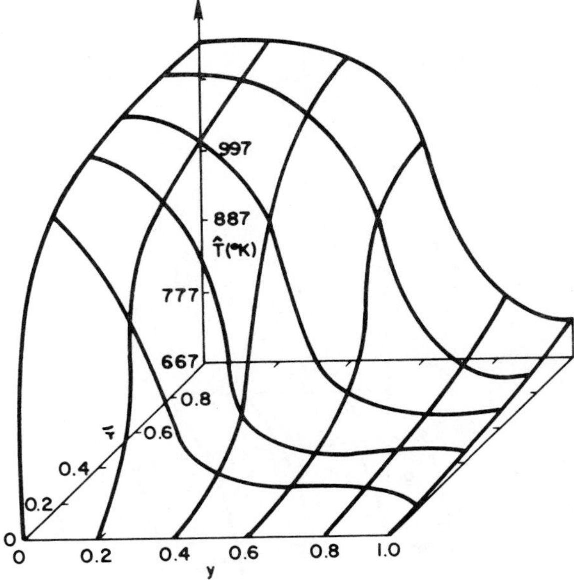

Figure 5. Particle temperature profile in a CF-Z reactor operating at standard conditions

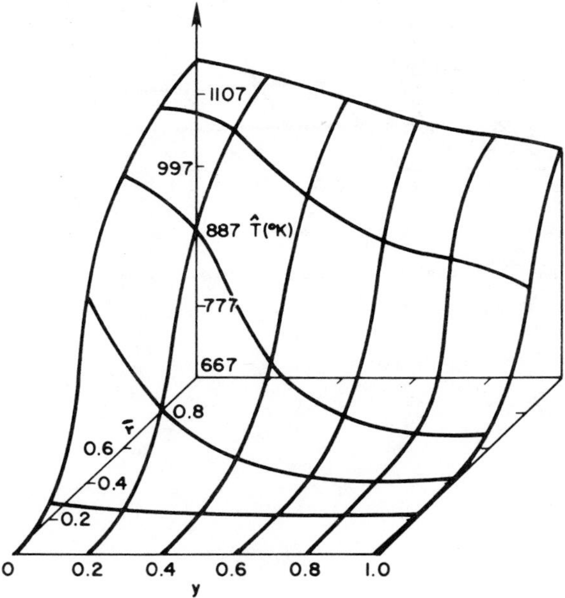

Figure 6. Particle temperature profile in a CF-π reactor at standard conditions

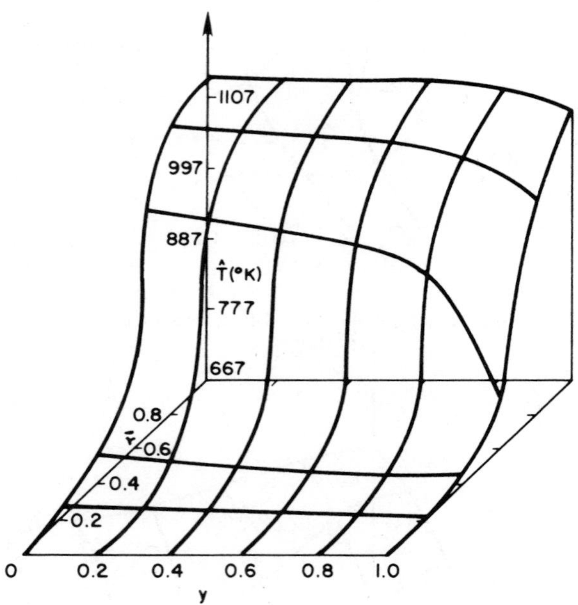

Figure 7. Particle temperature profile in a CF-π reactor ($R_o = 0.11$ *m*)

TABLE I

Total Conversion, χ (%)

Parameter Values*	CF-π	CP-Z	CF-Z	CP-π
A. "Standard"	75.4	41.8	50.	72.6
B. Perfect distribution	82.8	88.8	82.8	88.8
C. $\ell = 0.9$	69.4	43.1	47.7	66.7
D. $R_o = 0.11$ m	80.0	35.0	37.4	89.1
E. $\phi_1 = 0.1$	75.4	41.3	50.1	72.6
F. $w_{o_{c_E}} = 2.54$ m/s	97.4	96.2	94.4	98.3
G. Isothermal (1111 K, $w_{o_{c_E}} = 2.54$ m/s)	99.3	98.8	98.9	99.2
H. Isothermal (667 K, $w_{o_{c_E}} = 2.54$ m/s)	14.9	12.6	13.9	14.4
I. $r_2 = 0.25$ m. $T_o = 583$ K	18.7	15.9	19.1	15.7
J. $r_2 = 0.25$, $T_o = 583$ K (without axial flow)	20.3	19.0	20.2	19.0

*All parameter values are "standard" except those designated.

CF-π is preferable for ξ<1 and CP-π is superior for ξ<1. When maldistribution is absent, the effects of flow direction become evident, and for highly exothermic, irreversible reaction, CPRF is preferred. Based on this argument, CP-π would be the best flow configuration since it gives the best profile and also enjoys the advantageous effects of flow direction for highly exothermic reactions. In any case, the effects of all the factors at a certain reaction condition can be determined from the analysis presented here.

Legend of Symbols

b^*	friction factor in catalyst bed
B	dimensionless Heat of Reaction
c	dimensionless concentration
D	diffusivity
Da	Damkohler number
f	friction factor in the feed and outlet channels
ℓ	height of reactor
Le	Lewis number Pe_H/Pe_M
Pe	Peclet number
r	radial coordinate
r_1	inner radius of catalyst basket
r_2	outer radius of catalyst basket
R	outer radius of outer annulus
T_o,C_o	feed temperature and concentration
$\underline{v}$	dimensionless radial velocity
$\underline{w}$	dimensionless axial velocity
y	dimensionless catalyst bed dept
z	dimensionless axial coordinate, or strained y
θ	dimensionless temperature
ρ	aspect ratio r_1/r_2
ϕ	fractional free surface area of catalyst basket

Literature Cited

1. Dil'man, V.V.; Sergeer, S.P.; Genkin, V.S. Teor. Osn. Khim. Tekhn 1971, 5, 564.
2. Genkin, V.S.; Dil'man, V.V.; Sergeev, S.P. Int. Chem. Eng. 1973, 13, 24-8.
3. Kaye, L.A. AIChE 71st Annual Meeting, Paper No. 12E, Nov. 12-16, 1978, Miami Beach.
4. Raskin, A.Ya.; Sokolinskii, Yu. A.; Mukosei, V.I.; Aeror, M.E. Teor. Osn. Khim. Tekhn. 1968, 2, 220.
5. Raskin, A. Ya.; Sokolinskii, Yu. A.; Khim. Prom. 1969, 45, 520.
6. Hlavacek, V.; Kubicek, M. Chem. Eng. Sci. 1972, 27, 177.
7. Dudukovic, M.P.; Lamba, H.S. 80th AIChE National Meeting, Paper No. 576, 1975, Boston.

8. Ponzi, P.R.; Kaye, L.A. AIChE J. 1979, 25, No. 1, 100.
9. Calo, J.M. "Chemical Reaction Engineering - Houston"; V.W. Weekman, Jr., and Dan Luss, ACS Symposium Series 65, Washington, D.C., 1979, pp. 550-61.
10. Chang, H.-C.; Calo, J.M. Proceedings of the 1978 Summer Computer Simulation Conference Newport Beach, July, 1978, p.272.
11. Carberry, J. "Chemical and Catalytic Reaction Engineering"; Ch. 3, McGraw-Hill, 1976.
12. Ray, W.H.; Marek, M.; Elnashaie, S. Chem. Eng. Sci. 1972, 27, 1527.
13. Syceva, A.M.; Abrosimov, B.Z.; Mel'nikov, S.M.; Fadeev, I.G. Int. Chem. Eng. 1970, 10, No. 1, 66.
14. Vanderveen, J.W.; Luss, D.; Amundson, N.R. AIChE J. 1968, 14, No. 4, 636-43.
15. Rhee, H.-K.; Foley, D.; Amundson, W.R. Chem. Eng. Sci. 1973, 28, No. 2, 607.
16. Chang, H.-C. Ph.D. Thesis, Princeton University.

RECEIVED June 3, 1981.

17

Moving Bed Coal Gasifier Dynamics Using MOC and MOL Techniques

RICHARD STILLMAN

IBM Scientific Center, P.O. Box 10500, Palo Alto, CA 94304

The method of characteristics and both method of lines techniques, continuous-time discrete-space and continuous-space discrete-time, were used to solve the system of hyperbolic partial differential equations representing the dynamics of a moving bed coal gasifier with countercurrent gas-solid heat transfer. The adiabatic plug flow model considers 17 solids stream components, 10 gas stream components and 17 reactions. The kinetic and thermodynamic parameters were derived for a Wyoming subbituminous coal. The inherent numerical stiffness of the coupled gas-solids equations was handled by assuming that the gas stream achieved steady state values almost instantly. Calculated dynamic responses are shown for step changes in reactor pressure, blast temperature, steam flow rate, and coal moisture. Both steady state convergent and limit cycle responses were obtained.

The chemical industry is beginning to shift away from dependence on natural gas and petroleum to the use of coal as the basis for some of their hydrocarbon feed stocks. Examples can be found in ammonia manufacture, methanol production, acetic anhydride synthesis and synthetic gasoline production. Electric utilities are also looking at coal for use in combined cycle and fuel cell based power plants. A common first step in many of these plants is the gasification of coal, and since gasifier operation can impact the operation of other units in the plant, it is useful to predict the dynamic behavior of a coal gasifier reactor using a simulation model. The mathematical model can help to provide a better understanding of the complex dynamic behavior exhibited by the actual gasifier when subjected to simple and multiple disturbances. The extent of dynamic testing that can be performed on a simulation model, as well as the type of information which can be

0097-6156/81/0168-0331$09.00/0

obtained, often goes much beyond that which would be allowable or even possible in a given commercial installation.

Some work has already been done on the simulation of transient behavior of moving bed coal gasifiers. However, the analysis is not based on the use of a truly dynamic model but instead uses a steady state gasifier model plus a pseudo steady state approximation. For this type of approach, the time response of the gasifier to reactor input changes appears as a continuous sequence of new steady states.

Yoon, Wei and Denn (1, 2, 3) consider the time response of a gasifier to small changes in operating conditions such as might occur during normal operation of the reactor. They regard the time required to reach a new steady state, following a step change in operating conditions, as the most useful measure of transient response. For small step changes, they estimate this response time, and the changes in reactor variables during a transient, using a psuedo steady state technique. Their technique involves removal of the time variable from the system of dynamic equations by assuming that the space origin moves at the same velocity as the velocity of the thermal wave for the maximum bed temperature. They give detailed calculated results for small step changes in coal feed rates for both ash discharge and slagging gasifiers.

The gasifier modeling technique used by Hsieh, Ahner and Quentin (4) is based on the construction of a data space representing steady state reactor conditions using the University of Delaware steady state model described in Yoon, Wei and Denn (1). They made an analysis of the chemical reaction rates and thermal capacitance effects to develop the dynamic algorithms used to simulate the dynamic trends between the steady state points. Their dynamic responses are thus estimated by using the quasi steady state data bank, a linear interpolation routine and the derived dynamic algorithms. No actual details of their model are given. However, they do show a short 9 minute gasifier transient response for exit gas composition and temperature resulting from a ramp decrease in steam, air and coal feed rates. Daniel (5) has used a simplified method to develop a short time scale transient model for a moving bed gasifier.

Wei (6) presents a very brief discussion of coal gasification reactor dynamics. He describes the transient response to a small step change as a soft transience in which the movement from one steady state to another one nearby takes place as a wave through a series of pseudo steady states. He points out that the hard transience of start up and major upset in reactor operation are not well understood. One of the purposes of this paper is to increase this understanding.

Although the pseudo steady state approximation provides a useful tool for estimating some aspects of gasifier dynamics, it does not provide the means to examine the full range of dynamic behavior that one would expect to find for a gasifier. Therefore, a different approach has been taken here in that a nonlinear dynamic model representing moving bed gasifier dynamics is solved directly so that the global aspects of the dynamic behavior can be examined.

One further note, the University of Delaware gasifier model used in the pseudo steady state approximation assumes that the gas and solids temperatures are the same within the reactor. That assumption removes an important dynamic feedback effect between the countercurrent flowing gas and solids streams. This is particularly important when the burning zone moves up and down within the reactor in an oscillatory manner in response to a step change in operating conditions.

Process Description

A moving bed gasifier is a vertical reactor with countercurrent flowing gas and solids streams. Coal enters the top of the reactor and ash (and/or clinker or molten slag) is removed from the bottom. A mixture of steam and oxygen (or air) enters the bottom of the reactor and the raw product gas exits from the top.

An adiabatic steady state plug flow model has been developed by Stillman (7, 8) for this type of gasifier. For that model, the following sequence of physical and chemical events was assumed to take place in the reactor. Heat was extracted from the hot exiting gas by the entering solids stream so that the coal temperature was increased and the coal moisture was evaporated. A further increase in the coal temperature caused the coal volatile matter to be released, leaving a char. In the gasification zone, some of the char reacted with the carbon dioxide, water and hydrogen gas stream components. The oxygen in the feed gas burned all or almost all of the remaining char to provide the heat necessary to run this endothermic process. An ash residue was left from the combustion reaction. Some or all of this ash melted and then either solidified to form clinkers or else remained in a molten state, depending on the reactor operating conditions. The water gas shift reaction and the methanation reaction were also assumed to take place in the gas stream.

The model considered 17 components in the solids stream:

water, hydrogen, nitrogen, oxygen, carbon, sulfur, ash, slag, clinker, water(vs), hydrogen(vs), carbon dioxide(vs), carbon monoxide(vs), methane(vs), hydrogen sulfide(vs), ammonia(vs), tar(vs),

where (vs) indicates a volatile solid component. The gas stream had 10 components:

water, hydrogen, nitrogen, oxygen, carbon dioxide, carbon monoxide, methane, hydrogen sulfide, ammonia, tar.

A set of 17 reactions was written to simulate the reactor events and they included 1 reaction for drying, 8 parallel reactions for devolatilization, 5 reactions for gasification and 3 reactions for combustion. The kinetic and thermodynamic parameters for these reactions were derived for a Wyoming subbituminous coal.

The same model description will be used as the basis for deriving the gasifier dynamic model. All of the kinetic and thermodynamic parameters will be taken from Stillman (7, 8).

Dynamic Model

The continuity equations for mass and energy will be used to derive the hyperbolic partial differential equation model for the simulation of moving bed coal gasifier dynamics. Plug flow (no axial dispersion) and adiabatic (no radial gradients) operation will be assumed.

The mass balance dynamic equations for the solids stream are given by

$$\frac{\partial CjS}{\partial t} + \frac{\partial FjS}{\partial z} = \sum_{i} aij\ ri \qquad \begin{array}{l} i=1,2,\ldots,17 \\ j=1,2,\ldots,17 \end{array}$$

where CjS is the concentration of solids component j and FjS is the molar flux of solids component j defined as

$$FjS \equiv CjS\ uS \qquad j=1,2,\ldots,17.$$

The aij values are the stoichiometric coefficients for component j in reaction i, ri is the rate of reaction i, uS is the local solids stream velocity, t is real time, and z is distance measured from the bottom of the reactor.

The corresponding mass balance dynamic equations for the countercurrent flowing gas stream are given by

$$\frac{\partial CjG}{\partial t} - \frac{\partial FjG}{\partial z} = \sum_{i} aij\ ri \qquad \begin{array}{l} i=1,2,\ldots,17 \\ j=18,19,\ldots,27 \end{array}$$

where the molar flux of gas component j is defined as

$$FjG \equiv CjG\ uG \qquad j=18,19,\ldots,27.$$

The energy balance dynamic equation for the solids stream is

$$\frac{\partial \phi S}{\partial t} + \frac{\partial \psi S}{\partial z} = hGS\ AGS(TG - TS) - \sum_{i} ri\ \Delta Hi$$

$$i=1,2,\ldots,12,15,16,17$$

where the energy density and energy flux functions for the solids are defined by

$$\phi S \equiv cpS\ CS\ TS$$

and

$$\psi S \equiv \phi S \; uS,$$

respectively. The solids absolute temperature is denoted by TS and cpS is the molar heat capacity of the solids.

On the right-hand side of the solids energy balance, the first term represents the gas-solids heat transfer, and the second term is the heat gain and loss from the solids reactions. Exothermic reactions have a negative heat of reaction and endothermic reactions have a positive heat of reaction. The local gas-solids heat transfer coefficient is indicated by hGS, AGS is the local gas-solids heat transfer area, TG is the gas stream absolute temperature, and ΔHi are the heats of reaction.

The corresponding gas stream energy balance dynamic equation is

$$\frac{\partial \phi G}{\partial t} - \frac{\partial \psi G}{\partial z} = -hGS \; AGS(TG - TS) - \sum_{i} ri \; \Delta Hi \qquad i=13,14$$

where the energy density and energy flux functions for the gas are

$$\phi G \equiv cpG \; CG \; TG$$

and

$$\psi G \equiv \phi G \; uG,$$

respectively.

Gas and solids equations of state, and a reactor pressure equation, are needed to complete the definition of the dynamic model. Pressure drop as a linear function of coal bed height is used for the gasifier pressure equation and the ideal gas law is used for the gas equation of state. The solids equation of state is expressed in terms of both the bulk and raw densities. These equations are given in Stillman (7, 8).

The required initial conditions for the dynamic model are the temperature and flux profiles of the gas and the solids streams down the entire length of the gasifier at time zero.

FjS(t,z)	j=1,2,...,17	t=0	z=0,...,L
FjG(t,z)	j=18,19,...,27		
ψS(t,z)			
ψG(t,z)			
TS(t,z)			
TG(t,z).			

The boundary conditions needed for the model are the input molar fluxes and the temperatures of the solids and gas feed streams, and the inlet gas pressure. At the top of the reactor:

FjS(t,z)   j=1,2,...,17   t=t   z=L
TS(t,z).

At the bottom of the reactor:

FjG(t,z)   j=18,19,...,27   t=t   z=0
TG(t,z)
P(t,z).

This set of hyperbolic partial differential equations for the gasifier dynamic model represents an open or split boundary-value problem. Starting with the initial conditions within the reactor, we can use some type of marching procedure to solve the equations directly and to move the solution forward in time based on the specified boundary conditions for the inlet gas and inlet solids streams.

However, it is important to note that there is an inherent numerical stiffness in the coupled gas-solids equations because the gas stream moves through the reactor much more rapidly than the solids stream. In a typical example, while it only takes the gas about 7 seconds to move through the reactor, it takes the solids stream about a 1000 times longer.

Typical ratios of gas velocity to solids velocity are about 400, 4200, 1200, at the top of the reactor, in the burning zone, and at the bottom of the reactor, respectively. The solids and gas velocities represent the two characteristic directions for our hyperbolic system. If we plot these velocity curves on a reactor length versus time graph, the characteristic curves for the gas will be essentially horizontal in comparison to the solids stream characteristic because of the large gas to solids velocity ratios.

Making the assumption that the gas stream characteristic curves are indeed horizontal for all practical purposes, is equivalent to setting the time partial derivatives for the concentrations and the energy density equal to zero in the original system of partial differential equations for the gas. Using this approximation reduces the gas equations to a set of steady state equations.

Thus our final dynamic model for a moving bed coal gasifier consists of a set of hyperbolic partial differential equations for the solids stream coupled to a set of ordinary differential equations for the gas stream. Shampine and Gear (9) caution that, for systems containing elements with different time scales, removing stiffness by changing the model may be risky because it might be difficult to relate the solution of the modified model to that of the original model. In our case, no such difficulty was found. The steady state conditions predicted by the modified dynamic model were monitored by the steady state model (7, 8).

Distance Method of Lines

The distance method of lines (continuous-time discrete-space) technique is a straightforward way for obtaining the numerical solution of time dependent partial differential equations with one spatial variable. The original system of partial differential equations is transformed into a coupled system of time dependent ordinary differential equations by using spatial finite difference formulas to replace the spatial differentiation terms for a discrete set of spatial grid points. The number of ordinary differential equations produced by this operation is equal to the original number of partial differential equations multiplied by the number of grid points used. Thus, although we now have a larger number of equations to consider, usually the augmented system of ordinary differential equations is easier to solve numerically than the original smaller system of partial differential equations.

First order hyperbolic differential equations transmit discontinuities without dispersion or dissipation. Unfortunately, as Carver (10) and Carver and Hinds (11) point out, the use of spatial finite difference formulas introduces unwanted dispersion and spurious oscillation problems into the numerical solution of the differential equations. They suggest the use of upwind difference formulas as a way to diminish the oscillation problem. This follows directly from the concept of domain of influence. For hyperbolic systems, the domain of influence of a given variable is downstream from the point of reference, and therefore, a natural consequence is to use upstream difference formulas to estimate downstream conditions. When necessary, the unwanted dispersion problem can be reduced by using low order upwind difference formulas.

The Lagrange interpolation polynomial was used to develop the spatial finite difference formulas used for the distance method of lines calculation. For example, the two point polynomial for the solids flux variable F(t,z) can be expressed by

$$F(t,z) = \left(\frac{z - z(k)}{z(k-1) - z(k)}\right) F(t,k-1) + \left(\frac{z - z(k-1)}{z(k) - z(k-1)}\right) F(t,k)$$

where k represents the grid point index number. The index number increases in value from top to bottom of the reactor. If we take the partial derivative of the two point polynomial with respect to z at index point k, we obtain

$$\frac{\partial F(t,z)}{\partial z} = \frac{F(t,k) - F(t,k-1)}{\Delta z}$$

which is the two point upwind formula for the solids stream. The local grid spacing is indicated by Δz.

In a similar fashion, more accurate higher order formulas can be developed. The four point upwind formula is

$$\frac{\partial F(t,z)}{\partial z} = \frac{11F(t,k) - 18F(t,k-1) + 9F(t,k-2) - 2F(t,k-3)}{6\Delta z}.$$

The four point upwind biased formula is given by

$$\frac{\partial F(t,z)}{\partial z} = \frac{2F(t,k+1) + 3F(t,k) - 6F(t,k-1) + F(t,k-2)}{6\Delta z}$$

and the four point downwind biased formula is

$$\frac{\partial F(t,z)}{\partial z} = \frac{-F(t,k+2) + 6F(t,k+1) - 3F(t,k) - 2F(t,k-1)}{6\Delta z}.$$

Different combinations of spatial finite difference formulas were tried to determine the best set for our system of equations. The two point upwind formula was found to be best for the solids component molar fluxes. The low order formula was used because most of the gasifier reactions turn off abruptly when a component disappears and this creates sharp discontinuities. Higher order formulas tend to flatten out discontinuities, and in some cases, this causes material balances to be lost which then leads to numerical instability problems. Maintaining component material balance is an important aid to preserving numerical stability in the calculations. The low order formulas minimized these difficulties.

The four point upwind biased formula worked best for the solids stream energy flux calculation. Some downstream information was useful because of the countercurrent flow of the gas and solids streams. To keep the same order, the four point downwind biased formula was used at the top of the reactor and the four point upwind formula was used at the bottom.

Accuracy and calculation time are highly dependent on the number of spatial grid points used. More grid points give better accuracy but calculation time increases accordingly. To resolve this dilemma, a variable grid structure was used in the calculations. In the top part of the reactor, where the drying and devolatilization reactions were taking place, a coarse grid was used. In the bottom part of the reactor, where the gasification and combustion reactions were occurring, a finer mesh was used. A total of 82 grid points was used for the method of lines calculations. With the variable grid structure, the top third of the reactor had 13 nodes, the middle third had 21 nodes, and the bottom third had 48 nodes.

Grid spacing has no effect on the use of the two point upwind formula but it does effect the use of the four point formula. Therefore, grid reduction was done in a prescribed manner. For any one change, the grid spacing could only be cut in half and the grid change had to remain in effect for at least 3 node points.

This restriction allowed the coefficients for the four point upwind biased formula to sequence through the following values for one grid spacing change

2.0	3.0	-6.0	1.0	full step size
3.2	-1.5	-2.0	0.3	half step size
2.25	2.0	-4.5	0.25	half step size
2.0	3.0	-6.0	1.0	half step size

which is then repeated for each succeeding change.

The variables needed in the dynamic model calculations are the solids molar and energy fluxes. However, in the distance method of lines technique, when we replace the spatial differential terms by finite difference formulas, the time derivative in the remaining differential equation is in terms of either component concentration or energy density, which we do not want. Therefore, the following change was made in the distance method of lines model. Replacing CjS with its equivalent FjS/uS and taking the partial derivative with respect to time gives

$$\frac{\partial}{\partial t}\left(\frac{FjS}{uS}\right) = \frac{1}{uS}\frac{\partial FjS}{\partial t} - \frac{FjS}{uS\ uS}\frac{\partial uS}{\partial t} .$$

Neglecting the acceleration term, we have the approximation

$$\frac{\partial CjS}{\partial t} \simeq \frac{1}{uS}\frac{\partial FjS}{\partial t}$$

which gives us the desired flux variable in the model. A similar change was used to convert from energy density to energy flux.

Time Method of Lines

The time method of lines (continuous-space discrete-time) technique is a hybrid computer method for solving partial differential equations. However, in its standard form, the method gives poor results when calculating transient responses for hyperbolic equations. Modifications to the technique, such as the method of decomposition (12), the method of directional differences (13), and the method of characteristics (14) have been used to correct this problem on a hybrid computer. To make a comparison with the distance method of lines and the method of characteristics results, the technique was used by us in its standard form on a digital computer.

The original system of partial differential equations is transformed into a system of ordinary differential equations by replacing the time differential terms with time finite difference formulas. The number of equations in the new system is the same as the original number of equations. However, it is necessary to store intermediate results at spatial nodes for both current and previous time increments.

The Lagrange interpolation polynomial was again used to develop the finite difference formulas. To avoid additional iterations, only upwind differences were used. The two point upwind formula for the solids stream concentration variable at any location z within the reactor for time t is given by

$$\frac{\partial C(t,z)}{\partial t} = \frac{C(t,z) - C(t-1,z)}{\Delta t}$$

where Δt is the time increment. The three point upwind formula is

$$\frac{\partial C(t,z)}{\partial t} = \frac{3C(t,z) - 4C(t-1,z) + C(t-2,z)}{2\Delta t}$$

and the four point upwind formula is

$$\frac{\partial C(t,z)}{\partial t} = \frac{11C(t,z) - 18C(t-1,z) + 9C(t-2,z) - 2C(t-3,z)}{6\Delta t}.$$

The same 82 point variable grid structure was used in the time method of lines calculations as was used for the distance method of lines calculations. Also, the three and four point upwind formulas were found to attenuate the calculated step responses too much and they were discarded.

Method of Characteristics

The method of characteristics (15) is a natural way for solving hyperbolic partial differential equations. The technique is based on locating the characteristic propagation paths or directions for the partial differential equations and integrating the resulting ordinary differential equations along these directions. Thus, as with the method of lines, this technique transforms our problem from solving partial differential equations to solving ordinary differential equations.

For the gasifier dynamic model, the characteristic directions are given by the solids and gas stream velocities:

$$\frac{dz}{dt} = uS \qquad + \text{ direction}$$

$$\frac{dz}{dt} = -uG \qquad - \text{ direction.}$$

These two families of curves cross each other at a number of common nodes. However, the assumption was made earlier that only the steady state equations would be used for the gas stream calculations. Therefore, we only need to consider the application of the method of characteristics to the solids stream partial differential equations.

To use the technique for our system of equations, we first make the assumption that the solids stream velocity is piecewise constant for very small axial sections of the reactor, i.e., for the local integration step. The solids velocity still varies significantly within the gasifier, but its change is assumed piecewise rather than continuous.

If this is done, then the solids stream energy balance dynamic equations can be rewritten as

$$\frac{1}{uS}\frac{\partial \psi S}{\partial t} + \frac{\partial \psi S}{\partial z} = \text{RHS}$$

where RHS is the right-hand side of the original equation. Since solids velocity has been assumed piecewise constant, the total differential for the energy flux is given by

$$dt\,\frac{\partial \psi S}{\partial t} + dz\,\frac{\partial \psi S}{\partial z} = d\psi S.$$

Putting the above two equations into vector-matrix notation,

$$\begin{bmatrix} \frac{1}{uS} & 1 \\ dt & dz \end{bmatrix} \begin{bmatrix} \frac{\partial \psi S}{\partial t} \\ \frac{\partial \psi S}{\partial z} \end{bmatrix} = \begin{bmatrix} \text{RHS} \\ d\psi S \end{bmatrix}$$

we have a system of simultaneous linear algebraic equations in terms of the first partial derivatives. The characteristic solution for these equations is obtained when the determinant of the matrix vanishes. From linear equation theory, the determinant must then also vanish when the column vector on the right-hand side of the vector-matrix equation is substituted for either of the columns in the matrix on the left.

Substituting for the second column and setting the determinant to zero gives

$$\frac{1}{uS}\frac{d\psi S}{dt} = \text{RHS}.$$

Likewise, substituting for the first column gives

$$\frac{d\psi S}{dz} = \text{RHS}.$$

Both of these equations are ordinary differential equations for the energy flux along the solids velocity characteristic curve. Either one can be used, but we chose to implement the second equation for our method of characteristics dynamic model. A similar derivation can be done to develop the corresponding solids stream mass balance equations.

The integration of these equations is carried out for fixed time slices along the various characteristic curves within the reactor. Since the solids velocity varies down the reactor, this fixed time slicing gives the effect of a variable grid structure with a variable number of nodes. The location of the last node is determined by the bottom of the reactor rather than by the time slice. At any given time during a transient response calculation, nodes can be either added or subtracted to handle the changing conditions within the reactor.

A time slice of 1.5 minutes was used for most of the method of characteristics calculations. This time slice gave a total of 79 nodes for the base case initial condition steady state. With the variable solids velocity, the top third of the reactor had 13 nodes, the middle third had 15 nodes, and the bottom third had 51 nodes. This is very similar to the variable grid structure used in the method of lines calculations.

Numerical Considerations

As the gas and solids streams move through the gasifier, different reactions are slowly starting and abruptly stopping as components disappear at different locations in the various zones. While some reactions are proceeding vigorously, other reactions are just starting at very low rates. Extremely steep axial temperature and molar flux gradients are present in the burning zone area. Also, the zone locations are continually shifting up and down the reactor during a transient period. Thus , we would expect the gasifier differential equations to exhibit a high degree of numerical stiffness.

To examine the extent of this problem, an eigenvalue analysis was done for many different reactor transient and steady state time profiles. The range of stiffness ratios (absolute value of ratio of largest to smallest eigenvalue, real parts only) observed for the different reactor zones was as follows:

drying zone	10E5 - 10E10
devolatilization zone	10E2 - 10E8
gasification zone	10E1 - 10E4
burning zone	10E2 - 10E9
ash zone	10E0 - 10E1.

This indicates that the equations are extremely stiff with a constantly changing stiffness ratio throughout the reactor. Thus, the equations might be very stiff for a time, then moderately stiff for a while and then mildly stiff at various other locations within the reactor.

The eigenvalue mix was found to be very similar for all of

the profiles. On the average, 36.54% of the eigenvalues were negative real, 3.26% were positive real, 1.39% were negative complex, 0.16% were positive complex and the remaining 58.65% were zero. Also, the full Jacobian was singular at all locations within the reactor. With the presence of positive real eigenvalues and positive and negative complex eigenvalues, we would hope to find a few oscillatory transient responses, and indeed, some limit cycle responses were obtained. The ratio of the imaginary to the real part of the positive complex eigenvalue seemed to be the deciding factor between steady state convergent or limit cycle responses. This ratio was never observed to go above about 5 for the steady state results, but it went as high as 22 for the limit cycle runs.

Stiffness is not a problem for the transient part of the calculations since in the transient region integration step size is limited by accuracy rather than by stability (9). Nonstiff integration codes would be expected to perform better than stiff codes in this case. For the time method of lines and the method of characteristics, the differential equations are numerically in a transient state for the integration code even though the reactor is in a steady state condition. This is due to the reactions turning on and off at various locations as the integration proceeds along the reactor.

However, for the distance method of lines technique where integration is done at fixed distance nodes, stiffness could be a problem as we approach steady state conditions. Jacobian based stiff codes can not handle the integration in its present form because the Jacobian is always singular. In many sections of the reactor, the number of active differential equations is continually increasing or decreasing. Codes based on variable order multistep methods are very inefficient under these conditions because they must be restarted whenever a change occurs. This means that we would be using higher order code for some parts of the reactor and lower order code in the more rapidly changing sections. A fixed higher order method would probably be better. Also, for multistep methods, the integrator work- space has to be saved for each of the nodes.

Another possibility would be to use a multirate code which integrates each differential equation individually using different step sizes. Orailoglu (16) and Gear (17) discuss this approach but their procedure uses multistep Jacobian methods which are not efficient for our system of equations. What we need is a fixed order single step multirate method.

We finally decided to solve the problem by using a simple sifting procedure similar to the one used by Emanuel and Vale (18). Before the integrator was called, the derivatives were determined for the solids stream molar flux equations. Any flux equation that had a derivative below the value of the sift parameter was considered to be inactive at that time. The energy flux equation was always considered active. Only the active equations were then integrated for the given time step. This procedure was repeated for each node at each integrator time step. In this way the stiffness ratio could be reduced to less than 100 in the drying zone, less than 10 in the devolatilization zone and to about 1 in the gasification, burning, and ash zones, while still maintaining calculated results very close to those obtained for unsifted runs.

Various integration methods were tested on the dynamic model equations. They included an implicit iterative multistep method, an implicit Euler/modified Euler method, an implicit midpoint averaging method, and a modified divided difference form of the variable-order/variable-step Adams PECE formulas with local extrapolation. However, the best integrator for our system of equations turned out to be the variable-step fifth-order Runge-Kutta-Fehlberg method. This explicit method was used for all of the calculations presented here.

The initial conditions within the reactor needed to start the dynamic model calculations were established by using the steady state model (7, 8) to calculate a first estimate. This estimate was adjusted for use with the distance method of lines, time method of lines, and method of characteristics programs by running the individual programs to a steady state condition without changing input conditions.

The dynamic model calculations are done in a two phase process. In the first phase, the solids stream dynamic equations are integrated for the reactor while keeping gas stream conditions constant. When necessary, intermediate gas stream values are obtained by interpolation between storage nodes. These calculations proceed down the reactor for the distance and time method of lines, and back up the reactor for the method of characteristics. In the second phase, the gas stream steady state equations are integrated from the bottom to the top of the reactor while keeping the solids stream conditions constant. Intermediate solids stream values are obtained by interpolation between storage nodes.

The numerical stability requirement for the coupling of the gas-solids calculations in the distance method of lines model was estimated to be

$$uS\ \Delta t/\Delta z \leq 0.028.$$

This establishes the upper limit on the solids stream integration time step size for any specified grid spacing. Likewise, for the time method of lines model, the stability requirement was estimated as

$$uS\ \Delta t/\Delta z \geq 1.67$$

which sets the time grid size based on the specified storage node spacing.

An IBM 370/158 computer was used for all of the calculations. As would be expected, the calculation speed was different for each of the three dynamic models. The distance method of lines model ran at a speed of 0.23 times real time (4.3 times slower than real time) which was very slow. For short time transient calculations, the speed could be increased to real time speed, but long term numerical stability required the slower speed. The time method of lines model ran 3.92 times faster than real time. However, the calculated transient responses were incorrect and the model could not be used for that purpose. For a time slice of 1.5 minutes, the method of characteristics model ran 1.56 times faster than real time. If the time slice was increased to 5.0 minutes (fewer nodes), the speed increased to 4.75 times real time but the gas stream accuracy was reduced. Therefore, the 1.5 minute time slice was used for the calculations shown here.

Simulation Results

Based on the dynamic model presented in the previous sections, three computer programs were written to simulate moving bed gasifier dynamics using the method of characteristics and both method of lines techniques. Table I lists the ash (Lurgi) gasifier operating data for the base case initial conditions. Except for the multiple steady state runs, all of the step change response calculations were made using these initial condition values. The proximate, ultimate and simulation model analysis of the Roland seam subbituminous coal used in the calculations are given in Stillman (7).

Figure 1 shows the exit gas temperature time response to a step change in coal moisture from 34.67 to 27.00 wt % for the three dynamic models. This is a large step change involving a solids material wave moving through the reactor and it was intended to provide a severe test for the three methods.

The upper curve was calculated by the method of characteristics program and it exhibits a true limit cycle or sustained oscillation response (19). The middle curve was calculated by the distance method of lines program. The response is attenuated and stretched out. The final long term oscillations had random unequal periods and they were out of phase with the MOC results. The lower curve was calculated by the time method of lines program. The initial part of the response is similar to the DMOL results but then the temperature incorrectly levels out to a steady state condition. Thus, it was evident that the distance and time method of lines techniques were not as accurate as the method of characteristics procedure for calculating the gasifier step responses and they were discarded.

All of the remaining calculations were done by the MOC program. The solids temperature at the 0.3 m (1 ft) level will be used to display the step change responses. That location was chosen because it is initially slightly above the burning zone

Table I. Base Operating Data for Ash Discharge (Lurgi) Reactor

Reactor Bed Height	2.74 m (9.00 ft)
Reactor Internal Diameter	3.70 m (12.14 ft)
Exit Gas Pressure	2.84 MPa (28.0 atm)
Dry Coal Feed Rate	2067.7 kg/h-m-m (423.5 lb/h-ft-ft)
Coal Moisture Content	34.67 wt %
Dry Coal/Oxygen Ratio	2.80 wt/wt
Steam/Oxygen Ratio	8.20 mol/mol
Inlet Coal Temperature	78.0 °C (172.4 °F)
Inlet Gas Temperature	360.6 °C (681.0 °F)
Exit Gas Temperature	275.6 °C (528.1 °F)
Exit Solids Temperature	372.8 °C (703.0 °F)
Solids T at 0.3 m (1 ft)	912.6 °C (1674.6 °F)
Dry Exit Gas mol %	
Hydrogen	40.83
Carbon Dioxide	31.47
Carbon Monoxide	14.53
Methane	10.88
Nitrogen, Ammonia, Hydrogen Sulfide, Tar	2.29

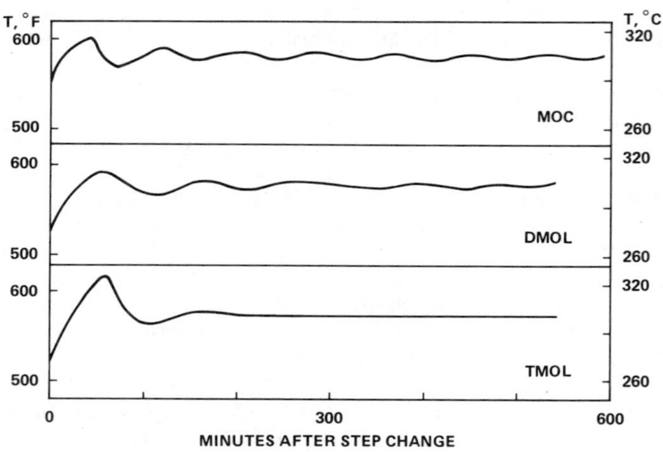

Figure 1. Exit gas temperature response, coal moisture reduced to 27.00 wt %

where large temperature fluctuations can occur when the burning zone shifts back and forth in the reactor.

Inlet Gas Temperature. A few step response runs were made for changes in inlet gas (blast) temperature. Only steady state convergent responses were observed and the final results are summarized in Table II. Figure 2 shows the solids temperature response when the inlet gas temperature is reduced from 360.6 °C (681.0 °F) to 305.0 °C (581.0 °F). This change puts less heat into the reactor which slows down the reactions in the gasifier above the burning zone. This causes the burning zone to shift down the reactor a short distance producing the rapid temperature drop exhibited in Figure 2.

Figure 3 shows the response when the inlet gas temperature is increased to 416.1 °C (781.0 °F). Since more heat is available in the gasifier, the reaction rates are increased, the burning zone shifts slowly up the reactor a short distance, and the solids temperature at the 0.3 m (1 ft) level gradually increases as the burning zone moves a little closer.

Coal Moisture. The results for changes in coal moisture content are given in Tables III, IV and V, and Figures 4 and 5. In addition to steady state convergent results, we now have underdamped (decaying oscillation) and limit cycle responses. Figure 4 shows the solids temperature response when the inlet coal moisture is reduced from 34.67 to 32.91 wt %. There is a 66 minute delay in the response which represents the transport time for the solids stream wave to reach the 0.3 m (1 ft) level. As the burning zone begins to shift up the reactor, the solids temperature peaks and starts back down during the next 50 minutes which is the time it takes the solids wave to move out of the gasifier. The solids temperature oscillates a few more times as the location of the burning zone shifts up and down before settling out to a steady state position slightly above its initial location.

The underdamped steady state results for reducing the coal moisture content to 31.05 wt % are given in the third data column of Table III. The solids temperature response starts out similar to the Figure 4 results except the temperature rises to a higher peak value of 1133 °C (2072 °F). Starting with an amplitude of about 50 °C (90 °F), the temperature response then oscillates in a decaying manner for the next 2300 minutes until it reaches a final steady state value. The final location of the maximum solids temperature is slightly below the 0.3 m (1 ft) level.

The limit cycle response caused by decreasing the coal moisture content to 27.00 wt % is shown in Figure 5. Again there is a steep solids temperature rise as the solids material wave reaches the 0.3 m (1 ft) level, but in this case, the decaying temperature oscillations only last about 200 minutes before the gasifier settles into a sustained oscillation mode with an esti-

Table II. Inlet Gas Temperature, Final Steady State Results

Inlet Gas Temperature	°C	305.0	388.3	416.1
	°F	581.0	731.0	781.0
Exit Gas Temperature	°C	269.4	276.7	277.6
	°F	517.0	530.0	531.7
Exit Solids Temperature	°C	330.9	398.1	424.3
	°F	627.7	748.6	795.7
Solids T at 0.3 m (1 ft)	°C	778.7	951.4	986.4
	°F	1433.6	1744.6	1807.5
Dry Exit Gas	mol %			
Hydrogen		40.92	40.82	40.82
Carbon Dioxide		32.08	31.45	31.43
Carbon Monoxide		13.83	14.56	14.59
Methane		10.88	10.88	10.87
Nitrogen, Ammonia, Hydrogen Sulfide, Tar		2.29	2.29	2.29

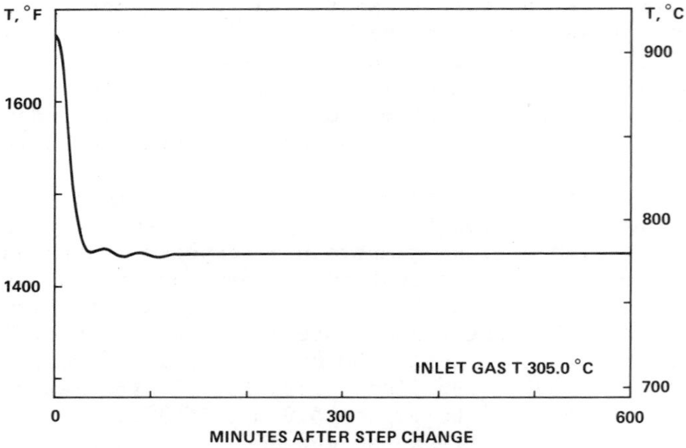

Figure 2. Solids temperature at 0.3 m, inlet gas temperature reduced to 305.0°C

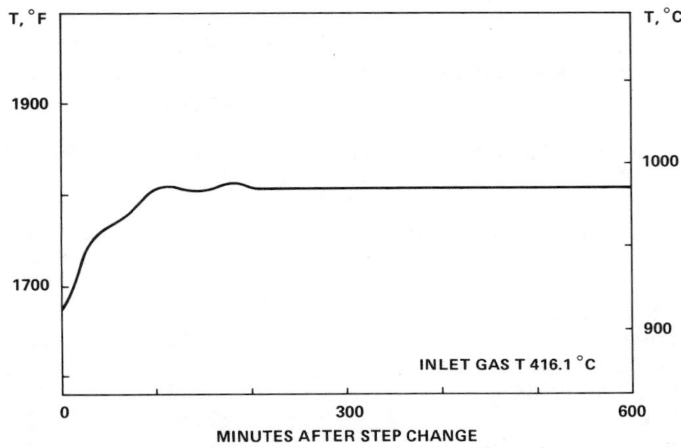

Figure 3. Solids temperature at 0.3 m, inlet gas temperature increased to 416.1°C

Table III. Coal Moisture Change, Final Steady State Results

Coal Moisture	wt %	33.80	32.91	31.05
Exit Gas Temperature	°C	276.0	281.2	290.4
	°F	528.8	538.1	554.8
Exit Solids Temperature	°C	375.1	372.1	368.1
	°F	707.1	701.7	694.6
Solids T at 0.3 m (1 ft)	°C	838.6	928.5	1112.9
	°F	1541.5	1703.3	2035.2
Dry Exit Gas	mol %			
Hydrogen		41.06	41.08	41.11
Carbon Dioxide		31.91	31.95	32.02
Carbon Monoxide		13.85	13.79	13.69
Methane		10.90	10.90	10.90
Nitrogen, Ammonia, Hydrogen Sulfide, Tar		2.28	2.28	2.28

Table IV. Coal Moisture 30.08 wt %, Limit Cycle Results

		Max	Min
Exit Gas Temperature	°C	297.4	291.7
	°F	567.3	557.1
Exit Solids Temperature	°C	367.3	366.4
	°F	693.1	691.6
Solids T at 0.3 m (1 ft)	°C	1126.1	1070.4
	°F	2058.9	1958.8
Dry Exit Gas	mol %		
Hydrogen		41.00	41.22
Carbon Dioxide		31.78	32.27
Carbon Monoxide		14.06	13.31
Methane		10.88	10.92
Nitrogen, Ammonia, Hydrogen Sulfide, Tar		2.28	2.28

Estimated Limit Cycle Period 81.4 minutes

Table V. Coal Moisture 27.00 wt %, Limit Cycle Results

		Max	Min
Exit Gas Temperature	°C	306.9	303.3
	°F	584.4	577.9
Exit Solids Temperature	°C	367.2	366.6
	°F	693.0	691.9
Solids T at 0.3 m (1 ft)	°C	1123.6	1095.8
	°F	2054.5	2004.5
Dry Exit Gas	mol %		
Hydrogen		41.25	41.36
Carbon Dioxide		32.30	32.57
Carbon Monoxide		13.27	12.86
Methane		10.91	10.94
Nitrogen, Ammonia, Hydrogen Sulfide, Tar		2.27	2.27

Estimated Limit Cycle Period 82.8 minutes

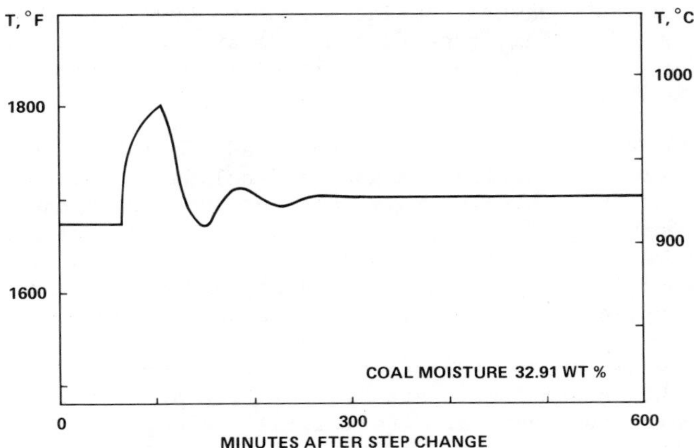

Figure 4. Solids temperature at 0.3 m, coal moisture reduced to 32.91 wt %

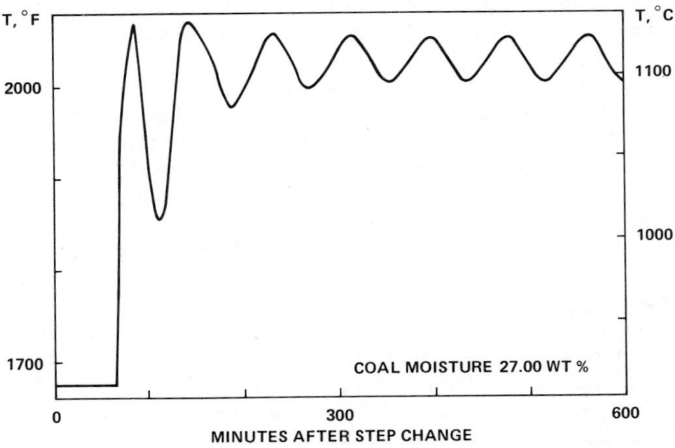

Figure 5. Solids temperature at 0.3 m, coal moisture reduced to 27.00 wt %

mated period of 82.8 minutes. The location of the solids peak burning zone temperature moves up and down in rhythm with the oscillations but always remains slightly above the 0.3 m (1 ft) level. The decaying type response means that the limit cycle surface, which encloses an equilibrium or steady state point, was approached from the outside direction.

Steam Flow. The steam/oxygen feed ratio is an important variable for controlling the gas and solids temperatures in the burning zone and changing this ratio has a significant effect on gasifier operation. The simulation results for steam flow rate step changes are given in Tables VI-X and Figures 6, 7 and 8. Besides having steady state convergent, underdamped steady state, and limit cycle responses, we also have an underdamped limit cycle response.

Figure 6 shows the steady state convergent results for reducing the steam flow rate by 1.875%. Since the oxygen flow rate remains constant at its initial value, this step change reduces the steam/oxygen molar ratio from 8.20 to 8.05. The response shows a small almost linear temperature rise for about 40 minutes, and then a heavily damped oscillatory drop to the final steady state condition. Even though reducing the steam flow rate raises the gas peak temperature, the increase in this run is not enough to overcome the effect of the reduced gas stream flow which puts less heat into the upper part of the reactor. This shifts the burning zone down the reactor and causes the final solids temperature at the 0.3 m (1 ft) level to drop by about 41 °C (74 °F).

The underdamped steady state results for a 2.25% decrease in steam feed rate are given in the third data column of Table VI. This step change reduces the steam/oxygen molar ratio to 8.02. The underdamped response starts out very similar to the Figure 6 results and appears to be headed for a steady state condition. However, beginning at about the 305 minute point, the burning zone begins to move back up the reactor. Thus, during the period of 305 to 365 minutes, the solids temperature at the 0.3 m (1 ft) level rises from 868 °C (1594 °F) to 962 °C (1763 °F). The response then settles back into a decaying oscillation mode until steady state conditions are reached 445 minutes later.

The underdamped limit cycle results for a 2.50% decrease in steam feed rate (8.00 steam/oxygen molar ratio) are given in Table VII. Once again, the response starts out like the Figure 6 response but then goes into a decaying limit cycle mode. However again, beginning at about the 340 minute point, the burning zone shifts back up the reactor. During the next 60 minutes, the solids temperature at the 0.3 m (1 ft) level goes from 860 °C (1580 °F) to 956 °C (1752 °F). The response then returns to the decaying limit cycle mode for the next 935 minutes until a very small stable limit cycle is reached.

The limit cycle responses for a 5.0% decrease and a 5.0%

Table VI. Steam Feed Change, Final Steady State Results

Steam Feed Rate Change	%	-1.25	-1.875	-2.25
Exit Gas Temperature	°C	273.2	271.9	271.3
	°F	523.8	521.5	520.3
Exit Solids Temperature	°C	374.1	374.8	372.4
	°F	705.3	706.7	702.4
Solids T at 0.3 m (1 ft)	°C	886.9	872.3	952.3
	°F	1628.5	1602.1	1746.2
Dry Exit Gas	mol %			
Hydrogen		40.66	40.58	40.53
Carbon Dioxide		31.36	31.31	31.28
Carbon Monoxide		14.74	14.84	14.90
Methane		10.94	10.97	10.99
Nitrogen, Ammonia, Hydrogen Sulfide, Tar		2.30	2.30	2.30

Table VII. Steam Feed Rate Change -2.5%, Limit Cycle Results

		Max	Min
Exit Gas Temperature	°C	271.0	270.8
	°F	519.9	519.5
Exit Solids Temperature	°C	372.8	372.7
	°F	703.0	702.8
Solids T at 0.3 m (1 ft)	°C	948.8	945.6
	°F	1739.8	1734.1
Dry Exit Gas	mol %		
Hydrogen		40.50	40.51
Carbon Dioxide		31.24	31.27
Carbon Monoxide		14.96	14.92
Methane		11.00	11.00
Nitrogen, Ammonia, Hydrogen Sulfide, Tar		2.30	2.30

Estimated Limit Cycle Period 82.0 minutes

Table VIII. Steam Feed Rate Change -3.75%, Limit Cycle Results

		Max	Min
Exit Gas Temperature	°C	273.5	267.7
	°F	524.3	513.8
Exit Solids Temperature	°C	375.2	372.6
	°F	707.3	702.6
Solids T at 0.3 m (1 ft)	°C	977.8	889.4
	°F	1792.1	1632.9
Dry Exit Gas	mol %		
Hydrogen		40.15	40.53
Carbon Dioxide		30.76	31.62
Carbon Monoxide		15.74	14.47
Methane		11.03	11.08
Nitrogen, Ammonia, Hydrogen Sulfide, Tar		2.32	2.30

Estimated Limit Cycle Period 82.6 minutes

Table IX. Steam Feed Rate Change -5.0%, Limit Cycle Results

		Max	Min
Exit Gas Temperature	°C	271.7	265.2
	°F	521.1	509.3
Exit Solids Temperature	°C	376.8	373.6
	°F	710.2	704.4
Solids T at 0.3 m (1 ft)	°C	958.7	860.1
	°F	1757.6	1580.2
Dry Exit Gas	mol %		
Hydrogen		39.92	40.37
Carbon Dioxide		30.52	31.53
Carbon Monoxide		16.15	14.64
Methane		11.08	11.15
Nitrogen, Ammonia, Hydrogen Sulfide, Tar		2.33	2.31

Estimated Limit Cycle Period 82.5 minutes

Table X. Steam Feed Rate Change +5.0%, Limit Cycle Results

		Max	Min
Exit Gas Temperature	°C	279.8	275.3
	°F	535.7	527.6
Exit Solids Temperature	°C	374.7	372.9
	°F	706.5	703.3
Solids T at 0.3 m (1 ft)	°C	844.5	802.6
	°F	1552.1	1476.7
Dry Exit Gas	mol %		
Hydrogen		41.46	41.66
Carbon Dioxide		31.93	32.34
Carbon Monoxide		13.68	13.05
Methane		10.66	10.69
Nitrogen, Ammonia, Hydrogen Sulfide, Tar		2.27	2.26

Estimated Limit Cycle Period 83.5 minutes

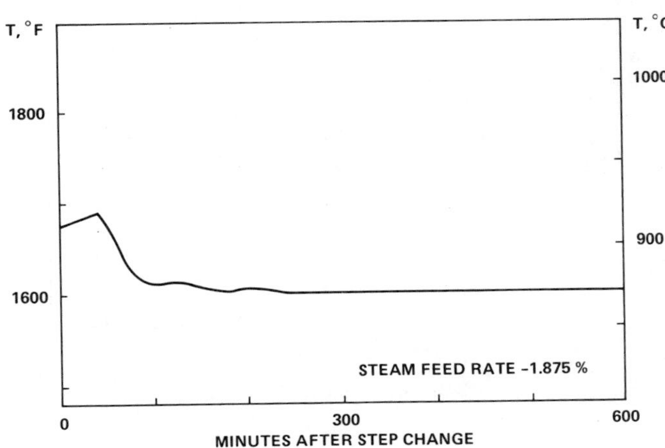

Figure 6. Solids temperature at 0.3 m, steam feed rate reduced 1.875 %

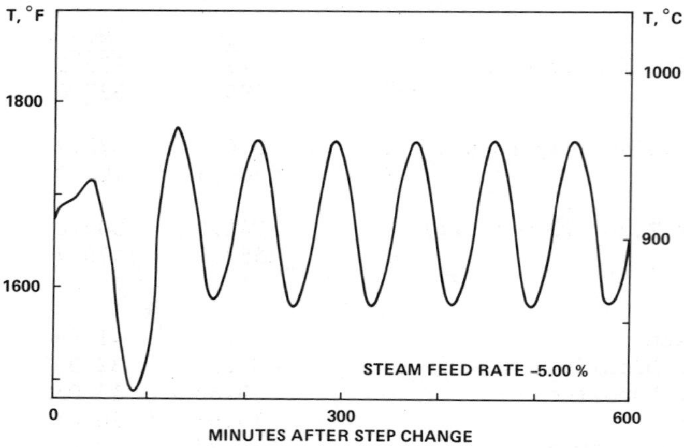

Figure 7. Solids temperature at 0.3 m, steam feed rate reduced 5.0 %

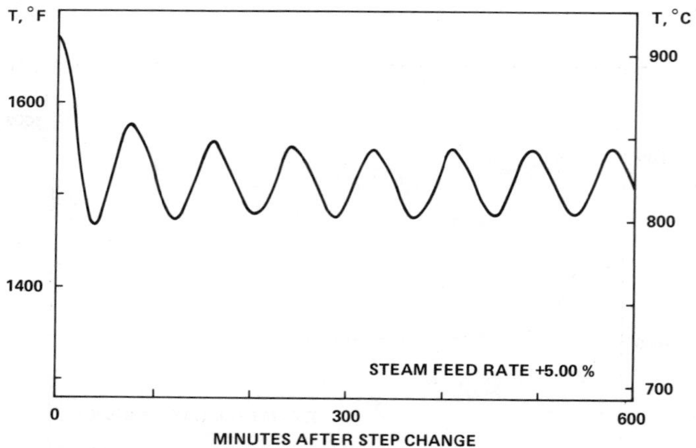

Figure 8. Solids temperature at 0.3 m, steam feed rate increased 5.0 %

increase in steam feed flow rate are given in Figures 7 and 8. For these runs, the steam/oxygen molar ratios are 7.79 and 8.61, respectively. Both of the curves show a classic decaying limit cycle form which indicates that the limit cycle surface was approached from the outside.

Exit Gas Pressure. Gasifier pressure has an important effect on the composition of the exit raw gas and on the operation of the reactor. At normal operating pressure, increases in reactor pressure increase the methane and carbon dioxide content of the raw gas and decrease the hydrogen and carbon monoxide content. Decreases in pressure have the opposite effect. The reactor pressure step change results are given in Figures 9 and 10, and Tables XI, XII and XIII.

Figure 9 shows the steady state convergent results for reducing the exit gas pressure from 2.84 MPa (28 atm) to 2.53 MPa (25 atm). The pressure decrease lowers the gas density but increases the gas velocity since the molar feed rate of the inlet gas remains constant. This puts less heat into the reactor which lowers the exit gas temperature and causes the burning zone to shift down the reactor a short distance. This is reflected in the Figure 9 solids temperature response which starts out with a short false temperature rise but then exhibits a rapid temperature drop to a final steady state condition (non-minimum phase response).

Figure 10 shows the limit cycle response produced when the exit gas pressure is increased to 3.45 MPa (34 atm). This change decreases the gas velocity and shifts the burning zone up the reactor a short distance where, in this case, it oscillates up and down with a period of about 82.6 minutes. In Figure 10, the solids temperature response at the 0.3 m (1 ft) level begins with a short false temperature drop and then quickly rises and goes into an expanding oscillation mode before settling into a stable limit cycle response. The expanding oscillations indicate that the limit cycle surface was approached from the inside. The Table XII calculation had a more pronounced expanding oscillation phase as it took 846 minutes to reach a stable limit cycle response.

No experimental gasifier data was found to verify any of the simulation model results. However, limit cycle responses have been experimentally observed for the pressure in combustion chambers and boilers (20). This lends credence to our calculated results since coal combustion is an important factor in gasifier operation.

Multiple Steady States. Not only can the nonlinearity of our dynamic model equations produce limit cycle responses, it can give rise to multiple steady state conditions in which the same set of operating parameters can produce different reactor profiles. By accident, three of these multiple steady state responses were obtained and they are summarized in Table XIV.

To check the validity of the pressure step change results,

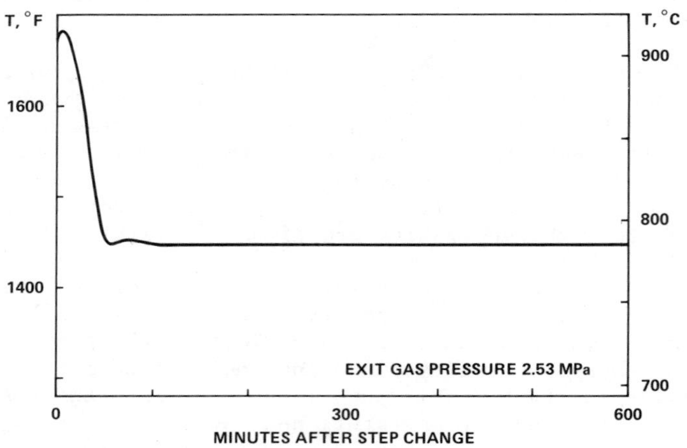

Figure 9. Solids temperature at 0.3 m, exit gas pressure reduced to 2.53 MPa

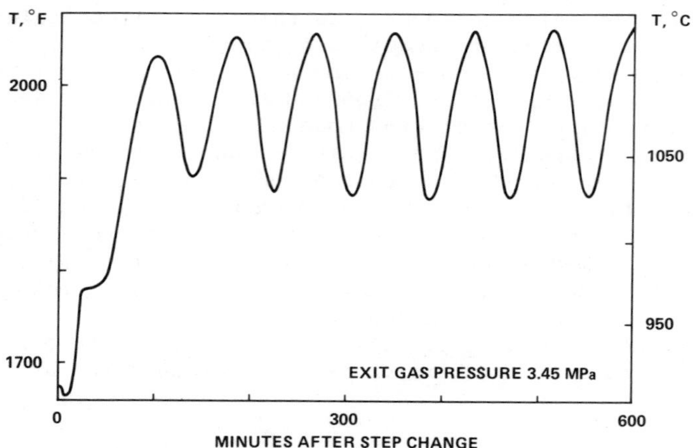

Figure 10. Solids temperature at 0.3 m, exit gas pressure increased to 3.45 MPa

Table XI. Exit Pressure Change, Final Steady State Results

Exit Gas Pressure	MPa	2.53	3.14
	atm	25.0	31.0
Exit Gas Temperature	°C	268.2	278.2
	°F	514.7	532.7
Exit Solids Temperature	°C	384.1	370.1
	°F	723.3	698.2
Solids T at 0.3 m (1 ft)	°C	785.2	992.8
	°F	1445.4	1819.0
Dry Exit Gas	mol %		
Hydrogen		41.23	40.44
Carbon Dioxide		31.00	32.30
Carbon Monoxide		15.00	13.58
Methane		10.50	11.37
Nitrogen, Ammonia, Hydrogen Sulfide, Tar		2.27	2.31

Table XII. Exit Pressure 3.24 MPa, Limit Cycle Results

		Max	Min
Exit Gas Temperature	°C	280.7	275.6
	°F	537.2	528.1
Exit Solids Temperature	°C	370.3	368.9
	°F	698.6	696.0
Solids T at 0.3 m (1 ft)	°C	1067.7	978.9
	°F	1953.9	1794.0
Dry Exit Gas	mol %		
Hydrogen		40.10	40.35
Carbon Dioxide		32.16	32.80
Carbon Monoxide		13.90	12.95
Methane		11.52	11.59
Nitrogen, Ammonia, Hydrogen Sulfide, Tar		2.32	2.31

Estimated Limit Cycle Period 82.7 minutes

Table XIII. Exit Pressure 3.45 MPa, Limit Cycle Results

		Max	Min
Exit Gas Temperature	°C	283.6	276.4
	°F	542.4	529.6
Exit Solids Temperature	°C	369.2	367.3
	°F	696.5	693.1
Solids T at 0.3 m (1 ft)	°C	1126.8	1026.6
	°F	2060.3	1879.9
Dry Exit Gas	mol %		
Hydrogen		39.66	39.98
Carbon Dioxide		32.51	33.46
Carbon Monoxide		13.61	12.21
Methane		11.88	12.03
Nitrogen, Ammonia, Hydrogen Sulfide, Tar		2.34	2.32

Estimated Limit Cycle Period 82.6 minutes

Table XIV. Exit Pressure, Multiple Steady State Results

Exit Gas Pressure	MPa	2.84*	2.84	2.84
	atm	28.0	28.0	28.0
Exit Gas Temperature	°C	275.6	275.7	270.9
	°F	528.1	528.2	519.7
Exit Solids Temperature	°C	372.0	370.4	379.0
	°F	701.6	698.8	714.2
Solids T at 0.3 m (1 ft)	°C	918.3	1002.6	782.7
	°F	1684.9	1836.6	1440.8
Dry Exit Gas	mol %			
Hydrogen		40.83	40.83	41.03
Carbon Dioxide		31.47	31.47	31.86
Carbon Monoxide		14.53	14.53	13.93
Methane		10.88	10.88	10.90
Nitrogen, Ammonia, Hydrogen Sulfide, Tar		2.29	2.29	2.28

*unstable steady state results

initial conditions different from the base case values were used for some of the transient calculations. For example, if the final steady state conditions produced by increasing the exit gas pressure from 2.84 MPa (28 atm) to 3.14 MPa (31 atm) are used as the initial conditions, and the exit gas pressure is stepped back to its original value of 2.84 MPa (28 atm), we obtain a steady state convergent response which returns to the original initial condition. Likewise, if we step the exit gas pressure up to 3.45 MPa (34 atm), we obtain the same limit cycle response that we obtained in stepping from 2.84 MPa (28 atm) to 3.45 MPa (34 atm).

Next we used the 3.45 MPa (34 atm) limit cycle results at the 810 minute point for our initial conditions. If we step the exit gas pressure down to 3.14 MPa (31 atm), we obtain the same under-damped steady state result that we obtained by stepping from 2.84 MPa (28 atm) to 3.14 MPa (31 atm). However, if we step the exit gas pressure down to 2.84 MPa (28 atm), we do not return to the original initial condition state. Instead, we obtain the multiple steady state result given in the third data column of Table XIV. The exit gas temperature, the solids temperature at the 0.3 m (1 ft) level, and the raw gas carbon monoxide content are lower than the original values. The exit solids temperature, and the hydrogen and carbon dioxide content of the raw gas are higher. This condition is caused by the burning zone not being able to move back up the reactor after its initial downward shift.

We next tried stepping the exit gas pressure down to 3.14 MPa (31 atm), and then after 270 minutes, stepping on down to 2.84 MPa (28 atm). These results are given in the first two data columns of Table XIV. After about 370 minutes, the reactor reached the spurious steady state conditions given in column one, which are very close to the original initial condition values. However, the reactor only remained there for about 20 minutes, and then spent the next 330 minutes moving to the multiple steady state result given in column two. The spurious steady state results represented an unstable multiple steady state condition. The column two results show that even though the solids temperature profile down the reactor is slightly different from the original profile, the exit gas composition remains the same (to two decimal places).

Finally, the exit gas pressure was stepped down from 3.45 MPa (34 atm) to 3.24 MPa (32 atm), 3.04 MPa (30 atm), and 2.84 MPa (28 atm) in 270 minute increments. This procedure gave a steady state convergent response which corresponded to the original initial condition.

Bifurcation. Bifurcation refers to the switching or branching from one type of response behavior to another as a parameter passes through a critical value. For us, when the parameter is below the critical value, our step change responses are in the steady state convergent region. When they are above the critical value, our responses are in the limit cycle region. As the parameter values approach the critical point, we enter a transition region. Just below the critical point, even though we still have steady state convergent responses, the steady state is only reached after a long period of decaying oscillation. The closer we approach the critical point, the longer the decaying oscillation time becomes. Just above the critical point, we have long periods of either decaying or expanding oscillations which finally result in stable limit cycle responses. Thus, the approach to the critical point from either above or below is asymptotic.

Table XV provides a summary of the coal moisture, steam feed rate and exit gas pressure transient response runs showing the time required to reach the given condition. It provides a rough estimate for the values of the bifurcation points for these runs. Thus, the bifurcation point for the coal moisture step change runs lies between 30.08 and 31.05 wt % moisture. For the steam feed rate changes, it lies between -2.25% and -2.50%, and for the exit gas pressure, it is bracketed by the 3.14 MPa (31 atm) and 3.24 MPa (32 atm) values.

Summary

The continuity equations for mass and energy were used to derive an adiabatic dynamic plug flow simulation model for a moving bed coal gasifier. The resulting set of hyperbolic partial differential equations represented a split boundary-value problem. The inherent numerical stiffness of the coupled gas-solids equations was handled by removing the time derivative from the gas stream equations. This converted the dynamic model to a set of partial differential equations for the solids stream coupled to a set of ordinary differential equations for the gas stream.

The method of characteristics, the distance method of lines (continuous-time discrete-space), and the time method of lines (continuous-space discrete-time) were used to solve the solids stream partial differential equations. Numerical stiffness was not considered a problem for the method of characteristics and time method of lines calculations. For the distance method of lines, a possible numerical stiffness problem was solved by using a simple sifting procedure. A variable-step fifth-order Runge-Kutta-Fehlberg method was used to integrate the differential equations for both the solids and the gas streams.

Step change dynamic response runs revealed that the distance and time method of lines techniques were not as accurate as the method of characteristics procedure for calculating gasifier transients. Therefore, these two techniques were discarded and the remaining calculations were all done using the method of characteristics.

The nonlinear dynamic gasifier model produced a wide variety of transient response types when subjected to step changes in operating conditions. However, stepping the inlet gas (blast) temperature up and down by as much as 55.6 °C (100 °F) only gave steady state convergent responses. Varying the feed coal moisture

Table XV. Summary of Bifurcation Response Time Data

		Minutes	Type
Coal Moisture wt %	33.80	375	Steady State
	32.91	480	Steady State
	31.05	2430	Decay Steady State
	30.08	324	Limit Cycle
	27.00	315	Limit Cycle
Steam Feed % Change	-1.25	270	Steady State
	-1.875	375	Steady State
	-2.25	810	Decay Steady State
	-2.50	1335	Decay Limit Cycle
	-3.75	345	Limit Cycle
	-5.00	300	Limit Cycle
	+5.00	210	Limit Cycle
Pressure MPa(atm)	2.53(25)	270	Steady State
	3.14(31)	1890	Decay Steady State
	3.24(32)	846	Expand Limit Cycle
	3.45(34)	351	Limit Cycle

content in the range of 34.67 to 27.00 wt % gave steady state convergent, underdamped (decaying oscillation), and limit cycle (sustained oscillation) responses. Increasing and decreasing the inlet steam flow rate by up to 5 % produced steady state convergent, underdamped steady state, limit cycle, and decaying oscillation phase limit cycle responses. Perturbing the exit gas pressure in the range of 2.53 MPa (25 atm) to 3.45 MPa (34 atm) gave non-minimum phase steady state convergent, underdamped steady state, limit cycle, and expanding ocillation phase limit cycle responses. Thus, in these calculations the limit cycle surface was approached from both the inside and the outside directions.

Changing the exit gas pressure also gave three multiple steady state responses in which the same set of operating parameters produced different reactor profiles. Finally, a rough estimate for the location of the bifurcation points was given for the coal moisture, steam feed rate, and exit gas pressure transient response runs.

Nomenclature

aij	stoichiometric coefficient component j in reaction i
AGS	local gas-solids heat transfer area/volume ratio
cpG	gas molar heat capacity
cpS	solids molar heat capacity
C	concentration variable
CG	total gas concentration
CS	total solids concentration
CjG	component j concentration in gas
CjS	component j concentration in solids
F	molar flux variable
FjG	gas molar flux for component j
FjS	solids molar flux for component j
hGS	local gas-solids heat transfer coefficient
ΔHi	heat of reaction i
i	gasifier reaction index number
j	gas/solids component index number
k	distance node index number
L	top of reactor node index number
P	absolute pressure
ri	rate of reaction i
t	real time
t	time node index number
Δt	time grid spacing
TG	gas absolute temperature
TS	solids absolute temperature
uG	local gas velocity
uS	local solids velocity
z	distance from reactor bottom
Δz	local distance grid spacing

Greek Letters

ϕG	gas energy density
ϕS	solids energy density
ψG	gas energy flux
ψS	solids energy flux

Literature Cited

1. Yoon, H.; Wei, J.; Denn, M. "Modeling and Analysis of Moving Bed Coal Gasifiers"; EPRI Report No. AF-590, Vol 2, 1978.
2. ________. "Transient Behavior of Moving Bed Coal Gasification Reactors"; 71st Annual AIChE Meeting, Miami, 1978.
3. ________. AIChE J. 1979, 25, (3), 429.
4. Hsieh, B. C. B.; Ahner, D. J.; Quentin, G. H. in Vogt, W. G.; Mickle, M. H., Eds.; "Modeling and Simulation, Vol 10, Pt 3, Energy and Environment"; ISA: Pittsburgh, 1979; p 825.
5. Daniel, K. J. "Transient Model of a Moving-Bed Coal Gasifier"; 88th National AIChE Meeting, Philadelphia, 1980.
6. Wei, J. "The Dynamics and Control of Coal Gasification Reactors"; Proceedings JACC, Vol II; ISA: Pittsburgh, 1978; p 39.
7. Stillman, R. IBM J. Res. Develop. 1979, 23, (3), 240.
8. ________. "Simulation of a Moving Bed Gasifier for a Western Coal Part 2: Numerical Data"; IBM Palo Alto Scientific Center Report No. G320-3382, 1979.
9. Shampine, L. F.; Gear, C. W. SIAM Review 1979, 21, (1), 1.
10. Carver, M. B. J. Comp. Physics 1980, 35, (1), 57.
11. Carver, M. B.; Hinds, H. W. Simulation 1978, 31, (2), 59.
12. Vichnevetsky, R. Simulation 1971, 16, (4), 168.
13. Buis, J. P. Simulation 1975, 25, (1), 1.
14. McAvoy, T. J. Simulation 1972, 18, (3), 91.
15. Abbott, M. B. "An Introduction to the Method of Characteristics"; American Elsevier: New York, 1966.
16. Orailoglu, A. "A Multirate Ordinary Differential Equation Integrator"; University of Illinois Department of Computer Science Report No. UIUCDCS-R-79-959, 1979.
17. Gear, C. W. "Automatic Multirate Methods for Ordinary Differential Equations"; University of Illinois Department of Computer Science Report No. UIUCDCS-R-80-1000, 1980.
18. Emanuel, G.; Vale, H. J. in Bahn, G. S., Ed.; "The Performance of High Temperature Systems, Vol 2"; Gordon and Breach: New York, 1969; p 497.
19. Bailey, J. E. in Lapidus, L.; Amundson, N. R., Eds.; "Chemical Reactor Theory: A Review"; Prentice-Hall: Englewood Cliffs, 1977; p 758.
20. Friedly, J. "Dynamic Behavior of Processes"; Prentice-Hall: Englewood Cliffs, 1972; p 516.

RECEIVED June 3, 1981.

18

Fixed Bed Reactors with Deactivating Catalysts

JAMES M. POMMERSHEIM and RAVINDRA S. DIXIT

Bucknell University, Lewisburg, PA 17837

Models are formulated and solved for the deactivation of catalysts by feed stream poisons. Catalyst deactivation effects are scaled from single pores, to pellets to catalyst beds. Design equations were presented for fixed bed reactors with catalyst pellets which deactivate by both pore-mouth (shell-progressive) and uniform (homogeneous) poisoning mechanisms. Conversion and production levels are predicted as a function of time and reactant Thiele modulus (h). Levels increased with increases in pellet deactivation times (or time constants). Levels were found to increase with Thiele moduli h for pore-mouth poisoning and decrease for uniform poisoning. An upper limit on bed production exists for pore-mouth and uniform irreversible poisoning, but not for uniform reversible poisoning, where at long times the production rate becomes constant. For uniform poisoning the interior of the catalyst acts as an internal guard-bed removing poison which could otherwise inhibit reaction near the pellet surface. This effect is most pronounced at higher h. Spherical and flat plate pellets gave substantially equivalent results.

In order to be able to predict the chemical production from a catalytic reactor, the kinetics of reaction must first be known. By applying the conservation equations to a single pore, the reaction rate for a catalyst pellet can be found. With a knowledge of this rate, the reactor design equations for a fixed bed of such pellets can be solved to predict conversion levels and chemical production. For solid catalytic reactors undergoing catalyst deactivation, the bed design equations must also incorporate the kinetics of deactivation and its effect on pore, pellet and bed transport rates. The overall models are often complex and unwieldy, although simplifying assumptions can be made for particular cases based on the degree of uncoupling which exists between

0097-6156/81/0168-0367$05.00/0

the deactivation and reaction processes. This type of approach can lead to insights about the overall deactivation process which may lie hidden in a more complicated analysis.

In pioneering work, Wheeler (1) showed the effect that two limiting but important modes of catalyst deactivation, pore-mouth and uniform (or homogeneous) poisoning, can have on the overall activity of a catalyst pore. In pore-mouth poisoning the catalyst has a strong chemical affinity for the poison precursor and poison will be strongly adsorbed on the catalyst surface. The outer part of the catalyst is poisoned first and a moving band of totally poisoned catalyst slowly moves into the unpoisoned catalyst interior. Catalyst fouling as a result of coke deposition can also result in a moving band or shell of deactivated catalyst. Pore mouth deactivation by poisoning or fouling is more likely when precursor molecules are large and the pores are narrow and long. These factors make the Thiele modulus for poison deposition, h_s, large. In such instances the poison precursor molecules will reside in the vicinity of the pore mouth longer and be more likely to lie down there.

On the other hand, uniform or homogeneous catalyst poisoning presumes that the poison precursor species has full access to the catalyst interior before deactivation begins. There is no diffusional resistance for this species. This will be more likely to occur when the pores are large, the catalyst pellets small, and the intrinsic deactivation rate is low. In addition smaller poison precursor molecules will be able to diffuse more rapidly into the catalyst interior. Here the Thiele modulus for poison laydown h_s will be small, and in the limit, zero.

Masamune and Smith (2) examined the problem of finding conversions in a fixed bed reactor containing a deactivating catalyst. Having obtained in detail the shape of the poisoning front in a single catalyst pellet, they indicated how these results can be used with the reactor design equations to numerically predict overall conversions. Olson (3) studied the time dependence of activity in a fixed-bed reactor. Wheeler and Robell (4) combined and condensed much of the previous theory. Their results predicted the decline in activity of a fixed bed reactor. They were successful in obtaining an analytical solution which had some degree of generality. Haynes (5) extended the work of Wheeler and Robell to include a factor to account for strong intraparticle diffusion resistance to the poison precursor. The general equations were simplified by making the assumption of shell-progressive poisoning, and dimensionless plots were obtained which showed the effect of the Thiele modulus, a dimensionless time, and the number of reaction transfer units on the activity and conversion in a fixed bed reactor. In a comprehensive review on catalyst deactivation, Butt (6) has summarized a number of experimental and theoretical studies dealing with deactivation in fixed bed reactors.

Pommersheim and Dixit (7) have developed models for poisoning occurring in the pores of flat plate and spherical catalyst pellets.

They considered deactivation to occur by either pore-mouth (shell-progressive) or uniform (homogeneous) poisoning and examined the effect these types of deactivation had on overall activity and production rates for a single catalyst pellet. Analytical solutions were obtained for the production per pore by considering the time dependence of activity. Their results will be used here as the basis for the development of models for deactivation in fixed bed reactors.

In the present work, solutions are presented for fixed-bed reactors subject to the following kinds of pellet poisoning:

(i) pore mouth poisoning for flat plate type pellets
(ii) homogeneous (uniform) reversible poisoning for flat plate pellets
(iii) homogeneous (uniform) reversible poisoning for spherical pellets

The following assumptions are made:

1) the bed is isothermal throughout
2) concentration gradients and activity variations in the radial direction are negligible
3) no change in moles upon reaction
4) concentrations of poison species are much less than concentrations of reactant species
5) the reaction is first order and irreversible
6) reactor pressure drop does not effect reaction kinetics or flow
7) no mass transfer film external to the pellets
8) plug flow
9) the change in activity with bed length is much slower than the changes in concentration with bed length.

The last assumption is referred to as the quasi-steady-state assumption. The fraction of the bed which is poisoned is a function of time only and not of bed length, reactor space time, or the concentration of the reactant A external to the pellets. At any given time the bed activity will be constant, and only one concentration of the poison precursor species S will exist in the bed. Such a situation will be more likely to occur when deactivation rates are low compared to reaction rates. Under this condition S will spread evenly throughout the bed. Within particles, however, concentration gradients of S may still exist depending on the poisoning mechanism and the pore and pellet properties.

Bed Concentration Profiles

Feed gas or liquid enters the bottom of the packed bed with concentration $(C_{Ao})_o$. Contained within the feed is a small amount of poison precursor S in concentration $(C_{So})_o$, such as lead in automotive exhaust gas or sulfur or metals (e.g., nickel and vanadium) contained in a petroleum feed-stock.

Consider a differential section of a fixed bed reactor which is packed with uniform sized catalyst particles. The plug flow design equation for this section is (8)

$$\frac{dW}{F_{Ao}} = \frac{dX_A}{-r'_A} \tag{1}$$

where F_{Ao} is the molar flow rate of A at the bed inlet, X_A is the (global) conversion of A, W is the weight of catalyst, and $-r'_A$ is the moles of A reacted per second per gram of catalyst.

Pore Mouth Poisoning: Flat Plate Pellets. For flat plate type pellets undergoing pore mouth poisoning, the moles of A reacting per pore is given by (7)

$$-r_A = \frac{\pi r^2 \bar{C}_{Ao} D_A h \tanh h(1-\alpha)}{L(1 + \alpha h \tanh h(1-\alpha))} \tag{2}$$

where $\bar{C}_{Ao}$ is the intraparticle concentration of A, r is the average pore radius, L is the pellet half width, D_A is the effective diffusivity of reactant in the pellet and h is the Thiele modulus for the reactant A and is given by $L\sqrt{k/D_A}$. k is the reaction velocity constant, based on pore volume, for the first order reaction A → R. $\bar{C}_{Ao}$ varies along the bed length, but the other model parameters will be constant.

The rate of reaction of A per unit weight of catalyst can be related to the rate of reaction of A per pore. Thus

$$-r'_A \frac{\text{moles of A}}{\text{sec. g cat.}} = -r_A \frac{\text{moles A}}{\text{sec. pore}} \quad \frac{\varepsilon_p}{\rho_p V_p} \tag{3}$$

where ε_p and ρ_p are the pellet porosity and density, respectively, and V_p is the volume of one pore. The design equation then becomes

$$-\frac{d\bar{C}_{Ao}}{dz} = \frac{V(1-\varepsilon)\varepsilon_p \bar{C}_{Ao} D_A h \tanh h(1-\alpha)}{HL^2 v_o(1 + \alpha h \tanh h(1-\alpha))} \tag{4}$$

where v_o is the volumetric flow rate, H and V are the height and volume of the reactor, respectively, and ε is the porosity of the fixed bed. Because of assumptions number 1), 3) and 6), v_o remains constant. Introducing dimensionless variables for concentration and distance:

$$\phi = \frac{\bar{C}_{Ao}}{(C_{Ao})_o} \qquad \text{and} \qquad \lambda = \frac{z}{H}$$

equation (4) becomes

$$\frac{d\phi}{d\lambda} = \frac{-V(1-\varepsilon)\varepsilon_p D_A}{v_o L^2} \frac{h \tanh h(1-\alpha)}{1 + \alpha h \tanh h(1-\alpha)} \phi \tag{5}$$

The boundary condition is

$$\phi = 1 \quad \text{at} \quad \lambda = 0$$

Because of assumption 9), α will be independent of λ, and equation (5) can be directly integrated to give

$$\phi = \exp \frac{-N_1 \lambda h \tanh h(1-\alpha)}{1 + \alpha h \tanh h(1-\alpha)} \tag{6}$$

where N_1 is a dimensionless number defined as

$$N_1 = \frac{V(1-\varepsilon)\varepsilon_p D_A}{v_o L^2} \tag{7}$$

V/v_o is the (superficial) reactor space time. N_1 represents the ratio of reactor space time to pellet diffusion time.

The variation of α with time follows a parabolic curve (7)

$$\alpha^2 = \frac{t}{\tau_1} = \Theta \tag{8}$$

where τ_1, the time for complete deactivation of the pellet, is equal to $\tau_1 = \omega_o L^2/rD_s(C_{So})_o$. ω_o is the intrinsic poison laydown per unit of pore surface (moles/cm^2), and D_s is the diffusivity of S. Because of the parabolic nature of equation (8), over thirty percent of the catalyst activity is already gone when $\Theta = 0.1$.

Figure 1 shows a plot of dimensionless concentration of reactant ϕ versus dimensionless time Θ, with $N_1\lambda$ fixed at unity. The Thiele modulus h is shown as a parameter. When both N_1 and λ are unity, the trace on the figure at any given modulus represents the way the output concentration from the fixed bed reactor varies with time. As time proceeds it increases towards the incoming concentration in the characteristic "S" shape shown. From this figure, it can be seen that for small values of h (less than about 0.1) there is little or no reaction and the concentration of A at the bed exit remains the same as the feed concentration.

At $t = \tau_1$ ($\Theta = 1$ in figure 1) each pore (and pellet) becomes completely deactivated. Since the bed deactivates uniformly, it will also completely lose its activity at this time. This is shown in figure 1 by the fact that the concentration of reactant coming from the bed becomes the incoming concentration $(C_{Ao})_o$ at $\Theta = 1$. The intercept of the curves (indicated by black circles at time zero) corresponds to the initial concentration of reactant in the bed, before deactivation begins. The total production from such a deactivating fixed bed corresponds graphically to the area located above each curve and below the line $\phi = 1$. The area out to $\Theta = 1$ would represent the total ultimate production.

In figure 1 the parameter N_1 is fixed. This implies that the diffusivity D_A is also fixed. Thus, increases in the Thiele modulus h, shown as a parameter in figure 1, are associated with increases in the reaction rate constant k. Higher reaction rates lead to greater conversions at any given time as evidenced by the lower values of the ordinate.

Increases in temperature raise the rate constant k and raise the production. This effect becomes less and less important as the temperature increases. Figure 1 shows that there is a limit curve at very high moduli ($h \to \infty$ and $\phi \to \exp[N_1\lambda/\sqrt{\Theta}]$), above which no further increase in production appears possible. However, some further increase can occur with increases in D_A (N_1 increases), although this may be offset if D_S also increases (τ_1 drops). The intrinsic rate of poison laydown ω_o may also change with temperature, but whether it rises or falls will depend on the specific system under consideration.

Uniform (homogeneous) Reversible Poisoning: Flat Plate Pellets. The rate of reaction for uniform or homogeneous reversible poisoning in flat plate pellets is given by (1)

$$-r_A = \frac{2\pi r\ L\ k_s\ \overline{C}_{Ao}\ \sqrt{1-\alpha}\ \tanh h\sqrt{1-\alpha}}{h} \tag{9}$$

where k_s is the surface (pore) reaction rate constant. The concentration profiles are given by

$$\phi = \exp \frac{-N_2\lambda\ \sqrt{1-\alpha}\ \tanh h\sqrt{1-\alpha}}{h} \tag{10}$$

where N_2 is a dimensionless number defined as

$$N_2 = \frac{2\pi r L k_s \varepsilon_p\ (1-\varepsilon)\ V}{V_p\ v_o} = \frac{2}{r}\ k_s\ \varepsilon_p\ (1-\varepsilon)\ \frac{V}{v_o} = k\varepsilon_p\ (1-\varepsilon)\ \frac{V}{v_o} \tag{11}$$

N_2 represents the ratio of reactor space time to reaction time. Comparison of equations (7) and (11) shows $N_2 \sim h^2 N_1$. α, the fraction of the pore (or pellet) poisoned, is given by (7)

$$\frac{\alpha}{\alpha_e} = 1 - \exp(-t/\tau) = 1-\exp(-\Theta') \tag{12}$$

τ is the time constant for poison laydown for homogeneous reversible poisoning and is given by: $\tau = \omega_\infty[k_d + k_a(C_{So})_o]$, where ω_∞ is the maximum value of ω when all the adsorption sites are fully occupied, k_d is the desorption rate constant and k_a the adsorption rate constant for poison laydown, Θ' is the dimensionless time t/τ, and α_e is the equilibrium fraction of active sites on the catalyst surface which are poisoned. When $\alpha_e = 1$ the poisoning is irreversible, and all catalytic sites eventually become deactivated. Unlike pore mouth poisoning τ is not a pore burn-out time but a true first order time constant for poisoning. Thus Θ' (unlike Θ) can assume values greater than unity since some catalyst activity is always present. At $\Theta' = 1$ only 63.2% of the poisonable catalytic sites on the surface have been deactivated, while at $\Theta' = 3$ it has risen to 95%.

Equations (10) and (11) were solved for different values of Θ', α_e, h and N_2. Figure 2 shows the variation of the dimensionless concentration $\emptyset$ as a function of the dimensionless time Θ' with h as a parameter. The product $N_2\lambda$ is fixed at unity, and the poisoning is irreversible ($\alpha_e = 1$). Unlike the pore mouth poisoning case, concentrations were found to increase with increasing h. Thus the conversion decreased with increases in the Thiele modulus. With N_2 fixed, the reaction rate k will be fixed. Increases in h are then associated with decreases in reactant diffusivity, D_A, or increases in pellet dimension, L. Reaction is then confined more to the periphery of the catalyst. Since poisoning is uniform, the inside of the catalyst will be able to sponge up poison which otherwise would inhibit reaction. The inside of the catalyst acts as an internal guard bed. The lower slope of the curves in figure 2 at higher values of h is attributable to this guard-bed action.

Uniform Reversible Poisoning: Spherical Pellets. The rate of reaction for uniform reversible poisoning in spherical pellets is given by (7)

$$-r_A = \frac{3\,k\,V_p\,\overline{C}_{Ao}}{h'^2}\left(-1 + h'\sqrt{1-\alpha}\,\coth h'\sqrt{1-\alpha}\right) \tag{13}$$

where

$$h' = R\sqrt{k/D_A} \tag{14}$$

h' is the Thiele modulus for spherical pellets of radius R.

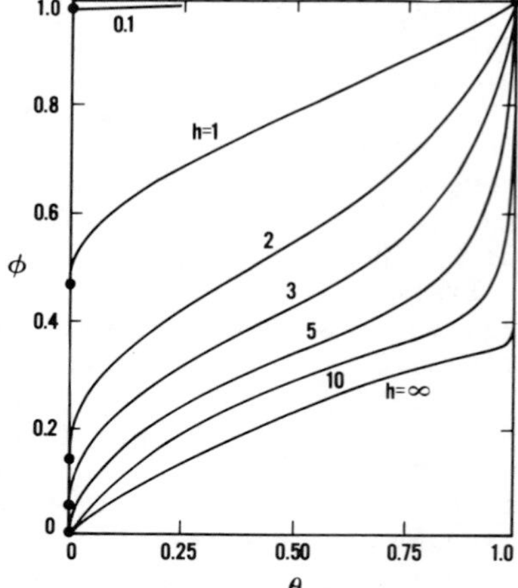

Figure 1. Dimensionless reactant concentration Φ vs. dimensionless time Θ for pore mouth poisoning: $N_1\lambda = 1$*; parameter,* h

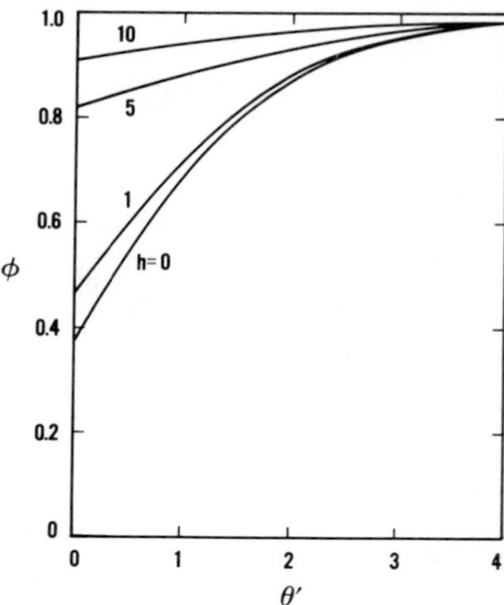

Figure 2. Dimensionless reactant concentration Φ vs. dimensionless time Θ′ for uniform poisoning: $N_2\lambda = 1$*;* $\alpha_e = 1$*; parameter,* h

The dimensionless concentration profiles are

$$\phi = \exp \frac{-N_2\lambda\,(-1 + h'\sqrt{1-\alpha}\,\coth h'\sqrt{1-\alpha})}{3\,h'^2} \tag{15}$$

α, the fraction poisoned, is given by equation (12). For spherical pellets as with uniform poisoning of flat plate pellets, increasing h increases the reduced concentration resulting in lower conversions at a fixed time. As h increases, the diffusional resistance for A increases, decreasing the conversion and increasing the reduced concentration. Equation (15) can be directly compared to equation (10) if the concept of an equivalent sphere (8) is used. For flat plate pellets of width 2L, the equivalent radius is 3L. Thus, equations (10) and (15) can be made analogous to one another by replacing h' by 3h. Calculations made on this basis indicated that ϕ values were slightly less (within 3%) for the spherical case compared to the flat plate case. When h = 0 both equations (10) and (15) reduce to $\phi = \exp[N_2\lambda(\alpha-1)]$. For both cases a decrease in the equilibrium surface coverage of poison α_e as well as an increase in the space time to reaction time ratio N_2 results in increased conversions (lower ϕ's). Higher values of N_2 give higher initial conversions and lower values of α_e give lower deactivation rates. For these conditions the analogous curves in figure 2 would begin lower and be flatter.

Production From A Fixed Bed Reactor.

The production from a fixed bed reactor can be found by integrating the instantaneous molar flow rate of product over the time of reactor operation. The production will be equal to the total consumption of reactant when the reaction has the same number of moles of product as reactant. The total production of product R for the reaction A → R is given by

$$N_R(t) = F_{Ao}\left[t - \int_0^t \frac{\overline{C}_{Ao}}{(C_{Ao})_o}\Big|_{z=H} dt\right] = F_{Ao}\left[t - \int_0^t \phi\,(\lambda = 1)\,dt\right] \tag{16}$$

In the absence of deactivation $N_R = F_{Ao}\,[1 - \phi(\alpha = 0)]t$; the production varying linearly with time.

Pore Mouth Poisoning: Flat Plate Pellets. For pore mouth poisoning of flat plate pellets, substitution of equation (6) into equation (16) yields

$$N_R(t) = 2\,F_{Ao}\,\tau_1\left[\frac{\Theta}{2} - \int_0^{\alpha(t)} \alpha \exp \frac{-N_1\,h \tanh h\,(1-\alpha)}{1 + \alpha h \tanh h(1-\alpha)}\,d\alpha\right] \tag{17}$$

for the production coming from a fixed bed reactor. The ultimate or final total production (at $t = \tau_1$, and $\alpha = 1$) is given by

$$N_R(1) = 2 F_{Ao} \tau_1 [1/2 - \int_0^1 \alpha \exp \frac{-N_1 h \tanh h(1-\alpha)}{1 + \alpha h \tanh h(1-\alpha)} d\alpha] \quad (18)$$

Figure 3 shows the dimensionless production (divided by the scale factor $2F_{Ao}\tau_1$) calculated using equation (17) and expressed as a function of dimensionless time Θ with h as a parameter. The value of $N_1\lambda$ is fixed at unity. The production rises with time towards its final value at complete deactivation ($\Theta = 1$), predicted by equation (18). Low values of h give low production, while high values give high production. This is in agreement with the increased reactivity at larger moduli. At very high values of h further increase in production is not gained by higher h's. The initial slope of the curves in figure 3 represents the rate of increase of production in the absence of deactivation. The straighter curves at the higher values of h indicate that deactivation is not as important at high moduli, and that during deactivation there is not as much lost production. Raising the deactivation time τ_1, will proportionally raise the production at any time.

Figure 4 shows a plot of the fraction of the final production $N_R/N_R(1)$ vs time Θ with h as a parameter. Lower values of h result in higher values of this fraction at any fixed time, indicating that deactivation has a more pronounced effect on production at lower Thiele moduli. At high h values the curves are relatively straight and appear to approach a common asymptote at very high moduli. This is consistent with equations (17) and (18).

Uniform or Homogeneous Reversible Poisoning: Flat Plate Pellets. Substituting equation (10) into equation (16), the total production for uniform reversible poisoning in flat plate pellets becomes

$$N_R(t) = 2F_{Ao} \tau [\frac{\Theta'}{2} - \int_{y*}^1 \frac{y}{y^2-1+\alpha_e} \exp \frac{-N_2 y \tanh hy}{h} dy] \quad (19)$$

where the lower limit on the integral is given by

$$y* = \sqrt{1 - \alpha_e [1 - \exp(-\Theta')]} \quad (20)$$

With $\alpha_e = 1$, equations (19) and (20) reduce to the case of uniform irreversible poisoning.

For uniform poisoning in flat plate pellets, figure 5 presents a plot of the production ratio $N_R/N_R(1)$ as a function of the dimensionless time Θ' with h as a parameter. The dimensionless group N_2 was set at unity, while the poisoning was irreversible

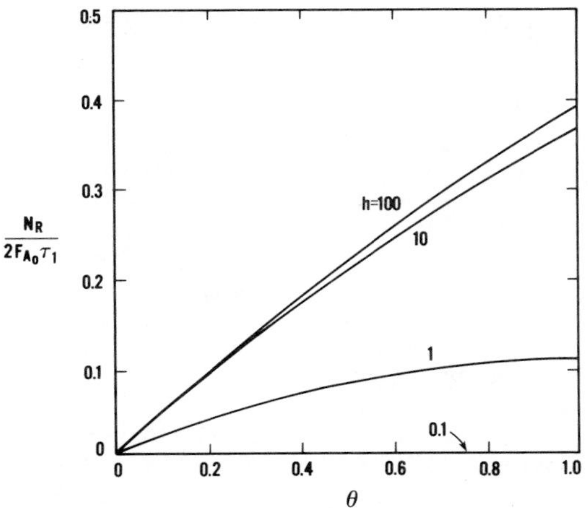

Figure 3. Dimensionless production $N_R/2F_{A_0}\tau_1$ *vs. dimensionless time* Θ *for pore mouth poisoning:* $N_1\lambda = 1$*; parameter:* h

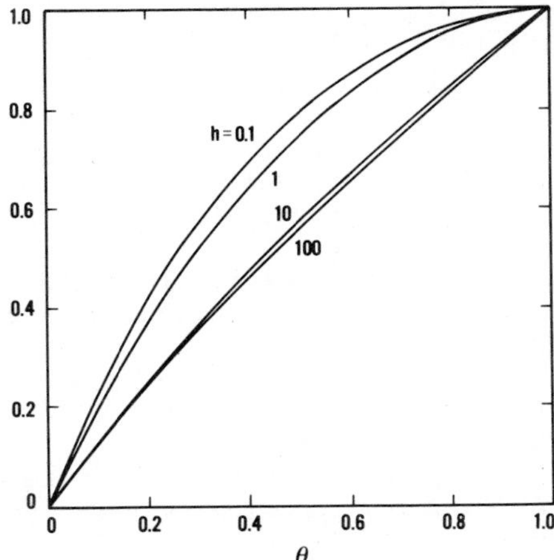

Figure 4. Fraction of final production $N_R/N_R(1)$ *vs. dimensionless time* Θ *for pore mouth poisoning:* $N_1\lambda = 1$*; parameter,* h

(α_e = 1). $N_R(1)$ represents the production at t = τ, the first-order time constant for deactivation. At this point 63.2% of the catalyst has become deactivated. As shown in figure 5 all curves pass through the common point Θ' = 1.

Both the production and the production ratio first rise rapidly with time and then level off at long times, generally only after the catalyst is almost completely deactivated. Thus the production is finite even though, unlike the pore mouth case, the deactivation, does not have a finite extinction time. For times less than τ, all curves are more or less common, regardless of the value of h. For times greater than τ, the production ratio levels off most rapidly for small moduli. At suitably small h, the production ratio becomes independent of h. This result can also be obtained theoretically by examining equation (17). As the value of the Thiele modulus is raised, the actual production N_R falls, as is indicated by equation (17) or by examining the relative areas above the curves in figure 2. Higher values of h are associated with severe diffusional limitations. The higher slope of the curve in figure 5 at the higher values of h is attributable to the guard bed action of the pellet interior. Because the interior of the catalyst sponges up poison, it is relatively more effective at higher moduli in raising the production ratio.

Equation (19) predicts that increases in α_e, the final fraction of the surface poisoned (as $\Theta' \to \infty$), will decrease the production of R. This effect is shown in figure 6 which presents a plot of the production ratio as a function of Θ' with the equilibrium fraction poisoned (α_e) shown as a parameter. $N_2\lambda$ and h were both set at unity. All curves cross at Θ' = 1. Comparing the curves for α_e = 0.1, 0.5 and 1.0, irreversible poisoning, α_e = 1.0, gives higher production ratios for times less than τ, but significantly lower ratios at greater times. At a low degree of poisoning, as indicated by α_e = 0.1, the production ratio curve is substantially linear. A catalyst which would not deactivate would have α_e = 0. On figure 6 this appears as a line of unit slope. The vertical distance between this line and any of the curves in figure 6 gives a measure of the production loss due to deactivation. Unlike the case of irreversible poisoning (α_e = 1), for reversible poisoning (α_e < 1) there is no upper limit on production. Note in figure 6 that only the curve for α_e = 1 becomes horizontal at longer times. With reversible poisoning at longer times Θ', $\alpha < \alpha_e$ and the production rate becomes constant. The catalyst is effectively functioning then with a reduced but constant activity, proportional to $(1 - \alpha_e)$.

Uniform Reversible Poisoning: Spherical Pellets. For uniform reversible poisoning in spherical pellets an expression for the instantaneous total production is obtained by substituting equation (15) into equation (16)

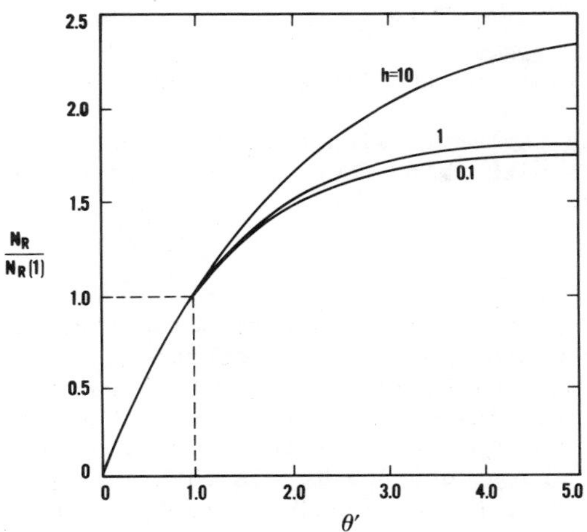

Figure 5. Production ratio $N_R/N_R(1)$ vs. dimensionless time Θ' for uniform poisoning: $N_2\lambda = 1$; $\alpha_e = 1$; parameter, h

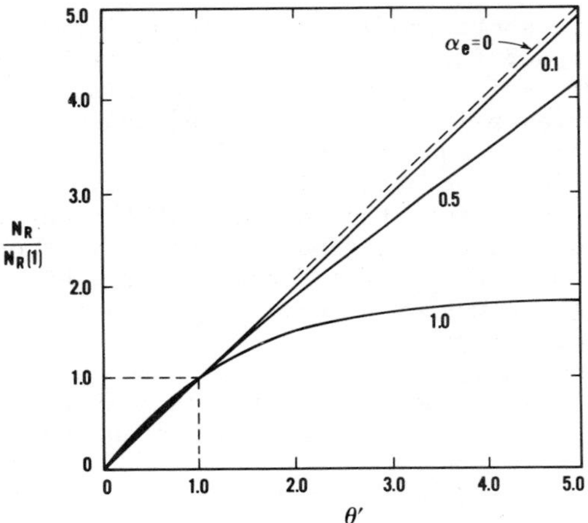

Figure 6. Production ratio $N_R/N_R(1)$ vs. dimensionless time Θ' for uniform poisoning: $N_2\lambda = 1$; h = 1; parameter, α_e

$$N_R(t) = 2F_{Ao}\,\tau\left[\frac{\Theta}{2} - \int_{y^*}^{1} \frac{y}{y^2-1+\alpha_e} \exp\frac{-N_2(-1 + h'y \coth h'y)}{3h'^2}\,dy\right] \quad (21)$$

where the lower limit on the integral is again given by equation (20).

Calculations performed with this equation gave substantially identical results with those found using equation (19) with h replaced by h'/3.

Conclusions

With either pore-mouth or uniform poisoning, fixed bed conversions and production levels are a strong function of the reactant Thiele modulus h, increasing with h for pore mouth poisoning and decreasing with h for uniform poisoning. These trends depend on the constancy of the dimensionless groups N_1 and N_2. Conversions and production levels both increased with increases in the pellet deactivation time (pore-mouth) or time constant for poison deposition (uniform).

For pore-mouth poisoning, deactivation has a more pronounced effect on production at lower moduli. At high values of h, further increases in h do not gain significant increases in production. Since pellet deactivation times decrease with increases in temperature, a best temperature may exist for maximum production.

For both pore mouth and irreversible uniform poisoning, an upper limit was found for the bed production of R, while with uniform reversible poisoning no such limit existed.

For uniform reversible poisoning, the production increased with a decrease in the equilibrium surface of the catalyst poisoned. At long times the production rate became constant.

For uniform poisoning, the interior of the catalyst acts as an internal guard-bed, removing poison which might otherwise inhibit reaction near the pellet surface. This effect is most pronounced at high values of h, where reaction is confined to the periphery of the catalyst.

Spherical and flat plate pellets give substantially equivalent conversions and production levels for uniform reversible poisoning when the Thiele moduli are put on an equivalent basis.

<u>Legend of Symbols.</u>

A	reactant species
$\overline{C}_{Ao}$	intrapellet concentration of A external to the pellets, moles/m^3
$(C_{Ao})_o$	concentration of A in the feed stream, moles/m^3
$(C_{so})_o$	concentration of S in feed, moles/m^3
D_A	effective diffusivity of A, m^2/s
D_s	effective diffusivity of S, m^2/s
F_{Ao}	molar flow rate of A at reactor inlet, moles/s
H	height of fixed bed, cm
h,h'	Thiele modulus for reactant A, L $\sqrt{k/D_A}$, or R $\sqrt{k/D_A}$, respectively, dimensionless
h_s	Thiele modulus for poison precursor
k	reaction rate constant, s^{-1}
k_a	adsorption rate constant, $m^4/mole^2$, s
k_d	desorption rate constant, m^2 / mole, s
k_s	reaction rate constant of A based on surface area, m/s
L	length of catalyst pore, half width of flat plate pellets, m
N_R	production of product R, moles
$N_R(1)$	production of R at Θ (or Θ') = 1
N_1	dimensionless group for pore mouth poisoning (equation 7)
N_2	dimensionless group for uniform reversible poisoning (equation 11)
r	average or mean pore radius, m
r_A	rate of reaction of A, mole / m^3,s
r'_A	rate of reaction, moles/g catalyst,s

R	radius of spherical pellets, m; product of reaction
t	time, s
V	volume of fixed bed, m^3
V_p	volume of single pore, m^3
v_o	(input) volumetric flow rate, m^3/s
W	weight of catalyst, g
y*	limit in integration, $\sqrt{1 - \alpha_e(-\exp -\Theta')}$, equation (20), dimensionless
z	distance along fixed bed, m

Greek Letters.

α	fraction of catalyst surface poisoned
α_e	equilibrium fraction of surface poisoned
ε	fixed bed porosity (interparticle)
ε_p	pellet porosity (intraparticle)
Θ, Θ'	dimensionless times, t/τ_1 and t/τ, respectively
λ	dimensionless distance, z/H
ρ_p	bulk density of catalyst pellet, g/m^3
τ	time constant for poison laydown, uniform reversible poisoning, s
τ_1	pellet deactivation time, pore-mouth poisoning, s
ϕ	dimensionless concentration in fixed bed, $\overline{C}_{Ao}/(C_{Ao})_o$
ω_o	intrinsic poison laydown per unit of pore surface, $moles/m^2$
ω_∞	maximum value of ω_o

Acknowledgments.

The authors are grateful to Ms. Niloofar Farhad of the department for her computational assistance and to the Bucknell University Computer Center for the generous use of its facilities. The author's also appreciate the invaluable aid of Mrs. Helen Mathias in typing the several versions of the manuscript.

Literature Cited.

1. Wheeler, A., "Catalysis," P. H. Emmett, Ed., Vol. II, Reinhold, New York, 1955.
2. Masamune, S.; Smith, J. M., "Performance of Fouled Catalyst Pellets," A.I.Ch.E. Journal 1966 12, 384-394.
3. Olson, J. H., "Rate of Poisoning in Fixed Bed Reactors," Ind. Eng. Chem. Fundam. 1968 10, (2).
4. Wheeler, A.; Robell, A. J., "Performance of Fixed-Bed Catalytic Reactors with Poison in the Feed," Journal of Catalysis 1979 13, 299-305.
5. Haynes, H. W. Jr., "Poisoning in Fixed Bed Reactors," Chem. Engr. Sci. 1970 25, 1615-1619.
6. Butt, J. B., "Catalyst Deactivation," in "Chemical Reaction Engineering, Advances in Chemistry Series 109," Gould, R. F. (ed.), American Chemical Society, Washington, D.C., 1972, 259-496.
7. Pommersheim, J. M.; Dixit, R. S.; "Models for Catalyst Deactivation," 70th Annual AIChE Meeting, New York, November 1977.
8. Levenspiel, O., "Chemical Reaction Engineering," 2nd Ed., Wiley, New York, 1972.

RECEIVED June 3, 1981.

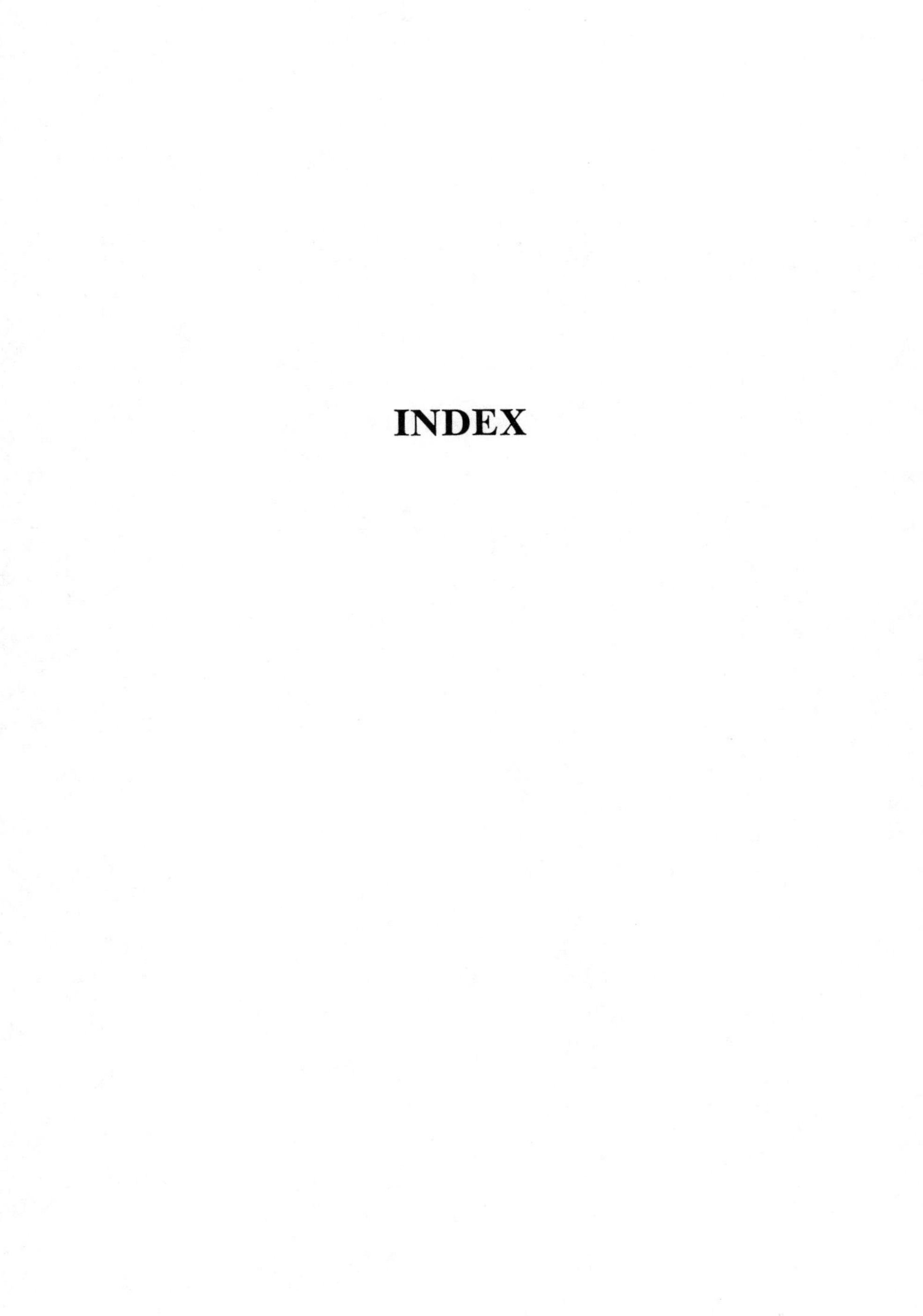

INDEX

INDEX

G

H

I

W

Z

Jacket design by Carol Conway.
Production by Robin Giroux and Cynthia E. Hale.

Elements typeset by Service Composition Co., Baltimore, MD.
The book was printed and bound by The Maple Press Co., York, PA.

PHYSICS OF THE ATOM

M. RUSSELL WEHR
Drexel Institute of Technology

JAMES A. RICHARDS, JR.
Community College of Philadelphia

PHYSICS OF THE ATOM

SECOND EDITION

ADDISON-WESLEY PUBLISHING COMPANY

READING, MASSACHUSETTS · PALO ALTO · LONDON · DON MILLS, ONTARIO

This book is in the **ADDISON-WESLEY SERIES IN PHYSICS**

 PRINTED IN THE UNITED STATES OF AMERICA. PUBLISHED SIMULTANEOUSLY IN CANADA. LIBRARY OF CONGRESS CATALOG CARD NO. 67-19432

Preface to the first edition

This book is designed to meet the modern need for a better understanding of the atomic age. It is an introduction suitable for any student with a background in college physics and mathematical competence at the level of calculus. Some parts of the book will be better appreciated by those with a knowledge of chemistry. This book is designed to be an extension of the introductory college physics course into the realm of atomic physics; it should give the student a proficiency in this field comparable to that he has in mechanics, heat, sound, light, and electricity.

The approach has been to deal with a logical series of topics leading to the practical conversion of mass-energy into kinetic energy. The development of modern physics is such a logical sequence of discoveries that we have had little difficulty in deciding what topics to include. The order of presentation is usually chronological. Mechanics can be presented in a from-simple-to-complex manner because mechanics is an organization of observations with which most of us are familiar. Atomic physics, on the other hand, involves concepts which are quite foreign to general experience, and it is both helpful and stimulating for the student to feel that he is growing in understanding as others grew before him. This book capitalizes on the fact that the story has a plot.

This plot is developed as simply and directly as possible. It is difficult to know how a new idea comes into being. Even the man who creates a new idea cannot fully describe that moment when the light dawned. But we, as "Monday-morning quarterbacks," can state the problem, present the relevant information, and, we hope, make the idea which originally required genius appear obvious. With informal discussion and many analogies the student may be led to understand new and difficult concepts. Although any serious student of atomic physics must go on to a more elaborate mathematical treatment, we feel that even at this level the most convincing argument is a careful mathematical development and the presentation of original data. We have not hesitated to use calculus where it is appropriate, and we have clearly indicated the weakness of our argument on those few occasions when the argument is indeed weak.

Many who use this book will not have access to a laboratory. Thus we have described a few projects most students can do for themselves.

The book is designed for a one-term course in either a liberal arts or an engineering curriculum. Basic physics is basic physics whether the student is motivated by curiosity or possible application. Our preference for the rationalized MKSA system of units is due to the fact that this system is rapidly becoming *the* system in engineering and physics, and because this system includes the units all scientists use to measure electrical quantities. Except in equations which include units peculiar to our subject, like the electron volt, our equations are valid in any consistent rationalized system of units.

The references at the end of each chapter include sources which have been helpful to us in preparing the text and to which students may *profitably* refer. Many of our references to original articles are to reprints of source materials. Our thought is that these compilations and translations are more convenient for students to use. In some cases we have made remarks designed to facilitate the student's choice of reference reading.

A few of the problems at the end of each chapter may involve mere substitution of numbers into formulas. In these cases the purpose is to give the student a feeling for the magnitudes of atomic quantities. Most of the problems, however, require thoughtful analysis. Some are stated in general terms instead of numbers. The order of the problems is the same as the order in which the theory is developed.

Appendix 1 is a chronology of discoveries important to the atomic view of nature. It was impossible to formulate any sharp criterion for selecting the items to be included, and so, in general, we have taken a broad view. The purpose of this chronology is to convey a sense of the development of a natural science. We think it will also serve to stress the international character of scientific achievement and to emphasize that many of the great scientists did their first important work early in life.

This textbook is dedicated to the students who will use it and to former students whose questions, comments, and reactions have largely determined our presentation. We gratefully acknowledge the assistance of Elliot H. Weinberg of North Dakota State College. We also wish to thank our colleagues at Drexel Institute of Technology, especially Henry S. C. Chen, Dennis H. Le Croissette, and Irvin A. Miller, for their constructive suggestions.

This book is neither a treatise nor a survey. It is a textbook which bridges the gap between classical physics and the present frontiers of physical investigation.

Philadelphia, Pa.
July 1959

M.R.W.
J.A.R., Jr.

Preface to the second edition

The principal changes in this edition are the rewriting of the chapter on waves and particles and the addition of a new chapter on matter waves, both prepared by I. A. Miller of Drexel Institute of Technology. Many sections have been revised and expanded and there are new sections on blackbody radiation, masers and lasers, relativity, and the Mössbauer effect. The discussion of radiation detectors has been updated and the chapter on high-energy physics has been enlarged. Some of the problems in the first edition have been retained without change, others have been rewritten, and over 160 new problems have been added.

This book is written for a one-semester course which meets three hours a week. The material can readily be covered in this time if the student has had an extensive course in introductory physics so that Chapters 1 and 2, and most of Chapter 3 can be gone over quickly or omitted entirely. For a short course, the following omissions are suggested: Sections 7–6 to 7–10 inclusive, Chapters 8 and 9, and Sections 12–4 to 12–12 inclusive.

We have sought to incorporate throughout pedagogical improvements suggested by students and teachers who have used this text. In addition to those whose help we acknowledged in the preface to the first edition, we especially wish to thank A. Caprecelatro, D. E. Charlton, E. M. Corson, T. T. Crow, E. I. Howell, A. P. Joblin, G. E. Jones, P. Kaczmarczik, A. Meador, M. W. Minkler, T. J. Parmley, L. E. Peterson, D. J. Prowse, F. W. Sears, F. O. Wooten, and M. W. Zemansky.

Although we have been sensitive to all suggestions, we have incorporated those which would retain and enhance the "spirit of natural philosophy" we have sought to achieve.

Philadelphia, Pa.
February 1967

M.R.W.
J.A.R., Jr.

Contents

CHAPTER 1

THE ATOMIC VIEW OF MATTER

1-1. INTRODUCTION

The ancient Greeks speculated about almost everything. Democritus, for example, theorized that not only matter but also the human soul consists of particles. His statements, made about twenty centuries before the advent of experimental science, can be regarded primarily as demonstrating the fertility of his imagination. Still, it would be a mistake to discredit Democritus completely, for although he was no student of atomic physics, he had characteristics which every student needs.

Because atomic physics is built from big ideas about very small things, its development often leads along paths which run counter to common sense. As we consider things and events that are orders of magnitude removed from everyday experience, the difficulty of understanding their nature increases. Our common sense enables us to understand the relationship between a brick and a house. Conceiving of the earth as round may involve a little uncommon sense, but for most people it presents no great difficulty. However, the relationship between water and a water molecule is more difficult. While we can see the earth, whether flat or round, we cannot see a water molecule even with the best of instruments. All of our information about single water molecules is of an indirect kind, yet it is a very unsophisticated chemist for whom the concept of a single water molecule is not a part of his common sense. As a man's knowledge expands, more and more facts assume the aspect of "common sense." Certain velocity relationships are common sense. To an observer in a moving car, the velocity of another moving car appears different than to an observer standing beside the highway. In fact, a very young child once observed when the car in which he was traveling was passed by another, "We are backing up from the car ahead." However, the statement made by Albert Einstein that the velocity of light is the same for all observers regardless of their own velocities is very uncommon sense. In a later chapter we will attempt to show that his statement is reasonable and can appropriately be incorporated into our common sense. The conflict between the earth's actual roundness and its apparent flatness is resolved conceptually, i.e., by imaginative understanding, with the realization that the earth is a very big sphere. Somewhat similarly, the apparent conflict between our statements about relative velocities is resolved conceptually with the realization that the velocity of light is a very large velocity. Democritus, who could propose an atomic theory in about 400 B.C., would have the courage and imagination to face the ideas that lie before us.

It is the business of philosophers to discuss the nature of reality. It is the business of physicists (once called natural philosophers) to discuss the nature of physical reality. Philosophy, therefore, includes all of physics and a lot more besides. It is natural, then, that physics should have a continuing influence on philosophy. As physical discovery is quickly put into engineering practice and made to bear on man's physical environment, so it also affects the formulation of philosophical theory and bears directly on man's outlook and interpretation of life.

The old or classical physics of Newton was extraordinarily successful in dealing with events observed in his day. Using methods he developed, it is simple to equate the earth's gravitational force on the moon to the centripetal force and obtain verifiable relationships about the behavior of the moon. The same methods can be extended to orbits which cannot be regarded as circular. In fact, three observations of a new comet enable astronomers to foretell with great accuracy the entire future behavior of the comet. Given a certain amount of specific data known as initial or boundary conditions, classical Newtonian mechanics enables us to determine future events in a large number of situations. It is easy to move a step further and argue that what Newton has demonstrated to be true often, is true always, and that given sufficient initial data and boundary conditions, laws may be found which show every future event to be determined. The motion of a falling leaf or the fluctuations in the price of peaches may be very complex phenomena. It may require tremendous amounts of data and the application of very complicated laws which we do not yet understand to be able to make predictions in these cases. The important philosophical consequence of classical mechanics was not that every problem had been solved, but that a point of view had been established. It was felt that each new discovery would fall into the Newtonian mechanistic framework. Philosophical questions like the following became more pressing. Do we humans make decisions which alter the course of our lives or are we, like the bodies of the solar system, acting according to a set of inflexible laws and in accordance with a set of boundary conditions? Are we free or is our apparent ability to make decisions an illusion? Is everything we do beyond our responsibility, having been determined at the time of creation? Although mechanistic philosophy is rather repulsive when applied to ourselves, we nevertheless lean heavily upon it in interpreting things that go on about us. Indeed, the whole argument over whether human behavior is influenced more by heredity or environment is based on the assumption that human behavior is determined by some combination of the two.

To the extent that this mechanistic philosophy is based on classical physics, it is due for revision. Upon examination of events that are either very large or very small, we find that classical physics begins to fail. When a new theory or a modified theory has had to be applied in order to describe experimental observations, it has often resulted that the new theory is very different from classical physics. The method of attack, the mathematical techniques, and the form of the solution are often quite different. At one point we shall show that the observations of natural phenomena are inherently *uncertain.* It becomes evident, then, that if some circumstance had led to the development of atomic physics before classical physics, the influence of atomic physics on philosophy would have been against mechanism rather than for it.

Atomic physics has given us electronics and all that that word implies, including radio, radar, television, computers, etc. Atomic physics has given us nuclear energy. The new physics is as successful with submicroscopic events as classical physics was with large-scale events. But it may be that the most important benefits that can result from the study of atomic physics are philosophical rather than technical.

1-2. CHEMICAL EVIDENCE FOR THE ATOMIC VIEW OF MATTER

The speculations of Democritus and of the Epicurean school, whose philosophy was based on atomism, were not the generally accepted views of matter during the Middle Ages and the Renaissance. The prevailing concepts were those of Aristotle and the Stoic philosophers, who held that space, matter, and so on were continuous, and that all matter was one primordial stuff which was the habitat of four elementary principles—hotness, coldness, dryness, and wetness. Different materials differed in the degree of content of these principles. The hope of changing the amount of these principles in the various kinds of matter was the basis of alchemy. Not until the development of quantitative chemistry in the last half of the eighteenth century did the experimental evidence needed for evaluating the conflicting speculations about the constitution of matter begin to appear.

Antoine Lavoisier of France was outstanding among the early chemists. He evolved the present concept of a chemical element as "the last point which analysis is capable of reaching"; and he concluded from his observations on combustion that matter was conserved in chemical reactions.

In 1799 the French chemist J. L. Proust stated the law of definite or constant proportions, which summed up the results of his studies of the substances formed when pairs of elements are combined. The law is: *in every sample of any compound substance, formed or decomposed, the proportions by weight of the constituent elements are always the same.* This statement actually defines chemical compounds, because it differentiates them from solutions, alloys, and other materials which do not have definite composition.

The principal credit for founding the modern atomic theory of matter goes to John Dalton, a teacher in Manchester, England. His concern with atoms seems to have originated with his speculations about the solubilities of gases in water and with his interest in meteorology, which led him to try to explain the fact that the atmosphere is a homogeneous mixture of gases. Eventually, he believed that an element is composed of atoms that are both *physically* and *chemically* identical, and that the atoms of different elements differ from one another. In a paper he read at a meeting of the Manchester Literary and Philosophical Society in 1803, Dalton gave the first indication of the quantitative aspect of his atomic theory. He said, "An enquiry into the relative weights of the ultimate particles of bodies is a subject, as far as I know, entirely new: I have lately been prosecuting this enquiry with remarkable success." This was followed by his work on the composition of such gases as methane (CH_4), ethylene (C_2H_4), carbon monoxide (CO), carbon dioxide (CO_2), and others which led him to propose the law of multiple proportions in 1804. This law states: *if substance A combines with substance B in two or more ways, forming substances C and D, then if mass A is held constant, the masses of B in the various products will be related in proportions which are the ratios of small integers.* The only plausible interpretation of this law is that when elementary substances combine, they do so as discrete entities or atoms. Dalton emphasized the importance of relative weights of atoms to serve as a guide in obtaining the composition of other substances, and stressed that a chemical symbol means not only the element but also a fixed mass of that element. The introduction of the concept of atomic weights (strictly, atomic masses) was Dalton's greatest contribution to the theory of chemistry, because it gave a precise quantitative basis to the older vague idea of atoms. This concept directed the attention of quantitative chemistry to the determination of the relative weights of atoms.

An important law pertaining to volumes of gases was announced by Gay-Lussac in 1808. He said that *if gas A combines with gas B to form gas C, all at the same temperature and pressure, then the ratios of the volumes of A,B, and C will all be ratios of simple integers.* Two examples of this law are (a) the combining of two volumes of hydrogen and one volume of oxygen to form two volumes of water vapor, and (b) the union of one volume of nitrogen and three volumes of hydrogen to produce two volumes of ammonia. The following are symbolic forms of these reactions:

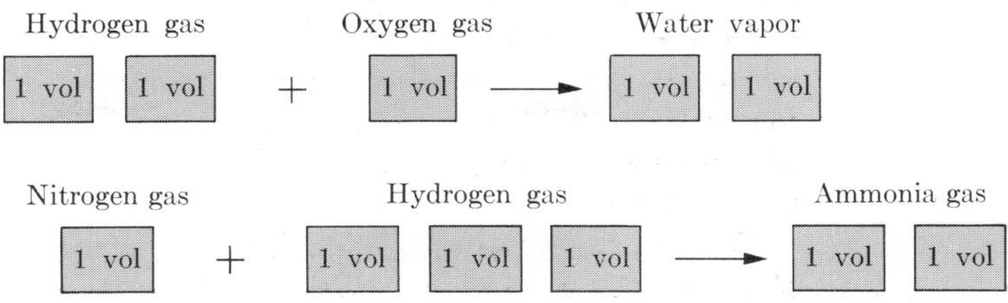

It is obvious that Gay-Lussac's law, like the law of multiple proportions, implies that the substances which participate in these reactions participate in discrete or corpuscular amounts. The ratio between the number of shoes worn to the number of people wearing them is almost an exact integer, namely two, showing that both people and shoes are discrete entities. The ratio of the number of tomatoes used per serving of tomato soup is quite a different kind of situation, and if the ratio is integral it is only by coincidence.

Gay-Lussac's law supported the work of Dalton, but it also raised difficult questions about the composition of an element in the gaseous state. In the case of the first reaction given, does each atom in the given oxygen gas divide to spread through the two volumes of water vapor? If so, the indivisibility of atoms must be abandoned. Or does each entity in the oxygen gas consist of a multiplicity of atoms? If so, how many atoms are grouped together? Similar questions can be raised about each of the gases in the two reactions given. It is evident that the numerical values of the relative weights of the atoms determined from these reactions will depend upon the answers to these questions.

In 1811, Avogadro, an Italian physicist, proposed the existence of different orders of small particles for the purpose of correlating the works of Dalton and Gay-Lussac. He postulated the existence of "elementary molecules" (atoms) as the smallest particles that can combine to form compounds, and the existence of "constituent molecules" (molecules of an element) and "integral molecules" (molecules of a compound) as the smallest particles of a body that can exist in the free state. He went on to state (without proof) a very important generalization, known as Avogadro's law, that *at the same temperature and pressure equal volumes of all gases contain the same number of molecules.* From this law and his concepts of atoms and molecules, Avogadro showed that the ammonia-producing reaction required that nitrogen gas consist of diatomic molecules and that oxygen must also be diatomic to account for the water-vapor reaction. He further concluded that water must consist of a union of two atoms of hydrogen and one atom of oxygen.

Unfortunately, the ideas advocated by Avogadro received little notice even when revived by Ampere in 1814. The notion that hydrogen and other gases were composed of diatomic molecules was ridiculed by Dalton and others, who would not conceive of a combination of atoms of the same kind. They asked, "If two hydrogen atoms in a container filled with this gas can cling together, why do not all cling together and condense to a liquid?" This is indeed a very good question. Science was not able to give a satisfactory answer until over a century later, after the introduction of the Bohr theory of the atom.

In the next two sections in this chapter we will describe some of the methods which were and still are used to determine the relative weights of atoms. The results obtained by the analytical chemists using these several methods during the first half of the nineteenth century were often contradictory. They frequently obtained different values for the atomic weight of the same element. By the 1850's inconsistencies were so numerous that many felt that the atomic theory of matter would have to be discarded. However, the contradictions were resolved in 1858 by the Italian chemist Cannizzaro, who had an intimate knowledge of the then known methods for determining atomic weights and a broad grasp of the whole field of chemistry. He showed that Avogadro really had provided a rational basis for finding atomic weights, and that the inconsistent results obtained by various experimenters resulted from a lack of clear distinction between atomic weights, equivalent weights, and molecular weights. The views of Canizzaro received the approval of the scientific world when they were adopted by the international conference on atomic weights which met in Karlsruhe, Germany, in 1860. This, then, is the year in which the fundamental ideas of modern chemistry were widely accepted.

1-3. MOLECULAR WEIGHTS

After Cannizzaro had clarified and established some of the basic definitions in chemistry, Avogadro's law opened the door to one of the methods for determining molecular weights. No one had any idea what a single molecule weighed, but once there was a way of isolating equal numbers of different kinds of molecules, the relative weights could be determined. The hydrogen molecule was found to be the lightest molecule, and the hydrogen atom proved to be the lightest atom. In 1815 Prout had proposed that the relative atomic weight of hydrogen be arbitrarily taken as one. On this basis most other light atoms and molecules had relative weights which were nearly integers. But, for reasons to be discussed later, it turned out that the atomic weights of many of the heavier atoms were not very nearly integers. Hydrogen appeared to be a poor basis for the system, and more nearly integral atomic weights for all atoms could be obtained by making a heavier atom the basis of the system. Originally, oxygen was chosen and its atomic weight was arbitrarily set at exactly 16. This was the basis for many years until, in 1961, it was replaced by setting the atomic weight of carbon at exactly 12. (This will be discussed further in Chapter 2 when we consider isotopes.) Under either plan the atomic weight of hydrogen is not exactly unity, although it is nearly so.

These relative molecular and atomic weights are all dimensionless ratios. If about four parts by weight of hydrogen were combined with 32 parts by weight of oxygen, about 36 parts by weight of water vapor can be formed, according to the familiar equation, $2H_2 + O_2 = 2H_2O$. It is convenient to arbitrarily select the gram as the unit of weight (in physics we should properly use the term *mass*, but common usage overrules). The gram molecular (or atomic) weight of a material is the mass in grams of the number of particles in a mole of the material. A *mole* is defined as the amount of substance containing the same number of atoms or molecules as the number of atoms in 12 grams of pure carbon whose atomic weight is exactly 12. The number of atoms in this amount of carbon is called Avogadro's number or the *Avogadro constant*, N_A, and it is of basic importance in physics and physical chemistry. These definitions emphasize that a mole is a fixed number of particles of a material and it is evident that the number of moles in an amount of a compound is the weight in *grams* of the compound divided by its formula (molecular) weight. (Note that the mole and the Avogadro constant have been defined in terms of 12 grams of carbon. This is the modern practice. However, these definitions do not always conform to the MKSA system of units. Therefore, we will on occasion use the kilomole, kmole, which is 1000 moles. Obviously a kilomole of carbon 12 has a mass of 12 kilograms.)

The value of the Avogadro constant was of relatively minor importance to chemistry in the early nineteenth century and its magnitude was not even estimated until Loschmidt did so in 1865. We will discuss Perrin's method of determining it later in this chapter. Here is an interesting case where knowing the existence of a number was more important than knowing its magnitude as, for example, in determining the relative weights of the atoms involved in the ammonia-producing hydrogen-nitrogen reaction previously described. The value of the Avogadro constant is almost inconceivably large; by modern measurements it is

$$6.02252 \pm 0.00028) \times 10^{23}$$

particles per mole. Only after the magnitude of Avogadro's number was known could the absolute mass of an atomic particle be computed.

It follows from Avogadro's law that the volume of a mole of a gas is the same for all gases. The normal volume of a perfect gas or the standard molar volume of an ideal gas, V_0, is the volume occupied by a mole of the gas at a pressure of 1 standard atmosphere and a temperature of 0°C. The value of V_0 is

$$(2.24136 \pm 0.00030)^* \times 10^{-2}\ \text{m}^3 \text{ per mole.}$$

1-4. ATOMIC WEIGHTS

Avogadro's law provided a systematic method for determining molecular weights, but a large amount of quantitative data on the formation of various compounds were required before the atomic weights of the known elements could be determined. The situation is somewhat like the following: Suppose that man *A* pays man *B* $1.00 in coin, using no coin smaller than quarters, and that we wish to determine how this is done. He may do this in any one of four ways:

(a) one $1.00 coin,
(b) two 50¢ coins,
(c) one 50¢ and two 25¢ coins,
(d) four 25¢ coins.

If man *B* now pays man *C* 25¢, possibilities (a) and (b) are eliminated, but there is still a doubt as to how the original transaction was made. By careful observation of further transactions of those who spend the original $1.00, it could be determined just what coins *A* must have had originally.

An aid to the solution of this puzzle was the empirical discovery by Dulong and Petit, in 1819, that for most elements in the solid state the atomic weight times the specific heat at constant volume is about 6 cal/mole · °K. The law of Dulong and Petit permits a rough independent determination of atomic weight by dividing this constant by the specific heat. We shall discuss the theoretical basis of this law in Chapter 9.

---

* If it seems strange that this and some other constants are given with the uncertainty expressed to more than one significant figure, refer to the article, "Probable Values of the General Physical Constants," by R. T. Birge, [*Phys. Rev. Suppl.* **1**, (1929) p. 6]. We note here only that the concepts of probable errors and significant figures do not correspond completely. If the probable error can be determined to less than 10 percent of itself, then more than one significant figure is required to express it.

Principal articles containing the values of various constants are: "Values for the Physical Constants Recommended by NAS-NRC," *Nat. Bur. Std. (U.S.) Tech. News Bull.*, **47** (1963), p. 175; "World Sets Atomic Definition of Time," *Nat. Bur. Std. (U.S.) Tech. News Bull.*, **48** (1964), p. 209; Mechtly, E. A., *The International System of Units*, NASA SP-7012. Washington, D.C.: U.S. Government Printing Office (1964); Cohen, E. R. and J. W. M. Du Mond, "Our Knowledge of the Fundamental Constants of Physics and Chemistry in 1965," *Rev. Mod. Phys.* **37** (1965), p. 537.

1-5. PERIODIC TABLE

Probably the most significant discovery in all chemistry, aside from the atomic nature of matter, was the periodic properties of the elements, now depicted in the familiar periodic table of the elements (Appendix 3). The chemical properties of this table are probably familiar to most readers of this book; the physical properties will be discussed later. The table was proposed independently by Meyer and by Mendeléev in 1869. Its usefulness lay both in its regularities and in its irregularities. One interesting irregularity in the original table was that in order to have the elements fall in positions consistent with their chemical properties, it was necessary to leave numerous spaces unoccupied. Mendeléev suggested that these spaces would be filled with as yet undiscovered elements. Using his table, he was able to describe in considerable detail the properties these elements could be expected to have when they were discovered. It was nearly one hundred years before all the predictions that Mendeléev made were fulfilled.

Reflect, for a moment, on the vast simplification that the chemical discoveries here outlined provide. Looking about us, we see innumerable kinds of materials. The atomic view indicates that these materials are of discrete kinds whose number, however large, is not infinite. The discovery of elements is a further simplification in that the many materials we encounter are shown to be composed of only about one hundred chemically distinguishable materials, many of which are rare. It turns out that even these elements are not a heterogeneous group but are subject to further classification into a periodic table. The problems of chemistry are many, however, it is easy to see that things are much simpler than might at first appear.

1-6. PHYSICAL EVIDENCE FOR THE ATOMIC VIEW OF MATTER

In our discussion thus far, all atomic properties have been inferred from studies of gross matter. In 1827 the English botanist Robert Brown observed that microscopic pollen grains suspended in water appear to dance about in random fashion. At first the phenomenon was ascribed to the motions of living matter. In time, however, it was found that any kind of fine particles suspended in a liquid performed such a perpetual dance. Eventually it was realized that the molecules of a liquid are in constant motion and that the suspended particles recoiled (Brownian movement), when hit by the molecules of the liquid. However, long before the equations for Brownian movement were derived early in the 20th century, the particles of matter were thought of as moving about in a random manner and undergoing frequent collisions. Such processes are decidedly in the domain of physics. How can the principles of mechanics be applied to molecular collisions?

The simplest state of matter to consider was a gas. The ideal gas law, for n moles of a gas is $pV = nRT$, where R is the universal gas constant per mole and p, V, and T are the pressure, volume, and temperature, respectively. This law was a well-established *empirical* relationship, and its derivation was one of the objectives of physics. The application of classical physics to the mechanics of gases is called the *kinetic theory of gases.* Although Daniel Bernoulli had some success in developing this theory as early as 1738, the principal contributions that led to its establishment were made between 1850 and 1900 by Clausius, Maxwell, Boltzmann, and Gibbs.

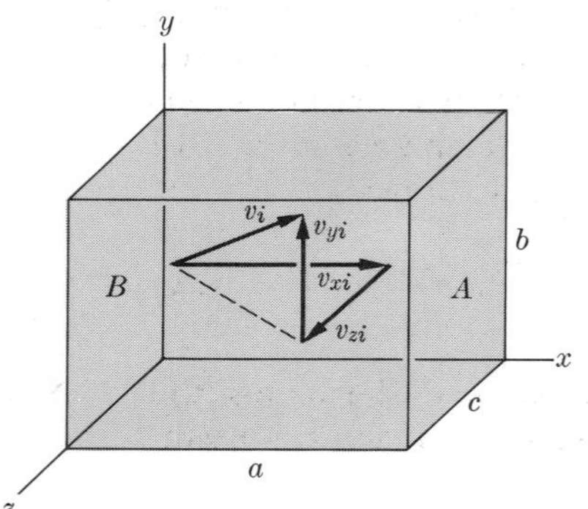

FIG. 1–1

1–7. KINETIC THEORY OF GASES. MOLAR HEAT CAPACITY

Early in our study of physics, we investigated the mechanics of bodies that can be regarded as particles. The study of extended bodies was treated by introducing certain averages, and the translational problem of extended bodies was solved by introducing the concept of a *center of mass* that moves as though it were a particle. The study of rotational properties of extended bodies was similarly facilitated by the introduction of another average property of the body, its *moment of inertia.* In the kinetic theory of gases, we assume that pressure, volume, temperature, etc., are *averages* of properties of all the molecules of a gas. Kinetic theory is a large and elegant subject. We can convey its spirit by deriving the ideal gas law and a few other relationships.

Consider a rectangular container, the edges of which are parallel to the X-, Y-, and Z-axes and have dimensions a, b, and c, as shown in Fig. 1–1. There are N identical particles, each of mass m, in the box moving in random directions with a wide range of speeds. These identical particles may be, but are not necessarily, atoms or molecules.

We assume that the particles are very small so collisions between them are rare compared to collisions with the plane walls of the container. Neglecting minor forces such as gravity and intermolecular forces, we shall consider that the only forces acting on the particle are those resulting from collisions with the walls. We number the particles $1, 2, \ldots, i, \ldots N$. Figure 1–1 shows the ith particle, whose velocity is v_i. This velocity may be broken into the rectangular components v_{xi}, v_{yi}, and v_{zi}, as shown. We assume collisions to be perfectly elastic such that when a particle strikes a wall the velocity component related to the axis that is perpendicular to that wall is reversed in direction but unchanged in magnitude. The other two velocity components remain unchanged. Thus if the particle strikes side A, the x-component of its momentum is changed from $+mv_{xi}$ to $-mv_{xi}$. The net change in the x-component of its momentum is

$$(-mv_{xi}) - (+mv_{xi}) = -2mv_{xi}.$$

Since collision with the wall causes the particle to change its momentum, the wall experiences an impulsive force. This impulsive force is unknown because we cannot estimate the time of contact in a meaningful way. Fortunately, it is not the impulsive force but the average force due to repeated hits that we seek. Since collisions with the top, bottom, far, and near sides have no effect on v_{xi}, and since collisions with ends A and B merely reverse the direction of v_{xi}, we see that the time interval between successive hits on side A is the total x-distance, $2a$, divided by the x-component of the velocity or $2a/v_{xi}$. By applying Newton's second law, we find the average force F_i of the wall on the ith particle to be

$$F_i = \frac{\Delta(mv)}{\Delta t} = \frac{-2mv_{xi}}{2a/v_{xi}} = -\frac{mv_{xi}^2}{a}. \tag{1–1}$$

This force is equal in magnitude and opposite in direction to the force of the particle on the wall and thus the particle produces an average pressure on side A given by

$$p_i = \frac{-F_i}{\text{area}} = \frac{-F_i}{bc} = \frac{mv_{xi}^2}{abc} = \frac{mv_{xi}^2}{V}, \tag{1–2}$$

where V is the volume of the container.

The pressure we have computed is due to but one, the ith, particle. The pressure due to each particle is computed the same way. Adding the pressures due to the N identical particles, we have

$$p = \sum_{i=1}^{N} p_i = \frac{m}{V} \sum_{i=1}^{N} v_{xi}^2. \tag{1–3}$$

To evaluate the sum on the right-hand side of the equation we note (see Fig. 1–1) that

$$v_i^2 = v_{xi}^2 + v_{yi}^2 + v_{zi}^2. \tag{1–4}$$

Since this equation holds for each of the particles, we can add the corresponding equations and obtain

$$\sum_{i=1}^{N} v_i^2 = \sum_{i=1}^{N} v_{xi}^2 + \sum_{i=1}^{N} v_{yi}^2 + \sum_{i=1}^{N} v_{zi}^2. \tag{1–5}$$

We now define the *mean square velocity*, $\overline{v^2}$, to be the average of the sum of the squares of the velocities; therefore

$$\overline{v^2} = \left(\sum_{i=1}^{N} v_i^2\right) \Big/ N. \tag{1–6}$$

Applying this definition to all terms in Eq. (1–5), we find that it becomes

$$\overline{v^2} = \overline{v_x^2} + \overline{v_y^2} + \overline{v_z^2}, \tag{1–7}$$

and, substituting terms from Eq. (1–6) into Eq. (1–3), we get

$$p = \frac{m}{V} N\overline{v_x^2}. \tag{1–8}$$

Since we assume these velocities to be completely random in direction and magnitude, the three mean square velocity components must be equal or $\overline{v_x^2} = \overline{v_y^2} = \overline{v_z^2}$. This assumption enables us to deduce from Eq. (1–7) that

$$\overline{v^2} = 3\overline{v_x^2} = 3\overline{v_y^2} = 3\overline{v_z^2}. \tag{1–9}$$

The square root of the quantity $\overline{v^2}$ is called the *root-mean-square speed, or velocity*, v_{rms}. Substituting $\overline{v_x^2}$ from Eq. (1–9) into Eq. (1–8), we get

$$p = \frac{Nm}{V} \frac{\overline{v^2}}{3}$$

or

$$pV = \tfrac{1}{3}Nm\overline{v^2}. \tag{1–10}$$

When the particles in the container are the molecules of a gas, then the number of moles n of the gas in the container equals the total number of molecules N in it divided by the number of molecules in a mole N_A, the Avogadro constant. Therefore we have $n = N/N_A$ or $N = nN_A$. Since the product of the mass of a molecule and the Avogadro constant is the molecular weight M, we can express the total mass of the gas in the box as $Nm = nN_Am = nM$. When this result is substituted in Eq. (1–10) it becomes

$$pV = \tfrac{1}{3}nM\overline{v^2}. \tag{1–11}$$

This is not the result we sought, $pV = nRT$, so we have as yet no justification for the many assumptions we have made. The result is interesting, however, because it contains the pV term, and the term $\frac{1}{3}nM\overline{v^2}$ has a familiar look. If we write

$$\tfrac{1}{3}nM\overline{v^2} = \tfrac{2}{3}(\tfrac{1}{2}nM\overline{v^2}), \tag{1–12}$$

the quantity in parentheses is clearly the total *translational* kinetic energy of the molecules. This energy must be the internal energy of the gas U because it was assumed that no forces of attraction or repulsion exist which could give rise to molecular potential energy. By combining Eqs. (1–11) and (1–12), we obtain

$$pV = \tfrac{2}{3}U. \tag{1–13}$$

If we could show that $U = \tfrac{3}{2}nRT$, then Eq. (1–13) would become $pV = nRT$, the ideal gas law.

TABLE 1–1 C_v OF GASES

Gas		C_v, cal/mole · °K
Helium,	He	3.00
Argon,	A	3.00
Mercury,	Hg	3.00
Hydrogen,	H_2	4.82
Oxygen,	O_2	4.97
Chlorine,	Cl_2	6.01
Ether,	$(C_2H_5)_2O$	30.8

By 1850 Joule had demonstrated that heat is a form of energy. When a gas is heated at constant volume, the heat energy supplied causes a temperature change that must increase the energy of the gas. If we tentatively accept the relation $U = \tfrac{3}{2}nRT$, then the change in internal energy with respect to temperature can be given by

$$dU/dT = \tfrac{3}{2}nR. \tag{1–14}$$

The change in internal energy with respect to temperature of one mole of an ideal gas at constant volume is called the *molar heat capacity* C_v. Therefore we obtain

$$C_v = \tfrac{3}{2}R, \tag{1–15}$$

and its value is

$$C_v = \frac{3}{2}R = \frac{3}{2} \times \frac{8.31\ \text{J}}{\text{mole} \cdot {}^\circ\text{K}} \times \frac{1\ \text{cal}}{4.18\ \text{J}} = 2.97\ \text{cal/mole} \cdot {}^\circ\text{K}.$$

The experimental values of C_v for several gases are given in Table 1–1. Note that three values agree very closely with the computed value but that the others are quite different. Both the agreements and the disagreements are interesting. The values which agree are those of monatomic gases and those that disagree are not. We shall have more to say about the apparent disagreements. The point here is to recall that in our discussion of kinetic theory we assumed that our molecules were isolated elastic spheres. We should not expect our result to apply to diatomic dumbbells or to complicated molecules.

1-8. EQUIPARTITION OF ENERGY

The agreement we have observed for monatomic molecules would not have been possible had there not been the number 3 in the expression $C_v = 3R/2$. Looking back over our derivation, we find that the 3 entered into the calculation from the statement $\overline{v^2} = 3\overline{v_x^2}$, that is, because the molecule was free to move in three-dimensional space. The expression for the average kinetic energy of translation of the molecules of a gas is composed of three equal parts, $R/2$ per degree of absolute temperature associated with each coordinate. The principle of *equipartition of energy* states that if a molecule can have energy associated with several coordinates, the average energy associated with each coordinate is the same. Because of this principle, the number of coordinates necessary to specify the position and configuration of a body is called the number of its *degrees of freedom.*

A monatomic molecule requires three coordinates to specify its position. A rigid diatomic molecule requires three position coordinates and two more are necessary to specify its configuration. If we assume the second atom is at a fixed distance from the first, its location is specified as being on a sphere with the first atom at its center. It requires but two additional coordinates to specify where on this sphere the second atom lies. Thus the addition of a second atom adds two degrees of freedom to the molecule. If our derivation for molar heat capacity had been based on diatomic instead of monatomic molecules, we would have obtained $5R/2$ instead of $3R/2$. We find that $5R/2 = 4.95$ cal/mole · °K, which agrees closely with the measured molar heat capacities of such diatomic molecules as hydrogen and oxygen, as shown in Table 1–1. Six coordinates are enough to specify the position of any *rigid* molecule, however complex, but if the molecules are composed of vibrating atoms, then the number of degrees of freedom may become very large. This accounts for the large molar heat capacity of ether. A fuller discussion of heat capacities requires the introduction of quantum theory but classical kinetic theory reveals much, both qualitatively and quantitatively.

With the help of independent data from molar heat capacities, we have found that the kinetic theory of matter provides a quantitative mechanical model for both the ideal gas law and molar heat capacities. A very useful relation can be obtained from one of the derived equations and the ideal gas law. Having shown that $pV = \frac{1}{3}nM\overline{v^2}$ and knowing that $pV = nRT$, we see that $\frac{1}{3}nM\overline{v^2} = nRT$. This may be written as

$$\boxed{\tfrac{1}{2}M\overline{v^2} = \tfrac{3}{2}RT.} \qquad (1\text{–}16)$$

This equation shows that the kinetic energy of translation of the molecules of a gas depends only on the absolute temperature, and that at a given temperature, the lighter molecules have the greater speeds. The root-mean-square speed of hydrogen molecules at room temperature is about 1800 m/s, or more than 1 mi/s.

1-9. MAXWELL'S SPEED DISTRIBUTION LAW

We have found that $\overline{v^2}$ can be computed from the temperature of a gas. The speed thus determined is one of the important average properties of a gas. But an average can be misleading. Consider first that a man who earns one million dollars a year is riding in a taxi, alone except for the driver. The average income of the occupants of the car is about $500,000. This true statement tends to imply that there are two wealthy men in the car. Similarly for a large group of people, a statement of the average income fails to disclose that some incomes are far from average. A fuller economic picture is conveyed if we know what fractions of the populace have incomes within certain ranges. We might find, for example, 1% below $1000, 5% between $1000 and $2000, etc. Such data determine what is called a distribution function.

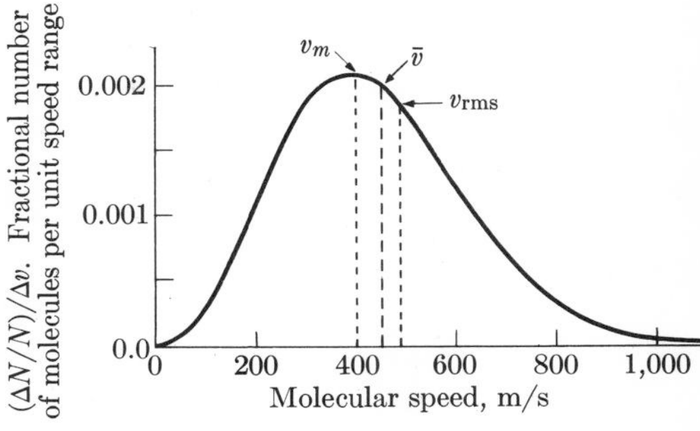

FIG. 1-2
Distribution of speeds for nitrogen molecules at 0°C.

Thus it is reasonable to ask how individual molecular speeds are distributed about the average. Maxwell solved the following problem: what fraction of the molecules of a gas have speeds between v and $v + \Delta v$? Maxwell's solution is too difficult to present here, but the result is

$$\frac{\Delta N}{N} = \frac{4}{\sqrt{\pi}}\left(\frac{M}{2RT}\right)^{3/2} v^2 e^{-Mv^2/2RT}\,\Delta v. \tag{1-17}$$

In this equation, $\Delta N/N$ is the fractional number of molecules that have speeds in the interval Δv, M is the molecular weight of the gas, T is its absolute temperature, and R is the gas constant. This Maxwellian distribution of speeds is shown graphically for a particular case in Fig. 1-2. The curve shows that the molecular speeds range from zero to infinity. The speed for which the curve is maximum is the *most probable speed*, v_m. If we were to pick molecules at random, this speed would be found most often. The average speed obtained from the pressure calculation was $\sqrt{\overline{v^2}}$, or the root-mean-square speed v_{rms}. It is larger than the most probable speed because it is based on the square of the speeds, which gives high speeds more relative importance than low ones. Intermediate between these is the *average speed* $\bar{v}$. The relations between the characteristic speeds of gas molecules are as follows:

$$v_{\text{rms}} = \sqrt{\frac{3RT}{M}},$$

$$v_m = \sqrt{\frac{2}{3}}\, v_{\text{rms}} = 0.817 v_{\text{rms}}, \tag{1–18}$$

$$\bar{v} = \sqrt{\frac{8}{3\pi}}\, v_{\text{rms}} = 0.921 v_{\text{rms}}.$$

EXAMPLE. Calculate (a) the root-mean-square speed of the molecules of nitrogen under standard conditions, and (b) the kinetic energy of translation of one of these molecules when it is moving with the most probable speed in a Maxwellian distribution. (The required physical data can be found in Appendixes 3, 7 and 8.)

Solution. (a) Using Eq. (1–16) and considering a kmole of the gas, we have

$$\tfrac{1}{3}M\overline{v^2} = RT,$$

or

$$\overline{v^2} = 3RT/M,$$

$$\overline{v^2} = 3 \times 8.31 \times 10^3 \frac{\text{J}}{{}^\circ\text{K} \cdot \text{kmole}} \times 273^\circ\text{K} \times \frac{1 \text{ kmole}}{28 \text{ kg}}$$

$$= 2.43 \times 10^5 \frac{\text{m}^2}{\text{s}^2};$$

$$v_{\text{rms}} = \sqrt{\overline{v^2}} = \sqrt{2.43 \times 10^5 \text{ m}^2/\text{s}^2} = 492 \text{ m/s}.$$

This result can also be obtained from Eq. (1–11):

$$pV = \tfrac{1}{3}nM\overline{v^2}$$

$$\overline{v^2} = \frac{3pV}{nM}$$

$$= 3 \times 1.013 \times 10^5 \frac{N}{\text{m}^2} \times 22.42 \frac{\text{m}^3}{\text{kmole}} \times \frac{1 \text{ kmole}}{1 \times 28 \text{ kg}}$$

$$= 2.43 \times 10^5 \frac{\text{m}^2}{\text{s}^2};$$

$$v_{\text{rms}} = \sqrt{\overline{v^2}} = \sqrt{2.43 \times 10^5 \text{ m}^2/\text{s}^2} = 492 \text{ m/s}.$$

(b)

$$E_k = \tfrac{1}{2}mv_m^2.$$

The mass of a nitrogen molecule can be calculated from its molecular weight and the Avogadro constant:

$$m = M/N_A$$

$$= 28 \frac{\text{kg}}{\text{kmole}} \times \frac{1 \text{ kmole}}{6.02 \times 10^{26} \text{ molecules}}$$

$$= 4.65 \times 10^{-26} \text{ kg/molecule}.$$

FIG. 1–3
Diagram of apparatus used by Zartman and Ko.

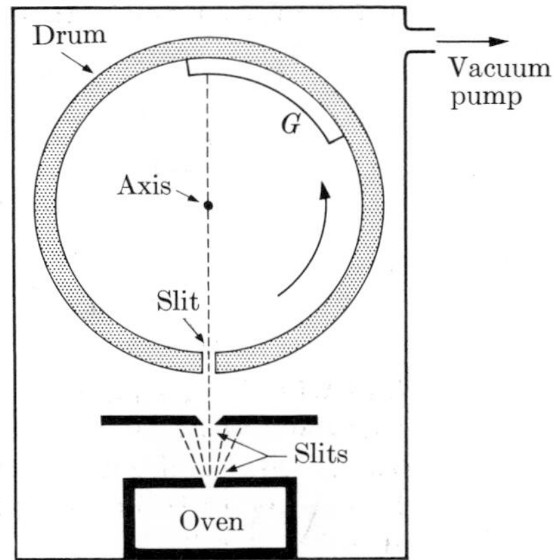

From Eq. (1–18) we have

$$v_m = \sqrt{\tfrac{2}{3}}\, v_{\text{rms}},$$

$$\therefore\quad E_k = \tfrac{1}{2} \times 4.65 \times 10^{-26}\ \text{kg} \times \tfrac{2}{3} \times 2.43 \times 10^5\ \text{m}^2/\text{s}^2 = 3.76 \times 10^{-21}\ \text{J}.$$

Maxwell's distribution of speeds was employed to calculate other gas properties and was indirectly verified in terms of these secondary properties. A direct experimental verification was obtained by Zartman and Ko in 1930. They used an oven, shown in Fig. 1–3, containing bismuth vapor at a known high temperature (827°C). Bismuth molecules streamed from a slit in the oven into an evacuated region above.* The beam was made unidirectional by another slit, which admitted only properly directed molecules. Above the slit was a cylindrical drum that could be rotated in the vacuum about a horizontal axis perpendicular to the paper. A slit along one side of the drum had to be in a particular position to enable the beam of molecules to enter it. When the drum was stationed so that the beam could enter, the beam moved along a diameter of the drum and was deposited on a glass plate G mounted on the inside surface of the drum opposite the slit. During the experiment, the drum was rotated at a constant angular velocity so that short bursts of molecules were admitted on each rotation. Because the speeds of the molecules varied, some crossed the diameter quickly and others took much more time, and since the drum was turning while the molecules were moving across it, they struck the glass plate at different places. Thus the distribution of speeds was translated by the apparatus into a distribution in space around the inside of the drum, as indicated by the variation in the darkening of the glass where the bismuth was deposited. The thickness of the deposit was measured optically, and comparison of the experimental distribution of speeds with Maxwell's theoretical distribution expression showed excellent agreement.

---

* Fast-moving molecules escape from the oven more often than slow ones. Computation shows that if the oven is at a temperature T, the root-mean-square speed of escaping molecules is the same as the root-mean-square speed within an oven at a higher temperature, $4T/3$.

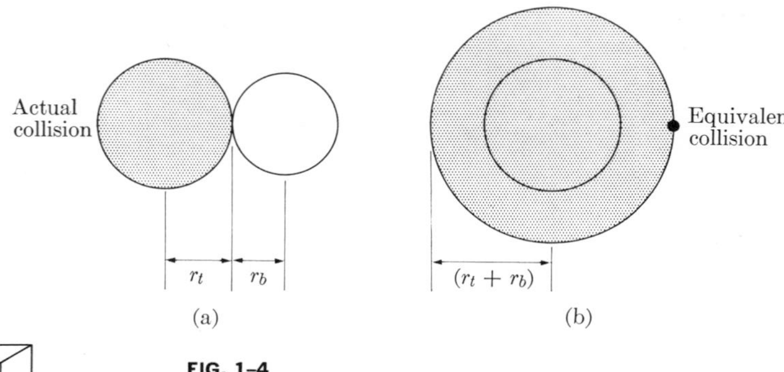

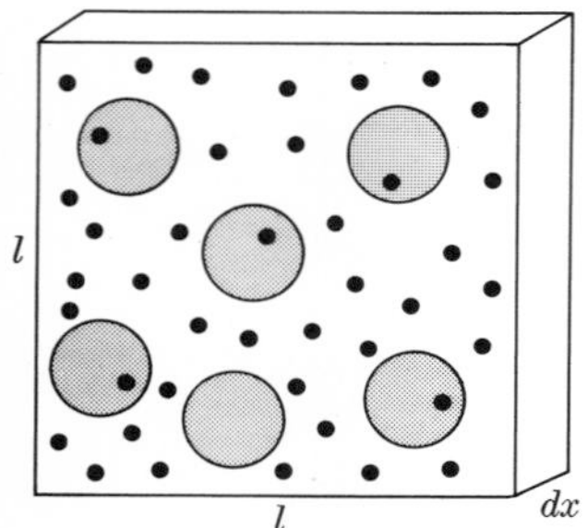

FIG. 1–4
For mathematical convenience, the actual collision depicted in part (a) may be represented by the equivalent collision shown in (b).

FIG. 1–5

1–10. COLLISION PROBABILITY. MEAN FREE PATH

If molecules were truly geometrical points, no collisions would take place between them. Actual molecules, however, are of finite size, and for the purposes of this discussion we are assuming that a molecule is a rigid, perfectly elastic sphere. A collision between two molecules is considered to take place whenever one molecule makes contact with another. Let us refer to one of the colliding molecules as the target molecule, of radius r_t, and to the other as the bullet molecule, of radius r_b. Then a collision occurs whenever the distance between the centers of the molecules becomes equal to the sum of their radii, $r_t + r_b$, as in Fig. 1–4(a).

When we are considering collisions of molecules of a given gas with other molecules of the same gas, the radii r_t and r_b are equal and there is no difference between target molecules and bullet molecules. In many instances, however, we wish to consider collisions between different kinds of particles, and so we shall speak of the target molecules as though they differed from the bullet molecules.

Since it is only the center-to-center distance that determines a collision, it does not matter whether the target is large and the bullet is small, or vice versa. We may therefore replace an actual collision with the equivalent collision shown in Fig. 1–4(b), in which the bullet molecule has been considered to shrink to a geometrical point and the target molecule to expand to a sphere of radius $r_t + r_b$.

Now consider a thin layer of material of dimensions l, l, and dx. The layer contains (equivalent) target molecules only, and to begin with we assume that these are at *rest*. We then imagine that a very large number N of bullet molecules are incident normally on the face of the layer like a blast of pellets from a shotgun, in such a way that they are distributed over the face. If the thickness of the layer is so small that no target molecule can *hide* behind another, the layer presents to the bullet molecules the appearance shown in Fig. 1–5, where the shaded circles represent the target molecules and the black dots the bullet molecules.

Most of the bullet molecules will pass through the layer, but some will collide with target molecules. The ratio of the number of collisions, dN, to the total number of bullet molecules, N, is equal to the ratio of the area presented by the target molecules to the total area presented by the layer:

$$\frac{dN}{N} = \frac{\text{target area}}{\text{total area}}.$$

The target area σ of a single (equivalent) molecule is

$$\sigma = \pi(r_t + r_b)^2. \tag{1–19}$$

This area is called the *microscopic collision cross section* of one (equivalent) molecule. The total target area is the product of this and the number of molecules in the layer. If there are n target molecules per unit volume, this number is $nl^2\,dx$, so the total target area is

$$n\sigma l^2\,dx.$$

The total area of the layer is l^2, so

$$\frac{dN}{N} = \frac{n\sigma l^2\,dx}{l^2} = n\sigma\,dx. \tag{1–20}$$

The quantity $n\sigma$ is called the *macroscopic cross section* of the group of (equivalent) molecules. Note that the unit of macroscopic cross section is reciprocal length, not area.

In the preceding equation the quantity dN/N is the fractional number of molecules that undergo collisions and therefore this ratio is simply the probability of a collision. (Strictly, this should have a negative sign because dN molecules are removed from the stream of bullets.) In the beginning of this discussion the cross section was thought of as an actual area presented by a target molecule, but this was soon replaced by an equivalent area. The microscopic cross section can now be defined in a more realistic way through Eq. (1–20), i.e., that the probability of an interaction between two molecules is directly proportional to σ.

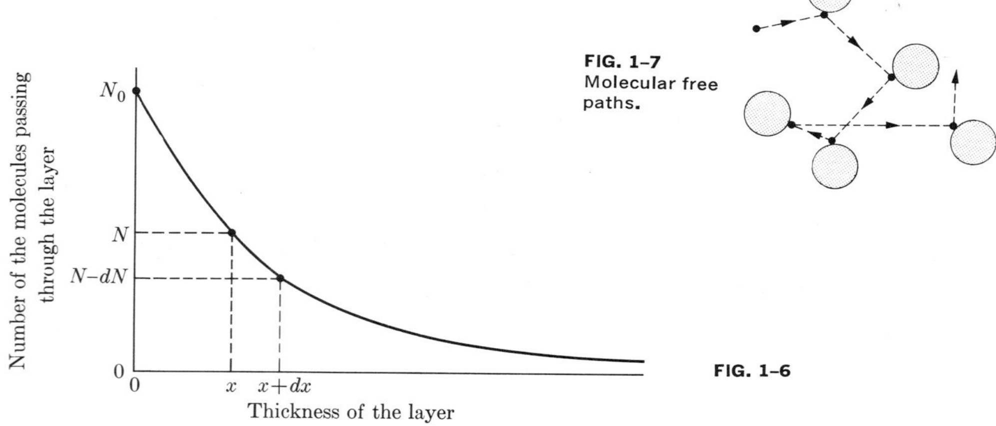

FIG. 1-7 Molecular free paths.

FIG. 1-6

If N_0 bullet molecules per unit area are incident normally on the face of a layer of material containing *stationary* molecules having the macroscopic cross section $n\sigma$, then N, the number transmitted per unit area through a finite thickness x, can be found by integrating Eq. (1–20). We then have

$$\int_{N_0}^{N} -\frac{dN}{N} = \int_0^x n\sigma\, dx, \tag{1–21}$$

and obtain

$$\ln \frac{N}{N_0} = -n\sigma x$$

or

$$N = N_0 e^{-n\sigma x}. \tag{1–22}$$

This exponential equation is plotted as a solid line in Fig. 1–6.

Let us next follow in imagination a single bullet molecule as it makes its way through a material along the zigzag path shown in Fig. 1–7. We wish to obtain an expression for the average distance traveled between collisions, known as the *mean free path, L*. This can be deduced from the results above by a type of reasoning that is common and useful in problems of this sort.

When molecules are passing through the thin layer of material in Fig. 1–5, the number removed from the beam by collisions is small compared with the original number, and we can say that N molecules have each traversed a thickness dx of material and that in the process a number dN of collisions have taken place. The total distance traveled by all of the N molecules is then $N\,dx$. We now make the hypothesis that the number of collisions made by a single molecule in traversing the same total distance $N\,dx$ is equal to the number of collisions made by N molecules, each traversing a distance dx. Then from Eq. (1–20), the total number of collisions made by the single molecule in a total path length $N\,dx$ is

$$dN = N\,n\sigma\,dx. \tag{1–23}$$

The mean free path of the molecule is equal to the total path length divided by the number of collisions, or

$$L = \frac{\text{total path length}}{\text{total number of collisions}}.$$

From the expressions above for the total path length and the total number of collisions, we have

$$L = \frac{N\,dx}{Nn\sigma\,dx}$$

or

$$\boxed{L = \frac{1}{n\sigma}.} \tag{1–24}$$

The concept of mean free path may be visualized by thinking of a man shooting a rifle aimlessly into a forest. Most of the bullets will hit trees, but some bullets will travel much farther than others. It is easy to see that the average distance the bullets go will depend inversely on both the denseness of the woods and the size of the trees.

The expression for the mean free path of a bullet molecule can also be found from Eq. (1–22). Obviously, L is the mean or average distance $\bar{x}$ traveled by the bullet molecules before collisions with the targets in a layer of material so thick that no molecule goes all the way through it. Therefore, what we seek is the numerical average of the distances traveled by the bullets as their number decreases from N_0 to 0. The required relation can be found by considering the curve in Fig. 1–6. Let N be the number of bullets that have traveled a distance x, and $N - dN$ those moving onward beyond $x + dx$; then dN is the number of bullets that collided with targets in the distance dx. Therefore, the combined ranges of the bullets in this group at the time of collision is $x\,dN$. The average distance traveled by a bullet will be the sum of the combined ranges of all the groups from N_0 to 0 divided by the total number of bullets. Mathematically, this is

$$\bar{x} = L = \frac{\int_{N_0}^{0} x\,dN}{\int_{N_0}^{0} dN}. \tag{1–25}$$

This is most easily integrated if dN in the numerator is replaced by its equivalent in terms of dx. From Eq. (1–22) we get

$$N = N_0 e^{-n\sigma x},$$

and the derivative of this is

$$dN = -N_0 n\sigma e^{-n\sigma x}\, dx. \tag{1–26}$$

The new limits are obtained from the conditions that when $N = N_0$, $x = 0$; and when $N = 0$, $x = \infty$. Substituting the expression for dN from Eq. (1–26) into the numerator of Eq. (1–25) and integrating the denominator, we have

$$\bar{x} = L = \frac{\int_0^\infty - xN_0 n\sigma e^{-n\sigma x}\, dx}{-N_0} = n\sigma \int_0^\infty xe^{-n\sigma x}\, dx. \tag{1–27}$$

We integrate this by parts, obtaining

$$L = \left[\frac{1 - x}{e^{n\sigma x}}\right]_0^\infty.$$

This is indeterminate at the upper limit but, by using l'Hospital's rule, we find the result is

$$L = \frac{1}{n\sigma}. \tag{1–28}$$

This is identical to the previous expression for the mean free path. (We will obtain equations of the same form as Eq. (1–22) in later chapters, especially in the discussions of x-rays and of radioactivity. Note that the derivation of Eq. (1–28) shows that the mean value, over the range 0 to ∞, of the variable in the exponent in Eq. (1–22) is equal to the negative of the reciprocal of the coefficient of that variable.)

In the above analysis, we assumed that the target molecules were at rest. This assumption is valid for a bullet molecule going through a solid, and we will meet this situation in a later chapter when we consider neutrons passing through a solid moderator in a nuclear reactor. If, however, we consider a gas in which both the target and the bullet molecules are moving randomly, the mean free path will decrease because now there are not only head-on collisions as before, but also "sideswipes" with targets moving across the line of travel of the bullet. It is found that the mean free path of a molecule of an ideal gas having a Maxwellian distribution of speeds is

$$\boxed{L = \frac{0.707}{n\sigma}.} \tag{1–29}$$

1-11. FARADAY'S LAW OF ELECTROLYSIS—SKEPTICISM

Another line of argument supporting the atomic view of matter came from the work of Faraday. In 1833 he observed that if the same electric charge is made to traverse different electrolytes, the masses of the materials deposited on the electrodes are proportional to the chemical equivalent weights of the materials. The quantity of electricity required to deposit a mole of univalent ions in electrolysis is called the Faraday constant, F, and is equal to 9.64870×10^4 coulombs. Like the law of multiple proportions proposed by Dalton, this also implied atomicity of matter. Faraday's law, however, brings electricity into the picture and implies that both electricity and matter are atomic.

We have traced a few highlights of the development of the atomic view of matter through most of the nineteenth century, but since no one had ever seen a molecule, the entire theory was still regarded with skepticism. Maxwell, who proposed the distribution of speeds already discussed, did his greatest work in electrical theory. It was he who found the relationship between electricity and light, and it is because of his work that we often call light "electromagnetic radiation." In his comprehensive book on electricity and magnetism (1873), after explaining Faraday's laws of electrolysis on the basis of the atomic theory of matter and electricity, Maxwell says, "It is extremely improbable that when we come to understand the true nature of electrolysis we shall retain in any form the theory of molecular charges, for then we shall have obtained a secure basis on which to form a true theory of electric currents and so become independent of these provisional theories."

As late as 1908 the physical chemist Wilhelm Ostwald and the physicist Ernst Mach opposed the atomic theory of matter. Their skepticism is an interesting question in epistemology. These scientists were unwilling to accept purely indirect evidence. Mach makes their position clear in the following analogy: A long elastic rod held in a vise may be made to execute slow, perceivable vibrations. If the rod is shortened, the vibrations become a blur in which individual motions of the rod cannot be followed. If the rod is shortened further, the blur may be visually unobservable but a tone is heard. If the rod is made so short that we no longer experience a physical sensation from its behavior, we may still think of it as vibrating when struck. This, according to Mach, is a safe extrapolation of our ideas because it proceeds from the *directly* observable to the *indirectly* observable. Those who were skeptical about the atomic theory objected to the fact that the evidence was *entirely indirect.* The experiments described in the next section provided the observable events which made the indirect evidence we have given acceptable to everyone.

1-12. PERRIN'S VERIFICATION OF THE ATOMIC VIEW OF MATTER

Credit for removing the remaining skepticism of atomic theory goes to the French physical chemist Perrin. He tested the hypothesis that the suspended particles which dance about in a stationary liquid in Brownian movement behave like large gas molecules. For his experiments, Perrin prepared a water suspension of particles which met a stringent set of requirements. They had to be large enough to be seen individually, but small enough to have an appreciable thermal motion which could be measured; and they had to be of known uniform size and mass. Further, the concentration of the particles in the suspension had to be so low that the force effects between them could be neglected. In short, the particles had to be directly observable and conform to the assumptions of the kinetic theory of gases. Perrin was able to obtain a suspension of particles that met these requirements by centrifuging a water mixture of powdered gamboge, a gum resin more dense than water. The centrifuge separated the particles according to size. After drawing off a portion of the mixture where the magnitude of the particle size was suitable for his experiments, he could centrifuge again and again until the size of the remaining particles was nearly uniform. Although gamboge is more dense than water, Perrin observed that the particles did not settle out of still water. They assumed a distribution in height, with more particles per unit volume near the bottom of the container than at the top. He measured this distribution as a function of height.

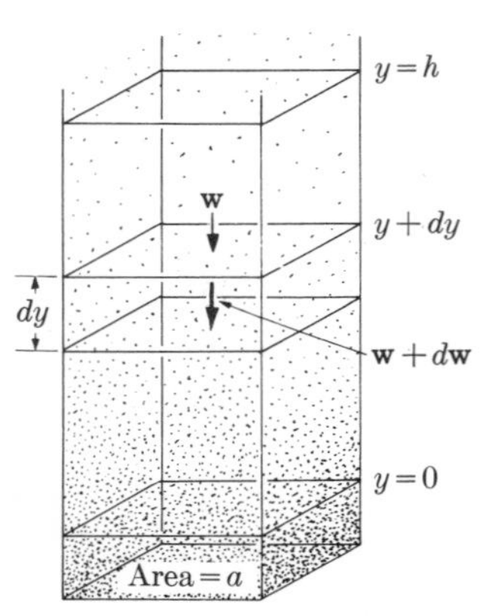

FIG. 1–8

To derive the distribution equation, consider a vertical column of gas (Fig. 1–8) which has a cross-sectional area a and which is at a uniform temperature. Through this column, let us take a horizontal slice of thickness dy. If the weight of the gas above this slice is w, then the weight of the gas above the bottom of the slice will be $w + dw = w + mgna\,dy$, where mg is the weight of a molecule, n is the average number of molecules per unit volume, and $a\,dy$ is the volume of the slice. (Note that we are employing an atomic view of the gas.) The difference of these weights per unit area, $mgn\,dy$ is the pressure difference due to the gas in the slice, that is, $dp = -mgn\,dy$. The minus sign denotes that the pressure decreases as the height increases. Since the molecules in the column of gas have weight, the number of molecules per unit volume at a low level is greater than at a higher one. Because of this difference in concentration, there is a corresponding difference in the number of molecular collisions per unit time at the two levels, and the column of gas comes to dynamic equilibrium. In this state, the weight of the molecules in any layer is just balanced by the net upward force caused by the difference between the number of molecular impacts per unit time on the lower and upper horizontal surfaces of the layer. We shall next obtain an expression for the difference in pressure between two levels produced by a difference in molecular concentrations.

The equation $pV = RT$ holds for one mole of any gas that can be regarded as ideal. According to atomic theory, the number of molecules in a mole is the Avogadro constant N_A. We can obtain an expression that contains the concentration of the molecules by dividing the ideal gas law equation by the Avogadro constant. Thus we have

$$\frac{pV}{N_A} = \frac{RT}{N_A} \quad \text{or} \quad p = \left(\frac{N_A}{V}\right)\left(\frac{R}{N_A}\right)T.$$

The quantity N_A/V is the molecular concentration n. (Note that here n is not the number of moles as earlier in this chapter.) This concentration is a function of the pressure and therefore, since R, N_A, and T are constant, $dp = (R/N_A)T\,dn$. When this expression for the difference of pressure due to the difference in molecular concentrations in the layer of height dy is equated to the difference due to the weight of the layer, we obtain

$$\frac{R}{N_A}T\,dn = -mgn\,dy,$$

or

$$\frac{dn}{n} = -\frac{N_A mg}{RT}\,dy. \tag{1–30}$$

The relation between the molecular concentration n_0 at $y = 0$ and n at $y = h$ can be found by integrating Eq. (1–30). We have

$$\int_{n_0}^{n} \frac{dn}{n} = \int_0^h -\frac{N_A mg}{RT}\,dy,$$

which gives

$$n = n_0 e^{-(N_A mgh/RT)}. \tag{1–31}$$

Equation (1–31) may also be written in the form

$$n = n_0 \exp\left[-\frac{N_A mgh}{RT}\right]. \tag{1–32}$$

Since the pressure of a gas is directly proportional to the number of molecules per unit volume, Eq. (1–32) can be rewritten in terms of pressures. The resulting equation is then called the *law of atmospheres*, since it gives the distribution in height of the pressure in a column of gas at constant temperature and subject to the force of gravity.

Equation (1–32) must be modified slightly to make it applicable to the study of Brownian movement. The effective weight of a particle suspended in a fluid is the resultant of its weight and Archimedes' buoyant force. The volume of a particle of mass m and density ρ is m/ρ and the mass of an equal volume of liquid having a density ρ' is $\rho' m/\rho$. Therefore, the buoyant force on this particle when submerged in the liquid is $mg\rho'/\rho$, and its effective weight becomes

$$mg - mg\frac{\rho'}{\rho} = mg\left(\frac{\rho - \rho'}{\rho}\right).$$

When we replace the actual weight mg in Eq. (1–32) by the effective weight, we obtain

$$n = n_0 \exp\left[-\frac{N_A mg(\rho - \rho')h}{\rho RT}\right]. \tag{1–33}$$

This is the equation for *sedimentation* equilibrium of a suspension as a result of Brownian movement.

Perrin measured a series of n's at a series of h's in a very dilute suspension. The results verified the sedimentation equation. Equally important, his measurements yielded a value for N_A, since all other quantities in the equation were known. The very existence of the Avogadro constant implies the correctness of the atomic theory.

It is interesting that Einstein, the greatest contributor to the development of modern physics, also had a part in the final establishment of the atomic theory. In 1905 he derived an equation which describes how a suspended particle should migrate in a random manner through a liquid. His expression involved the Avogadro constant, and Perrin's verification of the Einstein formula confirmed the value of this constant. The results showed that small particles suspended in a stationary liquid do move about in the manner predicted by the molecular-kinetic theory of gases.

Altogether, Perrin used four completely independent types of measurements, each of which was an observable verification of atomic theory and each of which gave a quantitative estimate of the Avogadro constant. Since the publication of his results in 1908, no one has seriously doubted the atomic theory of matter.

1–13. BOLTZMANN CONSTANT

Once N_A was determined, it was possible to evaluate the frequently recurring quantity R/N_A. This is the universal gas constant *per molecule*, and is known as the *Boltzmann constant, k*. Its value is 1.380×10^{-23} J/°K. Earlier in this chapter we computed the translational kinetic energy of the molecules of a gas from $\frac{1}{2}M\overline{v^2} = \frac{3}{2}RT$ (Eq. 1–16), where M and R are the mass and gas constant per mole, respectively. Replacing M by $N_A m$ and R by $N_A k$ in this equation, we obtain

$$\boxed{\tfrac{1}{2}m\overline{v^2} = \tfrac{3}{2}kT,} \tag{1–34}$$

where m and k are the mass and gas constant *per molecule*, respectively. Thus the average translational kinetic energy of any molecule of a gas depends only on its temperature (if indeed the word temperature has meaning for a single molecule), and the *energy per degree of freedom* is $kT/2$.

Let us now return to Eq. (1–31), the law of atmospheres. In this equation N_A/R equals $1/k$ and mgh is the gravitational potential energy E_p of a particle. Substituting in Eq. (1–31), we get

$$n = n_0 e^{-E_p/kT}. \tag{1–35}$$

This expression yields the relative number of particles per unit volume in two different energy states in terms of their difference of gravitational potential energy, E_p. It turns out that this same equation may be used for any case where the potential energy is a function of position. For two energy states of electrified particles, E_p would be the difference in their electrostatic potential energies. When all types or forms of energy are considered, we obtain the *Boltzmann distribution law*

$$\boxed{n = n_0 e^{-E/kT},} \tag{1–36}$$

where E is the difference in energy between two states. We will need this equation when we discuss the distribution of energy in a group of harmonic oscillators, in Chapter 3.

REFERENCES

American Institute of Physics, D. E. Gray, editor, *American Institute of Physics Handbook.* 2nd ed. New York: McGraw-Hill, 1963. A very comprehensive book.

Bridgman, P. W., *The Logic of Modern Physics.* New York: Macmillan, 1938.

Brown, T. B., editor, *The Lloyd William Taylor Manual of Advanced Undergraduate Experiments in Physics.* Reading, Mass.: Addison-Wesley, 1959. Discusses theory and apparatus required for numerous experiments.

Cajori, Florian, *A History of Physics.* New York: Macmillan, 1929.

Conant, James B., *On Understanding Science; An Historical Approach.* New Haven, Conn.: Yale University, 1947.

Dampier, William C., *A History of Science.* 3rd ed. New York: Macmillan, 1943. This is a history of the development of all of the natural sciences and their relations with philosophy and religion.

Einstein, Albert, *Investigation on the Theory of the Brownian Movement.* R. Fürth, translator. New York: Dover Publications, 1948.

Frank, Philipp, *Philosophy of Science.* Englewood Cliffs, N. J.: Prentice-Hall, 1957.

Hodgeman, C. D., editor, *Handbook of Chemistry and Physics.* 46th ed. Cleveland, Ohio: Chemical Rubber Publishing Co., 1965. A good reference book for chemical and physical data.

Holton, Gerald, *Introduction to Concepts and Theories in Physical Science.* Reading, Mass.: Addison-Wesley, 1952. This is an excellent historical account of how the scientific method has been employed in investigating the physical world. This book should be read by every student in the sciences.

JEANS, JAMES, *Physics and Philosophy*. New York: Maxmillan, 1943.

LAUE, MAX VON, *History of Physics*. R. E. Oesper, translator. New York: Academic Press, 1950. A short but comprehensive book.

LEICESTER, H. M. and H. S. KLICKSTEIN, *A Source Book in Chemistry*. New York: McGraw-Hill, 1952. A collection of the principal papers of the great men in chemistry. Every student of atomic theory should read the original works of Dalton, Avogadro, Gay-Lussac, and Cannizzaro.

PERRIN, JEAN, *Atoms*. D. L. Hammick, translator. London: Constable, 1923.

PLANCK, MAX, *The Philosophy of Physics*. New York: Norton, 1936.

SAMBURSKY, S., *The Physical World of the Greeks*. Morton Dagut, translator. New York: Macmillan, 1956.

SARTON, GEORGE, *A History of Science: Ancient Science through the Golden Age of Greece*. Cambridge, Mass.: Harvard University Press, 1952.

SEARS, F. W., *Thermodynamics*. 2nd ed. Reading, Mass.: Addison-Wesley, 1955. General treatment of the kinetic theory of gases in Chapters 11–15.

TAYLOR, F. SHERWOOD, *A Short History of Science and Scientific Thought*. New York: Norton, 1949.

ZEMANSKY, M. W., *Temperatures Very Low and Very High*, Momentum Book No. 6. Princeton, N. J.: Van Nostrand, 1954.

PROBLEMS

1–1. Describe one or more experiments which show that the earth is essentially spherical.

1–2. An early chemist wishes to determine the atomic weight of nitrogen. He assumes that the atomic weight of oxygen is exactly 16 and he prepares four oxides of nitrogen which are distinctly different compounds (see data below).

Nitrogen, parts by wt.	Oxygen, parts by wt.	Product oxide
86.3	197	*A*
500	285	*B*
300	343	*C*
108.2	186	*D*

(a) Show that these data demonstrate the law of multiple proportions. [*Hint:* First find the masses of one of these elements which unite with a unit mass of the other.] (b) From the above data the chemical formulas of the products cannot be determined completely but, by assuming that nature is simple, one may propose several possible sets of product formulas. Write out several possible sets of product formulas. (c) Calculate the atomic weight of nitrogen for each set.

1–3. The specific heats of the elements are given in Appendix 4. Choose eight of these which are solids at room temperature and which are distributed over the range of atomic weights. Calculate the molar heat capacity of each in cal/mole · °K. Plot these results against atomic weight as the abscissa.

1–4. (a) What would happen if the collisions between the molecules of a gas were not perfectly elastic? (b) What would happen to the shape of an air-filled spherical toy balloon if, on the average, the velocities of the gas molecules in one direction were to become greater than those in another?

1–5. A dart board with area A hangs on a vertical wall. Darts are thrown horizontally with a speed v and stick in the dart board. If the mass of each dart is m and r darts strike the board in each second, derive an equation for the average pressure p exerted on the board.

1–6. Compute the molar heat capacity at constant volume of a gas composed of molecules which are rigid three-dimensional structures.

1–7. Show that if ρ represents the density of a gas, then

$$v_{\text{rms}} = \sqrt{3p/\rho}.$$

1–8. Repeat the derivation of the ideal gas law assuming a mixture of particles having two different masses, m_1 and m_2. From this derivation generalize to the law of partial pressures which states that the pressure of a mixture of gases is the sum of the pressures due to each gas separately.

1–9. (a) Compute the arithmetic mean speed and the root-mean-square speed for each of the following distributions of the speeds of eight particles:

1) all eight have speeds of 10 m/s;
2) two have speeds of 3 m/s, four have speeds of 6 m/s, and two have speeds of 10 m/s;
3) one has a speed of 3 m/s, three have speeds of 6 m/s, and four have speeds of 10 m/s;
4) four are at rest and four have speeds of 10 m/s.

(b) In each case decide whether the shape of the graph of the speed distribution would be the same as that of the translational kinetic energy distribution assuming that each particle has the same mass.

1–10. The speed distribution function of a group of N particles is given by $dN_v = k\,dv$, where dN_v is the number of particles which have speeds between v and $v + dv$ and k is a constant. No particle has a speed greater than the value V and the range of speeds is from 0 to V. (a) Draw a graph of the distribution function, that is, plot (dN_v/dv) versus v. (b) Find the constant k in terms of N and V. (c) Compute the average and the root-mean-square speeds in terms of V. (d) What percent of the particles have speeds between the average speed and V? and between the root-mean-square speed and V?

1–11. The speed distribution function of a group of N particles is given by $dN_v = kv\,dv$ where dN_v is the number of particles which have speeds between v and $v + dv$ and k is a constant. No particle has a speed greater than V and the speeds range from 0 to V. (a) Draw a graph of the distribution function, that is, plot (dN_v/dv) versus v. (b) Find the constant k in terms of N and V. (c) Compute the average speed, the root-mean-square speed, and the most probable speed in terms of V. (d) What percent of the particles have speeds between the average speed and V? and between the root-mean-square speed and V?

1–12. (a) Show from Eq. (1–17) that the most probable speed in a Maxwellian distribution is given by $v_m = \sqrt{2RT/M}$, and (b) then show from Eq. (1–16) that $v_m = \sqrt{\frac{2}{3}}\, v_{\text{rms}}$. [*Hint:* Determine the condition for which the ordinate quantity in Fig. 1–2 is maximum.]

1–13. For which of the following gases do the molecules have (a) the highest most probable speed (b) the lowest most probable speed at a given temperature: CO, H_2, O_2, A, NO_2, Cl_2, He?

1–14. An object can escape from the surface of the earth if its speed is greater than $\sqrt{2gR}$, where g is the acceleration due to gravity and R is the radius of the earth. (a) Using a radius of 6.4×10^6 m, calculate this escape speed. (b) Explain why oxygen and nitrogen remain in the earth's atmosphere while hydrogen does not.

1–15. Show that the ratio of the rms speeds of the molecules in two different gases at the same temperature is inversely proportional to the square root of their masses.

1–16. The speed of propagation of a sound wave in air at 27°C is about 348 m/s. Find the ratio of this speed to the rms speed of nitrogen molecules at this temperature.

1–17. (a) At what temperature will the rms speed of oxygen molecules be twice their rms speed at 27°C? (b) At what temperature will the rms speed of nitrogen molecules equal the rms speed of oxygen molecules at 27°C?

1–18. Find the rms speed, the average speed, and the most probable speed of the molecules of gaseous hydrogen at a temperature of (a) 20°C and (b) 120°C.

1–19. Find the rms speed, the average speed, and the most probable speed of the molecules of gaseous hydrogen at a constant temperature of 20°C and a pressure of (a) 1 atm and (b) 100 atm.

1–20. (a) To what temperature would an ideal gas in which the "particles" are baseballs have to be heated so their rms speed in a Maxwellian distribution would equal that of a fast ball having a speed of 30.5 m/s? (The mass of a baseball is 144g.) (b) What is the microscopic cross section of these "particles"? (The circumference of a baseball is 23 cm.)

1–21. The drum of a Zartman-Ko apparatus, Fig. 1–3, has a radius of 8 cm and rotates at 6000 rpm. The oven contains mercury atoms at a temperature of 600°K. Two atoms of mercury, one with the most probable speed at oven temperature and the other with the rms speed at the same temperature, leave the oven and enter the rotating drum. These two atoms are then deposited on the glass plate at the far side of the drum. What is the separation of these two atoms on the glass plate?

1–22. A certain lecture hall measures 10m × 20m × 30m and the temperature in the hall is 0°C. (a) Calculate the total kinetic energy of the air assuming that the pressure is 1 atmosphere. (Note that the diatomic molecules, O_2 and N_2, have five degrees of freedom at this temperature.) (b) Calculate the force on the wall of the room which is 10m × 30m. (c) How many degrees of freedom would be involved in calculating the pressure used in part (b) from the kinetic theory of gases?

1–23. When the atoms in a deuterium gas have an average translational kinetic energy of 12×10^{-14} J, they can approach one another so closely that nuclear fusion will occur. (a) What is the speed of a deuterium atom having this kinetic energy? (b) To what temperature would the deuterium gas have to be heated so that the rms speed of the atoms would equal the speed in the preceding part? (Deuterium is hydrogen having an atomic weight of 2.014.)

1–24. (a) What is the total kinetic energy of translation of the atoms in 4 moles of helium at a temperature of 27°C? (b) What would be the answer for the same amount of any other ideal gas?

1–25. What is the total kinetic energy of the molecules in 2 moles of argon gas at a temperature of (a) 300°K? (b) 301°K? (c) Calculate the internal energy change of 2 moles of argon when the temperature is increased 1K°, using the specific heat at constant volume, and compare your answers with the change in kinetic energy.

1–26. The microscopic cross section for a certain bullet and particle is σ when they are electrically neutral. Would the value of σ increase or decrease if the bullet and particle carried electric charges (a) of like sign, (b) of unlike sign?

1–27. (a) Show that n, the number of molecules per unit volume of an ideal gas, is given by $n = pN_A/RT$ where N_A is the Avogadro constant. (b) Find the number of molecules in 1 m^3 of an ideal gas under standard conditions. (c) What is the number of molecules in 1 m^3 of an ideal gas at a pressure of two atmospheres and a temperature of 47°C?

1–28. The diameters of molecules can be computed from an equation derived by using the kinetic theory to explain the viscosity of a gas. The results show that the molecular diameters are about 2×10^{-10} m for all gases. Find (a) the microscopic cross section and (b) the macroscopic cross section of the molecules of hydrogen at a temperature of 20°C and a pressure of 1 atm.

1–29. Derive Eq. (1–28) from Eq. (1–27).

1–30. (a) If the pressure is kept constant, at what temperature will the mean free path of the molecules of a given mass of an ideal gas be twice that at 27°C? (b) If the temperature is kept constant, at what pressure in millimeters of mercury will the mean free path of the molecules of a given mass of an ideal gas be 1000 times greater than that at a pressure of 1 atm?

1–31. Find the mean free path for a Maxwellian distribution of speeds (Eq. 1–29) of the molecules of hydrogen gas when at a pressure of 1 atm and a temperature of (a) 20°C and (b) 120°C; and (c) when at a pressure of 100 atm and a temperature of 20°C. (Data for calculating the macroscopic cross section are given in Problem 1–28.) (d) How many collisions per second would a molecule that is always moving with the average speed in a Maxwellian distribution make in each of the preceding cases? (The time of contact during collisions is negligible.) (e) What is the ratio of the mean free path in part (a) to the wavelength of green light, $\lambda = 5500 \times 10^{-10}$ m?

1–32. A beam of bullet particles is incident normally on a layer of material containing stationary target particles. Find (a) the fraction of the incident beam transmitted and the fraction which experienced collisions in a layer whose thickness equals the mean free path, and (b) the thickness of the layer in terms of the mean free path required to reduce the transmitted beam to one-half the intensity of the incident beam.

1–33. Show that the kinetic energy of translation of a molecule having the most probable speed in a Maxwellian distribution is equal to kT.

1–34. A neutron is a fundamental particle. Like ordinary gas molecules, neutrons have a distribution of speeds, and this distribution is of prime importance in the theory of nuclear reactors. Quantitatively, a thermal neutron is usually defined as one having the most probable speed of a Maxwellian distribution at 20°C. Find (a) the kinetic energy of translation and (b) the speed of a thermal neutron. (c) A thermal neutron is sometimes called a "kT neutron." Why?

1–35. Assume that hydrogen atoms in the atmosphere of the sun obey the Maxwellian speed distribution. Calculate (a) the kinetic energy of one of these atoms moving with the most probable speed in the distribution given that the temperature is 6000°K, and (b) the speed of this atom.

1–36. Consider nitrogen molecules at a temperature of 0°C. (a) What is the average translational kinetic energy of a molecule? (b) What is the average rotational kinetic energy of a molecule? (c) What is the average total kinetic energy of a molecule?

1–37. (a) Show that when the Maxwellian speed distribution is expressed in terms of the mass m of a molecule the result is

$$\frac{\Delta N}{N} = \frac{4}{\sqrt{\pi}}\left(\frac{\mathrm{m}}{2kT}\right)^{3/2} v^2 e^{-mv^2/2kT}\,\Delta v.$$

(b) Assuming that the energy E of a molecule is only translational kinetic energy, show that the fractional number of molecules that have energies in the range ΔE is

$$\frac{\Delta N}{N} = \frac{2}{\sqrt{\pi}}\left(\frac{1}{kT}\right)^{3/2} E^{1/2} e^{-E/kT}\,\Delta E.$$

(c) From the energy distribution of part (b) show that the most probable energy is $\frac{1}{2}kT$. (d) What is the ratio of the average translational kinetic energy to the most probable translational kinetic energy?

1–38. In one of his experiments, using a water suspension of gamboge at 20°C, Perrin observed an average of 49 particles per unit area in a very shallow layer at one level and 14 particles per unit area at a level 60 microns higher. The density of the gamboge was 1.194 g/cm^3 and each particle was a spherical grain having a radius of 0.212 micron. (1 micron $= 10^{-6}$ m.) Find (a) the mass of each particle, (b) the Avogadro constant, and (c) the molecular weight of a particle if each grain is regarded as a single giant molecule. Use the results from parts (a) and (b) to calculate (c).

CHAPTER 2

THE ATOMIC VIEW OF ELECTRICITY

2-1. ELECTRICAL DISCHARGES

We have already considered how Faraday's law of electrolysis implies that both matter and electricity are atomic. In spite of Maxwell's skepticism, it is very difficult to explain the fact that the passage of one faraday of electricity through an electrolyte liberates or deposits an equivalent weight of a substance, except by assuming that both matter and electricity exist in units which preserve their identity throughout the process.

In order to learn more about "particles of electricity," we turn to another line of investigation and consider the passage of electricity through gases. Although Benjamin Franklin's very dangerous experiment with kite and key was not a particularly convincing one, it nevertheless led to the correct conclusion that lightning is the discharge of electricity through a gas (air). Every electric spark is an example of this process. Since sparks are one of the most dramatic electric effects, it is natural that they should have been a subject of early study.

The passage of electricity through gases is a very complicated process and a great deal has been learned from it. There are many ways in which the character of an electrical discharge can be altered, but here we shall direct our attention to the effect of gas pressure. A typical discharge tube is shown in Fig. 2–1. This system has a gauge which measures the gas pressure and a pumping system which varies the pressure. Electrodes are sealed into the ends of the tube so that an electric field can be established between them.

When the pressure in the tube is atmospheric, a very large electric field is required to produce a discharge (about 3×10^6 volts/m for air). The discharge is a violent spark as the gas suddenly changes from being an excellent insulator to being a good conductor. As the pressure is reduced, the discharges are more easily established (Fig. 2–2), until, at very low pressures, they again become difficult to start. Discharges start most easily at a pressure of about 2 mm of mercury (although this will depend

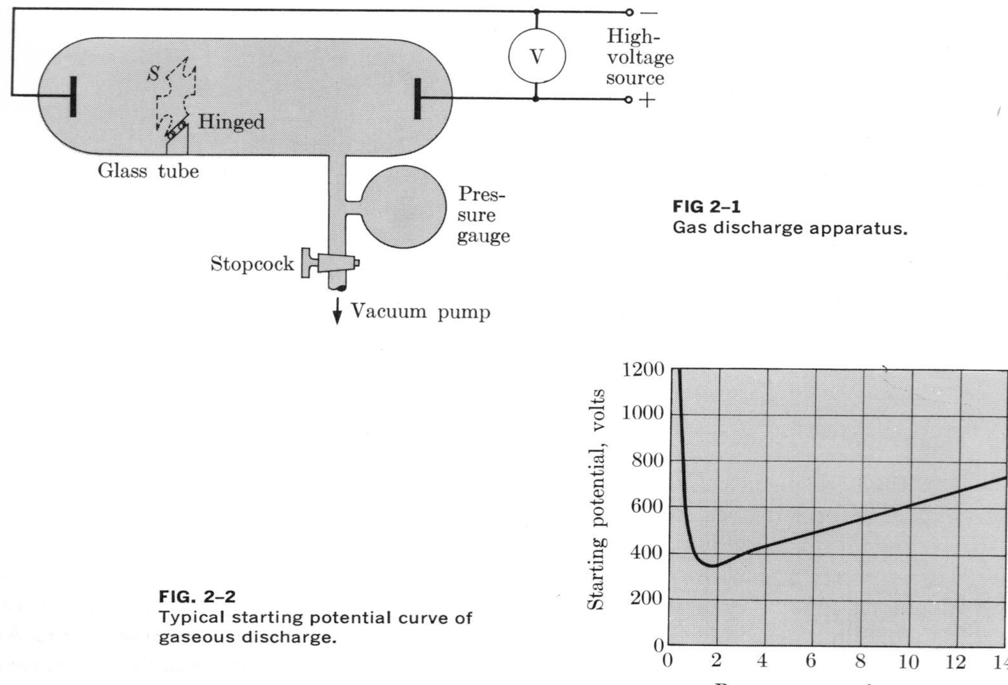

FIG 2–1
Gas discharge apparatus.

FIG. 2–2
Typical starting potential curve of gaseous discharge.

upon the kind of gas and the geometry of the electrodes). As the pressure is reduced, the discharge changes in character. With air in the tube, the bright spark changes to a purple glow filling the whole tube, and with neon, one obtains the red glow seen in many advertising signs. On further lowering of the pressure, the glow assumes a remarkable and complicated structure, with striations and dark spaces. At very low pressures the glow of the gas becomes dim and a new effect appears—the glass itself begins to glow. If the bulb has within it a device which is hinged so that it can be made to move into or out of the region between the electrodes by tipping the entire bulb (S in Fig. 2–1), then another effect may be seen. The greenish glow of the glass, which appears everywhere between the electrodes when the object S is out of the way, is partly obliterated when S is swung between the electrodes. If the object S has some distinctive shape, it may be seen clearly that it is casting a shadow. The shadow is on the side of S that is away from the negative electrode or cathode. If the cathode is small, the shadow is rather sharp. It is a simple deduction that the greenish glow is caused by some kind of rays from the cathode that cannot penetrate the obstruction S. These rays are called *cathode rays.* Many years ago it was observed that these rays could be deflected by both electric and magnetic fields, and the direction of these deflections showed that the rays were negatively charged.

Sir J. J. Thomson undertook a quantitative study of cathode rays in 1897. He was able to show that all cathode rays or corpuscles possess a common property. He showed that the ratio of their charge to their mass, q/m, was a constant. His measurements did not establish that all the rays have identical charges or identical masses, although this is the simplest interpretation of his results. He did, however, discover a unique characteristic of these rays and he is regarded as the discoverer of a fundamental particle of electricity, the electron.

2-2. CHARGED-PARTICLE BALLISTICS

Before discussing one of the methods by which q/m can be measured, let us review some basic facts of electricity and magnetism. When a particle having charge $+q$ is in a uniform electric field of intensity E, the particle experiences a force in the direction of the field, of magnitude

$$F = qE \quad [\text{in vector notation } \mathbf{F} = q\mathbf{E}]. \tag{2-1}$$*

If all other forces on the particle are negligible compared with this one, the particle will undergo uniformly accelerated motion, and we have

$$ma = qE. \tag{2-2}$$

Note that if the particle has a component of its velocity perpendicular to E, that component remains *unchanged*, whereas the component of velocity in the direction of E *changes* with an acceleration a.

If a particle having charge q moves in a uniform magnetic field where the magnetic induction is B with a velocity v which is perpendicular to B, the particle experiences a force F which is *perpendicular* to both v and B and has a magnitude given by

$$F = qvB \quad [\text{in vector notation } \mathbf{F} = q(\mathbf{v} \times \mathbf{B})]. \tag{2-3}$$

If all other forces on the particle are negligible compared with this one, the particle will change its direction of motion but not its speed. After an

* This equation and those following are valid in any consistent system of units. No conversion factors for units need be introduced provided *all* are electrostatic units, *all* are electromagnetic units, or *all* are meter-kilogram-second-ampere units. This will be true of all equations in this book except in cases where units peculiar to atomic physics, such as angstroms or electron volts, are specified.

infinitesimal change of direction, the particle has the same speed and is still moving perpendicular to B. Thus the particle experiences a force of constant magnitude and changing direction that causes it to move in a circular path in a plane perpendicular to the magnetic induction or flux density. Since the force required to maintain a mass m, with tangential velocity v, in a circular path of radius r is

$$F = mv^2/r,$$

we obtain

$$mv^2/r = qvB, \tag{2–4}$$

or

$$mv = qBr. \tag{2–5}$$

2–3. THOMSON'S MEASUREMENT OF q/m

We are now ready to consider how Thomson measured the ratio of charge to mass, q/m, for what he called "cathode corpuscles." His apparatus (Fig. 2–3) consisted of a highly evacuated glass tube into which several metal electrodes were sealed. Electrode C is the cathode from which the rays emerged. Electrode A is the anode, which was maintained at a high positive potential so that a discharge of cathode rays passed to it. Most of the rays hit A, but there was a small hole in A through which some of the rays passed. These rays were further restricted by an electrode A' in which there was another hole. Thus a narrow beam of the rays passed into the region between the two plates P and P'. After passing between the plates, the rays struck the end of the tube, where they caused fluorescent material at S to glow. (We will discuss how this happens when we consider fluorescence in Chapter 4.)

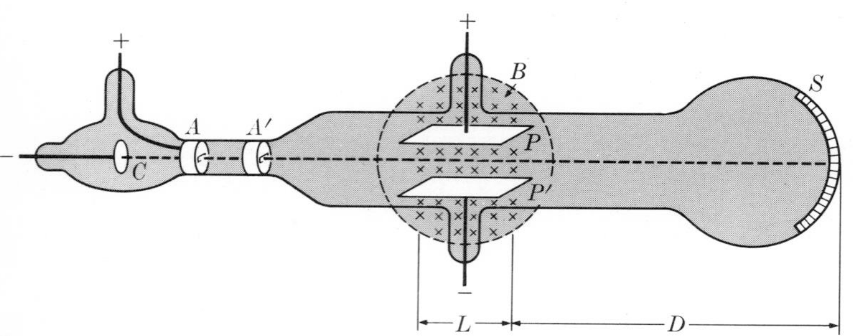

FIG. 2–3 Thomson's apparatus for measuring the ratio q/m for cathode rays.

The deflection plates P and P' were separated a known amount, so that when they were at a known difference of potential the electric field between them could be computed. We shall assume that the field was uniform for a distance L between the plates and zero outside them. When the upper plate P was made positive, the electric field deflected the negative cathode rays upward.

The trajectory of the cathode corpuscle is obtained in the same way that trajectories were found for projectiles in the gravitational field of force in mechanics. The only difference is that here the constant electric force is upward and is limited to the region between the plates. The gravitational force mg and the electrostatic force between two corpuscles are so small that they may be neglected.

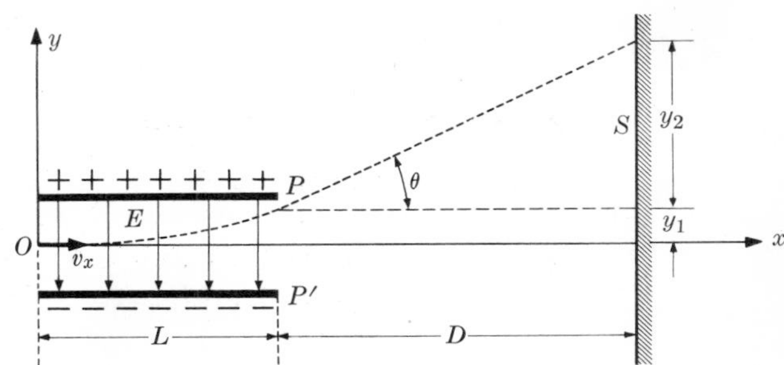

FIG. 2–4
Electrostatic deflection of cathode rays.

If, in Fig. 2–4, the cathode rays enter the region between the plates at the origin O with a velocity v_x, this velocity will continue to be the horizontal component of the velocity of the rays. The general equation for displacement in uniformly accelerated motion is

$$s = s_0 + v_0 t + \tfrac{1}{2}at^2. \tag{2–6}$$

Applying Eq. (2–6) to the horizontal direction, we obtain

$$x = v_x t. \tag{2–7}$$

Between the plates, however, the rays experience an upward acceleration

$$a_y = \frac{qE}{m}, \tag{2–8}$$

obtained from Eq. (2–2). In this case E is constant, since fringing of the electric field is neglected, and it is equal to the potential difference between the deflection plates divided by their separation. Hence the general displacement equation becomes

$$y = \frac{qEt^2}{2m}. \tag{2–9}$$

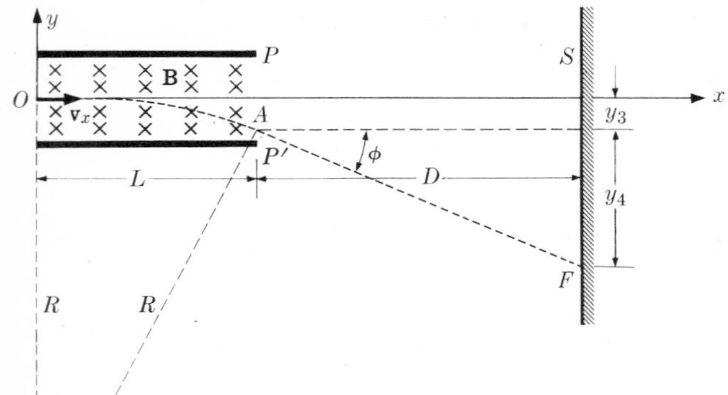

FIG. 2–5
Magnetic deflection of cathode particles.

Elimination of t between Eqs. (2–7) and (2–9), yields the equation for the parabolic trajectory,

$$y = \frac{qEx^2}{2mv_x^2}. \tag{2–10}$$

The quantity y_1, defined in Fig. 2–4, is the value of y when $x = L$.

Beyond the plates, the trajectory is a straight line because the charge is then moving in a field-free space. The value of y_2 is $D \tan\theta$, where D and θ are defined as in Fig. 2–4. The slope of this straight line is

$$\tan\theta = \left(\frac{dy}{dx}\right)_{x=L} = \left(\frac{qEx}{mv_x^2}\right)_{x=L} = \frac{qEL}{mv_x^2}. \tag{2–11}$$

The total deflection of the beam, y_E, is $y_1 + y_2$, so that

$$y_E = y_1 + y_2 = \frac{qEL^2}{2mv_x^2} + \frac{qELD}{mv_x^2} = \frac{qEL}{mv_x^2}\left(\frac{L}{2} + D\right). \tag{2–12}$$

If q/m is regarded as a single unknown, then there are two unknowns in this equation. The initial velocity of the rays, v_x, must be determined before q/m can be found. We need another equation involving the initial velocity v_x, so that this unknown velocity can be eliminated between the new equation and Eq. (2–12).

Thomson obtained another equation by applying a magnetic field perpendicular to both the cathode-corpuscle beam and the electric field. It is represented in Fig. 2–3 as being into the page and uniform everywhere within the x-marked area. Thus the electric and magnetic forces acted on the cathode rays in the same geometric space.

Figure 2–5 shows the situation when the magnetic field alone is present. The negatively charged rays experience a force that is initially downward. This force is not constant in direction, but is always normal to both the field and the direction of motion of the rays. Therefore the cathode corpuscles move in a circular path according to Eq. (2–5). The center of curvature of the trajectory is at C, and the radius of curvature of the path is

$$R = mv_x/qB, \tag{2–13}$$

where v_x is the initial velocity of the rays in the x-direction. Referred to the origin O, the equation of this circular path is

$$x^2 + (R + y)^2 = R^2. \tag{2–14}$$

Solving for R, we get

$$R = -\frac{x^2 + y^2}{2y} \approx -\frac{x^2}{2y}. \tag{2–15}$$

The approximation is good if the deflection is small compared with the distance the rays have moved into the magnetic field, that is, when $y^2 \ll x^2$.

Since the radius of curvature is difficult to measure, we eliminate R between Eqs. (2–13) and (2–15), and obtain

$$y = -\frac{qBx^2}{2mv_x}. \tag{2–16}$$

Therefore, for small deflections, the circular path may be approximated by the parabolic path of Eq. (2–16). The minus sign indicates that the curve is concave downward.

Just as in the electric case, we find that y_3 is the value of y for $x = L$. The rays again move in a straight line through the field-free region, so that

$$y_4 = D \tan \phi = D\left(\frac{dy}{dx}\right)_{x=L} = -\frac{D_qBL}{mv_x}. \tag{2–17}$$

For the total magnetic deflection y_B, we have

$$y_B = y_3 + y_4,$$

or

$$y_B = -\frac{q}{m}\left(\frac{BL^2}{2v_x} + \frac{BLD}{v_x}\right) = -\frac{qBL}{mv_x}\left(\frac{L}{2} + D\right). \tag{2–18}$$

Equation (2–18) is very similar to Eq. (2–12). It contains q/m and v_x together with measurable quantities, so that v_x can be eliminated and q/m found. It is interesting, however, to follow Thomson's procedure for determining v_x by considering the simultaneous application of the electric and the magnetic fields. If these are adjusted so that there is no deflection on the screen, then the force of the electric field on the charged particle is balanced by that of the magnetic field. For this condition of balance, we find from Eqs. (2–1) and (2–3) that

$$F = qE - qv_xB = 0, \tag{2–19}$$

or, in terms of v_x,

$$\boxed{v_x = \frac{E}{B}.} \quad ✓ \tag{2–20}$$

This result can also be derived from Eqs. (2–12) and (2–18). When y_E and y_B are equal and opposite, the resultant deflection of the cathode-ray beam is zero, and we then have

$$y_E + y_B = \frac{q}{m}\frac{EL}{v_x^2}\left(\frac{L}{2} + D\right) - \frac{q}{m}\frac{BL}{v_x}\left(\frac{L}{2} + D\right) = 0. \tag{2–21}$$

This expression is easily reduced to $E/v_x^2 = B/v_x$ or $v_x = E/B$.

For this particular ratio of the fields, the particle goes straight through both fields. It is undeflected, and therefore the measurement of v_x does not depend on the geometry of the tube. Since $y = 0$ at all times, the approximation in Eq. (2–15) is avoided. The velocity thus determined may be substituted into Eq. (2–12), which was derived without approximation.

Thomson measured q/m for cathode rays and found a unique value for this quantity which was independent of the cathode material and the residual gas in the tube. This independence indicated that cathode corpuscles are a common constituent of all matter. The modern accepted value of q/m is $(1.758796 \pm 0.000019) \times 10^{11}$ coulombs per kilogram. Thus Thomson is credited for the discovery of the first subatomic particle, the electron. Because it was shown later that electrons have a unique charge e, the quantity he measured is now denoted by e/m_e. He also found that the velocity of the electrons in the beam was about one-tenth the velocity of light, much larger than any previously measured material particle velocity.

It was fortunate that the electrons Thomson studied had nearly equal velocities. If this had not been the case, the spot on the end of his experimental tube would have been seriously smeared. Both Eqs. (2–12) and (2–18) include the electron velocity and if all electrons had not had the same velocity, each would have undergone a different deflection. Thomson could tell that the electrons had a uniform velocity when he observed an undeflected spot upon proper adjustment of **E** and **B**.

It is interesting to explore why the electrons in Thomson's apparatus had nearly uniform velocities. The electrons he studied came from the cathode and were accelerated toward the anode by a potential difference that we can call V. Since the energy required to separate the electrons from the cathode is negligibly small, the work done by the electric field on the charges went into kinetic energy.

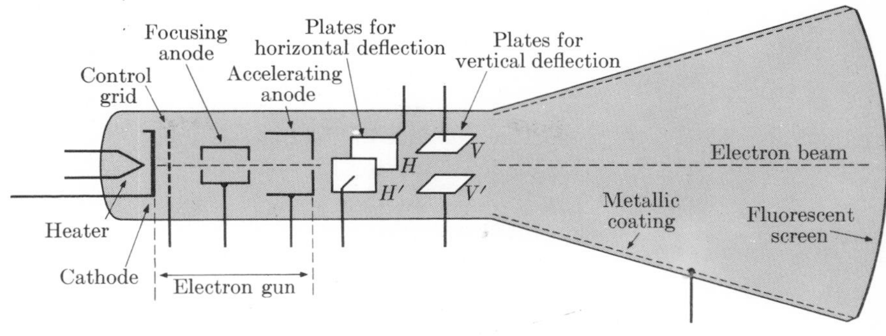

FIG. 2–6
Basic elements of a cathode-ray tube.

In general, from the law of conservation of energy, the change of kinetic energy plus the change of electrical potential energy of a charge as it goes from point 1 to point 2 must equal zero because no work is done by external forces. Therefore we have

$$\left(\tfrac{1}{2}mv_2^2 - \tfrac{1}{2}mv_1^2\right) + (qV_2 - qV_1) = 0, \tag{2–22}$$

or

$$\boxed{-q(V_2 - V_1) = \frac{m}{2}(v_2^2 - v_1^2).} \tag{2–23}$$

The quantity $V_2 - V_1$ is the potential of the second electrode relative to the first.

If we apply Eq. (2–23) to Thomson's experiment, noting that the charges are negative, the accelerating voltage $V_2 - V_1$ is V, the initial velocity v_1 is zero, and the final velocity v_2 is v_x, we obtain

$$qV = \frac{m}{2}v_x^2,$$

or

$$q/m = e/m_e = v_x^2/2V. \tag{2–24}$$

This is another equation relating e/m_e and v_x. It could have been used with Eq. (2–18) to give e/m_e. Thomson could have measured the potential difference between the cathode and anode and been spared either the electric or magnetic deflection of the beam in the vicinity of P and P'. Indeed, other methods of measuring e/m_e utilize this principle.

Cathode-ray tubes such as Thomson used have been developed into important modern electronic components. Electrostatic deflection of an electron beam is used in the cathode-ray tube of modern oscilloscopes. Such tubes usually have two sets of deflecting plates (Fig. 2–6), so that the electron beam can be deflected right and left as well as up and down. These tubes utilize the fact that the deflection is proportional to the electric field between the plates, as shown by Eq. (2–12). Television tubes, on the other hand, commonly utilize magnetic deflection to cause the beam to sweep over the face of the picture area.

Anyone can demonstrate for himself that electric and magnetic fields deflect electron beams. Holding a strong permanent magnet near the face of a television picture produces weird distortions. Rubbing the face of a picture tube or even the plastic protective window with wool, silk, or nylon will produce strong electric fields when the humidity is low. Neither the magnetic nor electric fields thus produced are uniform or perpendicular to the beam, and the deflections they produce are striking in their unpredictability.

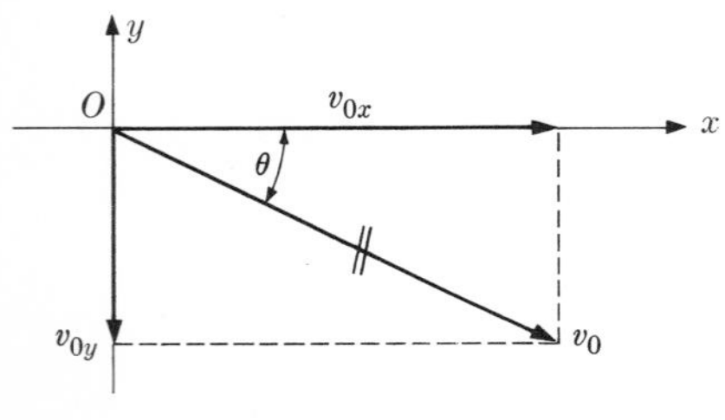

FIG. 2–7

EXAMPLE. (a) Find the speed an electron acquires if it is accelerated from rest through a potential difference of 565 volts.

This electron now moves in a vertical plane with its acquired speed and enters a region where there is a uniform electric field of 35 V/cm directed downward. Find (b) the coordinates of the electron 5×10^{-8} s after it passes through the point of entry along a course directed at an angle of 30° below the horizontal (Fig. 2–7), and (c) the direction of its velocity at this time.

Solution. (a) Recall Eq. (2–23):

$$-q(V_2 - V_1) = \frac{m}{2}(v_2^2 - v_1^2).$$

Since $v_1 = 0$, we obtain

$$v_2 = \sqrt{\frac{-2e(V_2 - V_1)}{m_e}} = \sqrt{-2 \times (-1.76 \times 10^{11}\ \text{C/kg}) \times 565\ \text{J/C}}$$

$$= \sqrt{1.99 \times 10^{14}\ \frac{\text{kg} \cdot \text{m}^2/\text{s}^2}{\text{kg}}} = 1.41 \times 10^7\ \text{m/s}.$$

(b) The field is uniform and directed downward, so there is a constant force on the electron directed upward along the positive y-axis. Therefore the acceleration of the electron is constant and its value is

$$a = \frac{F}{m} = \frac{Eq}{m} = \frac{Ee}{m_e}$$

$$= 35\ \frac{\text{J/C}}{\text{cm}} \times 10^2\ \frac{\text{cm}}{\text{m}} \times 1.76 \times 10^{11}\ \frac{\text{C}}{\text{kg}}$$

$$= 6.16 \times 10^{14}\ \frac{\text{m}}{\text{s}^2}.$$

Referred to the point of entry O as the origin of coordinates, the equations of motion for the electron are

$$x = v_0 \cos\theta\, t = 1.41 \times 10^7 \frac{\text{m}}{\text{s}} \times 0.866 \times 5 \times 10^{-8}\ \text{s}$$
$$= 0.61\ \text{m}$$

and

$$y = -v_0 \sin\theta\, t + \tfrac{1}{2}at^2$$
$$= -1.41 \times 10^7 \frac{\text{m}}{\text{s}} \times 0.500 \times 5 \times 10^{-8}\ \text{s}$$
$$+ \tfrac{1}{2} \times 6.16 \times 10^{14} \frac{\text{m}}{\text{s}^2} \times (5 \times 10^{-8}\ \text{s})^2$$
$$= -0.35\ \text{m} + 0.77\ \text{m} = 0.42\ \text{m}.$$

(c)

$$v_x = v_0 \cos\theta$$
$$= 1.41 \times 10^7 \frac{\text{m}}{\text{s}} \times 0.866 = 1.22 \times 10^7 \frac{\text{m}}{\text{s}}.$$
$$v_y = -v_0 \sin\theta + at$$
$$= -1.41 \times 10^7 \frac{\text{m}}{\text{s}} \times 0.500 + 6.16 \times 10^{14} \frac{\text{m}}{\text{s}^2} \times 5 \times 10^{-8}\ \text{s}$$
$$= -0.71 \times 10^7 \frac{\text{m}}{\text{s}} + 3.08 \times 10^7 \frac{\text{m}}{\text{s}}$$
$$= 2.37 \times 10^7 \frac{\text{m}}{\text{s}}.$$

Let ϕ_x be the direction of motion with respect to the x-axis; then

$$\tan\phi_x = \frac{v_y}{v_x} = \frac{2.37 \times 10^7\ \text{m/s}}{1.22 \times 10^7\ \text{m/s}} = 1.94,$$

and

$$\phi_x = 62^\circ\, 45'.$$

The direction of motion can also be obtained from the equation of the trajectory. Combining the x- and y-coordinate equations in part (b) by eliminating t gives

$$y = -x \tan\theta + \frac{1}{2} \frac{Ee}{m_e} \frac{x^2}{v_0^2 \cos^2\theta}.$$

It is left to the reader to show that the slope of this curve at $x = 0.61$ m is 1.94.

2-4. ELECTRONIC CHARGE

Although the measurement of e/m_e indicated the identity of electrons, another measurement was required before e and m_e could be known separately. This was first made with precision in 1909 by R. A. Millikan, who perfected a technique suggested by J. J. Thomson and H. A. Wilson.

Both the charge e and the mass m_e of an electron are incredibly small quantities. The mass of any body can be determined from the measurement of the force acting on it when it is accelerated. Even if a single electron could be isolated for study, no instrument could measure its mass directly. Similarly, the charge on a body can be determined by measuring the force it experiences in an electric field. This method does not require the isolation of a single electron and, since very intense electric fields can be created, a measurable force can be produced.

An experiment to measure e must be carried out with a body having so few charges that the change of one charge makes a noticeable difference. Since the experiment must be done with very little charge, the force the body experiences will be small even though a large electric field is utilized. If the force on the charged body is very small, then the body itself must be very light. The force of gravity is always with us, and if the small electric force is not to be masked by a large gravitational force, then the mass of the body must be both small and known. If the body is small enough that the electric force on its charges is of the same order of magnitude as the gravitational force it experiences, then it may be that the gravitational force will be a useful standard of comparison rather than an annoying handicap.

Millikan used a drop of oil as his test body. It was selected from a mist produced by an ordinary atomizer. The drop was so small that it could not be measured optically, but with a microscope it could be seen as a bright spot because it scattered light from an intense beam, like a minute dust particle in bright sunlight.

When such a drop falls under the influence of gravity, it is hindered by the air it passes through. The way in which the fall of a small spherical body is hindered by air had been described by Stokes, who found that such a body experienced a resisting force **R** proportional to its velocity, or

$$R = kv. \tag{2–25}$$

The proportionality constant k was found by Stokes to be

$$k = 6\pi\eta r, \tag{2–26}$$

where η is the coefficient of viscosity of the resisting medium and r is the radius of the body. (This law assumes that the resisting medium is homogeneous. A more complicated law must be used if the size of the body is of the same order of magnitude as the mean free path of the molecules of the medium.) This is a friction equation very different from that introduced in mechanics to describe the force between two sliding bodies. In that case we assumed that the friction force depended only on the nature of the sliding surfaces and the normal force pressing the surfaces together. Hence in mechanics we discussed a force which did not depend on the speed of the motion. In the problem of a box sliding against fric-

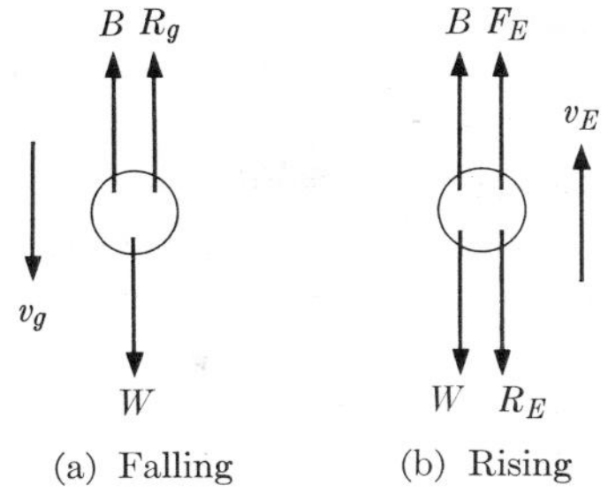

B R_g · B F_E · v_E · v_g · W · W R_E

(a) Falling (b) Rising

FIG. 2–8
Forces acting on an oil drop (equilibrium conditions).

tion down an inclined plane, the friction produced a constant force opposing the motion, but the acceleration was constant and the velocity increased continuously. A body subject to a frictional force like that given by Stokes' law will behave very differently.

A falling droplet of oil is acted on by its weight w, the buoyant force B of the air, and the resisting force $R = kv$ (Fig. 2–8). The resultant downward force F is

$$F = w - B - kv. \tag{2–27}$$

Initially, the velocity v is zero, the resisting force is zero, and the resultant downward force equals $w - B$. The drop therefore has an initial downward acceleration. As its downward velocity increases, the resisting force increases and eventually reaches a value such that the resultant force is zero. The drop then falls with a constant velocity called its *terminal velocity*, v_g. Since $F = 0$ when $v = v_g$, we have from Eq. (2–27),

$$w - B = kv_g. \tag{2–28}$$

Let ρ be the density of the oil and ρ_a the density of the air. Then

$$w = \tfrac{4}{3}\pi r^3 \rho g, \qquad B = \tfrac{4}{3}\pi r^3 \rho_a g, \tag{2–29}$$

and inserting the value of k from Eq. (2–26), we get

$$\tfrac{4}{3}\pi r^3(\rho - \rho_a)g = 6\pi\eta r v_g. \tag{2–30}$$

All of the quantities in this equation except r are known or measurable. We can therefore solve for the drop radius r and hence can express the proportionality constant k in terms of known or measurable quantities. The result is

$$k = 18\pi\left[\frac{\eta^3 v_g}{2g(\rho - \rho_a)}\right]^{1/2}. \tag{2–31}$$

In the experiment, the oil drop is situated between two horizontal plates where a known strong electric field may be directed upward or downward

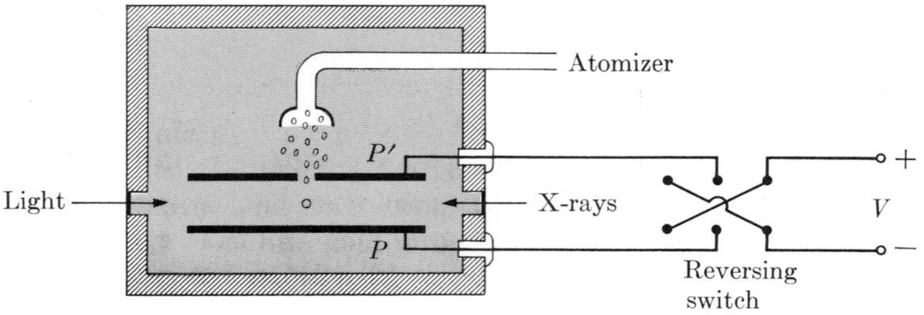

FIG. 2–9 Millikan's oil-drop experiment.

or may be turned off (Fig. 2–9). The droplet has a small electric charge q which may be minus or plus, depending on whether it has an excess or deficiency of electrons. The droplet gets this charge from rubbing against the nozzle of the atomizer and from encounters with stray charges left in the air by cosmic rays, or deliberately produced by x-rays or by bringing a radioactive material nearby. In the electric field the drop will experience a force $q\mathbf{E}$, which can always be directed upward by the proper choice of the direction of $\mathbf{E}$. The experimenter must manipulate $\mathbf{E}$ so that the drop rises and falls in the region between the plates but never touches either.

The microscope with which the drop's movements are followed is equipped with two horizontal hairlines whose separation represents a known distance along the vertical line in which the drop travels. By timing the trips of the drop over this known distance, the terminal velocities of the drop are found. The velocities of fall, $\mathbf{v}_g$, are all the same, since oil does not evaporate and therefore the weight of the drop is constant. The velocity of rise, however, depends on the charge q. The resultant force on the drop while it is rising is

$$F = qE + B - w - kv. \tag{2–32}$$

When the terminal velocity $\mathbf{v}_E$ is reached, the resultant force is zero, so

$$qE = w - B + kv_E. \tag{2–33}$$

But from Eq. (2–28), $w - B = kv_g$, so finally

$$q = \frac{k}{E}(v_g + v_E). \tag{2–34}$$

Since these terminal velocities are constant, they are relatively easy to measure.

Equation (2–34) permits the evaluation of q, the charge on the drop. In the oil-drop experiment, the value of v_g is determined for a particular drop with the electric field off, and a whole series of v_E's for the same drop is observed with the field on. If we knew that the electronic charge was unique and that there was only one charge on the drop, then Eq. (2–34) would give the value of this charge at once. Since the nature of the electronic charge was not known, Millikan repeated the experiment with many different charges on the drop. This provided a set of q's which he found to be integral multiples of one charge which he took to be the ultimate unit of charge, e. Thus he established the *law of multiple proportions* for electric charges and concluded from it that electricity must be atomic in character.

Millikan made observations on oil drops of different sizes and also on drops of mercury. In one instance a drop was watched continuously for eighteen hours. The sets of observations always gave the same value of the electronic charge or "atom" of electricity. The best modern determination of e is $(1.60210 \pm 0.00007) \times 10^{-19}$ coulomb.

2-5. MASS OF THE ELECTRON. AVOGADRO CONSTANT

Since e/m_e and e are now known, it is only simple arithmetic to find the mass of the electron to be

$$m_e = (9.1091 \pm 0.0004) \times 10^{-31} \text{ kg}.$$

Still another basic atomic constant may now be calculated with precision by using the value of the electronic charge. The Faraday constant is the amount of charge required to transport one atomic (molecular) weight of a univalent ion of a material through an electrolyte. Dividing the Faraday constant by e gives the number of electrons which have participated in this transport, or the Avogadro constant. The result agrees with Perrin's value, which had finally established the atomic view of matter.

2-6. POSITIVE RAYS

After the particle of negative electricity, the electron, had been identified, it was reasonable to ask about positive electricity. The search was made in a discharge tube very similar to that which disclosed cathode rays. In 1886, Goldstein observed that if the cathode of a discharge tube had slots in it, there appeared streaks of light in the gas on the side away from the anode. These channels of light, first called "canal rays," were easily shown to be due to charged particles. They moved in the direction of the electric field which was producing the discharge, and they were deflected by electric and magnetic fields in directions that proved that their charge was positive. Attempts were made to measure q/m, the ratio of the charge to the mass, of these *positive rays*. It was soon discovered that q/m for positive rays was much less than for electrons and that it depended on the kind of residual gas in the tube. The velocities of these positive rays were found to be nonuniform and much less than electron velocities.

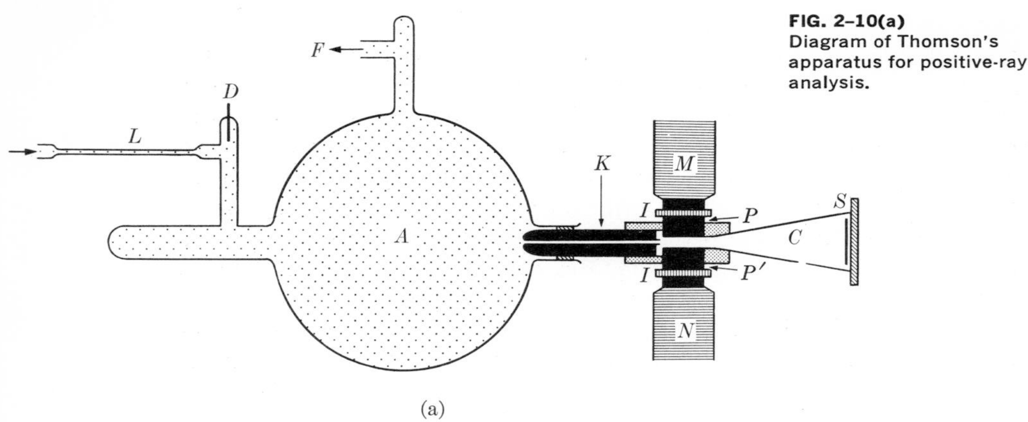

FIG. 2–10(a)
Diagram of Thomson's apparatus for positive-ray analysis.

Thomson devised a different method for measuring q/m of these positive rays having nonuniform velocities. Figure 2–10(a) shows the apparatus he used. The main discharge took place in the large bulb A at the left, where K is the cathode and D is the anode. The gas under study was slowly admitted through the tube at L and was simultaneously pumped out at F. Thus a very low gas pressure was maintained. Most of the positive rays produced in the bulb hit the cathode and heated it. The cathode had a "canal" through it, so that some of the positive rays passed into the right half of the apparatus. Just to the right of the cathode are M and N, the poles of an electromagnet. The pole pieces of this magnet were electrically insulated by sheets, I, so that the magnetic pole pieces could also be used as the plates of a capacitor for the establishment of an

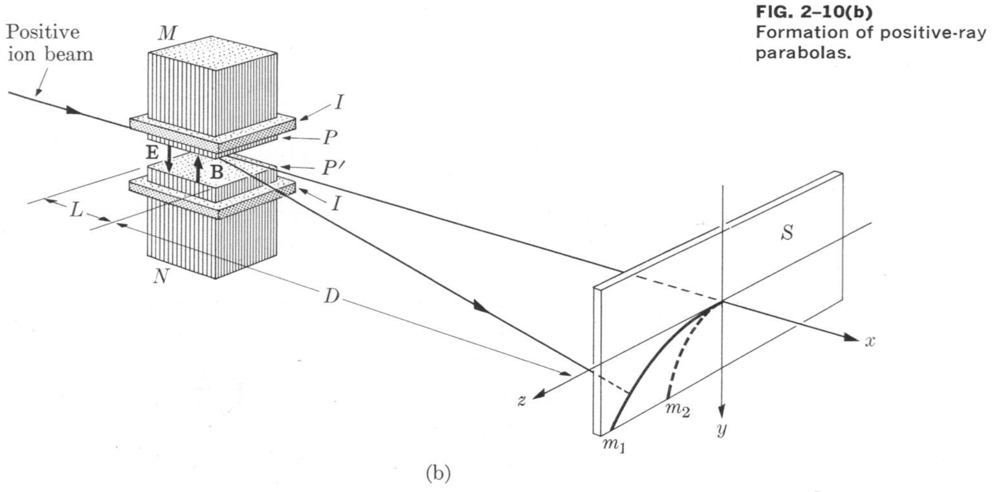

FIG. 2–10(b)
Formation of positive-ray parabolas.

electric field. With neither electric nor magnetic fields, the positive rays passed straight through the chamber C to the sensitive layer at S. This layer was either the emulsion on a photographic plate or a fluorescent screen. The beam was well defined because of the narrow tunnel in the cathode through which it had to pass. Instead of crossed fields as in the electron apparatus, this apparatus has its fields perpendicular to the rays but parallel to each other. The electric field is directed downward and the magnetic induction is upward, so that in Fig. 2–10(b) the electric force is toward the bottom of the page along the y-axis and the magnetic force is out of the page toward the reader along the z-axis.

Let a positively charged particle of unknown q/m enter the region between P and P' in Fig. 2–10(b) with an unknown velocity v_x along the x-axis. Then, according to Eq. (2–12), the deflection of the particle on the screen S due to the electric field is

$$y = \frac{qEL}{mv_x^2}\left(\frac{L}{2} + D\right), \qquad (2\text{–}35)$$

and, according Eq. (2–18), the deflection at S due to the magnetic field is

$$z = \frac{qBL}{mv_x}\left(\frac{L}{2} + D\right). \qquad (2\text{–}36)$$

These two equations together are the parametric equations of a parabola, where v_x is the parameter. Since v_x is different for different particles of the same type, the pattern on the screen is not a point but a locus of points. Elimination of v_x between these equations leads to

$$z^2 = \frac{q}{m}\frac{B^2L}{E}\left(\frac{L}{2} + D\right)y, \qquad (2\text{–}37)$$

which is the equation of a parabola.

Some actual parabolas obtained by Thomson's method are shown in Fig. 2–11. Examination of the figure reveals several things: Positive rays have distinct values of q/m, as is shown by the fact that the traces are clearly parabolas. That a single-experiment discloses several values of q/m is evident from the fact that there are several parabolas. It is apparent that the method is not capable of great precision because the parabolas are not sharp.

Thomson assumed that each particle of the positive rays carried a charge equal and opposite to the electronic charge, and he attributed the divergent parabolas to differences in mass. He assumed that the positive rays were positive because each had lost one electron. Thomson could identify particular parabolas with particular ions (charged atoms or molecules are called *ions*). Thus for atomic hydrogen, he could verify

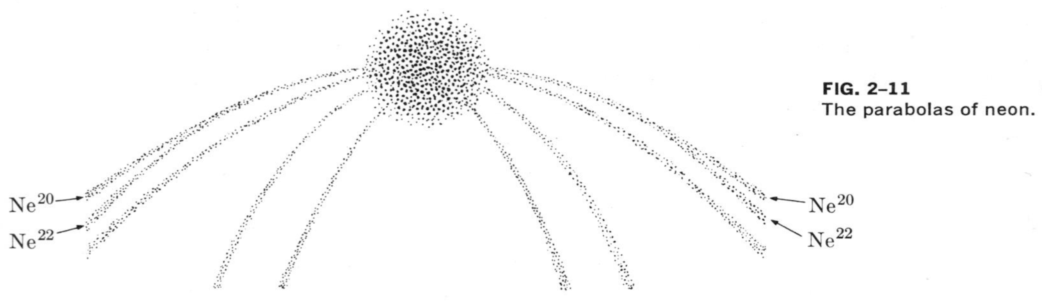

FIG. 2–11
The parabolas of neon.

that the q/m he measured was equal to the value one would expect from dividing the electronic charge by the mass per atom (the atomic weight of hydrogen divided by Avogadro's number). The reason that positive rays move more slowly than electrons and have lower values of q/m than electrons is now clear: the positive rays are much more massive. The largest q/m for positive rays is that for the lightest element, hydrogen. From the value of q/m it was found that the mass of the *hydrogen ion* or *proton* is 1836.2 times the mass of an electron. Electrons contribute only a small amount to the mass of material objects.

2–7. ISOTOPES

The most striking thing that was shown by the Thomson parabolas was that certain chemically pure gases had more than one value of q/m. Most notable was the case of neon, of atomic weight 20.2. Neon exhibited a parabola situated to correspond to a particle of atomic weight 20, but it also had a parabola which indicated an atomic weight of 22. Since the next heavier element, sodium, has an atomic weight of 23.0, efforts to explain away the unexpected value of q/m failed at first. Finally, it was concluded that there must be two kinds of neon, with different masses but chemically identical. The proof of this interpretation was given by Aston, one of Thomson's students.

Aston used a principle which we discussed in Chapter 1. We pointed out there that the average kinetic energy of a molecule in a gas is $3kT/2$. Different gas molecules mixed together in a container must be at the same temperature, and hence the average kinetic energy of each kind of molecule must be the same. If the two gases have different molecular masses, the lighter molecules must have the higher average velocity, and these will make more collisions per unit time with the walls of the container than the heavier molecules. Therefore if these molecules are allowed to diffuse through a porous plug from a container into another vessel, the lighter molecules will have a higher probability of passing through than the heavier, slower ones. Aston took chemically pure neon gas and passed part of it through such a plug. Since one such pass accomplishes only a slight separation, the process had to be repeated many times. He ended with two very small amounts of gas. One fraction had been through the plug many times and the other had been "left behind" many times. He measured the atomic weight of each fraction and found values of 20.15 for the former and 20.28 for the latter. The difference was not great, but it was enough to show that there are indeed two kinds of neon. Many other elements have since been shown to exist in forms which are chemically identical but different in mass. Such forms of an element are called *isotopes*. Thus Dalton's belief that all of the atoms of an element were physically identical in every way was not correct.

The discovery of isotopes solved several problems. It explained the two parabolas observed by Thomson. It also gave a logical explanation of the fact that the atomic weight of neon, 20.2, departs so far from an integral value. If chemical neon is a mixture of neon of atomic weight 20 and of neon of atomic weight 22, then there is some proportion of the two which will mix and have an average atomic weight of 20.2.

2-8. MASS SPECTROSCOPY

FIG. 2-12 Bainbridge's mass spectrometer, utilizing a velocity selector.

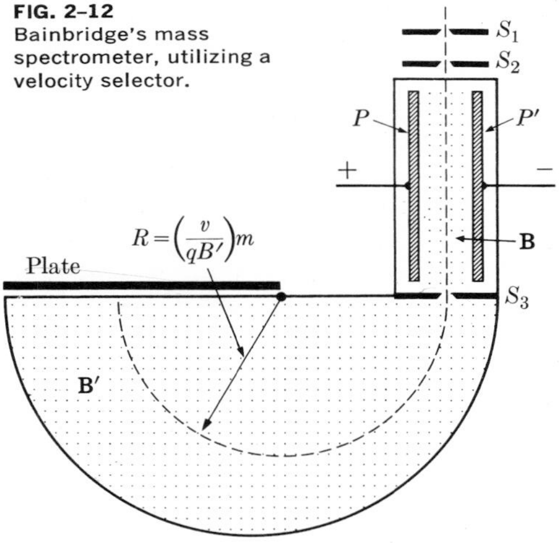

A detailed search for the isotopes of all the elements required a more precise technique. Aston built the first of many instruments called mass spectrographs in 1919. His instrument has a precision of one part in 10,000, and he found that many elements have isotopes. Rather than discuss his instrument, however, we shall describe an elegant one built by Bainbridge. The Bainbridge mass spectrograph (Fig. 2-12) has a source of ions (not shown) situated above S_1. The ions under study pass through slits S_1 and S_2 and move down into the electric field between the two plates P and P'. In the region of the electric field there is also a magnetic induction B, perpendicular to the paper. Thus the ions enter a region of crossed electric and magnetic fields like those used by Thomson to measure the velocity of electrons in his determination of e/m_e. Those ions whose velocity is E/B pass undeviated through this region, but ions with other velocities are stopped by the slit S_3. All ions which emerge from S_3 have the same velocity. The region of crossed fields is called a *velocity selector*. Below S_3 the ions enter a region where there is another magnetic field B', perpendicular to the page, but no electric field. Here the ions move in circular paths of radius R. From Eq. (2–5), we find that

$$m = \frac{qB'R}{v}. \tag{2–38}$$

Assuming equal charges on each ion, then, since B' and v are the same for all ions, we find that the masses of the ions are proportional to the radii of their paths. Ions of different isotopes are converged at different points on the photographic plate. The relative abundance of the isotopes is measured from the densities of the photographic images they produce. Figure 2–13 shows the mass spectrum of germanium. The numbers shown beside the isotope images are not exactly the relative masses or atomic weights of each of the atoms but the integers nearest the relative weights. The integer closest to the isotopic mass is called the *mass number* A, and isotopes are written with the mass number as a superscript to the chemical symbol. Thus the isotopes shown are written Ge^{70}, Ge^{72}, etc., or, in another notation, ^{70}Ge, ^{72}Ge, etc.

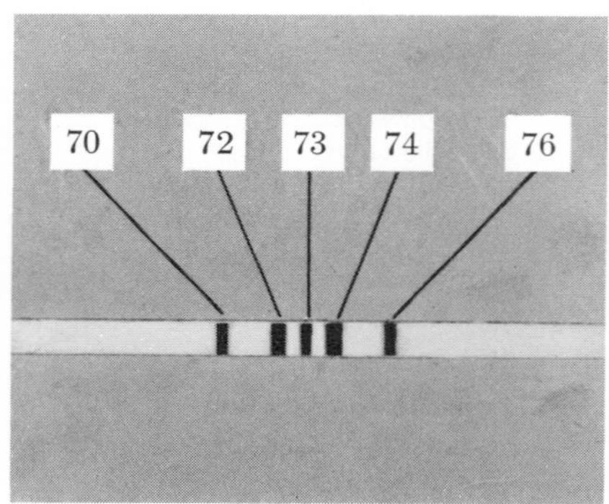

FIG. 2–13
The mass spectrum of germanium, showing the isotopes of mass numbers 70, 72, 73, 74, 76.

As in the case of neon, the discovery of the isotopes of the various elements largely accounted for the fact that many chemical atomic weights are not integers. If germanium has mass numbers 70, 72, 73, 74 and 76, it is no wonder that a mixture of isotopes of germanium has a chemical atomic weight of 72.6.

2–9. ISOTOPIC MASS. UNIFIED ATOMIC MASS UNIT

Before approximately 1930, both chemists and physicists had assigned an atomic weight of 16 to natural oxygen and thus used this as the basis for the scale of atomic weights. However, the discovery that natural oxygen is a mixture of three isotopes, O^{16}, O^{17}, and O^{18}, and a somewhat variable mixture at that, led physicists to assign the number 16 to the isotope of oxygen having the lowest mass, whereas chemists, not needing so precise a definition, continued to use 16 as the atomic weight of the mixture of the three isotopes of oxygen.

This dual basis led to two tables of relative weights of atoms which differed by about 275 ppm (parts per million). In addition, the Faraday constant, the Avogadro constant, and the gas constant had different values, depending on which basis was chosen. A great deal of confusion resulted from this situation. For example, values of the various constants given in tables could be misleading unless the scale upon which they were based was given.

Further confusion followed from the fact that although physicists were using the oxygen isotope 16 as the base of their system, mass spectroscopists had found it more convenient to use the carbon isotope 12 as a standard because this atom provides a series of reference points in mass spectrograms.

In 1957, the International Commission on Atomic Weights initiated steps to resolve the difficulties arising from the dual system. In 1961, the final action was taken which resulted in setting the mass of the most abundant isotope of carbon at 12 as the base of a common system for chemistry and physics.

The atomic weight of an element used by the chemist is the average value, weighted on the basis of abundance, of the isotopic masses, based on carbon 12, of the isotopes of the element. The atomic weight of an element varies slightly because of natural variations in the isotopic composition of the element. Some observed ranges are hydrogen, ± 0.00001; oxygen, ± 0.0001; and sulfur, ± 0.003. The chemist uses atomic weights because his samples of a material, even when small, contain a very large number of atoms and for his techniques isotopes are indistinguishable. If he were to use isotopic masses and consider all the possibilities, he would find about 1200 different values of the molecular weight of chemically pure hydrated zinc sulfate, $ZnSO_4 \cdot 6H_2O$.

In physics the relative abundance of isotopes, the separation of isotopes, and the properties of particular isotopes come in for detailed study. The relative weights of *neutral atoms* are called the *isotopic masses*. The unit of atomic masses, called the *unified atomic mass unit*, *u*, or *amu*, is by definition one-twelfth of the mass of C^{12}. Since the mass of an atom is its isotopic mass divided by the Avogadro constant, it follows from the definition that the u is equal to the reciprocal of the Avogadro constant, or $(1.66040 \pm 0.00008) \times 10^{-27}$ kg. Thus the discovery of isotopes not only accounted for those chemical atomic weights which were far from integral, but also provided a new basis in terms of which the isotopic masses are very near to being integers.

REFERENCES

ANDERSON, D. L., *The Discovery of the Electron*, Momentum Book No. 3. Princeton, N. J.: Van Nostrand, 1964.

ASTON, F. W., *Isotopes*. 2nd ed. London: Arnold, 1924. An account of the first precision determinations of isotopic masses.

BAILEY, P. T., "Discovery of the Electron," *Phys. Today* **19,** No. 7 (July, 1966), 12. Brief discussion of and complete references to the determinations of q/m for cathode rays prior to the work of J. J. Thomson.

GLASSTONE, SAMUEL, *Sourcebook on Atomic Energy*. 2nd ed. Princeton, N. J.: Van Nostrand, 1958. A comprehensive source of basic information in atomic and nuclear physics.

HARNWELL, G. P., and J. J. LIVINGOOD, *Experimental Atomic Physics*. New York: McGraw-Hill, 1933. An excellent book on experimental methods and relevant theory.

LOEB, LEONARD B., *Fundamental Processes of Electric Discharge in Gases*. New York: Wiley, 1939.

MAGIE, WILLIAM F., *A Source Book in Physics*. Cambridge, Mass.: Harvard University Press, 1963. A collection of the principal papers of the great men in physics. A valuable source for learning how some of the important discoveries were made.
Among the papers included are the following: "The Cathode Discharge," by William Crookes; "Laws of Electrolysis," by Michael Faraday; "The Canal Rays," by Eugen Goldstein; "The Cathode Discharge," by Johann W. Hittorf; "The Negative Charges in the Cathode Discharge," by Jean Perrin; "The Electron," by Joseph J. Thomson; "The Canal Rays," by Wilhelm Wien.

MILLIKAN, ROBERT A., *The Electron*. 2nd ed. Chicago, Ill.: University of Chicago, 1924. A detailed account of the determination of the electronic charge.

SHAMOS, M. H., editor, *Great Experiments in Physics*. New York: Holt, 1959. Includes the original accounts of 24 experiments that laid the foundation for modern physics, with annotations by the editor. An excellent book.

WHITE, R. S., "The Earth's Radiation Belts," *Phys. Today* **19,** 25 (October, 1966). A good summary of current knowledge of belts of charged particles trapped by the magnetic field of the earth. Contains an extensive list of references.

PROBLEMS

2–1. What is the ratio of the electric force on a charged particle in an electric field of 20 V/cm to the force of gravity on the particle if it is (a) an electron, (b) a proton? (c) Is the weight negligible compared with the electric force?

2–2. An electron moving in a vertical plane with a speed of 5.0×10^7 m/s enters a region where there is a uniform electric field of 20 V/cm directed upward. Find the electron's coordinates referred to the point of entry and the direction of its motion at a time 4×10^{-8} s later if it enters the field (a) horizontally, (b) at 37° above the horizontal, and (c) at 37° below the horizontal.

2–3. If the charged particle in Problem 2–2 were a proton instead of an electron, what must be the magnitude and direction of the electric field so that the answers for the proton would be the same as they were for the electron?

2–4. The dimensions of some parts of a typical commercial cathode-ray tube are given in Fig. 2–14. If electrons start from rest at the cathode, what is their velocity v_x at the origin O for an accelerating voltage of 1136 V between the anode and cathode?

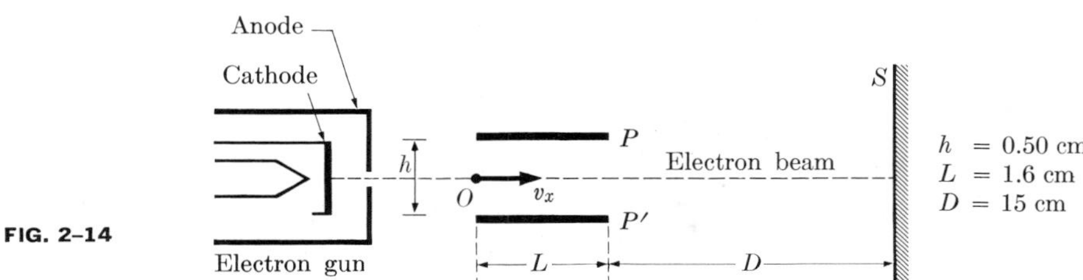

FIG. 2–14

2–5. Given that the potential difference between the deflecting plates P and P' is 50 V in the cathode-ray tube in Problem 2–4, (a) find the y-coordinate and the direction of motion of the electrons when $x = L$. (b) What is the total deflection on the screen S?

2–6. A large plane metal plate is mounted horizontally at a distance of 0.80 cm above another similar horizontal plate. They are charged to a potential difference of 40 V, the upper plate being positive. An electron is projected horizontally with a velocity of 10^6 m/s from a point O which is midway between the plates. (a) Find the x-coordinate of the point at which the electron strikes a plate. (b) Compute the tangent of the angle which gives the direction of the electron's motion as it strikes the plate. (c) What is the change in kinetic energy of the electron in going from O to the plate? (d) What would be the answer to Part (c) if the electron had no initial velocity at O?

2–7. An electron beam consists of electrons which have been accelerated from rest through a potential difference of 1000 V. The beam current is 50 mA. The beam is incident on a metal surface which stops the electrons. Calculate the force exerted on the metal plate by the electron beam.

2–8. A particle having a mass of 1 g carries a charge of -3×10^{-8} C. The particle is given an initial horizontal velocity in the plane of the paper of 6×10^4 m/s in the gravitational field of the earth. What is the magnitude and direction of the minimum magnetic induction that will keep the particle moving along the same straight line?

2–9. An electron is accelerated from rest through a potential difference of 50 V. The electron then enters a uniform electric field of intensity 3000 N/C, the electron moving in the same direction as the electric field lines. (a) How far into the field does the electron travel before it comes to rest? (b) How many seconds elapse between the time the electron enters the field and the instant it comes to rest? (c) Is it possible to use a magnetic field in place of an electric field to stop the electron?

2–10. (a) Through what potential difference would a deuterium ion have to be accelerated from rest in a vacuum so that it would have a speed of 8.47×10^6 m/s? (Refer to Problem 1–23.) (b) What would have to be the magnitude and direction of the smallest magnetic induction which would constrain the moving deuterium ion to a circular path in an evacuated toroidal tube 1 m in diameter?

2–11. A uniform electric field of intensity 40×10^4 V/m is perpendicular to a uniform magnetic field of flux density 2×10^{-2} T. (A tesla T is a weber per square meter, Wb/m^2.) An electron moving perpendicularly to both fields experiences no net force. (a) Show in a diagram the relative orientation of the electric field vector, the magnetic induction vector, and the velocity of the electron. (b) Calculate the speed of the electron. (c) What is the radius of the electron orbit when the electric field is removed?

2–12. An electron is accelerated from rest to a speed of 10^7 m/s by an electron gun. The electrons leave the gun and strike a screen 4 cm from the front of the gun. When a uniform magnetic field, perpendicular to the electron's velocity, is established between the gun and the screen a deflection of 2 cm is noted on the screen. (a) What is the accelerating potential in the gun? (b) What is the magnetic induction?

2–13. A charged particle enters the region between two very large parallel metal plates, the particle's velocity being parallel to the plates when the particle enters. The plates are separated a distance of 2 cm and the potential difference between the plates is 2000 V. When the particle has penetrated 5 cm into the space between the plates, it is found that the particle has been deflected 0.6 cm. If a perpendicular magnetic field of flux density 0.1 T is impressed simultaneously with the electric field, the particle is found to undergo no deflection at all. Calculate the charge to mass ratio for this particle.

2–14. A uniform magnetic field whose induction is 4×10^{-4} T acts on the electron beam over the distance L in the cathode-ray tube in Problem 2–4. The field is normal to the plane of the trajectory and directed outward. (a) Find the y-coordinate and the direction of motion of the electron when $x = L$. (b) What is the total deflection on the screen S? (c) What would be the deflection on the screen computed both from the approximate and the exact expressions for the radius of curvature if the magnetic induction extended the whole distance, $L + D$, from O to the screen?

2–15. The cathode-ray tube of Problem 2–4 is mounted horizontally and oriented so that the screen faces north. Given that the earth's magnetic induction is 5×10^{-5} T at a dip of 53°, find the magnitude and direction of the horizontal component and of the vertical component of the displacement of the electron beam with respect to the original position on the screen when the tube is turned 90° so that it faces west. Assume that the earth's field extends only over the distance $L + D$ and that the deflections are small.

2–16. Given a cathode-ray tube of the form shown in Problem 2–4, (a) show in general terms that the total deflection on the screen due to an electric field between P and P' is equal that which the electrons would undergo if they traveled along the axis from O to the point $x = L/2$ and then were deflected at the angle θ given in Eq. (2–11). (b) Would a similar relation hold if a magnetic field instead of an electric field extended over the length L?

2–17. A cathode-ray tube is placed in a uniform magnetic induction B with the axis of the tube parallel to the lines of force. If electrons emerging from the gun with a velocity v make an angle θ with the axis as they pass through the origin O, show (a) that their trajectory is a helix, (b) that they will touch the axis again at the time $t = 2\pi m_e/Be$, (c) that the coordinate of the point of touching is $x = 2\pi m_e v \cos\theta/Be$, and (d) that for small values of θ, the coordinate of the point of crossing or touching the axis is independent of θ. (e) The arrangement in this problem is called a magnetic lens or magnetic bottle. Why? (f) How do the trajectories of the electrons passing through the origin at an angle θ above the axis differ from those directed at an angle θ below the axis?

2–18. Electrons are accelerated through a potential difference of 1000 V in an electron gun and leave the narrow hole in the anode as a narrow diverging beam. What magnitude of axial magnetic induction is required to focus the beam on a screen 50 cm from the hole? [*Hint:* See Problem 2–17.]

2–19. What is the final velocity of an electron accelerated through a potential difference of 1136 V if it has an initial velocity of 10^7 m/s?

2–20. Two large, plane metal plates are mounted vertically 4 cm apart and charged to a potential difference of 200 V. (a) With what speed must an electron be projected horizontally from the positive plate so that it will arrive at the negative plate with a velocity of 10^7 m/s? (b) With what speed must it be projected from the positive plate at an angle of 37° above the horizontal so that the horizontal component of its velocity when arriving at the negative plate is 10^7 m/s? (c) What is the magnitude of the y-component of the velocity when arriving at the negative plate? (d) What is the electron's time of transit from one plate to the other in each case? (e) With what speed will the electron arrive at the negative plate if it is projected horizontally from the positive plate with a speed of 10^6 m/s?

2–21. Two positive ions having the same charge q but different masses, m_1 and m_2, are accelerated horizontally from rest through a potential difference V. They then enter a region where there is a uniform electric field E directed upward. (a) Show that if the ion beam entered the field along the x-axis, then the value of the y-coordinate for each ion at any time t is $y = Ex^2/4V$. (b) Can this arrangement be used for isotope separation?

2–22. Two positive ions having the same charge q but different masses, m_1 and m_2, are accelerated horizontally from rest through a potential difference V. They then enter a region where there is uniform magnetic induction B normal to the plane of the motion. (a) Show that if the beam entered the magnetic field along the x-axis, the value of the y-coordinate for each at any time t is

$$y = Bx^2(q/8mV)^{1/2}.$$

(b). Can this arrangement be used for isotope separation?

2–23. In a cathode-ray tube of the form shown in Problem 2–4 electrons are accelerated from rest through a potential difference V. These then enter a vertical electric field E which extends over the distance L, and also a magnetic field having flux density B normal to the paper over the distance $L + D$. If both fields are adjusted so that there is no deflection on the screen, show that

$$\frac{e}{m_e} = \frac{E^2L^2}{2VB^2}\frac{(L+2D)^2}{(L+D)^4}.$$

2–24. Particles with charge q and mass m are injected into a homogeneous magnetic field having induction B. When their velocities are initially perpendicular to the field, the particles travel in circular orbits. Derive an expression for the frequency of revolution of the particles and show that the frequency is independent of the velocity.

2–25. A charged oil drop falls 4.0 mm in 16.0 s at constant speed in air in the absence of an electric field. The relative density of the oil is 0.80, that of the air is 1.30×10^{-3}, and the viscosity of the air is 1.81×10^{-5} N · s/m². Find (a) the radius of the drop and (b) the mass of the drop. (c) If the drop carries one electronic unit of charge and is in an electric field of 2000 V/cm, what is the ratio of the force of the electric field on the drop to its weight?

2–26. Derive the expression for k in Eq. (2–31) from Eqs. (2–25), (2–29), and (2–30). What are the units of k in the mksA system?

2–27. When the oil drop in Problem 2–25 was in a constant electric field of 2000 V/cm, several different times of rise over the distance of 4.0 mm were observed. The measured times were 36.1, 11.5, 17.4, 7.55, and 23.9 s. Calculate (a) the velocity of fall under gravity, (b) the velocity of rise in each case, and (c) the sum of the velocity in part (a) and each velocity in part (b). (d) Show that the sums in part (c) are integral multiples (two significant figures) of some number and intepret this result. (e) Calculate the value of the electronic charge from these data.

2–28. Show that the electric field E necessary to raise an oil drop of mass m and charge q with a speed which is twice the speed of fall of the drop when there is no field is $E = 3mg/q$, given that the buoyant force of the air is negligible.

2–29. In an experiment to count and "weigh" atoms it is found that a current of 0.800 A flowing through a copper sulfate solution for 1800 s deposits 0.473 g of copper. The atomic weight of copper is 63.54, its valence is 2, and the electronic charge is 1.60×10^{-19} C. Using only the data *given in this problem*, find (a) the number of electronic charges carried by the ions which deposited as copper atoms, (b) the number of copper atoms deposited, (c) the mass of a copper atom, (d) the number of atoms in a gram-atomic weight of copper, (e) the number of electronic charges carried by a gram-equivalent weight of copper ions, (f) the number of coulombs required to deposit a gram-equivalent weight of copper, and (g) the mass of a hydrogen atom given that its atomic weight is 1.008.

2–30. What must be the direction of the electric field E and the magnetic induction B in Fig. 2–10(b) so that the segment of the positive-ion parabola will be in (a) the lower right quadrant, (b) the upper right quadrant, and (c) the upper left quadrant, as viewed from the right of the diagram?

2–31. (a) If the ion beam in Fig. 2–10(b) contains two types of ions having equal charges but different masses, which of the two parabolic segments will have those of greater mass? (b) If the masses are equal but the charges different, which segment will contain those having the larger charge?

2–32. For a particular parabola in Thomson's mass spectrograms, what physical quantity is different for the ions which land close to the origin than for those landing farther away? Why does this difference exist, since the accelerating voltage is the same for all the ions?

2–33. In a mass spectrometer, ions having the same charge q but different masses m_1 and m_2 are accelerated from rest through a potential difference V. A narrow beam of these ions then enters a magnetic field having a magnetic induction B which is perpendicular to the motion of the particles. (a) Derive a simple expression in terms of the *given* quantities for the *ratio* of the radii of the trajectories of the two types of ions in the magnetic field assuming the following:

Case 1. All m_1 ions have the same velocity, and all m_2 ions have the same velocity, which is not necessarily equal to that of the m_1 ions.

Case 2. There is a distribution of velocities in the ion beam but, before entering the magnetic field B, the beam passes through an effective velocity selector having an electric field E and magnetic induction B_1.
(b) In which of the preceding cases is the resolving power of the mass spectrometer greater?

2–34. If the electric field between the plates PP' in Fig. 2–12 is 100 V/cm and the magnetic induction in both magnetic fields is 0.2 T, (a) what is the speed of an ion which will go undeviated through the slit system? (b) Given that the source produces singly charged ions of the carbon isotopes C^{12} and C^{13}, find the distance between the center of the lines formed by them on the photographic plate. (Assume that the atomic weights of these isotopes are equal to their mass numbers.) (c) If the slit S_3 is 1 mm wide, will the images on the plate overlap?

2–35. A deuteron is an ionized hydrogen isotope with mass number 2. Protons and deuterons are accelerated through a potential difference of 150 V, pass through a small slit, and then enter a uniform magnetic field where the magnetic induction is 0.010 T. The field causes the particles to move in a circular path. What is the separation of the beams after completing a semicircle?

2–36. The text does not tell why in Bainbridge's mass spectrometer the ion beam is caused to execute a semicircle before striking the plate. In Fig. 2–12 the beam passing down through S_3 is slightly divergent. Use a compass (or a 25-cent coin) to show that divergent beams through S_3 that have a common radius of curvature are in best focus at the plane of the plate.

2–37. Copper has two isotopes whose masses are 62.9 and 64.9, respectively. What is the percent abundance of each in ordinary copper having an atomic weight of 63.5?

2–38. The isotopic mass of C^{12} on the former O^{16} scale was 12.003816. (a) By what percent must this value of the mass of C^{12} be reduced to make it exactly 12 (the C^{12} base)? (b) Calculate the isotopic mass of the former oxygen 16 base on the C^{12} scale.

2–39. Uranium hexafluoride, UF_6, is a gaseous compound at 100°C. Given a mixture of two such hexafluorides, one formed of the isotope U^{235} and the other of U^{238}, find the root-mean-square speed of each molecule. Comment on the possibility of separating these by a diffusion process.

CHAPTER 3

THE ATOMIC VIEW OF RADIATION

3–1. INTRODUCTION

All sources of light consist of matter which is excited in one way or another. The firefly excites his body matter by some obscure chemical process; the matter of the sun is excited by heat. But ever since Heinrich Hertz demonstrated the validity of Maxwell's theory of electromagnetic radiation, we have known that the ultimate source of radiation is an accelerated electric charge. We cannot begin the story of radiation with Maxwell, however, if we are to appreciate one of the most dramatic demonstrations of the scientific method.

3–2. PARTICLES OR WAVES

Certain Greeks of ancient times argued that since a blind man reaches out to feel his way about, the seeing man must reach out with his eyes. They thought of light as a kind of tentacle emitted by the eye yet retaining contact with the eye so that information about objects touched was conveyed to the mind. Such a view obviously fails to explain why a man cannot see at night unless there is an outside source of illumination.

It was realized long ago that light consists of something which goes out from certain "sources," bounces off objects, and may finally enter the eye. In the seventeenth century there were two views on the nature of the "something" that was bouncing about. Newton defended the premise that light consists of a stream of fast-moving elastic particles of very small mass. His view accounted for the law of reflection, which states that the angle of incidence is equal to the angle of reflection. (This is the way perfectly elastic balls bounce from the sidewalk.) He accounted for the law of refraction by arguing that when particles of light are very near any optically dense medium like glass, they are attracted to it, and this attraction increases the component of the velocity of light in a direction perpendicular to the surface. Thus, according to Newton, the light travels through the medium *faster* than it does in free space and has its direction altered *toward* the normal. Christian Huygens, on the other hand, supported the view that light consists of waves. The most impressive argument in his favor at that time was that two light beams can cross through each other without "colliding." He too explained reflection and refraction. His explanation of refraction was that when a wavefront penetrates an optically dense medium at an angle, the wave moves more *slowly*. This slowing of the wavefront causes the wave's direction of advance to be altered *toward* the normal.

3-3. ELECTRICITY AND LIGHT

After a century of neglect, the undulatory theory of light was revived by the versatile English scientist Thomas Young. In 1801, he showed that only the principle of interference of waves could explain the colors of thin films and of striated surfaces. During the next half century, further experimental work, especially by the French physicists Fresnel, Arago, Malus, Cornu, Fizeau, and Foucault, showed that the particle theory of light was not tenable. This work reached its culmination in 1864 when James Clerk Maxwell announced the results of his efforts to put the laws of electricity into good mathematical form. He had succeeded in this formulation and found in addition an important by-product: the laws could be combined into the mathematical form of the wave equation for electromagnetic waves. He showed, furthermore, that the velocity of these waves is the *velocity of light!* Thus in one dramatic move he put the theory of electricity in order and incorporated all optics into that theory.

Huygens' view completely displaced Newton's when Foucault found the velocity of light in an optically dense medium *less* than its velocity in free space, and Maxwell's theory was verified in 1888 when Hertz demonstrated that oscillating currents in an electric circuit can radiate energy through space to another similar circuit. Hertz used a circuit containing inductance and capacitance, hence capable of oscillating. Whenever a spark jumped across a gap in the active (transmitting) circuit, electromagnetic waves were radiated from the region in which the electric discharge occurred. (Modifications of this first transmitter were used for radio communication until the advent of vacuum tubes.) The passive or receiving circuit was a loop of wire containing a gap. When energy was transferred from one circuit to the other, sparks jumped across the receiver gap. Hertz's experiments showed that the radiation generated by electric circuits obeyed the known laws of optics. It thus appeared that the theory of light was in a satisfactory and elegant state.

Yet the last word on this subject had not been said. Hertz noted that the induced spark was more easily produced when the terminals of the receiving gap were illuminated by light from the sparks in the transmitter gap. This effect was studied more fully by one of Hertz's students, Hallwachs, who showed that a negatively charged clean plate of zinc loses its charge when illuminated by ultraviolet light. Thus Hertz's verification of Maxwell's wave theory of light led almost simultaneously to the discovery of the photoelectric effect which, as we shall see, led in turn to a profound reinterpretation of the wave theory of radiation.

3-4. ELECTRODYNAMICS

It is unfortunate that we do not have time to develop here the methods of Maxwell's electrodynamics. We can, however, develop qualitatively the idea that the ultimate source of radiation is an accelerated electric charge.

Every electric charge produces an electric field whose lines of force extend radially from the charge through all space. When this charge is in motion, a magnetic field is produced in accordance with Ampere's law and the magnetic field lines are circles concentric with the current. According to the viewpoint of Maxwell's field theory, it is the motion of the electric lines of force that sets up the magnetic field transverse to them. A steady electric current is accompanied by steady electric and magnetic fields; however, a varying current (i.e., one composed of accelerated charges) will produce changes in both fields associated with it. These changes are propagated outward from the accelerated charges through space with the speed of light. The acceleration of a charge produces a pulse of electromagnetic radiation which consists of an electric field component and a magnetic field component that are perpendicular to each other and to their direction of propagation. If the charge oscillates with a simple periodic motion, an electromagnetic wave like that in Fig. 3–1 will be produced.

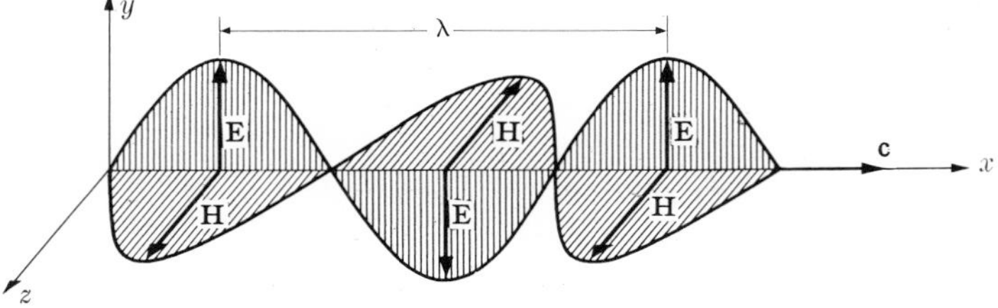

FIG. 3–1 A plane-polarized electromagnetic wave of wavelength λ showing the relation of the vectors **E**, **H**, and **c**.

The energy transmitted by the electromagnetic waves in a radiation field may be specified either in terms of the intensity or of the energy density of the wave motion. The intensity of the radiation is defined as the energy transmitted in unit time through a unit area normal to the direction of propagation of the waves. The mks unit of intensity is watts per square meter. The energy density or volume density of the radiation is defined as the amount of radiant energy in a unit volume of space. The MKSA unit of energy density is joules per cubic meter. It is evident that the energy density is equal to the intensity divided by the velocity of propagation of the wave. The term *energy density* is particularly useful in discussing the radiation within a heated enclosure.

Our discussion has implied a linear acceleration but, according to Maxwell's theory, radiation occurs whenever an electric charge is accelerated in any manner. For example, a charge moving with constant speed in a circular path will be a source of radiation. This case is equivalent to two mutually perpendicular simple periodic motions with equal amplitudes and frequencies, but with a phase difference of 90°.

Every radio transmission is a refined repetition of Hertz's original experiments verifying Maxwell's proposition. Instead of physical motion of charged bodies, a radio transmitter causes electrons to move back and forth in an antenna which is physically at rest. These accelerated electrons produce radiation. The electric field in this radiation will exert force on any charges it encounters. Thus the electrons in a receiving antenna respond to the radiation by being accelerated, and their motion constitutes an electric current. A modern radio receiver amplifies these currents and makes them easy to observe. Hertz had no amplifiers, so the emf induced in the receiver had to be large enough to create visible sparks.

3-5. THE UNITY OF RADIATION

From previous studies, the readers of this book are aware that the many forms of radiation, heat, light, radar, radio, etc., differ from one another in frequency but not in kind. The so-called "kinds" of radiation are characterized by the techniques used to produce and detect them; actually, they all travel through free space with the same velocity and should all be understood in terms of the same theory. The tremendous range of the electromagnetic spectrum is shown in Fig. 3–2 (photon energy, mentioned in the figure, will be discussed later in this chapter). The classical theory of Maxwell applies to all these radiations and all are due ultimately to the acceleration of electrical charges. Except for differences due to frequency, an observation made on one "kind" of radiation must also be true of all other kinds.

3-6. THERMAL RADIATION

Information about the nature of all radiation may be obtained from a study of any of the "kinds" of radiation. We now consider the radiation from heated bodies, since that investigation has proved particularly fruitful. We all know that a body will emit visible radiation if it is hot enough. A close relation between temperature and radiation is further implied by the fact that a white-hot body is hotter than a red-hot one. We might explore this matter further by passing the radiation from a hot body through some dispersive instrument such as a prism or grating spectrom-

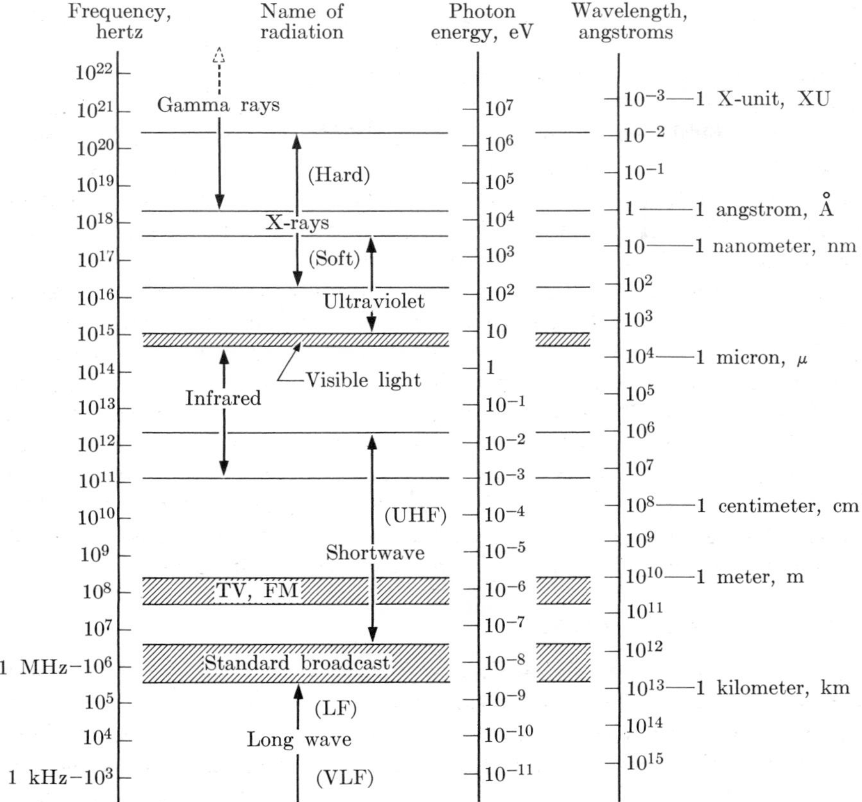

FIG. 3–2
The spectrum of electromagnetic radiation.

eter. If we measure the radiant energy emitted by a hot body (No. 1) for a whole series of radiant frequencies, we might obtain a graph similar to the dashed curve in Fig. 3–3. The ordinate of this curve is called the *monochromatic emittance*, W_λ, which is the amount of energy radiated per unit time per unit area of emitter in a wavelength range $d\lambda$; the abscissa is the wavelength rather than the frequency. Repeating the same experiment for another body (No. 2) of a different material but at the same temperature, we might now obtain the dotted curve of Fig. 3–3. It is clear from the figure that at most wavelengths the first body is a more efficient emitter at the given temperature than the second. Although the two curves differ, they have the same general character. They come to their highest points at about the same wavelength. Upon studying a great variety of substances all at the same temperature, we would obtain a great variety of emission

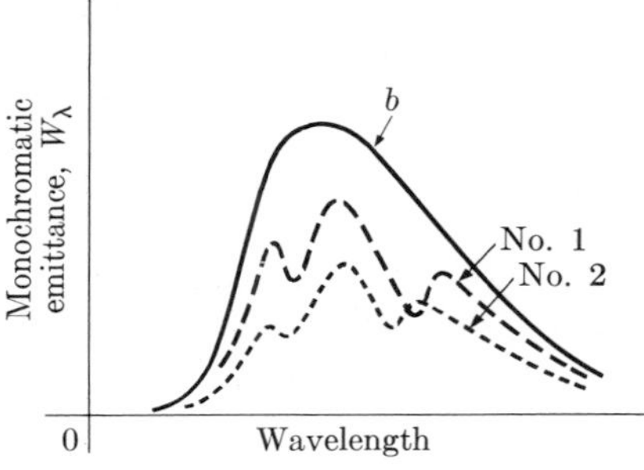

FIG. 3–3
The radiation spectrum of several hot bodies.

curves, but none of these would ever have a greater monochromatic emittance, $W_{\lambda b}$, than the envelope curve, shown as a solid line in Fig. 3–3. It appears that this curve may have a significance which does not depend on the nature of the emitting material. Let us attempt to find or make an emitter which has an emission curve identical to the solid curve of Fig. 3–3.

3–7. EMISSION AND ABSORPTION OF RADIATION

It may be wondered why it is, if the surfaces of all bodies are continually emitting radiant energy, that all bodies do not eventually radiate away all their internal energy and cool down to a temperature of absolute zero. The answer is that they would do so if energy were not supplied to them in some way. In the case of the filament of an electric lamp, energy is supplied electrically to make up for the energy radiated. As soon as this energy supply is cut off, these bodies do, in fact, cool down very quickly to room temperature. The reason that they do not cool further is that their surroundings (the walls, and other objects in the room) are also radiating, and some of this radiant energy is intercepted, absorbed, and converted into internal energy. The same thing is true of all other objects in the room—each is both emitting and absorbing radiant energy simultaneously. If any object is hotter than its surroundings, its rate of emission will exceed its rate of absorption. There will thus be a net loss of energy and the body will cool down unless heated by some other method. If a body is at a lower temperature than its surroundings, its rate of absorption will be larger than its rate of emission and its temperature will rise. When the body is at the same temperature as its surroundings, the two rates become equal, there is no net gain or loss of energy, and no change in temperature.

Figure 3–3 shows that the emittance of a surface W is different at different wavelengths and the paragraph above implies that it is greater for higher temperatures. To simplify our next discussion consider an infinitesimal band of wavelengths and several opaque bodies in thermal equilibrium with each other and their surroundings. Because the bodies are opaque they will not transmit radiation and therefore, in general, part of the incident radiation will be reflected and the remainder will be absorbed. The fraction absorbed, called the *absorptance* a, plus the fraction reflected, called the *reflectance* r, must be unity or, since the surfaces may be different,

$$a_1 + r_1 = 1, \qquad a_2 + r_2 = 1, \quad \text{etc.} \tag{3-1}$$

Since the bodies are in thermal equilibrium with their surroundings, they will be "bathed" in radiation of uniform intensity, I. If this is not obvious, it may be helpful to think of the bodies as being tiny specks near each other but too small to cast shadows on each other. We can now write the last sentence quantitatively. The total radiation in a time Δt from body No. 1, which has an area ΔA_1 and a radiant emittance W_1, is $W_1 \, \Delta A_1 \, \Delta t$. The absorption by the same body in the same time is $a_1 I \, \Delta A_1 \, \Delta t$. For the condition of thermal equilibrium to exist these must be equal. Therefore we have

$$W_1 \, \Delta A_1 \, \Delta t = a_1 I \, \Delta A_1 \, \Delta t, \tag{3-2}$$

and similarly for another body,

$$W_2 \, \Delta A_2 \, \Delta t = a_2 I \, \Delta A_2 \, \Delta t, \quad \text{etc.} \tag{3-3}$$

Dividing Eq. (3–2) by Eq. (3–3), we obtain

$$\frac{W_1}{W_2} = \frac{a_1}{a_2} \quad \text{or} \quad \frac{W_1}{a_1} = \frac{W_2}{a_2}, \quad \text{etc.} \tag{3-4}$$

Since the number of specks or kinds of surface has not been restricted, it becomes evident that W/a for any substance must be a constant (which may, of course, still depend on wavelength and temperature).

We have just proved that a body or surface which is a good emitter (high value of W) must be a good absorber (high value of a) and conversely. If we could find a perfect absorber, we would necessarily have found the best possible emitter, the graph of which is shown as b in Fig. 3–3.

3–8. BLACKBODY RADIATION

In acoustics, an open window is taken to be a perfect absorber of sound, since an open window reflects virtually no sound back into the room. In optics, there are few things darker than the keyhole of a windowless closet, since what little light gets into the closet bounces around against absorbing surfaces before it is redirected out the keyhole. Painting the inside of the closet black may increase the darkness of the keyhole, but the essential darkness of the hole is due to the geometry of the cavity rather than to the absorptivity of its surfaces. A small hole in a cavity of opaque material is the most perfect absorber of radiant energy man has found. Conversely, a small hole in a cavity is the most perfect emitter man has devised. We can

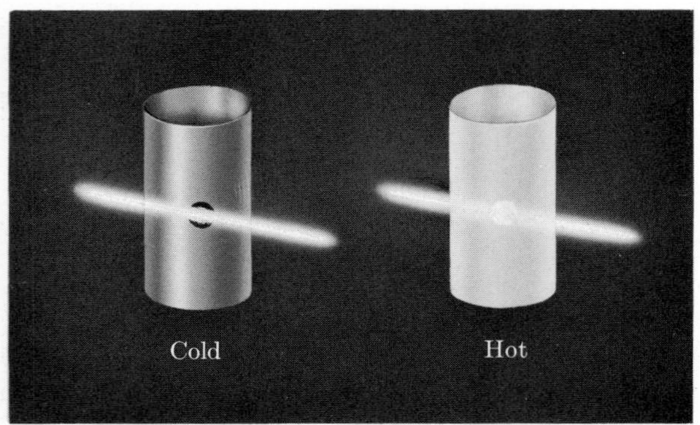

FIG. 3–4
Radiation from a hollow tungsten cylinder.

conclude this from the proof above or we can understand it more thoroughly from the following. If we look into a hole in a heated cavity, we can see the radiation from the inside wall just opposite to the hole. In addition, we see some radiation from other parts of the inside of the cavity which was directed toward the spot of wall we are looking at and is reflected to us by that spot. The absorption and emission of radiation by a hole in a hollow tungsten cylinder is shown in Fig. 3–4. The light streak across the center of the figure is an incandescent filament maintained at a constant color temperature for comparison purposes. When the cylinder is cold, the hole is darker than any other part and actually appears black, but when the cylinder is heated sufficiently, the hole is brighter than the body of the tube and matches the reference filament. Such a hollow absorber-emitter is called a blackbody. Since a is equal to unity for blackbody, from Eq. (3–4) we obtain

$$\frac{W_1}{a_1} = \frac{W_2}{a_2} = \frac{W_b}{1} = W_b. \tag{3–5}$$

This relation is called Kirchhoff's law of radiation: *The ratio of the radiant emittance of a surface to its absorptance is the same for all surfaces at a given*

TABLE 3–1

Initially emitted by M_1	Initially emitted by M_2
$\rightarrow W_1\,\Delta t$	$\leftarrow W_2\,\Delta t$
$\leftarrow (1-a_2)W_1\,\Delta t$	$\rightarrow (1-a_1)W_2\,\Delta t$
$\rightarrow (1-a_1)(1-a_2)W_1\,\Delta t$	$\leftarrow (1-a_1)(1-a_2)W_2\,\Delta t$
$\leftarrow (1-a_1)(1-a_2)^2W_1\,\Delta t$	$\rightarrow (1-a_1)^2(1-a_2)W_2\,\Delta t$
$\rightarrow (1-a_1)^2(1-a_2)^2W_1\,\Delta t$	$\leftarrow (1-a_1)^2(1-a_2)^2W_2\,\Delta t$
$\leftarrow (1-a_1)^2(1-a_2)^3W_1\,\Delta t$	$\rightarrow (1-a_1)^3(1-a_2)^2W_2\,\Delta t$
$+\cdots$	$+\cdots$

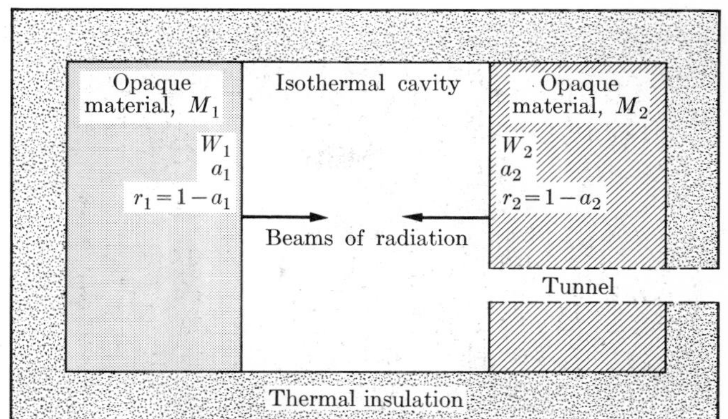

FIG. 3–5
Radiation traveling between two different materials

temperature and is equal to the radiant emittance of a blackbody at the same temperature.

We now discuss the multiple reflection situation of a radiant cavity quantitatively. Consider the radiation traveling back and forth in an isothermal cavity formed between two plane parallel sheets of different materials, as shown in Fig. 3–5. Let us now follow the history of the radiation emitted from a unit area of each face in a time interval Δt, which is just long enough to permit the radiation to travel across the space to the other face. It is reflected there with some loss of energy, returns to the first face, is reflected there with further loss, and so on. Table 3–1 gives the values and the direction of travel of the components in the radiation streams per unit area between the walls after several intervals of Δt have elapsed.

While these successive traverses of the initial radiation from each face are occurring, both faces continue to emit. Therefore, when the steady state has been reached, there are simultaneous columns of thermal radiation going back and forth in the space between the faces. The total radiation streaming in one direction, say to the right, is simply the sum of all the components in the direction $\rightarrow$. Therefore, the total effective emittance toward the right is given by

$$\begin{aligned} W_\mathrm{r}\,\Delta t = {} & W_1\,\Delta t + (1 - a_1)(1 - a_2)W_1\,\Delta t + (1 - a_1)^2(1 - a_2)^2 W_1\,\Delta t \\ & + \cdots + (1 - a_1)W_2\,\Delta t + (1 - a_1)^2(1 - a_2)W_2\,\Delta t \\ & + (1 - a_1)^3(1 - a_2)^2\,\Delta t + \cdots \end{aligned} \tag{3–6}$$

Let $x = (1 - a_1)(1 - a_2)$. Then substituting this in the preceding equation we obtain

$$W_\mathrm{r} = W_1[1 + x + x^2 + \cdots] + W_2(1 - a_1)[1 + x + x^2 + \cdots]. \tag{3–7}$$

The series within the pairs of brackets is a simple geometric progression whose limit, since $0 < x < 1$, is $1/(1 - x)$. In terms of the absorptances, this limit is

$$\frac{1}{1-x} = \frac{1}{1-(1-a_1)(1-a_2)} = \frac{1}{1-(1-a_1-a_2+a_1a_2)}$$
$$= \frac{1}{a_1+a_2-a_1a_2}. \tag{3-8}$$

Because the system is isothermal, we can obtain the following relations from Kirchhoff's radiation law (Eq. 3–5):

$$W_1 = a_1 W_b \quad \text{and} \quad W_2 = a_2 W_b. \tag{3-9}$$

Substituting the values from Eq. (3–8) and Eq. (3–9) in Eq. (3–7), we have

$$W_r = \frac{a_1W_b + a_2W_b(1-a_1)}{a_1+a_2-a_1a_2} = \frac{W_b(a_1+a_2-a_1a_2)}{a_1+a_2-a_1a_2} = W_b. \tag{3-10}$$

This equation shows that the radiation to the right (it could just as well have been in any other direction) is effectively radiated from the left surface as though from a blackbody. If a tunnel, which is so small it does not subtract a significant portion of the radiation in the cavity, is bored through the right-hand face, then *the leakage radiation will be blackbody radiation.* It is to be noted that the derivation contained no assumptions about either the nature of thermal radiation or of the kinds of surfaces inside the enclosure.

We now know how to make a blackbody and have achieved the goal we set at the end of Section 3–6.

The reader can demonstrate for himself that blackbodies can even be made from bright objects. A bundle of sewing needles held with their points directed toward the eye looks remarkably black. A pile of razor blades at least $\frac{1}{16}$-inch thick also looks black when viewed from the sharp side. In these cases the incident radiation is completely absorbed as a result of all the partial absorptions experienced at the many successive partial reflections it undergoes in traveling down into the relatively deep, narrow spaces between the needles or the blades.

We now return to the question of the spectrum of the radiation emitted by a hot body. If we take a blackbody as our sample, we can measure the emission from the hole as we did the material samples in getting the data for Fig. 3–3. This experiment shows that the emission of the blackbody gives at once the smooth solid curve of Fig. 3–3, which, unlike the other curves in the figure, is independent of the material used to make the emitter. This confirms what we might have suspected before, that the solid curve portrays a general characteristic of thermal radiation at a given temperature. A study of this curve should give information about radiation itself. With consideration of the material composing the cavity eliminated, the remaining important variable is the temperature of the radiation source. Mathematically, the total energy radiated per unit time per unit area of emitter is proportional to the area under the curve, and Stefan found empirically that this area is directly proportional to the fourth power of the absolute temperature, $W = \sigma T^4$. (Here σ is the Stefan-Boltzmann constant, not a cross section as in Chapter 1.) This is called the *Stefan-Boltzmann law* or the "fourth-power law." It was derived theoretically by Boltzmann, who used a thermodynamic argument. Wien found that as the temperature of any blackbody is changed the curve retains its general shape, but that the maximum of the curve shifts with temperature so that the wavelength of the most intense radiation is inversely proportional to the absolute temperature, or $\lambda_{max} = \text{const}/T$. This is a special case of *Wien's displacement law*, which states that at corresponding wavelengths the monochromatic energy density of the radiation in the cavity of a blackbody varies directly as the fifth power of the absolute temperature. The relation defining corresponding wavelengths at temperatures T_1 and T_2 is $\lambda_1 T_1 = \lambda_2 T_2$. The displacement law enables us to predict the entire curve at *any* temperature, given the entire curve at *one* particular temperature. Neither of these radiation laws, however, treats the basic problem of why the energy radiated from a blackbody has this particular wavelength distribution.

3-9. WIEN AND RAYLEIGH-JEANS LAWS

A comparison of the blackbody radiation curves of Fig. 3-6 and the Maxwell distribution of speeds in a gas shown in Fig. 1-2 shows a remarkable similarity. Wien noted this similarity and tried to fit a function such as Maxwell had derived for the speed distribution to the blackbody wavelength distribution. There is more than the similarity of the curves to justify this approach. If the molecules of the blackbody are thermally agitated, then their distribution of speeds may be somewhat like that derived by Maxwell.

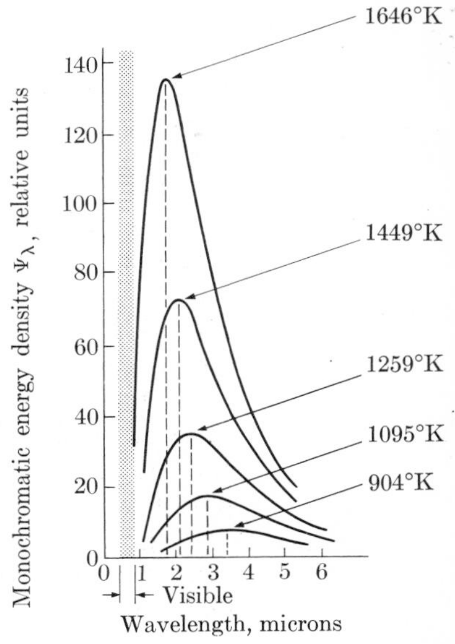

FIG. 3-6
The distribution of energy in the spectrum of the radiation from a blackbody at different temperatures.

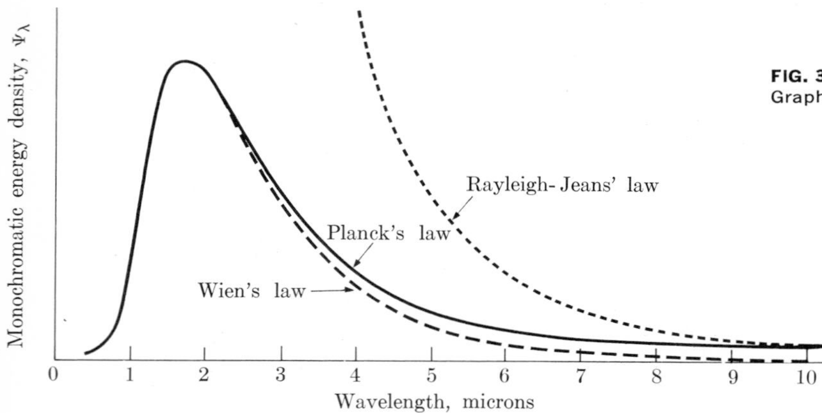

FIG. 3–7
Graphs of the radiation laws.

The accelerations of these molecules should be related to their velocities. These molecules contain charges which are therefore thermally accelerated, and we have shown that classical electrodynamics indicates that radiation results from accelerated charges. This argument is hardly rigorous, but it is a plausible explanation of the relationship between the similar curves. The expression that Wien obtained for the monochromatic energy density Ψ_λ within an isothermal blackbody enclosure in the wavelength range λ to $\lambda + d\lambda$ is

$$\Psi_\lambda = \frac{c_1 \lambda^{-5}}{e^{c_2/\lambda T}}\, d\lambda, \tag{3–11}$$

where λ is the wavelength, T is the absolute temperature, and e is the base of natural logarithms. This formula is essentially empirical and contains two adjustment constants c_1 and c_2, called the first and second radiation constants, respectively. Wien chose these constants so that the fit he obtained was rather good except at long wavelengths. The graph of Wien's law is shown dashed in Fig. 3–7. But a "pretty good fit" is not good enough, and a formula which is essentially empirical tells us nothing about the nature of radiation.

Lord Rayleigh set out to derive the radiation distribution law in a rigorous way. We shall not repeat his argument here except to mention that he concentrated attention on the radiation itself. He said that the electromagnetic radiation inside an isothermal blackbody cavity is reflected back and forth by the walls to form a system of standing waves for each frequency present. Thus these waves should resemble the standing waves on a violin string or, even more closely, the standing sound waves within an acoustic cavity. Just as a string can vibrate to produce a fundamental and a whole series of overtones, so there should be many modes of vibration present in the standing waves of radiation in the cavity space. It is a small sample of this radiation that streams out of a hole in a blackbody for spectrum analysis. Because the system is in thermal equilibrium, the radiation from the cavity absorbed by the interior walls must equal that emitted to the cavity by the atomic oscillators in the walls. Each mode of vibration introduces two degrees of freedom, one for the potential energy of the oscillator and one for its kinetic energy. This will be discussed in detail in Section 9–1. In Section 1–13 we found that the energy per degree of freedom was $\frac{1}{2}kT$, and thus each mode of vibration has a total energy kT associated with it.

We cannot reproduce here Rayleigh's involved derivation of the number of modes of vibration within the cavity. The result he obtained for the number of modes of vibration, dn_λ, per unit volume of space in the wavelength range λ to $\lambda + d\lambda$ is

$$dn_\lambda = 8\pi \frac{d\lambda}{\lambda^4}\cdot \tag{3–12}$$

If we accept this result, we need only multiply the energy per mode of vibration, kT, by the number of modes per unit volume of space within the wavelength interval $d\lambda$ to obtain the Rayleigh-Jeans law. The result is

$$\Psi_\lambda = 8\pi kT \frac{d\lambda}{\lambda^4}\cdot \tag{3–13}$$

This equation is also plotted in Fig. 3–7. At first glance, this law appears vastly inferior to Wien's. Although it fits well for long wavelengths, at short wavelengths or high frequencies it heads toward infinity in what has been dramatically called the "ultraviolet catastrophe." Theoretically, however, the Rayleigh law must be taken far more seriously than Wien's. It was derived rigorously on the basis of classical physics. It involves no arbitrary constants, and where it does fit the experimental curve, it fits exactly. Whereas the failure of the Wien law was "too bad," the failure of the Rayleigh law presented a crisis. It indicated that classical theory was unable to account for an important experimental observation. This was the situation to which Max Planck directed himself.

3-10. PLANCK'S LAW; EMISSION QUANTIZED

Planck's first step was essentially empirical. He found that by putting a mere minus one into the denominator of the Wien formula and by adjusting Wien's constants, he could get a formula which reduced to the Rayleigh formula at long wavelengths and which fitted the experimental curve everywhere. He knew that he had found a correct formula and that it should be derivable. Planck's position was a little like that of a student who has peeked at the answer in the back of the book and is now faced with the task of showing how that answer can be logically computed. Planck tried by every method he could conceive to derive this correct formula from classical physics. He was finally forced to conclude that there was no flaw in Rayleigh's derivation and that the flaw must lie in classical theory itself.

Planck had to eliminate the "ultraviolet catastrophe" which came into Rayleigh's derivation because of the assumption that the radiation standing waves had a fundamental and also an infinite number of harmonic modes of vibration. Each of these was assumed to have an average energy kT. Instead of taking the average energy per mode to be kT directly, Planck examined this matter in more detail.

The average we seek will be found by taking the sum of the products of the number of oscillators in each energy state and the energy of that state, and then dividing this sum by the total number of oscillators in all the states.

Let the energy which is associated with each of the degrees of freedom of an oscillator be some integral multiple m of a small unit of energy u. We can later make this as small as we choose and allow it to approach zero. (The method at this point is following a procedure used by Boltzmann, in 1877, to determine the distribution of kinetic energy among the molecules in a gas.) The number of oscillators having the energy mu, as given by the Boltzmann distribution law (Eq. 1–36), is

$$n_m = n_0 e^{-mu/kT}. \tag{3–14}$$

The energy contributed by the n_m oscillators is obviously

$$mun_m = mun_0 e^{-mu/kT}. \tag{3–15}$$

Therefore the average energy $\overline{w}$ of an oscillator is

$$\overline{w} = \frac{\sum_{m=0}^{\infty} mun_0 e^{-mu/kT}}{\sum_{m=0}^{\infty} n_0 e^{-mu/kT}}. \tag{3–16}$$

Since m is an integer, Eq. (3–16) becomes

$$\overline{w} = \frac{0 + ue^{-u/kT} + 2ue^{-2u/kT} + 3ue^{-3u/kT} + \cdots}{1 + e^{-u/kT} + e^{-2u/kT} + e^{-3u/kT} + \cdots}. \tag{3–17}$$

Let $x = e^{-u/kT}$. Then Eq. (3–17) can be written

$$\overline{w} = ux\frac{1 + 2x + 3x^2 + \cdots}{1 + x + x^2 + \cdots}. \tag{3–18}$$

The limits of these convergent series can be found by the usual methods (note that $x < 1$). The convergence limit of the series in the numerator is $1/(1 - x)^2$. This can be checked by expanding $(1 - x)^{-2}$ according to the binomial theorem. The denominator is a simple geometric progression converging to $1/(1 - x)$. Substituting these limits in Eq. (3–18), we have

$$\overline{w} = ux\frac{1/(1-x)^2}{1/(1-x)} = \frac{ux}{1-x} = \frac{u}{(1/x) - 1}. \tag{3–19}$$

When x is replaced by its equivalent, the result is

$$\overline{w} = \frac{u}{e^{u/kT} - 1}. \tag{3–20}$$

If we now multiply Eq. (3–20) by the number of modes of vibration in a unit volume of cavity space, from Eq. (3–12), we obtain the energy density in a wavelength range $d\lambda$,

$$\Psi_\lambda = \frac{8\pi}{\lambda^4}\frac{u}{e^{u/kT} - 1}\,d\lambda. \tag{3–21}$$

Recall that in this derivation the energy of an oscillator has been assumed to be an integer, m, times some small energy, u. Classical physics says the energy may have any value. This is equivalent to saying u may be exceedingly small and, in the limit, approach zero. If we set $u = 0$, Eq. (3–20) is indeterminate, 0/0. If we apply l'Hospital's rule, differentiating both numerator and denominator with respect to u before letting $u = 0$, we find that

$$\overline{w} = kT, \tag{3–22}$$

which is in complete agreement with Rayleigh's classical assumption. As we have seen, however, this assumption does not lead to the correct radiation law.

The relation given in Eq. (3–21) begins to look like Wien's law, Eq. (3–11), if u is not zero. Indeed, the denominators of these two equations become identical (except for the minus one) if a value of u is chosen so that the powers of the exponential terms are the same. To obtain this value of u, we let

$$\frac{c_2}{\lambda T} = \frac{u}{kT}, \tag{3–23}$$

or

$$u = \frac{c_2 k}{\lambda} = \frac{c_2 k}{c} f. \tag{3–24}$$

In this last equation, c is the free-space velocity of light and f is the frequency of the oscillator and therefore also the frequency of the radiation it emits. If we replace the constants (c_2k/c) by another constant h, we have

$$u = \frac{hc}{\lambda} = hf. \tag{3–25}$$

When the value of u from Eq. (3–25) is substituted in Eq. (3–21), we obtain Planck's law for the energy density of blackbody or cavity radiation. This law is

$$\Psi_\lambda = \frac{8\pi ch\lambda^{-5}}{e^{ch/\lambda kT} - 1}\, d\lambda. \tag{3–26}$$

This equation does agree with the experimental results. It is plotted in Fig. 3–7.

The new constant h is called the *Planck constant.* We have seen that it could be determined from Wien's constant c_2 but it can also be evaluated from the photoelectric effect discussed later in this chapter. Its value is $h = (6.6256 \pm 0.0005) \times 10^{-34}$ J · s. (Note that the units are those of angular momentum.)

Thus Planck was led to his startling, nonclassical assumption that the energy states of an oscillator must be an *integral* multiple of the product of the constant h and the frequency f of the electromagnetic radiation it emits. If E represents the smallest permissible energy change, Planck's famous quantum* equation is

$$\boxed{E = hf.} \tag{3–27}$$

Planck introduced the quantum concept in 1900, and it eventually led to the conclusion that radiation is not emitted in continuous amounts but in discrete bundles of energy each equal to hf. These bundles or packets of radiant energy are now called *quanta* or *photons*. This was the beginning of the atomic theory of radiation, which has grown to become the quantum theory. It is obvious, however, that quanta of radiation of different frequencies have different "sizes" (energies), and that they are atomic only in the sense that they are discrete. Planck thought at first that his *ad hoc*† hypothesis applied only to the oscillators and, possibly, to the emitted radiation in their immediate neighborhood and that, at most, it was a slight modification of Maxwell's theory of radiation. However, we shall see that he initiated a series of events which have changed our whole concept of the interaction of electromagnetic radiation with matter.

✓ EXAMPLE. What is the energy in a quantum of radiation having a wavelength of 5000 Å?

$$E = hf = \frac{hc}{\lambda} = 6.63 \times 10^{-34}\ \text{J}\cdot\text{s} \times 3 \times 10^{8}\ \frac{\text{m}}{\text{s}} \times \frac{1}{5000\ \text{Å}} \times \frac{1\ \text{Å}}{10^{-10}\ \text{m}}$$
$$= 3.98 \times 10^{-19}\ \text{J}.$$

---

* Quantum is the Latin word for *how much* or *how great.*

† *Ad hoc* means literally *to this,* and is used to describe a hypothesis which is applicable to but one (this) situation.

3-11. PHOTOELECTRIC EFFECT

We now turn from thermal radiation to another portion of the electromagnetic spectrum and consider an effect which is due to radiation of higher frequency. We mentioned earlier that even before the discovery of the electron Hallwachs observed that zinc irradiated with ultraviolet light lost negative charge. He proposed that somehow the radiation caused the zinc to eject negative charge. In 1899 Lenard showed that the radiation caused the metal to emit electrons.

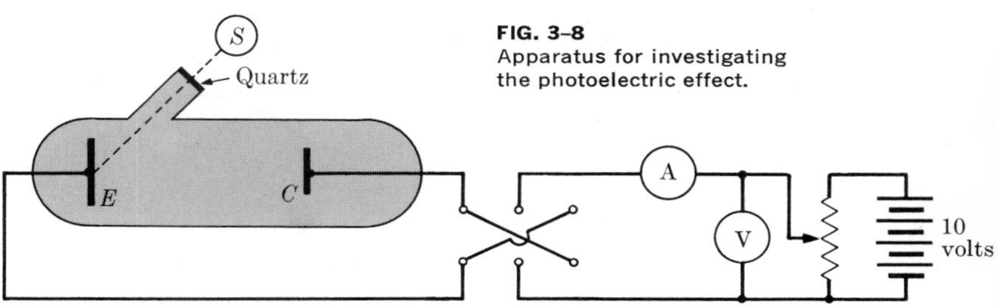

FIG. 3-8
Apparatus for investigating the photoelectric effect.

This phenomenon, called the photoelectric effect, can be studied in detail with the apparatus shown in Fig. 3-8. In this figure, S is a source of radiation of variable and known frequency f and intensity I, E is an emitting electrode of the material being studied, and C is a collecting electrode. Both electrodes are contained in an evacuated glass envelope with a quartz window that permits the passage of ultraviolet and visible light. The electric circuit allows the electrodes to be maintained at different known potentials and permits the measurement of any current between the electrodes. We first make the collecting electrode positive with respect to the emitting electrode, so that any electrons ejected will be quickly swept away from the emitter. About 10 volts is enough to do this but not enough to free electrons from the negative electrode by positive ion bombardment as was the case in the early cathode-ray tubes. If the tube is dark, no electrons are emitted and the microammeter indicates no current. If ultraviolet light is allowed to fall on the emitting electrode, electrons are liberated and the current is measured by the microammeter. It is found that the rate of electron emission is proportional to the light intensity. By holding the frequency f of the light and the accelerating potential V constant, we can obtain data like that represented in Fig. 3-9.

It is hardly surprising that if a little light liberates a few electrons, then more light liberates many. If we vary either the frequency of the light or the material irradiated, only the slope of the line changes.

We can now experiment by keeping the light intensity constant and varying the frequency. The graphs of these data are shown in Fig. 3-10, where A and B represent two different irradiated materials. The signifi-

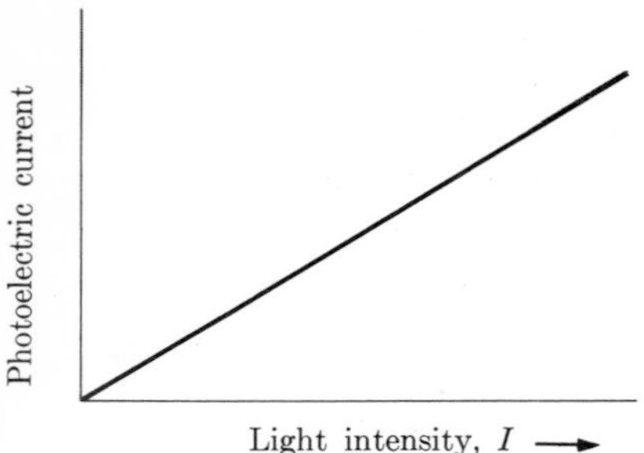

FIG. 3–9
Photoelectric current as a function of the intensity of the light. The frequency of the light and the accelerating potential are kept constant.

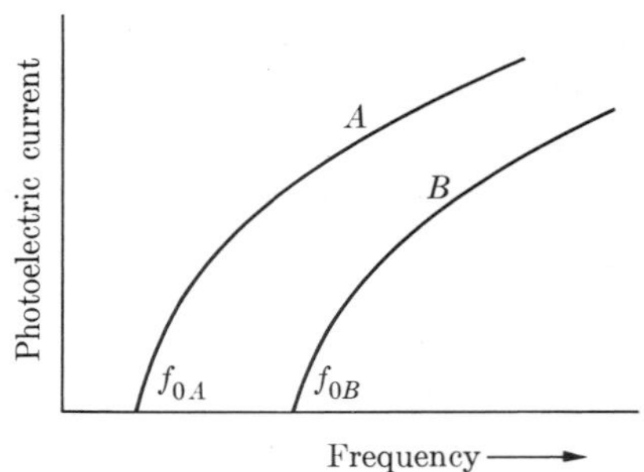

FIG. 3–10
Photoelectric current for two materials as a function of the frequency of the light. The intensity of the light and the accelerating potential are kept constant.

cant thing about these curves is that for every substance irradiated there is a limiting frequency below which no photoelectrons are produced. This frequency, called the *threshold frequency*, f_0, is a characteristic of the material irradiated. The wavelength of light corresponding to the threshold frequency is the *threshold wavelength*, λ_0. No photoelectrons are emitted for wavelengths greater than this.

The existence of a threshold frequency is difficult to explain on the basis of the wave theory of light. If we think of light as consisting of a pulsating electromagnetic field, we can imagine that that field is sometimes directed so as to tend to eject electrons from a metallic surface. We might even feel it reasonable that certain frequencies of light would resonate with the electrons of the metal so that, for a particular metal, there might be preferred light frequencies which would cause emission more efficiently. The striking thing about these data is that for each material there is a frequency below which no photoelectrons are emitted and above which they are emitted. This effect is independent of the intensity of the light.

In 1905 Einstein proposed a daring but simple explanation. He centered attention on the energy aspect of the situation. Whereas Planck had proposed that radiation was composed of energy bundles only in the neighborhood of the emitter, Einstein proposed that these energy bundles *preserve their identity throughout their life*. Instead of spreading out like water waves, Einstein conceived that the emitted energy bundle stays together, and carries an amount of energy equal to hf. For Einstein, the significance of the light frequency was not so much an indication of the frequency of a pulsating electric field as it was a measure of the energy of a bundle of light called a *photon*. His interpretation of the data of Fig. 3–10 would be that a quantum of light below the threshold frequency just does not have enough energy to remove an electron from the metal, but light above that frequency does.

The threshold frequency is dependent on the nature of the material irradiated because there is for each material a certain minimum energy necessary to liberate an electron. The *photoelectric work function* or *threshold energy*, W_0, of a material is the *minimum* energy required to free a photoelectron from that material.

In a third photoelectric experiment let us hold both the frequency and intensity of the light constant. The variable is the potential difference across the photoelectric cell. Starting with the collector at about 10 volts positive, we reduce this potential to zero and then run it negative until the photocurrent stops entirely. Curve I_1 of Fig. 3–11 shows the type of curve we might expect for this particular substance. This curve requires careful interpretation.

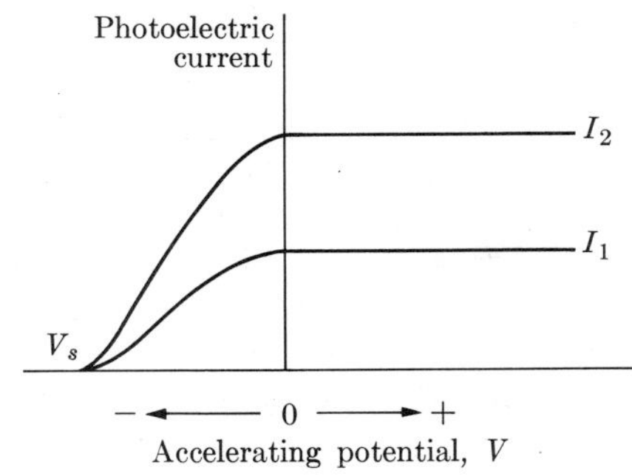

FIG. 3–11
Photoelectric current as a function of the accelerating potential for light of different intensities having a two-to-one ratio. The frequency of the light is constant.

When the potential difference across the tube is about 10 volts or more, *all* the emitted electrons travel across the tube. This stream of charges is called the saturation current, and it is obvious that an increase in the potential of the collector cannot cause an increase in current. As the accelerating potential is reduced from positive values through zero to negative values, the tube current reduces because of the applied retarding potential. Eventually this potential is large enough to stop the current completely.

The *stopping potential* V_s is the value of the retarding potential difference that is just sufficient to halt the *most energetic* photoelectron emitted. Therefore the product of the stopping potential and the electronic charge, $V_s e$, is equal to the maximum kinetic energy that an emitted electron can have. Since this stopping potential has a definite value, it indicates that the emitted electrons have a definite upper limit to their kinetic energy. Doubling the intensity of the light doubles the current at each potential, as in I_2 of Fig. 3–11, but the stopping potential is *independent* of the intensity.

If, however, the experiment is repeated with a series of different light frequencies, it is found that the stopping potential increases linearly with the frequency. This is best shown by plotting the stopping potential against the frequency, as shown in Fig. 3–12. Below the threshold frequency no electrons are emitted and the stopping potential is of course zero; however, as the frequency is increased above the threshold, the stopping potential increases linearly with the frequency.

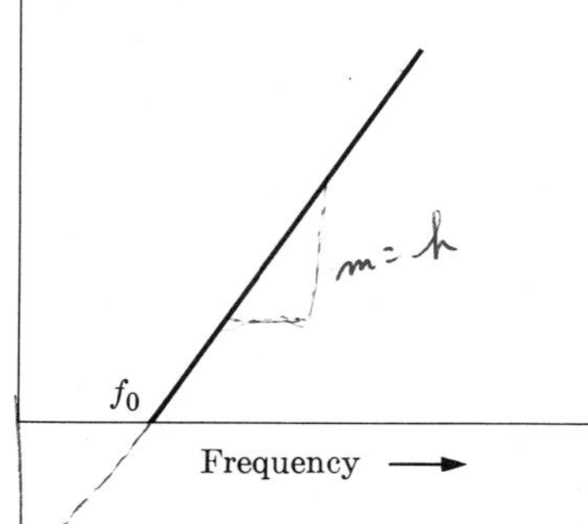

FIG. 3–12
Stopping potential as a function of the frequency of the light. Results are independent of intensity.

To see how fully the data of Fig. 3–12 confirm the Einstein photon interpretation of the photoelectric effect, we now interpret the graph of Fig. 3–12 as he would have. For light frequency between zero and the threshold frequency there are no photoelectrons produced, since the incident photons have less energy than the work function of the material. For light above the threshold frequency, photoelectrons are emitted. The energy of these emitted electrons may vary greatly. According to the Einstein view, however, there must be an upper limit to the energy of the emitted photoelectrons. No photoelectrons can have energy in excess of the energy of the incoming photon less the minimum energy to free an electron, the work function. Since the photon energy is proportional to frequency and since the stopping potential is a measure of the maximum kinetic energy of the emitted photoelectrons, the graph of stopping potential against frequency should be a straight line. This is precisely what Fig. 3–12 shows. The quantitative check of the Einstein interpretation is that the slope of the straight line provides a method of determining the way in which photon energy depends on photon frequency, which is the Planck constant.

A final and decisive experiment consists of making the irradiating light extremely weak. In this case, the number of photoelectrons is very small and special techniques are required to detect them. The significant result of this experiment is that dim light causes emission of photoelectrons which, however few, are emitted instantaneously and with the same maximum kinetic energy as for bright light of the same frequency.

According to the wave theory of light, pulsating electromagnetic fields spread out from their source. Dim light corresponds to waves of small amplitude and small energy. If dim light spreads over a surface, conservation of energy requires that either no photoelectrons should be emitted or the electrons must store energy over long periods of time before gathering enough energy to become free of the metal. The fact that high-energy photoelectrons appear immediately can be explained only by assuming that the light energy falls on the surface in concentrated bundles. According to Einstein, the dim light consists of a few photons each having energy depending only on the light frequency. This energy is not spread over the surface uniformly as required by the wave theory. A photon that is absorbed gives all its energy to one electron, and that electron will be emitted violently even though the number of such events is small.

If a man dives into a swimming pool, his energy is partly converted into waves which agitate other swimmers in the pool. If it were observed that when a man dives into a pool another swimmer is suddenly ejected from the pool onto a diving board, we would be forced to conclude that the energy provided by the diver did not spread out in an expanding wavefront, but was somehow transferred in concentrated form to the ejected swimmer. A swimming party held in such a superquantum pool would be a very odd affair compared with one in a classical pool.

To summarize Einstein's interpretation of the photoelectric effect, we equate the energy of an incident photon of frequency f to the sum of the work function of the emitter ($W_0 = hf_0$) and the maximum kinetic energy that an ejected photoelectron can acquire. We then have

$$hf = hf_0 + \tfrac{1}{2}m_e v_{\max}^2, \qquad \text{or} \qquad \tfrac{1}{2}m_e v_{\max}^2 = hf - hf_0,$$

or

$$\boxed{V_s e = hf - W_0.} \tag{3-28}$$

This is a linear equation. It is the equation of the graph in Fig. 3–12. It is evident that the slope of the curve V_s plotted as a function of f is equal to h/e, that f_0 is the f-axis intercept, and that W_0/e is the intercept of the extrapolated curve on the V_s-axis. Note that if $f < f_0$, v is imaginary. The physical meaning is that photoelectric emission is then impossible. Einstein's photoelectric equation was first verified with precision in 1916, eleven years after it was proposed, by Millikan, who made careful measurements of the photoemission from many different substances.

3–12. SUMMARY OF THE ATOMIC VIEW OF RADIATION

We introduced this chapter by outlining the disagreement between Newton and Huygens over the nature of light. We described how Maxwell strengthened Huygens' wave theory when he showed that electromagnetic waves were a consequence of the laws of electricity and magnetism. We reported that Hertz demonstrated that electric circuits could be made to produce the electromagnetic waves Maxwell predicted. But we also mentioned that Hertz observed that he could produce sparks more easily when his spark gap was illuminated. Thus Hertz's work, which supported Maxwell's theory, also contained the first observation of the photoelectric effect. Although the wave theory of light accounts beautifully for many optical phenomena, it fails to account for either the blackbody radiation or the photoelectric effect, where light appears to possess marked particle aspects. It is hardly satisfactory to regard light as a wave motion part of the time and a particle phenomenon at other times. We shall return to the question of the resolution of this conflict of viewpoint in Chapter 7. At this point it is clear, however, that the resolution of this paradox can never eliminate the idea that light is emitted and absorbed in bundles of energy called photons, and that now radiation must join matter and electricity in having a basically atomic character. The fact that radiant energy is quantized is a radical departure from classical physics and will require us to re-examine the whole energy concept from the quantum point of view. Establishing this fact has been the main business of this chapter. Before closing the chapter, however, we shall consider two related topics.

3–13. THE ELECTRON VOLT

In the Einstein equation, we measured the maximum kinetic energy of the emitted electrons by noting the potential energy difference (eV_s), which was equivalent to the electron kinetic energy. This method of determining and expressing electron energies is a particularly convenient one, and it suggests a new unit of energy. This new unit of energy is called the *electron volt*, eV, which is defined as the amount of energy equal to the change in energy of one electronic charge when it moves through a potential difference of one volt. Since the electron volt is an energy unit, it is in the same category as the foot-pound, the British thermal unit, and the kilowatt-hour.

Energies in joules can be converted to electron volts by dividing by $e_c = 1.60 \times 10^{-19}$. In this case e_c is *not a charge* but a *conversion factor* having the units of joules per electron volt. The Einstein equation, Eq. (3–28), is valid in any consistent system of units. If we choose mks units and divide Eq. (3–28) by the factor e_c, we obtain the same relation in electron volts:

$$E_{k(\max)}\ (\textit{numerically}\ \text{equal to}\ V_s) = \frac{hf}{e_c} - \frac{W_0}{e_c}. \qquad (3\text{–}29)$$

In words, this equation states that the maximum kinetic energy of a photoelectron in electron volts equals the energy of the photon in electron volts minus the work function in electron volts.

When the electron volt is too small a unit, it is convenient to use 10^3 eV = 1 keV (kilo-electron-volt), and 10^6 eV = 1 MeV (million-electron-volt). Other prefixes used to form multiples of units are listed in Appendix 7.

It is also useful to express photon energies in electron volts. In terms of photon frequency,

$$E = \frac{hf}{e} = 6.63 \times 10^{-34}\ \mathrm{J \cdot s} \times \frac{1\ \mathrm{eV}}{1.60 \times 10^{-19}\ \mathrm{J}} \times f$$
$$= 4.14 \times 10^{-15}\, f\ (\mathrm{s \cdot eV}). \tag{3–30}$$

Or, since $f = c/\lambda$, we get

$$E = 4.14 \times 10^{-15}\ \mathrm{s \cdot eV} \times 3 \times 10^8\ \frac{\mathrm{m}}{\mathrm{s}} \times \frac{1}{\lambda}$$
$$= \frac{1.24 \times 10^{-6}}{\lambda}\ (\mathrm{m \cdot eV}). \tag{3–31}$$

If the wavelength is expressed in angstroms instead of meters, we have

$$E = \frac{1.24 \times 10^{-6}}{\lambda}\ (\mathrm{m \cdot eV}) \times \frac{1\ \text{Å}}{10^{-10}\ \mathrm{m}},$$

or

$$\boxed{E = \frac{1.24 \times 10^{4}}{\lambda}\ (\mathrm{eV} \cdot \text{Å}).} \tag{3–32}$$

We shall use Eq. (3–32) *frequently.*

EXAMPLE. Light having a wavelength of 5000 Å falls on a material having a photoelectric work function of 1.90 eV. Find (a) the energy of the photon in eV, (b) the kinetic energy of the most energetic photoelectron in eV and in joules, and (c) the stopping potential.

Solution. (a) From Eq. (3–32),

$$E = \frac{1.24 \times 10^4\ \mathrm{eV} \cdot \text{Å}}{\lambda} = \frac{12400\ \mathrm{eV} \cdot \text{Å}}{5000\ \text{Å}} = 2.47\ \mathrm{eV}.$$

(b) The law of conservation of energy gives

Maximum kinetic energy = photon energy − work function

or

$$E_k = 2.47\ \mathrm{eV} - 1.90\ \mathrm{eV} = 0.57\ \mathrm{eV}.$$

Also

$$E_k = 0.57\ \mathrm{eV} \times \frac{1.60 \times 10^{-19}\ \mathrm{J}}{1\ \mathrm{eV}} = 9.11 \times 10^{-20}\ \mathrm{J}.$$

(c)

$$V_s = \frac{0.57\ \mathrm{eV}}{1\ \text{electronic charge}} = 0.57\ \mathrm{V}.$$

3-14. THERMIONIC EMISSION

We have already considered two ways in which electrons can be released from a metal. In Chapter 2 we discussed the discharge of electricity through gases. The electrons that participate in cold-cathode discharges are obtained from the cathode while it is bombarded by the positive ions produced in the residual gas in the tube. In this chapter we have considered another emission process, called photoelectric emission. There is still another kind of electron emission, which we shall now discuss briefly.

If a metal is heated, the thermal agitation of the matter may give electrons enough energy to exceed the work function of the material. Thus the space around a heated metal is found to contain many electrons. A study of this effect shows that the *thermionic work function* is very nearly the same as the photoelectric work function—a most satisfying result.

Since a fine wire can be heated easily by passing an electric current through it, thermionic emission is one of the most convenient electron sources. Most radio-type vacuum tubes use a heated cathode as their electron source. Thermal emission of negative charges from a hot wire in a vacuum was first observed by Edison in 1883 when he was making incandescent lamps, and such thermal emission is called the *Edison effect*. In 1899, J. J. Thomson showed that the thermions in this effect are electrons.

REFERENCES

ARONS, A. B., and M. B. PEPPARD, "Einstein's Proposal of the Photon Concept —A Translation of the *Annalen der Physik* Paper of 1905," *Am. J. Phy.* **33,** 367 (1965). This is the only English translation of the famous paper proposing the concept of photons and the photoelectric law.

GAMOW, GEORGE, *Thirty Years that Shook Physics.* Garden City, N. Y.: Doubleday, 1966. A lively account of the 20th century revolution in physics.

HERTZ, HEINRICH, *Electric Waves.* D. E. Jones, translator. London: Macmillan, 1893. Read pages 68–69 for the account of how Hertz's careful observations revealed an unexpected phenomenon and led to the "accidental" discovery of the photoelectric effect.

HUGHES, A. L., and L. A. DUBRIDGE, *Photoelectric Phenomena.* New York: McGraw-Hill, 1932. A comprehensive treatise.

KLEIN, M. J., *The Natural Philosopher,* Vol. 2. New York: Blaisdell, 1963. This contains a critical analysis and the historical background of Einstein's first paper on photons.

MAGIE, WILLIAM F., *A Source Book in Physics.* Cambridge, Mass.: Harvard University Press, 1963. This contains reprints of the following papers: "Electric Discharge by Light," by Wilhelm Hallwachs; "Temperature Radiation," by Josef Stefan.

Richtmyer, F. K., E. H. Kennard, and T. Lauristen, *Introduction to Modern Physics*. 5th ed. New York: McGraw-Hill, 1955. Chapter 1, brief history of the evolution of physics; Chapter 3, photoelectric and thermionic effects; and Chapter 4, origin of the quantum theory.

Simon, Ivan, *Infrared Radiation*, Momentum Book No. 12. Princeton, N. J.: Van Nostrand, 1966.

PROBLEMS

3–1. The cavity of a blackbody radiator is in the shape of a cube measuring 2 cm on a side and the blackbody is at a temperature of 1500°K. (a) Calculate the number of modes of vibration per unit volume in the cavity in the wavelength band between 4995 Å and 5005 Å. (b) What is the total radiant energy in the cavity in this wavelength band?

3–2. Show that the area under the Planck radiation curve is proportional to the fourth power of the absolute temperature.

3–3. At the surface of the earth a 1-cm^2 area oriented at right angles to the sun's rays receives about 0.13 J of radiant energy each second. Assume that the sun is a blackbody radiator. What is the surface temperature of the sun? (The radius of the sun is about 7×10^8 m and the earth is about 1.49×10^8 km from the sun.)

3–4. A tungsten sphere 0.5 cm in radius is suspended within a large evacuated enclosure whose walls are at 300°K. Tungsten is not a blackbody but has an average emissive power that is 0.35 that of a blackbody. What power input is required to maintain the sphere at a temperature of 3000°K, if heat conduction along the supports is neglected?

3–5. Find the percent change in the total energy radiated per unit time by a blackbody if the temperature of the blackbody is increased by (a) 100%, (b) 10%, (c) 1%, (d) 0.1%.

3–6. Show that the Planck radiation law is consistent with the special case of Wien's displacement law, as discussed in the text material.

3–7. At what wavelength does the maximum intensity of the radiation from a blackbody occur at a temperature of 300°K? 1000°K? 6000°K?

3–8. Predict from the curves in Fig. 3–6 the sequence of colors which would be seen if a piece of iron were heated from 20°C to 2500°C in a dark room.

3–9. Radiant energy with wavelengths longer than visible radiation (infrared region) is often called "heat radiation." (a) Can shorter wavelengths heat a body on which the radiation falls? (b) To what extent is it proper to call infrared rays heat rays? (See Fig. 3–6.)

3–10. If 5% of the energy supplied to an incandescent light bulb is radiated as visible light, how many visible quanta are emitted per second by a 100-watt bulb? Assume the wavelength of all the visible light to be 5600 Å.

3–11. In order that an object be visible to the naked eye, the intensity of light entering the eye from the object must be at least 1.5×10^{-11} J/m$^2 \cdot$ s. What is the minimum rate at which photons must enter the eye so that an object is visible, given that the diameter of the pupil is 0.7 cm? Assume a wavelength of 5600 Å.

3–12. The directions of emission of photons from a source of radiation are random. According to the wave theory, the intensity of radiation from a point source varies inversely as the square of the distance from the source. Show that the number of photons from a point source passing out through a unit area is also given by an inverse square law.

3–13. Show that Planck's radiation law, Eq. (3–26), reduces to Wien's law for short wavelengths, and to the Rayleigh-Jeans' law for long ones. [*Hint:* Express the exponential term as a series to obtain the second of these laws.]

3–14. Show that the energy density of blackbody radiation expressed in terms of frequency is

$$\Psi_f = \frac{8\pi h f^3}{c^3(e^{hf/kT} - 1)}\, df.$$

3–15. Since it has been shown that matter, electricity, and radiant energy are discrete or "atomic" in character, is it strictly correct to write their differentials dm, dq, and dE, respectively? Explain.

3–16. What is the energy in eV of a photon having a wavelength of 912 Å?

3–17. Show that Planck's constant h has the same physical units as angular momentum.

3–18. A mass of 10 g hangs from a spring with a force constant of 25 N/m. Assume this oscillator is quantized just as the radiation oscillators are. (a) What is the minimum energy that can be supplied to the mass? (b) If the mass at rest absorbs the minimum energy of part (a), what is the resulting amplitude? (c) How many quanta must the mass absorb in order to have an amplitude of 10 cm?

3–19. Particles of a certain system can have energies of E, $2E$, or $3E$, where $E = 0.025$ eV. (a) What are the ratios of the number of particles in each of the upper states to the number of particles in the lowest state when the system is in equilibrium at 290°K? (b) What is the average energy of a particle in the equilibrium distribution of these states?

3–20. (a) In Problem 1–34 it was found that a thermal neutron has a velocity of 2200 m/s. What is a thermal neutron's kinetic energy in eV? (b) What is the increase in kinetic energy in eV of each water molecule in a stream which goes down a 450-ft (137.1 m) waterfall?

3–21. The fissioning of a U^{235} atom yields 200 MeV of energy. How many such atoms must fission to provide an amount of energy equal to that required to lift a mosquito one inch? The mass of an average mosquito is 0.90 mg.

3–22. (a) Calculate the energy released in eV per atom or molecule in the complete combustion of each of the compounds listed at the top of page 87. (b) The fissioning of a U^{235} atom releases 200 MeV of energy. What is the ratio of this energy to that released per molecule of the compounds in the first part of this problem? (c) What is the ratio of the energy released per unit mass of uranium to the energy released per unit mass of each of the compounds?

Compound	Atomic or molecular weight	Heat of combustion
1. Coal	12 (carbon)	14,000 Btu/lb (7,780 kcal/kg)
2. Ethyl alcohol	46	327.6 kcal/mole
3. Starch	162 (assumed)	4,179 kcal/kg
4. Trinitrotoluene (TNT)	227	821 kcal/mole
5. Gasoline	100 (n-heptane)	20,750 Btu/lb (11,530 kcal/kg)

3–23. At 200 MeV per fission per atom of U^{235} find, on the basis of the energy released in complete combustion (data given in Problem 3–22), (a) the number of grams of uranium needed to furnish a megawatt-day (24 h) of energy, (b) the number of tons of coal equivalent to 1 lb of uranium, and (c) the number of pounds of uranium which equal the blast effect of a megaton of TNT. [*Note:* Only about 30% of the energy of combustion of TNT goes into its blast or explosive effect. Assume that the blast efficiency of the uranium is 100%.] It is very useful in nuclear engineering to remember the relations or conversion factors calculated in parts (a) and (b).

3–24. The visible light from a 40-watt incandescent bulb is incident normally on a potassium surface 50 cm from the bulb. (a) How long will it take a potassium atom to absorb 2.0 eV of energy which is its photoelectric work function? Consider the bulb a point source which radiates 7.5% of the input power as visible light. Assume that the absorbing area of the potassium atom is equivalent to a circular disk having a diameter of 5.0 Å. (b) How long would it take the potassium atom to absorb 2.0 eV of energy if it is illuminated by full moonlight? The illumination at 50 cm from a 40-watt bulb is about 740 times greater than full moonlight.

3–25. The light-sensitive compound on most photographic films is silver bromide, AgBr. We will assume that a film is exposed when the light energy absorbed dissociates this molecule into its atoms. (The actual process is more complex, but the quantitative result does not differ greatly.)

The energy of or heat of dissociation of AgBr is 23.9 kcal/mole. Find (a) the energy in eV, (b) the wavelength, and (c) the frequency of the photon which is just able to dissociate a molecule of silver bromide. (d) What is the energy in eV of a quantum of radiation having a frequency of 100 MHz? (e) Explain the fact that light from a firefly can expose a photographic film, whereas the radiation from a TV station transmitting 50,000 watts at 100 MHz cannot. (f) Will photographic films stored in a light-tight container be ruined (exposed) by the radio waves constantly passing through them? Explain. (g) After a class in which this problem was discussed one student came to the instructor to remark "But I have photographed TV towers." Comment!

3–26. When a certain photoelectric surface is illuminated with light of different wavelengths, the following stopping potentials are observed:

λ, A	3660	4050	4360	4920	5460	5790
V_s, V	1.48	1.15	0.93	0.62	0.36	0.24

Plot the stopping potential as ordinate against the frequency of the light as abscissa. Determine (a) the threshold frequency, (b) the threshold wavelength, (c) the photoelectric work function of the material, and (d) the value of the Planck constant h (the value of e being known).

3–27. The photoelectric work function of potassium is 2.0 eV. If light having a wavelength of 3600 Å falls on potassium, find (a) the stopping potential, (b) the kinetic energy in eV of the most energetic electrons ejected, and (c) the velocities of these electrons.

3–28. What will be the change in the stopping potential for photoelectrons emitted from a surface if the wavelength of the incident light is reduced from 4000 Å to 3980 Å? (Assume that the decrease in wavelength is so small that it may be considered a differential.)

3–29. The threshold wavelength for photoelectric emission from a certain material is 6525 Å. Find the stopping potential when the material is irradiated with (a) light having a wavelength of 4000 Å, and (b) with light having twice the frequency and three times the intensity of that in the previous part. (c) If a material having double the work function were used, what would then be the answers to parts (a) and (b)?

3–30. Light of wavelength 4000 Å liberates photoelectrons from a certain metal. The photoelectrons now enter a uniform magnetic field having an induction of 10^{-4} T. The electrons move normal to the field lines so that they travel circular paths. The largest circular path has a radius of 5.14 cm. Find the work function for the metal.

3–31. A surface is irradiated with monochromatic light of variable wavelength. Above a wavelength of 5000 Å, no photoelectrons are emitted from the surface. With an unknown wavelength, a stopping potential of 3 V is necessary to eliminate the photoelectric current. What is the unknown wavelength?

3–32. In a certain radio tube the electron current from the tungsten cathode to the plate is 200 mA. The thermionic work function of tungsten is 4.5 eV. (a) At what rate must energy be supplied, in watts, to get the electrons outside of the cathode? (b) Assuming a Maxwellian distribution of the velocities of the electron "gas" in tungsten, what must the temperature of the cathode be if the energy of the electron with the most probable velocity is to equal 4.5 eV? Does the answer seem reasonable? (c) What assumptions were made about the molecules of a gas when the kinetic theory equations were derived in Chapter 1? To what extent are these fulfilled by the electron "gas" in this problem?

CHAPTER 4

THE ATOMIC MODELS OF RUTHERFORD AND BOHR

4–1. INTRODUCTION

We have traced how matter, electricity, and radiation came to be regarded as atomic in character. We have established the existence of some elementary particles that are more fundamental than the chemical elements. Electrons, for example, are common to all elements and are a common building block of all matter. Our discussion of positive rays and mass spectroscopy showed that matter also has positive constituents which are much more massive than electrons. Thomson, who made the first quantitative measurements on electrons and positive rays, assumed that a normal chemical atom consists of a mixture of constituents. This mixture came to be called the "plum-pudding" atomic model: the atom was regarded as a heavy positive sphere of charge seasoned with enough electron plums to make it electrically neutral.

4–2. THE RUTHERFORD NUCLEAR ATOM

A very different atomic model was indicated by an experiment performed by Rutherford in 1911.

We shall discuss radioactivity at some length in Chapter 10, but in order to be able to discuss the Rutherford experiments a few observations need to be made now. Certain atoms are unstable and fly apart of their own accord. The nature of these disintegrations depends on the element that is disintegrating, but in every case the fragments ejected consist either of electrons, here called beta rays, or of doubly ionized helium atoms, called alpha particles. These disintegration fragments usually are ejected with high energies from a radioactive substance and are often accompanied by very short-wavelength photon radiation called gamma rays. Radium, for example, is an excellent source of high-energy alpha particles. These alpha particles can travel through a few centimeters of air before they are stopped, and in a vacuum they travel long distances without losing energy. When they strike certain materials, they cause visible fluorescent light flashes.*

Rutherford studied how these alpha particles from radium were absorbed by matter. He found they were absorbed by sheets of metal a few hundredths of a millimeter thick, but that they could readily pass through gold foil several ten-thousandths of a millimeter thick. Rutherford's apparatus is shown schematically in Fig. 4–1. Radium was placed in a cavity at the end of a narrow tunnel in a lead block. Alpha particles were emitted in random directions by this source and the lead absorbed all except those emitted along the axis of the tunnel. In this way Rutherford obtained a collimated beam of alpha particles which streamed toward the gold foil. Particles that went through the foil produced flashes on the fluorescent screen.

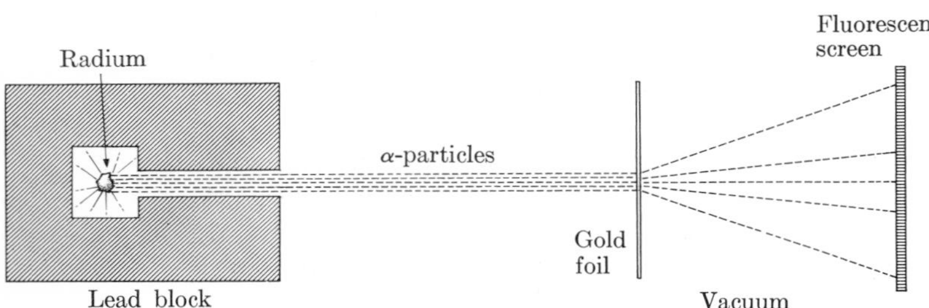

FIG. 4–1 Schematic diagram of Rutherford's alpha-particle scattering apparatus.

Many of the alpha particles did go straight through the foil or were deflected only by very small amounts. Amazingly, however, some alpha particles were deflected through very large angles. A few even returned to the side of the gold foil from which they came. Rutherford's astonishment at this is evident in his comment, "It was quite the most incredible event that has ever happened to me in my life. It was almost as incredible as if you fired a 15-inch shell at a piece of tissue paper and it came back and

* Radiolight watch dials are painted with a paste containing fluorescent material and a trace of radium. Under a microscope the glow of the dial can be seen to be a multitide of flashes, called scintillations, which remind one of the twinkling of the stars on a summer night. The effect may be seen through a 4-power magnifier, but it is better to use two or three times this magnification. The light is less likely to appear continuous if there is very little radioactive material on the dial. Observations must be made in a completely darkened room and it may be necessary to wait about five minutes for the eyes to become dark-adapted. This time delay will also permit any phosphorescence to fade and eventually die out. Both fluorescence and phosphorescence will be discussed in more detail in Section 4–11.

hit you." This observed scattering through large angles was contrary to predictions based on the Thomson model of the atom. Only small deflections were expected for the following reasons. The effect of electrons can be neglected. Although there are electrostatic forces between the electrons and the alpha particles, it is the electrons, rather than the much more massive alpha particles, that are appreciably deflected. The deflection of the alpha particle is small, even when it passes near or through a positive sphere of charge. The particle experiences only a small resultant force because it is repelled in many directions by other nearby charges. The haphazard deflections of an alpha particle as it passes through the foil nearly cancel, and therefore the net deviation is nearly zero. Thus the plum-pudding model provides no mechanism to account for large deflections. Rutherford replaced the plum-pudding model by another which correctly explained his experimental results. He assumed that the positive, massive part of the atom was concentrated in a very small volume at the center of the atom. This core, now called the *nucleus*, is surrounded by a cloud of electrons, which makes the entire atom electrically neutral. This model accounts for alpha-particle scattering in the following way. The forces between electrons and the alpha particles may be neglected as before, but now an alpha particle passing near the center of an atom is subject to an increasingly large Coulomb repulsion as the separation of the two particles decreases. Because the atom is mostly empty space, many of the alpha particles go through the foil with practically no deviation. But an alpha particle that passes close to a nucleus experiences a very large force exerted by the massive, positive core, and is deflected through a large angle in a single encounter.

The significant interaction, then, was between the doubly charged alpha particle ($+2e$) and the positive core having an integral number Z of positive charges of electronic magnitude ($+Ze$.) Rutherford assumed that these two charges acted on each other with a coulomb force which in this case was repulsive. The equation for this force in rationalized units is

$$F = \frac{1}{4\pi\epsilon_0}\frac{2e \cdot Ze}{r^2} = \frac{2Ze^2}{4\pi\epsilon_0 r^2}, \tag{4-1}$$

where ϵ_0 is the permittivity of free space ($1/4\pi\epsilon_0 = 9 \times 10^9$ MKSA units).

This force is inversely proportional to the square of the distance between the bodies. In advanced mechanics it is shown that such a force always results in an orbit that is a conic section. When the force is attractive, like the gravitational force between the sun and planets of the solar system, the orbits may be parabolas, ellipses, hyperbolas, or, in special cases, circles or straight lines. When the inverse square force is repulsive, however, the conic section must be the far branch of a hyperbola or its degenerate form, a straight line.

Of course the alpha particle and the atom core of the scattering metallic foil both experience the force given by Eq. (4–1). But since the gold atom is much more massive than the alpha particle, we shall assume that the gold atom remains fixed.

Consider, now, an alpha particle aimed directly at the core of an atom of a metal. The repelling force tends to slow the alpha particle, and since this force becomes greater as the particle nears the core, the alpha particle must finally stop. After this momentary pause, the alpha particle will be accelerated away from the atom along the same line by which it approached. In this case the path of the particle is a straight line. Obviously, this case is rare, since it requires that the alpha particle move directly toward the atom core.

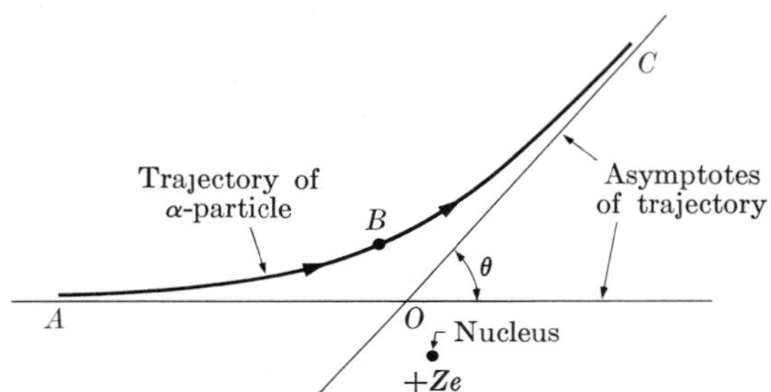

FIG. 4–2
Scattering of an alpha particle by the Rutherford nuclear atom.

In general, the alpha particle will not be perfectly aimed. If an alpha particle is aimed so as to pass near the core (Fig. 4–2), then it will begin its approach at A, moving parallel to a line which is one asymptote of the hyperbolic orbit. Because the repulsive force is radial whereas the particle's motion is not, the alpha particle is forced to move away from this asymptote of its trajectory. At its point of nearest approach, B, the velocity of the alpha particle is entirely tangential, and beyond this point the path straightens out, becoming parallel to another asymptote through C. The alpha particle is scattered through the angle θ between these two asymptotes. In the rare case of perfect aim, θ is 180°, whereas for the most common case of a wide miss, it is 0°. Between these extremes the angle of scattering depends on the initial speed of the alpha particle, the charge on the core, and the aiming of the alpha particle.

In a given experimental situation the first two of these variables are constant, while the third is impossible to measure or control. But the very randomness of the aim makes it possible to treat this variable statistically, and Rutherford could compute the angular distribution of the alpha particles emerging from the foil.

In 1913, Geiger and Marsden carried out an exhaustive test of the Rutherford theory, using gold and silver foils. The agreement between the theory and their experimental results was excellent. Their technique was not very sensitive to the core charge Z, but they estimated that it was about half the atomic weight. Later, Chadwick made more refined measurements with copper, silver, and platinum foils. He showed that the integer Z is very near the atomic number. In Chapter 6 we shall discuss the work of Moseley, who showed that the number of core charges is *exactly* the atomic number.

The Rutherford theory assumes that alpha particles are scattered by stationary point charges. This theory also gives a way of estimating the size of the "point." The required equation for the special case of an alpha particle aimed directly at the nucleus is readily derived from the law of conservation of energy. Let M be the mass of the alpha particle, $+2e$ its charge, and v its initial velocity. At its distance of closest approach d to the core charge $+Ze$, the particle is momentarily at rest, so that all of its kinetic energy must then have been converted into electrical potential energy. Thus we have

$$\tfrac{1}{2}Mv^2 = \frac{1}{4\pi\epsilon_0}\frac{2e \cdot Ze}{d}, \tag{4–2}$$

or

$$d = \frac{Ze^2}{\pi\epsilon_0 M v^2}. \tag{4–3}$$

For the metals studied, the "radius" of the nucleus, d, is about 10^{-14} m. This is far less than the radius of the space the atom occupies as computed from the density of the metal and the Avogadro constant or from the kinetic theory of matter. This atomic "radius" is about 10^{-10} m. Furthermore, the distance of nearest approach is only an upper limit on the size of the core.

Since aluminum is a light metal with a rather low atomic number, aluminum should repel alpha particles less readily than the heavier metals do. Equation (4–3) shows that for smaller Z the distance of nearest approach is less. For aluminum, d can be as small as 0.8×10^{-14} m. But it has been found that the Rutherford formulas do not fit perfectly for aluminum. The deviations can be explained by assuming that at very small distances from the force center, the force of repulsion is actually less than that computed from Coulomb's law. This suggests that a very short-range attractive force is competing with the repulsive electric force. This force cannot be gravitational. Not only is the gravitational force many magnitudes too small, but also it is an inverse square force which varies with distance, exactly as the Coulomb force does. This new force must depend on distance more strongly than does an inverse square force. The size of the region where this new short-range force is significant, compared with the Coulomb force, can be investigated with very high-energy particles. When such particles are sent through thin foils, the analysis of the distribution of the scattering angles leads to the discovery that the radius R of the nucleus (assumed spherical) is proportional to the cube root of the mass number of the scattering atoms. The relation is:

$$R = R_0 A^{1/3}, \tag{4-4}$$

where R_0 is equal to about 1.4×10^{-15} m. The value of R_0 is somewhat different for different incident particles—alpha particles, protons, or electrons—but it is always of the order of 10^{-15} m. A unit of length that is sometimes used in giving nuclear dimensions is the *fermi* f which is 10^{-15} meters. Equation (4–4) shows that the cube of the nuclear radius is proportional to A; thus the nuclear volume is proportional to the mass number. Therefore, since the density of a substance is its mass divided by its volume, the density of the nucleus is about the same for all atoms.

The success of Rutherford's theory of alpha-particle scattering gives him the distinction of having discovered the *atomic nucleus*. He probed matter with alpha particles and found it mostly empty space. He found the most massive part of the atom concentrated in a region of density about 10^{17} kg/m^3 (relative density about 10^{14}). This massive part contains positive particles which must repel one another with large electric forces. But nuclear stability and the fact that the Coulomb law is not accurate for short distances indicate that there is some entirely *new short-range force* that binds nuclear particles together. The Rutherford model raised many questions about how atoms are held apart and how solid bodies can retain their rigid structure. The answers to these questions come from the structure of the electronic "mist" outside the nucleus. To study this electron arrangement, we next direct our attention to atomic spectra.

4-3. SPECTRA

Most readers of this book have studied light and know that spectrographs are instruments which analyze light according to its distribution of frequency or color. These instruments always have an entrance slit and lens, a dispersive component which may be a prism or grating, and an optical system that forms an image of the slit on a detector that is usually a photographic plate. The instrument forms an image of the slit for each frequency of light present, so that light which is continuous in its frequency distribution forms a wide image which is a continuous succession of slit images. Light which is discontinuous in frequency distribution forms a discrete set of slit images that are called spectral lines.

In the earlier study of light, attention was particularly directed toward the theory of operation of spectrographs, but it was pointed out at that time that the light from any element in gaseous form produces a discontinuous line spectrum. Each element has its own characteristic frequency distribution or spectrum, so that each element can be identified by the light that it emits. The most dramatic instance of such identification occurred when the element helium was "discovered" in the spectrum of the sun before it was chemically isolated here on earth.

Our present interest in spectroscopy goes far deeper than an interest in instruments or a technique for analysis. Our concern lies in what the light emitted by an element can tell about the structure of that element. Our situation is like that of a man from Mars who might attempt to learn the structure of a piano by analyzing the sounds it can make.

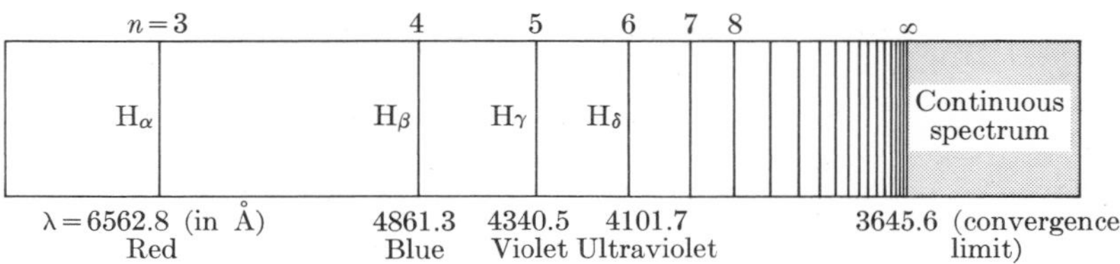

FIG. 4–3
Diagram of the Balmer series of atomic hydrogen. (The wavelengths are the values in air.)

4–4. THE HYDROGEN SPECTRUM

The obvious place to start the study of spectra is with the spectrum of hydrogen. It is not surprising that this lightest element has the simplest spectrum and probably the simplest structure. The hydrogen spectrum is shown in Fig. 4–3. The regularity of the spectral lines is immediately evident, and it appears obvious that there is some interrelationship among them. In 1884, a Swiss high-school mathematics teacher by the name of Balmer took the wavelengths of these lines as a problem in numbers. He set out to find a formula which would show their interrelation. He hit upon a formula which could be made to give these wavelengths very precisely. The Balmer formula is

$$\lambda \text{ (angstroms)} = \frac{3645.6n^2}{n^2 - 4}. \tag{4–5}$$

Each different wavelength is obtained by putting into the formula different values of the running integers n, which are $n = 3$, $n = 4$, $n = 5$, etc.

The success of the Balmer formula led Rydberg to attempt a formulation which would apply to heavier elements. He proposed an equation of the form

$$\bar{f} = \frac{1}{\lambda} = A - \frac{R}{(n + \alpha)^2}, \tag{4–6}$$

where $\bar{f}$ is the *wave number*,* R is the *Rydberg constant* which is equal to $1.09737 \times 10^7\ \text{m}^{-1}$, and n is a running integer. A and α are adjustment constants which depend on the element and the part of the spectrum or spectral series to which the formula is applied. Rydberg found that this formula, which can be regarded as a generalization of the Balmer formula, could be fitted to many spectral series, and further that the value of R was nearly the same when the formula was applied to different elements.

In 1908, Ritz noted that the wave numbers of many spectral lines are the differences between the wave numbers of other spectral lines, and that the A term of the Rydberg formula was really a particular value of a term, like the second term of the Rydberg formula. Using this "combination principle," Ritz rewrote the Rydberg formula as

$$\bar{f} = \frac{R}{(m + \beta)^2} - \frac{R}{(n + \alpha)^2}, \tag{4–7}$$

where α and β are adjustment constants which depend on the element. For different spectral series of a given element, m takes on different integral values. The different lines within a series are computed by changing the running integer n. It is easily shown that $\alpha = \beta = 0$ and $m = 2$, Eq. (4–7) reduces to the Balmer formula for hydrogen.

In the same year, 1908, Paschen found another hydrogen series of lines in the infrared region to which Eq. (4–7) could be fitted by making $\alpha = \beta = 0$, $m = 3$, and $n = 4, 5, 6$, etc. Thus, both the then-known hydrogen series could be represented by

$$\bar{f} = R\left(\frac{1}{m^2} - \frac{1}{n^2}\right). \tag{4–8}$$

This gives the Balmer series when $m = 2$ and $n = 3, 4, 5$, etc., and correctly predicts the Paschen series for $m = 3$ and $n = 4, 5, 6$, etc.

---

* The wave number is the number of waves per unit length, that is, reciprocal wavelength. One might suppose that the logical quantity to use for the reciprocal form of wavelength would be the frequency, c/λ. In spectroscopy, the wave number is used because in order to compute the frequency without losing the remarkable precision of wavelength measurements, it would be necessary to know the velocity of light to an equal precision. Wave numbers can be computed without knowing the velocity of light and so they retain all the accuracy of spectroscopic wavelength measurements. However, the wave number is not an absolute constant for a given spectral line because its wavelength depends upon the index of refraction of the medium in which the measurements are made. The wavelength in air is corrected to vacuum by means of the relation $\lambda_{\text{vac}} = \mu\lambda_{\text{air}}$, where μ is the index of refraction of air for the particular wavelength. In the visible region, λ_{vac} is approximately 2.5×10^{-2} percent greater than λ_{air}.

4–5. THE BOHR MODEL AND THEORY OF THE ATOM

Equation (4–8) represented the entire known hydrogen spectrum with great precision, but it was an empirical formula. At this point we have a correct but underived formula for the hydrogen spectrum. In 1913, Niels Bohr succeeded in deriving this important relation.

Bohr extended Rutherford's model of the atom. He retained the small core or nucleus of the atom and proposed, in addition, that there were electrons moving in orbits around the nucleus. In the case of hydrogen, Bohr proposed that the nucleus consisted of one proton with one electron revolving about it. This is a planetary model of the atom where the heavy positive nucleus is like the sun and the light, negative electron is like the planet earth. In this model, hydrogen is a tiny, one-planet solar system with the gravitational force of the solar system replaced by the electrostatic force of attraction between the oppositely charged particles. The general equations for the gravitational force and the electrostatic force are, respectively,

$$F = G\frac{MM'}{r^2} \quad \text{and} \quad F = \frac{1}{4\pi\epsilon_0}\frac{qq'}{r^2}. \tag{4–9}$$

Both forces are inversely proportional to the square of the distance between the particles. The planets of the solar system have elliptical orbits which are nearly circular. Bohr assumed that the planetary electron of hydrogen moves in a circular orbit, which makes the analysis of the classical aspects of the problem straightforward. Let v be the tangential speed of a mass M' that is revolving around a very large mass M in a circular orbit of radius r. Revolution occurs around the center of mass of the system which, in effect, is at the center of the large, massive body. The centripetal force acting on M' is the gravitational force of attraction due to M. Thus we have

$$F = G\frac{MM'}{r^2} = M'a = \frac{M'v^2}{r}, \tag{4–10}$$

from which we obtain

$$v^2 = \frac{GM}{r}. \tag{4–11}$$

In Bohr's model of the atom, an electron of charge e, mass m_e, and tangential speed v revolves in a circular orbit of radius r around a massive nucleus having a positive charge Ze. In this case, too, the center of the oribit is essentially at the center of the heavy nucleus. The centripetal force acting on the orbiting electron is the electrostatic force of attraction of the nuclear charge, and therefore the force equation is

$$F = \frac{1}{4\pi\epsilon_0}\frac{Ze \cdot e}{r^2} = m_e a = \frac{m_e v^2}{r}. \tag{4–12}$$

From this equation we find that

$$v^2 = \frac{Ze^2}{4\pi\epsilon_0 m_e r}. \tag{4–13}$$

(For hydrogen, the atomic number Z equals one. We include Z for generality.) Each of the equations (4–11) and (4–13) provides a relationship between the variables v and r. If one is known, the other can be found. In the gravitational case, any pair of values of v and of r which satisfy Eq. (4–11) may actually occur. In the electrical case, classical physics imposes no limitation on the number of solutions there can be for Eq. (4–13). For the case of the hydrogen atom, Bohr introduced a restrictive condition which is known as the first Bohr postulate. He assumed that not all the possible orbits that can be computed from Eq. (4–13) are found in hydrogen. Bohr's *first postulate* is that *only those orbits occur for which the angular momenta of the planetary electron are integral multiples of* $h/2\pi$, that is, $nh/2\pi$.* Here n is any integer and h is Planck's constant. Bohr's first postulate introduces the integer idea that appears in the Ritz formula and also introduces Planck's constant, which we have seen plays an important role in the atomic view of radiation. Stated mathematically, this first postulate is

$$I\omega = \frac{nh}{2\pi}, \tag{4–14}$$

where I is the moment of inertia of the electron about the center, ω is its angular velocity, and $n = 1, 2, 3, \ldots.$ For the revolving electron, the quantity $I\omega = mr^2\omega = mvr$. From this we obtain an equation that expresses Bohr's first postulate in a very useful form:

$$\boxed{m_e vr = \frac{nh}{2\pi}.} \tag{4–15}$$

The product $m_e vr$ is also called the moment of momentum of the electron.

The orbiting electron in hydrogen must simultaneously satisfy the conditions expressed by Eqs. (4–13) and (4–15). After eliminating v between these two equations, we find that the orbits which exist or are "permitted"

---

* The 2π in this expression has no particular physical significance. This factor could have appeared in quantum theory, although this was not pointed out in our discussion of it. Planck said the energy of a photon is $E = hf$. If, instead of using the frequency in cycles per second, he had used the angular frequency, $\omega = 2\pi f$, in radians per second, then his constant would have been $h/2\pi$, a quantity often written $\hbar$ and read "h-bar." In terms of $\hbar$, the energy of a photon is $E = \hbar\omega$, and the angular momentum of Bohr's first postulate is $n\hbar$. (Similarly, $\lambda\!\!\!^{-}$ means $\lambda/2\pi$.)

in the hydrogen atom are only those that have radii

$$r = \frac{\epsilon_0 h^2 n^2}{\pi m_e Z e^2}. \tag{4-16}$$

Since we must next consider the energy of the planetary electron and since it may seem odd that it proves convenient to consider the electron energy as being negative, we now review some basic energy concepts. The energy concept is useful in calculations only when there is an exchange of energy. Fundamentally, every energy calculation is the result of an integration that always involves either an initial and a final state (evaluation of a definite integral) or an arbitrary constant (evaluation of an indefinite integral). As a matter of convenience we arbitrarily assign a certain energy to a particular state. Thus, in considering kinetic energy, we usually say that a body at rest has no kinetic energy. A man on a moving train has no kinetic energy relative to himself, but an observer on the ground regards the man on the train as moving and having kinetic energy. Each is correct in terms of *his* aribtrary definition of zero energy. This arbitrary choice of reference level is more familiar in the case of potential energy. When we say the gravitational potential energy of a mass m is mgy, it is necessary to state what is meant by $y = 0$.

In discussing the energy of a planetary electron, we shall use the usual convention of field theory, namely, that the electron has no energy when it is at rest infinitely far from its nucleus. Because an electron can do work as it moves nearer the positive nucleus, it loses electric potential energy. Since the electron starts from rest at infinity with zero energy, its potential energy must become more negative as it approaches the nucleus.

To obtain an expression for the electric potential energy of an electron, we note that the potential at a point which is at a distance r from a nucleus having a charge Ze is

$$V = \frac{1}{4\pi\epsilon_0}\frac{Ze}{r}.$$

The potential energy of a negative electronic charge at this point is $E_p = V(-e)$ or

$$E_p = -\frac{Ze^2}{4\pi\epsilon_0 r}. \tag{4-17}$$

Note that the potential energy of the electron in this case is zero at infinity and negative elsewhere. We can use Eq. (4–13) to find its kinetic energy:

$$E_k = \frac{1}{2} m_e v^2 = \frac{Ze^2}{8\pi\epsilon_0 r}. \tag{4-18}$$

The total energy of the planetary electron is the sum of the potential and kinetic energies:

$$E = E_k + E_p = \frac{Ze^2}{8\pi\epsilon_0 r} - \frac{Ze^2}{4\pi\epsilon_0 r} = -\frac{Ze^2}{8\pi\epsilon_0 r}. \tag{4-19}$$

We have computed the total energy as a function of r. But we have seen that r can have only those values given by Eq. (4–16). Using this equation to eliminate r, we find that

$$\boxed{E_n = -\frac{m_e e^4 Z^2}{8\epsilon_0^2 h^2 n^2},} \tag{4-20}$$

where $n = 1, 2, 3, \ldots$ for the energy states that it is possible for the electron* to have. The integer n is called the *total* or *principal* quantum number and it can have any of the series of values, $1, 2, 3, \ldots$. The values of n determine the energies of the states. When n is large, the energy is large, that is, less negative than for small integers. The energy required to remove an electron from a particular state to infinity is called the *binding energy* of that state. It is numerically equal to E_n.

We now consider how Bohr used this set of energies to account for the hydrogen spectrum. In Chapter 3 we described how classical electrodynamics predicts that energy will be radiated whenever a charged particle is accelerated. We were careful to point out that the acceleration could be due to a change of direction of motion as well as due to a change of speed. According to classical theory, an orbital electron should radiate energy because of its centripetal acceleration. In order to preserve his atomic model of planetary electron orbits, Bohr had to devise a theory which would violate this classical prediction since, according to it, any electron that separated from the nucleus would soon radiate away its energy and fall back into the nucleus. Bohr's second break with classical physics is contained in his *second postulate*, which states that *no electron radiates energy so long as it remains in one of the orbital energy states; and that radiation*

---

* In calling E the energy of the electron, we are not precise. We have assumed that the electron does all the moving while the nucleus remains at rest. Since the mass M of the proton is 1836 times the mass m_e of the electron, the latter has most of the kinetic energy of the atomic system. A detailed treatment would require us to consider the movement of all particles about their common center of mass. There is a theorem in mechanics which states that in a two-body problem such as this, the motion of one body may be neglected if the mass of the other body is taken to be the *"reduced mass,"* which is the product over the sum of the two masses, $m_e M/(m_e + M) = m_e/(1 + m_e/M)$. If, in Eq. (4–20) and elsewhere, we *replace* the electron mass m_e by the reduced mass, then our equations correctly describe the atomic system *as a whole.*

occurs only when an electron goes from a higher energy state to a lower one, the energy of the quantum of radiation, hf, being equal to the energy difference of the states. Let the quantum number $n = n_2$ represent a higher energy state and $n = n_1$ represent a lower energy state ($n_1 < n_2$), then the second Bohr postulate can be written as

$$\boxed{hf = E_{n_2} - E_{n_1}.} \tag{4–21}$$

Substituting for the energies from Eq. (4–20), we have for the frequency of the emitted radiation

$$\boxed{f = \frac{m_e e^4 Z^2}{8\epsilon_0^2 h^3}\left(\frac{1}{n_1^2} - \frac{1}{n_2^2}\right)} \tag{4–22}$$

or, in terms of the wave number,

$$\bar{f} = \frac{1}{\lambda} = \frac{f}{c} = \frac{m_e e^4 Z^2}{8\epsilon_0^2 h^3 c}\left(\frac{1}{n_1^2} - \frac{1}{n_2^2}\right) \tag{4–23}$$

where c is the speed of light in a vacuum. Comparing Eq. (4–23) with Eq. (4–8) shows that both have the same form.

Equally impressive is the fact that the constant factor of the Bohr formula is the Rydberg constant, R. Again comparing Eqs. (4–8) and (4–23), we find that, since $Z = 1$ for hydrogen,

$$R = \frac{m_e e^4}{8\epsilon_0^2 h^3 c} = 1.0973731 \times 10^7 \text{ m}^{-1}. \tag{4–24}$$

The R given here is R_∞, which would be correct if the mass of the nucleus were infinite compared with the mass of an electron. If the motion of the nucleus is taken into account, m_e must be replaced by the reduced mass. Therefore, in general, $R = R_\infty/(1 + m_e/M)$. This accounts for the slight variation of R from element to element noted by Rydberg. It is a triumph of the Bohr model and theory that the slight differences between the spectra of ordinary hydrogen and its isotope, heavy hydrogen (deuterium), can be attributed to the influence of the nuclear mass. In fact, heavy hydrogen was discovered spectroscopically by Urey in 1932.

The Bohr formula gives the Balmer series for $n_1 = 2$ and the Paschen series for $n_1 = 3$, as we have seen before. But the Bohr theory places no restrictions on n_1 and his result suggested that there might be additional hydrogen series not yet found experimentally. In 1916 Lyman found a series in the far ultraviolet, in 1922 Brackett found a new series in the infrared, and in 1924 Pfund located another in the same region. Table 4–1 summarizes the five hydrogen series.

TABLE 4–1
THE SPECTRAL SERIES OF HYDROGEN

Values of n_1	Name of series	Values of n_2
1	Lyman	2, 3, 4, etc.
2	Balmer	3, 4, 5, etc.
3	Paschen	4, 5, 6, etc.
4	Brackett	5, 6, 7, etc.
5	Pfund	6, 7, 8, etc.

4–6. EVALUATION OF THE BOHR THEORY OF THE ATOM

Bohr's successful derivation of the Rydberg-Ritz formula opened the door to an understanding of the extra-nuclear structure of atoms which is now virtually complete. The student may feel that anyone can get the right answer to a problem if he is permitted to write his own postulates and break the rules of the game—especially when he knows the answer before he starts. Upon reflection, however, it is remarkable that Bohr could get the right answer without being more arbitrary than he was. There is no harm in introducing a new constant into physics as Planck did, but it is a great achievement to find that an empirical constant like the Rydberg constant is expressible in terms of basic constants already known. It is true that Bohr did work backwards in order to know that in his first postulate he should let the angular momentum be $nh/2\pi$. The impressive thing is that he found something so simple. There was no need to set the angular momentum to n times an arbitrary number. In view of later developments, it is surprising that Bohr could retain as much classical theory as he did.

4–7. ENERGY LEVELS

Bohr's second postulate that there are only certain discrete energy levels in the hydrogen atom is especially important because it has been found to have wide application throughout atomic physics. Energy levels are most conveniently expressed in electron volts. Equation (4–20) gives the energy states of a one-electron atom in basic units. If E_n is in joules, it may be converted into electron volts by dividing by the conversion factor e_c, 1.60×10^{-19} J/eV. For this case we have

$$E\ (\text{eV}) = \frac{E_n}{e_c} = -\frac{1}{e_c} \cdot \frac{m_e e^4 Z^2}{8\epsilon_0^2 h^2} \cdot \frac{1}{n^2}. \tag{4–25}$$

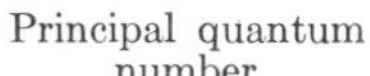

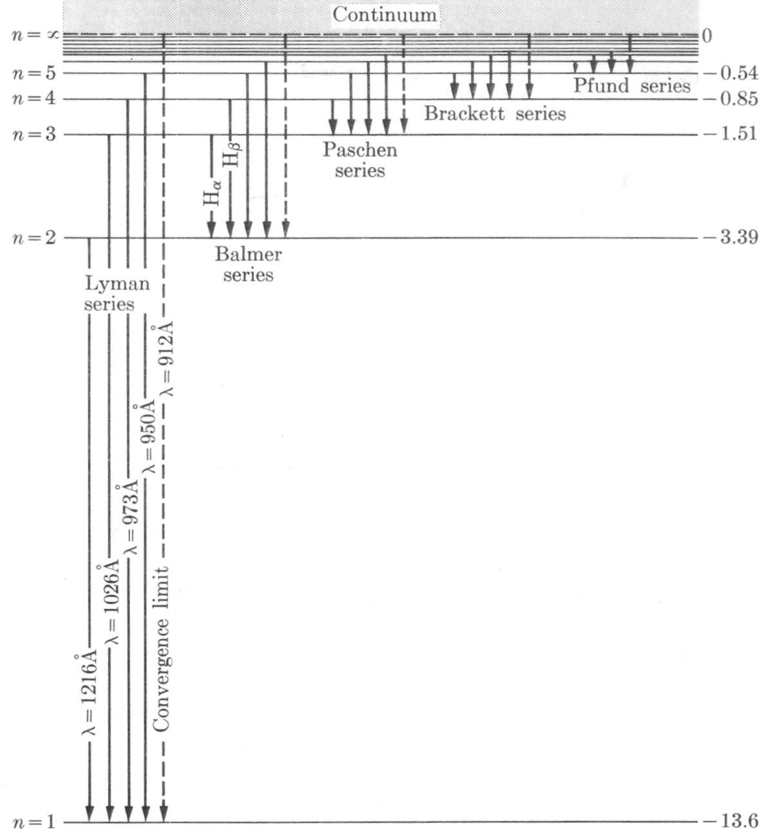

FIG. 4–4
Energy-level diagram of the hydrogen atom.

Substituting the values of the constants in this equation and letting $Z = 1$, we find that the energy levels of the hydrogen atom are given by

$$E \text{ (eV)} = -\frac{13.6}{n^2}. \tag{4–26}$$

These energy levels can now be represented graphically as shown in Fig. 4–4. The quantum numbers are shown at the left and the corresponding energies of hydrogen in electron volts are given at the right. In this array of energies, the higher (less negative) energies are at the top while the lower (more negative) are toward the bottom. In a normal unexcited hydrogen atom, the electron is in its lowest energy state at the bottom, with $n = 1$. An electron in this *ground state* is stable and moves in this orbit continuously without emitting or absorbing energy. The "excitement" begins when the electron absorbs energy in some way. There are a variety of ways in which this may be brought about. If the hydrogen is in an electric discharge, a free electron which has been accelerated by the electric field may hit* it. If the hydrogen is heated, the electron may be excited by a thermal-motion collision. If the hydrogen is illuminated, it may absorb energy from a photon. Suppose the electron in hydrogen absorbs about 20 eV of energy in one of these ways. This is enough energy to lift the electron to $n = \infty$ (13.6 eV) with 6.4 eV left over. In this case, the electron is made entirely free of its home nucleus and is given 6.4 eV of kinetic energy besides. If the electron absorbs just 13.6 eV, it is merely freed from its home nucleus and drifts about with only its thermal kinetic energy. In either of these cases the remaining nucleus is an ion. If the energy of a bombarding electron is less than that required for ionization but equal to or greater than that needed to raise an electron in an atom to one of its permitted energy levels, then the atomic electron will absorb just enough energy to put it into some higher energy state. After the bombarding electron has transferred enough energy to the atom to excite it, the electron will leave the encounter, carrying away any excess as kinetic energy. The *excitation energy* of any level in electron volts is *numerically* equal to the *excitation potential* of that state in volts.

After excitation, the atomic electron returns to its normal state. If it was excited to $n = 4$, it may jump from 4 to 1 in one step. It may also go 4, 2, 1 or 4, 3, 1 or 4, 3, 2, 1. In each step of the return trip, the electron must lose an amount of energy equal to the difference of the energy levels. The only mechanism available for this energy loss is through the emission of electromagnetic radiation. Thus, in Fig. 4–4 we have represented graphically the second Bohr postulate, given in Eq. (4–21). When we see the light from a hydrogen discharge, we are "seeing" the electrons go from excited states to lower states.

The electron transitions which end on $n = 1$ constitute the Lyman series, on $n = 2$, the Balmer series, on $n = 3$, the Paschen series, etc. From the energy-level diagram we can see that the Lyman transitions involve the largest changes of energy, produce the highest frequencies, and provide the "bluest" (ultraviolet) light.

The shaded region at the top of Fig. 4–4 represents the fact that electrons at infinity may have kinetic energy, so that their energy there is not zero but positive. If the electron of hydrogen is completely removed from its

* We use the word "hit" loosely. We saw in the discussion of alpha-particle scattering that a collision between charged bodies does not involve physical contact in the usual sense.

nucleus, then one of these electrons at infinity having *any* energy may fall into any one of the energy levels. Such an electron undergoes a change of energy equal to its energy at infinity minus the negative energy of the level to which it falls. The "double negative" in the last sentence enables us to conclude that the energy radiated by such a transition is the sum of the electron's kinetic energy at infinity and that involved in the transition from $n = \infty$ to the final level. The energy radiated will have a value greater than that involved in the transition from $n = \infty$ to the final level. Since there is a wide distribution of the energies among the electrons at infinity, there is a continuous spectrum below the short-wavelength convergence limit of any series.

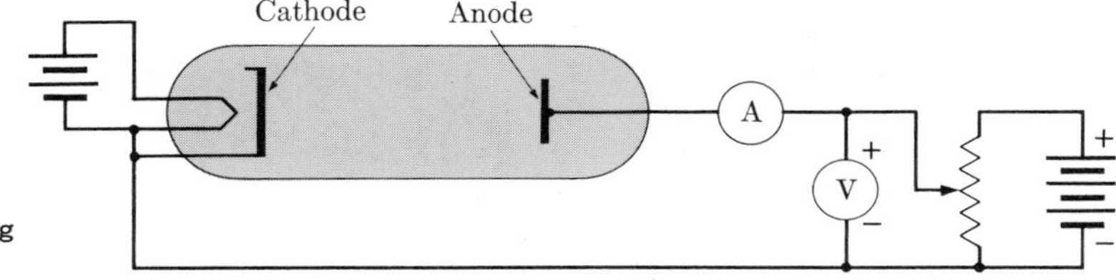

FIG. 4–5
Apparatus for determining ionization potential.

4–8. IONIZATION POTENTIALS

Confirmation of the energy level concept is convincingly given by a consideration of ionization potential. Consider first a radio-type tube, Fig. 4–5, which contains only a filament-heated cathode and a plate as anode. When the plate is positive with respect to the cathode, electrons will move across the tube to the anode. This current is limited by two factors. First, the number of electrons emitted per unit area from a cathode depends upon cathode composition and temperature. In the remainder of this discussion we shall assume that the cathode is operated hot enough so that the tube current is not significantly limited by cathode emission. The second factor is the effect of the electrons in the region between the electrodes upon those emerging from the cathode. The concentration of negative charge in the inter-electrode region is called *space charge*, and it lowers the potential in the vicinity of the emitting surface. Indeed, the potential in this region usually falls below the potential of the emitter. Thus, although the plate or accelerating potential is still positive, this decrease in potential due to space charge will reduce the electron current because the potential barrier will turn back electrons emitted with low kinetic energy. (The electrons in thermionic emission have a distribution of speeds similar to that of the molecules of a gas.) However, those high-energy electrons that get beyond this barrier caused by space charge arrive at the plate with the same energy they would have had if the space

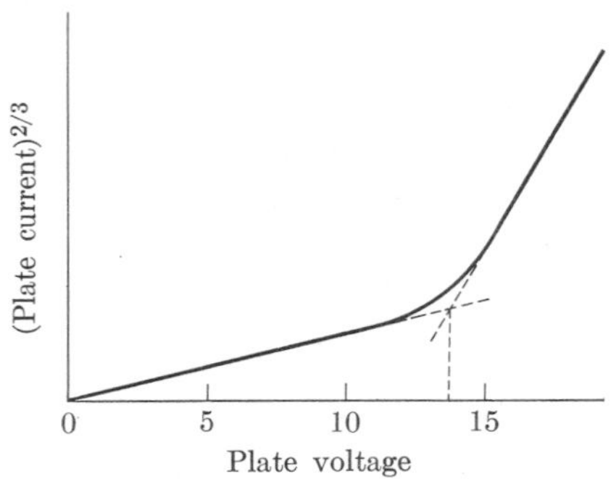

FIG. 4–6
Plate current in hydrogen-filled tube.

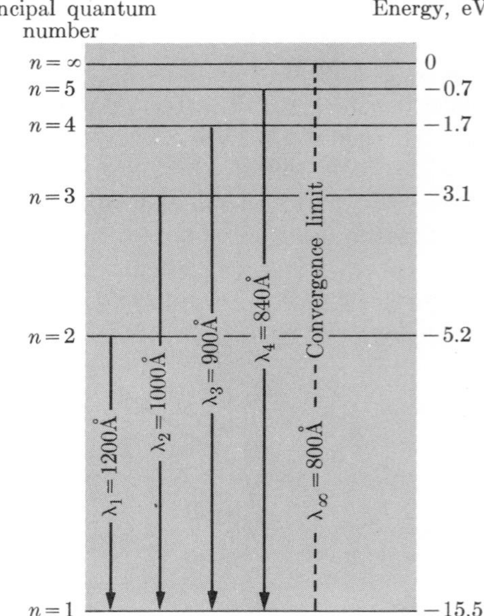

FIG. 4–7

charge had been absent, since the total potential difference V between the electrodes is independent of the space charge. The space charge limited current is not a linear function of the potential difference as in the case of an ohmic resistor; it is found to be proportional to the three-halves power of the potential difference. This is known as the Child-Langmuir Law. In mathematical form, it is $I = kV^{3/2}$. The value of k depends upon the geometry of the tube and the volume density of the charges between the electrodes.

The ionization potential of a gas is determined by introducing some of the gas into a tube such as that shown in Fig. 4–5, and then measuring the plate current as a function of plate voltage. As the potential difference is increased, it is found that above some particular value of the potential the current increases much more rapidly than it does below that value, as shown in Fig. 4–6. When the plate potential reaches this critical value, the electrons arriving at the anode have acquired enough energy to knock electrons off the atoms of the gas close to this electrode. The positive ions produced when the voltage equals or exceeds the critical value neutralize some of the negative space charge. Thus ionization causes a marked increase in the tube current. If the gas under study is hydrogen, the ionization potential is found to be 13.6 volts. It is a remarkable confirmation of the energy-level idea that there should be excellent agreement for the ionization potential as measured by the very different techniques of spectroscopy and electronics. Since the electrons are emitted from the cathode with an initial velocity distribution, some of them acquire enough energy to produce ionization at lower accelerating potentials than others. This accounts for the curved section joining the two straight-line portions of the graph in Fig. 4–6. Except for this short curved part, the break in the line is quite abrupt. This sudden change of slope would not occur if several low-energy electrons could combine in their "efforts" to ionize the atom of hydrogen.

EXAMPLE. Consider a *hypothetical* one-electron atom which we assume does not have the hydrogen energy levels, but which obeys Bohr's second postulate. The wavelengths of the first four lines of the spectral series terminating on $n = 1$ are: 1200 Å, 1000 Å, 900 Å, and 840 Å. The short wavelength limit of this series is 800 Å. (a) Find the values of the first five energy levels of this atom in eV, and construct the energy-level diagram. (b) What is the ionization potential? (c) What is the wavelength of the line emitted for the transition from the energy level $n = 3$ to $n = 2$? (d) What is the minimum energy that must be supplied to the electron in the ground state so that it can make the transition in part (c)?

Solution. (a) The energy-level transitions that produce the lines of the series are shown in Fig. 4–7. The energy in a quantum of radiation of given wavelength is found from Eq. (3–32). Thus the energy for the first line is

$$E = \frac{1.24 \times 10^4}{\lambda} (\text{eV} \cdot \text{Å}) = \frac{1.24 \times 10^4}{1200 \text{ A}} (\text{eV} \cdot \text{Å}) = 10.3 \text{ eV}.$$

Similarly, the energies for the other wavelengths are found to be 12.4 eV, 13.8 eV, 14.8 eV, and 15.5 eV, respectively.

Because the energy in the quantum radiated when the electron goes from the zero energy level ($n = \infty$) down to the lowest level ($n = 1$) is 15.5 eV, the energy of the ground state is $E_1 = -15.5$ eV. The difference in energy between the levels $n = 2$ and $n = 1$ is 10.3 eV. Therefore the energy of the state $n = 2$ is found from $E_2 - E_1 = 10.3$ eV, or

$$E_2 = E_1 + 10.3 \text{ eV} = -15.5 \text{ eV} + 10.3 \text{ eV} = -5.2 \text{ eV}.$$

In the same way, we find that

$$E_3 = -3.1 \text{ eV}, \qquad E_4 = -1.7 \text{ eV}, \qquad \text{and} \qquad E_5 = -0.7 \text{ eV}.$$

(b) 15.5 volts.

(c) From Eq. (3–32) we have

$$\lambda = \frac{1.24 \times 10^4}{E_3 - E_2} (\text{eV} \cdot \text{Å}) = \frac{1.24 \times 10^4}{(-3.1 \text{ eV}) - (-5.2 \text{ eV})} (\text{eV} \cdot \text{Å})$$

$$= \frac{1.24 \times 10^4}{2.1} \text{ Å} = 5900 \text{ Å}.$$

(d) $E_3 - E_1 = (-3.1 \text{ eV}) - (-15.5 \text{ eV}) = 12.4$ eV.

4–9. RESONANCE POTENTIALS

The experiment just described was set up to measure the potential through which electrons must be accelerated before they can lift orbital electrons from their lowest energy state (ground state) to infinity. This ionization was detected by an increase in the current through the tube. But before the bombarding electrons have enough energy to take the atom apart by removing an electron, they have enough energy to lift an electron to an excited state. Orbital electrons in the gas can be transferred from their lowest energy state to any of the higher states. As implied earlier, the quantum conditions which require an orbital electron to emit only certain frequencies as radiation apply also to the absorption process. These electrons can absorb only energies represented by transitions between energy levels. If an orbital electron is hit by a bombarding electron with insufficient energy to produce an energy transition, the orbital electron absorbs no energy from the bombarding electron and *the collision is perfectly elastic.* If the orbital electron is hit by a high-energy bombarding electron, then the orbital electron can absorb energy by making a transition. This leaves the bombarding electron with that much less energy. *Such a collision is inelastic,* since the bombarding electron is left with less energy than it had before the collision. Such an inelastic collision puts the orbital electron in one of the excited states, and hence it can radiate energy in returning to a lower state.

Consider again the ionization experiment. As the potential difference across the tube is slowly increased, the electrons from the heated cathode are accelerated to higher and higher velocities. At low speeds, these electrons make completely elastic collisions with the electrons of the gas, so that they are deviated but not slowed by the collision process. As the potential difference across the tube is increased, however, a potential is reached where energy can be transferred to an orbital electron. For the purpose of discussion consider hydrogen, which has an ionization potential of 13.6 volts. A look at Fig. 4–4 discloses that the least amount of energy the orbital electron in the ground state can absorb is 13.6 eV − 3.4 eV or 10.2 eV. A bombarding electron with 10.2 eV of energy can "resonate" with hydrogen and transfer its energy to the hydrogen. This produces no ionization, so the current through the tube is not changed, but after making such collisions, the bombarding electrons proceed more slowly and the hydrogen shows its "excitement" by radiating. The hydrogen will not glow visibly because this resonance radiation is one line of the Lyman series, which is in the ultraviolet region; however, ultraviolet spectroscopy confirms that the radiation is there.

In order to demonstrate the resonance phenomenon electronically, we need a more elaborate tube, such as that used by Franck and Hertz, who first performed the experiment in 1913 using mercury vapor as the gas.

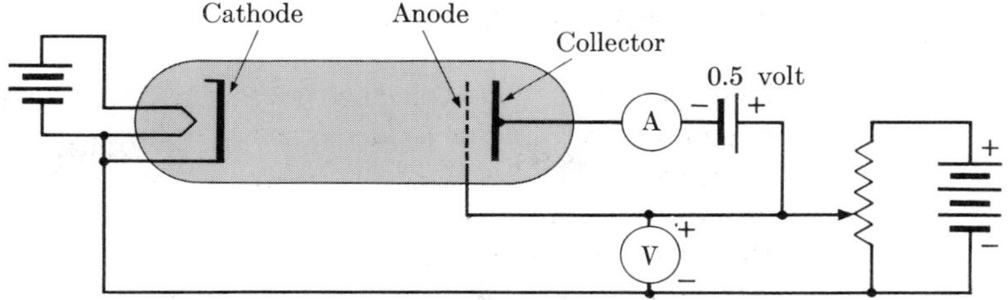

FIG. 4–8
Apparatus for determining resonance potential of gas

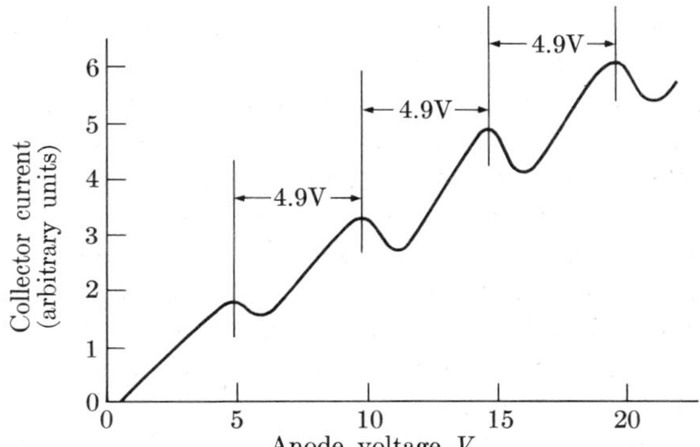

FIG. 4–9
Resonance potential curve for mercury.

The principal parts of such a tube are shown schematically in Fig. 4–8. From the standpoint of electronics, the effect of resonance is that the bombarding electrons are slowed down, so we need a device which will measure the energy of the bombarding electrons after they have made collisions. Suppose that the anode of the ionization tube is perforated or made of wire mesh. In this case, some of the bombarding electrons will pass through the electrode rather than hit it. We now need to know the energy with which the bombarding electrons arrive at the anode.

In our consideration of the photoelectric effect, we measured the energy of photoelectrons by making them move against the force action of an electric field, and the energy of the photoelectrons was given by the stopping potential. Here we use much the same technique and insert into the tube another electrode beyond the anode. This collector electrode is maintained less positive than the anode, say 0.5 V, so that any electrons that pass through the anode will be slowed by the field between the electrodes. Electrons which reach this last electrode must have passed the anode with an energy of at least 0.5 eV.

The experimental procedure consists of measuring the collector current as a function of the anode potential with respect to the cathode, and typical results are shown in Fig. 4–9. From $V = 0$ to $V = 0.5$ V, there is no collector current, since no electrons can reach the anode if they have less than 0.5 eV of energy. Above $V = 0.5$ V, the collector current rises because the number of electrons having at least this minimum energy increases. When V reaches the resonance potential of the gas, the collector current begins to decrease because some of the bombarding electrons are slowed by inelastic collisions with orbital electrons in the gas. The current rises again as V is further increased since, in the stronger field, bombarding electrons can make inelastic collisions early and still undergo enough acceleration to surmount the 0.5-V barrier.

The second dip is not due to a new energy transition, because very few electrons ever get enough energy to excite the next transition. The second dip occurs at twice the resonance potential and is caused by the bombarding electrons suffering two inelastic collisions of the same kind. Thus each peak of the curve signifies more collisions and each peak is an integral multiple of the resonance potential. The separation of successive resonance peaks is 4.9 V for mercury. (Since mercury ionizes at 10.4 V, the third resonance peak will be masked by other effects if the tube shown in Fig. 4–8 is used. Actually, a more complicated tube which differentiates the resonance and ionization effects is used. This important experimental detail does not in any way alter the principle discussed in this section. Mercury vapor is chosen because it is monatomic whereas hydrogen is ordinarily a diatomic molecule whose dissociation energy is 4.5 eV. If the gas were hydrogen, most of the bombarding electrons would give up their energy to excite molecular energy states and also to dissociate the molecules after the accelerating potential reached 4.5 V. This complex situation would conceal the effects of atomic hydrogen in this type of experiment.)

The data from the Franck-Hertz experiment corroborate the spectrographic observations and further support the concept of discrete energy levels.

4–10. PHOTON ABSORPTION

In the previous two sections we have discussed how an atom can absorb discrete amounts of energy from bombarding electrons. An atom may also absorb energy from photons, but there is an important difference. Absorbed photons disappear entirely. A photon with more energy than the ionization energy of an atom can always be absorbed because the excess energy will appear as kinetic energy of the photoelectron. A photon with less than the ionization energy cannot be absorbed unless its energy is equal to one of the excitation energies of the absorbing atom. (Some of these statements about photon absorption will be slightly modified when we discuss the scattering of high-energy x-ray quanta in Chapter 6.) Consider hydrogen again. Ordinarily, the probability of finding a hydrogen atom in an excited state is very small; therefore we *assume* that it is always in its ground state. Thus hydrogen can *only* absorb photons whose wavelengths correspond to those emitted in the far ultraviolet, the Lyman series. Hydrogen atoms are therefore transparent to visible and infrared light. If we pass radiation of all wavelengths through hydrogen and analyze the transmitted light by means of a spectrograph, we find the transmitted intensity *reduced* for the Lyman wavelengths. Such a spectrum, having a bright background and *dark lines*, is called an *absorption spectrum*. Because the atoms that have been excited by the absorption of radiation re-emit photons in *random* directions upon returning to the ground state, there is a decrease of intensity along the direction of the transmitted radiation. The absorption lines observed are actually very faint bright lines that appear dark by contrast.

There are certain advantages to studying absorption spectra. For many atoms, as for hydrogen, the absorption spectrum is simpler than the emission spectrum. For hydrogen, one can usually observe absorption for only one of the five emission series. This one series is sufficient to establish the entire energy-level diagram, and these levels permit an explanation of all the emission series.

The determination of elements on the sun is a dramatic example of absorption spectroscopy. The sun is a hot body which emits a continuous spectrum of photons. As these photons pass through the outer atmosphere of the sun, wavelengths which are characteristic of the gases present are absorbed. Thus the continuous spectrum of the light from the sun is crossed with (relatively) dark lines which were first observed by Fraunhofer in 1815. Most of the Fraunhofer lines correspond to the wavelengths of elements found on the earth. The absorption lines of the Balmer series of hydrogen are especially prominent in the spectrum of the sun. The Balmer lines of hydrogen and the visible-region lines of other elements have rather long wavelengths and can be absorbed *only* by excited atoms of the respective elements. These higher energy states are produced in the following way. Although the gaseous atmosphere of the sun is cooler than its surface, the temperature of the gas is still so high that a large number of atoms have sufficient kinetic energy to excite many other atoms by collision. Thus these atoms are raised above the lowest energy level to states where they can absorb wavelengths longer than the ultraviolet. For many years one set of Fraunhofer lines in the visible region could not be associated with any known element. It was presumed to be due to a new "sun element" which was appropriately named helium. This hypothesis was confirmed when helium was isolated on the earth and its emission spectrum was found to correspond with the previously unidentified Fraunhofer lines.

Many of the spectral series that are characteristic of molecular structure are in the infrared region. Absorption spectroscopy is the only feasible way of investigating the structure of those molecules that would be dissociated by being excited in an electric arc in an attempt to produce emission lines.

Molecules have the three degrees of freedom of translation at very low temperatures, and additional degrees of freedom due to rotation and vibration at higher temperatures. These additional degrees of freedom produce many quantized energy states. Since the energy differences between these states is small, the emitted wavelengths associated with them are long and so, for the most part, they lie in the infrared or beyond. Molecular spectra have many lines and are very complex. The rotational and vibrational states are characteristic of the molecule. These states differ even in the case of isomers, which are molecules having the same atomic composition but different arrangement or structure.

By means of molecular spectra it is possible to measure the frequencies of rotations and vibrations of the atoms that join together to form compounds. In this way we can learn about moments of inertia, spring constants, and the angles such as those between the lines joining the three atoms in H_2O. But the mechanical motions of atoms in molecules are much slower than the activity of atomic electrons, and few molecular studies can be carried out with radiation in the visible region. Progress in molecular research has led from the visible region into the infrared, the microwave (radar), and the radio region. An interesting application of one of the results of this work (to be discussed in Section 4–12) is the use of one of the molecular frequencies of ammonia as the "pendulum" for controlling a so-called "atomic" clock. Such clocks have very high precision. The problems of molecular structure are still under active study, especially in the field of microwave spectroscopy.

4–11. FLUORESCENCE AND PHOSPHORESCENCE

Another application of the energy-level concept is in the explanation of fluorescence. The fluorescent lamps used for modern lighting work in the following way. The electric discharge within the lamp is through mercury vapor. The spectrum of mercury has some lines in the visible region, but most of the emitted radiation is concentrated in a line in the ultraviolet. If the tube were made of quartz, which can transmit ultraviolet light, this radiation would be able to get out of the tube, where it could be used for air sterilization or to produce "sunburn." But a clear tube is a poor source of visible light. Therefore the inside of a fluorescent lamp is coated with a material which absorbs the invisible ultraviolet light. Thus the atoms of this fluorescent material become excited. If the excited electrons fell back to their normal state in one step, they would re-emit the ultraviolet light, which would still be invisible. But the excited electrons return to their normal state in more than one step. Each step produces radiation of less energy than the original excitation, so that the energy of the ultraviolet light is converted into visible light. The various tints that different lamps have are controlled by the nature of the fluorescent material used.

Some materials have what are called *metastable* states. When an electron is excited into one of these states, it does not return to its normal state at once, but may remain excited for an appreciable time. Such materials have a persistent light, called *phosphorescence,* which may last several hours after all external excitation is removed. These materials are sometimes used on the screens of cathode-ray tubes, and they are sometimes used to make light switches glow, so that they may be found in the dark. Most fluorescent tubes have some phosphorescence. It may be observed in a dark room a few minutes after the light is turned off.

4-12. MASERS AND LASERS

We now have the background to understand a whole group of new devices called *masers* [*m*olecular (formerly microwave) *a*mplification by *s*timulated *e*mission of *r*adiation]. We shall discuss the underlying principle of these devices and then consider their functional and practical differences.

In our discussion of the Bohr model of the hydrogen atom we made the too-simple assumption that *all* atoms are normally in the lowest or ground state. This is not *quite* true. If we assume that we have monatomic hydrogen gas at room temperature (chemically impossible), then it is easy to calculate the relative numbers of atoms in higher energy states. Thermal collisions will raise some atoms to higher energy states just as thermal collisions cause "air" molecules to rise physically high in the atmosphere. In Chapter 1 we derived the law of atmospheres, Eq. (1–32), which we generalized to the Boltzmann distribution law, Eq. (1–36). To compare the numbers of atoms at two different energies in thermal equilibrium we write this distribution law as

$$n_2/n_1 = \exp[-(E_2 - E_1)/kT]. \tag{4–27}$$

For hydrogen at room temperature this becomes

$$\frac{n_2}{n_1} = \exp\left(-\left\{\frac{-[-3.39 - (-13.6)]e_c}{k \cdot 293}\right\}\right),$$

where the energy levels have been introduced in electron volts and room temperature has been taken to be 20°C or 293°K. Since $e_c = 1.60 \times 10^{-19}$ J/eV and k is Boltzmann's constant ($= 1.38 \times 10^{-23}$ J/°K), we find that

$$\frac{n_2}{n_1} = \exp\left[-\left(\frac{10.2 \times 1.60 \times 10^{-19}}{1.38 \times 10^{-23} \times 293}\right)\right] = e^{-404} = 10^{-176},$$

which is incredibly small. This calculation shows that we were *almost* correct in stating that all atoms are in the ground state. Note that if the temperature had been much higher (as in the atmosphere of the sun,

6000°K) the proportion of atoms in the excited state would then have been significant, $n_2/n_1 = 10^{-8.6}$. Note too that the fraction of excited atoms would have been great, even at room temperature, if the separation of the energy levels had been much less. Note especially that e^0 equals 1 so that no matter how small the energy level difference or how high the temperature, the number of atoms in the higher state cannot exceed the number in the lower state (this whole discussion assumes the hydrogen to be in thermal equilibrium).

Having explained that some atoms (however few) may "normally" be in an excited state, we now state an additional fact. Whereas we have explained previously that a 10.2 eV photon can raise an hydrogen atom from the $n = 1$ state to the $n = 2$ state with the absorption of the photon, we now state that a 10.2 eV photon interacting with an excited hydrogen atom can cause a transition from the $n = 2$ state to the $n = 1$ state *with the emission of an additional photon* which must have the same energy. When 10.2 eV photons pass through hydrogen, absorption of photons is obviously the observed effect. But in a *few* instances the reverse effect of *stimulated emission* takes place. For future reference it is important to mention that stimulated photons and the stimulating photon are coherent (that is, in phase with one another).

Lasers and masers are devices in which these stimulated photons are cleverly enhanced so they outnumber absorbed photons.

Neither lasers nor masers can be readily made with hydrogen, but we will develop the idea one step more before we leave our consideration of this familiar substance.

Suppose that a chamber of atomic hydrogen were subjected to *very* intense radiation of 10.2 eV photons. Absorptions would tend to increase the number of atoms in the excited state. However, the more atoms there are in the excited state, the more likelihood there is of stimulated emission to the ground state. The probability of one process is intrinsically equal to that of the other, so that if the number of excited atoms became equal to the number in the lower state, the number of stimulated emissions would become equal to the number of absorptions. Thus radiation disturbs thermal equilibrium. We have calculated that hydrogen "in the dark" has only a very small fraction of its atoms excited. Hydrogen in intense radiation of appropriate energy will tend to have equal numbers of atoms in the higher and lower states, although the tendency toward thermal equilibrium always prevents the number in the higher state from quite equaling the number in the lower.

To illustrate this, let us consider a box containing grains of rice not more than one layer deep. Grains on the bottom of the box correspond to hydrogen atoms in the ground state. If we shake the box, some grains hit the top, which correspond to the atoms in the excited state. Shaking the box is intrinsically as likely to bang a grain down as up, but gravity (as in the law of atmospheres) corresponds to thermal equilibrium and tends to keep the grains at the bottom. If the box is shaken violently, the number of grains at the top and bottom tend to become equal, but there are always more at the bottom than at the top. Thus mere intense radiation can never cause more stimulated emission than absorption. If we could invent a way to establish a condition in which more atoms were in the higher state, then stimulated emission would exceed absorption. We would get out more than we "put in," and amplification would result.

Several ingenious ways of accomplishing this have been found. In 1955 Gordon, Zeiger, and Townes operated the first maser. They used ammonia molecules instead of the hydrogen we have been discussing. The ammonia molecule (NH_3) has energy states like hydrogen, but instead of the excited state being 10.2 eV above the ground state, it is excited by a mere 10^{-4} eV and thus at 20°C we have

$$\frac{n_2}{n_1} = \exp - \frac{10^{-4} \times 1.60 \times 10^{-19}}{1.38 \times 10^{-23} \times 293} = \exp - \frac{1}{252} \approx 1.$$

In thermal equilibrium the relative fraction of excited atoms is nearly unity. There are "naturally" almost as many molecules excited as unexcited. The remarkable achievement of Townes and his associates was their invention of a molecule sorter. They passed ammonia through a jet into a vacuum. The stream of molecules passed through an electrode structure where, because of their differing electrical properties, excited molecules were deflected one way and unexcited molecules were deflected the other. The excited molecules were allowed to pass on into a chamber. Now, the energy of the excited state is 10^{-4} eV which corresponds to a wavelength of 1.24 cm and a frequency of 23,870 MHz. This is in the microwave region. By making the chamber of proper size it became a resonant cavity for any radiation produced. If one molecule went from its excited state to its ground state, it was likely to stimulate another, and since initially all atoms were excited, this process could quickly build up to significant proportions. As excited molecules were converted to absorbing molecules, new excited molecules were introduced to sustain the production of photons. Thus the cavity "sang" with microwave radiation like a sounding organ pipe. It was an oscillator.

The maser just described has many remarkable properties. One of these is its stability. The frequency of the oscillations is determined by the nature of the ammonia molecule itself. One excited ammonia molecule is indistinguishable from another. They all cause radiation of the same frequency. By electronic techniques, this oscillator can be made to govern the frequency of slower oscillators. A succession of such frequency reductions provides a low frequency oscillator which ticks like a clock but which has all the stability built into ammonia molecules. Such a clock is called an *atomic clock* and its precision is such that it will neither gain nor lose more

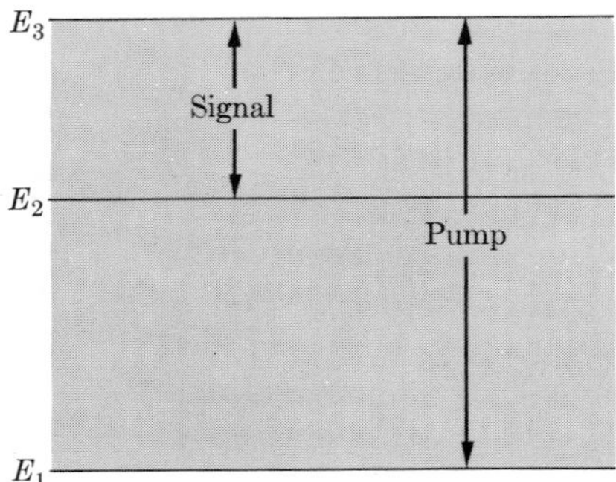

FIG. 4–10

than one second in 1000 years! Whereas this gas maser is superb as a stable oscillator, another type, the *solid state maser*, makes a better amplifier.

The most common solid state maser is made from ruby. A ruby is basically clear alumina made red by a small concentration of chromium. It is the chromium "impurity" that is the active atom of the maser. When the ruby is in a steady magnetic field, the chromium acquires energy states, three of which are represented schematically in Fig. 4–10. In thermal equilibrium the number of atoms in the three states obeys Boltzmann's law, as shown graphically in Fig. 4–11 (a). The ruby material is irradiated with photons from an external source whose frequency (energy) corresponds to the energy difference, $E_3 - E_1$. This causes absorption transitions from E_1 to the *metastable* state E_3 and stimulated transitions from E_3 to E_1. Although these latter transitions are stimulated, they themselves do not account for the maser action. The effect of these transitions is, as we have seen, to make the concentration of chromium atoms in states E_1 and E_3 tend to become equal. The concentration of atoms in E_2 remains substantially unchanged, as shown in Fig. 4–11 (b). This *optical pumping* of atoms from E_1 to E_3 causes a population inversion between states E_2 and E_3, which is enhanced because E_3 is a metastable state of chromium. If the ruby material is now exposed to photons whose energy is $E_3 - E_2$, transitions will occur both up and down between these two levels. The significant fact is that with this inverted population of these states, there are more stimulated emissions than there are absorptions. Amplification results.

It is true that signal frequency photons (with power in the microwatt range) tend to make the populations of states 2 and 3 become equal, but the overwhelming pumping between states 1 and 3 (milliwatts) keeps state 3 more populated than state 2.

Spontaneous transitions from state 3 to state 2 do occur, as you would expect from our discussion of the Bohr theory of hydrogen. Photons thus

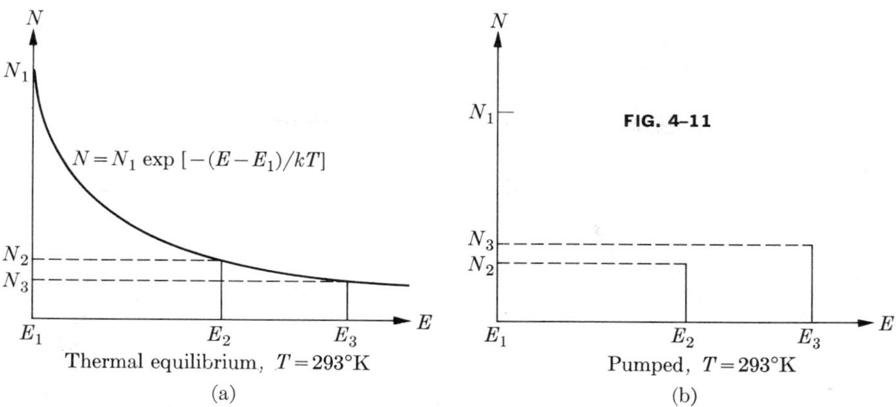

FIG. 4-11

produced are random in origin and constitute the ultimate defect of all amplifiers, noise. But these random transitions are so remarkably rare that this amplifier has less noise and can amplify weaker signals than amplifiers of any other type. One application of maser amplifiers is in detecting and measuring extremely weak microwave signals from outer space.

The frequency of microwaves that this maser can amplify is determined by the energy difference $E_3 - E_2$. This difference can be varied by changing the magnetic field in which the ruby material is placed. Thus, unlike the frequency stability of the ammonia maser, the solid state maser has versatility as an amplifier.

Ruby masers are always operated at low temperatures—frequently at the temperature of liquid helium, 4.2°K. Comparing Fig. 4–12(a) and (b) to the earlier figures shows how chilling enhances the inversion between states 2 and 3.

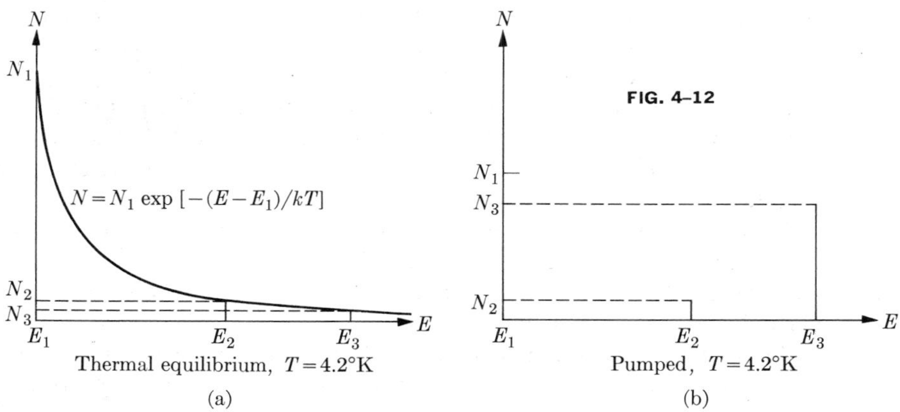

FIG. 4-12

Lasers (*L*ight *a*mplification by *s*timulated *e*mission of *r*adiation) are, in principle, no different from the maser just described. A material is used which has an energy diagram like that shown in Fig. 4–10, except that the energy differences are much greater, and thus the wavelengths are in the visible region instead of in the microwave region. The pumping is done with light and the laser produces light. Of course the pumping light must be "bluer" than the laser light. Like the ammonia maser, lasers are usually used as oscillators, light sources. The laser crystal is a cylinder with optically parallel ends, one of which is fully silvered and the other is partly silvered. The pumping light is admitted through the sides, and the laser light emerges from the partly transmitting end. Laser light traveling off axis is quickly lost by absorption at the sides, but the beam along the axis builds up to a high intensity as portions of the beam go back and forth between the reflecting surfaces insuring encounter with as many excited atoms as possible. We mentioned near the beginning of this discussion that stimulated radiation is in phase with stimulating photons. As photons stimulate more and more photons, there results a cascade of photons, all of which are in phase (coherent) and almost all of which move parallel to the crystal axis. As a consequence, laser light forms a beam with almost no "spread." If a laser produces a spot of light one centimeter in diameter on a nearby wall, the spot will be very little larger on a distant building.

4–13. MANY-ELECTRON ATOMS

We have stated that the Bohr theory is successful only for one-electron atoms, which pretty well limits quantitative application of the theory to hydrogen, heavy hydrogen, and ionized helium.

In our solar system, the various planets interact with one another only very slightly. The motion of the planets is understood by first considering that each planet experiences only the gravitational force of the sun. When these problems are solved for each planet, the weak planet-to-planet forces are treated as minor perturbations. The solar analogy is fruitful in understanding hydrogen, since hydrogen has but one electron. When an atom has two or more electrons, the interelectron interactions are not small and the simple planetary model breaks down. But one can sense what is in fact found: that a multiplicity of energy levels exists in these cases and so there are many lines in the spectra. Furthermore, the whole treatment so far has been of an *isolated* atom, that is, atoms in the gaseous state. In the liquid and solid states, the interactions between the electronic systems of the closely packed atoms introduce so many additional energy levels that the spectrum is now essentially continuous. Thus if the electrodes for an electric arc are copper rods, the spectrum of the light from the incandescent rods is continuous, whereas that from the gaseous copper ions in the arc between the electrodes is a line spectrum.

For a logical treatment of atomic structure, we should next turn to wave mechanics, which is a form of quantum mechanics. These theoretical techniques permit a more sophisticated understanding of both hydrogen and the heavier elements. But, historically, electronic structure was developed empirically before it was developed theoretically. Because our interest is directed inward toward the nucleus rather than outward to the more complex electronic structure of the heavier elements and molecules, our treatment of these subjects will be somewhat empirical and somewhat superficial.

4–14. QUANTUM NUMBERS

We have computed the spectrum of hydrogen in terms of but one quantum number n, called the principal quantum number. Measurements of extreme precision on hydrogen and of less precision on other elements show that the energy levels have fine structure (spectroscopes used in elementary courses in light are often able to separate the sodium doublet). If the substance under study is placed in a magnetic or an electric field, the energy levels are further subdivided. These are called the *Zeeman* and *Stark* effects, respectively. It turns out, therefore, that additional quantum numbers are needed in order to specify in detail the energy levels of atomic electrons. It is found that four quantum numbers are required.

The Bohr atomic model assumes that the electron orbits are circular. In planetary motion the circular orbit is possible but it is a very special case of an ellipse. An obvious extension of the Bohr model was to allow for the possibility of elliptical orbits. This extension was made by Sommerfeld in 1915.

Bohr's first postulate was that the angular momentum of an orbital electron is an integral multiple of $(h/2\pi)$. Although this integer n is associated only with angular momentum in the case of circular orbits, this association must be modified when elliptical orbits are considered because both the distance of the electron from the nucleus and the angular position of it change during a revolution. Just as the circular path is one extreme of an elliptical orbit, so a straight line is another. Mathematically, at least, we can consider a linear orbit where the electron passes directly through the nucleus. Such an orbit will have an energy associated with it, although it has no angular momentum. We can imagine a whole set of orbits as having the same energy but ranging from the straight-line orbit with no angular momentum to a circular orbit for which the angular momentum is maximum.

Sommerfeld's argument led *ultimately* to the retention of the principal or total quantum number, n, associated with the mean separation of the electron and proton as in our Eq. (4–16), and to the introduction of an orbital quantum number to characterize the angular momentum. The *orbital* quantum number, l, specifies the *number of units of angular momentum* ($h/2\pi$) associated with an electron in a given orbit. This quantity can be represented vectorially as shown in Fig. 4–13 by a straight line which is parallel to the axis of rotation and whose positive sense is given by the direction of advance of a right-hand screw rotated with the motion. It was found that the integer l could take on only positive values from zero to $(n - 1)$. Thus the electron in the smallest orbit ($n = 1$) will have no angular momentum, since for it l must equal zero. For a larger orbit with $n = 3$, there are three possible orbital shapes or eccentricities: a straight line through the nucleus without angular momentum for $l = 0$, an ellipse with an angular momentum for $l = 1$, and a "rounder" ellipse with more angular momentum for $l = 2$. It may seem mechanically absurd to think of an electron passing through the nucleus as required for $l = 0$, but this was reasonable in the wave mechanics which had been introduced when this scheme was completed.

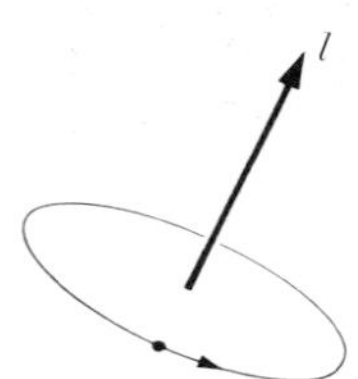

FIG. 4–13
Vector representation of orbital angular momentum.

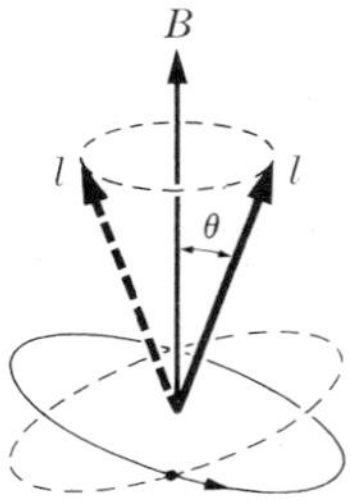

FIG. 4–14
Larmor precession of an electron orbit in a magnetic field.

The orbital motion of an electron is equivalent to a current in a loop of wire, so each orbit has a magnetic moment. The magnetic moment vector is normal to the plane of the orbit, but *antiparallel* to l because the electron's charge is negative. When the atom is placed in an external magnetic field, each electronic orbit will be subject to a torque which tends to make the l-vector parallel to the field. These vectors are shown in Fig. 4–14, where θ is the angle between l and the external magnetic induction, B. Because of the righting torque of the field on the revolving system, the l-vector will precess about the field for the same general reasons that account for the motion of a spinning top. This motion of the orbit, called the *Larmor precession,* introduces additional energy states into the atomic system. The amount of energy depends upon the precessional velocity and this depends upon the righting torque which, in turn, depends upon θ. If all values of θ could occur, then there would be an infinite number of new energy states. The Zeeman effect shows, however, only a few additional lines, not a continuous spectrum. Therefore only a few values of θ are "allowed" and these are those for which $l \cos \theta$, the projection of the l-vector on the direction of the magnetic induction, is an integer.

This introduces another quantum number called the *orbital magnetic* quantum number, m or m_l. It can have any integral value from $-l$ to $+l$ inclusive. Since m_l is limited to whole numbers, it means that the component of the orbital angular momentum along the magnetic field is restricted to *integral multiples* of $(h/2\pi)$. This is shown in Fig. 4–15 for $l = 2$. The "allowed" values of θ are given by $m_l = l \cos \theta$. Because of the restrictions on the orientation of the electron orbits, they are said to be space quantized. Space quantization of atoms was first verified in 1921 by Stern and Gerlach by observing the deflection of a narrow beam of atoms in silver vapor in an inhomogeneous magnetic field.

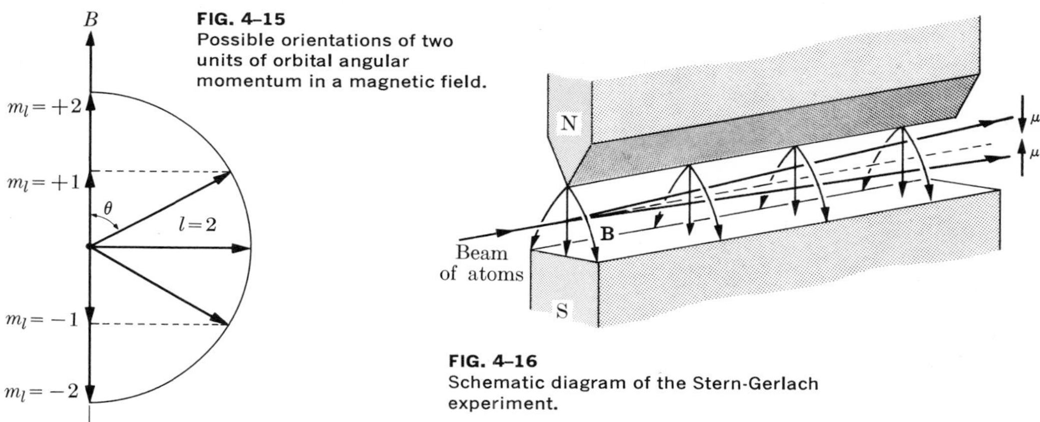

FIG. 4–15
Possible orientations of two units of orbital angular momentum in a magnetic field.

FIG. 4–16
Schematic diagram of the Stern-Gerlach experiment.

To discuss the Stern-Gerlach experiment, we return to the discussion earlier in this section and note that an orbiting electron is equivalent to a current loop whose magnetic moment vector μ is antiparallel to its angular momentum vector l. When the loop is in a uniform magnetic field, it experiences only a torque which produces the Larmor precession already mentioned. In a nonuniform field, however, in addition to the torque action there is a resultant force which gives the loop motion of translation. The magnitude of the force depends on the nonuniformity of the magnetic induction B and on the inclination of the magnet moment vector to the field, as specified by the angle θ in Fig. 4–14. The direction of the translational force is either toward or away from the region of increased magnetic field intensity, depending upon the direction of the component of μ along a field line. The force is maximum in one direction when $\theta = 0°$, maximum in the opposite direction when $\theta = 180°$, and zero when $\theta = 90°$. In the Stern-Gerlach experiment, a beam of silver atoms was projected through a nonuniform magnetic field, as shown in Fig. 4–16. Classically, one would expect the beam to be "smeared" out in the vertical direction, because the atomic current loops would be randomly oriented. However, it was found that the beam split into two distinct parts with nothing between them. This means that the magnetic moment in the case of the silver atom can only be aligned with or against the field. Similar experiments with beams of atoms of other elements also show space quantization of the atoms in a magnetic field.

In order to account for the fine structure of some spectral lines, Uhlenbeck and Goudsmit introduced a fourth quantum number in 1925. This is the electron spin quantum number, s or m_s. This spin of the electron about its own axis as it revolves about the nucleus is analogous to the rotation of the earth as it moves along its orbit around the sun. The spin momentum has the *numerical value* $\frac{1}{2}(h/2\pi)$. The rotating electron also has a magnetic moment, and spectroscopic observations show that the spin vector is capable of orientation only in either of two ways, parallel or antiparallel to the surrounding magnetic field. Therefore m_s can have only two values, $+\frac{1}{2}$ or $-\frac{1}{2}$. In the absence of a magnetic field, there is no unique assignment to the direction of m_l or m_s.

TABLE 4–2

VALUES OF QUANTUM NUMBERS

Principal	$n = 1, 2, 3, 4, \ldots$
Orbital	$l = 0, 1, 2, 3, \ldots, (n-1)$
Orbital magnetic	$m_l = -l, (-l+1), \ldots, (l-1), l$
Spin	$m_s = -\frac{1}{2}, +\frac{1}{2}$

The order in which the four quantum numbers have been discussed gives, in general, the order of their importance in determining the energy of any particular level. The restrictions on the values of these quantum numbers are summarized in Table 4–2.

The elements whose spectra most closely resemble hydrogen are the alkali metals, lithium, sodium, and potassium. Like hydrogen, these elements display series of lines. These series were given the descriptive names, *sharp*, *principal*, *diffuse*, and *fundamental*. When it was learned that the sharp series resulted from electron transitions to the $l = 0$ state, and the principal series resulted from transitions to the $l = 1$ state, etc., electrons in the $l = 0, 1, 2, 3, 4, 5, \ldots$ states came to be described as being in $s, p, d, f, g, \ldots$ (continuing alphabetically) states. This notation is extended by writing the value of n for a state before the letter, and if an atom has more than one electron with particular values of n and l, this number is written as a superscript of the letter. Thus if an atom has six electrons for which $n = 3$ and $l = 2$ the atom is said to have $3p^6$ electrons.

Taking four quantum numbers into consideration greatly increases the complexity of an energy-level diagram and the spectrum from which it is deduced. But the increase in the complexity of the spectrum is less than might at first be expected because many of the energy transitions that appear possible *do not occur*. There are certain rules, called *selection rules*, which specify between which energy levels an electron does move and, in consequence, which energy differences do appear as radiation. For example, one selection rule is that an atom can undergo no energy transition unless $\Delta l = \pm 1$. Thus $4s$ to $2p$, $2p$ to $1s$, and $5d$ to $2p$ are observed transitions. On the other hand, $4s$ to $1s$, $3d$ to $1s$, and $4f$ to $2p$ transitions are not observed.

4–15. PAULI EXCLUSION PRINCIPLE

Another generalization of utmost importance which was empirically arrived at is known as the *Pauli exclusion principle.* It states that, in any one atom, no two electrons have identical sets of quantum numbers. The electron in hydrogen does not demonstrate this rule since there is but one electron in this element, but the Pauli principle provides a vital key to the electron structure of the heavier elements. This principle, together with the *usual* energy importance of the quantum numbers, enables us to specify the states of the electrons in unexcited atoms—at least near the beginning of the periodic table.

The electron in hydrogen may have any energy state listed in Table 4–2. When this electron is in its lowest energy state, however, two of the quantum numbers are $n = 1$ and $l = 0$, and it is then in the $1s$-state. It is obvious from Table 4–2 that $m_l = 0$ whenever the electron is in any s-state. Helium has two electrons, one of which has those quantum numbers assigned to hydrogen. The second electron has the same quantum numbers except that its spin has a sign opposite to that of the first. Therefore, these two $1s^2$ electrons have different quantum numbers, as required by the Pauli exclusion principle.

Lithium has three electrons, two of which have the quantum numbers already assigned. These two $1s^2$ electrons constitute a helium core. The third electron must have considerably more energy than either of the first two. We have exploited both permitted variations of spin. The next lowest energy quantum number, m_l, cannot be increased without increasing l, but since l must remain less than n, which thus far has been 1, we see that we must now increase the most important quantum number n. Thus the lowest set of quantum numbers that can be assigned to the third lithium electron require it to be in the $2s$-state. The $2s$-state has $l = 0$ which requires that $m_l = 0$. But since the spin may be either $+\frac{1}{2}$ or $-\frac{1}{2}$, we find that the fourth electron, which forms beryllium, may also be a $2s$ electron. The electron configuration of beryllium is written $1s^2 2s^2$.

TABLE 4–3

Element	Z	First ionization potential, V	Atom in the ground state				
			Quantum number of last added electron		Electron configuration		Outermost shell occupied
			n	l			
H	1	13.6	1	0	$1s$		K
He	2	24.6	1	0	$1s^2$		K
Li	3	5.39	2	0	Helium core, 2 electrons	$2s$	L
Be	4	9.32	2	0		$2s^2$	L
B	5	8.30	2	1		$2s^2\ 2p$	L
C	6	11.3	2	1		$2s^2\ 2p^2$	L
N	7	14.5	2	1		$2s^2\ 2p^3$	L
O	8	13.6	2	1		$2s^2\ 2p^4$	L
F	9	17.4	2	1		$2s^2\ 2p^5$	L
Ne	10	21.6	2	1		$2s^2\ 2p^6$	L
Na	11	5.14	3	0	Neon core, 10 electrons	$3s$	M
Mg	12	7.64	3	0		$3s^2$	M
Al	13	5.98	3	1		$3s^2\ 3p$	M
Si	14	8.15	3	1		$3s^2\ 3p^2$	M
P	15	10.6	3	1		$3s^2\ 3p^3$	M
S	16	10.4	3	1		$3s^2\ 3p^4$	M
Cl	17	13.0	3	1		$3s^2\ 3p^5$	M
Ar	18	15.8	3	1		$3s^2\ 3p^6$	M
K	19	4.34	4	0	Argon core	$4s$	N

Boron has $Z = 5$ and has five electrons. The first four states "used up" all four possible s-states with $n = 2$. But since $n = 2$, then l may equal 1, and the fifth electron is a $2p$ electron. When $l = 1$, then m_l may be $-1, 0$, or 1, and each value of m_l may have spins $\pm\frac{1}{2}$ so there may be six electrons in the $2p$-state. The states described account for the configurations through neon, whose 10 electrons have the configuration $1s^2 2s^2 2p^6$. With neon we have exploited every possible state with electrons for which $n = 1$ or 2. The eleventh electron in sodium must therefore have $n = 3$, and its configuration is $1s^2 2s^2 2p^6 3s$. The $3s$-state will also accommodate magnesium, but the last electrons added for aluminum through argon must be $3p$ electrons. Table 4–3 summarizes how the Pauli principle enables us to establish atomic electron configurations through potassium where irregularities set in because the l quantum number becomes more energy-important than n, so that it is "cheaper" to increase n rather than l. Although our scheme becomes more complex with potassium, it is possible to extend the table through all the irregularities and assign quantum numbers to all the electrons of the unexcited elements. Short as Table 4–3 is, it permits a qualitative understanding of many properties of the atoms.

Table 4–3 shows a distinct periodicity for those elements with $n = 1$, 2, 3. The $n = 1$ group consists of chemically active hydrogen and chemically inactive helium. The $n = 2$ and $n = 3$ groups each begin with active alkalies and end with inert gases. Silicon and carbon have very similar chemistry and we note that, except for n, their most energetic electrons have identical quantum numbers. These groups constitute the rows of the periodic table, and Table 4–3 accounts for the first three rows with two, eight, and eight elements, respectively.

4–16. ELECTRON SHELLS AND CHEMICAL ACTIVITY

From the preceding discussion and from a study of Table 4–3, it is evident that whenever an element is reached for which the vector sum of the magnetic quantum numbers and the sum of the spin numbers are each equal to zero, we have a very stable electron configuration. This is the case for each of the noble gases. Whenever these sums are zero, we say that we have completed an *electron grouping or shell.* The innermost grouping is called the *K-shell* and is complete for helium, which has two K-shell electrons. As we go on adding electrons to the structure (and, of course, an equal number of protons to the nucleus), the grouping takes place farther out than before, in the region called the *L-shell.* This is filled when we arrive at neon. This element has a helium core surrounded by eight electrons in the completed L-shell. Continuing the addition of electrons brings us to argon, which has a neon core surrounded by eight electrons in the completed *M-shell.* In a similar way, the *N-, O-, and P-shells* become populated as we go on in the periodic table. In general, none of these outer shells is filled completely before some electrons enter one or more other shells which are not yet fully populated. The last column of Table 4–3 shows the outermost of each unexcited atom. We will need the electron-shell concept in our discussion of x-rays in Chapter 6.

Atoms having completed shells are very stable. This conclusion is supported by their chemical inactivity, but there is substantiating physical evidence too. The stability of the atomic system can be judged from the first ionization potential, which is a measure of the energy required to remove an electron from it. (The second ionization potential refers to the removal of the second electron and so on.) Table 4–3 shows that this first ionization potential rises to its greatest value, 24.6 V, for helium and then drops sharply to 5.39 V for lithium. This means that it is difficult to disrupt the completed K-shell, but easy to remove a lone electron in the L-shell. The required potential increases as we go from element to element until it again reaches a maximum in neon. After that, it drops once more to a low value but gradually increases until another maximum is reached in argon. This pattern is repeated periodically as one goes on through the periodic table of elements. Since the L-shell electrons of neon are somewhat screened from the attraction of the nucleus by the helium core, the ionization potential of neon is less than that of helium. Similarly, the M-shell electrons of argon are partially screened by the neon core. You probably have noticed that the potentials do not increase uniformly in going from beryllium to neon or from sodium to argon. Indeed, some are out of order on the basis of increasing numerical values. It is beyond the scope of this book to discuss these apparent anomalies, but the explanation can be found in more exhaustive books on atomic structure.

The stability of chemical compounds can be inferred from the ionization potentials in Table 4–3. From this it is evident that fluorine does not give up an electron readily; rather, it seeks an additional electron to complete its L-shell quota and thus attain the stable neon configuration. It is electron hungry and, therefore, readily becomes a negative ion in solution. On the other hand, sodium has a lone electron in the M-shell which is easily removed. If it lost this one, it would have the stable neon structure. Because of this excess electron, sodium readily becomes a positive ion in solution. Thus one expects, and in fact finds, sodium fluoride to be a very stable compound because each atom provides what the other one needs. Now consider carbon, which is halfway between helium and neon. It could revert to the helium arrangement by "giving up" electrons. This is what it does when it forms carbon dioxide, CO_2. On the other hand, it could approach the neon state by "taking on" electrons. This is what happens when it unites with beryllium to form beryllium carbide, Be_2C. In general, elements located just before a noble gas in the periodic table form strongly electronegative ions; those which follow are strongly electropositive; and those in the middle range between two of the noble gases are electronegative in their reactions with some elements but electropositive in others.

4–17. MOLECULES

The tendency for electrons to pair off within an atom so that the vector sum of their quantum numbers is zero has a strong bearing on the tendency of atoms to combine into compounds. A single, unexcited hydrogen atom has one electron with a spin of either $+\frac{1}{2}$ or $-\frac{1}{2}$. If this atom is near another hydrogen atom, the situation becomes more complex. If both atoms have their spin vectors parallel and if they should combine, the total electron spin of the molecule would be -1. Since this is in opposition to the observed tendency, we know these atoms will repel each other and not combine. If, however, the spins are antiparallel, the two atoms attract each other, forming a molecule whose total electron spin is zero. Therefore, we normally find hydrogen as the molecule H_2. Thus we have a stable molecule of an element when the resultant of the electron spins of its constituent atoms is zero. This is the answer to the objection, stated in Chapter 1, of Dalton and others to Avogadro's view that the molecules of hydrogen, nitrogen, oxygen, and many other gases are diatomic.

Helium, on the other hand, has two electrons with oppositely directed spins. Since its two electrons are paired, it has no tendency to react with atoms of its own or other kinds. To make helium chemically active, we would have to reverse the spin of one of its electrons. But we have seen that the Pauli principle requires an increase in the principal quantum number n before the spin can be changed. This requires a large energy which might be supplied by heating. Thermal agitation might excite the helium until it was chemically active, but the hot helium atom is not near other excited helium atoms long enough to react. Any that do react have so much energy that the molecule is unstable and quickly dissociates. Thus helium has no compounds and is found only in the atomic state.

In carbon, the six electrons are all paired off and carbon is inert at room temperature. The eight oxygen electrons are similarly paired off (though oxygen easily forms O_2, since its last electron can be reversed if m_l becomes unity). If we hold a match to paper, the carbon in the paper is easily excited. Its electrons become unpaired so that it reacts with oxygen. If there is not much oxygen available, the product is carbon monoxide, CO. In this case, one electron of each atom reverses and the two unpaired electrons of each atom attract each other. In CO, each atom has a valence of two. If there is an abundance of oxygen, the reaction product is CO_2. In this case, two carbon electrons reverse, causing all four outer electrons to be unpaired. Each two of these four attracts an oxygen atom to the one carbon atom. In this case, carbon has a valence of four while oxygen still has a valence of two.

Although examples of this kind can be cited almost without limit, it must be admitted that we have only scratched the surface of the problem of molecular structure. But in showing that chemical forces between atoms are determined by their electronic structure, we have suggested an answer to the question raised by Rutherford's nuclear model of the atom. A rigid body is not a cluster of plum-pudding atoms stuck together with glue. A rigid body is mostly empty space. Most of the mass is concentrated in tiny positive nuclei, each of which is surrounded by electrons. These electrons are not uniquely associated with their home nuclei but are shared with neighboring nuclei of the same or different kinds. Thus electronic interactions provide the "glue" which holds matter together, while electrostatic repulsion keeps the nuclei apart.

4–18. THE STATUS OF BOHR'S MODEL AND THEORY OF THE ATOM

In this chapter we have seen how Rutherford's discovery of the nucleus led Bohr to propose a model for the hydrogen atom and devise a theory which accounted for the spectral lines that hydrogen emits. Unfortunately, the Bohr model of the atom suffers from serious defects. In the first place it applies only to one-electron atoms, principally hydrogen. But even for this simple element, Sommerfeld found it necessary to postulate elliptical orbits and to consider the relativistic mass change (see Chapter 5) of the revolving electron in order to account for the details in the fine structure of the spectral lines. Attempts to apply Bohr theory to heavier elements were moderately successful in a few cases. The next heavier element, helium, has two electrons, but when one of these is removed by ionization, the remaining ion has but one electron and is hydrogen-like. The alkaline metals share with hydrogen the same column in the periodic table and their spectra are also hydrogen-like when ionized until only one orbital electron remains. But the Bohr model is not a general solution to the problem of atomic structure.

Furthermore, the Bohr model does not answer all the questions we can raise about hydrogen. A complete analysis should not only reveal the frequency of the emitted light, but also the *relative intensity* of the different spectral lines. In Chapter 8 we will discuss wave mechanics, which has replaced the Bohr analysis. However, we need not be dismayed with Bohr's work simply because it has been replaced by something better. Road maps are far less accurate than topographical survey maps. But on a cross-country trip, road maps are a better guide because the survey maps are cumbersome and contain details that are often irrelevant. The Bohr theory of hydrogen is like a road map which we use even when we know a more precise map exists.

Despite its shortcomings, the Bohr model was a conceptual "breakthrough" that facilitated many empirical observations about the electronic structure of the heavier atoms to which Bohr's work did not apply directly. We shall find Bohr's conceptual scheme very useful in describing the x-ray spectra of heavy elements. Although wave mechanics has replaced the Bohr model, wave mechanics confirms and builds upon the energy-level concept that Bohr introduced.

REFERENCES

BALMER, JOHANN J., "The Hydrogen Spectral Series," in MAGIE, W. F., *A Source Book in Physics*, pp. 360–365. Cambridge, Mass.: Harvard University Press, 1935.

BEYER, ROBERT T., *Foundations of Nuclear Physics*. New York: Dover, 1949. A collection of facsimiles of thirteen fundamental studies as they were originally reported in the scientific journals. Included is "The Scattering of α and β Particles by Matter and the Structure of the Atom" by Rutherford.

BOHR, NIELS, *The Theory of Spectra and Atomic Constitution*. 2nd ed. Cambridge, England: University Press, 1924. Bohr's own account of the stages in the development of his theory of atomic structure.

CANDLER, CHRIS, *Atomic Spectra and the Vector Model*. 2nd ed. Princeton, N. J.: Van Nostrand, 1964.

CHRISTY, R. W., and AGNAR PYTTE, *The Structure of Matter*. New York: Benjamin, 1965.

FLINT, GRAHAM, "Determination of Laser Hazards," *Proceedings of the First Conference on Laser Safety (1966)*, sponsored by the Martin Company, Orlando, Fla.

FRIEDMAN, F. L. and L. SARTORI, *The Classical Atom*. Reading, Mass.: Addison-Wesley, 1965. An excellent account of the experiments and subsequent theory of the atom prior to quantum mechanics.

GOLDENBERG, H. M., D. KLEPPNER, and N. F. RAMSEY, "Atomic Hydrogen Maser," *Phys. Rev. Letters* **5**, 361 (1960).

HERZBERG, GERHARD, *Atomic Spectra and Atomic Structure*. 2nd ed. J. W. T. Spinks, translator. New York: Dover, 1944. A good basic book.

HUND, FRIEDRICK, "Paths to Quantum Theory Historically Viewed," *Phys. Today* **19**, No. 8, 23 (August, 1966).

KAPLAN, IRVING, *Nuclear Physics*. 2nd ed. Reading, Mass.: Addison-Wesley, 1962. Chapters 3 and 7.

KING, G. W., *Spectroscopy and Molecular Structure*. New York: Holt, Rinehart, and Winston, 1964.

LENGYEL, B. A., *Lasers: Generation of Light by Stimulated Emission*. New York: Wiley, 1962.

MOOS, H. W., *Masers and Optical Pumping*, Resource Letter MOP-1 and Selected Reprints. New York: American Institute of Physics, 1965.

PAULING, LINUS, *The Nature of the Chemical Bond*. 3rd ed. Ithaca, N. Y.: Cornell University Press, 1960.

RICHTMYER, F. K., E. H. KENNARD, and T. LAURITSEN, *Introduction to Modern Physics*. 5th ed. New York: McGraw-Hill, 1955. Chapter 6.

ZORN, J. C., *Experiments with Molecular Beams*, Molecular Beams, Vol. 1 and *Atomic and Molecular Beam Spectroscopy*, Molecular Beams, Vol. 2, Resource Letter MB-1 and Selected Reprints. New York: American Institute of Physics, 1965.

PROBLEMS

4–1. Derive the equation for the radii of the Bohr orbits, Eq. (4–16), from Eqs. (4–12) and (4–14).

4–2. Show that the tangential speed of an electron in its orbit is

$$v = \frac{Ze^2}{2\epsilon_0 nh},$$

and its angular speed is

$$\omega = \frac{\pi m Z^2 e^4}{2\epsilon_0^2 n^3 h^3}.$$

4–3. (a) Calculate the radii of the first, second, and third "permitted" electron orbits in hydrogen in angstroms. (b) What is the diameter of the hydrogen atom in the ground state?

4–4. The Bohr model for hydrogen shows that the orbital electron can only be found at certain fixed distances from the proton, the larger radii corresponding to higher quantum numbers. Assume that the electron in a hydrogen atom moves outward to larger radii. Which of the following quantities increase and which decrease: angular momentum, total energy, potential energy, kinetic energy, frequency of rotation?

4–5. A particle of mass m moves in a circular orbit of radius r under the influence of a force kr directed toward the center (k is a constant). Assuming that Bohr's postulates apply to this system, derive the equation for (a) the radii of the permissible orbits and (b) the energies of these orbits in terms of the quantum number n. (c) Show that the frequency radiated when the particle makes a transition from one orbit to the adjacent orbit is the same as the frequency of the circular motion.

4–6. Suppose one has a tank of atomic hydrogen at a temperature of 20°C and a pressure of 1 atm. If each of the atoms and also the mean free path were expanded until the diameter of each nucleus is 1 inch, what would then be (a) the radius of the smallest electron orbit in feet, and (b) the average distance traveled between collisions in miles? (Assume that the nuclear diameter is 3 fermis, and that the mean free path is 9×10^{-5} cm. This is four times the mean free path calculated for molecular hydrogen in Problem 1–17.) (c) What would be the mean free path on this expanded scale if the gas pressure were reduced to 1 mm of mercury? (d) If the gas were liquefied so that the atoms "touched," would there still be open space in the world of hydrogen atoms?

4–7. Calculate the binding energy of the electron in hydrogen in joules and in eV when $n = 1, 2, 3$, and infinity.

4–8. Show that the average translational kinetic energy of a molecule of a gas at 27°C is much less than the energy required to raise a hydrogen atom from its ground state to its first excited state.

4–9. An alpha particle having a kinetic energy of 7.68 MeV is projected directly toward the nucleus of a copper atom. What is their distance of closest approach? The mass of an alpha particle is four times that of the proton and the atomic number of copper is 29.

4–10. Evaluate the Rydberg constant for hydrogen from atomic constants, assuming a nucleus of infinite mass.

4–11. Rearrange and alter the Balmer formula, Eq. (4–5), so that the left-hand side is wave number in reciprocal meters. Show that the result agrees with Eq. (4–23) when the Rydberg constant is substituted in the latter equation and when $n_1 = 2$ and $n_2 = n$.

4–12. Calculate (a) the frequency, (b) the wavelength, and (c) the wave number of the H_β line of the Balmer series of hydrogen. This line is emitted in the transition from $n_2 = 4$ to $n_1 = 2$. Assume that the nucleus has infinite mass.

4–13. Using the reduced mass equations, calculate the wavelength of the H_β line (see Problem 4–12) in the Balmer series for the three isotopes of hydrogen: H^1; deuterium, H^2; tritium, H^3.

4–14. An atom of tungsten has all of its electrons removed except one. (a) Calculate the ground-state energy for this one remaining electron. (b) Calculate the energy and wavelength of the radiation emitted when this electron makes a downward transition from $n = 2$ to $n = 1$. (c) In what portion of the electromagnetic spectrum is this photon?

4–15. (a) Calculate the first three energy levels for the electron in Li^{++}. (b) What is the ionization potential of Li^{++}? (c) What is the first resonance potential for Li^{++}?

4–16. The Rydberg constants (reduced mass form) for hydrogen and singly ionized helium are 10967757.6 m^{-1} and 10972226.3 m^{-1} respectively. For the nuclei of these atoms,

$$M_{\text{He}} = 3.9726 M_{\text{H}}.$$

From the given data, calculate the ratio of the mass of the proton to that of the electron to four significant figures.

4–17. Calculate the short wavelength limit of each of the series listed in Table 4–1, and find the energy of the quantum in eV for each.

4–18. Some of the energy levels of a *hypothetical* one-electron atom (not hydrogen) are listed in the table below.

n	1	2	3	4	5	∞
E_n, eV	−15.60	−5.30	−3.08	−1.45	−0.80	0

Draw the energy level diagram and find (a) the ionization potential, (b) the short wavelength limit of the series terminating on $n = 2$, (c) the excitation potential for the state $n = 3$, and (d) the wave number of the photon emitted when the atomic system goes from the energy state $n = 3$ to the ground state. (e) What is the minimum energy that an electron will have after interacting with this atom in the ground state if the initial kinetic energy of the electron was (1) 6 eV, (2) 11 eV?

4–19. For a certain *hypothetical* one-electron atom the wavelengths in angstroms for the spectral lines for transitions originating on $n = p$ and terminating on $n = 1$ are given by

$$\lambda = \frac{1500p^2}{p^2 - 1}, \qquad p = 2, 3, 4, \ldots$$

(a) What are the least energetic and most energetic photons in this series? (b) Construct an energy level diagram for this element showing the energies of the lowest three levels. (c) What is the ionization potential of this element?

4–20. (a) What is the least amount of energy in eV that must be given to a hydrogen atom so that it can emit the H_β line (see Problem 4–12 and Fig. 4–4) in the Balmer series? (b) How many different possibilities of spectral line emission are there for this atom when the electron goes from $n = 4$ to the ground state?

4–21. In a certain gas discharge tube containing hydrogen atoms, electrons acquire a maximum kinetic energy of 13 eV. What are the wavelengths of all the lines which can be radiated?

4–22. The energy levels in eV of a *hypothetical* one-electron atom (not hydrogen) are given by

$$E_n = -18.0/n^2, \qquad \text{where} \qquad n = 1, 2, 3, \ldots$$

(a) Compute the four lowest energy levels and construct the energy-level diagram. (b) What is the excitation potential of the state $n = 2$? (c) What wavelengths in angstroms can be emitted when these atoms in the ground state are bombarded by electrons which have been accelerated through a potential difference of 16.2 V? (d) If these atoms are in the ground state, can they absorb radiation having a wavelength of 2000 Å? (e) What is the photoelectric threshold wavelength of this atom?

4–23. The frequencies f of the spectral lines emitted by a certain *hypothetical* one-electron atom (not hydrogen) are given by the relation:

$$f = 864 \times 10^{12} \left(\frac{1}{n_1^2} - \frac{1}{n_2^2}\right) Hz$$

where the n's are the principal quantum numbers. (a) Find the wavelengths in angstroms of the first three lines of the series terminating on the ground state. (b) What is the photoelectric threshold wavelength of this atom? (c) Construct the energy level diagram. Give the values of the energies in eV of the first four levels and show, with labeled arrows, the transitions which cause the emission of the wavelengths in parts (a) and (b). (d) What is the binding energy of the electron when it is in the state $n = 3$? (e) State and clearly explain the possible interactions when a large number of these hypothetical atoms in the ground state are bombarded by: (1) a beam of electrons having 2.90 eV of kinetic energy; (2) a beam of 2.90 eV photons.

4–24. Compute the magnitude of (a) the longest wavelength in the Lyman series and (b) the shortest wavelength in the Balmer series. (c) From these wavelengths determine the ionization potential of hydrogen.

4–25. The first ionization potential of helium is 24.6 V. (a) How much energy in eV and in joules must be supplied to ionize it? (b) To what temperature would an atmosphere of helium have to be heated so that an atom of it moving with the most probable speed in a Maxwellian distribution would have just enough kinetic energy of translation to ionize another helium atom by collision?

4–26. Neglecting reduced mass corrections, show that (a) the short-wavelength limit of the Lyman series ($n = 1$) of hydrogen is the same as that of the Balmer series ($n = 2$) of singly ionized helium, and (b) that the wavelength of the first line of this helium series is 1.35 times the wavelength of the first line of the Lyman series of hydrogen. (c) How much energy must be given to singly ionized helium in the ground state so that the first line in the Paschen series will be emitted?

4–27. (a) Show that the frequency of revolution of an electron in its circular orbit in the Bohr model of the atom is $f = mZ^2e^4/4\epsilon_0^2n^3h^3$. (b) Show that when n is very large, the frequency of revolution equals the radiated frequency calculated from Eq. (4–22) for a transition from $n_2 = n + 1$ to $n_1 = n$. (This problem illustrates Bohr's *correspondence principle,* which is often used as a check on quantum calculations. When n is small, quantum physics gives results which are very different from those of classical physics. When n is large, the differences are not significant and the two methods then "correspond.")

4–28. A 10-kg satellite circles the earth once every 2 h in an orbit having a radius of 8000 km. (a) Assuming that Bohr's angular momentum postulate applies to satellites just as it does to an electron in the hydrogen atom, find the quantum number of the orbit of the satellite. (b) Show from Bohr's first postulate and Newton's law of gravitation that the radius of an earth-satellite orbit is directly proportional to the square of the quantum number, $r = kn^2$, where k is the constant of proportionality. (c) Using the result from part (b), find the distance between the orbit of the satellite in this problem and its next "allowed" orbit. (d) Comment on the possibility of observing the separation of the two adjacent orbits. (e) Do quantized and classical orbits correspond for this satellite? Which is the "correct" method for calculating the orbits?

4–29. On the average, an atom will exist in an excited state for a shake, 10^{-8} s, before it makes a "downward" transition and emits a photon. Assuming that the electron in hydrogen is in the state $n = 2$, how many revolutions about the nucleus are made before the electron "jumps" to the ground state?

4–30. What volume, in cubic miles, of *atomic* hydrogen at a pressure of 1 atmosphere and a temperature of 20°C is required so that, according to the Boltzmann distribution, it will contain one atom in the energy state $n = 2$?

4–31. (a) What would the temperature of the atmosphere of the sun have to be so that thermal agitation alone would put one millionth of the hydrogen atoms there into the necessary energy state for the absorption of the Balmer series wavelengths, that is, into the state $n = 2$? (b) What other means would be available in the atmosphere of the sun for pumping up hydrogen to this higher energy state? (c) Compare the answers to parts (a) and (b) with the results obtained in Problem 1–35 in which the temperature given is about the actual value for the atmosphere of the sun.

4–32. (a) Show that the magnetic moment of a circular Bohr orbit in hydrogen is given by $n(h/4\pi)(e/m_e)$. (The magnetic moment of a current-carrying loop of wire is equal to the product of the current and the area bounded by the loop.) (b) Calculate the magnetic moment of the orbit in hydrogen for which $n = 1$. (This particular value of the magnetic moment is called the Bohr magneton, μ_B.)

4–33. How can you account for the fact that lithium and sodium have similar chemical properties?

4–34. (a) Show that the maximum number of electrons that can be accommodated in a shell specified by the quantum number n is $2n^2$. (b) How many elements would there be if the electronic shells through $n = 7$ were completely occupied?

$$N = \sum_{l=0}^{n-1} 2(2l+1) = 2 + 4\sum_{l=1}^{n-1} l = 2 + 4\frac{(n-1)n}{2} + 2(n-$$

CHAPTER 5

RELATIVITY

5-1. IMPORTANCE OF VIEWPOINT

What one observes often depends on one's viewpoint. An election will be interpreted differently by Democrats and Republicans. A change in college rules may be viewed differently by students and administration. What scientists observe also depends on their viewpoint. Consider the interpretations of physical events that might be made by a person of high I.Q. born and brought up on a merry-go-round. He would experience a force somewhat like the force of gravity. It would be down at the center of rotation, and, because of the centrifugal component, it would be directed down and out at points away from the center. Our observer could learn to live happily with this force. He might even seek to write a quantitative description of it and express its nature in terms of mathematical equations. Though we express the force of gravity we experience as $F = mg$ (always down), our observer would have a harder time of it, since his force is not constant in either magnitude or direction. If our merry-go-round observer were to have the genius to devise a whole system of mechanics, the mechanics, he would devise would not be the mechanics of Newton. Newton's mechanics, which incorporates the concept of inertia introduced by Galileo, includes the law that a moving body subject to no forces will continue to move with a constant speed in a constant direction. But an airplane which flew "straight" over the merry-go-round would appear to move in a curved path to our special observer.

The situation described appears far-fetched, but the earth turns, and we were all born and brought up on a merry-go-round! The acceleration due to gravity is complicated by the rotation of the earth, and there are many other effects, like cyclonic storms, which are a direct result of the rotation. Newton knew of this complication. The earth turns rather slowly compared with a regular merry-go-round, and the effects of the earth's rotation are small enough that most of the time we can neglect them. But, strictly speaking, Newton's laws just do not hold precisely for an observer who is earthbound. For Newton's laws to be valid precisely, we must observe events from what is called an *inertial frame* or a Galilean-Newtonian coordinate system. Such a system is one which has no acceleration relative to a system of coordinates known to be fixed absolutely. This raises the question: What, *if anything*, constitutes a fixed system?

The whole structure of classical physics, then, is based on the assumption that we interpret all events as they would be interpreted by an observer whose viewpoint is in an inertial reference frame. The man on the merry-go-round would have a physics far from Newtonian. By stepping onto the earth he would find his physics more nearly Newtonian. The genius of Newton is, in part, that although he never could step off the earth physically, he did step off the earth mentally. He interpreted events as though he had no acceleration. Because of this shift in his viewpoint, he was able to write his laws of mechanics in the particularly simple form that he did.

But Newton never really knew to where he projected himself. He excluded the earth as a vantage point because the earth not only rotates on its axis but revolves about the sun. The sun offered possibilities, but even the sun moves and is probably accelerated through space. The stellar constellations were named by the ancients and the stability of their arrangement led to their being called the "fixed" stars. Yet it would be the strangest of coincidences if the "fixed" stars really were fixed. Any motion looks small when seen from far enough away, and from modern astronomical measurement we know that the constellations are slowly changing.

It would seem, however, that if we locate a frame of reference so that it is fixed relative to the stars, this will be sufficiently steady for Newton's laws to serve well enough for every practical purpose. Such a frame is good enough for the practical men who want to fly aircraft, build rocket engines, or communicate via television. But for those whose primary concern is the understanding of the nature of things and whose goal is the discovery of truth, uncertainty about the frame of reference represented a serious flaw in the logical structure of classical physics. *It is an amazing fact that concern over this seemingly minor point led ultimately to the discovery of how to release the energy of the nucleus of the atom.*

5-2. THE SEARCH FOR A FRAME OF REFERENCE—THE ETHER

The search for something more fixed than the stars went something like this. As we pointed out in Chapter 3, James Clerk Maxwell demonstrated that electricity and light are related phenomena. Starting with known properties of electricity and magnetism, Maxwell derived equations which are identical in form to the equations which describe many wave phenomena. He could demonstrate, furthermore, that the velocity of the waves he discovered was the same as the velocity of light. He could derive many other properties of light, and it was soon accepted that he had put the wave theory of light on a firm foundation. In this theory, light is an electromagnetic wave motion.

Every wave motion has something that "waves." Sound waves have air and water waves have water. Surely, it was argued, light waves must involve the waving of something even in free space. No one knew what it was, but it was given the name "luminiferous ether."

Light passes through many kinds of materials. It passes through relatively heavy materials like glass and it passes through the nearly perfect vacuum that must lie between the stars and the earth. Thus ether must permeate all of space. Light is a transverse wave motion. This comes out of Maxwell's theory and from many experimental observations, particularly those on polarized light. This implies that the ether is a solid. Transverse waves involve shear forces and can occur only in solids which can support shear. Sound waves in air must be longitudinal because of this fact. Furthermore, the ether must be a rigid solid. The propagation velocity of mechanical waves in various materials depends on the elastic constants of the material. These are much greater for steel than for air. The very great velocity of light thus implies that the ether must have a very large shear modulus. It is rather hard on the imagination to suppose that all space is filled with this rigid solid and that all material objects move through this solid without resistance, yet it was supposed to exist. However fanciful it may seem to us, physicists felt that this ether might be just the thing to which to attach a Newtonian coordinate system. It was conceived that Newton's laws would hold exactly for an observer moving without acceleration *relative to the ether*.

If the ether is assumed to be at rest, then the interesting question is: How fast are we moving through the ether? Since all speculations about the ether stem from its properties as a medium for carrying light, an optical experiment is indicated. It is not hard to compute how sensitive the apparatus must be in order to measure the ether drift. Assuming, for the sake of argument, that the sun has no ether drift, the velocity of the earth through the ether must be its orbital velocity. If the sun has an ether drift, then the drift of the earth will be even greater than its orbital velocity at some seasons. Knowing that the radius of the earth's orbit is about 93 million miles, we can find the orbital velocity to be about 18.5 mi/s. By performing the experiment at the best season of the year, we know that we should be able to find an ether drift of at least 18.5 mi/s. The velocity of light is 186,000 mi/s. Great as our orbital velocity is, it is only about 10^{-4} times the velocity of light; so it is evident that a very sensitive instrument is required.

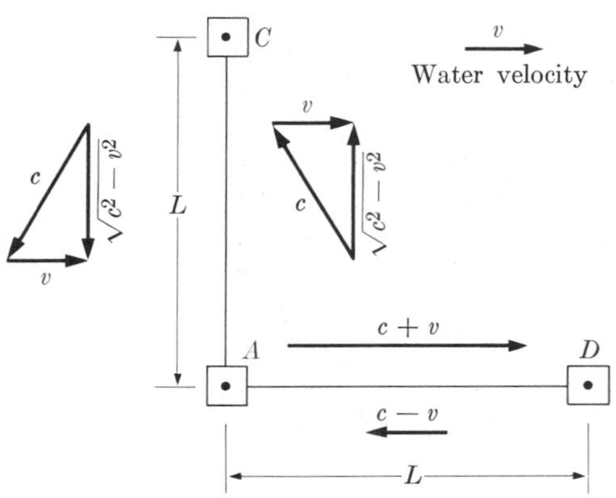

FIG. 5–1
Swimming analogy to the Michelson-Morley experiment.

5–3. THE MICHELSON INTERFEROMETER

A device of sufficient sensitivity was made and used in the United States by Michelson and Morley in 1887. The principle of their apparatus is brought out by the following analogy. Suppose two equally fast swimmers undertake a race in a river between floats anchored to the river bed. Two equal courses, each having a total length $2L$, are laid out from the starting point, float A, as shown in Fig. 5–1. One course is AD, parallel to the flow of the river relative to the earth, and the other is AC, perpendicular to it. How will the times compare if each of the swimmers goes out and back on his course? Let the speed of each swimmer relative to the water be c, and let the water drift or velocity with respect to the earth be v. When the swimmer on the parallel course goes downstream, his velocity will add to that of the water, giving him a resultant velocity of $(c + v)$ with respect to the earth. The time required for him to swim the distance L from A to D is $L/(c + v)$. On his return, he must overcome the water drift. His net velocity then is $(c - v)$, and his return time is $L/(c - v)$. His total time is the sum of these two times. This is seen to depend upon the velocity of the water, and is given by

$$t_{\parallel} = \frac{L}{c+v} + \frac{L}{c-v} = \frac{2Lc}{c^2 - v^2}. \qquad (5\text{–}1)$$

The other swimmer, going perpendicular to the water drift, spends the same time on each half of his trip, but he must head upstream if he is not to be carried away by the current. The component of his velocity that carries him toward his goal is $\sqrt{c^2 - v^2}$ with respect to the earth. The total time for his trip also depends on the water drift, and is

$$t_{\perp} = \frac{2L}{\sqrt{c^2 - v^2}}. \qquad (5\text{–}2)$$

To see how these two times compare, we divide the parallel course time, Eq. (5–1), by the perpendicular course time, Eq. (5–2), and obtain

$$\frac{t_{\parallel}}{t_{\perp}} = \frac{2Lc}{c^2 - v^2} \cdot \frac{\sqrt{c^2 - v^2}}{2L} = \frac{1}{\sqrt{1 - (v^2/c^2)}}. \qquad (5\text{–}3)$$

In still water $v = 0$, the ratio of the times is unity, and the race is a tie, as we would expect. In slowly moving water, the ratio is greater than unity and the swimmer on the perpendicular course wins; or, put differently, if the swimmers are stroking in phase when they leave float A, they will be out of phase when they return to it. If the velocity of the river increases to nearly that of the swimmers, then the ratio tends toward infinity. If the river velocity exceeds the swimmer velocity, the entire analysis breaks down. The ratio becomes imaginary and both swimmers are swept off the course by the current. The point is that, by observing the race, the water velocity relative to the anchored floats can be measured.

The optical equivalent of the above situation is to have a race between two light rays over identical courses, one parallel and one perpendicular to the ether drift. The instrument used, called a Michelson interferometer, is shown schematically in Fig. 5–2.

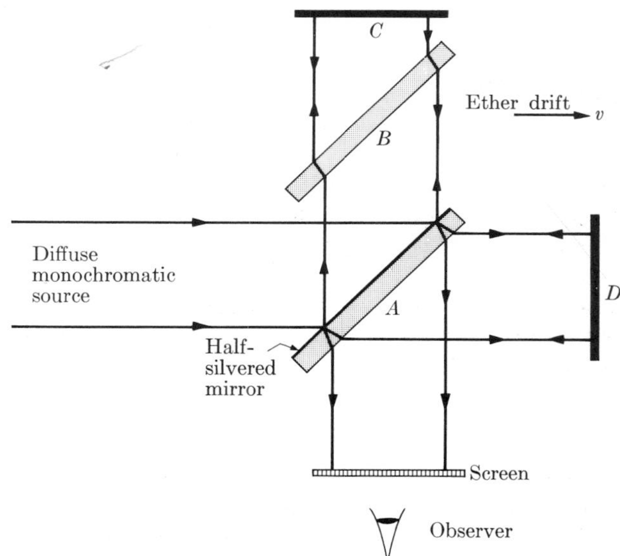

FIG. 5–2 The Michelson interferometer.

Light enters the apparatus from the source at the left. At A it strikes a glass mirror which has a half-silvered surface. Half the light is reflected up toward B and C, while the other half refracts at both surfaces of A and emerges parallel to the original beam and goes on to D. Both C and D are full-silvered, front-surface mirrors which turn their beams back toward A. The beam from C is partly reflected at A, but part of that beam refracts through A and goes to the observer. The beam from D partially refracts through A and is lost, but part of that beam is also reflected toward the observer. The plate of glass at B has the same thickness and inclination as that at A, so that the two light paths from source to observer pass through the same number of glass thicknesses. If the light from the slit did not diverge and remained very narrow in going through the apparatus, the observer would see a line of light. The brightness of this line would depend on the difference in the optical length* of the two light paths. If these differed by any whole number of wavelengths of the light (including zero), the line would be bright. If the paths differed by an odd number of half-wavelengths, then the line would be dark. Between these extremes every brightness gradation would be observed. In practice, the light does diverge in the apparatus, and there are a great many slightly different paths being traversed simultaneously. Consequently, the observer does not see one line but a multiplicity of lines. The loci of points where the paths differ by whole wavelengths are bright, and where the paths differ by an odd number of half-wavelengths there is darkness. Thus, as one path length is varied, the observer sees fringes, like the teeth of a comb, move across the field, rather than a single line becoming lighter and darker. It is fortunate that the optical system works as it does, since it is easier for the eye to detect differences in position than differences in intensity.

The precision of this device is remarkable. If yellow light from sodium is used, the wavelength is 5.893×10^{-7} m. Moving the mirror C away from A one-half this distance will increase one path length by a whole wavelength and cause the pattern to move an amount equal to the separation of two adjacent dark lines. If we can estimate to hundredths of fringes, then the smallest detectable motion is only 2.9×10^{-9} m. Upon moving a mirror one-thousandth of an inch, 86 fringes would go by. (One way of defining a meter is to say that it is the length of 1,650,763.73 wavelengths in vacuum of the orange-red line of krypton-86.)

The similarity between the Michelson interferometer and the swimming race should be evident. Light corresponds to the swimmers and has the free-space velocity, c, with respect to its ether medium. The ether drift corresponds to the water current drift and has the velocity v with respect to the earth. Just as we could learn about the river flow by seeing the outcome of the swimmers' race, so we wish to measure the ether drift by conducting a "light race" over equal paths parallel and perpendicular to the ether drift.

Suppose that instead of taking the ratio of the times for the two paths of the river race we now take their difference; then

$$\Delta t = \frac{2Lc}{c^2 - v^2} - \frac{2L}{\sqrt{c^2 - v^2}} = \frac{2L}{c}\left[\left(1 - \frac{v^2}{c^2}\right)^{-1} - \left(1 - \frac{v^2}{c^2}\right)^{-1/2}\right]. \tag{5–4}$$

---

* Two paths have the same optical length if light traverses both in the same time. The optical lengths of the interferometer paths can be changed by changing their physical length, by changing the index of refraction of the region through which the light passes, or, if the swimming analogy applies, by moving the apparatus relative to the light-carrying medium.

Using the first two terms of the binomial expansion, we have, to a good approximation if $v \ll c$, that

$$\Delta t = \frac{2L}{c}\left[\left(1 + \frac{v^2}{c^2}\right) - \left(1 + \frac{v^2}{2c^2}\right)\right] = \frac{Lv^2}{c^3}. \tag{5–5}$$

In the interferometer, the time difference should appear as a fringe shift from the position the fringes would have if there were no ether drift. The distance light moves in a time Δt is $d = c\,\Delta t$ and if this distance represents n waves of wavelength λ, then $d = n\lambda$. Therefore the fringe shift would be

$$\boxed{n = \frac{Lv^2}{\lambda c^2}.} \tag{5–6}$$

Thus if the light race is carried out with light of speed c and wavelength λ in an interferometer whose arms are of length L, one of which is parallel to the ether drift of velocity v, then Eq. (5–6) gives the number of fringes that should be displaced because of the motion of the earth through the ether compared with their positions if the earth were *at rest* in the ether.

5–4. THE MICHELSON–MORLEY EXPERIMENT

The apparatus used was large and had its effective arm length increased to about 10 m by using additional mirrors to fold up the path. The entire apparatus was floated on mercury so that it could be rotated at constant speed without introducing strains that would deform the apparatus. *Rotation was necessary* in order to make the fringes shift, and by rotating through 90°, first one arm and then the other could be made parallel to the drift, thereby *doubling* the fringe displacement of Eq. (5–6). We can now estimate whether this instrument should be sensitive enough to detect the ether drift. Recall that at some time of the year the ether drift v was expected to be at least the orbital velocity of the earth, which is about $10^{-4}\,c$. Thus we expect v/c to be at least 10^{-4}. Using light of wavelength 5×10^{-7} m, the computed shift is $\Delta n = 0.4$ fringe. Michelson and Morley estimated that they could detect a shift of one-hundredth of a fringe. Sensitivity to spare!

Measurements were made over an extended period of time at all seasons of the year, but no significant fringe shift was observed. Thinking that the earth might drag a little ether along with it just as a boat carries a thin layer of water when it glides, Michelson and Morley took the entire apparatus to a mountain laboratory in search of a site which would project into the drifting ether. Again a diligent search failed to measure an ether drift. The experiment "failed."

Few experimental "failures" have been more stimulating than this. The negative result of the Michelson-Morley experiment presented a challenge to explain its failure. Fitzgerald and Lorentz proposed an *ad hoc* explanation. They pointed out that there might be an interaction between the ether and objects moving relative to it, such that the object became shorter in all its dimensions parallel to the relative velocity. Recall that in the flowing-river analogy the ratio of the times of the swimmers in Eq. (5–3) was

$$\frac{1}{\sqrt{1 - v^2/c^2}}.$$

If the route parallel to the flow had been shorter by this factor, then the ratio of the times would have been one and the race would have been a tie. A similar shortening of the parallel interferometer arm would account for the tie race Michelson and Morley always observed. The shortening could never be measured because any rule used to measure it would also be moving relative to the ether and would shorten also. Whether you accept the Fitzgerald-Lorentz contraction hypothesis or not, the Michelson-Morley experiment indicates that all observers who measure the velocity of light will get the same result regardless of their own velocity through space.

5–5. THE CONSTANT SPEED OF LIGHT

Speed trials of cars, boats, and airplanes are never official unless they are made in the following way. The record contender must drive his craft in opposite directions over a measured course and the speed attained is calculated to be the double distance divided by the total time spent between markers. This technique is used to make any wind, water, or other conditions which may be helpful in one direction be cancelled out by their hindrance on the reverse trip. Measurements of the speed of light are similarly made by timing a flash of light as it goes to a distant mountain and returns. This technique is rather good and, *to a first approximation*, the influence of a moving medium does cancel out. But the cancellation is not perfect and *to a second approximation*, the effect of a moving medium does *not* cancel. This becomes obvious if the medium moves faster than the speed under test. In this case, the test cannot be made, since the thing tested is carried away. The fact that Fizeau, in France, and others obtained consistent results for the velocity of light at a variety of times, places, and in different directions was in itself evidence that the speed of the earth through the ether (if any) was small compared with the velocity of light. It is highly significant that the Michelson-Morley interferometer was sensitive enough to detect the second-order term. Referring to Eq. (5–5), you will note that the first terms (the ones) cancel out and the significant result remains only because the second terms do not cancel. The Michelson-Morley experiment was sensitive to the second-order terms because, instead of trying to measure the times of transit of light through their apparatus, they measured the *difference* of times. Michelson and Morley found that the speed of the earth through space made *no difference* in the speed of light relative to them. The inference is clear either that the earth moves in some way through the ether space more slowly than it moves about the sun, or that *all observers must find that their motion through space makes no difference in the speed of light relative to them.*

The consideration of electromagnetic phenomena led Einstein to conclude, apparently without knowledge of the Michelson-Morley experiment, that the speed of light does not depend upon the motion of the observer and that there is no preferred reference frame for the laws of physics. We quote the beginning of his first paper on relativity which is titled "On the Electrodynamics of Moving Bodies."*

"It is known that Maxwell's electrodynamics—as usually understood at the present time—when applied to moving bodies leads to asymmetries which do not appear to be inherent in the phenomena. Take, for example, the reciprocal electrodynamic action of a magnet and a conductor. The observable phenomenon here depends only on the relative motion of the conductor and the magnet, whereas the customary view draws a sharp distinction between the two cases in which either the one or the other of these bodies is in motion. For if the magnet is in motion and the conductor at rest, there arises in the neighborhood of the magnet an electric field with a certain definite energy, producing a current at the places where parts of the conductor are situated. But if the magnet is stationary and the conductor in motion, no electric field arises in the neighborhood of the magnet. In the conductor, however, we find an electromotive force, to which in itself there is no corresponding energy, but which gives rise—assuming equality of relative motion in the two cases discussed—to electric currents of the same path and intensity as those produced by the electric forces in the former case.

"Examples of this sort, together with the unsuccessful attempts to discover any motion of the earth relatively to the 'light medium,' suggest that the phenomena of electrodynamics as well as of mechanics possess no properties corresponding to the idea of absolute rest. They suggest rather that, as has already been shown to the first order of small quantities, the same laws of electrodynamics and optics will be valid for all frames of reference for which the equations of mechanics hold good. We will raise this conjecture (the purport of which will hereafter be called the 'Principle of

---

* Excerpt from A. Einstein, "Zur Electrodynamik bewegter Körper," *Annalen der Physik* **17,** 891 (1905), translated by W. Perrett and G. B. Jeffery, in *The Principle of Relativity* by Einstein, Lorentz, Minkowski, and Weyl. Published by Dover Publications, Inc., New York 14, N. Y. and reprinted through permission of the publisher.

Relativity') to the status of a postulate, and also introduce another postulate, which is only apparently irreconcilable with the former, namely, that light is always propagated in empty space with a definite velocity c which is independent of the state of motion of the emitting body. These two postulates suffice for the attainment of a simple and consistent theory of the electrodynamics of moving bodies based on Maxwell's theory for stationary bodies. The introduction of a 'luminiferous ether' will prove to be superfluous inasmuch as the view here to be developed will not require an 'absolutely stationary space' provided with special properties, nor assign a velocity-vector to a point of the empty space in which electromagnetic processes take place.

"The theory to be developed is based—like all electrodynamics—on the kinematics of the rigid body, since the assertions of any such theory have to do with the relationships between rigid bodies (systems of coordinates), clocks, and electromagnetic processes. Insufficient consideration of this circumstance lies at the root of the difficulties which the electrodynamics of moving bodies at present encounters."

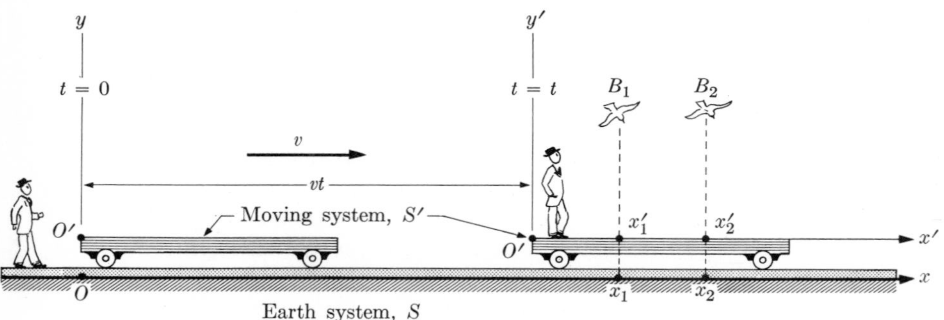

FIG. 5–3
Coordinate systems S and S'.

5–6. CLASSICAL RELATIVITY

Let us first consider the pre-Einstein relativity of physical quantities in classical or Galilean-Newtonian physics, and ask how events in one system, S, moving with constant linear velocity, appear from another system S', also moving with constant linear velocity.

No generality will be lost if one of the systems, say S, is regarded as being at rest, and the other, S', as moving with a uniform velocity v. Our problem, then, is like that of comparing the observations of a man on the ground with those of a man on a uniformly moving train.

In Fig. 5–3 the earth observer considers himself to be at the origin O of his system S and he chooses an x-axis parallel to the track. The train observer in S' likewise measures distances from himself at O' and chooses his x'-axis parallel to the track. Let us measure time from the moment when the two observers are exactly opposite each other. Suppose that at some later time t each observer decides to measure the separation of two birds, B_1 and B_2, which happen to be hovering over the track. The observer at O' observes that the positions of B_1 and B_2 are the small distances x'_1 and x'_2. The observer at O finds the positions of B_1 and B_2 are the larger distances x_1 and x_2. The observer at O can compute the O' observations by noting that the observer at O' has moved a distance vt, with the result

$$x'_1 = x_1 - vt \qquad \text{and} \qquad x'_2 = x_2 - vt. \tag{5-7}$$

The observer on the train (who may think the train is at rest with the earth moving under it) can account for the difference in their observations by observing that O has drifted away from him a distance vt. He obtains

$$x_1 = x'_1 + vt \qquad \text{and} \qquad x_2 = x'_2 + vt. \tag{5-8}$$

These are transformation equations in that they transform observations from one system to the other. By solving either set of equations for the separation of the birds, we get

$$x_2 - x_1 = x'_2 - x'_1 \qquad \text{or} \qquad \Delta x = \Delta x'. \tag{5-9}$$

This shows that the two observers agree on how far apart the two birds are. Note that the relative velocity v of the observers need not be known. Similarly, the time t since the two observers were opposite each other need not be known, but all observations must be *simultaneous*—even if the birds cooperate by staying the same distance apart.

What we have just shown is that *distance* or length is an *invariant* quantity when transformed from one Newtonian coordinate system to another.

We now compare results by the same two observers for a velocity measurement. Suppose that one bird flies from position B_1 at time t_1 to position B_2 at time t_2. The transformation equations of the positions of the bird are of the same form except for the difference of time, thus

$$x'_1 = x_1 - vt_1, \qquad x'_2 = x_2 - vt_2. \tag{5-10}$$

To solve for the average observed velocity u', we take the difference between the two transformation equations and divide by the time interval, obtaining

$$\begin{aligned} x'_2 - x'_1 &= (x_2 - vt_2) - (x_1 - vt_1) \\ &= (x_2 - x_1) - v(t_2 - t_1), \end{aligned} \tag{5-11}$$

or

$$\Delta x' = \Delta x - v\,\Delta t. \tag{5-12}$$

Therefore the velocity relation is

$$u' = \frac{\Delta x'}{\Delta t} = \frac{\Delta x}{\Delta t} - v = u - v. \tag{5-13}$$

Thus the *velocities* measured by the two observers are *not* the same. They are *not invariant* under a transformation between Galilean-Newtonian coordinate systems.

Suppose the bird again cooperates and swoops over the train with different velocities at positions 1 and 2. Each observer could measure the velocity at each position. Using the result just derived, we find that the velocity transformation equations are

$$u'_1 = u_1 - v \qquad \text{and} \qquad u'_2 = u_2 - v. \tag{5-14}$$

If the observers also measured the time Δt it took the bird to get from one position to the other, they could solve for the average acceleration. Taking the difference between the velocity transformations gives

$$u'_2 - u'_1 = u_2 - u_1. \tag{5-15}$$

Dividing by Δt, we have $a' = a$, which shows that velocity difference and *acceleration* are *invariant* under transformation between Newtonian inertial frames of reference.

The transformation equations we have just derived formally are fairly obvious. The most complicated result, the *variant* velocity transformation, is used without formal proof in the study of impact and Doppler effect, and was used in our recent discussion of the Michelson-Morley experiment. In fact, the problem presented by the negative result of that experiment was that, contrary to the Newtonian transformation, observers *must* find the velocity of light the *same* whether the observers are moving or not.

The classical transformation equations just derived apply only for reference frames moving with *constant linear velocity* with respect to one another. It is their failure in cases involving uniform motion that leads us into Einstein's special theory of relativity. We have not treated the classical transformation equations which deal with accelerated or rotating frames of reference, since these are much more complicated.

In the derivations above we have tacitly made the *classical assumption* that time intervals are the same for all observers. Actually, there is no a priori reason* for assuming that time or any other physical quantity is invariant under a transformation of coordinates. Whether or not an assumption is correct is determined solely by the experimental verification of the results predicted with its aid. We will find later in this chapter that

* An a priori reason is one deduced from previous assumptions or known causes.

the invariance of time interval will have to be abandoned, when the invariance of the free-space velocity of light is assumed.

5-7. GENERAL AND SPECIAL THEORIES OF RELATIVITY

Relativity is divided into two branches. One is called the general theory of relativity and the other is the special, or restricted, theory of relativity. In the general theory, proposed in 1915, Einstein treats the class of problems that arise when one frame of reference is accelerated relative to another, particularly in the phenomena of gravitation. In the special theory, proposed in 1905, Einstein treats problems that arise when one frame of reference moves with a *constant linear velocity* relative to another. The general theory is quite difficult. Fortunately, the special theory is sufficient for discussing most atomic and nuclear phenomena. It contains much "uncommon sense" and many ideas which tease the imagination. But the mathematics is simple enough that we can derive several interesting relationships which are basic to an understanding of the material which follows.

5-8. EINSTEINIAN RELATIVITY

The Michelson-Morley experiment was carried out to measure the velocity of the earth through the ether in the hope that the ether would provide a fixed frame of reference relative to which Newton's laws would hold exactly. Einstein assumed that all experiments designed to locate a fixed frame of reference would fail. Since he assumed that a fixed frame of reference could never be found, he went to the other extreme and postulated that the laws of physics should be so stated that they apply relative to any frame of reference.

If the laws of physics took different forms for different moving observers, then the different forms of the laws would provide a method for determining the nature of an observer's motion. This would take us back to the Newtonian view that there is a unique frame of reference for which the laws of physics are Newton's laws. Going back to our man on the merry-go-round, it is hard to see how the laws of motion could ever seem the same to him as they seem to a man standing on the earth. We have pointed out, however, that this is a problem in general relativity. For special or restricted relativity, we need only face the problem of making the laws of physics take the same form for all observers *translating* relative to each other with *constant* velocity. Limiting ourselves to special relativity, we may state its two postulates as follows:

(1) *The laws of physics apply equally well for any two observers moving with constant linear velocity relative to each other or, in other words, the observations on one reference frame are not preferred above those on any other.*

(2) *All observers must find the same value of the free-space velocity of light regardless of any motion they may have.*

5-9. RELATIVISTIC SPACE-TIME TRANSFORMATION EQUATIONS

The equations we are about to derive were first obtained by Lorentz as he successively refined electromagnetic theory to conform with the results of experiments. As stated earlier, both Fitzgerald and Lorentz obtained an ad hoc explanation of the negative result of the Michelson-Morley experiment by supposing that the interferometer shortened along its velocity vector through the ether. But we shall derive the relativistic transformation equations as consequences of Einstein's two postulates.

We return to our two observers at O and O' in systems S and S'. We ask the observer at O' to go back and pass the observer at O again. This time—just as O is opposite O'—we fire a photographic flashbulb. Thus at time $t = 0$, each observer is at the source of a spherical light wave. If we endow each observer with a supernatural power so that each can "see" the light spread out into space, then *each* observer must feel that *he* is at the center of the growing sphere of light. This must be the case, since we are now imposing the condition that the velocity of light is the same for all observers even if the velocity of an object is not.

The equation of the expanding sphere seen by the observer at O is

$$x^2 + y^2 + z^2 = c^2t^2, \tag{5-16}$$

where c is the velocity of light. Similarly, the observer at O' writes for his equation of the sphere

$$x'^2 + y'^2 + z'^2 = c^2t'^2. \tag{5-17}$$

Clearly, the relativistic transformation equations must be different from the classical ones if both observers are *each* to seem to be at the center of the same sphere.

As we consider what the new transformation equations are to be, we find that we are somewhat limited. These new equations must be linear. Any quadratic equation has two solutions and higher-order equations have more solutions. Surely any observations from the system S must have a unique interpretation in the system S'. There must be a "one-to-one" correspondence between what each observer "sees." The transformation equations must be linear in the space coordinates and in the times. Since the classical transformation equations were found to be $x' = x - vt$ and $x = x' + vt'$, let us here assume the next simplest linear equations,

$$x' = k(x - vt) \qquad \text{and} \qquad x = k'(x' + vt'), \tag{5-18}$$

where the k's are transformation quantities to be determined. Note that our notation t and t' admits the possibility that time intervals may not be identical for the two observers. Note further that k and k' must be equal if, as required by Einstein's first postulate, there is to be no preferred reference system. Hereafter we shall let k represent both k and k'.

We now seek to transform Eq. (5–17) so that the observer at O' "sees" what the observer at O would see. Mathematically, this requires that we eliminate x', y', z', and t' by using the assumed transformation equations. The coordinates y, y', z, and z' are perpendicular to the relative velocity of the observers, so that $y' = y$ and $z' = z$.* We have assumed a transformation equation for x', but we need some manipulation before we can eliminate t'. To get t' in terms of unprimed quantities, we work with our two assumed transformation equations in Eq. (5–18). Solving the second for t' and using the first to eliminate x', we obtain

$$t' = k\left[\frac{x}{v}\left(\frac{1}{k^2} - 1\right) + t\right]. \tag{5–19}$$

We can now transform the observation from O' into the observation from O by eliminating the primed quantities from Eq. (5–17). We obtain

$$k^2(x - vt)^2 + y^2 + z^2 = c^2k^2\left[\frac{x}{v}\left(\frac{1}{k^2} - 1\right) + t\right]^2. \tag{5–20}$$

Upon expanding and collecting terms, we have

$$\left[k^2 - \frac{c^2k^2}{v^2}\left(\frac{1}{k^2} - 1\right)^2\right]x^2 - \left[2vk^2 + \frac{2c^2k^2}{v}\left(\frac{1}{k^2} - 1\right)\right]xt$$
$$+ y^2 + z^2 = [c^2k^2 - v^2k^2]t^2. \tag{5–21}$$

This equation represents the expanding sphere of light as seen by the observer at O' whose observations have been interpreted (transformed) by the observer at O. This equation must be identical to the direct observation of the expanding sphere as seen from O, namely Eq. (5–16). For this to be the case, it is necessary that the quantities within brackets in Eq. (5–21) equal 1, 0, and c^2, respectively.†

---

* If this conclusion seems hasty, one can write $y' = k_y(y - v_yt)$. Since the only relative velocity is along the x-axis, we have $v_y = 0$. This leaves $y' = k_y(y)$, which we could carry into our subsequent derivation. If we did so, it would be obvious immediately that $k_y = 1$. A similar argument applies to the z-transformation. In order to make these simplifications in our derivations, we assumed the relative velocity to be parallel to the x-axis.

† In the $x = 0$ plane, the sphere becomes a circle, so that the sphere of Eq. (5–16) degenerates to $y^2 + z^2 = c^2t^2$ and the same sphere described by Eq. (5–21) degenerates to $y^2 + z^2 = [c^2k^2 - v^2k^2]t^2$. Since these equations describe the same circle, they must be identical, and to be identical, the quantity within the brackets must equal c^2.

Since the quantity within the third pair of brackets appears the simplest, we set it equal to c^2, obtaining

$$c^2 = c^2k^2 - v^2k^2 = k^2(c^2 - v^2). \tag{5-22}$$

From this we get

$$k^2 = \frac{c^2}{c^2 - v^2}$$

or

$$\boxed{k = \frac{1}{\sqrt{1 - v^2/c^2}}\cdot} \tag{5-23}$$

We choose the positive value of the square root, so that relativistic and classical physics correspond at low velocities.

Setting the quantity in the second pair of brackets in Eq. (5–21) equal to zero and solving for k, we again get the same result. This is encouraging. It appears possible that the transformation we have assumed can indeed allow each observer to consider himself to be at the center of the expanding sphere of light. The final check is to determine whether the value of k we have obtained makes the coefficient of x^2 become unity. Upon substituting for k, we find that the quantity within the first pair of brackets becomes

$$\begin{aligned} k^2 - \frac{c^2k^2}{v^2}\left(\frac{1}{k^2} - 1\right)^2 &= \frac{c^2}{c^2 - v^2} - \frac{c^2c^2}{v^2(c^2 - v^2)}\left(\frac{c^2 - v^2}{c^2} - 1\right)^2 \\ &= \frac{c^2}{c^2 - v^2} - \frac{c^4}{v^2(c^2 - v^2)}\left(\frac{c^2 - v^2 - c^2}{c^2}\right)^2 \\ &= \frac{c^2}{c^2 - v^2} - \frac{v^2}{c^2 - v^2} = 1. \end{aligned}$$

The consistency of these results means that mathematically it is possible to devise linear transformation equations which permit the transformation of one velocity to be invariant. We have chosen that one velocity to be the free-space velocity of light by assuming our two observers to be "watching" an expanding light wave. We need only modify the classical transformation equations by the inclusion of $k = 1/\sqrt{1 - v^2/c^2}$ and note that time, which was the same for all Newtonian observers, must now be transformed along with the space coordinates. Knowing k, we can simplify Eq. (5–19) to

$$t' = k\left[t - \frac{vx}{c^2}\right]\cdot \tag{5-24}$$

We can now summarize both classical and relativistic transformation equations for two observers having a relative velocity v parallel to the

x-axis as follows:

$$\begin{array}{ll} \text{Galileo-Newton} & \text{Lorentz-Einstein} \\ \text{(classical)} & \text{(relativistic)} \\ x' = x - vt & x' = k(x - vt) \\ y' = y & y' = y \\ z' = z & z' = z \\ t' = t & t' = k\left(t - \dfrac{vx}{c^2}\right), \end{array} \tag{5-25}$$

where

$$k = \frac{1}{\sqrt{1 - v^2/c^2}}.$$

Note that if $v \ll c$, then k is nearly 1 and the relativistic transformation equations reduce to the Newtonian forms. Thus Newtonian physics can be regarded as a special case of relativistic physics. Recall that the tremendous velocity of the earth about the sun is 18 mi/s, but that this is still only one ten-thousandth of the velocity of light. Thus the relativity correction for the observation of positions from the earth compared with observations from the sun is only about 5×10^{-7} %. Relativity makes no significant difference for "ordinary" engineering applications.

The relativistic transformation equations, however, point the way to important philosophical advances. The portion of special relativity we have developed applies only to observers not accelerated relative to each other. Note, however, that neither of these observers has a *preferred* viewpoint. If the transformation equations for primed quantities in terms of unprimed quantities are solved for the unprimed in terms of the primed ones, we obtain the *inverse transformations.* The resulting forms are the same except that the sign of their relative velocity, v, changes. [Thus $x = k(x' + vt')$.] This one difference is expected, since if O' moves north relative to O, then O must move south relative to O'. A step has been taken toward the goal of general relativity, namely, that the laws of physics shall take on the same form for *all* observers.

5–10. LENGTH CONTRACTION

With the aid of the Lorentz-Einstein transformation equations we can now "explain" the Michelson-Morley experiment. The *ad hoc* factor of Fitzgerald and Lorentz is the factor k, which has been derived from the basic relativistic assumption that all observers must get the same result if they measure the speed of light. Under relativity, space is not a rigid fixed thing, but changes in size depending on the motion of the observer. This is very different from Newton's view. He said, "Absolute space, in its own nature, without regard to anything external, remains always similar and immovable."

Let us now consider a rod placed in the O' or primed reference frame, (Fig. 5–3) with its length parallel to the x-axis. Let the coordinate of the left end be x'_1 and that of the right end be x'_2. The length of the rod as measured by an observer at rest with respect to the rod is called its *proper length.* Its value is $(x'_2 - x'_1)$ in this case. What is the length of the rod measured by an observer in the O, or unprimed, frame when the primed frame is moving in the positive x-direction with speed v? From Eq. (5–25) we find that

$$x'_1 = k(x_1 - vt_1) \qquad \text{and} \qquad x'_2 = k(x_2 - vt_2).$$

Subtracting the first of these expressions from the second, we obtain

$$(x'_2 - x'_1) = k[(x_2 - x_1) - v(t_2 - t_1)]. \tag{5–26}$$

This equation shows that the measured distance between the ends of the rod in the O-frame can have many different values which depend upon the choice of t_1 and t_2, the times when the ends of the rod are observed. Because of this, we define the length of a moving rod as the measured distance between its ends obtained when the two ends are observed simultaneously. Then $t_1 = t_2$ and Eq. (5–26) reduces to

$$(x'_2 - x'_1) = k(x_2 - x_1) \tag{5–27}$$

or

$$L' = kL = (1 - v^2/c^2)^{-1/2}L.$$

Since k is always greater than unity, $(x_2 - x_1)$ will always be less than the proper length, $(x'_2 - x'_1)$, and therefore it is said that the rod has contracted. We now see that the Fitzgerald-Lorentz contraction discussed in Section 5–4 is mathematically the same as the relativistic length contraction given by Eq. (5–27). However, the two equations are based on significantly different concepts. In the Fitzgerald-Lorentz contraction, v is the speed of the rod relative to the ether, whereas in the relativistic equation, it is the speed of the rod relative to an observer.

Assuming sufficient visual acuity, could one see that a moving body is contracted in the direction of its motion? It turns out that one could not if the moving body subtends a small angle at the observer. The situation here is not the same as that when we obtained Eq. (5–27), the length contraction equation. When we see an object we have a retinal image produced by light quanta which arrive simultaneously at the retina from different points of the object. Therefore these light quanta could not have been emitted by every point on the object at the same time. The points farther away from the observer must have emitted their part of the image earlier than the closer points. In this case we cannot let $t_1 = t_2$ in Eq. (5–26) and obtain Eq. (5–27). When the differences in the times of emission are taken into account, it is found that a moving body which subtends a small angle at the observer will appear to have undergone a rotation but not a contraction. It is not at all difficult to derive this result, but it is too long for inclusion here. The method of arriving at this interesting conclusion is given in a paper by Terrell and summarized and discussed in one by Weisskopf. For the case when the moving object subtends a large angle at the observer, then, under suitable conditions which are stated in a paper by Scott and Viner, the length contraction will be visible. The reader is urged to study these three articles. The references to them are given at the end of this chapter.

5-11. TIME DILATION AND CAUSAL SEQUENCE

The classical concept of time is contained in Newton's statement that "Absolute, true, and mathematical time, of itself and from its own nature, flows equably without relation to anything external." There is implicit in this that information can be transmitted at velocities so enormous compared with the relative velocity of observers on different reference frames that no question about defining simultaneity arises. When, however, we assume that light signals are the fastest possible way of communicating, we can expect to encounter difficulties in describing the motion of objects having velocities approaching that of light.

How is the time interval between two events in one reference frame related to the interval between events as observed from another frame moving with respect to the first? Again, as in Fig. 5–3, let the x-axes of the frames be parallel and the relative velocity of the origins O and O' be v. Consider an event which occurs at time t_1 at place x_1 in the unprimed frame and a later event in the same frame at t_2, x_2. The transformations to t', obtained from Eq. (5–25), are

$$t_1' = k\left(t_1 - \frac{vx_1}{c^2}\right) \qquad \text{and} \qquad t_2' = k\left(t_2 - \frac{vx_2}{c^2}\right) \cdot$$

By taking the difference between these expressions, we find the time interval in the primed frame to be

$$(t_2' - t_1') = k\left[(t_2 - t_1) - \frac{v}{c^2}(x_2 - x_1)\right] \cdot \tag{5–28}$$

To discuss this equation, let us start by considering the case when both events in the O-frame occur at the same place, that is, $x_1 = x_2$. The time interval between two such events measured in the coordinate frame in which the two events occur at the same place is called the *proper time* or local time. In the case under discussion, $(t_2 - t_1)$ is the proper time and Eq. (5–28) becomes

$$(t_2' - t_1') = k(t_2 - t_1). \tag{5–29}$$

Since k is greater than unity, the time interval between two events measured by the observer in the O'-frame will be longer than the proper time, which is the value of the time interval between the same two events obtained by the observer in the O-frame. This is *time dilation* and we say that the "clock" in the O-frame, the moving clock, runs more slowly than the one in the O'-frame. It is to be noted that although $x_1 = x_2$, the coordinates x'_1 and x'_2 are not the same because of the relative displacement of the reference frames during the time $t_2 - t_1$. If two events had occurred at the same place in the primed system at times t'_1 and t'_2, then the inverse transformation shows that the unprimed system would also have reported time dilation given by

$$t_2 - t_1 = k(t'_2 - t'_1).$$

Returning to Eq. (5–29), we see that if the events in the unprimed frame are simultaneous, that is, $t_2 - t_1 = 0$, then we also have $t'_2 - t'_1 = 0$. Therefore two observers will agree on the simultaneity of two events if they occur simultaneously at the *same* place in either system of coordinates. Finally, note that if the speed of light were infinite, then $k = 1$ and the time intervals in each frame would be equal or absolute, as Newton assumed.

Direct confirmation of time dilation is found in an experiment with mu-mesons. These are subatomic particles which can be created in the laboratory by high-energy particle accelerators. These particles are unstable and it is found that they disintegrate at an exponential rate such that one-half of them remain unchanged after 3.1×10^{-6} s, measured in the reference frame in which the mesons are at rest. These particles are also formed high in the earth's atmosphere by cosmic ray bombardment and are projected towards the earth's surface with a very high velocity. Consider a beam of these particles traveling down toward the earth at a speed of 0.9c. (For this speed, $k = [1 - (0.9c/c)^2]^{-1/2} = [0.19]^{-1/2} = 2.3$.) This beam would traverse a distance of 840 m in 3.1×10^{-6} s. Therefore one would expect that if a meson counter were placed 840 m above another one in the atmosphere, then the count observed at the lower level would be half of that at the upper one. But it is found to be significantly greater than one half of the count at the upper level. The error in the prediction arose because the earth-bound observer used the proper half-life instead of the dilated one in his computations. The dilated half-life of the mesons, calculated from Eq. (5–29), is $2.3 \times 3.1 \times 10^{-6}$ s $= 7.2 \times 10^{-6}$ s. Therefore the earth observer will obtain a higher count at a lower level than he first expected because this dilated half-life is much longer than 3.1×10^{-6} s, the time of flight between the counters. Let the earth observer increase the distance between counters to 1920 m, the product of the dilated half-life and the relative velocity of the reference frames. This observer will now find the lower level count to be half of that at the higher level. What does the observer riding along with the mesons report as the ratio of the counts? From his point of view, the separation of the counters is not 1920 m but, because of length contraction, is 840 m. This contracted length divided by $0.9c$ gives his time of flight between counters to be 3.1×10^{-6} s, which is equal to the proper half-life. Therefore, the observer in the meson frame will also find the lower-level count to be half of the upper-level one. Thus, both observers do agree upon the relative number of mesons that survived the trip between the counters, although the basic data each used in his calculations were quite different.

Let us now discuss the transformed time interval when the two events in the unprimed system do *not* occur at the same place. In this case the value of $t_2' - t_1'$ is given by Eq. (5–28). If the quantity within the brackets in this equation is equal to zero, then the two events are simultaneous in the primed frame. If the quantity is positive, then events are observed in the same order in the primed frame as in the unprimed one. If it is negative, then the two events in O should appear in the reverse order from O'. To find the mathematical condition for a reversal, we will rewrite Eq. (5–28) as follows:

$$t_2' - t_1' = k(t_2 - t_1)\left[1 - \frac{v}{c}\frac{x_2 - x_1}{c(t_2 - t_1)}\right].$$

This shows that the quantity within brackets will be negative only if

$$\frac{x_2 - x_1}{c(t_2 - t_1)} > \frac{c}{v}.$$

This inequality can be realized only if the distance between x_2 and x_1 is greater than the distance traveled by light in the time $t_2 - t_1$. But if the separation of the two places were that large, then a light signal leaving event 1 could not reach event 2 soon enough to cause event 2. Thus, if two events appear in different time order to two observers, one event cannot be the cause of the other. Therefore, we must conclude that cause and effect will never appear in the reverse order to different observers.

Is time *really* dilated and is the length of a moving rod *really* contracted in the direction of its motion? The answers depend upon what is meant by *really*. In the physical sciences what is real is what is measured. Only through measurement can one obtain the information needed for assigning properties to a clock, to a rod, to an atom and so forth. Time dilation and length contraction are real in this sense. Proper time and proper length have nothing of an absolute nature about them. Time, length, area, volume and other quantities are all relations between an observed body and the observer.

5-12. THE RELATIVISTIC VELOCITY TRANSFORMATION

To illustrate further how the transformation equations of space and time are used, let us repeat a calculation we made classically. We found the classical velocity transformation for observers with relative velocity v to be $u' = u - v$. To get the corresponding relativistic expression, we use the defining equations

$$u'_x = \frac{dx'}{dt'} \quad \text{and} \quad u_x = \frac{dx}{dt}.$$

We now express u' in terms of the differentials of unprimed quantities obtained from the relativistic transformation equations. The result is

$$u'_x = \frac{dx'}{dt'} = \frac{k(dx - v\,dt)}{k(dt - v\,dx/c^2)}$$

$$= \frac{dx/dt - v}{1 - (v/c^2)(dx/dt)},$$

or

$$\boxed{u'_x = \frac{u_x - v}{1 - u_x v/c^2}.} \tag{5-30}$$

In a similar way we find the velocity components parallel to the other two axes to be

$$u'_y = \frac{u_y}{k(1 - u_x v/c^2)} \quad \text{and} \quad u'_z = \frac{u_z}{k(1 - u_x v/c^2)}. \tag{5-31}$$

Note that the velocity components which are transverse to the relative motion of the reference frames depend upon the relative motion but that transverse distances do not.

Equation (5-30) is the relativistic rule for transforming velocities when the observed velocity is *parallel* to the relative velocity of the observers. We see that as in the classical case, the velocity transformation is not invariant. Consider, however, what happens when the velocity under observation is the velocity of light, c. If one observer measures the velocity of light and gets $u_x = c$, what will another observer moving relative to the first obtain?

The equation yields

$$u'_x = \frac{u_x - v}{1 - (u_x v/c^2)} = \frac{c - v}{1 - (cv/c^2)} = \frac{c - v}{(c - v)/c} = c.$$

Thus, regardless of their own relative velocity v, any two observers will agree that the velocity of light is c. In relativity, the velocity of light is *invariant*. This result should not surprise us, since it was used as a basic assumption for the derivation of the transformation equations. It is one example of the "uncommon" sense we promised to discuss at the very beginning of this book.

Although the result just derived follows logically from the assumptions of relativity and therefore should not surprise us, a further comment may be helpful. Suppose observer O measures the distance light travels during an interval of time and computes the velocity of light, obtaining c. Let another observer O' moving relative to O perform the same experiment. Intuitively we feel he cannot get the same result, c. But if O and O' both have "rubber" rulers and "defective" clocks, we see that their results may have any values—including c. The transformation equations of the special theory of relativity are a description of how "rubber" rulers and "defective" clocks vary so that the velocity of light can be unique, so that velocities small compared with c behave classically, and so that intermediate cases lead to no contradictions.

If this were the end of the matter, there is no doubt relativity would have been forgotten or remembered only as a fantastic speculation of a fertile mind. No one would make such a tremendous break with traditional thought merely to "explain" the Michelson-Morley experiment. Before it can be taken seriously, the inductive structure of relativity or any other theory must be subjected to deductive verification. If relativity is to be "true" or useful, it must correctly foretell some unforeseen events that can be subjected to experimental tests.

There are several such verifications of relativity. Some come from general relativity, which we are not treating here, but the most striking and important deductive consequence of relativity comes from the special relativity which we are treating. This is the mass transformation equation. To introduce mass into the picture, we next consider an impact situation.

5-13. RELATIVISTIC MASS TRANSFORMATION

Consider two basketballs which are spherical, perfectly elastic, and have identical masses when compared by an observer at rest relative to them. We give one ball to an observer O' on a railroad train, moving relative to the ground with a constant translational velocity v. We give the other ball to an observer O on the ground at a distance d from the railroad track. The observers have not yet passed each other and we tell each that he is to throw his ball with a certain velocity in a direction perpendicular to the track in such a way that the basketballs bounce perfectly off each other just at the moment when the two observers pass each other. (Deflections due to gravity are irrelevant to this discussion.)

Assuming sufficient skill, this experiment could be carried out and fully understood on the basis of classical mechanics. Each observer sees his ball hit and return to him just as though it had bounced from a perfectly elastic

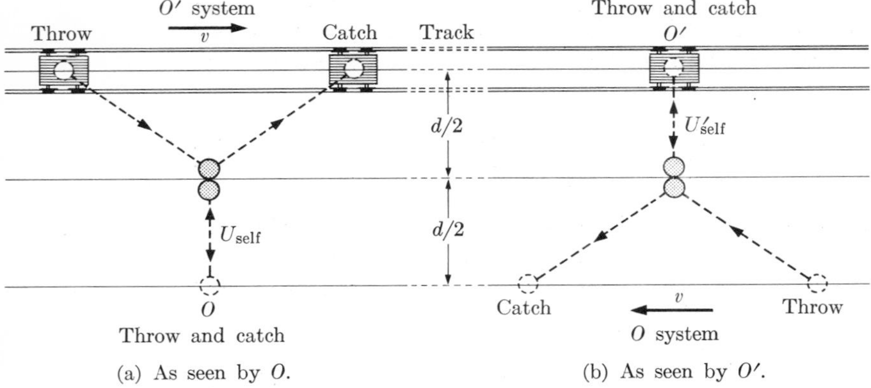

FIG. 5–4
Perfectly elastic collision of two identical basketballs as viewed from systems in relative motion.

wall with kinetic energy conserved. Each observer must anticipate the moment of passing, and the movements of the balls are as shown in Fig. 5–4.

In the analysis that follows we will consider four velocities. U'_{self} is the velocity O' observes he has given to his ball, and U_{self} is the velocity O observes he has given to his ball. By a condition of the experiment, $U'_{\text{self}} = U_{\text{self}}$. The velocity that observer O assigns to the ball thrown by O' will be called U'_{other}, and the velocity O' assigns to the ball thrown by O will be U_{other}.

If we now treat the situation just described from the standpoint of Einsteinian relativity, a new idea emerges. Distances perpendicular to the relative velocity of the balls are unaffected by relativity considerations, since we found that $y = y'$ and $z = z'$. Thus both observers agree that their basketballs move to and from the impact point a total distance d. But the two observers have different kinds of time. Observer O' was told how fast he should throw his ball. He throws at time t'_1 and catches at time t'_2, so that in his opinion

$$U'_{\text{self}} = \frac{d}{t'_2 - t'_1}. \tag{5–32}$$

Observer O disagrees with observer O' as to when and where he threw and caught his ball. Observer O says O' threw the ball from x_1 at the time $t_1 = k[t'_1 + (vx'_1/c^2)]$ and caught it at x_2 at the time $t_2 = k[t'_2 + (vx'_2/c^2)]$. Observer O computes the velocity with which O' threw his ball to be U'_{other}, where

$$U'_{\text{other}} = \frac{d}{t_1 - t_2} = \frac{d}{k[(t'_2 - t'_1) + (v/c^2)(x'_2 - x'_1)]}. \tag{5–32a}$$

In this expression, the quantity $(x_2' - x_1')$ is the amount observer O' must displace himself in order to catch the ball. Since we stated that the ball comes directly back to each observer, this quantity is zero, and Eq. (5–32a) becomes

$$U'_{\text{other}} = \frac{d}{k(t_2' - t_1')} = \frac{U'_{\text{self}}}{k}. \tag{5–33}$$

Since k is always greater than unity, it appears to observer O that observer O' has thrown his ball slower than he was told to. A similar calculation would show that observer O' would think observer O had thrown *his* ball slower than he was told to, or

$$U_{\text{other}} = \frac{U_{\text{self}}}{k}. \tag{5–33a}$$

We are now faced with a dilemma. If we assume that the masses remain the same as when they were compared without relative velocity, then conservation of momentum is violated. If we keep conservation of momentum, then the relative motion which made a difference in the observed velocities of the basketballs also resulted in an observed difference in their masses. In relativity it is assumed that *momentum is conserved.* Therefore observer O, who sees his ball change in momentum an amount $2m_{\text{self}}U_{\text{self}}$, equates this change to the change he observes for the other ball. He writes

$$2m_{\text{self}}U_{\text{self}} = 2m_{\text{other}}U'_{\text{other}}. \tag{5–34}$$

When the value of U'_{other} from Eq. (5–33) is substituted in Eq. (5–34), it becomes

$$2m_{\text{self}}U_{\text{self}} = 2m_{\text{other}}\frac{U'_{\text{self}}}{k}, \tag{5–35}$$

which, since the throws were equal such that $U_{\text{self}} = U'_{\text{self}}$, reduces to

$$m_{\text{other}} = km_{\text{self}}. \tag{5–36}$$

Thus observer O concludes that the masses of the balls are not equal and that there must be a transformation equation for mass in addition to those for the other quantities we have discussed.

In this derivation we have considered an especially simple situation in order to get our result as quickly as possible. We have computed the observed change in mass by considering the effect of a relative x-velocity on observed y-velocities. We could neglect the changes in mass each observer produced in his own ball when he threw it because both observers threw in the same way ($U'_{\text{self}} = U_{\text{self}}$). A general derivation would have led to a result of the same form, namely

$$\boxed{m = km_0 = \frac{m_0}{\sqrt{1 - (v^2/c^2)}},} \tag{5-37}$$

where m_0 (called the rest mass) is the mass of a body at rest relative to the measurer, and m (called the moving, or relativistic, mass) is the observed mass of the body when it has a velocity v relative to the measurer. The system in which the observer is located is usually called the laboratory system.

This result, which has been deduced from the space transformation equations, can be tested experimentally. What is required is the measurement of the mass of a body moving at great speed relative to an observer. J. J. Thomson was startled when he computed the speed of electrons in his cathode-ray tubes. These speeds were greater than any previously measured by man and, by modern methods, can be made significant compared with the free-space velocity of light, c. Furthermore, an e/m_e experiment is a measure of m_e if we assume that the charge e of the particle is constant. In principle, then, J. J. Thomson provided a technique for the testing of the mass-transformation equation. The first adequate test and verification was made in 1908 by Bucherer, using electrons from a radioactive source, and since then the mass-transformation equation has been confirmed innumerable times. It is now a cornerstone of atomic physics.

5–14. RELATIVISTIC MASS-ENERGY EQUIVALENCE

The derived and experimentally verified mass-transformation equation has important consequences. Whereas in classical physics we considered that the kinetic energy of a body could be increased only by increasing its velocity, we now must take mass variation into account. Suppose we increase the kinetic energy of a body an amount dE_k by exerting a force through a distance. The change of kinetic energy equals the work done, or

$$dE_k = F ds.$$

Since we cannot regard mass as constant in relativity, we write Newton's second law in its more general form, force equals the time rate of change of momentum:

$$F = \frac{d(mv)}{dt}.$$

The change of kinetic energy then is

$$dE_k = \frac{d(mv)}{dt} ds. \tag{5-38}$$

But since $ds/dt = v$, we can write the preceding equation as

$$dE_k = v\,d(mv) = v^2\,dm + mv\,dv. \tag{5-39}$$

This expression can be simplified by using the mass-transformation equation, Eq. (5–37), which after squaring and rearranging becomes

$$m^2c^2 = m^2v^2 + m_0^2c^2. \tag{5-40}$$

We now differentiate, noting that both m_0 and c are constant, and obtain

$$2mc^2\,dm = 2mv^2\,dm + 2m^2v\,dv. \tag{5-41}$$

When $2m$ is divided out of this equation, the right side becomes identical with our expression for dE_k in Eq. (5–39), so we have, finally,

$$\boxed{dE_k = c^2\,dm.} \tag{5-42}$$

This famous equation shows that in relativity a change in kinetic energy can be expressed in terms of mass as the variable. This equation is valid for a change of energy in any form whatever, although it was derived here for a change of kinetic energy.

Since the kinetic energy is zero when $v = 0$, then it is also zero when $m = m_0$. Therefore we integrate, and obtain

$$E_k = \int_0^{E_k} dE_k = c^2 \int_{m_0}^{m} dm = c^2(m - m_0)$$

or

$$\boxed{E_k = mc^2 - m_0c^2.} \tag{5-43}$$

This is the relativistic expression for kinetic energy. The classical relation $E_k = \frac{1}{2}mv^2$ does *not* give the correct value of the kinetic energy even when the relativistic values of the mass and the velocity are used.

Like the other transformation equations, this expression for E_k should reduce to the classical expression for $v \ll c$. If we transform m in Eq. (5–43), we obtain

$$E_k = \frac{m_0c^2}{\sqrt{1 - (v^2/c^2)}} - m_0c^2. \tag{5-44}$$

Now, since $v \ll c$, we can say

$$\frac{1}{\sqrt{1 - (v^2/c^2)}} = \left(1 - \frac{v^2}{c^2}\right)^{-1/2} = \left(1 + \frac{v^2}{2c^2} + \frac{3}{8}\frac{v^4}{c^4} + \cdots\right), \tag{5-45}$$

where we have carried three terms of the binomial expansion. We then have, from Eq. (5–44),

$$E_k = m_0c^2\left(1 + \frac{v^2}{2c^2} + \frac{3}{8}\frac{v^4}{c^4} + \cdots - 1\right)$$
$$= \frac{1}{2}\,m_0v^2 + \frac{3}{8}\,m_0\,\frac{v^4}{c^2} + \cdots \qquad (5\text{–}46)$$

The first term of this is the classical expression for the kinetic energy. Obviously, it is the only significant term for low velocities.

Equation (5–42) suggests a broader interpretation of Eq. (5–43). Equation (5–42) implies that, in addition to the familiar forms of energy, such as kinetic, potential, and internal, there is yet another kind, *mass-energy*. Just as Joule's constant relates heat to energy by telling us the number of joules per calorie, so Eq. (5–42) relates mass to energy by telling us the number of joules per kilogram. Just as heat is energy and energy can be observed as heat, so mass is energy and energy can be observed as mass. According to this view, Eq. (5–43) can be written as

$$E = mc^2 = m_0c^2 + E_k, \qquad (5\text{–}47)$$

where E and mc^2 are the *total* energy, m_0c^2 is the *rest mass-energy*, and E_k is the kinetic energy of a body.

We can now obtain an important relation between the total mass-energy E of a body and its momentum p from Eq. (5–40). If we multiply through this equation by c^2 and make the appropriate substitutions from $E = mc^2$ and $p = mv$, we get

$$\boxed{E^2 = p^2c^2 + m_0^2c^4.} \qquad (5\text{–}48)$$

This equation will be used several times in later chapters.

The equivalence of mass and energy brings a satisfying consequence. The broad principle of conservation of energy now takes unto itself another broad conservation principle, the conservation of mass. The identity of mass and energy resulting in a unified concept of mass-energy is certainly the most "practical" consequence of relativity. We shall see in Chapter 11 how nuclear reactions illustrate the fact that neither mass nor energy as conceived classically is conserved separately. In relativity we see that it is mass-energy that is conserved. In any interaction the *total mass, m* (which includes any kinetic energy in mass units), is *the same before and after* the interaction. Similarly, in any interaction, the *total energy E* (which includes any rest mass in energy units), is *the same before and after* the interaction. We are now witnessing the growth of the whole new field of nuclear engineering which is based on E equals mc^2.

Most of the mass-energy we use comes from the sun, where rest mass-energy is converted into thermal mass-energy. Today mankind is converting rest mass-energy into thermal mass-energy here on the earth. We shall see that nuclear fission and fusion are the techniques for this conversion.

5-15. THE UPPER LIMIT OF VELOCITY

An examination of Eq. (5–37), the mass transformation equation, shows that for $v = c$, $m = \infty$. Thus, as the velocity of a body increases toward the free-space velocity of light, the mass of the body increases toward infinity. An infinite mass moving at a snail's pace would have infinite energy. Since it is absurd for any body with finite rest mass m_0 to have infinite energy, we must conclude that it is impossible for such a body to move with the free-space velocity of light. Thus in relativistic mechanics the free-space velocity of light is the greatest velocity that can be given to any material particle. If a tiny particle is given equal increments of energy, the first increments increase the velocity significantly and the mass insignificantly. But as the particle gains more and more energy, the velocity changes become less as the velocity approaches the velocity of light. As the velocity changes become insignificant, the mass changes become significant.

Putting the same argument another way, we have two equations for the kinetic energy of a body, both of which are correct relativistic expressions. Equation (5–44) expresses E_k in terms of m_0 and v. For v small compared with c, this is the convenient formula, especially since it reduces to the first term of Eq. (5–46), the classical expression. The other equation is $E_k = c^2(m - m_0)$, Eq. (5–43). This equation expresses E_k in terms of m_0 and the variable moving mass, m. When the velocity is near c, the mass is the more convenient variable. We can express E_k in terms of either m or v, since we have the mass-variation equation, Eq. (5–37), relating m and v.

Since we argued that the free-space velocity of light is an upper limit for material particles, it is fair to ask whether any velocities equal or exceed this velocity. We have been careful in this discussion to emphasize that the limit is the *free-space* velocity of light. Light itself often travels more slowly than its free-space velocity. The index of refraction of a medium is the ratio of the free-space velocity of light to the actual velocity in the medium. In water, for example, a particle may travel faster than light *in that medium.* An example of this will be discussed when we consider Cherenkov radiation in Chapter 12.

Photons in free space obviously travel with the free-space velocity of light. We have found that photons have energy hf, and since we have shown that mass and energy are equivalent we can conclude that each photon has a mass hf/c^2. Here, then, is a mass particle actually traveling with the free-space velocity of light. But there is no contradiction here. A photon has no rest mass m_0. If we stop a photon to measure its mass, all its mass-energy is transferred to something else and the photon no longer exists. Perhaps the main difference between a "material" particle and a photon particle is that one has rest mass while the other does not.

There are times when one may encounter velocities greater than the free-space velocity of light, but on these occasions the thing moving is a mathematical function rather than a physical reality. The transmission of a "dot" in radio telegraphy may be regarded as the resultant of the transmission of several frequencies having different velocities. These velocities, called phase velocities, may exceed the free-space velocity of light. The motion of the resultant "dot" is called a group velocity, and it can never exceed the free-space velocity of light. (These two types of velocities associated with waves will be discussed in Chapter 7.) We may state with complete generality that no observer can find the *velocity of mass-energy in any form to exceed the free-space velocity of light.*

5-16. EXAMPLES OF RELATIVISTIC CALCULATIONS

(1) The basic mass unit of nuclear physics is the unified atomic mass unit, u, which is 1/12 of the rest mass of carbon-12. One u = 1.66×10^{-27} kg, and this is nearly the proton's mass. In relativity, this mass-energy may be expressed in energy units as

$$\begin{aligned} E = m_0c^2 &= 1.66 \times 10^{-27}\ \text{kg} \times (3.00 \times 10^8)^2\ \frac{\text{m}^2}{\text{s}^2} \\ &= 1.49 \times 10^{-10}\ \text{J}. \end{aligned}$$

Or 1 u may be expressed in eV and MeV:

$$\begin{aligned} 1\ \text{u} = \frac{E}{e_c} &= 1.49 \times 10^{-10}\ \text{J} \times \frac{1\ \text{eV}}{1.60 \times 10^{-19}\ \text{J}} \\ &= 9.31 \times 10^8\ \text{eV} = 931\ \text{MeV}. \end{aligned}$$

We have now expressed the relativistic rest mass-energy of one u in five ways:

$$\begin{aligned} 1\ \text{u} &= \tfrac{1}{12}\text{C}^{12} = 1.66 \times 10^{-27}\ \text{kg} = 1.49 \times 10^{-10}\ \text{J} \\ &= 9.31 \times 10^8\ \text{eV} = 931\ \text{MeV}. \end{aligned}$$

This same quantity could also be expressed in any other units of mass or energy—slugs, calories, grams, kwh, ft·lb, etc. We could have made similar calculations with any quantity of mass or energy. We chose the u as an important example. It is very useful to *remember* that 1 u = 931 MeV. (To six significant figures, 1 u equals 931.441 MeV on the C^{12} isotopic mass scale.)

(2) Consider next the velocity that one u would have if it has a kinetic energy equal to twice its rest mass-energy. It would be very difficult to give a particle with a mass of one u so much energy, since it would be about equivalent to accelerating a proton through 2×931 megavolts, but the situation provides a good example of calculating velocity relativistically.

This kinetic energy can be expressed as 2 u, 3.32×10^{-27} kg, 2.98×10^{-10} J, 1.86×10^{9} eV, or 1.86 GeV.

The total or moving mass of the particle is the sum of its rest mass and kinetic energy. Thus its moving mass, m, is $1 + 2 = 3$ u, 4.47×10^{-10}J, etc.

We now have everything necessary to calculate the relativistic velocity. The mass-velocity equation, Eq. (5–37), solved for the velocity is

$$v = c\sqrt{1 - (m_0/m)^2}.$$

We may substitute for the rest and moving masses in any consistent units. The velocity will have the same units as those used for c. Using the masses in u, we have

$$v = c\sqrt{1 - (\tfrac{1}{3})^2} = 0.942c = 2.83 \times 10^8 \,\frac{\text{m}}{\text{s}}.$$

This is a very large velocity, and we may expect that a classical calculation will be seriously in error. To show this, we next calculate v on the basis of classical physics. We know that the kinetic energy is $E_k = 2.98 \times 10^{-10}$ J, and that the mass is $m = 1.66 \times 10^{-27}$ kg. Solving $E_k = \frac{1}{2}mv^2$ for v, we have

$$v = \sqrt{\frac{2E_k}{m}} = \sqrt{2 \times 2.98 \times 10^{-10}\text{ J} \times \frac{1}{1.66 \times 10^{-27}\text{ kg}}}$$

$$= 6.0 \times 10^8 \,\frac{\text{m}}{\text{s}} = 2.0c.$$

Not only is this result very different from the relativistic one, but it is also greater than the velocity of light. Whenever the two methods differ significantly, the relativistic method must be used.

(3) We next seek the kinetic energy of a 1-slug earth satellite traveling with a velocity of 18,600 mi/h. Since the velocity of light in English units is 186,000 mi/s, the ratio of the satellite's velocity to the velocity of light, v/c, is

$$\frac{v}{c} = 18{,}600 \,\frac{\text{mi}}{\text{h}} \times \frac{1\text{ h}}{3600\text{ s}} \times \frac{1\text{ s}}{186{,}000\text{ mi}} = \frac{1}{36{,}000}.$$

Strictly speaking, we should compute the satellite's moving mass from Eq. (5–37), $m = m_0/\sqrt{1 - (v/c)^2}$, and then the kinetic energy from Eq. (5–47), $E_k = (m - m_0)c^2$. But the satellite's velocity is so small compared with c that slide rule calculations would make the use of Eq.

(5–46), $E_k = \frac{1}{2}m_0v^2 + \frac{3}{8}m_0(v^4/c^2)\ldots$, much more convenient. In this example the second term on the right is negligible. This shows that a classical calculation is sufficient for a body moving so slowly.

(4) The labor of calculating the numerical results in problems in special relativity may often be reduced by a trigonometric substitution. To illustrate this, let us find the relativistic velocity of an electron that has been accelerated from rest through a potential difference of 120,000 V.

If we find the final moving mass of the electron, the velocity can be obtained from the equation

$$m = \frac{m_e}{\sqrt{1-(v^2/c^2)}}\cdot$$

The moving mass is

$$\begin{aligned} m &= m_e + \Delta m = m_e + \frac{\Delta E_k}{c^2} = m_e + \frac{qV}{c^2} \\ &= 9.11 \times 10^{-31}\ \text{kg} + 1.60 \times 10^{-19}\ \text{C} \\ &\quad \times 120{,}000\ \frac{\text{J}}{\text{C}} \times \frac{\text{s}^2}{(3 \times 10^8)^2\ \text{m}^2} \\ &= (9.11 + 2.13) \times 10^{-31}\ \text{kg} = 11.24 \times 10^{-31}\ \text{kg}. \end{aligned}$$

Rather than substitute the values of m and m_e in the mass-variation equation and perform the long operations of squaring, taking a difference, and obtaining a square root, let us substitute $\sin\theta = v/c$. Then we have

$$\sqrt{1-(v^2/c^2)} = \sqrt{1-\sin^2\theta} = \cos\theta,$$

and the mass-variation equation becomes

$$\cos\theta = \frac{m_e}{m}\cdot$$

Upon substituting the values of the mass, we get

$$\cos\theta = \frac{m_e}{m} = \frac{9.11 \times 10^{-31}\ \text{kg}}{11.24 \times 10^{-31}\ \text{kg}} = 0.810.$$

From trigonometric tables we then find that

$$\theta = 35°53' \qquad \text{and} \qquad \sin\theta = 0.586.$$

Therefore, the result sought is

$$\sin\theta = \frac{v}{c} = 0.586$$

or

$$v = 0.586c = 0.586 \times 3 \times 10^8\ \frac{\text{m}}{\text{s}} = 1.76 \times 10^8\ \frac{\text{m}}{\text{s}}\cdot$$

5-17. PAIR PRODUCTION

Certainly one of the most dramatic instances of the relativistic change of energy from one form to another is in the phenomenon known as pair production. *Pair production* is the process in which a photon becomes an electron and a new positive particle. This new particle was anticipated by Dirac in 1928 from his relativistic wave-mechanics theory of the energy of an electron, and was experimentally observed by Anderson in 1932.

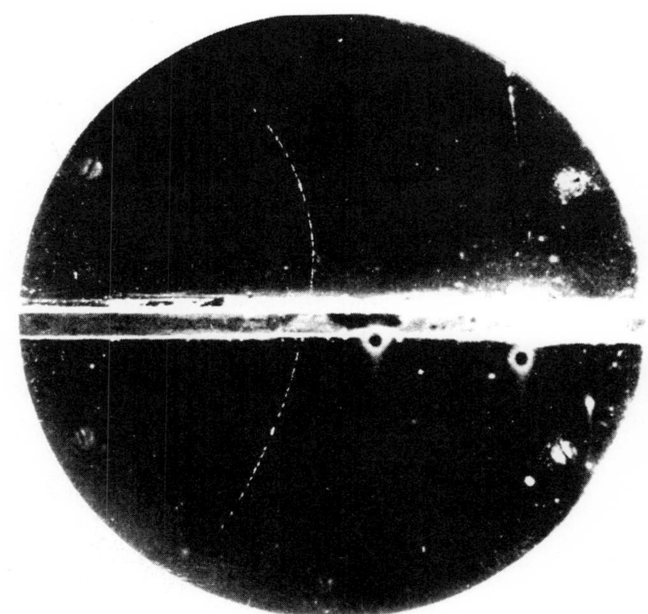

FIG. 5-5
Cloud-chamber track of a positive electron (positron) in a magnetic field. (Courtesy of C. D. Anderson, California Institute of Technology)

The discovery was made in the course of cosmic-ray research with a cloud chamber. We shall discuss cloud chambers in Chapter 9. All we need say now is that a cloud chamber is a device which makes the paths of charged particles visible and subject to photographic recording. Figure 5-5 is Anderson's most famous cloud-chamber photograph. It shows a charged particle moving with high speed and with its path curved by a magnetic field directed into the paper. The particle is seen to traverse a sheet of lead 6 mm thick in the middle of the chamber. From the beady nature of the track it was established that the particle was electronlike. Since the lead sheet could have only slowed the particle, its direction of motion must have been from the region of low curvature to the region of high curvature. From the observed direction of curvature and the known direction of the motion and of the magnetic field, it was concluded that the

FIG. 5–6
Cloud-chamber tracks of three electron-positron pairs in a magnetic field. Three gamma-ray photons entering at the top materialize into pairs within a lead sheet. The coiled tracks are due to low-energy photoelectrons ejected from the lead. (Courtesy of Radiation Laboratory, University of California)

charge of the particle was positive. On the basis of this picture and others similar to it, Anderson announced the discovery of a new particle which he called the *positron* or positive electron.* The positron is just like the electron except for the sign of its electric charge.

The source of the positron is not visible in Fig. 5–5 but several sources are apparent in Fig. 5–6. Note the three sets of diverging tracks leaving the lower side of the lead plate. These are tracks originating at the lead but with no corresponding tracks coming into the lead on the upper side. Evidently these pairs of particles are not produced by other charged particles but by something else that cannot produce tracks, namely, photons. The curvature of the paths is due to a magnetic field normal to the plane of the paper. Since the tracks of the particles of the pair are oppositely curved, the individuals must have opposite charges. In each case one particle is a positron and the other an ordinary electron. A quantum of radiant energy coming from above the lead sheet has changed into matter forming a *pair* consisting of a positron and an electron. In the process of *pair production* we have the *materialization* of energy.

It is easy to show that the rest mass of an electron is equivalent to 0.51 MeV. Thus, to create a pair, a photon must have at least enough energy to convert into two electron masses or 1.02 MeV. A photon could never convert into just one electron or positron, since this would violate the law of conservation of electric charge. The photons which produced the pairs shown in Fig. 5–6 must each have had more than 1.02 MeV of energy, since these pairs were not only created but also were given considerable kinetic energy.

A photon cannot produce a pair just anywhere. The process must take place in an intense field such as that close to the nucleus of an atom. Since a nucleus must be involved in pair production, conservation of energy and of momentum become a bit complex. If the nucleus involved is massive, it can carry away its share of momentum without taking away an appreciable amount of kinetic energy. In this case pair production can

* At this same time it was proposed that the electron be called the *negatron* or negative electron. These names have never come into general use.

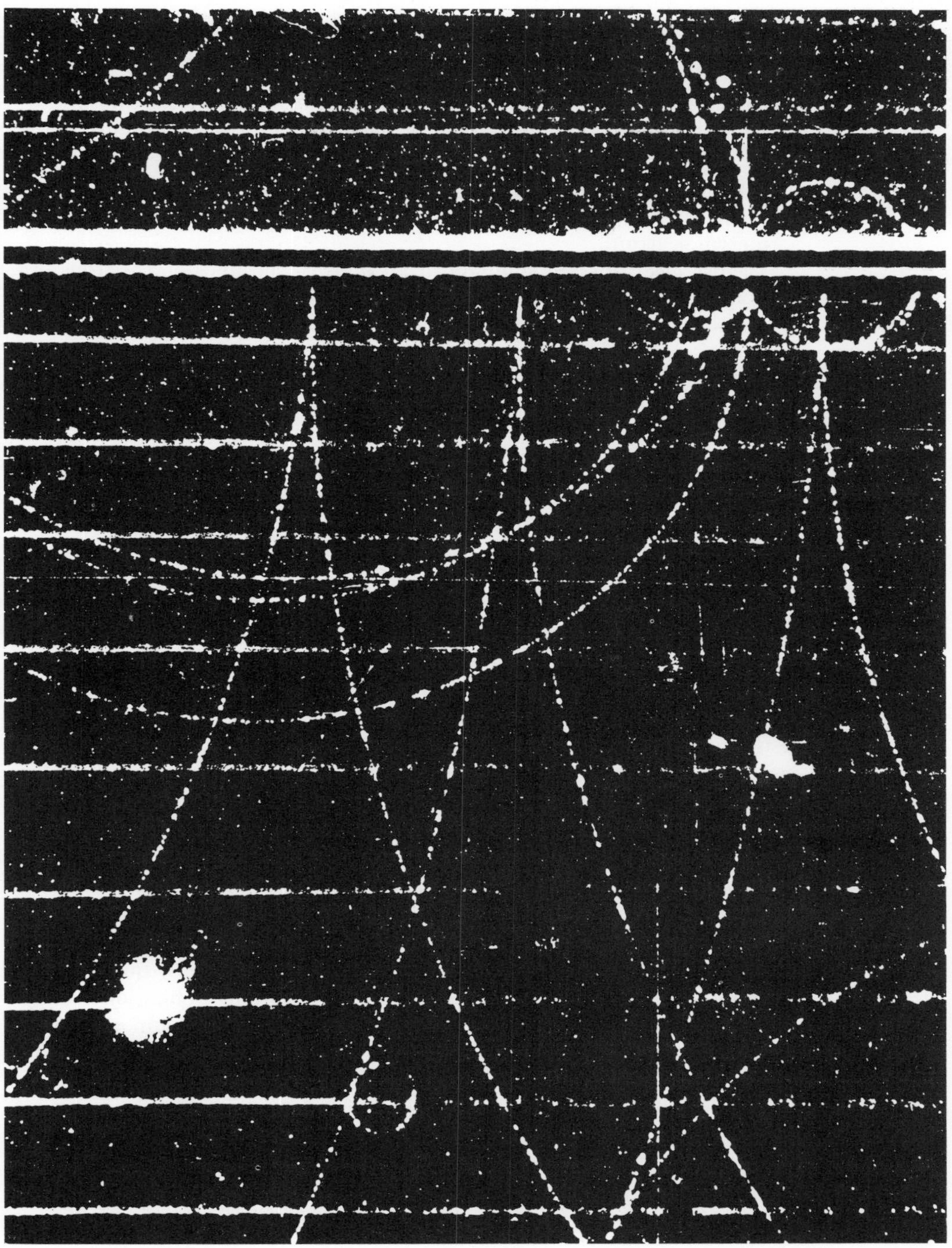

occur if the energy of the photon is just a little above the mass-required minimum of 1.02 MeV. If the pair is produced near a light nucleus, on the other hand, the light nucleus carries away appreciable kinetic energy at the expense of the originating photon and thus this photon must have considerably more than 1.02 MeV of energy. For 20-MeV photons, the nuclei of the air are about as effective as those of lead in "catalyzing" pair production. Since pair production occurs near the nucleus of an atom, the repelled positron emerges with a somewhat higher velocity than the attracted electron.

Because electrons are common in all matter, the lives of positrons are very short. Once a positron has lost most of its kinetic energy in passing near charged particles, it will move quickly toward an attracting electron. Before union occurs, however, these particles sometimes revolve momentarily about their center of mass in a semi-stable configuration called *positronium.*

Positronium is a non-nuclear "element" with an average existence of less than 10^{-7} s. Despite its short life, its spectrum has been measured. These measurements show that positronium is hydrogenlike, as expected. The frequencies of its spectral lines are about half those of the corresponding hydrogen lines because the reduced mass of the electron in positronium is about half its value in hydrogen.

When the electron and positron finally join, they annihilate each other and become electromagnetic radiation. The production of *annihilation radiation* is the converse of the materialization of energy. Just as pair production requires a photon of at least 1.02 MeV of energy, so annihilation produces photons whose energies add up to at least 1.02 MeV. Annihilation resulting in two photons is by far the most common, but the three-photon case has been observed. This rare event is discussed in advanced books. The laws of conservation of mass-energy, of momentum, and of charge hold in all these processes.

Pair production is not limited to the positron and electron. In 1955 the *antiproton* was produced in experiments with high-energy particles from accelerators. The antiproton and the proton have equal charges of opposite sign and equal masses. It requires a minimum of 1.88 GeV of energy to produce a proton-antiproton pair. The antiproton is usually quickly captured by a proton. The subsequent proton-antiproton annihilation usually produces a number of subatomic particles. The positronium analog of this pair has not been observed.

The positron and the antiproton are often called *antiparticles.* They may be thought of as plane mirror images of the electron and proton, respectively. Since the discovery of these pairs, there has been speculation about the possibility of the existence of *antihydrogen.* This would have an antiproton nucleus and an orbital positron. One can indeed imagine the existence of the *antimatter* form of each of the elements. No such mirror-image forms have been found, but perhaps other galaxies are composed of them. Note that the term "antimatter" does not mean negative mass.

The processes of pair production and annihilation are both inconceivable without relativity. These are cases where an appreciable amount of radiant energy without rest mass and particles with appreciable rest mass are alternately materialized and destroyed.

REFERENCES

AMAR, HENRI, "New Geometric Representation of the Lorentz Transformation," *Am. J. Phys.* **23,** 487 (1955).

BARKER, E. F., "Energy Transformations and the Conservation of Mass," *Am. J. Phys.* **14,** 309 (1946). This is an excellent discussion of the mass-energy relation.

EDDINGTON, A. S., *Space, Time, and Gravitation.* Cambridge, England: University Press, 1920. A very readable discussion of the special and general theories of relativity, including the four-dimensional space-time continuum.

EINSTEIN, ALBERT, *Relativity, the Special and General Theory.* R. W. Lawson, translator. New York: Holt, 1920. A simple, lucid account of relativity.

EINSTEIN, ALBERT, *Out of My Later Years.* New York: Philosophical Library, 1950. Essays on a number of subjects.

FRANK, PHILIPP, *Einstein, His Life and Times.* New York: Knopf, 1947. This is a good biography.

FRISCH, D. H. and J. H. SMITH, "Measurement of the Relativistic Time Dilation Using μ-Mesons," *Am. J. Phys.* **31,** 342 (1963).

GAMOW, GEORGE, *Mr. Tompkins in Wonderland.* New York: Macmillan, 1944. This delightful fantasy is an account of life in a strange, relativistic world where the velocity of light is only ten miles per hour.

HOLTON, GERALD, *Special Theory of Relativity,* Resource Letter SRT-1 and Selected Reprints. New York: American Institute of Physics, 1963. Contains, among others, the article by Terrell and the one by Weisskopf on the invisibility of length contraction.

KATZ, ROBERT, *An Introduction to the Special Theory of Relativity,* Momentum Book No. 9. Princeton, N. J.: Van Nostrand, 1964.

MICHELSON, A. A., *Studies in Optics.* Chicago, Ill.: University of Chicago Press, 1927. This contains descriptions of famous experiments to determine the velocity of light and to measure the ether drift.

MICHELSON, A. A. and E. W. MORLEY, "The Michelson-Morley Experiment," in Magie, W. F., *A Source Book in Physics.* Cambridge, Mass.: Harvard University Press, 1963.

SCOTT, G. D. and M. R. VINER, "The Geometrical Appearance of Large Objects Moving at Relativistic Speeds," *Am. J. Phys.* **33,** 534 (1965).

SEARS, F. W. and R. W. BREHME, *Relativity*. Reading, Mass.: Addison-Wesley (in press).

SHANKLAND, R. S., "Conversations with Albert Einstein," *Am. J. Phys.* **31,** 47 (1963). Contains very interesting historical material on relativity.

SHANKLAND, R. S., "Michelson-Morley Experiment," *Am. J. Phys.* **32,** 16 (1964). A historical account.

SLEPIAN, JOSEPH, "Magnetic Speedometer for Aircraft," *Elec. Eng.* **68,** 308 and 449, 1949.

TAYLOR, E. F. and J. A. WHEELER, *Spacetime Physics*. San Francisco: Freeman, 1966. An introductory textbook in relativity.

TERRELL, JAMES, "Invisibility of the Lorentz Contraction," *Phys. Rev.* **116,** 1041 (1959).

WEBSTER, D. L., "Relativity and Parallel Wires," *Am. J. Phys.* **29,** 841 (1961).

WEISSKOPF, V. F., "The Visual Appearance of Rapidly Moving Objects," *Phys. Today* **13,** 24 (1960).

WHITTAKER, EDMUND T., *A History of the Theories of Aether and Electricity*. Revised ed. London: Thomas Nelson and Sons. Vol 1, 1951; Vol. 2, 1953. Comprehensive treatment of the development of various concepts in these fields.

PROBLEMS

5–1. A person who knows Galilean-Newtonian mechanics leaves his earth-bound laboratory and establishes an isolated one in a closed, over-the-road trailer which a truck can pull along a level highway without noise or vibration. Is it possible for him to perform at least one experiment in his new laboratory system to determine whether the trailer has (a) a linear acceleration, (b) a radial acceleration, (c) a constant linear velocity, or no velocity? Describe an experiment he might perform in each case.

5–2. An experimenter who is skilled in Newtonian mechanics moves from a laboratory in an astronomical observatory to one in a cave which is shut off from all contact with the outside world. Could he perform any experiment in his new laboratory system which would determine (a) whether the earth is rotating and (b) whether it is moving with constant linear velocity? Describe an experiment he might perform in each case.

5–3. (a) The ends of a rectangular loop of wire are connected to a sensitive galvanometer. If the whole system (loop and galvanometer) is then moved with constant velocity in a uniform magnetic field where the lines of induction are normal to the plane of the loop, will the galvanometer deflect? Explain. (b) It is proposed to make a ground-speed indicator for an airplane in the following way: A galvanometer on the instrument panel is to be connected to the ends of a wire stretched between the wing tips (or simply to the tips themselves for a metal airplane). It is believed that there will be an induced emf in the wire that is directly proportional to the speed of "cutting" the vertical component of the magnetic field of the earth when in level flight. Will the proposed indicator work? Explain. (Assume that the magnetic field is constant.) (c) Is it possible to measure the speed of an automobile relative to the earth from the fact that it is moving in the earth's magnetic field? Explain.

5–4. A person who is unaware of Newtonian mechanics and who has always lived in a closed trailer is told that a plumb line points in the direction of the force of gravity. (a) In what direction will he think this force acts if, as viewed by an observer on the earth, the trailer has always been moving at a constant speed around a horizontal circular track? (b) Does a plumb line on your trailer system, the earth, point in the direction of the gravitational pull of the earth? (c) Does it point in the direction of the acceleration due to gravity?

5–5. A super-searchlight which is rotating at 120 rev/m throws its beam on a screen 50,000 mi away. What is (a) the sweep speed of the beam across the screen, (b) the speed of a photon from the searchlight to the screen, and (c) the speed with which any particular photon sweeps across the screen? (d) Do any of these results violate the second postulate of the special theory of relativity? (The value of c is 186,000 mi/s.)

5–6. Cathode-ray tubes have been made in which the sweep speed or trace speed of the electron beam across the fluorescent screen exceeds the free-space velocity of light. Does this require that the velocity of any one of the electrons exceed the limiting relativistic velocity? Explain.

5–7. Show from Eq. (5–25) that the relativistic equations to transform space and time to system S from observations on S' are $x = k(x' + vt')$ and

$$t = k[t' + (vx'/c^2)],$$

respectively.

5–8. The coordinates of an event are x, y, z, and t in the unprimed frame and x', y', z', and t' in the primed frame. Suppose a quantity ds^2 is defined by $ds^2 = dx^2 + dy^2 + dz^2 - c^2\,dt^2$. Show that this quantity is invariant under a relativistic transformation of coordinates, that is, that $ds^2 = ds'^2$.

5–9. Consider two events which when observed from the S system occur at different points x_1 and x_2 but at the same time t_0. Show that these two events are not simultaneous to an observer in the S'-system, which is moving with a velocity v parallel to the x-axis, and that the time difference in S' can vary from $-\infty$ to $+\infty$.

5–10. The relative speed of S and S' is $0.98c$. The following statement is made by S': "At noon a red light flashed at my origin and a blue light flashed 10 seconds later at $x' = 9 \times 10^8$ meters, $y' = 0$, $z' = 0$." What is the temporal separation of the two flashes as measured by S?

5–11. A man in a super rocket travels between two markers on the ground placed 90 m apart in 5×10^{-7} s at constant speed. This time is measured by a ground observer. (a) How far apart does the man in the rocket judge the markers to be? (b) What time interval does his own watch show elapsed? (c) What is his estimate of the speed of the markers relative to him? (d) Compare this with the ground observer's judgment of the rocket's speed.

15 sec

5–12. An observer A is on a rocket ship which passes the earth at a speed of 1.8×10^8 m/s. An observer B on earth sets his watch so that it reads zero, the same as A's watch when the ship passes. (a) If observer B looks at A's watch through a telescope, what time does B see on A's watch when B's watch reads 30 s? (b) If A observes B's watch, what time does he see when his watch reads 30 s? (Note that the transit time of the light signal must be taken into account.)

5–13. A subatomic particle with a mean proper lifetime of 3.0 μs is formed in a high-energy accelerator and projected through the laboratory with a speed of $0.8c$. (a) What is the mean lifetime of the particle in the laboratory system? (b) How far does the particle travel, on the average, from the point of formation in the laboratory before disintegrating? (c) What is distance in part (b) as observed from the reference frame in which the particle is at rest?

5–14. Space ship A is traveling with a speed of $0.8c$ with respect to the earth. Space ship B, traveling on a parallel course, passes A with a relative speed of $0.5c$. What is the speed of B with respect to the earth when calculated (a) classically, and (b) relativistically?

5–15. (a) With respect to the laboratory, ball A rolls eastward with a constant speed of 2 m/s and ball B rolls westward with a constant speed of 2 m/s. What is the relative velocity of A with respect to B calculated relativistically? (b) With respect to the laboratory, the electrons in accelerator A are projected toward the east with a speed of 2×10^8 m/s and electrons in accelerator B are projected toward the west with a speed of 2×10^8 m/s. What is the relative velocity of the A electrons with respect to the B electrons calculated relativistically?

5–16. Given a source of light in the primed frame which is moving parallel to the x-axis with a velocity v relative to the unprimed frame. (a) A beam of radiation from the source is emitted in the $x'y'$-plane at an angle θ' with the x'-axis. Transform the velocity components of the beam, $u'_x = c \cos \theta'$ and $u_y = c \sin \theta'$, to the unprimed frame and find the resultant of the transformed components. (b) Show that the angle θ made by the beam with the x-axis is given by

$$\cos \theta = \frac{\cos \theta' + (v/c)}{1 + (v/c) \cos \theta'}.$$

(c) The axis of a cone of light in the primed frame is parallel to the x-axis. Given that the half angle of the cone is 60° and that $v = 0.8c$, find the half angle of the cone in the unprimed frame. Why would the result in a problem like this be called the "headlight effect"?

5–17. A stick of length L_0 is at rest in the S-system and is oriented at an angle θ with respect to the x-axis. What is the length of the stick and its orientation angle as viewed by an observer in the S'-system, which is moving with the velocity v parallel to the x-axis?

5–18. In elementary optics, some derivations of the equation for the Doppler effect for light contain a step in which the velocity of the light source is added vectorially to the velocity of the light emitted. Relativistically, the velocity of light is independent of that of the source. How would you explain the existence of the Doppler effect on the basis of the special theory of relativity? [Consider time intervals (frequency), etc.]

5–19. Show that the rest mass of an electron is equivalent to 0.511 MeV.

5–20. An electron in a certain x-ray tube is accelerated from rest through a potential difference of 180,000 V in going from the cathode to the anode. When it arrives at the anode what is its (a) kinetic energy in eV, (b) its relativistic mass, (c) its relativistic velocity, and (d) the value of e/m_e? (e) What is the velocity of the electron calculated classically?

5–21. A kilogram of water is heated from 0°C to 100°C. (a) What is its increase in mass due to its increase in thermal energy? (b) What is the ratio of this increase to the initial mass? (c) Could this mass increase be measured? (d) How much mass is lost by a kilogram of water at 0°C when it freezes to ice at 0°C?

5–22. A doubly ionized helium atom is accelerated from rest through a potential difference of 6×10^8 volts in a linear accelerator. Find (a) its kinetic energy in MeV, (b) its relativistic mass in kg, and (c) its relativistic velocity. (Assume that the rest mass of a helium atom is equal to four times that of a hydrogen atom.)

5–23. Show that the expression $E_k = \frac{1}{2}mv^2$ does *not* give the relativistic value of the kinetic energy of a body even if the relativistic mass is used. [*Hint:* Substitute m from Eq. (5–37) into the given relation for kinetic energy, expand by the binomial theorem, and compare the result with Eq. (5–46).]

5–24. Calculate, relativistically, the amount of work in MeV that must be done (a) to bring an electron from rest to a velocity of $0.4c$, and (b) to increase its velocity from $0.4c$ to $0.8c$. (c) What is the ratio of the kinetic energy of the electron at the velocity of $0.8c$ to that of $0.4c$ when computed from (1) relativistic values, and (2) from classical values?

5–25. Plot the ratio of m/m_0 of a particle as a function of v/c for the following values of the independent variable: 0.0, 0.2, 0.4, 0.6, 0.8, 0.9, and 0.99.

5–26. An airplane has a speed of 500 mi/hr with respect to the earth. (a) For earth-bound observers, what is the fractional change in the length of the airplane since take-off due to its speed? (b) What is the fractional change in mass?

5–27. What is the measured length of a meter stick moving parallel to its length when its mass is found to be twice its rest mass?

5–28. A particle moves at a speed such that its kinetic energy just equals its rest mass energy. What is the speed of the particle?

5–29. At the surface of the earth the radiation intensity from the sun is

$$0.140 \text{ J/cm}^2 \cdot \text{s}.$$

The present mass of the sun is 2×10^{30} kg and the sun is 149×10^6 km from the earth. Calculate the percent loss in the mass of the sun in one hour.

5–30. (a) Show that the speed of a particle whose total energy is E is

$$v = c\left[1 - \left(\frac{m_0c^2}{E}\right)^2\right]^{1/2}.$$

(b) Show that the speed of a particle whose momentum is p is

$$v = \frac{pc}{(p^2 + m_0^2c^2)^{1/2}}.$$

5–31. An electron having a kinetic energy of 0.50 MeV enters a region which has a uniform magnetic induction of 5×10^{-3} T normal to the motion of the electron. Calculate the radius of the trajectory using classical physics and then repeat the calculation using relativistic physics.

5–32. Some high-energy accelerating devices produce electrons with kinetic energies as high as 10^9 eV. (a) What is the ratio of the mass of the electron at this energy to the electron rest mass? (b) What is the speed of this high-energy electron?

5–33. A certain type of charged particle accelerator fails to work when the mass of the accelerated particle becomes 25% greater than its rest mass. (a) What is the highest kinetic energy this accelerator can give to (1) a proton and (2) an electron? (b) What is the relativistic velocity of the proton and the electron in part (a)?

5–34. Assuming that calculations must be made relativistically instead of classically when the results from the two methods differ by more than 1%, find the potential difference through which a charged particle must be accelerated from rest in a vacuum so that its relativistic or moving mass exceeds its rest mass by 1% of the rest mass if the particle is (a) an electron and (b) a proton.

5–35. Assuming that calculations must be made relativistically instead of classically when the results from the two methods differ by more than 1%, find the potential difference through which a charged particle must be accelerated from rest in a vacuum so that its classical velocity is 1% more than its relativistic velocity if the particle is (a) an electron and (b) a proton.

5–36. When the radioactive isotope cobalt-60 disintegrates, it emits two gamma-ray photons and one 0.31-MeV beta particle (electron). (a) What is the relativistic velocity of the ejected beta particle? (b) If the cobalt-60 is immersed in water, is the velocity of ejection calculated in part (a) less than the velocity of light in water, which has an average index of refraction of 1.33?

5–37. What is the energy equivalent in joules and in kilowatt-hours of 1 kg of uranium, or of wood, or of sand?

5–38. The fissioning of an atom of uranium-235 releases 200 MeV of energy. What percent is this fission energy of the total which would have been available if all the mass of the uranium atom had appeared as energy?

5–39. In a famous experiment on nuclear reactions performed by Cockcroft and Walton a stationary target composed of the isotope lithium-7 was bombarded with 0.70-MeV hydrogen ions. It was found that these elements then combine and change to two alpha particles, each having a kinetic energy of 9.0 MeV. (An alpha particle is doubly ionized helium-4.) (a) Calculate and compare the total mass, in u, of the initial particles with the total mass of the final particles. (The rest masses of the neutral atoms are given in Appendix 5. Use the conversion factor 1 u/931 MeV. Assume that all data are precise enough to warrant answers to three significant figures.) (b) Calculate and compare the total energy (includes rest mass energy) in MeV of the initial particles with the total energy of the final particles. (c) Show that the difference between the sum of the kinetic energies of the final particles and the sum of the kinetic energies of the initial particles is equal to the energy equivalent of the difference between the sum of the rest masses of the initial particles and the sum of the rest masses of the final particles. (d) Using the results of the preceding parts, discuss the law of conservation of mass and the law of conservation of energy in relativity. In which part might one say that mass has been "transformed" or "changed" into energy? Why might this be said?

5–40. How much work must be done in terms of the rest mass energy, m_0c^2, to increase the velocity of a particle (a) from rest to $0.90c$, and (b) from $0.90c$ to $0.99c$? (c) What is the increase in mass in terms of m_0 for each case?

5–41. A gamma-ray photon having a wavelength of 0.0045 Å materializes into an electron-positron pair in the neighborhood of a heavy nucleus. (a) What is the total kinetic energy of the pair in MeV immediately after being produced? (b) Could two 0.75-MeV quanta which are passing through lead materialize into an electron-positron pair? Explain.

5–42. An electron and a positron which have negligible velocities combine to produce two-photon annihilation radiation. (a) What is the energy of each photon? (b) What will be the relative direction of motion of these photons? Explain.

5–43. A 2.90-MeV quantum of radiation, passing through lead, materializes into an electron-positron pair. Given that these particles have equal kinetic energies, find (a) the relativistic mass, and (b) the relativistic velocity of each. Neglect the recoil energy of the lead atom. (c) What is the direction of and the smallest value of the magnetic induction which will cause each of these particles to have a trajectory with a radius of curvature of 12 cm?

5–44. If an electron-positron pair is produced near a lead nucleus so that the positron has 11 MeV of energy and the electron has 10 MeV of energy, how far from the center of the lead nucleus was the pair produced?

5–45. Positronium is a hydrogen-like revolving system composed of a positron and an electron. (a) Show that the energies of the Bohr orbits are one-half those of hydrogen. (b) How do the frequencies of the emitted spectra compare with those of hydrogen?

CHAPTER 6

X-RAYS

6-1. DISCOVERY

In Chapter 2 we reported how the study of electric discharges through gases at low pressure led Thomson to the discovery of the electron and of isotopes. Let us now consider an earlier discovery made in 1895 in connection with electric discharges: Roentgen's discovery of x-rays. We quote a translation of Roentgen's words:*

"If the discharge of a fairly large induction-coil be made to pass through a Hittorf vacuum-tube, or through a Lenard tube, a Crookes tube, or other similar apparatus which has been sufficiently exhausted, the tube being covered with thin, black card-board which fits it with tolerable closeness, and if the whole apparatus be placed in a completely darkened room, there is observed at each discharge a bright illumination of a paper screen covered with barium planto-cyanide, placed in the vicinity of the induction-coil, the fluorescence thus produced being entirely independent of the fact whether the coated or the plain surface is turned towards the discharge-tube . . .

"The most striking feature of this phenomenon is the fact that an active agent here passes through a black card-board envelope which is opaque to the visible and the ultra-violet rays of the sun or of the electric arc; an agent, too, which has the power of producing active fluorescence . . .

"We soon discover that all bodies are transparent to this agent, though in very different degrees. I proceed to give a few examples: Paper is very transparent; behind a bound book of about one thousand pages I saw the fluorescent screen light up brightly, the printers' ink offering scarcely a noticeable hindrance. In the same way the fluorescence appeared behind a double pack of cards; a single card held between the apparatus and the screen being almost unnoticeable to the eye. A single sheet of tin-foil is

---

* Reprinted by permission of the publishers from William Francis Magie, *Source Book in Physics*. Cambridge, Mass.: Harvard University Press, 1935.

also scarcely perceptible; it is only after several layers have been placed over one another that their shadow is distinctly seen on the screen. Thick blocks of wood are also transparent, pine boards two or three centimeters thick absorbing only slightly. A plate of aluminum about fifteen millimetres thick, though it enfeebled the action seriously, did not cause the fluorescence to disappear entirely."

The history of science has many instances of "accidental" discovery and of these the discovery of x-rays is a prime example. Columbus' discovery of America was another instance where a man looking for one thing found another. Although accidents can happen to anyone, "accidental" discovery of truth seems to be reserved to those observers who have earned the right to be lucky. Such luck falls on the deserving few whose courage, patience, insight, and objectivity enable them to take advantage of the "breaks of the game." There is a word for this process that is better than the word "accident." It is "serendipity"—the process or art of taking advantage of the unexpected.

Of all the discoveries made by man, there is probably none that attracted public attention more quickly than the discovery of x-rays. The fact that the rays permitted one to "see" through opaque objects was sensational "tabloid material" and there was great consternation lest, by their use, fully-dressed people might be made to appear unclothed. When such speculation died down, however, there was wide appreciation of the value of x-rays in setting broken bones, and the rays were quickly put to this use.

6–2. PRODUCTION OF X-RAYS

As Roentgen said in his paper, the observed x-rays came from low-pressure gas discharge tubes like those already described in our discussion of cathode rays. We quote further from Roentgen's original paper published in 1895:* "According to experiments especially designed to test the question, it is certain that the spot on the wall of the discharge-tube which fluoresces the strongest is to be considered as the main centre from which the x-rays radiate in all directions. The x-rays proceed from that spot where, according to the data obtained by different investigators, the cathode rays strike the glass wall. If the cathode rays within the discharge-apparatus are deflected by means of a magnet, it is observed that the x-rays proceed from another spot—namely, from that which is the new terminus of the cathode rays."

Modern x-ray tubes still produce x-rays by causing cathode rays to strike a solid target, but the techniques have been greatly refined. In a modern Coolidge tube, the source of the cathode-ray electrons is a heated filament. There is no need for a residual gas to be ionized, and so modern

* Magie, op. cit.

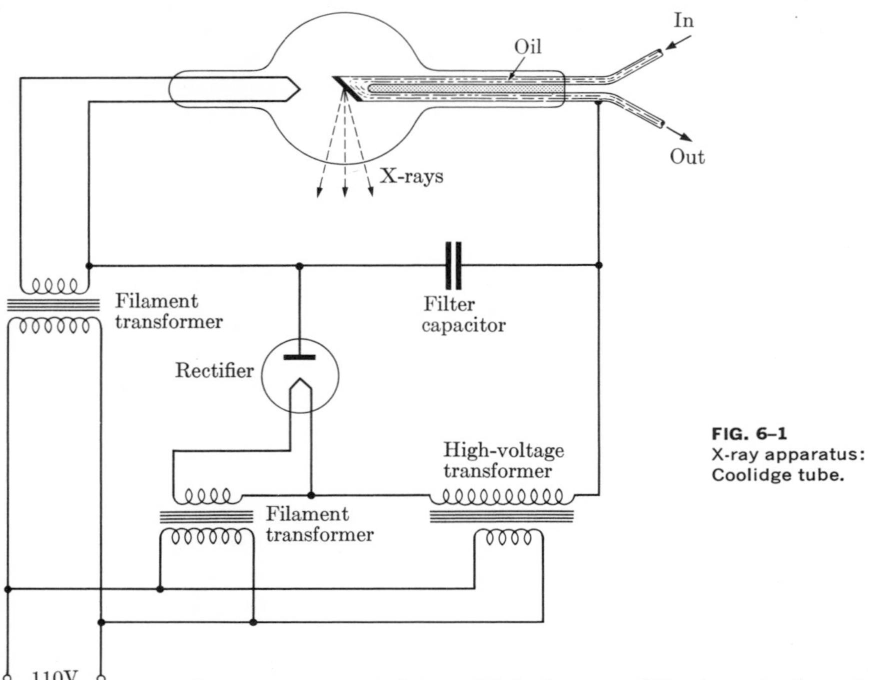

FIG. 6–1
X-ray apparatus:
Coolidge tube.

tubes are evacuated to a high degree. The target of modern tubes is a metal having a high melting point and a high atomic number. In the production of x-rays, a large amount of heat is generated in the target and so it is usually made hollow to permit cooling water or oil to be circulated through it. In addition to the tube itself, the other major part of the apparatus is a source of high potential to accelerate the cathode-ray electrons. Originally, this was provided by an induction coil (spark coil), but the common technique today is to use step-up transformers which operate at power-line frequency alternating current. If a.c. is applied to the tube, cathode rays and, consequently, x-rays are produced during only one-half of the cycle and the potential across the tube is constantly changing. Most x-ray machines rectify and filter the high alternating potential so that there is a steady high voltage across the tube. The potential used depends on the ultimate use of the x-rays, but it usually ranges between ten thousand and a million volts. A schematic diagram of the essential features of an x-ray machine is shown in Fig. 6–1.

6–3. THE NATURE OF X-RAYS. X-RAY DIFFRACTION

Roentgen failed to determine the nature of x-rays. He eliminated some possibilities and he suggested what they might be, but his own uncertainty is shown by the fact that he chose the name x-rays.

Upon finding that x-rays are not deflected by electric or magnetic fields on the one hand or perceptibly refracted or diffracted on the other, Roentgen concluded that the rays were neither charged particles nor light of any ordinary sort. Tongue-in-cheek, he proposed that x-rays might be the "longitudinal component" of light, in analogy with the fact that acoustic vibrations in solids have both a longitudinal and a transverse part. There is some appeal in the idea that light might have a component which, being longitudinal, might be "slimmer" than the transverse part and therefore penetrate solids more easily than the "fatter" transverse waves. But Maxwell's wave theory of light, which was then at its height, accounted for light being a transverse wave motion, and an effort was made to identify x-rays with this more familiar phenomenon.

In 1912, Max von Laue conceived of a way of testing the idea that x-rays might be light of very short wavelength. Recall the diffraction grating formula,

$$n\lambda = d \sin \theta, \tag{6–1}$$

where n is the spectrum order (1, 2, 3, . . .), λ is the wavelength, d is the grating space, and θ is the angle of diffraction. We can see that if λ is very small compared with d, the diffraction angle must be very small unless the order n is large. Since the intensity in high orders is very weak, we can see that assuming λ to be very small can account for the observed failure of gratings to produce measurable diffraction for x-rays.

The obvious remedy is to make gratings with much finer rulings, although we shall see in Section 6–12 that there is another remedy. But the art of making gratings was already in a high state of mechanical perfection. To improve gratings by reducing their grating space by a few orders of magnitude would have required better materials for both the ruling machine and the grating material itself. However, the basic granularity of matter imposes limitations. This dilemma led Laue to the idea of taking advantage of the very granularity that stood in his way. It was felt that the regular shapes of crystals, with their plane cleavage surfaces and well-defined edges, must mean that the atoms are regularly arranged throughout their structure. Laue thought that the atoms of a single crystal might provide the grating needed for the diffration of x-rays. At Laue's suggestion, Friedrich and Knipping directed a narrow beam (pencil) of x-rays at a crystal and set a photographic plate beyond it. The result was a picture like that shown in Fig. 6–2. (Most of the x-rays go directly through the crystal and strike the plate at its center. To prevent gross overexposure at this point it is usual to fasten a disk of lead over the center of the plate. When the photograph shown in Fig. 6–2 was taken, a lead disk masked the center for all but the last second of a 40-min exposure.) Laue knew he had met with success when he observed the complicated but symmetrical pattern on the plate. These spots could only be due to diffraction from the

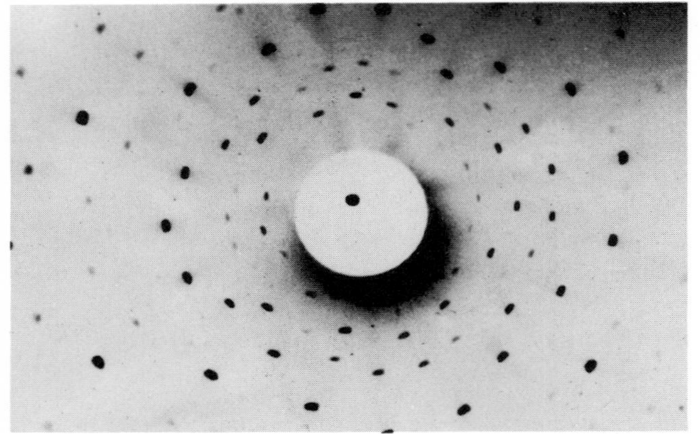

FIG. 6–2
Laue diffraction of NaCl taken with radiation from a tungsten-target tube operated at 60 kV. The dark patch below the center disk was caused by scattered x-rays.

atoms of the crystal. [Strictly, the structural units of such crystals as NaCl and KCl are ions, not neutral atoms. These are called *ionic crystals* because they are held together by strong electrostatic forces acting between charged particles, e.g., (Na^+) (Cl^-) and (K^+) (Cl^-).]

This diffraction pattern is hardly like that produced by a man-made grating. In optical spectroscopy, the grating consists of parallel lines in a plane. You can appreciate that its diffraction would become more complex if a grating were turned ninety degrees and then ruled again with a second set of lines perpendicular to the first. If many such gratings made of glass were stacked behind one another, the diffraction pattern would be further complicated. Since a crystal consists of a regular array of atoms in three-dimensional space, it produces an intricate pattern.

Later we shall discuss qualitatively why a Laue pattern looks as it does. Our immediate point, however, is that by diffracting x-rays Laue demonstrated that x-rays can be combined destructively—the sole, unique experimental criterion for wave motion. At the same time, Laue patterns established that x-rays have very short wavelengths, and confirmed the supposition that crystals have their atoms arranged in a regular structure.

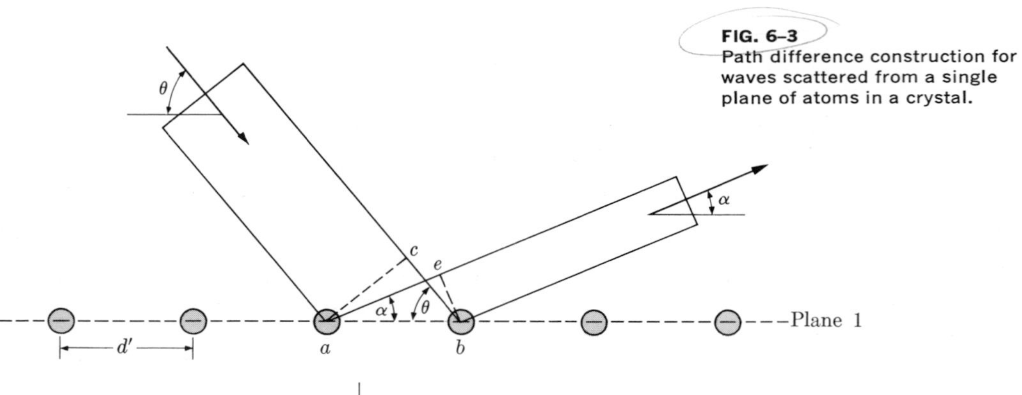

FIG. 6–3
Path difference construction for waves scattered from a single plane of atoms in a crystal.

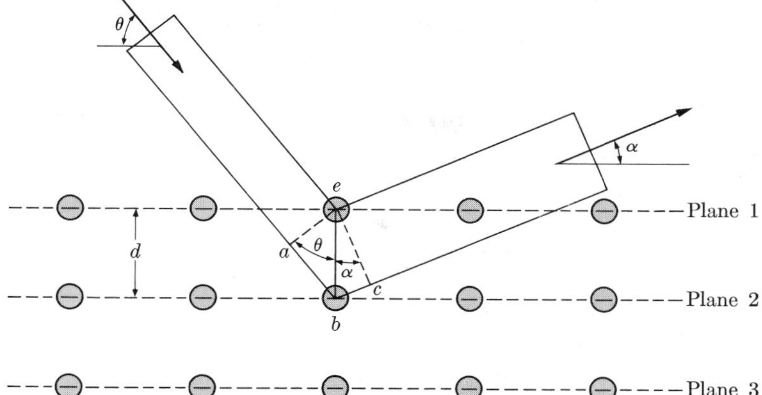

FIG. 6–4
Path difference construction for waves scattered by successive planes of atoms in a crystal.

Late in 1912, shortly after the Laue experiment, William L. Bragg devised another technique for diffracting x-rays. Instead of observing the effect created by passing the rays through a crystal, Bragg considered how x-rays are scattered by the atoms in the crystal lattice. Consider an x-ray wavefront incident on a surface row of atoms in a crystal plane, as shown in Fig. 6–3. Each atom becomes a source of scattered x-radiation. In general, the scattered x-rays from all the atoms in the crystal combine destructively as they fall on top of one another in a random manner. If certain conditions are met, however, constructive interference will occur at a few places. One of the relations which must be satisfied for reinforcement can be derived with the aid of Fig. 6–3. In this figure d' is the distance between adjacent atoms, θ is the angle* between the incident rays and a row of atoms in the surface plane, and α is the angle between scattered rays and the surface plane. To obtain the path difference between rays from adjacent atoms, we construct the lines $\overline{ac}$ and $\overline{be}$ perpendicular to the incident and scattered rays, respectively. This path difference is obviously $\overline{ae} - \overline{cb}$, and, for reinforcement, this difference must equal some integral multiple of the x-ray wavelength, or $m\lambda$. Therefore, we have $\overline{ae} - \overline{cb} = m\lambda$, which can be written as

$$d' \cos \alpha - d' \cos \theta = m\lambda. \tag{6–2}$$

The other relation which must be satisfied for maximum reinforcement is that the scattered rays from successive planes of atoms meet in phase. Referring to Fig. 6–4, in which d is the distance between successive planes,

* Note that whereas in optics it is usual to measure angles of incidence and reflection between the rays and the normal to the surface, the Bragg angle is measured between the ray and a crystal plane.

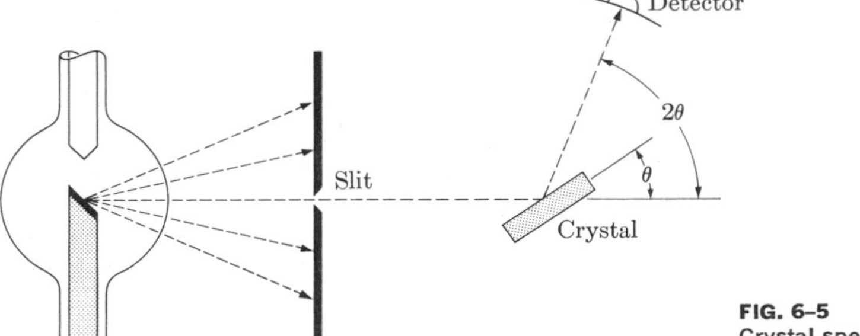

FIG. 6–5
Crystal spectrometer.

we note that rays scattered from the second plane travel a greater distance than those from the first plane. In order that the rays from successive planes reinforce one another, it is necessary that these additional distances shall be some integral multiple of the wavelength, $n\lambda$. If we draw the line $\overline{ea}$ normal to the incident rays and the line $\overline{ec}$ normal to the scattered rays, we see that the length of the path for a ray from the second plane exceeds that for a ray from the first plane by the sum of the distances $\overline{ab}$ and $\overline{bc}$. Therefore the rays from the two planes will interfere constructively when $\overline{ab} + \overline{bc} = n\lambda$, which is equivalent to

$$d \sin \theta + d \sin \alpha = n\lambda. \tag{6–3}$$

In general, the conditions imposed by Eqs. (6–2) and (6–3) cannot be satisfied simultaneously without considering the scattered wavelets from atoms which are in the various layers but not in the plane of incidence. Both conditions are met, however, in the special case where $\theta = \alpha$. Then Eq. (6–2) reduces to zero and Eq. (6–3) becomes

$$n\lambda = 2d \sin \theta, \tag{6–4}$$

where n is the order of the spectrum. When $\theta = \alpha$, we have precisely the condition of regular optical reflection. Because of this, Bragg scattering is usually called Bragg "reflection." This is actually a misnomer, but we shall follow common usage. The planes of atoms in the crystal which are responsible for Bragg reflection are called *Bragg planes.* Let us now summarize the conditions for constructive interference of x-rays scattered from Bragg planes. The *first condition* is that the angle the incident beam makes with the planes must equal that made by the reflected beam; and the *second condition* is that the reflections from the several Bragg planes must meet in phase, that is, must satisfy the relation, $n\lambda = 2d \sin \theta$.

The Bragg technique of using crystals as diffraction gratings for an x-ray spectrometer is shown in Fig. 6–5. The x-rays coming from the tube

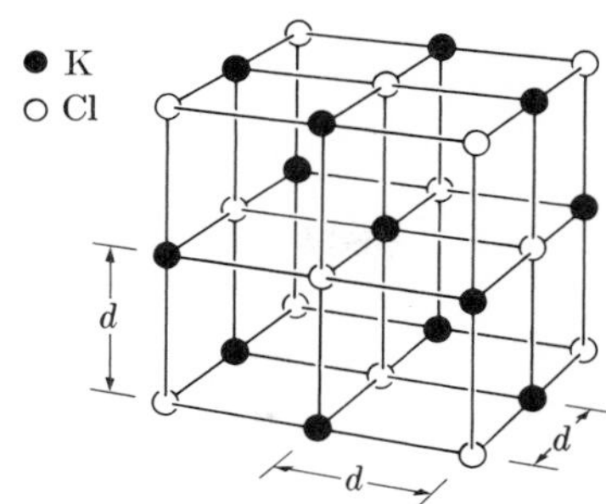

FIG. 6–6
Cubic lattice of sylvite, KCl.

at the left are restricted to a narrow beam or collimated by a lead sheet which absorbs all rays except those which pass through a small slit. These rays fall on a crystal which can be rotated about an axis parallel to the slit and perpendicular to the plane of the figure. The angle θ between the crystal planes and the original x-ray direction can be measured. Bragg reflection can take place only in the direction 2θ from the original x-ray direction. Whether or not the rays actually reinforce one another in this direction depends on whether the second Bragg condition is fulfilled. Since d is a fixed crystal property, the measurements of those angles for which reflections do occur provide a measure of the x-ray wavelengths that are present. In order that spectra measured in this way may be quantitative, the crystal spacing d must be known.

The regular arrangement of the space positions of the atoms in a crystal is called a *lattice array*. The basic interplanar distance or *principal grating space, d,* is shown for the cubic crystal KCl in Fig. 6–6. Note that d is the distance between *adjacent* atoms. The *unit cell* of a lattice is the smallest block or geometric figure of a crystal which is repeated again and again to form the lattice structure. The length of the side of a unit cell is the distance between atoms of the *same* kind. This is equal to $2d$ for the cubic crystal in Fig. 6–6. The length of the side of the unit cell is equal to the basic distance d only in the case of a simple cubic structure in which all of the atoms are of the same kind. Some pure metals form crystals of this type. These basic interatomic distances can be computed from knowledge of the molecular weight of the crystalline compound, Avogadro's number, the density of the material, and its crystalline form. For the cubic type the general procedure is to calculate the number of atoms per unit volume, multiply this by the volume associated with each atom, and equate the product to unity. This procedure is illustrated in the example that follows.

EXAMPLE. KCl (sylvite) is a cubic crystal having a density of 1.98 g/cm^3. (a) Find the distance between adjacent atoms in this crystal, and (b) find the distance from one atom to the next one of the same kind.

Solution. (a) Molecular weight of KCl = 39.10 + 35.45 = 74.55.

$$\text{Mass of KCl molecule} = 74.6\,\frac{\text{g}}{\text{mole}} \times \frac{1\text{ mole}}{6.02 \times 10^{23}\text{ molecules}}$$

$$= 12.4 \times 10^{-23}\text{ g}$$

$$\frac{\text{No. KCl molecules}}{\text{unit volume}} = 1.98\,\frac{\text{g}}{\text{cm}^3} \times \frac{\text{molecule}}{12.4 \times 10^{-23}\text{ g}}$$

$$= 1.60 \times 10^{22}\,\frac{\text{molecules}}{\text{cm}^3}.$$

Since KCl is diatomic,

$$\frac{\text{No. of atoms}}{\text{unit volume}} = \frac{2\text{ atoms}}{\text{molecule}} \times 1.60 \times 10^{22}\,\frac{\text{molecules}}{\text{cm}^3}$$

$$= 3.20 \times 10^{22}\,\frac{\text{atoms}}{\text{cm}^3}.$$

Let d be the distance, measured along the edge of the cube, between adjacent atoms in the crystal and let n be the number of atoms along the edge of a 1-cm cube. Then the length of an edge is nd, and the volume of this unit cube is n^3d^3. However, n^3 is the number of atoms in 1 cm^3; therefore

$$3.20 \times 10^{22}\,\frac{\text{atoms}}{\text{cm}^3} \times d^3 = 1$$

or

$$d = (3.20 \times 10^{22})^{-1/3}\text{ cm} = 3.14 \times 10^{-8}\text{ cm}$$

$$= 3.14\text{ Å} = 3140\text{ XU.*}$$

This is the principal grating space, or basic interplanar distance, of KCl.

(b) The distance between two atoms of the same kind is twice the above value, or 6.28 Å. This is the length of the side of a unit cell of KCl.

The calculation of the grating space is simple only for a cubic crystal, and the distance thus computed is only the *basic* or *principal* interplanar distance. Within a crystal there are many crystal planes from which Bragg reflection can result. Consider the many possibilities presented by the two-dimensional situation of a marching band on a football field, Fig. 6–7. The basic distance between any adjacent members of the band might be 5 ft. But as the band marches past, it is evident that there are

---

* X-ray wavelengths are commonly less than an angstrom, 10^{-10} m, and are often expressed in X-units, abbreviated XU. The XU is about 10^{-13} m. Its true definition is not based on the meter but on the interatomic distance or lattice constant of rock salt, NaCl. When the XU was first used, the lattice constant of rock salt was taken to be 2.8140 Å or 2814.0 XU. High-precision measurements have revised the rock salt lattice constant so that we now have 1000.00 XU = 1.00202 Å.

many lines through the band which resemble a three-dimensional array of parallel planes. These sets of parallel lines or planes are separated from one another by different amounts which can be computed from the basic interatomic distance. Reflections from some sets of planes are more intense than those from some other sets. Intensity variations are introduced by the differences in the nature of the planes in crystals composed of more than one type of atom. Elements with high atomic numbers scatter radiation more effectively than those with low atomic numbers. If the x-marked band members in Fig. 6–7 represent one kind of atom and the dot-marked ones another kind, we note that some planes contain only x-marked scatterers, some only the dot-marked type, and some have a mixed population. (It should be kept in mind that the lines in Fig. 6–7 are actually analogous only to the traces of the crystal planes in the plane of the paper. These planes of the crystal are not necessarily normal to the paper. It is evident from a study of Fig. 6–6, for example, that a plane containing only potassium atoms is not perpendicular to any of the faces of the cube.)

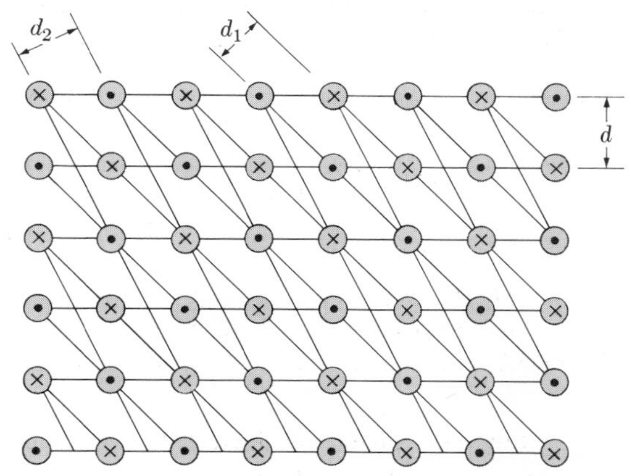

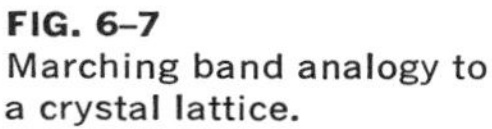
FIG. 6–7
Marching band analogy to a crystal lattice.

We are now in a position to appreciate both why the Bragg method of x-ray spectroscopy is simple and why Laue pictures are complex. In the Bragg method, the crystal can be set so that one set of populated planes is reflecting a beam of x-rays, and the three-dimensional crystal is used in a two-dimensional manner. In the Laue method, a pencil of x-rays is made to pierce the crystal perpendicular to a set of planes of the crystal. Most of the spots are due to diffraction from sparsely populated planes that happen to be situated so that both of the Bragg conditions for reflection are satisfied. The Bragg method is better for the study of the wavelengths of x-rays, but the Laue technique is very useful for the study of crystals. It is a tedious but rewarding task to start with a Laue

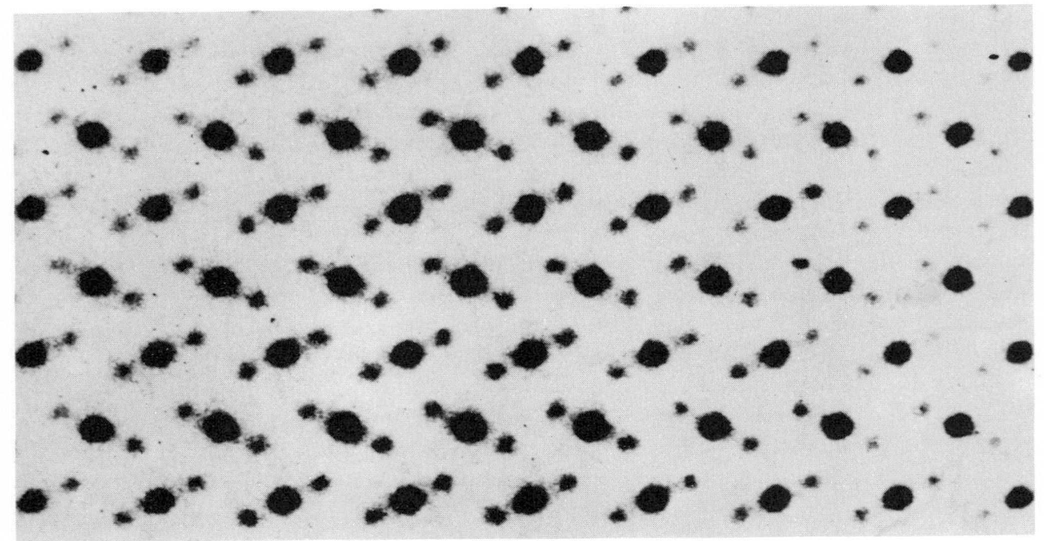

FIG. 6–8
Atoms in marcasite, FeS^2, looking along the crystallographic c-axis magnified 4.5 million times. The larger spots are iron atoms with 26 electrons each; the smaller dots are sulfur atoms with 16 electrons each. The regular array of the atoms in the crystal is quite evident. This remarkable photograph was taken with a two-wavelength microscope. (Courtesy of M. J. Burger, Massachusetts Institute of Technology)

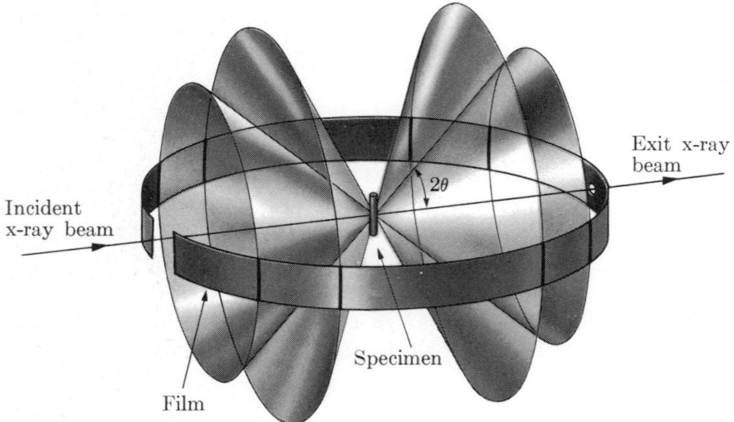

FIG. 6–9
Relation of film to specimen and incident beam in the Debye-Scherrer powder diffraction method. X-rays forming the cone having a half-angle of 2θ were reflected from the same kinds of planes, all oriented at an angle θ to the incident beam. Other cones are formed in a similar way.

pattern and work back to the geometry of the array of atoms that must have produced it. An ingenious type of microscope employing both x-rays and visible light has been devised for crystal study. With its aid an enormous, useful magnification can be obtained, so that one can "look" into the lattice structure, as shown in Fig. 6–8.

We shall not dwell on the subject of crystallography, but there is another x-ray crystallographic technique we must mention in passing. The Bragg and Laue methods require single crystals large enough for study. Many materials that are basically crystalline cannot be obtained as large single crystals, but these materials may be studied be means of the powder technique. One form of this technique, the Debye-Scherrer method, requires that the substance be ground and powdered, so that it can be assumed that all the tiny crystals have random orientation. When a pencil of x-rays is passed through this powder, a series of rings is formed on a photographic film. Each ring is the intersection of the film plane and a cone of rays. Each cone is the locus of rays for which some set of crystal planes is so oriented that both Bragg conditions of reflection are fulfilled. The formation of a powder diffraction pattern is shown schematically in Fig. 6–9, and Fig. 6–10 is the pattern for NaCl made with the x-radiation from copper.

Another form of powder technique is used to study the orientation of crystals in an extruded or drawn material like wire. If the crystallites in the wire actually are random in their orientation, a true powder pattern of circles results. But if the crystallites are somewhat oriented, then each circle has nonuniform intensity and the pattern on the film tends toward a Laue pattern. The degree of orientation can be determined from such pictures.

Laue patterns of metals or other materials rather opaque to x-rays are sometimes obtained by studying the x-rays scattered back from the side of the material nearer the x-ray source. This technique is used, for example, to study the crystalline structure of steel.

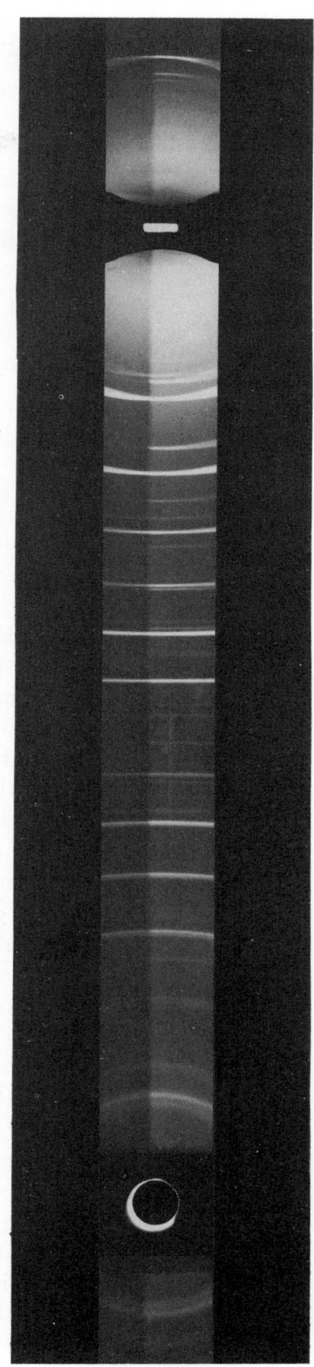

FIG. 6–10
Powder diffraction pattern of NaCl made with the K_α and K_β wavelengths of copper. During exposure, the left portion of the film was covered with nickel foil. This filtered out the K_β line.

X-rays of a single wavelength are often useful. One way of obtaining such monochromatic rays is to use x-rays that have undergone a single-crystal Bragg reflection before they are used for powder studies. Later, we shall discuss filters which strongly absorb all but one of the wavelengths in certain x-ray spectra.

Since a single crystal is used for Laue patterns, there are only a very small number of planes which fulfill the Bragg conditions for constructive interference for any one x-ray wavelength. Therefore, to obtain many Laue spots requires the use of an incident beam containing many wavelengths. On the other hand, the minute crystals in a powder have random orientations, so that many planes are available for the production of interference maxima even when monochromatic x-rays are used.

6-4. MECHANISM OF X-RAY PRODUCTION

Roentgen reported in his original paper that x-rays are produced where cathode rays strike some material object. Since we now know that cathode rays are high-velocity electrons, we can restate Roentgen's observation by saying that x-rays are produced when high-velocity electrons hit something. To explore the mechanism of x-ray production, let us ask what happens when electrons strike solid matter.

Many of the electrons that strike matter do nothing spectacular at all. Most of them undergo glancing collisions with the particles of the matter, and in the course of these collisions, the electrons lose their energy a little at a time and thus merely increase the average kinetic energy of the particles in the material. The result is that the temperature of the target material is increased. It is found that most of the energy of the electron beam goes into heating the target.

Some of the bombarding electrons make solid hits and lose most or all of their energy in just one collision. These electrons are rapidly decelerated. We have pointed out that radiation results when a charged body is accelerated. Therefore when an electron loses a large amount of energy by being decelerated, an energetic pulse of electromagnetic radiation is produced. This is an *inverse photoelectric effect* in which an electron produces a photon. In Chapter 3 we found that photons of a given energy produce photoelectrons with a certain maximum energy. Here we find that electrons of a given energy produce x-ray photons with a certain maximum energy. Both effects confirm the quantum view of radiation. According to classical electromagnetic theory, there is no lower limit to the wavelength of the radiation that a moving electron can produce when it is stopped suddenly. But there is a quantum limit. Given that an electron has been accelerated through a difference of potential of V volts, we can use Eq. (3–32), which Duane and Hunt showed was valid for x-rays, to compute the wavelength of the resulting radiation in angstroms if the

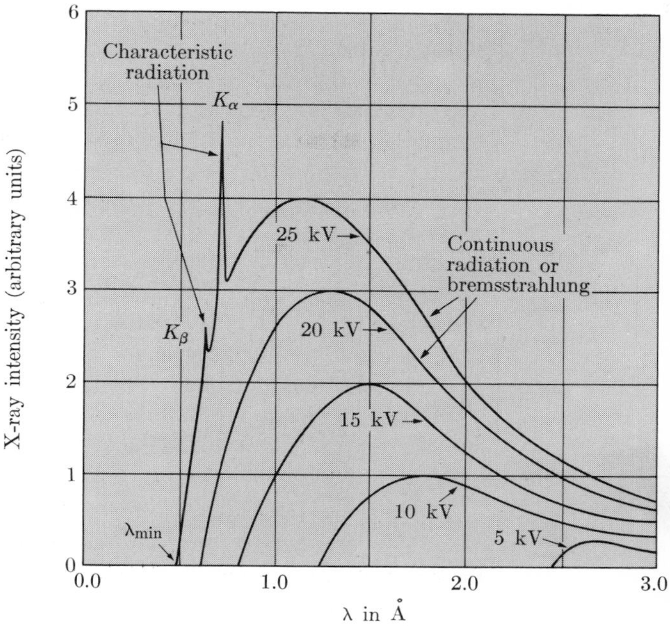

FIG. 6–11
X-ray spectrum of molybdenum as a function of the applied voltage. Line widths are not to scale. The K-series excitation potential of this element is 20.1 kV. The shortest L-series wavelength is 4.9 Å.

electron loses all its energy in a single encounter. In this case the energy loss of the electron in eV is *numerically* equal to the accelerating potential in volts. Therefore we have

$$\lambda_{\min}\,(\text{Å}) = \frac{12400}{V}. \tag{6–5}$$

This expression gives the minimum wavelength, since no electron can lose more energy than it has, and there will be a continuous distribution of radiation toward longer wavelengths because there are all sorts of collisions, from direct hits to glancing ones. Thus glancing collisions account for the *continuous spectrum* of x-rays from any target material and also for the inefficiency of the conversion of electron energy into x-ray energy. The Germans named this continuous radiation *bremsstrahlung*, meaning literally "braking radiation." This is a highly descriptive term, since it refers to the radiation that results from the braking or stopping of charged particles. Some continuous spectra are shown in Fig. 6–11.

Looking at the collision process more closely, however, we find there is another very important kind of collision energy exchange. The bombarding electron may also give energy to electrons bound to the target atoms. If these atomic electrons are freed from their home atoms, ions are produced. Since x-ray producing electrons have energies of the order of many thousands of electron volts, it is very easy for them to produce ions by removing outer electrons. X-ray producing electrons may also have enough energy to produce ions by removing inner electrons from the atom, even down to the innermost or K-shell. Such an ion has a low-energy hole in its electronic structure, and this vacancy is promptly filled when one of its electrons in a higher energy state falls to this low-energy level. When an electron falls into a low-energy level, it releases its energy as radiation, just as in the Bohr theory. Although the energy required to ionize an atom by removing an outer electron is much less than 100 eV, the energy required to ionize by removing an inner electron may be as high as 120,000 eV. When an outer electron falls into such a vacancy, it will radiate a photon of this energy. Such photons are in the x-ray region and have wavelengths which are fractions of angstroms. This mechanism, which accounts for a significant part of x-ray production, produces x-rays having particular wavelengths which are *characteristic* of the target material.

6-5. X-RAY ENERGY LEVELS

In our discussion of the periodic table we noted that a heavy element has two electrons with $n = 1$, eight with $n = 2$, etc. The Pauli exclusion principle requires that no two electrons can have identical quantum numbers and thus no two electrons can have identically the same energy. But n is the principal quantum number and it has the most to do with the energy of the electron. Thus the two electrons with $n = 1$ have energies which differ only slightly. The electrons for which $n = 2$ have much more energy than those for which $n = 1$, but they differ among themselves very little. In discussing x-rays, it is often convenient to ignore the minor energy differences of those electrons having the same n and concentrate on the large differences associated with different values of n. In the discussion of shell structure in Section 4–16 we said that those electrons having $n = 1$ constitute the K-shell electrons. Those electrons with $n = 2$ form the L-shell. The heaviest elements have seven populated shells, K, L, M, N, O, P, and Q. However, the N-, O-, P-, and Q-shells have energies which give rise to the optical spectrum and their energies are *negligible* so far as x-rays are concerned.

With these simplifications we can at once make an energy-level diagram which is typical of all the heavy elements, Fig. 6–12. It is important now to point out the similarities and differences between this energy-level diagram and that of hydrogen.

The one electron in hydrogen is normally in the level for which $n = 1$, and the higher levels in hydrogen are vacant unless the atom is excited. The x-ray diagram represents a typical heavy element having an atomic number of about 40. This element has many electrons. In general, the

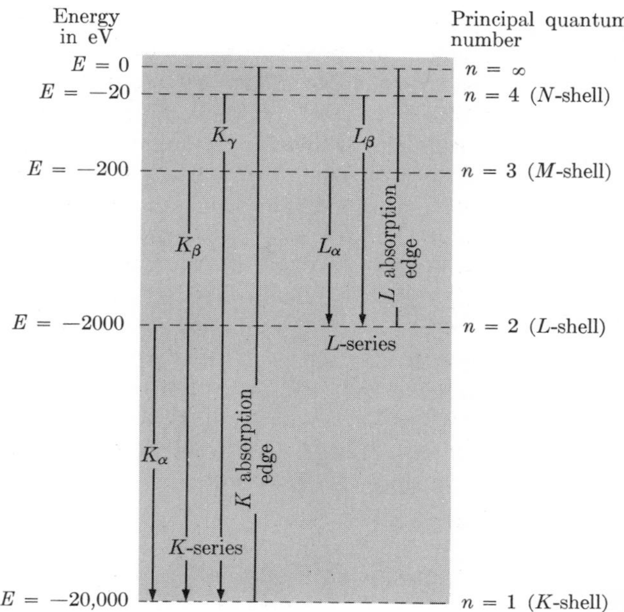

FIG. 6–12
Hypothetical x-ray energy-level diagram.

levels of the heavy elements are occupied by 2, 8, 18, 32, etc., electrons. A hydrogen atom may be excited from the state $n = 1$ to the state $n = 3$. But in a heavy element, a K-shell electron, $n = 1$, cannot be excited to the states $n = 2$ or $n = 3$, since these L- and M-shells are normally filled.

The electron in hydrogen is bound to its nucleus by only one proton with a charge of plus e. The inner electrons of a heavy element are bound by electrostatic forces due to a nucleus with a large charge Ze. This difference in nulear charges leads to vast differences in the energies of the various levels. Whereas the level having $n = 1$ in hydrogen has an energy of only -13.6 eV, in a heavy element the level for which $n = 1$ has an energy of the order of minus thousands of electron volts. It is $-20{,}000$ eV in our hypothetical example.

For both light and heavy elements, each level except K is actually a narrow group of levels within which we are ignoring the energy range. Indeed, in x-ray studies, we frequently ignore the energy difference between the energy of the N-shell ($n = 4$) and the energy of a free electron ($n = \infty$). This leads to an interesting result: all heavy elements have x-ray spectra which are similar in character.

Suppose the hypothetical substance whose energy-level diagram is shown in Fig. 6–12 were the target of an x-ray tube across which there is a potential difference of 50,000 volts; then all the cathode-ray electrons strike this target with 50,000 electron volts of energy. Most of these electrons fritter their energy into heat. Some cathode-ray electrons undergo collisions which produce radiation quanta having energy equal to the energy loss in each collision. These cause x-rays with a *continuous* spread of wavelengths down to 0.248 Å, as determined by the exciting voltage. Some cathode-ray electrons undergo collisions which cause the removal of target electrons from the various target shells. If an outer, N-shell electron is knocked out, the optical spectrum of the target material will be excited. Light will be emitted as electrons fall into the N-shell vacancy. This is a negligible effect so far as x-rays are concerned, and the light cannot be seen, since it is usually produced at a slight depth within the target which is opaque to visible light. If an M-shell electron is removed, the same remarks apply except that the radiation is in the far ultraviolet region. If an L-shell electron is knocked out, we have a more interesting situation. An L vacancy can be filled in two important ways: an electron can fall into the L-shell from outside the atom (or from the N-shell—we ignore the difference), producing an x-ray of wavelength 6.20 Å, or the vacancy in the L-shell can be filled from the M-shell. This energy difference is 1800 eV, and causes an x-ray of wavelength 6.89 Å. (In the latter case, the vacancy in the M-shell will also be filled, by the production of a far ultraviolet photon.) These x-radiations of the target are called the L_β and L_α radiations, respectively. Since it is more probable that the L-shell will be filled from the nearer M-shell, the less energetic or longer wavelength L_α radiation is more intense than the L_β. If the bombarding electrons had less than 2000 eV of energy, they would not be able to remove electrons from the L-shell and *neither* the L_α *nor* L_β radiation could result.

We have assumed, however, that the bombarding electrons have 50,000 eV of energy, so that they can not only remove electrons from the L-shell but also from the K-shell. If a K vacancy is filled from the L-shell, the radiation would have wavelength 0.689 Å and would be called K_α radiation. (The consequent L vacancy would then result in L_α or L_β radiation.) If the K vacancy is filled from the M-shell, there will be more energetic radiation, K_β, of wavelength 0.626 Å. Or if the K vacancy is filled from the N-shell (or from infinity), there will be slightly more energetic radiation, K_γ, of wavelength 0.620 Å. These K-series radiations are listed in the order of their relative probability. The K radiations of this example are quite energetic or "hard," whereas the L-series radiations are in the "soft" x-ray region.

In this example we have used a hypothetical substance in order to have simple numbers to work with, but what we have done is typical of all x-ray spectra. If we ignore the small energy differences of the electrons within a given shell, all elements produce x-ray spectra of similar character and with very few lines. The shortest wavelength associated with a shell is called the *absorption edge wavelength* of that shell, for reasons given later in this chapter. The corresponding energy is equal to that needed to remove an electron from the shell to infinity.

6-6. X-RAY SPECTRA OF THE ELEMENTS. ATOMIC NUMBER

In 1913, Moseley used a large number of elements as x-ray tube targets and found that the K_α radiation of each element was distinct from that of any other element. This result not only provides a unique way of identifying the various elements, but also is of theoretical significance. Specifically, Moseley's law states that the frequencies of corresponding x-ray spectral lines, such as K_α's, as we go from element to element, may be represented by an equation of the form

$$\boxed{\sqrt{f} = a(Z - b),} \tag{6-6}$$

where a and b depend on the particular line and Z is the atomic number of the element.

The original ordering of the elements in the periodic table was on the basis of their atomic weights. Initially, atomic numbers were mere ordinal numbers specifying where the element lay in a list based on these weights. These atomic numbers had no more physical significance than the house numbers that identify the houses on a given street. Moseley found that if he arranged the wavelengths of the K_α lines in the order of atomic weights, these wavelengths formed a remarkably regular progression. The sequence was not perfect, however, since he found both gaps and wavelengths out of order. He attributed the gaps in the series to undiscovered elements and proposed that there should be a unique correlation between the wavelength series and atomic number. His words were, "We have here a proof that there is in the atom a fundamental quantity, which increases by regular steps as we pass from one element to the next. This quantity can only be the charge on the central positive nucleus." In Moseley's day, nickel with an atomic weight of 58.69 was listed ahead of cobalt with an atomic weight of 58.94. There was some chemical evidence that the order of these two elements in the periodic table should be reversed, and Moseley demonstrated this by showing that the atomic number of cobalt is 27 and that of nickel is 28.

Moseley's work was announced in 1913, the same year that Bohr accounted for the hydrogen spectrum. In the case of hydrogen, Bohr came to the same conclusion that Rutherford had reached from alpha-particle scattering and that Moseley verified for all elements. Recall that in the Bohr model the charge of the nucleus was basic to his analysis and the mass of the nucleus played a minor role. Note how nicely Bohr's idea of electrical forces accounts for the Moseley observation of the regular progression of the K radiations. The electron in hydrogen is a K-shell electron held to the nucleus by the attraction of one proton. The two K electrons of helium are held by two protons and thus their binding energies are greater (more negative) than that of the electron in hydrogen. The two K electrons of a heavy element are bound by the attraction of many protons. As we move from one element to the next, the energy to remove a K electron, the K binding energy, increases in a regular way.

We can now combine the work of Bohr and Moseley in an interesting and quantitative way that strengthens both their views. From Eq. (4–22) for the frequency of the spectral lines of hydrogen, we have

$$f = \left(\frac{m_e e^4}{8\epsilon_0^2 h^3}\right) Z^2 \left(\frac{1}{n_1^2} - \frac{1}{n_2^2}\right),$$

or

$$\sqrt{f} = \left[\left(\frac{m_e e^4}{8\epsilon_0^2 h^3}\right)\left(\frac{1}{n_1^2} - \frac{1}{n_2^2}\right)\right]^{1/2} Z. \tag{6–7}$$

This bears some resemblance to Moseley's law, so we are encouraged to go further. Suppose we remove one of the two K-shell electrons of an element to infinity. The remaining outer electrons will be attracted to the nucleus not by the nuclear charge Ze as Bohr assumed, but by a charge $(Ze - e)$. One may think of the remaining K electron as "screening" the outer electrons from the full nuclear attraction. Thus the b term of Moseley's law is a nuclear screening constant and it equals one for the K-series lines. Since the K_α line is a transition from $n_2 = 2$ to $n_1 = 1$, we have everything we need to attempt a Bohr theory calculation of an x-ray K_α line. Choosing molybdenum ($Z = 42$), we have

$$f = \frac{me^4{}_e}{8\epsilon_0^2 h^3}\left(\frac{1}{1^2} - \frac{1}{2^2}\right)(42 - 1)^2.$$

Expressed in energy units, we obtain 17.2 keV for this line, which is in remarkable agreement with the experimental value of 17.4 keV.

6–7. X-RAY ABSORPTION

The most spectacular property of x-rays is their ability to penetrate materials that are opaque to less energetic radiation. The basic mechanism of absorption of ultraviolet, visible, and infrared radiation is the transfer of photon energy to the electronic, vibrational, and rotational energy states of the material doing the absorbing. For an x-ray photon, whose energy is orders of magnitude greater than visible radiation, these mechanisms are trivial because a high-energy photon has a small probability of interacting in a comparatively low-energy process. High-energy x-ray photons are more likely to interact with electrons in the K- or L-shells. Since the probability of an x-ray interacting with the many loosely bound electrons in an absorber is small and since there are relatively few tightly bound electrons, high-energy (short-wavelength or hard) x-rays have remarkable penetrating ability.

6-8. INTENSITY MEASUREMENTS

To discuss the penetrating ability of x-rays further, we must first consider the techniques for measuring x-ray intensities.

Roentgen first detected x-rays by the fluorescence they produce in certain materials. This effect can be used for quantitative intensity measurements if it is coupled with an objective measurement of the fluorescent light produced. Roentgen also observed that x-rays blacken a photographic plate, and this may be used to measure the intensity if a densitometer is used to measure the blackening. A third method was suggested by Roentgen's experiments on the conductivity of air due to x-rays. The x-rays ionize the air, and this effect can be measured quantitatively with an ionization chamber. There are many forms of ionization chambers.

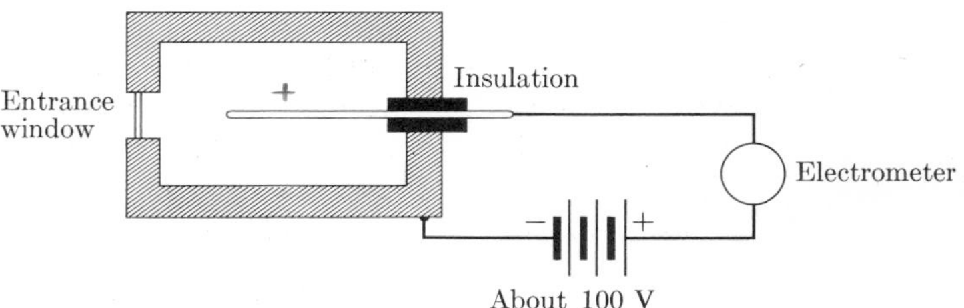

FIG. 6–13
Ionization chamber.

One of these is shown schematically in Fig. 6–13. The chamber itself is a metallic box into which an electrode is inserted through an insulating plug. This electrode is usually maintained at a positive potential, so that any free electrons within the chamber will be attracted to it. The potential difference is chosen low enough so that the dielectric strength of the air prevents a discharge, and high enough so that the charges are collected before the electrons recombine with positive ions. When x-rays enter the chamber and produce ions, the collected charges constitute a small current which is very nearly proportional to the intensity of the x-rays. This current is amplified and measured by an electrometer, an instrument which is very convenient because it is sensitive and responds quickly to changes in x-ray intensity.

6-9. ABSORPTION COEFFICIENTS

An ionization chamber can be used to study the penetrating ability of x-rays with apparatus like that shown in Fig. 6–14. X-rays are collimated by slits, rendered monochromatic by a Bragg reflection, and passed through a material under study; then their intensity is measured by the ionization chamber. If we vary the thickness of the absorbing material, a plot of transmitted intensity against thickness looks like that in Fig. 6–15.

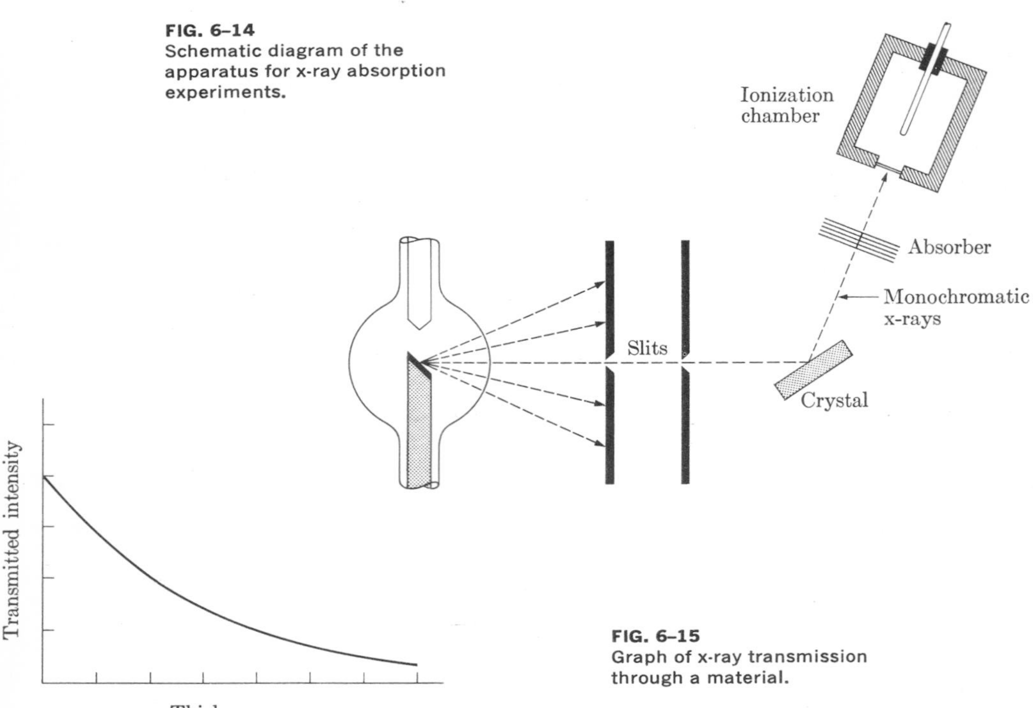

FIG. 6–14
Schematic diagram of the apparatus for x-ray absorption experiments.

FIG. 6–15
Graph of x-ray transmission through a material.

This is an exponential decay curve which can be derived by assuming that a small thickness of absorber, dx, reduces the intensity of the beam by an amount $-dI$ proportional to both the intensity I and the thickness dx. This assumption leads to the differential equation

$$dI = -\mu I \, dx, \tag{6-8}$$

where μ is a constant of proportionality. If the intensity of the beam which is incident on a layer of material of thickness x is I_0, then I, the intensity of the *transmitted* beam, can be found by integrating Eq. (6–8). We have

$$\int_{I_0}^{I} \frac{dI}{I} = \int_{0}^{x} -\mu \, dx, \tag{6–9}$$

which gives

$$\ln I - \ln I_0 = -\mu x$$

or

$$\boxed{I = I_0 e^{-\mu x}.} \tag{6–10}$$

This is the equation of the curve in Fig. 6–15.

The quantity μ is usually called the *linear absorption coefficient*, although *macroscopic absorption coefficient* and *linear attenuation coefficient* are also used. When Eq. (6–8) is solved for μ, we see that the linear absorption coefficient is equal to the fractional decrease in intensity of the radiation per unit thickness of the absorber. Thus it has the dimensions of reciprocal length. The value of this coefficient depends on the x-ray wavelength and the material used.

It is interesting to express the absorption properties of a material in another way. If we let $I = I_0/2$, we can solve for the corresponding value of x. Since this gives the thickness of absorber that reduces the intensity of the transmitted beam to one-half of that of the incident beam, it is called the *half-value layer*, T, or hvl. Thus, from Eq. (6–10), we have

$$\ln 2 = \mu T \tag{6–11}$$

or, if we use logarithms to the base 10,

$$2.30 \log 2 = \mu T,$$

or

$$\boxed{T \text{ (hvl)} = \frac{0.693}{\mu}.} \tag{6–12}$$

Although the equations we have derived are correct for a *narrow* beam of x-rays, they must be modified before they are applied to the general case of shielding against dangerous radiation. The absorber in the apparatus shown in Fig. 6–14 reduces the intensity of the beam both by "soaking up" x-rays and by scattering them in a new direction. Thus when a broad beam of x-rays passes through an extended absorber, x-ray photons removed from the beam at one point may be scattered into it again at another point. When these photons have less than 200 keV of energy, a rule-of-thumb method for correcting for the scattered rays is to assume that the source is one and one-half times as intense as it is measured to be. This requires increasing the absorber thickness originally calculated by about 0.6 of a half-value layer. For energies above 1 MeV, two other effects, which will be discussed later in this chapter, also become significant in preventing a rapid exponential decrease in the intensity of the transmitted beam. The *"buildup factor"* due to all the processes may then be such that the transmitted intensity I is four times the value calculated from Eq. (6–10).

The value of μ depends on the x-ray wavelength and on the absorbing material. The basic variation of μ with λ is that μ is very nearly proportional to λ^3. But this basic variation is profoundly influenced by the nature of the absorbing material. If the x-ray wavelengths are so long that they can excite only outer electrons of the absorbing material, then interaction between x-rays and the K, L, or M electrons of the absorber must be elastic. But if the x-rays have shorter wavelengths, they may be able to eject M, L, or even K electrons. The various absorption mechanisms are additive, and each varies approximately as λ^3. The variation of μ with the wavelength of a particular material is shown in Fig. 6–16. In this figure λ_K, λ_L, and λ_M are absorption edge wavelengths.

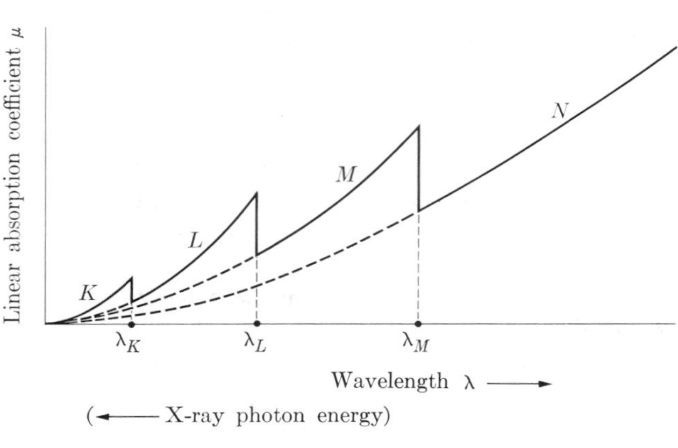

FIG. 6–16
Simplified graph of the linear absorption coefficient of a material as a function of x-ray wavelength. Fine structure of the absorption edges is not shown.

X-rays of short wavelength are very penetrating despite the fact that they can be absorbed by ejecting electrons from any shell of the absorber, that is, by the photoelectric effect. If the wavelength is longer than the K absorption limit of the absorber, the x-rays are unable to ionize the absorber from the K level and one absorbing level is eliminated. At this wavelength, there is an abrupt change in the absorption coefficient, called an absorption edge. The absorption coefficient again rises with increasing wavelength until suddenly the x-rays are unable to ionize from the L-shell and there is another absorption edge. The contribution to μ for each mechanism varies approximately as λ^3 for each absorbing level and the discontinuities occur when the number of absorption mechanisms changes.

Figure 6–16 is a simplified diagram, since it shows but one absorption edge for each shell. The fact is that there is one K edge, there are three L edges, and five M edges. This fine structure of the variation of absorption coefficient with wavelength comes about because the two K electrons have practically the same energy, but the L and M electrons fall into three and five energy subgroups, respectively. These energy levels of a substance are more easily observed in absorption than in emission and were ignored when we discussed x-ray emission spectra. One might be tempted to assign absorption edges to electron transitions within the atom, such as from K to L. These transitions cannot occur in absorption, however, since there is normally no vacancy in the L-shell into which a K electron can go. For this reason, an atom which absorbs energy from an x-ray beam must absorb enough energy to lift some electron *entirely free* of the atom. If an absorbing atom is ionized by havng one of its K electrons removed, that atom then has a K-shell vacancy and so it will emit one or more of its characteristic wavelengths as it returns to the ground state. Since this secondary emission can be radiated in any direction, the effect of the process is the removal of an x-ray photon from the direction of the original beam. The characteristic radiations emitted as a result of absorbing photons of high energy are often called *fluorescence* x-rays.

Just as there is a regular progression of x-ray emission wavelengths as we vary the atomic number of the target material, there is also a regular progression in the absorption edges of different elements. The shifting of x-ray spectra with atomic number provides an important technique for obtaining essentially monochromatic x-rays. Suppose we have an x-ray tube with a copper target. The K-series of the copper consists of an intense K_α line, and weak K_β and K_γ radiations, listed in the order of decreasing wavelength and of increasing energy. Nickel is a metal whose atomic number is one less than that of copper. Its energy levels are less negative than those of copper, and its emission lines and absorption edges are at slightly longer wavelengths than those of copper. The K_γ wavelength of nickel is almost identical with its λ_K absorption edge, and it falls between the strong K_α and the weaker K_β and K_γ emission lines of copper. Figure 6–17 shows the K-series emission spectrum of copper and the linear absorption coefficient of nickel as a function of wavelength. Since the minimum of the absorption curve is at a slightly shorter wavelength than one of the K lines but at a longer wavelength than the others, nickel filters out most of the radiation from the copper except its strong K_α line. There are several sets of elements such that by using one as a filter for the emission of the other, nearly monochromatic x-rays are obtained. The filtering effect of nickel for copper radiation was also shown in Fig. 6–10.

We have described how the linear absorption coefficient depends on the wavelength of the x-rays. We now consider how it depends on the absorbing material.

—— Intensity of x-rays from copper
- - - - Linear absorption coefficient of nickel

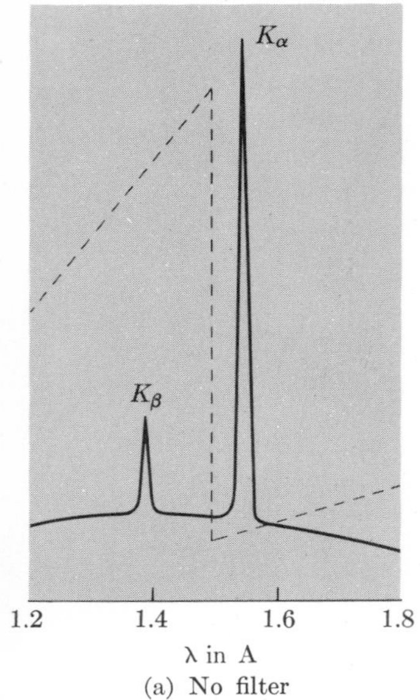

(a) No filter

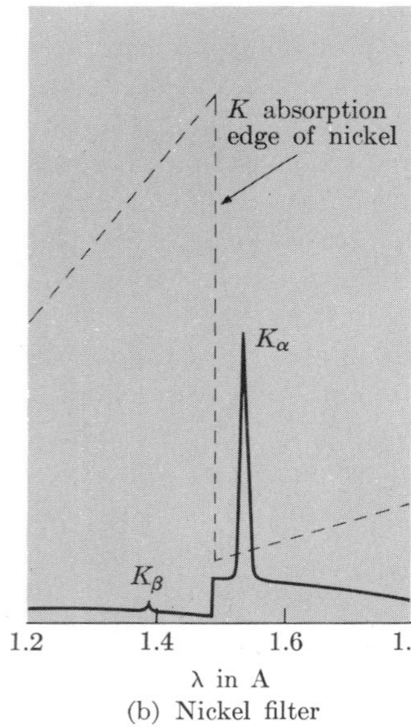

(b) Nickel filter

FIG. 6–17 Comparison of the spectrum of x-radiation from copper (a) before and (b) after passing through a nickel filter.

The absorption coefficient depends strongly on the density of the absorbing material which, of course, changes greatly if the material goes from gaseous to solid states.

We can write

$$\mu x = \frac{\mu}{\rho} x\rho, \tag{6–13}$$

where ρ is the mass density of the material. The quantity μ/ρ is called the *mass absorption coefficient*, μ_m. Its units are of the form of area divided by mass. Dimensional analysis shows that $(x\rho)$ is a mass per unit area, m_a. It is the mass of a sheet or slab of the absorber which has the thickness x and a unit of surface area normal to the incident x-ray beam. In these terms, the exponent of Eq. (6–10) becomes $-\mu_m m_a$ and we have

$$\boxed{I = I_0 e^{-\mu_m m_a}.} \tag{6–14}$$

Although the mass absorption coefficient varies from material to material far less than the linear coefficient does, it is still far from constant. Materials with large atomic numbers absorb x-rays more readily than the lighter elements, so that a certain mass per unit area of lead is more effective than the same mass per unit area of, say, carbon. In general, μ_m varies approximately as the atomic number cubed. Combining this empirical fact with the dependence on λ stated earlier, we can say, approximately, that

$$\mu_m = k\lambda^3 Z^3, \tag{6–15}$$

where k is nearly constant so long as λ does not vary through the absorption edges of the materials being compared.

Concrete is a shielding material which is widely used at the present time. It has about the same mass absorption coefficient as aluminum, but this is often increased by adding barytes to the mix or by using ground limonite instead of sand, and steel punchings instead of gravel. Looking ahead to the problem of shielding against gamma rays, we note that we then have radiations which are much more energetic than the x-rays usually used and which have much shorter wavelengths than the K absorption edge of any element. In the 1-MeV range, the absorbing properties of the heavy elements depend more on the number of electrons per unit mass than upon the electronic structure. For a given attenuation at a given photon energy, the mass per unit area required does not vary widely from one material to another. For 1-MeV radiation, the weight per square foot of five half-value layers of water is 112 lb, of concrete is 138 lb, of steel is 134 lb, and of lead is 110 lb. These change to other sets of values at other energies.

EXAMPLE. The mass absorption coefficient μ_m of aluminum for x-rays having a wavelength of 0.33 Å is 0.61 cm^2/g. The mass density of aluminum is 2.70 g/cm^3. Find (a) the linear absorption coefficient of aluminum, (b) the half-value layer, and (c) the thickness of absorber needed to reduce the intensity of the beam to $\frac{1}{24}$th of its incident value.

Solution. (a) $\mu_m = \mu/\rho$, or

$$\mu = \rho\mu_m = 2.70 \frac{\text{g}}{\text{cm}^3} \times 0.61 \frac{\text{cm}^2}{\text{g}} = 1.65 \text{ cm}^{-1}.$$

(b) According to Eq. (6–12),

$$T = \frac{0.693}{\mu} = \frac{0.693}{1.65 \text{ cm}^{-1}} = 0.42 \text{ cm}.$$

(c) From Eq. (6–10), we have $I/I_0 = e^{-\mu x}$. Substituting the values in this example, we obtain

$$\tfrac{1}{24} = e^{-1.65x}, \qquad \text{or} \qquad 24 = e^{1.65x}.$$

Therefore $\ln 24 = 1.65x$, or, if we use logarithms to the base 10, $2.30 \log 24 = 1.65x$, then

$$x = \frac{2.30 \times 1.38}{1.65} = 1.93 \text{ cm}.$$

There is an alternative method. Let n be the number of half-value layers needed to reduce the intensity the required amount, then

$$\left(\tfrac{1}{2}\right)^n = \tfrac{1}{24}.$$

Therefore $n \log 2 = \log 24$, or $0.301n = 1.38$, or $n = 4.6$. The thickness of this many half-value layers is $x = 4.6 \times 0.42 \text{ cm} = 1.93 \text{ cm}$.

6-10. COMPTON SCATTERING

The mechanism of absorption that we have considered is the photoelectric effect, in which the x-rays lose energy by ejecting electrons from the absorbing material. There is another effect which is weaker than the photoelectric effect in the usual range of x-ray energies but which becomes more important than the photoelectric effect at high x-ray energies (about 0.1 MeV and above). This is Compton scattering. A scattered x-ray is not really absorbed, since it does not lose a very large fraction of its energy. But, as mentioned earlier, the apparatus of Fig. 6–14 is sensitive to scattering, since an x-ray photon deflected out of the beam will fail to reach the detector, just as if it had been absorbed completely.

Planck introduced the idea that radiation must be emitted in bundles of energy, although he believed that once emitted, the energy spread in waves. Einstein extended the Planck idea to the absorption of radiation in his explanation of the photoelectric effect. He added the assumption that once a quantum of energy was radiated, it preserved its identity as a photon until it was finally absorbed. In the chapter on relativity we showed that mass and energy are identical. Since photons have energy, they must have mass. This developing concept that photons are true particles throughout their life comes to its climax in the Compton effect.

Compton assumed that the photons had the very "mechanical" property of momentum and solved the problem of impact of a photon and a material particle by means of relativistic mechanics.

Let us consider a material particle which is initially at the origin and at rest relative to the coordinate frame shown in Fig. 6–18. (The initial motion of the particle is negligible compared with the other velocities in the following analysis.) Initially, it has no momentum p_1 and its energy E_1 is only its rest-mass energy, m_0c^2. This particle is then hit by a photon moving along the x-axis. This photon brings to the impact an energy

$$E = hf = \frac{hc}{\lambda}\cdot \tag{6–16}$$

The photon has no rest mass, but its moving mass is equal to its energy, divided by c^2, or

$$m_{ph} = \frac{E}{c^2} = \frac{h}{c\lambda}\cdot \tag{6–17}$$

FIG. 6–18
Compton scattering.

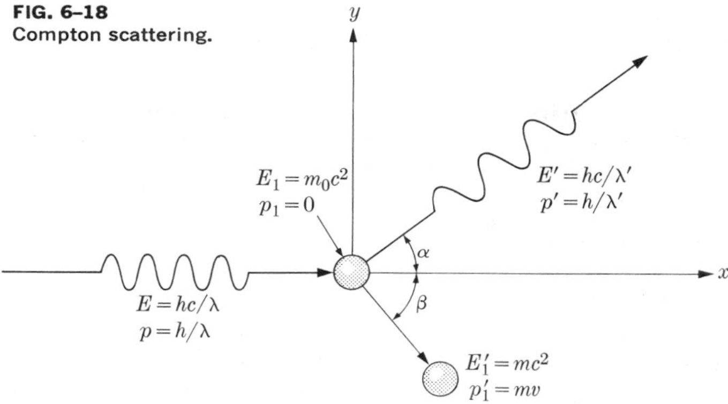

The momentum of the photon is its mass times its velocity, which is c. Thus its momentum p, which is all in the x-direction, is

$$p = \frac{h}{c\lambda}\, c = \frac{h}{\lambda}\cdot \tag{6–18}$$

We assume that the impact is slightly glancing, so that the motion of the material particle is directed below the x-axis, while the photon is deflected above the x-axis. The impact gives the material particle kinetic energy, so that its total energy becomes mc^2. The material particle also acquires momentum, mv, which is directed at an angle β to the x-axis. The mass of the particle in these statements is the relativistic moving mass.

The photon leaves the impact with its energy changed to hc/λ' and its momentum changed in magnitude and direction to h/λ' at an angle α to the x-axis. The above statements are summarized in Fig. 6–18.

We may now use the laws of conservation of mass-energy and of momentum to describe the impact situation. If we *neglect* any binding energy that the particle may have and apply the law of conservation of mass-energy, then the kinetic energy of the ejected particle is given by

$$E_k = \frac{hc}{\lambda} - \frac{hc}{\lambda'} = mc^2 - m_0c^2. \tag{6–19}$$

For the conservation of the x-component of momentum, we may write

$$\frac{h}{\lambda} = \frac{h}{\lambda'}\cos\alpha + mv\cos\beta, \tag{6–20}$$

and for the y-component,

$$0 = \frac{h}{\lambda'}\sin\alpha - mv\sin\beta. \tag{6–21}$$

The experimental test of the theory we are developing consists in measuring the wavelength λ' of the deflected photons as a function of their angle of deflection α. The initial photon wavelength λ is known. Since the material particles are usually electrons, m_0 is also known. We find, then, that the three equations (6–19) through (6–21) involve five variables, m, v, λ', α, and β, of which three, m, v, and β, relating to the electron, will be eliminated. This requires an additional equation which is the relativistic interdependence of m and v. From Eq. (5–37), we have

$$m = \frac{m_0}{\sqrt{1 - (v^2/c^2)}},$$

or

$$m^2v^2 = c^2(m^2 - m_0^2). \tag{6–22}$$

We first eliminate β between Eqs. (6–20) and (6–21) by isolating the terms containing β and squaring both equations. We thus obtain

$$m^2v^2 \cos^2 \beta = h^2 \left(\frac{1}{\lambda^2} - \frac{2 \cos \alpha}{\lambda\lambda'} + \frac{\cos^2 \alpha}{\lambda'^2}\right) \tag{6–23}$$

and

$$m^2v^2 \sin^2 \beta = h^2 \left(\frac{\sin^2 \alpha}{\lambda'^2}\right). \tag{6–24}$$

Adding these equations and using $\sin^2 x + \cos^2 x = 1$, we find that

$$m^2v^2 = h^2 \left(\frac{1}{\lambda^2} - \frac{2 \cos \alpha}{\lambda\lambda'} + \frac{1}{\lambda'^2}\right). \tag{6–25}$$

We next eliminate v between Eqs. (6–25) and (6–22), obtaining

$$c^2(m^2 - m_0^2) = h^2 \left(\frac{1}{\lambda^2} - \frac{2 \cos \alpha}{\lambda\lambda'} + \frac{1}{\lambda'^2}\right). \tag{6–26}$$

Solving Eq. (6–19) for m and squaring yields

$$m^2 = m_0^2 + \frac{2m_0h}{c}\left(\frac{1}{\lambda} - \frac{1}{\lambda'}\right) + \frac{h^2}{c^2}\left(\frac{1}{\lambda} - \frac{1}{\lambda'}\right)^2. \tag{6–27}$$

We eliminate m by substituting from Eq. (6–27) into Eq. (6–26). This gives

$$2m_0hc\left(\frac{1}{\lambda} - \frac{1}{\lambda'}\right) + h^2\left(\frac{1}{\lambda^2} - \frac{2}{\lambda\lambda'} + \frac{1}{\lambda'^2}\right) = h^2\left(\frac{1}{\lambda^2} - \frac{2 \cos \alpha}{\lambda\lambda'} + \frac{1}{\lambda'^2}\right). \tag{6–28}$$

By canceling terms and multiplying both sides of the equation by $\lambda\lambda'/h$, we find that Eq. (6–28) simplifies to

$$m_0c(\lambda' - \lambda) - h = -h \cos \alpha. \tag{6–29}$$

We have finally

$$\boxed{\Delta\lambda = \lambda' - \lambda = \frac{h}{m_0c}(1 - \cos \alpha).} \tag{6–30}$$

When the particle involved in Compton scattering is an electron, the first term on the right-hand side of Eq. (6–30) becomes (h/m_ec). This quantity is called the Compton wavelength of the electron.

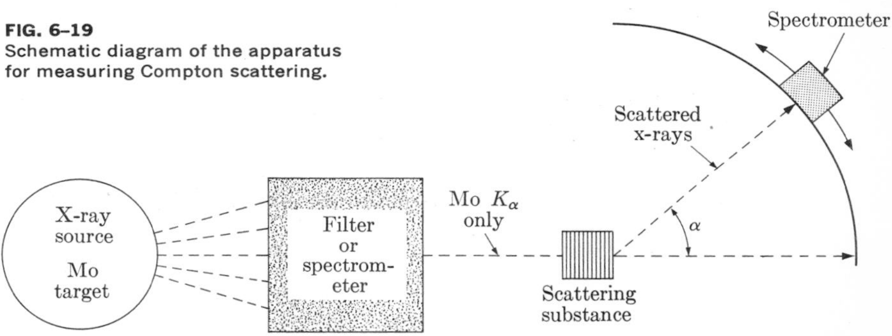

FIG. 6–19
Schematic diagram of the apparatus for measuring Compton scattering.

The experimental test of Eq. (6–30) can be made with an apparatus such as that schematically represented in Fig. 6–19. Monochromatic x-rays are scattered through an angle α and the spectrum of the scattered radiation is measured. The measuring spectrometer must be free to swing in an arc about the scatterer, so that the scattered wavelength can be measured as a function of the scattering angle α. Compton's results for the molybdenum K_α line scattered by carbon for four values of α are shown in Fig. 6–20. The solid vertical lines correspond to the primary wavelength, λ, and the dashed ones to the calculated modified wavelength λ'. The electrons which, when hit, modify the photon wavelength, must suffer a change of energy, as required by our derivation. Such electrons must be either free or loosely bound. If the photon encounters a tightly bound electron, the whole atom is involved in the collision, and the value of m_0 in Eq. (6–30) then becomes the mass of the atom rather than the mass of an electron. Since the mass of the atom is several thousand times the electronic mass, the Compton wavelength shift for it is too small to detect. These scattered photons comprise the unmodified peaks shown in Fig. 6–20.

For very high-energy photons most of the atomic electrons appear free and a large fraction suffer a wavelength shift. For low-energy x-rays most electrons appear bound unless the scattering atom has a low atomic number.

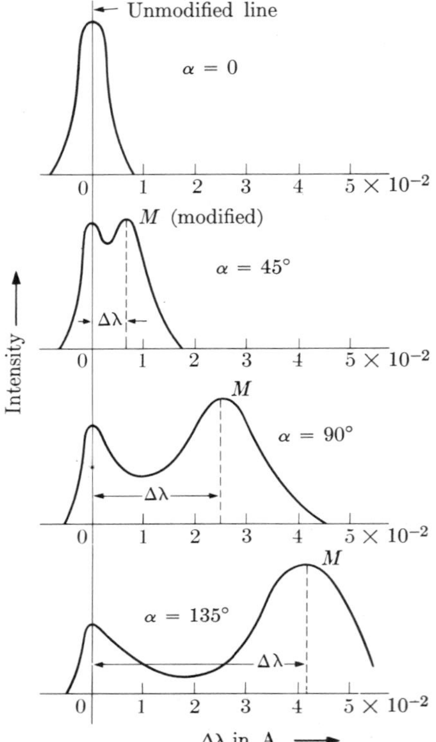

FIG. 6–20
Wavelength displacement of the modified line, M, of the $K\alpha$-radiation of molybdenum scattered from carbon, as a function of scattering angle.

For *visible photons*, all the electrons appear bound and there is *no* observed Compton shift. Although we have just seen that the likelihood of a Compton wavelength shift depends on the photon energy and the scattering material, the amount of the wavelength shift depends only on the scattering angle. This aspect of Compton scattering has been tested, and essentially the same change in wavelength was found for x-rays scattered from fifteen different elements.

In Fig. 6–20, the modified peak is broader than the unmodified one: The reason for this difference does not appear in the derivation because we assumed that the electron doing the scattering was initially at rest. Motion of the target electron fully accounts for the fact that, for a given angle, some photons had their energy changed a little more or less than what would be expected for stationary electrons.

Not only does the Compton effect contribute to our understanding of absorption coefficients, its theory also provides a good check of relativity and extends the particle concept of photons. With photons behaving like billiard balls, we are nearly back to Newton's particle theory of light. In the next chapter we shall synthesize the divergent wave and particle theories of radiation.

6–11. ABSORPTION BY PAIR PRODUCTION

The third mechanism of x-ray absorption is pair production, which was discussed in Section 5–16. Although it cannot occur if the photon has less than about 1 MeV of energy, above 5 MeV pair production in lead is a more important process than either photoelectric absorption or Compton scattering. The radiation arising from the annihilation of the pair contributes to the buildup factor in the absorber.

The three mechanisms of x-ray absorption are summarized in Fig. 6–21. The contribution of each of these to the linear absorption coefficient in

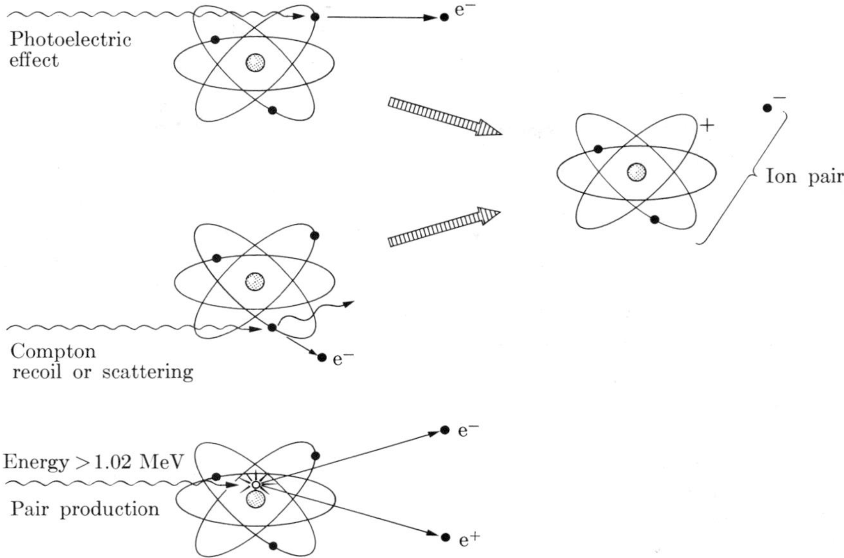

FIG. 6–21
Summary of x-ray and gamma-ray interactions with matter.

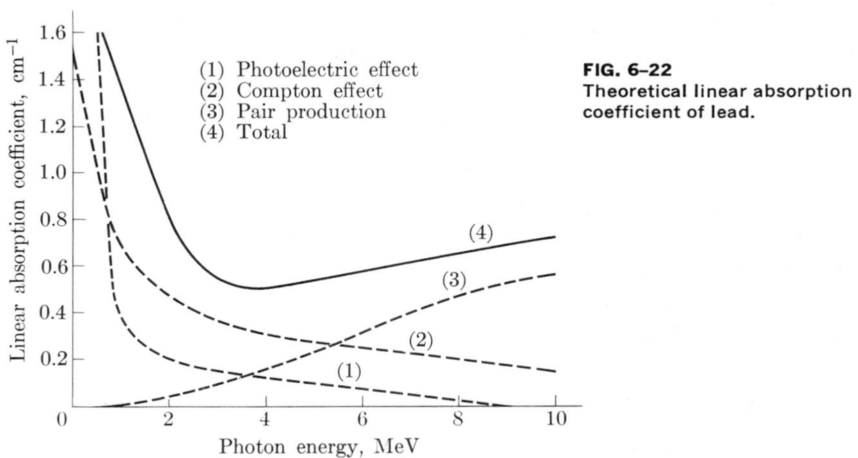

FIG. 6–22
Theoretical linear absorption coefficient of lead.

lead for photons of different energies is shown in Fig. 6–22. The slope of the photoelectric variation in Fig. 6–16 is positive because the abscissa is the wavelength. In Fig. 6–22 the abscissa is energy and thus this slope is negative. The discontinuities of Fig. 6–16 do not show on Fig. 6–22 because the scales are very different.

6-12. DIFFRACTION WITH RULED GRATINGS

X-rays were first diffracted with crystals but eventually a method was also devised for diffracting them with man-made gratings. This method is an interesting and important example of how physical research often builds crosslinks that reinforce the entire structure of our ideas.

We stated earlier that x-rays were not diffracted by man-made gratings because the wavelength of x-rays is much less than the separation of the closest lines man can rule with the granular materials at his disposal. Compton found a way around this difficulty which was certainly ingenious. He used a grating in an unusual way. The index of refraction of glass for x-rays is slightly *less* than one. Thus when x-rays go from air or vacuum into glass, they are passing from an *optically* dense medium into one less dense, and the rays are refracted away from the normal. Recall that whenever light goes from a more dense to a less dense medium, there is a critical angle, and that if the rays approaching the interface make an angle greater than the critical angle, total reflection results. X-rays striking glass at "grazing incidence" will not enter the glass at all, but will be totally reflected. If, now, the glass surface is a ruled grating, these rulings will appear closer together from a grazing angle than when viewed from a point normal to the grating surface. Thus the very geometrical situation which makes the bands between the rulings of the grating capable of reflecting also makes the apparent grating space much less. Compton found that with his ingenious arrangement he could diffract x-rays by measurable amounts.

The important result of this work was that Compton could measure x-ray wavelengths absolutely. The grating space was determined by a screw whose pitch could be measured by counting the threads in a length measured by an ordinary length standard. Therefore, the grating space was known from very direct elementary measurement. All prior x-ray wavelength measurements had been made in terms of a calculated crystal spacing which depended on the Avogadro constant, and the Avogadro constant depended in turn upon the charge of the electron. When high precision measurements were made, it was found that the wavelengths measured by a grating differed from those measured by crystal diffraction by about 0.3%. Since this was an absolute measurement, it was concluded that the crystal spacings that had been used formerly were in error. Shiba, a Japanese physicist, traced this back through the Avogadro constant to the charge of the electron and concluded that Millikan's value was in error. Shiba pointed out that the error lay in the viscosity of air used in the calculations in the oil-drop method. Once the new value of the charge was accepted, all physical constants based on that charge were revised. It is interesting that an experiment in x-ray spectroscopy should provide a technique for measuring the basic unit of electrical charge.

6-13. RADIATION UNITS

Some of the energy absorbed from ionizing radiation which passes through matter damages the medium by causing molecular changes or by altering the crystalline form. The amount of damage produced depends upon the nature of the absorbing material, the energy of the photon, and the intensity of the radiation. The effects are large in complex organic molecules and so x-rays are injurious to living tissue. To study this quantitatively, it is necessary to define a unit for amount of radiation absorbed.

One of the units of the amount of x-radiation is the *roentgen*, R. The roentgen is the amount of x- or gamma radiation that produces in 1 cm^3 of *air* under standard conditions ions carrying $1/(3 \times 10^9)$ coulomb of charge of either positive or negative sign.* Air was chosen for the ionized medium because its mass absorption coefficient is nearly the same as that of water and of body tissue over a considerable range of wavelengths. The *milliroentgen*, mR, is a thousandth of a roentgen. In radiobiology these units are usually called the *exposure* units for radiation delivered to a specified area or volume of the body. Note that the units of amount of radiation do not depend upon time, and they are not a measure of the energy flux in a beam of radiation.

The *exposure rate* is measured by the quantity of radiation delivered per unit time. Some of the units are the *roentgen per hour*, R/h, the *milliroentgen per hour*, mR/h, and so on.

Absorbed radiation frees photoelectrons and Compton recoil electrons. These secondary particles or corpuscles will also ionize the air as they move toward the collecting electrodes in an ionization chamber. The charges carried by these ion pairs produced by the secondaries must be counted as a part of the quantity of charge to be measured. [An *ion pair*, ip, is composed of the positive ion and the negative ion (usually an electron) produced when a neutral particle is ionized. The magnitude of the charge on each member of the pair is necessarily the same. Each ion of an ion pair produced in air is singly charged.] Although all the secondary particles originate in 1 cm^3 of air where the absorption occurred, the measured charge comes from the whole volume through which these secondaries range.

When a large number of ions are formed in air, it is found that the *average* energy required to produce an ion pair is 33.7 eV. This is much greater than the minimum energy needed to ionize either oxygen or

---

* The definition adopted by the International Commission on Radiological Units and Measurements is as follows: "The roentgen is that quantity of x- or gamma radiation such that the associated corpuscular emission per 0.001293 g of dry air produces, in air, ions carrying 1 esu of quantity of electricity of either sign." This mass of air has a volume of 1 cm^3 at 0°C and a pressure of 760 mm of mercury, and 1 esu of charge is equal to $1/(3 \times 10^9)$ C.

nitrogen. The value is larger because some photons eject inner electrons whose binding energies are much greater than the minimum ionization energy. If we take the product of the number of ion pairs in air corresponding to a roentgen and the average energy required to produce such a pair, we find that the roentgen is equivalent to 0.112 ergs of energy absorbed per cubic centimeter of standard air or 86.9 ergs per gram of air. When a beam of x-rays enters another medium where the atoms have different atomic numbers, the absorption of energy changes. Thus it was found that exposure to a dose of 1R of x-rays results in the absorption of *about* 97 ergs per gram of soft body tissue. The *Rep* (roentgen equivalent physical) is that dose of ionizing radiation which results in the absorption of 97 ergs per gram of body tissue. This unit is *not* very *definite* because of the variations in the composition of the body. However, a definite unit is the *rad* (radiation absorbed dose), which is 100 ergs (10^{-5} J) of *absorbed* energy per gram of *any* absorbing material. The *millirad*, mrad, is a thousandth of a rad. Additional units will be discussed in the chapter on natural radioactivity.

Radiation damage to the human body depends upon the absorbed dose, the exposure rate, and the part of the body exposed. The safe limit for those exposed to radiation over the *whole* body during their working day is now set at 250 mR/week. This is an exposure rate of 6.25 mR/h based on a 40-hour week. Up to a few years ago, the safe tolerance level was thought to be twice this amount. It is likely that it will be lowered again soon. Although the absorbed dose due to long exposure to low-level radiation is large, the resulting direct damage is negligible because the body has time to repair the injury. The effects of acute radiation exposure over the *whole* body are about as follows: 20–50 R, some blood changes; 100–250 R, severe illness but recovery within 6 months; 400 R, fatal to 50% of the persons affected (this is called the median lethal dose, MLD or LD-50); and 600 R, fatal to all.

The mechanism of tissue destruction is not completely understood. When radiation is absorbed by the various complex organic molecules in the body, an electron is either raised to a higher energy level within the molecule or removed altogether. One might expect that after the electron returns, all would be normal again, as in the case of hydrogen and other single atoms. However, during the time the molecule is in the excited state, its constituent atoms sometimes rearrange themselves. Then, although the system is again neutral after the electron's return, it is no longer the same molecule. A reaction which occurs in some cases is that two hydroxyl radicals, OH, combine to form hydrogen peroxide, H_2O_2. The molecular situation is somewhat analogous to a high tower which a child builds with small wooden blocks. If we "ionize" the tower by pulling out a block near the base, it is evident that the tower will not rebuild itself when we return the "electron" to the heap. It is unlikely that any substantial change would have occurred if only the topmost block had been removed and then returned. Destruction of a body cell depends upon which electron in the molecular structure absorbs the radiation energy.

In the human body, the hands and feet can receive a much larger dose of radiation without permanent injury than any other part. However, the genes in the cells of the body are readily damaged. Injury to those in the reproductive cells is particularly serious because it gives rise to mutations in the generations which follow. These mutations are almost always adverse and the process is irreversible. There is no safe lower limit of radiation when considering the inheritance of genetic damage. It has been said that "a little radiation is a little bad, and a lot is a lot bad." This means that the probability of absorption by any one gene or group of genes is less for a small dose than for a large one. However, the damage per quantum of radiation absorbed is the same in both cases. It is estimated that the rate of mutation will show significant increase if the exposure is more than 10 R during their reproductive lifetime—a period of about 30 years. During that length of time, one will receive about 4 R from cosmic rays and from the radioactive materials which are found in low concentrations everywhere. Any exposure to x-rays adds to the accumulated dose. Remember that the damage discussed here is transmitted to the generations to come. A whole body exposure of 10 R over a period of three decades would not harm the parent.

The things considered in this section lie in the field of radiological physics. As we advance into the nuclear age, the solution of an increasing number of problems of public health will become the responsibility of persons trained in this area of science.

REFERENCES

BUERGER, M. J., "Generalized Microscopy and the Two-Wave-Length Microscope," *J. Appl. Phys.* **21**, 909 (1950).

CLARK, GEORGE L., *Applied X-Rays*. 4th ed. New York: McGraw-Hill, 1955.

COMPTON, A. H. and S. K. ALLISON, *X-Rays in Theory and Experiment*. 2nd ed. Princeton, N. J.: Van Nostrand, 1935. A comprehensive treatise.

CULLITY, B. D., *Elements of X-Ray Diffraction*. Reading, Mass.: Addison-Wesley, 1956. Good discussion of x-rays and crystal structure.

JAUNCEY, G. E. M., "Birth and Early Infancy of X-Rays," *Am. J. Phys.* **13**, 362 (1945). In addition to Roentgen's account of his discovery, this article contains a number of sensational accounts of x-rays from the newspapers of 1896.

MEYER, CHARLES F., *The Diffraction of Light, X-Rays, and Material Particles*. Chicago, Ill.: University of Chicago, 1934. A treatise.

National Bureau of Standards, Washington, D. C.: U. S. Government Printing Office.

Handbook 51, Radiological Monitoring Methods and Instruments, 1962.

Handbook 76, X-Ray Protection up to Three Million Volts, 1961.

Handbook 84, Radiation Quantities and Units, 1962.

Handbook 85, Physical Aspects of Irradiation, 1964.

Richards, Horace C., "A. W. Goodspeed: A Pioneer in Radiobiology," *Am. J. Phys.* **11,** 342 (1943).

Roentgen, Wilhelm C., "The Roentgen Rays," in Magie, W. F., *A Sourcebook in Physics.* Cambridge, Mass.: Harvard University Press, 1963.

Sproull, Wayne T., *X-Rays in Practice.* New York: McGraw-Hill, 1946. Contains theory and industrial practice.

Watson, E. C., "The Discovery of X-Rays," *Am. J. Phys.* **13,** 281 (1945).

PROBLEMS

6–1. Potassium iodide, KI, is a cubic crystal which has a density of 3.13 g/cm^3. Find (a) the basic interplanar distance, and (b) the length of a side of the unit cell.

6–2. The spacing between the principal planes in a crystal of NaCl is 2.820 Å. It is found that a first-order Bragg reflection of a monochromatic beam of x-rays occurs at an angle of 10°. (a) What is the wavelength of the x-rays in Å and in XU, and (b) at what angle would a second-order reflection occur?

6–3. What is the shortest wavelength that can be emitted by the sudden stopping of an electron when it strikes (a) the screen of a TV tube operating at 10,000 V and (b) the plate of a high-power radio transmitter tube operating at 30,000 V? (c) Determine the spectral region in which these wavelengths lie by referring to Fig. 3–2.

6–4. (a) At what potential difference must an x-ray tube operate to produce x-rays with a minimum wavelength of 1 Å? 0.01 Å? (b) What is the maximum frequency of the x-rays produced in a tube operating at 20 kV? at 60 kV?

6–5. An x-ray tube is operating at 150,000 V and 10 mA. (a) If only 1% of the electric power supplied is converted into x-rays, at what rate is the target being heated in calories per second? (b) If the target weighs 300 g and has a specific heat of 0.035 cal/g°C, at what average rate would its temperature rise if there were no thermal losses? (c) What must be the physical properties of a practical target material? What would be some suitable target elements?

6–6. Since the index of refraction of all materials for x-rays is very close to unity, it is impractical to make lenses for them. Therefore all radiographs (x-ray photographs) are necessarily shadowgraphs. (a) What requirement is imposed on the target area from which the radiation originates so that the pictures will be sharply defined? (b) What is the smallest size film with respect to the size of the object that can be used in radiographing it?

6–7. X-rays from a certain cobalt target tube are composed of the strong K series of cobalt and weak K lines due to impurities. The wavelengths of the K_α lines are 1.785 Å for cobalt, and 2.285 Å and 1.537 Å for the impurities. (a) Using Moseley's law, calculate the atomic number of each of the two impurities. (b) What elements are they? (For the K-series, $b = 1$ in Moseley's law.)

6–8. The K absorption edge of tungsten is 0.178 Å and the average wavelengths of the K-series lines are $K_\alpha = 0.210$ Å, $K_\beta = 0.184$ Å, and $K_\gamma = 0.179$ Å. (a) Construct the x-ray energy-level diagram of tungsten. (b) What is the least energy required to excite the L-series? (c) What is the wavelength of the L_α line? (d) If a 100-keV electron struck the tungsten target in a tube, what is the shortest x-ray wavelength it could produce, and (e) what is the shortest wavelength characteristic of tungsten that could be emitted?

6–9. The radiation from an x-ray tube operated at 80 kV is incident on a sheet of tungsten. (a) Using the energy levels characteristic of tungsten calculated in Problem 6–8, find the maximum kinetic energies of the electrons ejected from the K-, L-, and M-shells, in eV. (b) What is the range of kinetic energies of the electrons ejected by 80 keV photons from the shells between M and infinity?

6–10. A thin sheet of nickel is placed successively in a beam of x-rays from cobalt, then in one from copper, and finally in one from zinc. Discuss the effect of each of these beams on the nickel atoms and show that the filtered radiation from copper is essentially monochromatic. X-ray data for these elements are given in the accompanying table. Consider only K-shell absorption.

Element	Emission wavelengths, Å		
	K_α	K_β	K_{absorb}
Co	1.79	1.62	1.61
Ni	1.66	1.49	1.48
Cu	1.54	1.39	1.38
Zn	1.43	1.29	1.28

6–11. (a) How does the linear absorption coefficient vary with the atomic number of an absorber? (b) Before an examination of the gastrointestinal tract of a patient is made with x-rays, the patient drinks a suspension of barium sulfate, $BaSO_4$. Why must this be done? (c) Some fluids in the human body can be followed by the use of x-rays if a solution of potassium iodide, KI, is introduced into the region to be studied. Why must this be done?

6–12. How many half-value layers of a material are necessary to reduce the intensity of an x-ray beam to (a) $\frac{1}{16}$, (b) $\frac{1}{80}$, and (c) $\frac{1}{200}$ of its incident value?

6–13. The half-value layer of steel for 1.5-MeV radiation is 0.5 in. (a) Plot the curve of transmitted intensity against thickness in half-value layers for a sheet of steel 3 in. thick when the intensity of the incident beam is 200 units. (b) Using the same coordinate scales as in part (a), plot the radiation that has been absorbed against the thickness. (c) Most of the radiation absorbed appears as heat. Assuming low thermal conductivity so that the temperature gradient in the steel is approximately the slope of the absorption curve, discuss the variation of this gradient within the sheet. (d) Will the sheet have thermal stresses?

6–14. For 0.2-Å x-rays, the mass absorption coefficients in cm^2/g for several metals are: aluminum, 0.270; copper, 1.55; and lead, 4.90. (a) What is the half-value thickness of each for a narrow beam of x-rays? (b) What thickness of each is required to reduce the intensity of the transmitted beam to $\frac{1}{32}$ of its incident value? (c) If the "buildup" due to scattering and other processes for a broad beam of radiation is equivalent to making the incident beam 1.5 times its actual intensity, what thickness of each material is then needed to obtain an intensity reduction to $\frac{1}{32}$?

6–15. Show that the tenth-value layer of an absorber equals $2.30/\mu$.

6–16. The linear absorption coefficient of a certain material for x-rays having a wavelength of 1 Å is $\mu_1 = 3.0\ cm^{-1}$, and for those having a wavelength of 2 Å is $\mu_2 = 15\ cm^{-1}$. If a narrow beam of x-rays containing equal intensities of 1-Å and 2-Å radiation is incident on the absorber, for what thickness of the material will the ratio of the intensities of the transmitted rays be 4 to 3.

6–17. The linear absorption coefficients of aluminum and copper are $0.693\ cm^{-1}$ and $13.9\ cm^{-1}$, respectively, for the K_α line from tungsten. (a) What percentage of the intensity of this line will pass through a 5-mm plate of aluminum? of copper? (b) What thickness of aluminum is equivalent to the absorption of 5 mm of copper for this wavelength?

6–18. Observations made from very high altitude balloons and from rockets that go to the "top" of the earth's atmosphere show that the sun and certain stars emit x-rays (x-ray astronomy). Why does visible light from these sources reach the surface of the earth whereas the x-rays do not?

6–19. Calculate the fraction of an x-ray beam transmitted by mercury for which $\mu_m = 3.0\ cm^2/g$ when passing through: (a) 0.1 mm of liquid mercury, (b) 10 cm of a solution of 0.2 mole of $Hg(NO_3)_2$ dissolved in a liter of water, and (c) 80-cm column of mercury vapor at 500°C and 1 atmosphere pressure.

6–20. Show that when the exponent μx in $I = I_0 e^{-\mu x}$ is equal to or less than 0.1, then, with an error of less than 1%, (a) the transmitted radiation equals $I_0(1 - \mu x)$, and (b) the absorbed radiation equals $I_0 \mu x$. (*Hint:* Expand the exponential term into a series.)

6–21. For 2-MeV x- or gamma rays, the half-value layers of some commonly used shielding materials are: water, 5.9 in; ordinary concrete (relative density, 2.6), 2.3 in; iron, 0.80 in; and lead, 0.53 in. Find (a) the mass absorption coefficient of each in ft^2/lb, and (b) the weight per unit area in lb/ft^2 of each material for shielding walls which are 5 half-value layers thick.

6–22. Three quanta, each having 2 MeV of energy, are absorbed in a material. One of the quanta is absorbed by the photoelectric process, another is involved in a Compton scattering, and the third is involved in pair production. (a) Discuss the possible methods by which the scattered photon and each of the charged particles produced can be involved in energy interchanges until all the original photon energy is reduced to thermal energy. (b) Will all this final thermal energy be freed at the point where absorption occurred originally?

6–23. (a) Show that the mean free path of an x-ray photon in a solid is $1/\mu$. [*Hint:* Refer to the derivation of Eq. (1–28).] (b) When a part of the human body, say the arm, is overexposed to a beam of soft (long wavelength) x-rays, a "burn" appears on the skin on the side of incidence, but for rather hard (short wavelength) x-rays, the "burn" usually occurs only on the skin farthest from the incident side. Explain.

6–24. The K_α line of molybdenum (wavelength, 0.712 Å) undergoes a Compton collision in carbon. What is the wavelength change of the line scattered at 90° if the scattering particle is (a) an electron, and (b) the whole carbon atom? (c) What is the scattered wavelength in each case?

6–25. Each of three quanta of radiation undergoes a 90° Compton scattering in a block of graphite. Assuming that the ejected electron had no binding energy in carbon, compute the fractional change in wavelength ($\Delta\lambda/\lambda$) for the scattered radiation when the incident quantum is: (a) a gamma ray from cobalt,

$$\lambda = 1.06 \times 10^{-2}\ \text{Å};$$

(b) K_α x-rays from molybdenum, $\lambda = 0.712$ Å; and (c) visible light, $\lambda = 5000$ Å. (d) The photoelectric work function of carbon is actually 4.0 eV. If this is taken into account, what would be the answers to the preceding parts? (e) Comment on the feasibility of resolving the two waves in a beam composed of the incident and scattered radiation in each of the preceding cases.

6–26. An x-ray quantum having a wavelength of 0.15 Å undergoes a Compton collision and is scattered through an angle of 37°. (a) What are the energies of the incident and scattered photons and of the ejected electron? (b) What is the magnitude of the momentum of each photon? (c) Find the momentum of the electron both graphically and analytically using the values found in the preceding part.

6–27. Using the data and results of Problem 6–23 for Compton scattering of the K_α line of molybdenum by carbon, find the energy of the recoil particle given that it is (a) an electron and (b) the whole carbon atom. (c) What is the direction of motion of the recoil particle in each case with respect to the direction of incidence of the photon?

6–28. A photon of energy E undergoes a Compton collision with a free particle of rest mass m_0. (a) Show that the maximum recoil kinetic energy of the particle is

$$E_{k_{\max}} = \frac{E^2}{E + m_0c^2/2}.$$

(b) What is the maximum energy which can be transferred to a free electron by Compton collision of a photon of violet light ($\lambda = 4000$ Å)? (c) Could violet light eject electrons from a metal by a Compton collision?

6–29. When a hydrogen atom goes from the state $n = 2$ to the state $n = 1$, the energy of the atom decreases by 10.2 eV. In Chapter 4 it was stated that all of this energy is radiated as a single photon. Actually, the energy of the photon must be slightly less than 10.2 eV, since a small part of the energy is needed to provide the kinetic energy of the recoiling atom. (a) What is the recoil momentum of the hydrogen atom? (b) What fraction of the 10.2 eV is taken by the recoil of the atom?

6–30. (a) Radiation of wavelength λ incident normally on a certain mirror is reflected with no loss of energy. Consider a beam of radiation to be a stream of photons. Show that the pressure exerted by a beam on the mirror is $2I/c$, where I is the incident intensity. (b) Calculate the pressure exerted by a beam having an intensity of 0.6W/m^2.

6–31. A photon of wavelength λ undergoes Compton scattering from a free electron as shown in Fig. 6–18. Prove that, regardless of the energy of the incoming photon, the scattered photon cannot undergo pair production if α is greater than 60°.

6–32. The apparatus for a Compton scattering experiment is arranged so that the scattered photon and the recoil electron are detected only when their paths are at right angles to one another. Using Fig. 6–18, show that under these conditions (a) the scattered wavelength is given by $\lambda' = \lambda/\cos\alpha$ and (b) the energy of the scattered photon is m_0c^2.

6–33. A photon of energy 1.92 MeV undergoes pair production in the vicinity of a lead nucleus. The created particles have the same speed, and both travel in the direction of the original photon. Calculate the recoil momentum of the lead nucleus.

6–34. Use the conservation laws to show that it is impossible for pair production to take place in free space. Assume that the velocity v of each of the particles of the pair is parallel to the direction of motion of the incident photon.

6–35. Show that the roentgen is equivalent to (a) 2.083×10^9 ion pairs per cm^3 of standard air, and (b) 1.611×10^{12} ion pairs per g of air. The density of air is 1.293 g/liter at standard conditions.

6–36. Given that it requires 33.7 eV of energy to produce an ion pair in air, show that the roentgen is equivalent to an energy absorption of (a) 7.02×10^4 MeV per cm^3 of standard air, (b) 5.42×10^7 MeV per g of air, (c) 86.9×10^{-7} J per g of air, and (d) 2.07×10^{-6} cal per g of air, and 0.87 rads.

6–37. An x-ray beam having an exposure rate of 480 mR/h produces ions in 6 cm^3 of standard air in an ionization chamber. What is the saturated ionization current in amperes?

6–38. A common form of exposure meter or dosimeter is a cylindrical capacitor filled with a gas. One model worn by personnel working in the vicinity of x-ray equipment has an absorption equivalent to 6 cm^3 of standard air. (a) If the electrodes are charged to a difference of potential of 400 V, what must be the resistance of the electrode insulation so that the charge leakage in 8 h shall not exceed 10% of an assumed safe tolerance level of 50 mR/8-h day? (Assume that the voltage remains constant.) (b) What would be the effect on the accuracy of the dosimeter if (1) there are finger streaks on the insulator, and (2) the relative humidity is high?

CHAPTER 7

WAVES AND PARTICLES

7-1. WAVE-PARTICLE DUALITY OF LIGHT

Electromagnetic radiation, which includes visible light, infrared and ultraviolet radiation, and x-rays, was shown to be a wave motion by interference experiments. This type of experiment, which involves constructive and destructive interference, is considered to be a test for the existence of waves since it requires the presence of two waves at the same position at the same time, whereas it is impossible to have two particles occupy the same position at the same time. On the other hand, the experimental results for blackbody radiation, photoelectric effect, and x-ray absorption are explained by considering the radiation to appear as a stream of particles which are, for example, absorbed one at a time. It appears that light must be considered as a wave in some experiments and as a particle in others. This does not occur randomly, however; the experiments can be sorted into the following two types. Those which require the wave nature are ones which may be called propagation experiments. An important part of their explanation is the consideration of the path or paths traveled by the light, as in interference experiments where a path difference is determined. The experiments which require the particle nature may be called interaction experiments. The radiation interacts with matter to produce a resultant absorption or scattering.

This dual nature of light was not readily accepted. The main reason for this is the apparently contradictory aspects of the two natures. A wave is specified by a frequency f, wavelength λ, phase velocity u, amplitude A, and intensity I. These are not all independent. Thus the velocity, frequency, and wavelength are related by $u = f\lambda$. A wave is necessarily spread out and occupies a relatively large region of space. Actually, a sinusoidal wave with a unique wavelength would have to have infinite length, otherwise there would be a change in wavelength at the ends of the wave. A particle, on the other hand, is specified by a mass m, velocity v, momentum p, and energy E. The characteristic which seems in conflict with a wave is that a particle occupies a definite position in space. In order for a particle to be at a definite position, it must be very small. It is difficult to accept the conflicting ideas that light is a wave which is spread out over space and also a particle which is at a point in space. This acceptance is necessary, however, to explain all the results of the experiments which can be performed with light. (We use the word light to include the entire electromagnetic spectrum.)

We do have connections between the wave and particle characteristics of light. As indicated in Chapter 3, Planck related the energy of the photon, E, and the frequency of the wave, f, by

$$E = hf. \tag{7–1}$$

In Chapter 6 we saw how Compton used the relation

$$p = h/\lambda, \tag{7–2}$$

between the momentum of the photon, p, and the wavelength of the wave, λ. In addition, the intensity of the wave is related to the rate at which photons pass through a unit area. Consequently, the particle characteristics of light can be found from the wave characteristics, even though the concepts of wave and particle appear to contradict each other. We shall attempt to resolve this contradiction in Section 7–6.

7–2. THE DE BROGLIE HYPOTHESIS

The dual nature of light, made necessary by experimental results, was extended by de Broglie in 1924. He felt that nature was symmetrical and the dual nature of light should be matched by a dual nature of matter. His argument was that if light can act like a wave sometimes and like a particle at other times, then things like electrons, which were considered particles, should also act like waves at times.

To specify the wave properties, de Broglie proposed that the relation between the momentum and the wavelength of a photon is a general one, applying to photons and material particles alike. Since the momentum of a material particle is its mass m times its velocity v, the de Broglie wavelength is

$$\boxed{\lambda = \frac{h}{mv}.} \tag{7–3}$$

These proposed waves were not electromagnetic waves but were a new kind of wave, which were called matter waves or *pilot* waves. The word *pilot* implies that these waves pilot or guide the particle. When de Broglie published his hypothesis, it was not supported by any experimental evidence. His only real argument was his intuitive feeling that nature must be symmetrical.

At this point we may consider the fact that the wavelength is not sufficient to completely specify a wave. The frequency or velocity must also be known. We choose to define the frequency of the matter wave by extending the photon analogy, $f = E/h$, and by using the relativistic energy expression, $E = mc^2$. Since the phase velocity of any wave motion is $u = f\lambda$, we find the velocity of the waves associated with a particle to be $u = f\lambda = (E/h)(h/mv)$. Thus we have

$$u = \frac{mc^2}{h}\frac{h}{mv} = \frac{c^2}{v}. \tag{7–4}$$

In this equation, v is the speed of the material particle, which must be *less* than the speed of light, c; thus u is *greater* than the speed of light. The speed of the mass-energy of the particle does not exceed the speed of light in free-space, but the phase velocity of its associated waves does. This result does not conflict with the concepts of relativity, since the speed of light is a limiting speed only for mass-energy. In the chapter on relativity we warned that we would encounter situations where we would use speeds greater than c.

7–3. BOHR'S FIRST POSTULATE

The theoretical implications of the de Broglie wavelength of matter are interesting. By making a very plausible assumption, we can relate this wavelength to the Bohr model of the atom.

If a stretched string, fastened at each end, is caused to vibrate, the disturbances move along the string in both directions, are reflected at the ends, and come back on one another. In general, this disturbance of the string causes a complex motion that makes the string blurred everywhere. If the exciting frequency, the length of the string, or the string tension is varied, stationary waves can be produced. Displacement loops and nodes appear. Instead of running up and down the string, the waves appear to stand still with transverse activity at the loops. The condition for standing waves is that the length of the string be an integral number of half wave-lengths of the disturbance, since there is a node at each end.

If we were to form a circular loop with the string, the disturbances would move around the loop in both directions and there would be no reflections. The condition for standing waves for this configuration is that the length of the loop (the circumference of the circle) be an integral number of whole wavelengths of the disturbance. Thus, the condition for circular standing waves is

$$2\pi r = n\lambda,$$

where r is the radius of the circle and n is an integer.

Let us now assume that the Bohr orbits correspond to standing electron waves, analogous to the circular loop of string. If we use the above condition with the de Broglie wavelength, we have

$$2\pi r = \frac{nh}{m_e v}. \tag{7-5}$$

Recalling that the angular momentum of a particle moving in a circular orbit is $m_e vr$, we find that the angular momentum of the Bohr electron is

$$m_e vr = n\frac{h}{2\pi}, \tag{7-6}$$

which is precisely Bohr's first postulate. This discussion should not be interpreted as deriving Bohr's postulate, for essentially what we have done is replace Bohr's postulate with de Broglie's postulate.

7-4. MATTER REFRACTION

Einstein pointed out that if the de Broglie hypothesis is valid, then it should be possible to diffract electrons. Schroedinger felt that if de Broglie were right, then the waves associated with matter should suffer refraction.

Light usually travels in straight lines, but when it goes from one medium to another it is refracted. Refractive bending is due to changes in the velocity of propagation of light, which is low in a medium of high refractive index. We are aware of abrupt refractions such as occur at the surfaces of a lens. However, the refractions at layers of air of differing temperatures are quite gradual. When the summer sun beats on a blacktop highway, the layers of air next to the road are expanded. Although the refractive index of air is very close to unity, the hot air over the highway has an index even nearer unity. Light coming down to the highway enters a region of lower index and is refracted away from the normal because of the refractive index gradient. If the light approaches this region of index gradient at a grazing angle, total reflection causes the familiar mirage effect. In this case, the bending of the path of light by refraction is not abrupt. The light rays form a continuous curve.

Schroedinger felt that the continuous curved paths of material objects might be such a continuous refraction of the associated matter waves. To see how this works out, we consider the following example. The parabolic flight of a baseball is a simple case of a material body moving in a curved path. At the top of the flight the path is concave downward, the instantaneous velocity v is horizontal, and the path may be regarded momentarily as the arc of a circle of radius of curvature R. The acceleration of the ball is the centripetal acceleration, v^2/R, which is the acceleration due to gravity, g. Equating these two expressions for the acceleration, we can solve for the radius of curvature of the path:

$$R = \frac{v^2}{g}. \tag{7–7}$$

To express this result in terms of energy, we note that the total energy E of the baseball is constant and that the gravitational potential energy is $E_p = m_0gy$, where y is the height of the ball above the ground. The velocity of the ball is so small that we may use classical physics. Since the kinetic energy must be $(E - E_p)$, we have

$$E_k = \tfrac{1}{2}m_0v^2 = E - E_p$$

or

$$v^2 = \frac{2(E - E_p)}{m_0}, \tag{7–8}$$

so that R, in terms of energy, becomes

$$R = \frac{2(E - E_p)}{m_0g}. \tag{7–9}$$

Having found R by classical, nonrelativistic means, we next attempt to solve the same problem by considering the refraction of matter waves.

We apply Eq. (7–8) and $u = c^2/v$ to determine the phase velocity of the associated waves of the baseball, and obtain

$$u = c^2\sqrt{m_0/2(E - E_p)} = c^2\sqrt{m_0/2(E - m_0gy)}, \tag{7–10}$$

where $E_p = m_0gy$. It is evident that u is a function of the height y, and that u increases when y increases. This means that the higher the ball, the less is the "refractive index" of the space in which the associated waves move. Let us again compute the radius of curvature that this "refraction" imparts to the trajectory of the ball.

We assume, as before, that for an instant dt the ball moves on the arc of a circle of radius R. As in the study of optical refraction, we shall talk in terms of an infinitesimal wavefront of width dy perpendicular to the "ray." In Fig. 7–1 the phase waves move at a height y above the earth, and they curve concave downward because the top of the wavefront moves faster than the bottom. From the figure we find that

$$d\theta = \frac{u\,dt}{R} = \frac{(u + du)\,dt}{R + dy}. \tag{7–11}$$

To find du, we first put Eq. (7–10) in the form

$$u(E - m_0gy)^{1/2} = c^2\left(\frac{m_0}{2}\right)^{1/2}$$

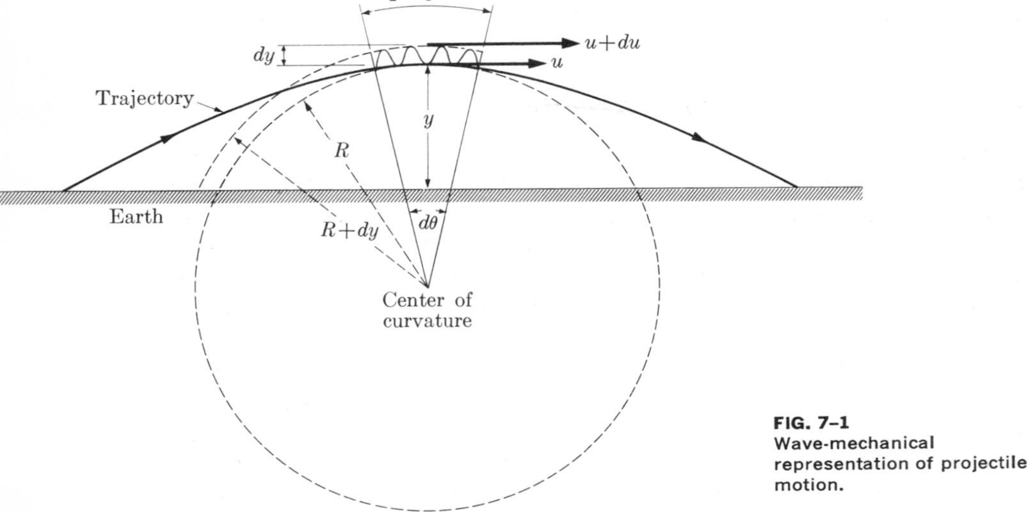

FIG. 7–1
Wave-mechanical representation of projectile motion.

and then differentiate u with respect to y. The result reduces to

$$du = \frac{1}{2}\frac{m_0 g}{E - m_0 g y} \cdot u \cdot dy. \tag{7–12}$$

Substituting this expression for du in Eq. (7–11), we get a relation which simplifies to

$$m_0 g R = 2E - 2m_0 g y. \tag{7–13}$$

Substituting from $E_p = m_0 g y$, we finally have

$$R = \frac{2(E - E_p)}{m_0 g}. \tag{7–14}$$

Agreement between Eqs. (7–9) and (7–14) supports Schroedinger's idea that it may be possible to devise a system of mechanics in which the paths of rays of matter waves replace the classical Newtonian trajectories.

In 1926 Schroedinger announced his system of wave mechanics. (We shall consider wave mechanics in Chapter 8.) In analogy to Maxwell's wave equation which completely describes the wave properties of light, Schroedinger wrote the wave equation which completely describes the wave nature of matter. At first physicists were not excited by the speculations of de Broglie, Schroedinger, and others. There was no real experimental evidence, such as the observation of interference effects that provided the most convincing proof of the wave nature of light. However, in 1925 Elsasser deduced from de Broglie's theory that a beam of electrons diffracted by a crystal should show interference phenomena. This prediction eventually led to the experimental verification of the wave nature of matter.

7-5. THE DAVISSON AND GERMER EXPERIMENT

In 1927 Davisson and Germer were studying the scattering of electrons by nickel. Their technique was reminiscent of both Rutherford alpha-particle scattering and Compton x-ray scattering. They directed a beam of electrons onto a block of nickel and measured the intensity of the electrons as they scattered from the nickel in different directions. In the course of the experiment, their vacuum system broke accidentally and had to be repaired.

When the vacuum system broke the nickel target was at a high temperature, and the air caused the nickel to acquire a heavy coat of oxide. To remove the oxide from the block of nickel, Davisson and Germer reduced the oxide slowly in a high-temperature oven. When their apparatus was reassembled, they began to get very different results. Whereas the number of scattered electrons had previously become continuously less as the scattering angle increased, they now found that the number of electrons went through maxima and minima. *The electrons were being diffracted.* Using the familiar techniques of x-ray diffraction by crystals, Davisson and Germer computed the wavelength their electrons must have, and they found that this wavelength agreed with the de Broglie formula.

The prolonged heating to clean the nickel block had caused it to become a single crystal, and the electron diffraction pattern was completely analogous to x-ray diffraction by the Laue technique. This experiment verified the de Broglie hypothesis and indicated that material particles have wave properties.

To consider the Davisson and Germer experiment in more detail, we show their apparatus schematically in Fig. 7–2. At the right is an electron gun that provides a collimated beam of electrons whose energy is known from the accelerating potential. These electrons were scattered by the nickel target, which could be rotated about an axis perpendicular to the

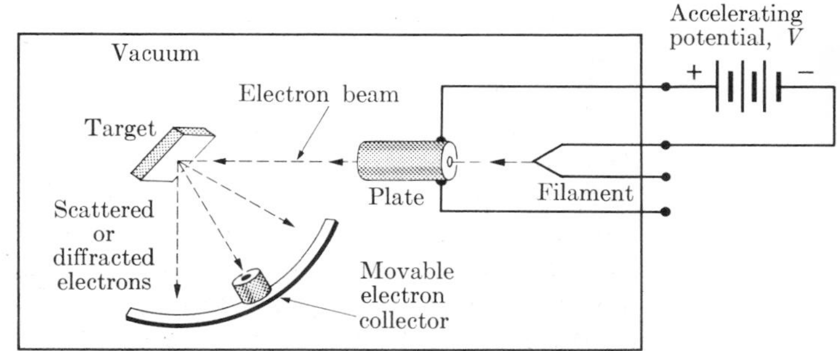

FIG. 7–2
Schematic diagram of the Davisson-Germer electron diffraction apparatus.

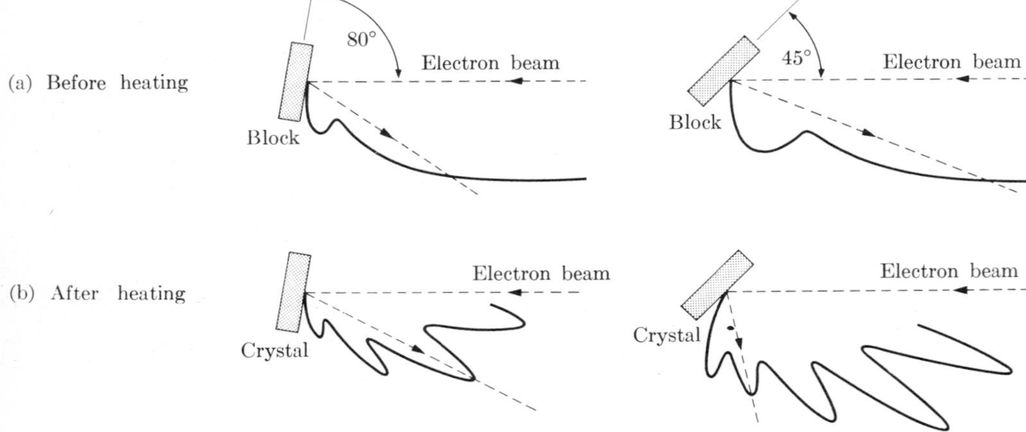

FIG. 7–3
(a) Electron scattering from a block of nickel, and (b) electron diffraction from a single crystal of nickel.

page. The movable electron collector could be swung about the same axis as the target, so that it could receive the electrons coming from the target in any direction included in the plane of the diagram. Figure 7–3 shows the results obtained for two target orientations both before and after the target was heat-treated. The wavelength of the electrons observed experimentally agreed well with the de Broglie wavelength which could be computed from the accelerating potential, V. Since Davisson and Germer used 75-eV electrons, they could obtain the electron velocity from the classical expression

$$E_k = Ve = \frac{m_e v^2}{2}, \tag{7–15}$$

and substitute into the de Broglie relation, obtaining

$$\lambda = \frac{h}{m_e v} = \frac{h}{\sqrt{2Vem_e}}. \tag{7–16}$$

The de Broglie hypothesis was further verified in Germany when Estermann and Stern diffracted helium atoms from a lithium fluoride crystal, and in the United States when Johnson diffracted hydrogen from the same kind of crystal. G. P. Thomson, son of J. J. Thomson, obtained excellent powder diffraction patterns by sending a collimated beam of electrons through very thin sheets of various metals. Figure 7–4(a) shows an electron diffraction pattern of aluminum. For comparison an x-ray

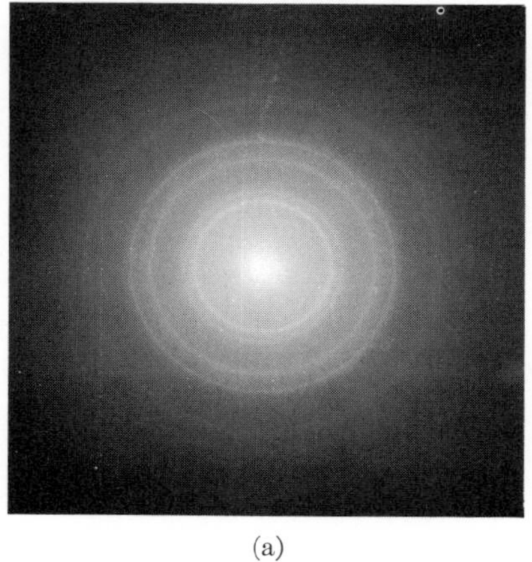

(a)

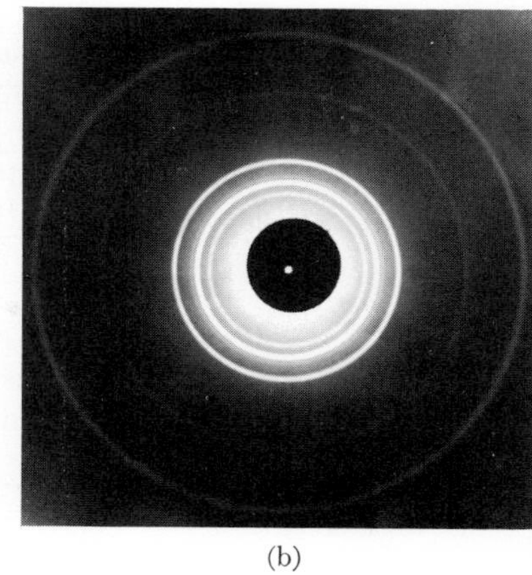

(b)

FIG. 7–4
(a) Electron diffraction pattern of aluminum, using electrons accelerated through 8000 volts. (b) X-ray diffraction pattern of aluminum, using *K*-radiation from copper. The wavelengths and geometry are different for the two photographs. A lead disk was placed over the center of the x-ray film for most of the exposure.

diffraction pattern of aluminum is shown in part (b) of the figure. The effective wavelengths for the two patterns are different but the similarities are evident. Davisson and Germer, using electrons instead of x-rays, repeated Compton's experiment of diffracting soft x-rays from a man-made grating at grazing incidence.

With the complete verification of the de Broglie hypothesis, we have arrived at a point where our atomic world has strange aspects. Compton scattering showed that waves have particle aspects and the de Broglie hypothesis shows that particles have wave characteristics. Another result of all these experiments is contained in Bohr's principle of *complementarity*, which states that in no single experiment does a photon show both wave and particle properties, nor does a particle simultaneously show both particle and wave properties. In Section 7–6 we will attempt to resolve this duality more seriously than Eddington did when he humorously suggested that the primordial entity is really a "wavicle."

7–6. WAVE GROUPS

Although the interference effects compel us to accept the fact that, like light, material particles have the dual nature of waves and particles, it is still difficult to see how the wave extended over a region of space does not conflict with the small particle located at a specific position. This apparent conflict can be resolved by the formation of what is called a wave group or wave packet. In essence this is a wave which extends over a very limited region of space. It is formed by applying the principle of superposition to many waves. When many waves travel through the same medium, the resultant wave is the summation (in our case it will be algebraic) of all the waves. In acoustics such a summation of two waves of slightly different frequencies produces beats, which are fluctuations in the amplitude of the resultant wave. We shall see that when more and more frequencies are added, the regions where the amplitude is large get smaller.

To simplify the discussion of the superposition we will use a one-dimensional wave; that is, a wave whose displacement, y, depends on the position along a line, x, and the time, t. A sinusoidal wave with amplitude A may be written as

$$y = A \sin 2\pi \left(\frac{x}{\lambda} - ft\right) = A \sin (kx - \omega t), \tag{7-17}$$

where we have introduced k, the propagation number which equals $2\pi/\lambda$, and the angular frequency $\omega = 2\pi f$. Even with one-dimensional conditions it is difficult to draw the wave, since there are two independent variables. What we usually do is hold one fixed and vary the other. Thus we either draw the displacement as a function of position at some fixed time (equivalent to a snapshot) or the displacement as a function of time at some fixed position. In the following development, we will use time as the variable. The results would be the same using the position as the variable if t were replaced by x and ω were replaced by k.

The time dependence of a sinusoidal wave with amplitude A can be considered as a component of a vector of length A which is rotating with a constant angular velocity ω, as shown in Fig. 7–5(a). Such a rotating vector is called a *phasor*. Since the sum of the components of many vectors is the component of the vector sum, we can get the time dependence of the sum of many waves by adding many phasors. Let us add n waves with angular velocities between ω_1 and ω_n. To simplify the addition we choose waves with the same amplitude and angular velocities which vary uniformly, so that $\omega_2 - \omega_1 = \omega_3 - \omega_2 = \cdots = \omega_n - \omega_{n-1} = \delta\omega$. Since there are $n - 1$ intervals between ω_1 and ω_n, this interval is

$$\delta\omega = \frac{\omega_n - \omega_1}{n - 1}.$$

If all the phasors are parallel at $t = 0$, they will be spread out uniformly at time t, as indicated in Fig. 7–5(b). The angle between adjacent phasors will be given by

$$\delta\theta = (\delta\omega)t = \left(\frac{\omega_n - \omega_1}{n - 1}\right) t. \tag{7-18}$$

The magnitude of the vector sum is the amplitude of the resulting wave, while the projection of the resultant on the vertical axis is the instantaneous

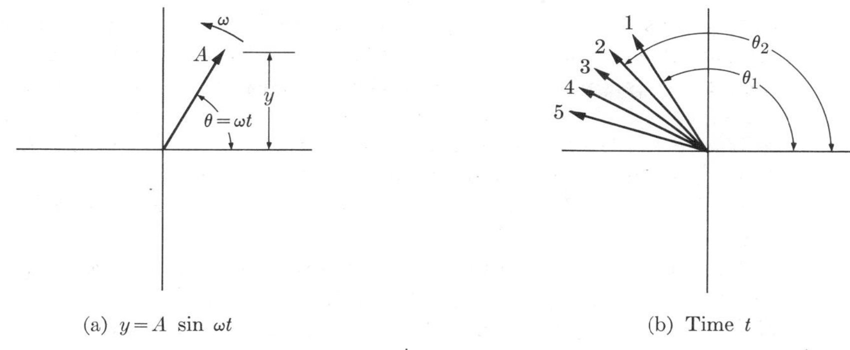

(a) $y = A \sin \omega t$ (b) Time t

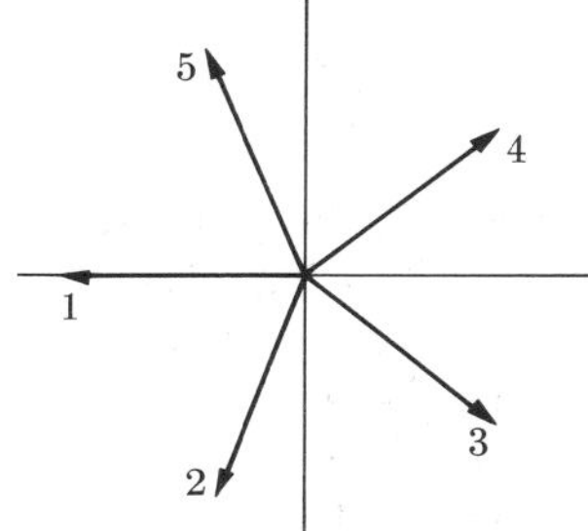

(c) Time t_1, resultant zero

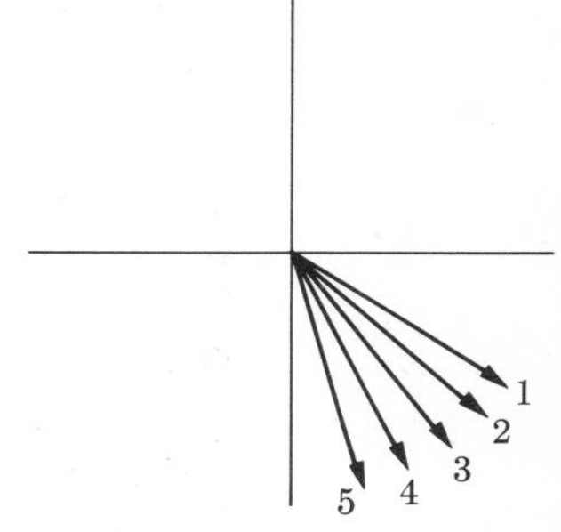

(d) Just before time t_2

FIG. 7-5
(a) Single phasor. (b) Five phasors a short time after $t = 0$. (c) Five phasors with zero resultant. (d) Five phasors a short time before they become parallel.

displacement. We assume that $\delta\omega$ is so small that the amplitude does not change very much during one revolution. The amplitude represents the envelope of the oscillations and we will determine the general shape of this envelope. As the phasors fan out from $t = 0$, the magnitude of the resultant decreases to zero. This first zero occurs when the fan covers the entire angle 2π and the angle between adjacent phasors is $2\pi/n$, as shown in Fig. 7-5(c) for five phasors. If we call the time for the first zero t_1, Eq. (7-18) becomes

$$\frac{2\pi}{n} = \left(\frac{\omega_n - \omega_1}{n - 1}\right) t_1,$$

which can be solved for t_1 to give

$$t_1 = \left(\frac{2\pi}{\omega_n - \omega_1}\right) \frac{(n - 1)}{n}. \tag{7-19}$$

For a while after t_1 the magnitude of the resultant will be small, since the phasors will still be spread out. Eventually, however, the second phasor will catch up to the first and at this time all the other phasors will also be aligned with the first. The resultant will be the same magnitude that it was when $t = 0$. Fig. 7-5(d) shows the phasors just before this occurs. If

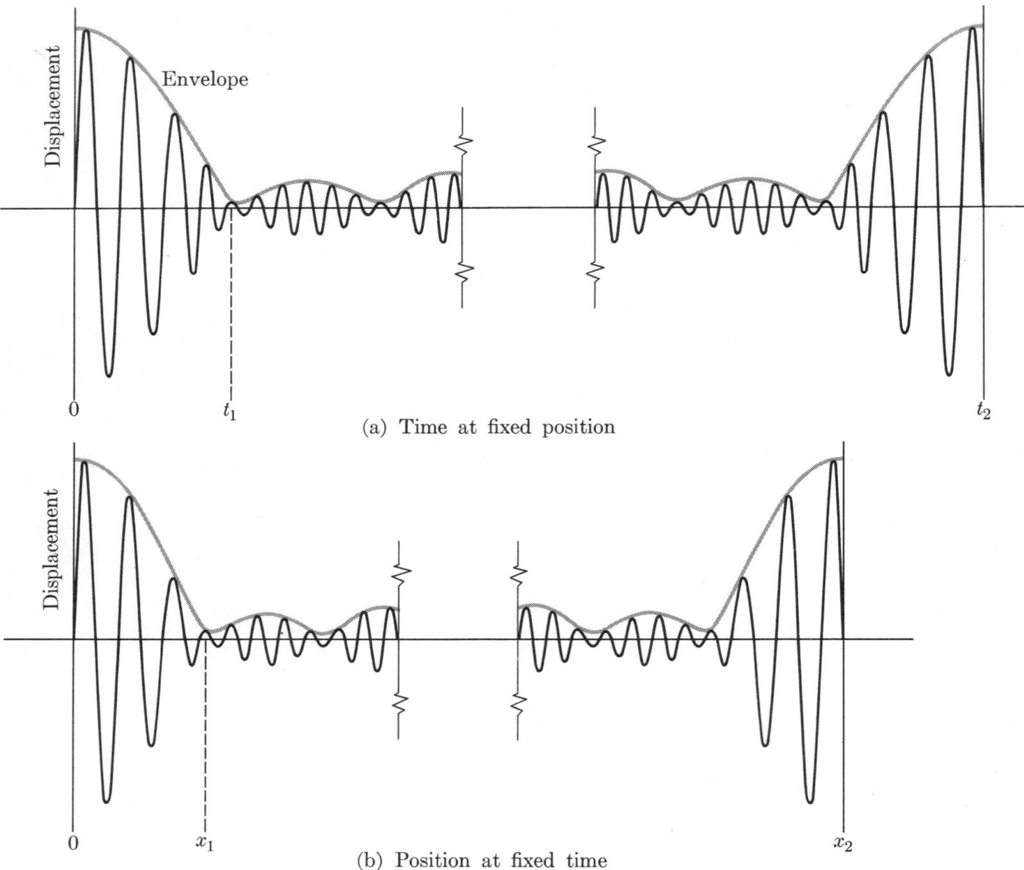

FIG. 7–6
A resultant wave produced by summation of many waves with different frequencies; (a) as a function of time, (b) as a function of position.

we call the time when they are aligned t_2, the angle between adjacent phasors is 2π and Eq. (7–18) becomes

$$2\pi = \left(\frac{\omega_n - \omega_1}{n - 1}\right) t_2,$$

which gives for t_2:

$$t_2 = \left(\frac{2\pi}{\omega_n - \omega_1}\right)(n - 1). \tag{7–20}$$

After time t_2 has elapsed, the envelope will repeat the shape it took between $t = 0$ and $t = t_2$, so that we have the magnitude of the sum of the phasors as a function of time.

The quantitative variation of the envelope of the resulting wave will depend on the number of waves added together; however, the preceding qualitative discussion gives us some idea of how the resultant wave will appear. This is shown in Fig. 7–6(a). The wave has a large amplitude only in limited regions with sections of relatively low amplitude between. The regions of large amplitude are called the *wave groups* and the individual waves which are added together are called the *phase waves.* Remember that we chose to consider time as the variable. If we had varied x at a fixed time, the analysis would have been the same with ω replaced by k. The resultant wave as a function of position is shown in Fig. 7–6(b).

Although we have managed to produce wave groups, there are too many of them to reasonably represent a particle. For the wave to appear like a particle there should be only one group. This can be accomplished if we can get the time between the group maxima, t_2, to be infinite. A consideration of Eq. (7–20) shows that this can be done by having n become infinite. A *single* wave group can be formed by the addition of an infinite number of waves. Note that this does not mean that the frequency difference in the denominator of Eq. (7–20) must be infinite. An infinite number of frequencies can be contained in a finite frequency interval. The "size" of the group in time is $2t_1$, which we call the *time duration* of the group. For an infinite number of waves Eq. (7–19) for t_1 becomes

$$t_1(n = \infty) = \frac{2\pi}{\omega_n - \omega_1}.$$

If we call $\omega_n - \omega_1$ the frequency spread $\Delta\omega$, we have

$$2t_1 = \Delta t = 4\pi/\Delta\omega; \tag{7–21}$$

the time duration of the wave group is determined by the frequency spread. In the same way the spatial size of the group, Δx, is given by

$$\Delta x = 4\pi/\Delta k; \tag{7–22}$$

the spread of propagation numbers determines the size of the group.

Although we have been able to construct a wave group which has the particle-like characteristic of extending over a very limited region of space, there is still the question of the propagation velocity of the group. Each of the phase waves has a phase velocity $u = f\lambda = \omega/k$. In a nondispersive medium all the phase waves have the same velocity and consequently the group will move with this velocity. In a dispersive medium the phase velocities differ and the question of the group velocity is more complicated.

Let us consider two of the phase waves, which can be written as

$$y_1 = A \cos (k_1 x - \omega_1 t)$$

and

$$y_2 = A \cos (k_2 x - \omega_2 t),$$

where $u_1 = \omega_1/k_1$ and $u_2 = \omega_2/k_2$. The functions have been chosen such that the two waves are in phase and maximum at $x = 0$ and $t = 0$. The addition of these two waves can be done by using the trigonometric identity

$$\cos a + \cos b = 2 \cos \tfrac{1}{2}(a - b) \cos \tfrac{1}{2}(a + b).$$

The resultant wave becomes

$$\begin{aligned} y &= y_1 + y_2 \\ &= 2A \cos \tfrac{1}{2}[(k_2 - k_1)x - (\omega_2 - \omega_1)t] \cos \tfrac{1}{2}[(k_1 + k_2)x - (\omega_1 + \omega_2)t]. \end{aligned}$$

If the propagation numbers and frequencies differ only by differential amounts, we have

$$k_2 - k_1 = dk, \qquad \omega_2 - \omega_1 = d\omega,$$

$$\tfrac{1}{2}(k_1 + k_2) \doteq k_1 \quad \text{and} \quad \tfrac{1}{2}(\omega_1 + \omega_2) \doteq \omega_1.$$

The resultant wave is then

$$y = 2A \cos \tfrac{1}{2}[dk\, x - d\omega\, t] \cos (k_1 x - \omega_1 t). \tag{7–23}$$

The second cosine function is the original wave. The coefficient of this cosine can be considered to be an amplitude which varies with x and t. This variation of amplitude is called the *modulation* of the wave and has a long wavelength (π/dk) and a low frequency $(d\omega/\pi)$. From the product of these we can get the velocity of propagation of the modulation, which is the group velocity v_g:

$$v_g = \left(\frac{\pi}{dk}\right)\left(\frac{d\omega}{\pi}\right) = \frac{d\omega}{dk}\cdot \tag{7–24}$$

The group velocity depends on the way in which the frequency varies with the propagation number. Since we have considered only two phase waves, it is possible that v_g will differ for different pairs of phase waves; that is, the group velocity may be a function of frequency. When many waves are added, the derivative of Eq. (7–24) is evaluated for the central frequency of those used in the summation.

There is a simple and instructive demonstration experiment illustrating these results that can be performed with a pair of pocket combs which have slightly different tooth spacings. We have found that a set of black combs is best—one with 12 and the other with about 14 teeth to the inch. If the combs are held in front of a light surface (or on an overhead projector) and superimposed with their teeth parallel, light and dark regions can be observed. These bands are like acoustical "beats" and the number of bands per inch is the difference in the number of teeth in the combs, per inch. The teeth of the two combs represent the phase waves and the light bands of constructive interference are the wave groups. By adding many more combs with different tooth spacings one behind the other, we could form an arrangement such that only one constructive region remained.

If we now slide one comb over the other in a direction perpendicular to the teeth, we observe that the bands move with a velocity different from that of the comb motion. Groups move faster or slower than the phase teeth and have a direction with or against the comb motion, depending on which comb is moved. It is much better to experience this than to read a description.

If the wave group is to represent a particle, then in some way it is necessary that the speed of the group and the speed of the particle be the same. If these speeds differed, the particle would soon be in a region where the amplitude of the wave is negligible and the wave would not give a useful indication of the position of the particle. Using the mass-energy relation from relativity, we can find the particle velocity in terms of its momentum and energy:

$$v = p/m = pc^2/mc^2 = pc^2/E.$$

To get the group velocity in the same terms we use $p = h/\lambda$ and $E = hf$, which are written as

$$k = \frac{2\pi p}{h} \quad \text{and} \quad \omega = \frac{2\pi E}{h}. \tag{7–25}$$

The group velocity can then be written as

$$v_g = \frac{d\omega}{dk} = \frac{dE}{dp}. \tag{7–26}$$

To evaluate this we need the energy-momentum relationship from relativity, Eq. (5–48);

$$E^2 = p^2c^2 + m_0^2c^4.$$

Differentiating this, we obtain

$$2E\,dE = 2pc^2\,dp.$$

The group velocity becomes

$$v_g = \frac{dE}{dp} = \frac{pc^2}{E}. \tag{7–27}$$

This is the same as the above expression for the particle velocity and we see that the choices that were made for the frequency and wavelength to be associated with the particle also lead to the satisfying result that the wave group and the particle have the same speed. Thus it is possible to have a wave motion which has the particle characteristic of being in a small region of space and which will move with the particle's speed.

7–7. WAVE-PARTICLE DUALITY

The wave-group idea is an attempt to resolve the conflict contained in the fact that some explanations of experiments require the use of waves while others require the use of particles. This interpretation means that an electron is a group of matter waves and a photon is a group of electromagnetic waves, that is, there are only waves in nature but they form wave packets which we call particles. As indicated in Section 7–4, the trajectories of the particles are due to the refractions of the waves. Macroscopic objects consist of many of these wave groups and the interactions between objects are simply interactions of waves.

Although this "reality" of the waves seems reasonable, there are some difficulties associated with it. When a wave is incident on a boundary between two media, it generally splits into a reflected wave and a refracted wave. The incident wave group becomes two groups. If the wave group is to be an electron, where the boundary would be produced by a change in electric potential energy, we know that the electron does not split. It is difficult to see how the two wave groups, one reflected and one refracted, can be the single electron. In addition, the Coulomb force, which is stated for point charges and acts between electrons, cannot be handled conveniently with this interpretation.

Another possible interpretation was presented by de Broglie. In this the particles are "real" and the associated waves are pilot waves which guide the particle. These waves are abstract quantities which are to be looked upon as probability waves. The amplitude of these probability waves at a certain position is a measure of the probability of the particle being at that position. The pilot waves are abstract quantities and the word "wave" is used as it is, for instance, in the phrases "wave of enthusiasm" and "crime wave." Since the waves are not directly observable, there is no necessity for a medium. This interpretation removes the difficulty of the wave group concept when the wave is partially reflected and partially refracted. For the probability waves, the amplitude of the reflected wave determines the probability that the particle reflects from the boundary and the amplitude of the refracted wave determines the probability that the particle penetrates the boundary. A single particle will do *only* one or the other, but a large number of particles will divide according to the probabilities.

This pilot wave interpretation, which means that a light beam is a stream of photons, also runs into difficulty. Suppose that we pass a light beam through a pair of slits separated by a small distance, as in the interference experiments of Young. The interference pattern produced on a screen consists of alternate light and dark bands. The positions of these bands are determined by the differences of the paths of the two waves proceeding from the slits to the screen. The alternate light and dark bands represent alternate large and small probabilities of a photon arriving at the respective positions on the screen. If one of the slits is covered, the interference pattern changes, which means the probabilities of where the photon will arrive at the screen changes. For example, the photon may now have a large probability of arriving at a position where the previous probability was very small. If the photons are real, however, they *must* go through one slit or the other. Consequently, as the photon goes through one slit, its motion is influenced by the other slit and the photon is able to know whether the slit through which it did not pass is open or not. This gives some "intelligence" to the particles which does not seem reasonable. We will return to this problem in Section 7–9.

The resolution of the conflict between waves and particles lies in an appraisal of what we mean by a wave and a particle. Both of these terms, when applied to the fundamental entities, are abstractions of the human mind which are arrived at by extrapolation from the macroscopic world of grains of sand and waves on strings. The following is another very clever and useful trick of the human mind. A hollow rubber ball has its center of gravity at its center. Discussion of most, but not all, of the motion of the ball can be greatly facilitated by regarding the ball as a point mass with all its mass at the center of gravity. The center of gravity has no objective reality, and if someone cuts the ball open, points to the center and says, "Ha! You see, there is no mass there," we reply that the center of gravity makes a poor description of what is at the center of the ball, but that it continues to be useful in describing the motion of the ball. No one description of the ball can ever completely represent what the reality of the ball is. In the same way the particle description and the wave description are each incomplete in attempting to describe physical reality.

The mistake of those who say that interference shows that light *is* a wave phenomenon is a verbal mistake that is made every day. We point to a map and say this *is* the United States. What we mean is that this diagram on a piece of paper is a scale representation of many of the physical and political features of the United States. We know that the real United States cannot be folded, rolled up, or burned. We know that the states are not different in color, only a few square inches in area, and completely flat. The map is a clever, useful, elegant model invented by the human mind. More may be learned about some aspects of the United States in one hour of map study than in a lifetime of looking at the real United States, We do not scoff at maps because they are unreal, we admire them as useful descriptions.

Both the wave and the particle are models we have constructed in attempting to describe matter. Quite naturally, we do not expect either model to give a complete description. Some properties, such as interference, are contained in one model, the wave, while other properties, such as mass, are contained in the other model, the particle. The two models complement each other in that together they give a description of matter. Thus we should say that the electrons are waves *and* particles, not waves *or* particles. The same statement can be made about electromagnetic radiation. During an experiment the particular model which is used is determined by the apparatus used.

Even though we admit that waves and particles are not "real," it is very awkward to talk about experimental procedures and results in such a way as to indicate this. Consequently, we will still make statements which seem to imply that particles exist; for example, we say that the intensity of the wave is a measure of the probability of the location of the particle. This is for convenience only. Both the wave and the particle are incomplete models and both are necessary for a description of all the properties of matter which are experimentally determined. With this interpretation, there is no conflict in the dual nature of matter or electromagnetic radiation.

7-8. THE HEISENBERG UNCERTAINTY PRINCIPLE

An important consequence of the wave-particle duality can be developed from the wave group analysis of Section 7-6. In an attempt to get a wave that had a limited extent, we added many waves to form a wave group. If we correlate the wave model to the particle model by assuming that the amplitude of the group measures the probability of the particle being at that position, we see that there is still no certainty in knowing the location of the particle. It could be anywhere in the group.

To decrease the uncertainty in location of the particle, we have to reduce the size of the group, Δx. If we rearrange Eq. (7-22) into

$$\Delta x \, \Delta k = 4\pi, \tag{7-28}$$

the size of the group can be reduced by increasing the spread of propagation numbers, Δk. It appears that we may eliminate the uncertainty in position of the particle by using an infinite spread of propagation numbers. We see from $k = 2\pi p/h$ that the momentum of the particle is determined by the propagation number of the wave. If we use an infinite spread of propagation numbers, we will have an infinite spread in the momentum of the particle. When we decrease the uncertainty of the particle's position, we increase the uncertainty of the particle's momentum. If we put Eq. (7-28) in terms of momentum, we have

$$\Delta x \, \Delta p = 2h.$$

The coefficient of h in this expression was obtained through our simplification of using waves with the same amplitudes and with a uniform spread of frequencies. A different choice of these would give a different number for this coefficient. Consequently, the equation is written as

$$\Delta x \, \Delta p_x \simeq h. \tag{7-29}$$

The subscript is added to the momentum to indicate that it is the momentum associated with the x-displacement. Equation (7–29) is interpreted as indicating that the uncertainty in position of the particle times the uncertainty in the associated momentum is approximately the Planck constant. The words *associated momentum* are used because in the three dimensional case there are also momentum components in the y- and z-directions and with the x-position we use the x-component of momentum. There are equivalent equations for the other directions:

$$\Delta y \, \Delta p_y \simeq h, \tag{7-30}$$

$$\Delta z \, \Delta p_z \simeq h. \tag{7-31}$$

These uncertainties are involved with the nature of matter and are not the same as the uncertainties introduced by the limited precision of some measuring device. In a practical experiment the uncertainties introduced by the equipment will usually be much larger than the ones associated with the wave-particle duality.

We can perform the same analysis with Eq. (7–21), $\Delta t = 4\pi/\Delta\omega$, where the frequency spread determined the time duration of the group as it passed a given position. Using $E = hf = h\omega/2\pi$, we obtain

$$\Delta t \, \Delta E \simeq h. \tag{7-32}$$

This relationship can be interpreted as meaning that the uncertainty of the energy of the particle is dependent on the time interval used, since the particle must be observed for a time Δt to be certain that the particle has passed the point of observation. In a broader sense, if an energy measurement is performed in a time interval Δt, there will be a corresponding minimum uncertainty in the energy given by Eq. (7–32).

This result throws light on a question not considered in the Bohr theory of atomic energy states. An electron in an atom spends most of its time in its unexcited state, and since in this state Δt is large, then ΔE is small and the energy of the electron is well determined. Although the time electrons spend in excited states is usually rather short, it is highly variable. We said in Chapter 4 that phosphorescence is due to some energy states being metastable, in which case Δt may be hours. The longer an electron remains in a particular state, the better is its energy in that state determined. This means that the energy difference between the excited state and the ground state has less uncertainty and there is little spread in the radiated frequency; that is, the spectral line is sharper. Experimentally, the transitions from short-lived excited states to the ground state result in broad intense spectral lines, while the transitions from metastable states result in sharp weak spectral lines. The fact that weak spectral lines are sharper than intense lines agrees with these uncertainties in energies.

These equations concerning uncertainties, that is, Eqs. (7–29), (7–30), (7–31), and (7–32) are variant statements of the *Heisenberg Uncertainty Principle.* We look upon these statements as indicative of the inherent nature of the physical world. Thus it is impossible for us to know an exact position and an exact momentum of a photon or an electron. A precise knowledge of one can be obtained only at the expense of the precision of the other. Frequently, the uncertainty principle is introduced as being due to the act of measurement. For example, when the position of an electron is determined, the measurement introduces the uncertainty in position and momentum because of an interaction between the apparatus and the electron. This implies that the electron had an exact position and momentum before the measurement. We prefer the interpretation that the uncertainties are inherent in the nature of the things we do experiments with. Admittedly, this choice cannot be made on the basis of any physical result. Physics deals with the results of measurements and either interpretation leads to the same uncertainties in measured quantities.

As anticipated in Chapter 1, another philosophical idea affected by the uncertainty principle was that of causality. According to classical theory, the path of a particle is determined by its initial position and momentum and by the forces acting on it. If the forces between the particles of the universe were known and if it were possible to measure, at a given time, the exact position and momentum of every particle, then the past and future positions and momenta of every particle could be calculated. The past and future are completely determined by the information known at an instant of time. The uncertainty principle indicates that we cannot know an exact position and momentum for each particle; we can only determine what the particles will probably do. For macroscopic objects the uncertainties are so small that the probable motion does not differ significantly from classical motion. However, for photons and electrons, the classical prediction of their motion gives a poor idea of the experimental results.

7–9. THE DOUBLE-SLIT EXPERIMENT

To crystallize the ideas we have been discussing about the wave-particle duality and to emphasize the limits on our ability to describe the behavior of the basic entities of the physical world, we take the double-slit diffrac-

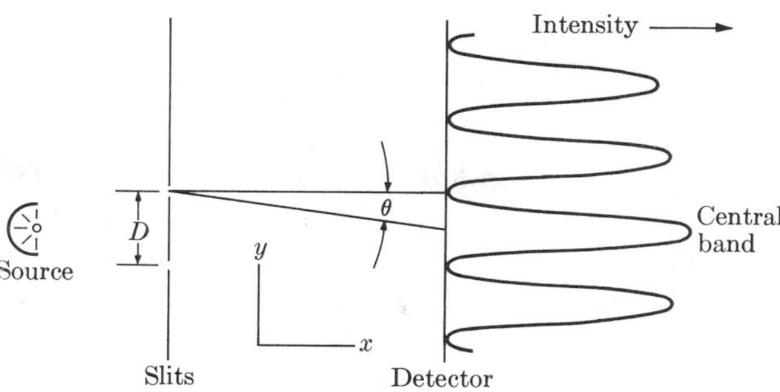

FIG. 7–7
The double-slit diffraction experiment.

tion experiment mentioned in Section 7–7 for an example. A schematic diagram of the experiment is shown in Fig. 7–7. The source emits a beam which is incident on two slits separated by a distance D. The beam could be electromagnetic, that is, a stream of photons, or a beam of cathode rays, that is, a stream of electrons. The following analysis applies to either, but for convenience we shall consider the stream to be photons. At a distance from the slits which is large, compared to D, we have a detector, which could be a photographic film or a series of photoelectric cells.

If the detector is a film, it will show, after exposure, the characteristic interference pattern indicated by the curve labeled intensity on the figure. The positions of the maxima and minima are found by considering the superposition of the wave from one slit with the other; that is, the interference pattern is a prediction of the wave model. The angular separation of the central maximum and the adjacent minimum is determined from

$$D \sin \theta = \lambda/2 = h/2p_x,$$

where λ is the wavelength of the light and p_x is the associated momentum of the photon.

When angle θ is small, it becomes

$$\theta = h/2p_x D, \tag{7–33}$$

and θ is the angle between any adjacent maximum and minimum.

If the detector is a series of photoelectric cells, the light will release photoelectrons and the maxima will indicate those cells where many electrons are emitted. Each photoelectron represents the arrival at the detector of a photon and thus more photons strike the detectors around the maxima in the pattern than around the minima. The pattern gives very little information pertaining to a single photon but only the effect produced by many photons. The correlation between the wave model and the particle model is statistical. The maxima of the wave pattern represent regions of greatest probability for the arrival of a photon.

If one of the slits is covered during the experiment, the resulting single-slit pattern will differ from the preceding one. If a cover is placed over the bottom slit for the first half of the experiment and then the cover is placed over the top slit for the second half, the total exposure at the detector is different than when both slits are uncovered. The positions on the screen, where the photon will most likely strike, change. If we consider the light beam to be a stream of photons, a single photon must pass through either one slit or the other. The fact that the pattern is different for two slits than one slit means that the motion of the photon is influenced by the other slit. It appears that the photon, as it passes through one slit, "knows" whether the other slit is open. Thus the particle model seems to lead to an absurd result.

An analysis of the experiment shows that no information was obtained experimentally about the passage of a photon through either slit. Consequently a statement about such passage is meaningless. This emphasizes that physics deals with predictions for experimental results. The definition of every physical quantity *must* contain, explicitly or implicitly, the operation necessary to measure the physical quantity. It makes *no sense* to talk about results which cannot be observed. If we want to know through which slit the photon passes, we must revise the experiment so that we can determine this. To detect the passage of a photon we place many small particles just to the right of the slits. After passing through a slit, the photon will collide with one of these particles. Observation of the recoil of the small particle will allow us to determine the slit used by the photon. In this ideal experiment the only uncertainties introduced will be those predicted by the uncertainty principle; we will assume that the precision achieved in constructing the apparatus was so high that there are no uncertainties in dimensions, such as the separation of the slits.

To be certain of the slit from which the small particle recoiled, the uncertainty in its y-position must be much smaller than the separation of the slits; we must have

$$\Delta y \ll D.$$

There will be an exchange of momentum during the collision, but there is an uncertainty in the amount since we don't know the details of the collision. The uncertainty in the y-component of the momentum of the photon, Δp_y, must not be great enough to cause it to deviate from the interference pattern. This means that we must have

$$\Delta p_y/p_x \ll \theta = h/2p_xD,$$

or

$$\Delta p_y \ll h/2D.$$

Since momentum is conserved, Δp_y is also the uncertainty in the y-component of the momentum of the small particle. If we multiply the uncertainty in the y-position of the small particle with the uncertainty in its y-component of the momentum, we obtain

$$\Delta y \, \Delta p_y \ll Dh/2D = h/2. \tag{7–34}$$

If we compare this with the prediction of the uncertainty principle, $\Delta y \, \Delta p_y \simeq h$, we see that the small uncertainties necessary in Eq. (7–34) are not possible. If Δy is small enough that we can determine through which slit the photon passes, Δp_y is so large that we do not get the interference pattern at the detector. *We cannot detect both the particle nature, as indicated by the collision with the small particle, and the wave nature, as indicated by the interference pattern.* This example also illustrates how the uncertainty principle is used to resolve apparent conflicts between the wave model and the particle model.

7–10. SUMMARY

The wave-particle duality of matter is substantiated by experiment. Neither the wave model nor the particle model is sufficient alone to include all of the properties of the physical world. The two models are synthesized to complement each other in providing a description for all experimental results. This synthesis is contained in the relations between the wave properties of wavelength and frequency and the particle properties of momentum and energy. The predictions of the two models seem to be correlated statistically; a large wave amplitude means a large probability of the particle being at that position. For material particles like electrons, the particle model has been developed so that, for example, we can calculate electron trajectories in magnetic fields. The corresponding wave model for material particles requires further elaboration. Although we can find the associated wavelength and frequency, we have no method for calculating the amplitude of the "electron wave." Our experience with previous wave motions, such as sound waves, indicates that a wave equation is needed. The solution of this equation would give the amplitude of the wave, which we can correlate with the particle model. A *complete* acceptance of the dual nature of matter requires a further development of the wave model.

REFERENCES

BORN, MAX, *Physics in My Generation.* London and New York: Pergamon Press, 1956. Thought-provoking essays on relativity, wave mechanics, and other fields of modern physics.

CARRUTHERS, P., *Quantum and Statistical Aspects of Light,* Resource Letter QSL-1 and Selected Reprints. New York: American Institute of Physics, 1963.

DE BROGLIE, LOUIS, *Matter and Light.* W. H. Johnston, translator. New York: Dover Publications, 1946.

GAMOW, GEORGE, *Mr. Tompkins Explores the Atom..* New York: Macmillan, 1944. A fantasy describing life in a wave-mechanical atomic world.

MARGENAU, HENRY, *The Nature of Physical Reality.* New York: McGraw-Hill, 1950. Comprehensive and thorough.

MARGENAU, H. and G. M. MURPHY, *The Mathematics of Physics and Chemistry.* Princeton, N. J.: Van Nostrand, 1943.

ROSSI, B., *Optics.* Reading, Mass.: Addison-Wesley, 1957.

SMITH, M. S., *Modern Physics.* New York: Wiley, 1960.

PROBLEMS

7–1. Show that relativistically the wavelength associated with a particle having a rest mass m_0 and a velocity v is

$$\lambda = \frac{h(1 - v^2/c^2)^{1/2}}{m_0 v}.$$

7–2. Show that the de Broglie wavelength in angstroms for an electron accelerated from rest through a potential difference V in volts is (a) classically, $\lambda = 12.27/V^{1/2}$, and (b) relativistically,

$$\lambda = \frac{12.27}{V^{1/2}}\left(\frac{Ve}{2m_e c^2} + 1\right)^{-1/2}.$$

7–3. What is the de Broglie wavelength of the waves associated with an electron which has been accelerated from rest through a potential difference of (a) 100 V? (b) 8000 V?

7–4. Show that the de Broglie wavelength for a particle of mass m moving with the rms speed of a Maxwellian distribution at temperature T is $\lambda = h/(3mkT)^{1/2}$, where k is Boltzmann's constant.

7–5. Calculate the de Broglie wavelength for a neutron moving with the rms speed of a Maxwellian distribution at (a) 20°C. (b) 100°C. (c) At what angle will these neutrons undergo a first-order Bragg reflection from the principal planes of MgO, which is a cubic crystal with a principal grating space of 2.101 Å?

7–6. What is the de Broglie wavelength (a) for a 22-caliber rifle bullet having a mass of 1.1 g and a speed of 3×10^4 cm/s, and (b) for a 75-kg (165-lb) student ambling along at 0.5 m/s? (c) What would have to be the grating space of a plane grating which would diffract the "particles" in parts (a) and (b) through an angle of 30° in the first order assuming normal incidence? (d) Discuss the feasibility of such diffraction.

7–7. An alpha particle (doubly ionized helium) is ejected from the nucleus of a radium atom with 5.78 MeV of kinetic energy. (a) What is the de Broglie wavelength of this particle? (b) How does this wavelength compare with the nuclear diameter, which is about 2×10^{-14} m?

7–8. Using the rotating vector idea of Section 7–6, add four waves with equal amplitudes and different frequencies to obtain the resultant amplitude as a function of time. Assume that $\omega_1 = 2\pi$ rad/s and $\delta\omega = \pi$ rad/s. Plot the amplitude as a function of time for every $\frac{1}{8}$ s and show that the time between groups is 2 s and the time duration of the group is 1 s.

7–9. (a) By actual experiment with two combs, determine whether the group velocity of the waves is greater or less than their phase velocity when the "long-wavelength" comb is moved faster than the "short-wavelength" one. (b) Which is the greater, the group velocity or the phase velocity, (1) in a beam of white light in glass, and (2) in a beam of sound waves having different frequencies in air? (c) Except in free space, the group and the phase velocity of light are identical only for ideally monochromatic light. Is it possible to obtain such light? (Consider Doppler effect in the source of light, filters, finite slit widths of monochromators, etc.)

7–10. A wave group is formed by the addition of an infinite number of waves. The ratio of the angular frequency and propagation number is the phase velocity. (a) Given that the phase velocity is constant show that the group velocity is equal to the phase velocity. (b) Given that the phase velocity is proportional to the propagation number, show that the group velocity is twice the phase velocity.

7–11. Show that the group velocity v_g can be obtained from the phase velocity u by the relation

$$v_g = u - \lambda \frac{du}{d\lambda}.$$

7–12. Under proper conditions the velocity of water waves having the wavelength λ is given by $u = (g\lambda/2\pi)^{1/2}$, where g is the acceleration due to gravity. Show that the group velocity for such waves is one half of the phase velocity.

7–13. (a) Show from Eqs. (7–3) and (7–4) that the phase velocity in free space of the de Broglie waves associated with a moving particle having a rest mass m_0 is given by

$$u = c\sqrt{1 + \left(\frac{m_0 c}{h}\lambda\right)^2}.$$

(b) According to this equation, which wavelengths will have the greater phase velocity, the long ones or the short ones? Will each exceed c? (c) Does the equation indicate that there is dispersion of de Broglie waves in free space? (d) By referring to the experiment in Problem 7–9, show that the answer to part (b) of this problem involves no contradiction in saying that the velocity of a particle is less than c and that the phase velocity of its associated waves is greater than c.

7–14. (a) What is the momentum of a photon of wavelength 0.02 Å. (b) What is the momentum of an electron which has the same total energy as a 0.02 Å photon? (c) What is the de Broglie wavelength of the electron in part (b)?

7–15. Compute the minimum uncertainty in the location of a 2-g mass moving with a speed of 1.5 m/s and the minimum uncertainty in the location of an electron moving with a speed of 0.5×10^8 m/s, given that the uncertainty in the momentum is $\Delta p = 10^{-3}p$ for both.

7–16. Assume that the uncertainty in the location of a particle is equal to its de Broglie wavelength. Show that the uncertainty in its velocity is equal to its velocity.

7–17. The electron in a hydrogen atom moves into the excited state $n = 2$ and remains there for 10^{-8} s before making a downward transition to the ground state. Calculate the uncertainty of the energy in the state $n = 2$. Is this a significant correction to the Bohr theory prediction of —3.39 eV?

7–18. We wish to simultaneously measure the wavelength and position of a photon. Assume that our measurement of wavelength gives $\lambda = 6000$ Å and that our equipment allows an accuracy of one part in a million in the measurement of λ. What is the minimum uncertainty in the position of the photon?

CHAPTER 8

MATTER WAVES

8-1. THE CLASSICAL WAVE EQUATION

The subject of matter waves cannot be treated meaningfully without employing far greater mathematical sophistication than has been necessary heretofore. The reader will encounter functional notation, operators, partial differential equations, probability, and the so-called complex algebra. Although we attempt to explain each mathematical step, our purpose here is to use mathematics rather than to teach it. We hope that if some readers of this book cannot follow every step in the argument, the flavor and elegance of the wave mechanics will, nevertheless, be conveyed.

A development of the wave nature of matter requires that a wave equation be used to find the displacement of the matter wave, which will be a function of position and time. Before the equation for matter waves is developed, let us consider the classical wave equation. In particular, we will consider the equation for transverse waves in a string. This means there will be one position coordinate, x. Since the displacement, y, is a function of two independent variables, x and t, a differentiation of y with respect to x must be a partial differentiation; that is, one for which t is held constant. When y is differentiated with respect to t, x is held constant; thus we expect a partial differential equation for y.

The phase velocity of a traveling wave in the string is determined by the elastic properties of the string. If the string is uniform, this velocity, u, is constant and the wave travels along the string without change of shape. An observer moving parallel to the string with velocity u will see this stationary shape. If the position coordinate used by the moving observer is x', he will say that the displacement of the string is a function of x' but not of t, or that

$$y = f(x').$$

The coordinate x' is related to the original coordinate x by the Galilean-Newtonian transformation from Chapter 5: $x' = x - ut$. Consequently, for an observer at rest with respect to the string, the displacement is the same function of $(x - ut)$ or

$$y = f(x - ut).$$

The sinusoidal wave of Eq. (7–17) was simply a particular function of $(x - ut)$. Since there can also be a traveling wave moving in the opposite direction, this functional relation can be generalized to

$$y = f(x \pm ut). \tag{8–1}$$

(The use of the relativistic transformation would not change this result.)

We can use Eq. (8–1) to obtain the partial differential equation for y. Some partial derivatives of y with respect to x are

$$\frac{\partial y}{\partial x} = \frac{\partial f}{\partial x} = \frac{\partial f}{\partial(x \pm ut)} \frac{\partial(x \pm ut)}{\partial x} = \frac{\partial f}{\partial(x \pm ut)}$$

and

$$\frac{\partial^2 y}{\partial x^2} = \frac{\partial}{\partial x}\left[\frac{\partial f}{\partial(x \pm ut)}\right] = \frac{\partial^2 f}{\partial(x \pm ut)^2} \frac{\partial(x \pm ut)}{\partial x} = \frac{\partial^2 f}{\partial(x \pm ut)^2}.$$

Differentiating with respect to t, we have

$$\frac{\partial y}{\partial t} = \frac{\partial f}{\partial t} = \frac{\partial f}{\partial(x \pm ut)} \frac{\partial(x \pm ut)}{\partial t} = \frac{\partial f}{\partial(x \pm ut)}(\pm u) = \pm u \frac{\partial f}{\partial(x \pm ut)}$$

and

$$\frac{\partial^2 y}{\partial t^2} = \pm u \frac{\partial^2 f}{\partial(x \pm ut)^2} \frac{\partial(x \pm ut)}{\partial t} = u^2 \frac{\partial^2 f}{\partial(x \pm ut)^2}.$$

A second-order partial differential equation, which includes both directions of the velocity, can be formed from the two second-order derivatives obtained above:

$$\frac{\partial^2 y}{\partial x^2} = \frac{1}{u^2} \frac{\partial^2 y}{\partial t^2}. \tag{8–2}$$

This wave equation can also be derived by considering the forces acting on a differential element of the string and its resulting acceleration. The solutions of Eq. (8–2) are the possible displacements of the string. A *particular solution* is determined by the initial displacement of the string in conjunction with any imposed boundary conditions. Thus, if the string is fixed at two points, the displacements at these two points must always be zero. The particular solution is then the usual standing wave. Even though we obtained Eq. (8–2) from a traveling wave, we see there are other types of waves possible. Since the equation is linear, any linear combination of solutions will also be a solution. This implies the *principle of superposition* and gives the usual method of forming a standing wave by a linear combination of traveling waves moving in opposite directions.

A review of the development of Eq. (8–2) will show that use of a string was not vital to the argument. The equation is called the classical wave equation and applies to the one-dimensional wave motion of strings, air columns, etc. Many of the properties which are associated with such waves are a result of the form of Eq. (8–2). We seek the corresponding equation for the waves which are associated with material particles.

8–2. THE SCHROEDINGER EQUATION

Since the concept of matter waves is not a result of previous physical theories, it is *impossible* to derive the corresponding wave equation for a particle. We are dealing with a new field of physics and cannot expect the basis for this field to be dependent on the bases of other fields. We are using the wave model, however, and the equation is developed in a manner analogous to that used for other waves.

The dependent variable of a matter wave is called the wave function and is denoted by Ψ (called capital psi). For one-dimensional problems the wave function is a function of the coordinates x and t. Analogous to the classical wave, we may expect that for a traveling matter wave, Ψ will be a function of $(x - ut)$. Since the velocity of the wave, u, is the ratio of the circular frequency and the propagation number, $u = \omega/k$, the wave function can be written as a function of $(kx - \omega t)$. From Eq. (7–25) we recall that the propagation number of the wave is proportional to the momentum of the particle and that the angular frequency is proportional to the energy; therefore we can write

$$\Psi = f\left[\frac{2\pi}{h}(px - Et)\right] = f\left(\frac{px - Et}{\hbar}\right),$$

where $\hbar$ (called h-bar) stands for $h/2\pi$. We may select as our function a sine or cosine wave. A more general wave would be a sum of a sine and cosine wave. By taking a certain combination of sine and cosine, we can use de Moivre's theorem* to express Ψ in exponential form:

$$\Psi = A \exp\left[\frac{i}{\hbar}(px - Et)\right]. \qquad (8\text{–}3)$$

This exponential form is very convenient when derivatives are to be found.

To obtain a differential equation for Ψ, we seek a combination of partial derivatives for which Ψ is a solution. We assume that the energy and momentum of the particle are constant. If Eq. (8–3) is differentiated with

* De Moivre's theorem is $e^{i\theta} = \cos\theta + i\sin\theta$, where i is the imaginary number $\sqrt{-1}$. This has the properties $i^2 = -1$ and $i^{-1} = -i$.

respect to x, we obtain

$$\frac{\partial \Psi}{\partial x} = A \frac{ip}{\hbar} \exp\left[\frac{i}{\hbar}(px - Et)\right] \qquad \text{or} \qquad \frac{\partial \Psi}{\partial x} = \frac{ip}{\hbar}\Psi,$$

which can be rearranged as

$$p\Psi = \frac{\hbar}{i}\frac{\partial \Psi}{\partial x} = -i\hbar \frac{\partial \Psi}{\partial x}. \tag{8–4}$$

This relationship is interpreted as meaning that the momentum p may be replaced by the differential operator $-i\hbar(\partial/\partial x)$, which operates on the wave function. Differentiation of Eq. (8–3) with respect to t gives

$$\frac{\partial \Psi}{\partial t} = -\frac{iE}{\hbar} A \exp\left[\frac{i}{\hbar}(px - Et)\right] = -\frac{iE}{\hbar}\Psi,$$

which is rearranged to give

$$E\Psi = i\hbar \frac{\partial \Psi}{\partial t}. \tag{8–5}$$

Thus the differential operator $i\hbar(\partial/\partial t)$, when operating on Ψ, is equivalent to the energy. Although Eqs. (8–4) and (8–5) were obtained for constant momentum and energy, we assume that they are valid even when p and E are not constant.

The partial derivatives with respect to x and t are connected by means of the relation between the energy and the momentum. The classical expression for the kinetic energy, in terms of the momentum, is

$$E_k = \frac{mv^2}{2} = \frac{(mv)^2}{2m} = \frac{p^2}{2m}, \tag{8–6}$$

where m is the rest mass of the particle. The total energy and momentum are related by the expression for energy conservation of a particle, which is

$$\frac{p^2}{2m} + V = E, \tag{8–7}$$

where V is the potential energy, which is a function of x. This step restricts us to a nonrelativistic wave equation. To obtain the wave equation, Eq. (8–7) is multiplied by Ψ and the differential operators are substituted for p and E. When this is done, we obtain

$$\boxed{-\frac{\hbar^2}{2m}\frac{\partial^2 \Psi}{\partial x^2} + V\Psi = i\hbar \frac{\partial \Psi}{\partial t}.} \tag{8–8}$$

This is the Schroedinger equation for one-dimensional matter waves. It was obtained with the use of analogies to other wave equations and solutions. Since the wave aspects of particles are not a result of previous properties, this equation cannot be derived. The test of the Schroedinger equation is whether its predictions agree with experimental results. To apply this equation to a particular problem, we need to know the particle mass and the potential function.

Let us compare the Schroedinger equation with the classical wave equation,

$$\frac{\partial^2 y}{\partial x^2} = \frac{1}{u^2}\frac{\partial^2 y}{\partial t^2}.$$

Both equations are linear partial differential equations. This means that the principle of superposition is valid and therefore that a sum of solutions is also a solution. This is necessary in order to have interference phenomena. Equation (8–8) differs from the classical wave equation by having a term with no derivative, $V\Psi$, and having a first-order time derivative. In addition the Schroedinger equation contains i, which makes it a complex equation. This means that Ψ is a complex function, but of real variables x and t. Because the two wave equations have these differences, we cannot expect the analogy between them to be maintained. In particular, the complex nature of Ψ indicates that Ψ cannot be observed like the displacement of a string and also means that we should not look for a medium which transmits our matter waves. There is no "ether" for matter waves, as opposed to the existence of, say, water for water waves.

8–3. INTERPRETATION OF THE WAVE FUNCTION

As with other wave motions, the wave equation we have obtained must be supplemented by additional conditions in order that the solutions have physical significance. These conditions are that Ψ and $\partial\Psi/\partial x$ must be single-valued, finite, and continuous. For brevity we will call a function that satisfies these conditions a *well-behaved function*. A comparison with the string shows that these conditions are reasonable. At some specific values of x and t there must be only one value of the displacement, and it cannot be infinite. Because the energy of the string is proportional to the square of the amplitude, an infinite amplitude would require infinite energy. The need for continuity of the displacement of the string is obvious. The fact that similar conditions are placed on $\partial\Psi/\partial x$ is also reasonable and eliminates the possibility of a kink in the string.

Now that we have a wave equation with its associated conditions requiring it to be well-behaved, we must decide on an interpretation of Ψ. The discussion in Chapter 7 indicates that the intensity of the wave should provide a prediction of the probable location of the particle. Since the position coordinate x varies continuously, we will have a probability density as we did in the Maxwellian speed distribution in Chapter 1. We shall consider the probability of the particle being located in an interval between x and $x + dx$. In general, the probability is time-dependent and the probability for this location is written as $P(x, t)\,dx$, where $P(x, t)$ is the *probability distribution function*. This probability must be real, whereas the solutions to the Schroedinger equation are complex; consequently Ψ cannot directly be used to give the probability. To obtain a real quantity we postulate that the probability that a particle is located within the interval between x and $x + dx$ is

$$P(x, t)\,dx = \Psi^*(x, t)\Psi(x, t)\,dx, \tag{8–9}$$

where Ψ^* (called psi-star) is the *complex conjugate* of Ψ. For the functions that we will deal with, the complex conjugate can be obtained by reversing the sign of any imaginary, i, that may be in the function. The product $\Psi^*\Psi$ is real; it is also written as $|\Psi|^2$, where the bars mean absolute value. This is similar to other waves where the intensity is proportional to the square of the amplitude. The probability that the particle is in a finite interval between x_1 and x_2 is found by summing the probabilities, as indicated by the integral:

$$\text{probability between } x_1 \text{ and } x_2 = \int_{x_1}^{x_2} \Psi^*\Psi\,dx. \tag{8–10}$$

This probability may be time dependent, since the wave function is time dependent. The interpretation of $\Psi^*\Psi$ as a probability density leads to another restriction on the wave function. The particle must be *somewhere* in the entire range of x; therefore the probability of the particle being in this range must be unity, which means that

$$\int_{-\infty}^{\infty} \Psi^*\Psi\,dx = 1. \tag{8–11}$$

When this restriction has been satisfied, the solution is said to be *normalized*.

A well-behaved and normalized solution of the Schroedinger equation for a specific system determines the probability distribution function for the system. This probability density is not directly observed experimentally for a single particle. To show this and to introduce some of the uses of probabilities, we consider a discrete system; that is, a system with a discrete number of possible results, each with a certain probability. Suppose that we use a loaded coin for which the probability of obtaining a head is twice that of obtaining a tail when the coin is tossed. If this probability distribution is normalized, the probability of a head is $\frac{2}{3}$ and the probability of a tail is $\frac{1}{3}$. The problem is to determine how these probabilities will affect experiments performed with the coin. We certainly do not expect to obtain a tail on every third toss. We would not be surprised if we didn't obtain four tails in twelve tosses. As the number of tosses increases, however, we expect that approximately one-third of them will be tails, and that the approximation will improve with the number of tosses. A quantitative test of the approximation is to give each possibility a numerical value and then to calculate the average number obtained in a series of tosses. If we let x represent the value of a toss and we give a tail the value of one and a head the value of two, the average of N tosses is

$$\overline{x} = \frac{\sum_{i=1}^{2} x_i N_i}{\sum_{i=1}^{2} N_i} = \frac{(1)N_1 + (2)N_2}{N_1 + N_2},$$

where N_1 is the number of tails obtained and N_2 is the number of heads, so that $N_1 + N_2 = N$. As N increases, we expect this average value to approach the average value calculated on the assumption that the results coincide with the prediction of the probabilities. We shall call this the *expectation value* and denote it by $\langle x \rangle$. The expectation value is

$$\langle x \rangle = \frac{\sum_{i=1}^{2} x_i P_i}{\sum_{i=1}^{2} P_i} = \sum_{i=1}^{2} x_i P_i,$$

where P_i is the probability of obtaining the value x_i and $\sum_{i=1}^{2} P_i = 1$ when the probabilities are normalized. For our coin $P_1 = \frac{1}{3}$ and $P_2 = \frac{2}{3}$, so that we have

$$\langle x \rangle = (1)(\tfrac{1}{3}) + (2)(\tfrac{2}{3}) = \tfrac{5}{3}.$$

We expect the average value of a large number of tosses to be very close to $\frac{5}{3}$. (Note that the expectation value cannot be observed experimentally in a single toss of the coin.) A knowledge of P_i also allows us to calculate the expectation value of any function of x, in the same manner:

$$\langle f(x) \rangle = \sum_{i=1}^{2} f(x_i) P_i.$$

As an example, let us calculate the expectation value of x^2 for the coin. We have

$$\langle x^2 \rangle = \sum_{i=1}^{2} (x_i)^2 P_i = (1)^2(\tfrac{1}{3}) + (2)^2(\tfrac{2}{3}) = 3.$$

Note that $\langle x^2 \rangle$ for this system is not equal to the square of $\langle x \rangle$.

It is entirely possible to encounter probability distributions which have the same $\langle x \rangle$ but different $\langle x^2 \rangle$. The value of $\langle x^2 \rangle - \langle x \rangle^2$ indicates the shape of the distribution—low values indicate a sharp distribution and large values indicate a broad one. A distribution which has no spread is one for which the probability of a certain value is one and the probabilities of all other values are zero. If the value x_j has $P_j = 1$, we have

$$\langle x \rangle = x_j \qquad \text{and} \qquad \langle x^2 \rangle = x_j^2.$$

Thus for a distribution which permits only one value to be observed, we see that

$$\langle x^2 \rangle = \langle x \rangle^2. \tag{8-12}$$

This equation is usually used as a test to see if the probability distribution corresponds to a system where only one value of the variable is observed.

When there is a continuous range of possible values, we must use a probability distribution function, and the probability of obtaining a value between x and $x + dx$ is given by $P(x)\,dx$. In the expressions for expectation values the sums are replaced by integrals and we have

$$\langle x \rangle = \int_{-\infty}^{\infty} xP(x)\,dx$$

and

$$\langle f(x) \rangle = \int_{-\infty}^{\infty} f(x)P(x)\,dx.$$

This applies to the situation where we have a solution to the Schroedinger equation with $P(x, t) = \Psi^*\Psi$. When calculating expectation values with this probability distribution, we choose to write the function between Ψ^* and Ψ. The expressions for expectation values become

$$\langle x \rangle = \int_{-\infty}^{\infty} \Psi^* x \Psi\,dx \tag{8-13}$$

and

$$\langle f(x) \rangle = \int_{-\infty}^{\infty} \Psi^* f(x) \Psi\,dx. \tag{8-14}$$

The wave function does not give the position of the particle but only permits us to compute the expectation value of the position. For the wave group constructed in Chapter 7, this corresponds to the center of the group and is the average value of the position after many measurements have been taken. When we say that the wave function gives a complete description of the particle, we mean that we can calculate the expectation value of any physical quantity that can be measured. Such quantities are called *physical observables.* A necessary consequence of this is that expectation values of physical observables must be real, since physical measurements give real numbers.

Two physical observables of interest, in addition to position, are momentum and energy. The expectation value of the momentum of a particle is given by

$$\langle p \rangle = \int_{-\infty}^{\infty} \Psi^* p \Psi\,dx.$$

This integration cannot be performed unless the momentum is expressed as a function of x. We already used a function of x to replace the momentum in our development of the Schroedinger equation. According to Eq. (8–4) we replace p with the operator $-i\hbar(\partial/\partial x)$. Our decision to place the variable between Ψ^* and Ψ is due to the fact that we must use operators. For the momentum the partial differentiation is performed on Ψ only. The expectation value of momentum is

$$\langle p \rangle = \int_{-\infty}^{\infty} \Psi^* \left(-i\hbar \frac{\partial}{\partial x}\right) \Psi \, dx. \tag{8–15}$$

For the energy we have the operator from Eq. (8–5); thus the expectation value of the energy is

$$\langle E \rangle = \int_{-\infty}^{\infty} \Psi^* \left(i\hbar \frac{\partial}{\partial t}\right) \Psi \, dx. \tag{8–16}$$

If we were interested in other physical observables, such as angular momentum, we would have to select equivalent operators for them, either by analogy or as some combination of previous operators. All of these expectation values may be dependent on time but they are not dependent on position.

Let us review the use of the Schroedinger equation. The system in which the particle is contained is described by a potential function $V(x, t)$. For this potential a well-behaved solution of the partial differential equation is found. This solution Ψ, when normalized, gives the probability distribution function $\Psi^*\Psi$ from which the expectation value of any physical observable can be determined. The expectation values are related to the average values obtained from experiments on the system. Since we can predict all experimental results from Ψ, we say that it provides a complete description of the motion. If the expectation value of the square of a physical observable is equal to the square of the expectation value of that observable, a measurement of the observable will always yield the expectation value.* A wave function which yields this result represents a state of the system which can be said to have a definite value of the observable. If this is not true, there is a spread in the distribution and the system does not have a definite value of the observable.

8–4. THE TIME-INDEPENDENT EQUATION

Before we consider a specific potential function we shall try a standard technique for solving partial-differential equations. The technique is to see if there is a solution $\Psi(x, t)$ which is a product of a function of x, say

* Strictly, the condition is that $\langle x^n \rangle$ must equal $\langle x \rangle^n$ for all values of n. In practice the value $n = 2$ is used.

$\psi(x)$ (called small psi), and a function of t, say $\phi(t)$. We assume that the solution can be written as

$$\Psi(x, t) = \psi(x)\phi(t) \tag{8–17}$$

and insert this in Eq. (8–8) to obtain

$$-\frac{\hbar^2}{2m}\frac{\partial^2}{\partial x^2}(\psi\phi) + V\psi\phi = i\hbar\frac{\partial}{\partial t}(\psi\phi).$$

In each differentiation the function of the other variable can be placed in front of the derivative and then the partial derivative can be replaced by an ordinary derivative, since it operates on a function of the variable only. This gives

$$-\frac{\hbar^2}{2m}\phi\frac{d^2\psi}{dx^2} + V\psi\phi = i\hbar\psi\frac{d\phi}{dt}\cdot$$

This equation is divided by $\Psi = \psi\phi$ and we have

$$-\frac{\hbar^2}{2m}\frac{1}{\psi}\frac{d^2\psi}{dx^2} + V = i\hbar\frac{1}{\phi}\frac{d\phi}{dt}\cdot \tag{8–18}$$

We now impose the restriction that the potential energy V is a function of x only. This restriction means that the entire left-hand side of Eq. (8–18) is a function of x only while the right-hand side is a function of t only. Since x and t are independent variables, the only way a function of x can always equal a function of t is to have both functions equal to a constant. Otherwise, the two functions would vary independently and even if they were equal for certain values of x and t, they would not be equal for all values of x and t. The constant that each side of Eq. (8–18) must equal is called the *separation constant* W and we obtain two equations:

$$-\frac{\hbar^2}{2m}\frac{1}{\psi}\frac{d^2\psi}{dx^2} + V = W \qquad \text{and} \qquad i\hbar\frac{1}{\phi}\frac{d\phi}{dt} = W.$$

We have separated the partial differential equation into two ordinary differential equations which can be written as

$$\boxed{-\frac{\hbar^2}{2m}\frac{d^2\psi}{dx^2} + V\psi = W\psi,} \tag{8–19}$$

called the *time-independent* Schroedinger equation, and

$$\boxed{i\hbar\frac{d\phi}{dt} = W\phi.} \tag{8–20}$$

Thus letting $\Psi = \psi\phi$ enables us to obtain two ordinary differential equations to be solved.

Since we restricted the potential to be a function of x, it does not appear in Eq. (8–20); therefore we can solve for the time dependence of the product solution without specifying the potential. If Eq. (8–20) is written as

$$\frac{d\phi}{dt} = -\frac{iW}{\hbar}\phi,$$

it is similar to Eq. (6–8) for the absorption of x-rays. The solution is

$$\phi = e^{-iWt/\hbar}. \tag{8–21}$$

Ordinarily, there would be a constant before the exponential. Since ϕ is multiplied by ψ, we can absorb this constant into ψ. Equation (8–21) gives the time dependence for all product solutions, $\psi\phi$, and we have

$$\Psi = \psi e^{-iWt/\hbar}. \tag{8–22}$$

Because of the similarity to sinusoidal waves, ψ is called the amplitude and the differential equation for ψ, Eq. (8–19), is also called the *amplitude equation.*

Since ϕ must be well-behaved, the separation constant W is restricted to a real number. If W were complex, the exponential would have a real part and would approach infinity as t approached infinity. To see what the separation constant represents we calculate the expectation value of the energy from Eq. (8–16) and obtain

$$\begin{aligned}\langle E\rangle &= \int_{-\infty}^{\infty} \Psi^*\left(i\hbar\frac{\partial}{\partial t}\right)\Psi\,dx \\ &= \int_{-\infty}^{\infty} \psi^* e^{iWt/\hbar}\left(i\hbar\frac{\partial}{\partial t}\right)\psi e^{-iWt/\hbar}\,dx.\end{aligned}$$

Performing the differentiation, we have

$$\langle E\rangle = \int_{-\infty}^{\infty} \psi^* e^{iWt/\hbar} W \psi e^{-iWt/\hbar}\,dx = W\int_{-\infty}^{\infty} \Psi^*\Psi\,dx.$$

Since the wave function is normalized, the last integral is unity and we have

$$\langle E\rangle = W.$$

Using the fact that the square of an operator means the operator applied successively, we can also obtain

$$\langle E^2\rangle = W^2,$$

which means that $\langle E^2 \rangle = \langle E \rangle^2$. Since this is the requirement for a probability distribution to yield a single value of the observable, the value W will always be obtained from a measurement of the energy of the system. (This statement cannot be made just from the fact that $\langle E \rangle = W$.) *A product solution corresponds to a state of the system with a definite energy W.* Values of the separation constant for which the solutions are well-behaved are the possible energies of the system. The probability distribution function for these solutions is independent of time, since we have

$$\begin{aligned} P &= \Psi^*\Psi = \psi^* e^{iWt/\hbar}\psi e^{-iWt/\hbar} \\ &= \psi^*\psi. \end{aligned}$$

This means that the expectation value of any function of x will be constant. States which have this time-independent probability distribution are called *stationary states.* The wave function of these stationary states has the exponential time dependence of Eq. (8–21). The position-dependent part of the wave function must be found from the amplitude equation, which requires knowledge of the potential distribution. Since the amplitude equation has only real terms in it, we shall be able to find real solutions for ψ. This is convenient because it will permit us to graph the solutions and thus obtain a picture of the amplitude as well as the probability distribution.

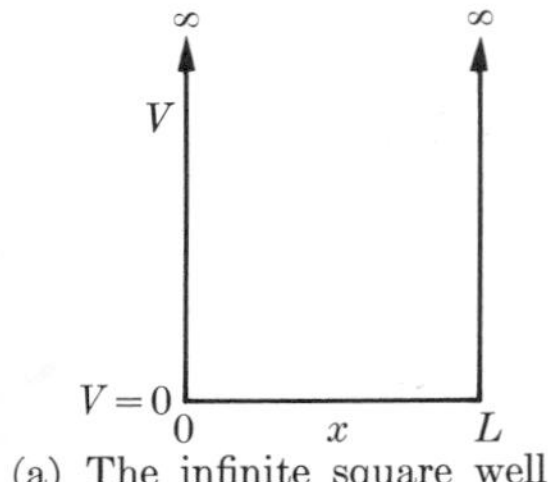

(a) The infinite square well

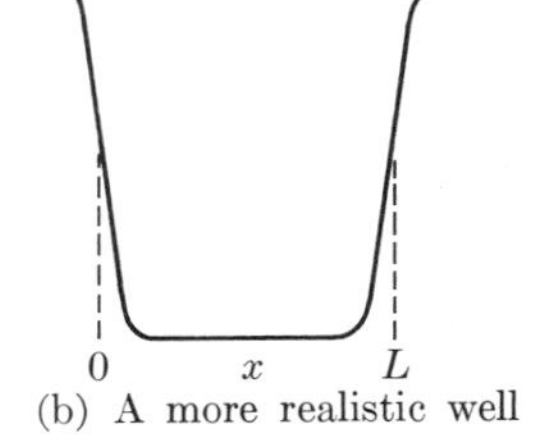

(b) A more realistic well

FIG. 8–1
One-dimensional potential energy distributions.

8–5. THE INFINITE SQUARE WELL

In order to show some of the characteristic features of the solutions to the amplitude equation without the complication of a difficult differential equation, we choose a simple potential distribution which is an idealization similar to the rigid walls of mechanics. The particle is restricted to a region ranging from $x = 0$ to $x = L$, where the potential is constant and most conveniently taken to be zero. Outside this region the potential is very large and we shall simplify our problem by taking it to be infinite. This potential distribution is shown in Fig. 8–1(a). It is called a square well because the potential rises vertically from the horizontal potential inside the well. This is an approximation for a more realistic potential, such as that shown in part (b) of the figure. Our task is to solve the time-independent Schroedinger equation for the amplitude function ψ for the infinite square well.

The potential walls of the well divide the problem into three parts: two outside the well and one inside the well. For the regions outside the well the infinite potential causes the second term of the amplitude equation,

$$-\frac{\hbar^2}{2m}\frac{d^2\psi}{dx^2} + V\psi = W\psi,$$

to be infinite unless $\psi = 0$. If this term is infinite, the equation can be satisfied in these regions only if the first term is also infinite. A function with an infinite second derivative is not a well-behaved function; therefore the solution for $x < 0$ and $x > L$ must be $\psi = 0$. This zero probability of being outside the well corresponds to the classical rigid wall where the particle cannot penetrate the wall. The continuity condition requires that the solution inside the well must be zero at the walls. There will be discontinuity of the slope at the walls, but this is due to the practical impossibility of the potential we are using.

The amplitude equation inside the well, where the potential is zero, is

$$-\frac{\hbar^2}{2m}\frac{d^2\psi}{dx^2} = W\psi. \tag{8–23}$$

The boundary conditions at the walls are

$$\psi = 0 \qquad \text{at} \qquad x = 0, L. \tag{8–24}$$

The differential equation can be written as

$$\frac{d^2\psi}{dx^2} = -k^2\psi, \tag{8–25}$$

where

$$k^2 = \frac{2mW}{\hbar^2}. \tag{8–26}$$

The general solution of Eq. (8–25) is

$$\psi = A \sin kx + B \cos kx.$$

The boundary condition at $x = 0$ gives

$$0 = A \sin (0) + B \cos (0) \qquad \text{or} \qquad B = 0.$$

At the other wall, $x = L$, we have

$$0 = A \sin kL.$$

Except for the trivial case when $A = 0$ (and thus $\psi = 0$), this equation is satisfied for the values of k given by

$$kL = \pi n, \qquad n = 1, 2, 3, \ldots,$$

or

$$k = \pi n/L, \qquad n = 1, 2, 3, \ldots \tag{8–27}$$

From Eq. (8–26) these discrete values of k lead to discrete values of the energy:

$$W_n = \frac{\hbar^2 \pi^2 n^2}{2mL^2}, \qquad n = 1, 2, 3, \ldots, \tag{8–28}$$

where we have used the integer n, called the quantum number, as a subscript on W to indicate this discrete character. The only constant to be determined is A, for which we use the normalization condition, Eq. (8–11):

$$\int_0^L A^2 \sin^2 \left(\frac{n\pi x}{L}\right) dx = 1.$$

For all values of n the value of A is $(2/L)^{1/2}$. The final solutions are

$$\psi_n = \sqrt{2/L} \sin (n\pi x/L), \qquad n = 1, 2, 3, \ldots \tag{8–29}$$

The first three energy levels and amplitude functions are shown in Fig. 8–2.

These results for the infinite square well show one of the most significant features of wave mechanical solutions. When a particle is in a bounded system, there are discrete energy levels. These quantized energy levels enter the problem naturally and are a result of the application of the condition that the solutions of the Schroedinger equation be well-behaved. Classically, a particle inside a one-dimensional box with rigid walls could have any positive energy. This disagreement between wave and classical mechanics is also evident in the probability distribution functions. These are shown in Fig. 8–3 for the lowest three states.

Classically, the particle passes back and forth between the walls and it has an equal probability of being found anywhere between $x = 0$ and $x = L$. The classical probability function is constant with a value of $1/L$. The quantum functions show peaks of magnitude $2/L$ and valleys where the probability is very small. The number of peaks is equal to the quantum number, n. As n gets very large, the number of peaks becomes large and they get closer together. The average of the distribution is $1/L$ and when n is very large, the distribution approaches the classical distribution. This is an example of the correspondence principle introduced in Problem 4–27. The discrete character of the energy levels disappears also as n gets large, since the energy difference when n changes by one becomes very small compared to the energy.

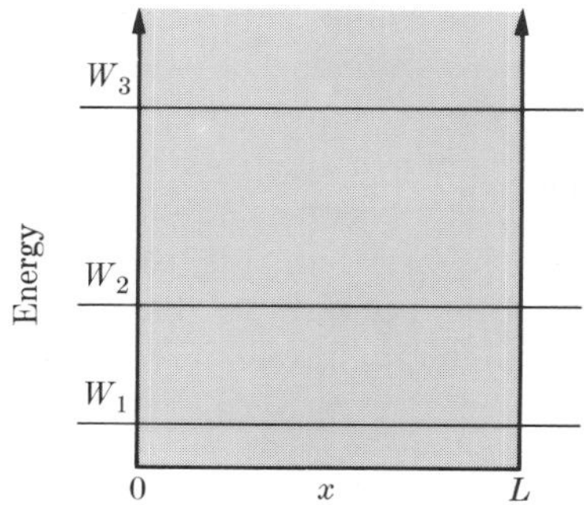

FIG. 8-2
Discrete energy states for the infinite square well.

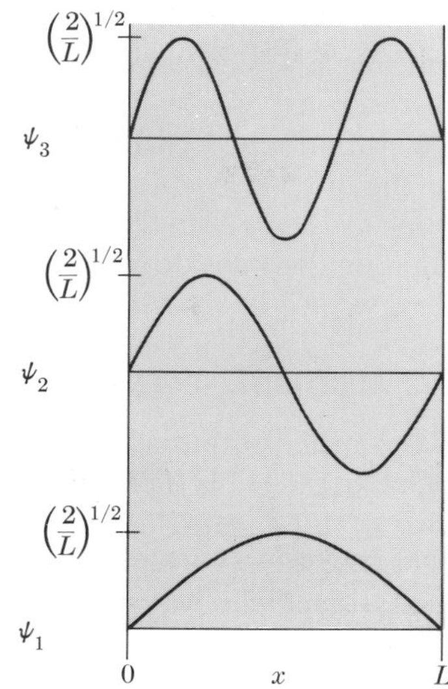

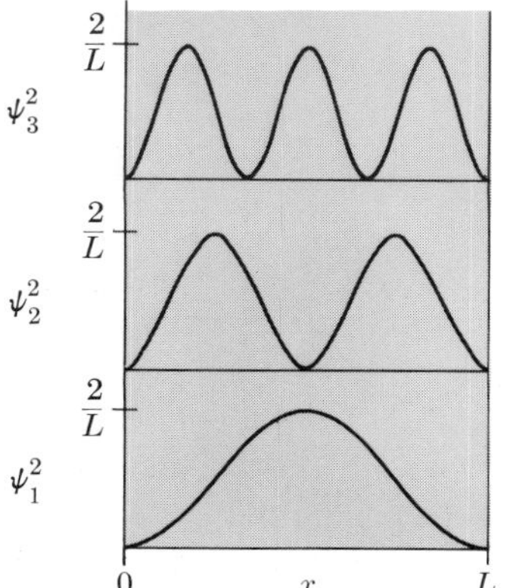

FIG. 8-3
Probability distribution functions for the infinite square well.

The expectation value of x for all states of the infinite well is $L/2$, which can be seen from the symmetry of the probability distributions about $x = L/2$. The expectation value of x^2 is *not* $L^2/4$. This is just an indication of the many experimental values possible in a measurement of the position of the particle. The expectation value of the momentum p is zero for all states. The expectation value of p^2 is not zero and again this indicates that there is more than one possible value for a measurement of the momentum of the particle. These results show that the solutions of the Schroedinger equation are not in conflict with the Heisenberg uncertainty principle. The fact that these states correspond to a definite energy does not conflict with $\Delta E\, \Delta t \simeq h$, since the wave functions represent stationary states; that is, $\Delta E = 0$ and $\Delta t = \infty$.

If the square well has a finite potential at the walls, the qualitative features are the same, except that ψ is not zero outside the walls but decreases exponentially as it "penetrates" the walls. Discrete energy levels are still obtained, although there are a finite number of them. This limit on the number occurs because eventually the energy becomes equal to the value of the potential at the walls. For energies higher than this, the wave function does not decrease outside the walls and the particle can have any energy. We have what is called a *continuum* of energy states.

8-6. ADDITIONAL POTENTIAL DISTRIBUTIONS

When other potential distributions are considered, there are very few for which exact analytic solutions of the Schroedinger equation can be found. Two of these are the harmonic oscillator and the Coulomb potential of electrostatics. We shall not solve the amplitude equation for these but we shall discuss the wave-mechanical solutions and compare them with older theories.

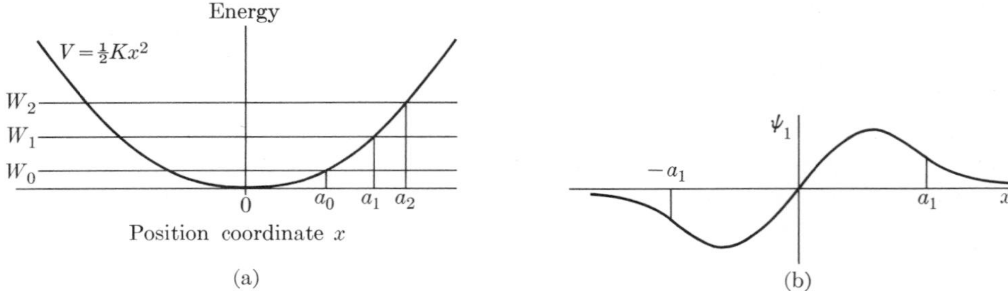

FIG. 8-4
(a) Potential distribution and energy levels for the harmonic oscillator. (b) The solution of the amplitude equation for $n = 1$.

The simple harmonic oscillator is a very important system in classical mechanics. When there are oscillations about an equilibrium position, many systems at least approximate a simple harmonic oscillator. The potential for the harmonic oscillator is $V = Kx^2/2$, where K is the force constant of the system. The amplitude equation is

$$-\frac{\hbar^2}{2m}\frac{d^2\psi}{dx^2} + \frac{1}{2}Kx^2\psi = W\psi.$$

As with the square well, this equation has well-behaved solutions only for discrete values of W and they can be denoted by a quantum number, $n = 0, 1, 2, \ldots$ The potential distribution is shown with the three lowest energy levels in Fig. 8-4(a). The solution for $n = 1$ is shown in part (b) of the figure. The displacements where the energy W_n is equal to the potential energy V are labeled a_n and are the classical turning points. According to classical mechanics, this is where the particle stops and reverses its motion; these points also indicate the amplitude of the sinusoidal motion. The solution for ψ_1 (which must be squared to give the probability distribution) shows that there is a probability of the particle being *beyond* the classical turning point.

The classical treatment permits the oscillator to have any energy. The wave-mechanical treatment quantizes the energy. In terms of the quantum number n, the energies are given by

$$W_n = (n + \tfrac{1}{2})hf, \tag{8-30}$$

with f being the frequency of oscillation:

$$f = (1/2\pi)\sqrt{K/m}.$$

For a harmonic oscillator the energy levels are uniformly spaced and separated by hf. This is the smallest permissible energy change which Planck introduced to explain blackbody radiation, as was discussed in Chapter 3. The only difference between the energy levels predicted by the Schroedinger equation and those of Planck is the term $\frac{1}{2}$ in Eq. (8–30). This does not affect the energy spacing but it does mean that the lowest energy level is $hf/2$ instead of zero. This is necessary in order that the Heisenberg uncertainty principle not be violated. If the oscillator had zero energy, it would have zero position and zero momentum, which are not simultaneously possible.

If, for the sake of comparison, we choose to give the classical solution only those energies permitted by the quantum solution, we get a set of amplitudes $a_0, a_1, a_2, \ldots$, where the subscripts are the quantum numbers. These amplitudes are shown as dashed vertical lines in Fig. 8–5. The curves which are drawn concave upward show the relative probability, on a classical basis, of finding the moving mass at any distance from its equilibrium position. These curves are concave upward because the mass moves through the equilibrium position with maximum velocity and spends most of its time near one end of the path or the other.

The shaded areas represent the wave-mechanical solution for the probable location of the mass for several energies. In the lowest energy state, with $n = 0$, the classical and quantum solutions violently disagree. Instead of the equilibrium position being least likely, it turns out to be the most likely in the new mechanics.

As the energy becomes greater, the classical probability retains its form, but the number of peaks in the quantum solution increases. This number is always $(n + 1)$. As the number of these peaks increases, their maxima begin to resemble the classical curve. Projecting this tendency to large

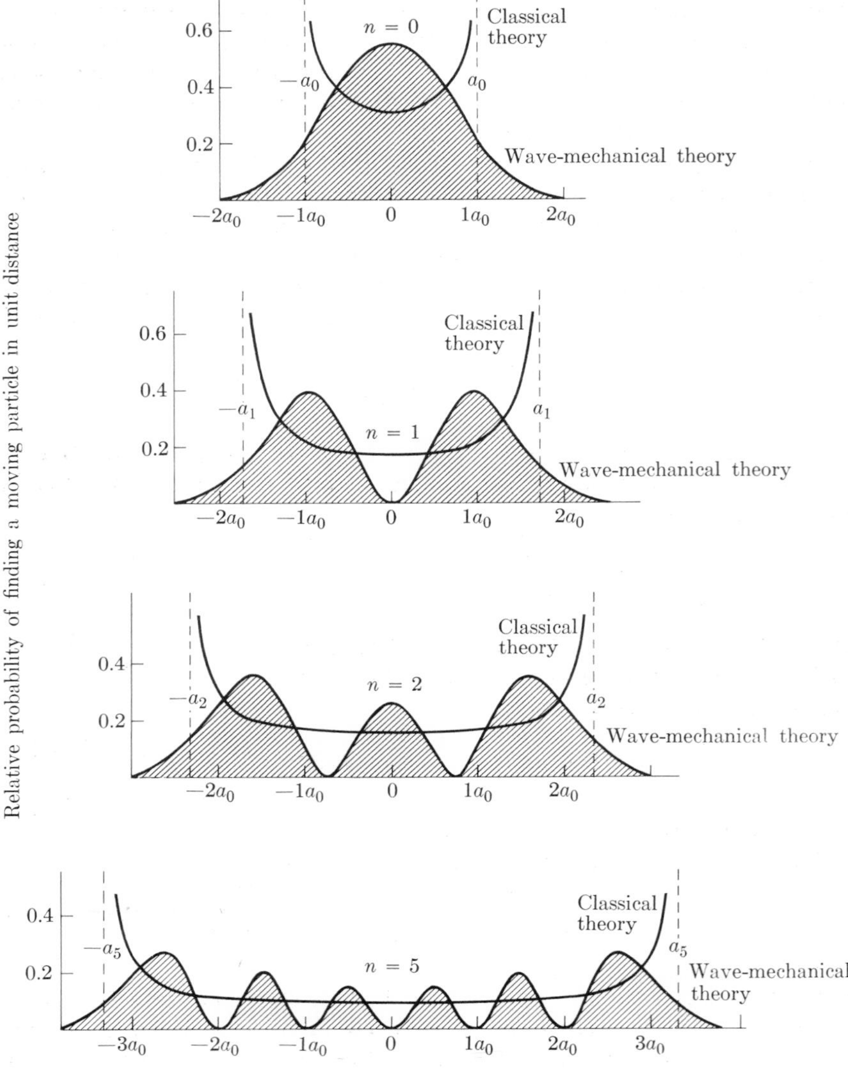

FIG. 8–5
Each diagram shows the relative probability of finding a harmonic oscillator at various displacements both classically and wave-mechanically. Four different energies are shown, corresponding to quantum numbers $n = 0, 1, 2$, and 5. The classical amplitudes are a_2 for $n = 2$, etc. Thus 2_{a_0} is twice the classical amplitude for $n = 0$. (Reprinted with permission from Blackwood, Osgood, and Ruark, *An Outline of Atomic Physics*, 3rd ed. New York: J. Wiley, 1955.

values of n, we see that the classical curve and the average of the quantum curve merge and finally become experimentally indistinguishable. The probability of finding the particle outside the classical turning point, as represented by the shaded area outside a_n in Fig. 8–5, decreases as n increases and becomes negligibly small for very large n. This agreement between the classical theory and the quantum theory for large n is in accord with the correspondence principle.

Another system that can be solved analytically is that of a charge moving in the Coulomb potential of another charge. This corresponds to the hydrogen atom, where the electron has a potential energy due to the presence of the proton, which we assume remains fixed. This is a three-dimensional problem, however, and we must extend the theory. By analogy with the classical wave equation, the amplitude equation becomes*

$$-\frac{\hbar^2}{2m_e}\left(\frac{\partial^2\psi}{\partial x^2}+\frac{\partial^2\psi}{\partial y^2}+\frac{\partial^2\psi}{\partial z^2}\right)+V\psi = W\psi. \tag{8–31}$$

The partial derivatives are used in Eq. (8–31) because ψ is a function of x, y, and z. This combination of partial derivatives is also written in operator form as

$$\frac{\partial^2}{\partial x^2}+\frac{\partial^2}{\partial y^2}+\frac{\partial^2}{\partial z^2} = \nabla^2,$$

where ∇^2 is called the Laplacian and is read "del-squared." The normalization condition becomes a volume integral:

$$\int \psi^*\psi \, dv = 1, \tag{8–32}$$

where the integration is performed over all space.

The potential energy for the hydrogen atom is

$$V = -\frac{1}{4\pi\epsilon_0}\frac{e^2}{r}.$$

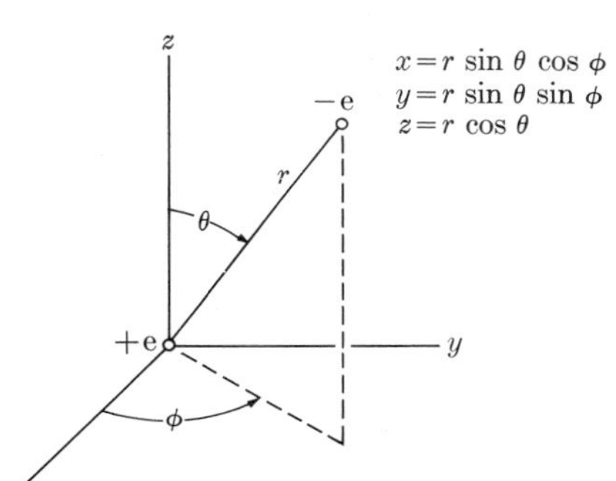

FIG. 8–6
The spherical polar coordinate system.

For this potential the problem is simplified by using spherical polar coordinates r, θ, and ϕ. This coordinate system and its relation to the xyz-system are shown in Fig. 8–6. The differential operator ∇^2, written in terms of derivatives with respect to r, θ, and ϕ, is

$$\nabla^2 = \frac{\partial^2}{\partial r^2}+\frac{2}{r}\frac{\partial}{\partial r}+\frac{1}{r^2 \sin\theta}\frac{\partial}{\partial\theta}\left(\sin\theta\frac{\partial}{\partial\theta}\right)+\frac{1}{r^2\sin^2\theta}\frac{\partial^2}{\partial\phi^2}.$$

---

* Since the potential is independent of time, the Schroedinger equation will still separate and the time-dependence is given by $e^{-iWt/\hbar}$.

The resulting partial differential equation is separated by assuming a product solution:

$$\psi(r, \theta, \phi) = R(r)\Theta(\theta)\Phi(\phi), \tag{8-33}$$

where the functional dependencies are indicated. This separation of the equation results in three ordinary differential equations, one for each coordinate. These equations are not easily solved, so we will merely present some of the results. The well-behaved solutions of the three equations are characterized by three quantum numbers. These three numbers are the n, l and m_l, which were discussed in Section 4–14, with m_l coming from the ϕ equation, l coming from the θ equation, and n coming from the r equation. The spin quantum number, s, is not obtained from the Schroedinger equation, which is applied only to nonrelativistic systems, since we used the classical expression for the kinetic energy. The spin quantum number does appear in the relativistic quantum mechanics, which was developed by Dirac in 1928.

As with the infinite well and the harmonic oscillator, the energies of the hydrogen atom are quantized; however, they depend only on the quantum number n and are given by

$$W_n = -\frac{m_e e^4}{8\epsilon_0^2 h^2 n^2}, \qquad n = 1, 2, 3, \ldots \tag{8-34}$$

These are the energies which were derived from Bohr's postulates in Chapter 4. Instead of Bohr's circular orbits, we have a stationary probability distribution that is a function of r, θ, and ϕ and is characterized by three quantum numbers. In general, we see that for every independent coordinate of a bound system the well-behaved conditions introduce a quantum number. These quantum numbers specify the wave function of the system and one of the permissible discrete energy levels.

8-7. SUMMARY

With the help of an analogy to the classical wave equation we have developed a matter wave equation. This equation, with the associated boundary conditions, has solutions which yield probability distribution functions and quantized energy levels. It is the agreement between these predictions and experimental results which is the sole justification for the Schroedinger equation. For instance, we assumed a product solution of the time-dependent equation in order to separate the equation. At that point we were not assured that these mathematical solutions were also physical solutions. The fact that the ensuing energy levels agree with experimental results shows that these are physical solutions. In particular, this wave theory for matter predicts the correct energy levels for the harmonic oscillator and the hydrogen atom. These two systems were originally quantized in two entirely different ways. The Schroedinger theory is seen to be applicable to any system which can be specified by a potential energy. Although there are only a few systems which can be solved exactly, approximation techniques have been developed which are used to find the solutions for a great many problems.

There is a major flaw in the development of the wave theory of matter presented here: there must be time-dependent probability distributions. If there were only time-independent probabilities, there would be no transitions from one state to another. Generally, every experiment is concerned with causing a transition between two states. The radiation from an excited atom originates from a transition to a lower energy level. Even though the stationary state solutions which we have found are physical solutions, they are not the only solutions. We will not go into the details but merely state that the theory has been developed to include the time-dependent features necessary for an understanding of transition probabilities. We recall that these transition probabilities, which determine the intensities of the discrete wavelengths emitted by atoms, were lacking in the Bohr theory of the atom.

There is some conflict between previous classical ideas and those of the theory of matter waves. For example, the concept of a classical trajectory of a particle must be discarded and replaced by a probability distribution spread over a large region of space. It may be useful at this point to consider the purpose and limitations of a physical theory. A theory is developed to answer questions. In physics the questions for which answers are to be provided by the theory must be able to be answered also by experiment. It is the agreement between the theoretical prediction and the experimental result which determines the status of the theory. (The experimental result, properly interpreted, is always correct.) A question which cannot be tested experimentally has no meaning. As an illustration of this, the wave theory of matter predicts the probability of finding a particle in a certain region. If an experiment is performed to determine whether the particle is there, the result will be that it is or it isn't; only the average result of many such experiments must agree with the predicted probability. Nothing is said about where the particle was before the experiment was performed and no questions concerning a previous position can be answered. In order to make predictions, theories use models (such as the particle and the wave that we have been using) and abstract concepts which are not themselves directly measured (such as Ψ). It is the correctness of the predictions that justifies the models and the abstract concepts.

It is also not disturbing when theories developed for one field of physics do not apply in other fields. We expect that, as physics is extended to new regions, new theories will be developed. At the present time the quantum theory, by including the previous classical theory through the correspondence principle, has successfully predicted the experimental results for a wider range of problems than any other theory.

REFERENCES

BOHM, DAVID, *Quantum Theory.* New York: Prentice-Hall, 1951.

BRIDGMAN, P. W., *The Logic of Modern Physics.* New York: Macmillan, 1949. This classic discusses the significance of measurable and unmeasurable quantities; it also discusses meaningless questions.

EISBERG, ROBERT M., *Fundamentals of Modern Physics.* New York: Wiley, 1961.

PARK, DAVID, *Introduction to the Quantum Theory.* New York: McGraw-Hill, 1964.

PAULING, LINUS and E. N. WILSON, *Introduction to Quantum Mechanics.* New York: McGraw-Hill, 1935.

SHERWIN, CHALMERS W., *Introduction to Quantum Mechanics.* New York: Holt, Rinehart, and Winston, 1959.

PROBLEMS

8–1. The possible values obtained from a measurement of a discrete variable, called x, are 1, 2, 3 and 4. (a) Given that the respective probabilities are $\frac{1}{4}$, $\frac{1}{4}$, $\frac{1}{4}$ and $\frac{1}{4}$, calculate the expectation values of x and x^2. (b) Given that the respective probabilities are $\frac{1}{12}$, $\frac{5}{12}$, $\frac{5}{12}$ and $\frac{1}{12}$, calculate the expectation values of x and x^2. (c) Compare the results for parts (a) and (b).

8–2. Show that the normalizing constant A in the solutions for the infinite square well of width L is $(2/L)^{1/2}$ for all values of the quantum number n.

8–3. An electron is bounded by a potential which is approximated by an infinite square well with a width of 2×10^{-8} cm. Calculate the lowest three permissible energies of the electron.

8–4. An infinite square well has a width of 2.00 Å. What is the fractional change in the lowest two permissible energies of an electron in this well if the width is increased by (a) 0.02 Å and (b) 2.00 Å? (c) What is the fractional change in the two lowest energy levels if a proton replaces the electron in this well?

8–5. A particle of mass m is in an infinite square well of width L in a state specified by the quantum number n. Use Eq. (8–29) to show (a) that the expectation value of x is $L/2$ and that the expectation value of x^2 is

$$\frac{L^2}{3}\left(1 - \frac{3}{2\pi^2 n^2}\right)$$

and (b) that the expectation value of p is zero and the expectation value of p^2 is $(\hbar\pi n/L)^2$.

8–6. A steel ball with a mass of 30 g is bouncing back and forth between rigid walls. The walls are 2 m apart and the ball has a velocity of 0.3 m/s. (a) Find the kinetic energy of the ball. (b) Assuming that this system corresponds to a particle in an infinite square well, find the quantum number of the state of the steel ball. (c) How much additional energy must the ball acquire to reach the next higher state?

8–7. The classical probability distribution function for a harmonic oscillator with amplitude a is $P(x) = A/(a^2 - x^2)^{1/2}$, where x lies between $-a$ and $+a$. (a) Determine the constant A by normalizing the distribution. (b) Calculate the expectation values of x and x^2.

8–8. The wave function which is the solution of the amplitude equation for the lowest energy state of a harmonic oscillator is $\psi(x) = Be^{-x^2/2a^2}$, where a is the corresponding classical amplitude. (a) Determine the constant B by normalizing the wave function. (b) Calculate the expectation values of x and x^2. (c) Calculate the expectation values of p and p^2.

8–9. Assume that the uncertainty in position for a harmonic oscillator is $\Delta x = (\langle x^2 \rangle - \langle x \rangle^2)^{1/2}$ and that the uncertainty in momentum is

$$\Delta p = (\langle p^2 \rangle - \langle p \rangle^2)^{1/2}.$$

Use the results of Problem 8–8 to show that these uncertainties for the lowest state are consistent with the Heisenberg uncertainty principle.

8–10. (a) Show that the function given below is a wave function of the amplitude equation for the harmonic oscillator. [*Hint:* This means that when the kinetic and potential energy operators are applied to this function, the result is the original function multiplied by a constant.] (b) What is the energy E of the system described by this wave function?

The function is

$$\psi(x) = Axe^{-\sqrt{mK}x^2/2\hbar},$$

where A is a constant, x is the displacement, m is the mass of the oscillator, and K is the force constant.

8–11. A 40-g mass suspended from a helical spring is oscillating vertically with an amplitude of 3 cm and a period of $\pi/5$ s. (a) Find the energy of the oscillating mass. (b) Assuming that mechanical energy can be quantized according to $E = hf$, how many quanta of energy does this mass have? (c) What is the distance between the peaks of the probability curve for locating the oscillating mass along its path? (d) Could this distance be observed?

CHAPTER 9

THE ATOMIC VIEW OF SOLIDS

9–1. INTRODUCTION

The solid state of matter has always been of interest to physicists, but there have been important advances in this field since quantum mechanics has been applied to it. In the last decade or so, solid-state physics has come to the forefront as one of the most fruitful fields of research.

Whereas the classical study of the solid state was primarily concerned with the measurement of the macroscopic properties of solids—mechanical, optical, thermal, and electrical—the modern study of the solid state is primarily concerned with the microscopic properties.

We have already discussed some solid-state physics. In Chapter 1 we presented the law of Dulong and Petit and showed how the specific heat capacities of solids assisted in the determination of the atomic weights. In Section 4–16 we discussed the interatomic forces which account for molecular structure and we suggested that a projection of these arguments could account for the structure of crystals. In this chapter we will show how some of the properties of solids were "explained" by classical theory and then we will go on to see how the introduction of quantum mechanics led to better explanations. Transistors are among the practical devices which have come from our deepening understanding of the solid state.

9–2. CLASSICAL ATOMISTIC APPROACH TO MOLAR HEAT CAPACITY OF SOLIDS

In Sections 1–7 and 1–8 we developed the classical kinetic theory of gases. In particular, we found that the molar heat capacity of a gas could be expressed as

$$C_v = \frac{fR}{2}, \tag{9–1}$$

where f is the number of degrees of freedom—three for a monatomic gas, five for a diatomic gas, etc.

We can extend the kinetic theory of matter to account for the molar heat capacities of solids by generalizing our definition of the number of degrees of freedom over that given in Section 1–8. In vibrating systems the number of degrees of freedom is *not* equal to the number of coordinates necessary to specify the positions of the particles.

Consider a mass m that is acted on by a system of springs so that it moves linearly with simple periodic motion of amplitude A and angular frequency ω. If the position varies as $\cos \omega t$, the velocity varies as $\sin \omega t$. The expression for the potential energy is $E_p = \frac{1}{2}mA^2\omega^2 \cos^2 \omega t$, and the kinetic energy is $E_k = \frac{1}{2}mA^2\omega^2 \sin^2 \omega t$. The total energy of the oscillator is the sum of these two expressions. It is constant and equal to the common coefficient of the squared trigonometric functions. This constant value could be expected from the law of conservation of energy. The graphs of the sine and cosine functions have the same shape; they differ only in phase. Thus the time average of the sine squared over a cycle equals the time average of the cosine squared. From this we may conclude that the average value of the kinetic energy of the oscillating body equals the average value of its potential energy, and the total energy is "partitioned" equally between these two types.

Since the average energy of a body or a mechanical system is distributed equally among the various ways in which it may have energy, it remains only to state the new rule for determining the number of these ways. Although in mechanics the number of degrees of freedom is taken to be the number of space coordinates necessary to completely specify the positions of the parts of a system, a more general rule used in statistical mechanics is the following: Write the complete energy expression for the system in terms of position coordinates and velocities, using any coordinate system that may be convenient. Count the number of variables in this expression which appear to the *second power*. This number is the number of statistical degrees of freedom. This count will include all kinetic energy terms and all elastic potential energy terms, but it will exclude such terms as gravitational potential energy, since this kind of energy depends on the height to the first power. We now apply this new rule to a solid.

A solid consists of a collection of atoms each of which is more or less bound, but which are, nevertheless, able to vibrate. Although each atom has its position completely specified by three coordinates, application of the rule for determining the number of degrees of freedom shows that there are six for each. The total energy of an atom in a solid has the form

$$E = \tfrac{1}{2}m(v_x^2 + v_y^2 + v_z^2) + \tfrac{1}{2}k(x^2 + y^2 + z^2) + mgy, \qquad (9\text{–}2)$$

where k is the force constant. The velocity terms represent the translational kinetic energy and have the same form regardless of the state of matter. The middle terms represent the elastic potential energy of an oscillator. These depend on the displacement of the atom from its equilibrium position. The last term represents the gravitational potential energy. This energy expression contains six variables which appear to the second power.

Since the energy is shared equally by these six degrees of freedom, we should expect the molar heat capacity at constant volume to be

$$C_v = 6R/2 = 3R = 5.97 \text{ cal/mole} \cdot {}^\circ\text{K}, \tag{9–3}$$

where $R = 1.99$ cal/mole · °K. According to this theory, the value of C_v should be the same for all solids. This is a derivation of the law of Dulong and Petit, which was rather well verified in Problem 1–3. Although it is pleasing to have extended the kinetic theory to solids with apparent success, we have led ourselves into a trap. According to the argument above, the specific heat of a solid should be independent of temperature, whereas in reality the specific heats decrease when the temperature is lowered. Furthermore, the theory above is inconsistent with the classical view of the theory of electrical conductivity.

9–3. CLASSICAL THEORY OF ELECTRON GAS IN SOLIDS

The classical theory of insulators is that in these materials the electrons are bound to their molecules, which are in turn tied to fixed locations in the solid. Conductors, on the other hand, are assumed to have atoms whose outer electrons are free to migrate from atom to atom throughout the solid crystalline structure. These electrons are assumed to behave within a conductor like an ideal gas. This assumption leads to the conclusion that the number of degrees of freedom in a conductor should be nine: three for the kinetic energy of the molecules, three for the elastic potential energy of the molecules, and three for the kinetic energy of the electrons. We shall explain the inconsistency between the nine degrees of freedom that every conductor should have classically and the experimentally verified six degrees of freedom of the law of Dulong and Petit after seeing where we are led by the "electron gas" theory as it applies to conduction.

In order to calculate the resistivity of the metal, we assume that the electron gas is thermally agitated and has a root-mean-square velocity which may be obtained by solving Eq. (1–34) for v. We get

$$v_{\text{rms}} = \sqrt{\frac{3kT}{m_e}}. \tag{9–4}$$

If the electrons have a mean free path L, then the average time $\bar{t}$ between collisions is

$$\bar{t} = \frac{L}{v_{\text{rms}}} = L\sqrt{\frac{m_e}{3kT}}. \tag{9–5}$$

Suppose that the electrons are in a bar of cross-sectional area A and length l, across which there is a potential difference V. This potential difference causes an average electric field V/l which exerts a force on each electron. The force on the electron is

$$F = eE = \frac{eV}{l}. \tag{9-6}$$

This force accelerates the electrons an amount

$$a = \frac{e}{m_e}\frac{V}{l}. \tag{9-7}$$

They accelerate for an average time $\bar{t}$ and acquire an average drift velocity $\overline{v_d}$ given by

$$\overline{v_d} = \frac{eV}{2m_e l} L \sqrt{\frac{m_e}{3kT}}. \tag{9-8}$$

This drift velocity is along the bar and is very small compared with the random $v_{\rm rms}$. Indeed, $\overline{v_d}$ is so small that each interval between thermal collisions begins a new acceleration "from rest" and the transport of electrons proceeds at an average velocity of $\overline{v_d}$.

If the concentration of free valence electrons is n per unit volume, then the number which pass across a plane perpendicular to the axis of the bar per unit time is $nA\overline{v_d}$, and the electrons constitute a current given by

$$I = enA\overline{v_d} = \frac{e^2nLA}{2m_e l}\sqrt{\frac{m_e}{3kT}}\,V. \tag{9-9}$$

Comparison with Ohm's law shows that this development gives the resistance of the bar as

$$R = \frac{2l}{e^2nLA}\sqrt{3kTm_e}. \tag{9-10}$$

Since in terms of the specific resistance we may write $R = \rho(l/A)$, we see that

$$\rho = \frac{2\sqrt{3kTm_e}}{e^2nL}, \tag{9-11}$$

and the electrical conductivity is

$$\sigma = \frac{1}{\rho} = \frac{e^2nL}{2\sqrt{3kTm_e}}. \tag{9-12}$$

It is interesting to find that for some pure metals, assuming that n is the number of valence electrons per unit volume and that L is the interatomic distance, quite good agreement is reached at room temperature.

Unfortunately for this theory, it is well known that the resistivity of most metals is proportional to the absolute temperature over a wide range, so that this agreement is largely concidental. Still worse, we have already pointed out that this theory would assign a molar heat capacity of $3\dot{R}/2$ to the electron gas, while in fact the molar heat capacity of metals is quite readily explainable on the basis of translational and vibrational energy of the molecules alone. Good conductors of heat are also good conductors of electricity. The proportionality of these two kinds of conduction is called the *Wiedemann-Franz* relationship, and it strongly implies that the two types of conduction have the same mechanism. If the motion of electron gas accounts for electrical conductivity and therefore thermal conductivity, it is paradoxical that the thermal motions of the electrons do not contribute to the specific heat of the material.

9–4. QUANTUM THEORY OF ELECTRONS IN A SOLID

In our efforts to account for the thermal and electrical properties of conducting solids by assuming an "ideal electron gas," we have implied that the character of the gas was in every respect like an ideal molecular gas except that the "molecules" are electrons. In the kinetic theory of gases we assumed that the molecules were not subject to any forces except during collisions. This is not true of an "ideal electron gas" within a solid. Whenever an electron passes near a positive nucleus it is electrostatically attracted to it, and the electrons are always in this electric field. Even though the electrons may be "free" to migrate, they are still associated with atoms. In Chapter 4 we found that the electrons associated with atoms can exist only in certain permitted energy levels. Just as the electrons in an atom may have only the energies permitted by that atom, so the electrons in a solid may have only the energies permitted by that solid. Whereas the molecules in a gas may have *any* energy, the electrons in a solid have their energies quantized.

In Chapter 4 we found that if an atom has more than one electron, no two of these electrons may have identical quantum numbers. This is the Pauli exclusion principle, and it imposes a restriction on the number of electrons that can occupy any one energy state. Thus we saw that even in an unexcited atom most of the electrons are forced to have energies well above that of the lowest energy state.

Sommerfeld, Fermi, and Dirac assumed that the electrons in a solid must have their energies quantized and that the electrons in a solid must obey the Pauli exclusion principle. Whereas the maximum number of electrons in any atom is about 10^2, the number of valence electrons per cubic centimeter of solid metal exceeds 10^{20}. If the Pauli principle applies to a solid over any appreciable region, it implies that although a few of the valence electrons may have very small energies, most of the electrons must occupy energy states extending to high energies. Following these assumptions, Fermi and Dirac derived an expression for the distribution of the energies and the speeds of the electrons in the electron gas in a solid. This Fermi-Dirac distribution is in marked contrast to the Maxwell distribution of gas molecules because the energies of the gas molecules are not quantized. Compare Fig. 9–1 with Fig. 1–2. In Fig. 9–1 we note that there are few electrons in the region of low speed and low energy, and that in accord with the Pauli principle most of them are forced to have high speed and high energy. A quantitative comparison of the Maxwell and Fermi-Dirac distributions may be made by stating that those electrons at the high-speed edge or maximum of the Fermi-Dirac distribution have speeds which would correspond to the most probable speed they would have at about 30,000°K in a Maxwellian gas if the electron energies were not quantized. Contrary to the classical view, this statement applies even when the solid is at absolute zero.

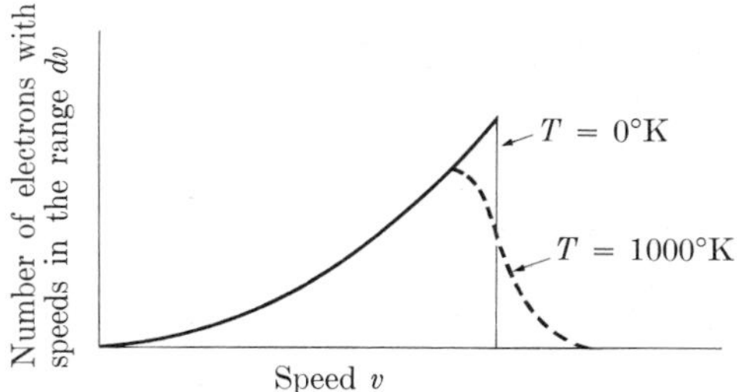

FIG. 9–1
Distribution of speeds in an electron gas according to Fermi-Dirac statistics.

Another important difference between these two distributions is their sensitivity to temperature. Whereas the Maxwell distribution retains its shape but is "stretched" toward higher speeds at higher temperatures, the Fermi-Dirac distribution is nearly independent of the temperature of the metal. Figure 9–1 suggests how it changes from 0°K to 1000°K.

We are now in a position to resolve two of the difficulties we have encountered in classical theory. First, when a conductor is heated the electrons contribute a negligible amount to the molar heat capacity. The electrons in the "Fermi gas" may indeed absorb a little energy when a conductor is heated, but those that are free to absorb energy are so few that the amount they absorb is a negligible fraction of the energy absorbed by the molecules of the crystal. Second, the high energy of the electrons at the maximum of the Fermi-Dirac distribution does not prevent them from acquiring a low-energy drift velocity in the presence of either an electric field or a thermal gradient. Thus the electrons contribute insignificantly to the specific heat, but supply a common mechanism for electrical and thermal conductivity.

9-5. WAVE-MECHANICAL TREATMENT OF ELECTRICAL CONDUCTIVITY

We have used the Fermi-Dirac distribution of electron speeds to resolve two of the classical difficulties, but we still must account for the temperature coefficient of resistance.

The electrons that contribute to electrical conductivity are those at the high-speed edge of the Fermi-Dirac distribution. These have much higher root-mean-square velocities than would be implied by Eq. (9–4). We can use this equation if, instead of the absolute temperature T of the conductor, we use the "equivalent" temperature of the maximum of the Fermi distribution. We have already mentioned that this "equivalent" temperature is very high and almost independent of the physical temperature of the conductor. Thus "T" in our earlier derivation takes on a new meaning and is no longer a significant variable.*

Since we have argued that the equivalent T is essentially constant, it is difficult to see how the resistivity and conductivity, Eqs. (9–11) and (9–12), can be temperature-dependent. The least well defined quantity in these equations is now L, the mean free path, which we supposed earlier to be the interatomic distance. The question is whether we can redefine L so as to restore to the resistivity the magnitude and temperature dependence we know it has experimentally.

Since the value of the equivalent T is much greater than room temperature, the required value of L must be greater than the interatomic distance if ρ and σ are to be consistent with experiment. Furthermore, if L were inversely proportional to the actual absolute temperature of the sample, the resistivity in Eq. (9–11) would vary with temperature, as it should. A proper proof that L varies inversely with temperature is complicated and not particularly enlightening. We therefore substitute a qualitative argument.

A vibrating atom presents a larger cross section than one at rest. The area increase is proportional to the square of the amplitude, just as the area of a circle is proportional to the square of its radius. The amplitude of an oscillator varies as the square root of its energy, and the average energy of an atomic oscillator is proportional to the temperature. Thus the increment in cross-sectional area is proportional to the temperature. If the atoms in a crystal had a negligible area when at rest, this argument could explain the temperature coefficient of resistance. At absolute zero the atom would have no significant cross section and electrons could streak through a conductor without encountering any resistance at all. Since, experimentally, the resistance of a conductor does tend to zero as the temperature approaches absolute zero, we must now justify the statement that nonvibrating atoms have a cross section for electron collisions that really is negligible.

---

* This will be discussed further under "kinetic temperature" in Section 12–14.

This justification comes about most easily by considering the wave nature of the electrons and by regarding the scattering of electrons to be analogous to the scattering of light. We are all familiar with the scattering of a beam of light by dust particles in air or by colloidal suspensions in a liquid. Even in the absence of particles, molecules of air will themselves contribute to the scattering, thus producing the blue color of the sky and the reddish tint of the sun when low on the horizon. If the widely dispersed molecules of the gaseous air or water vapor molecules can cause such appreciable scattering, what would be expected of the same substance, say water, in the liquid form? Since the density of the liquid is about 1000 times greater than that of the gas, it is surprising to find the actual scattering of the liquid is less than 50 times as great.* As a matter of fact, it is not the particles themselves, but the variation in density of groups of particles that contributes to the scattered light. Thus regularly arranged crystalline solids, whose appearance from a macroscopic view approaches a continuous medium, are strikingly free of scattering. If sunlight is focused first on a clean cover glass, such as is used for microscope slides, and then on a freshly cleaved flake of mica, the strong scattering by the first, in contrast to the lack of scattering by the second, can be easily observed if both are viewed against a very dark background.

We are thus able to support our original contention that it is not the cross-sectional area of the undisturbed atoms which must be considered in calculating the collision frequency and the mean free path of the electron in the metal, but rather that area which exhibits the characteristic random thermal fluctuations associated with the vibrational motion of the crystal lattice.

An alternative explanation of this interesting lack of scattering of electrons by regular arrays of atoms may be seen by considering the Bragg reflection conditions which we introduced in our discussion of x-rays:

$$n\lambda = 2d \sin \theta. \tag{9–13}$$

If we apply this formula to the waves associated with electrons within a crystal, λ is the de Broglie wavelength of the electrons, n is an integer, d is the basic crystal spacing, and θ is the reflection angle (see Fig. 6–4). The de Broglie wavelengths of electrons at the high-speed edge of the Fermi-Dirac distribution is large compared with the crystal spacing. Since $\sin \theta$ cannot exceed unity, the only possible value for n is zero and no lateral scattering by diffraction can take place. Thus the electrons should move undeflected through the crystal, and the electrical resistance should be zero.

This conclusion is justified if the crystal is absolutely pure and its atoms are in perfect order, as at absolute zero. If the crystal contains impurities

---

* R. W. Wood, *Physical Optics*. 3rd ed., p. 428. New York: The Macmillan Co., 1936.

or structural imperfections, these "flaws" will be separated by some multiple of the crystal spacing. If d in the Bragg formula is interpreted as the distance between imperfections then it may become greater than λ, and n and θ may become greater than zero. Since these spacings are irregular, incoherent scattering will be observed, rather than ordinary interference diffraction patterns. This scattering of electrons hinders their progress through the crystal and causes resistance. In this interpretation the temperature coefficient of resistance arises from the fact that thermal agitation introduces random structural irregularities.

In summary, we have shown that the linear relationship between the resistance of a conductor and its absolute temperature can be justified. The classical derivation contained a temperature to the wrong power. We reinterpreted T on the basis of the Fermi-Dirac distribution of electron velocities and found it to be a large constant rather than the absolute temperature. Having removed the temperature dependence altogether, we next reintroduced it by showing that the electron mean free path, L, varies inversely with temperature.

9-6. ELECTRIC POTENTIALS AT A CRYSTAL BOUNDARY

Ordinary gas molecules are usually confined by some sort of container, but the atmosphere is confined to the earth by the earth's gravitational field, which in the large scale is by no means negligible. The most probable reason the moon has no atmosphere is that its gravitational field is too weak to hold one. What is the container that tends to confine electrons within a metal? For a metal, the field which contains the electrons is electrical. A piece of metal is ordinarily uncharged, but if an electron escapes, the piece becomes positive. The positive piece then has an electric field which attracts the electron back to itself. Electric fields are so strong that electrons which would otherwise get away "gravitate" back to the metal. Thus if an electron is near the surface and its motion is directed so as to escape, it encounters a *potential barrier* which ordinarily prevents the escape. Of course if the electron has sufficient energy, it may surmount the potential barrier and become free of the metal. We have discussed both the photoelectric effect, in which electrons are given excess energy by photons, and thermoelectric emission, where the excess energy comes from thermal excitation. In both cases we found that there was a definite threshold energy that electrons must be given to enable them to surmount the potential barrier, and we called this energy the work function.

9-7. ENERGY BANDS IN SOLIDS

If we turn our attention from the boundary to the body of a metal, we find that the electric potential is not uniform from a microscopic point of view. Although the valence electrons in which we are interested are shielded from the positive nuclei by inner, bound electrons, nevertheless the valence electrons are attracted to the nuclei. These attractions cause potential variations. Figure 9–2 shows the nature of the potential variation in a crystal along a line that intersects the surface of the crystal.

While our discussion so far has apparently been able to explain several properties of metals on the basis that the electrons have a Fermi-Dirac distribution, we still have no theoretical basis for distinguishing the metals from other classes of solids. In this section we shall introduce refinements in the theory to account for the variation in electrical conductivity which distinguishes insulators and semiconductors from conductors.

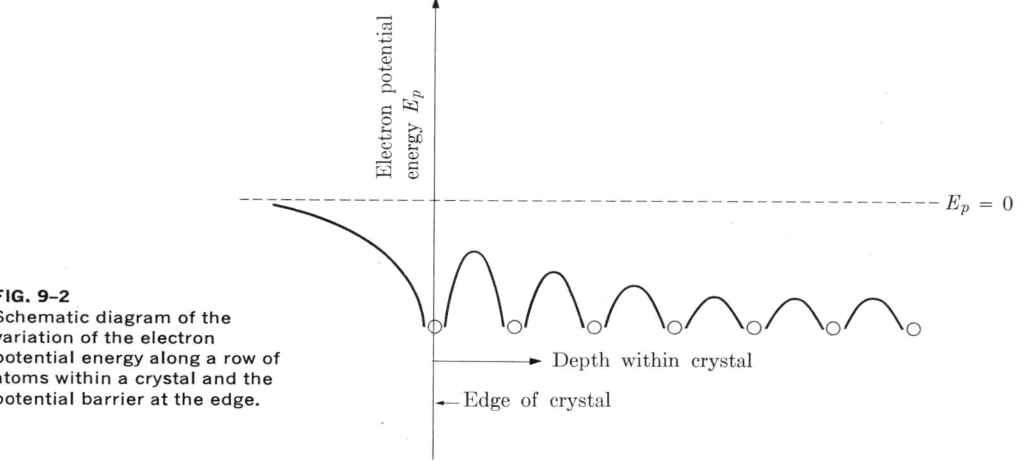

FIG. 9–2
Schematic diagram of the variation of the electron potential energy along a row of atoms within a crystal and the potential barrier at the edge.

The permissible energy levels of the electrons in a crystal can be found with the Schroedinger equation in Chapter 8. The potential distribution, such as that shown in Fig. 8–2, is put in the equation and solutions are determined which are well-behaved. The crystal is three-dimensional and the actual distribution of the electric potential depends on the specific crystal. If a representative simplified potential is used, the form of the solutions to the Schroedinger equation and the energy levels can be found. When the well-behaved conditions are used, the permissible energy levels are found to be collected in what are called *energy bands.* Within these energy bands the discrete energy levels are so close that they appear to be continuous. These bands are separated by regions, called *forbidden bands,* where there are no permissible energies.

Since these energy levels are for the entire crystal, they are to be occupied by a large number of electrons. When the Pauli principle is applied to these electrons, the limit on the number of electrons which may occupy a single energy level means that a large number of these levels must be filled. The implications of this result will be discussed in Section 9–8.

There is an analogy to this situation which could be stated in either electrical or mechanical terms. We choose the mechanical case, and consider a group of identical simple pendulums, each of which will oscillate with the same frequency f. If two of these pendulums are hung from a flexible common support, they become coupled oscillators and the system has *two* natural frequencies, $f + \delta$ and $f - \delta$, a little above and below the natural frequencies of the isolated pendulums. If the entire group of pendulums is suspended from the same flexible support, the system has as many *different* frequencies as there are pendulums. Thus the system has a *band* of frequencies. The identical electron energies of isolated atoms correspond to the identical frequencies of the isolated pendulums. As the atoms become coupled in the solid, the band of electron energies corresponds to the band of frequencies of the coupled pendulum system.

The following argument is more sophisticated, in that it is closer to the rigorous treatment. It is based on Bragg reflection of de Broglie waves, first discussed in Section 9–5. Whether an electron is moving through the crystal or not, its de Broglie waves extend over a considerable region of the crystal. Therefore we must consider the possibility of internal diffraction effects. Because of many partial reflections in successive layers within the crystal, some of these reflections become "total," and because of angular variations, the wavelengths and energies of these back "reflected" waves have an energy spread. Thus certain bands of electron energy are permitted and other bands are forbidden. Again we have a qualitative argument which accounts for the possibility that some electron energies can occur while others cannot.

We repeat that although these energy bands appear continuous, they are actually collections of closely spaced energy levels. The arguments we are about to present depend on the fact that each permitted band can hold only a certain number of electrons.

9–8. ELECTRICAL CLASSIFICATION OF SOLIDS

The energy bands we have just discussed provide an especially simple and satisfying explanation of the electrical classification of solids. If the highest energy band which has some electrons is unfilled, as in Fig. 9–3(a), these electrons may be excited from a lower to a higher energy level within the band. The separation of the levels within the band is so slight that the electrons are readily able to be accelerated to higher energies by weak electric fields. A material with such an electronic band structure is a conductor. All metals fall into this class, and it is this class we have been discussing in this chapter.

Most nonmetals are insulators. For insulators the highest occupied band is always completely filled. Although one might suppose that this case should be rare, it is actually the most common. Not only is the highest

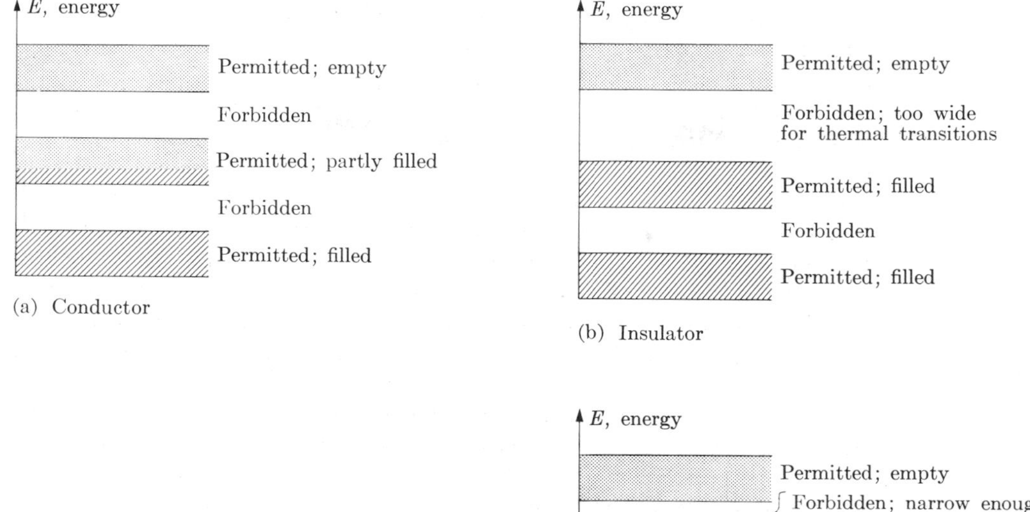

FIG. 9–3
Schematic energy bands (a) of a conductor, (b) of an insulator, and (c) of a pure intrinsic semiconductor at absolute zero.

occupied band filled, as shown in Fig. 9–3(b), but also the unoccupied forbidden band just above the filled band is wide on the energy scale. Since thermal agitation is unable to lift electrons from the filled band to the next higher permitted band, the electrons cannot absorb small amounts of energy due to electric fields or thermal agitation. Therefore these electrons are "fixed," and the material must be both a thermal and electrical insulator. In an intense electric field, of course, some electrons may be pulled into the higher permitted band, where they may migrate conductively. This is dielectric breakdown and results in permanent damage to a solid dielectric.

Semiconductors comprise a third class of materials. The difference between semiconductors and insulators is that the gap between the band which is filled at absolute zero, called the *valence band*, and the higher permitted band is narrow in a semiconductor, as shown in Fig. 9–3(c). In this case thermal energy at room temperature is sufficient to raise some electrons from the valence band to the higher permitted band, called the *conduction band*. Electrons in the conduction band are free to transport electrical charge. Their number per unit volume is the n in Eqs. (9–9) through (9–12). For a semiconductor, n increases with temperature, and it is easy to see that if n increases with temperature faster than L decreases, the electrical conductivity of a semiconductor will increase with temperature instead of decreasing as it does for conductors.

When electrons are transferred to the conduction band from the valence band, the valence band is no longer filled, and there are then energy states within the valence band available to the valence electrons. Valence electrons can then move successively, like automobiles in a traffic jam. When an opening appears, one car moves to fill it, leaving an opening behind. This is repeated over and over, so that the net effect is that the physical motion of the cars is forward while the gap between them moves backward. In the electrical case, the vacancy created by the removal of an electron is a small positively charged region called a *hole*. When a nearby electron moves into a positive hole, it leaves another positive hole at its original location, and so on. Thus there is a net transport of positive charge which is called *hole migration*. Holes are less mobile than electrons, and therefore hole currents are usually less than electron currents. Collectively, electrons and holes are called *carriers*. There is a dynamic equilibrium between mutual cancellation and thermal creation of conduction electrons and holes.

While amorphous materials have no proper place in our discussion, there is nevertheless an interesting demonstration that may be performed with a glass rod to illustrate the possibility of an insulator becoming a conductor at a high temperature. If two leads are connected to the ends of a thick glass rod, with suitable resistance in series with the rod, and connection is made to an ordinary 110-volt line, no charge will flow, of course. But application of a Bunsen burner flame to the midpoint of the rod will soon heat the glass to the point where it becomes conducting. If the burner is then shut off, the I^2R heating alone will sustain the temperature of the glass, so that it finally glows brightly and begins to melt.

In actual insulators, there are usually sufficient imperfections and chemical impurities in the lattice so that additional intermediate conducting levels exist. One would certainly expect, then, that such an irregularly arranged solid as glass would possess a large proportion of such localized conducting regions and hence exhibit the behavior described above.

Detailed examination of the known crystalline structure of various solids, plus study of the electronic structure of individual atoms in the solid, have made it possible to predict fairly well what the band structures of various materials should be, and hence which particular arrangements should be conductors, insulators, or semiconductors. Details of these calculations can be found in some of the references given at the end of this chapter.

We conclude this portion of our discussion with a summary of the differences between conductors, insulators, and semiconductors. A *conductor* is a solid with a large number of current carriers, a number independent of temperature. An *insulator* contains very few carriers at ordinary temperatures. A *semiconductor* contains relatively few carriers at low temperatures but a rather large number at higher temperatures. For semiconductors the actual dependence of resistance upon temperature is a result of two opposite effects. First, the increase in scattering of the electron wave with temperature tends to diminish the conductivity. This tends to cancel the second effect, which is due to increase in the number of carriers. At room temperature semiconductors may have either positive or negative resistance-temperature slopes.

9-9. SEMICONDUCTOR RADIATION DETECTORS

We have indicated that the energy to create conduction electrons and holes can be supplied from thermal energy. Another way to supply this energy is from high energy radiation such as x-rays. A photoelectron produced within the solid becomes a conduction electron and a conduction hole is left in the source atom. If the semiconductor is in an electric field, the electrons and holes can migrate and constitute a measurable current. This behavior of the semiconductor material is in close analogy with the gas in an ionization chamber. But since the concentration of atoms in a solid semiconductor radiation detector is much greater than in a gas, the semiconductor detector can be much smaller than an ionization detector. Other advantages of solid state detectors will be discussed in Chapter 10.

9-10. IMPURITY SEMICONDUCTORS

The characteristics of *pure* semiconductors, called *intrinsic* semiconductors, can be changed in very important ways by the introduction of trace amounts of impurities. Intrinsic semiconductors to which impurities have been added are said to be *doped.* Consider the intrinsic semiconductor, germanium, doped with antimony. Each germanium atom has four valence electrons, and therefore each atom has four neighboring atoms bonded to it. Atoms of antimony will fit into this structure, but they have five valence electrons of which four participate in bonding to neighbor atoms. The fifth electron is superfluous to the structure and is therefore loosely bound to the antimony atom. Since thermal energy is sufficient to cause some of the germanium electrons to leave their valence bonds and jump to the conduction band, it is easy to see that the fifth antimony electron is even more easily excited into this conduction band. Thus practically every antimony atom introduced into the germanium lattice contributes a conduction electron without creating a positive hole. Of course, each antimony atom has become a positive ion, but this ion is tied into the lattice structure so that it cannot contribute to conduction. Thus in addition to the electrons and holes intrinsically available in germanium, the addition of antimony greatly increases the number of conduction electrons. In this case antimony is called a *donor* impurity and it makes the germanium an *n-type* (*n* is for negative) semiconductor.

Gallium, on the other hand, has three valence electrons. If it is introduced into germanium, it can supply only three of the four electrons necessary to fit into the germanium lattice. Since thermal energy is sufficient to excite some bonded germanium electrons into the conduction band, it is easy to see that thermal excitation is sufficient to cause germanium valence electrons to complete the lattice structure by attaching themselves to the gallium without leaving the valence band. This causes the gallium to become a fixed negative ion and it leaves an electron hole in the valence band. Thus gallium is an *acceptor* which, at room temperature, causes as many positive holes as there are gallium impurity atoms. The acceptor, gallium, makes germanium a *p-type* (*p* for positive) semiconductor.

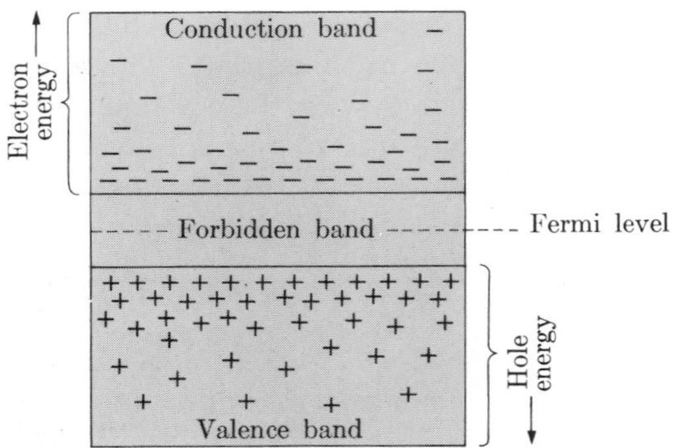

FIG. 9–4
Schematic energy-level diagram of an intrinsic semiconductor at room temperature.

9–11. FERMI LEVELS

The concept of the *Fermi level* is very useful in describing what happens when *n*- and *p*-type semiconductors are brought together. The Fermi level is a characteristic energy of a material.

We shall define the Fermi level in a rather qualitative way since a proper definition requires concepts beyond the level of this discussion. Our qualitative definition is that the Fermi level is the energy which corresponds to the "center of gravity" of the conduction electrons and holes "weighted" according to their energies. In an intrinsic semiconductor there is an equal number of electrons and holes. Furthermore, both have energy distributions which fall off almost exponentially from the band edges, in close analogy to the Boltzmann distribution given by Eq. (1–36). Thus there is a decreasing concentration of electrons at energies farther above the bottom of the conduction band and there is a decreasing concentration of holes at energies farther below the top of the valence band. Figure 9–4 is a schematic representation of this situation, which is presented without proof. An examination of the figure shows that the "center of gravity" of

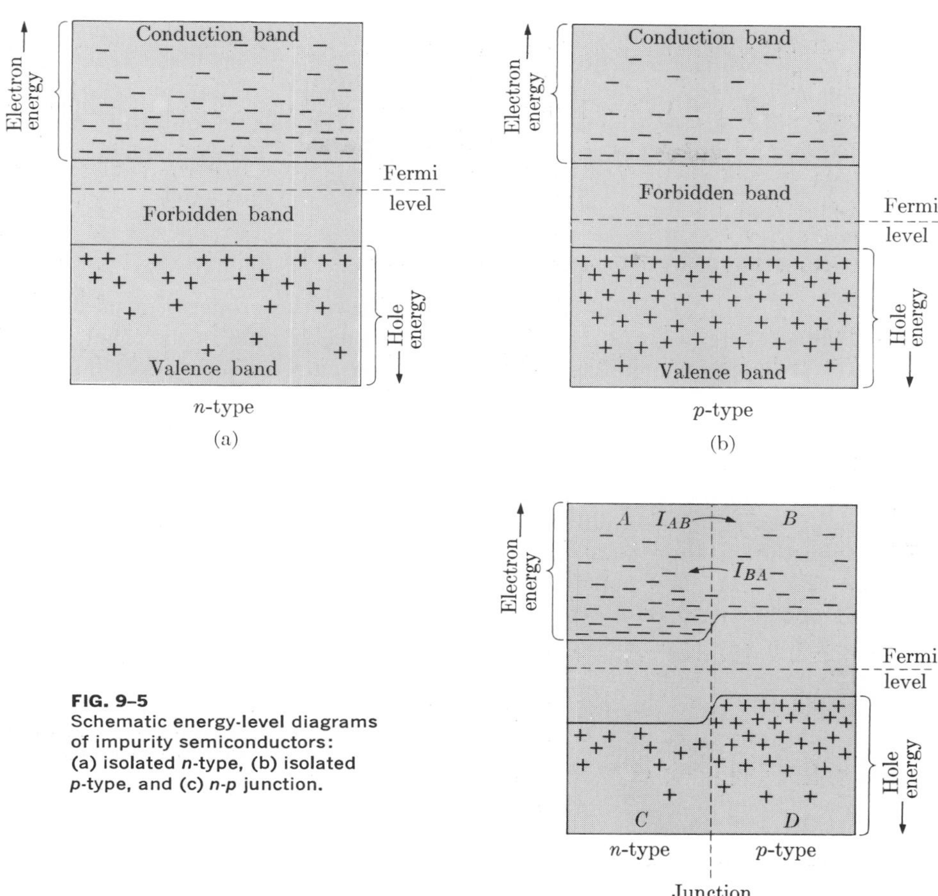

FIG. 9–5
Schematic energy-level diagrams of impurity semiconductors: (a) isolated *n*-type, (b) isolated *p*-type, and (c) *n*-*p* junction.

the pluses and minuses lies in the exact middle of the forbidden gap, where the Fermi level is drawn. Note that the Fermi level is an abstraction. Just as a hollow body can have a center of gravity where there is no matter, so a material can have a Fermi level at an energy forbidden to all electrons.

Fermi levels have two important characteristics. First, if an intrinsic semiconductor has a donor impurity added to it, it becomes *n*-type and has more conduction electrons than holes, Fig. 9–5(a). This moves the "center of gravity" up so that the Fermi level is above the middle of the forbidden band. Similarly, the addition of acceptors converts the intrinsic semiconductors into a *p*-type, with more holes than conducting electrons (Fig. 9–5b). In this case the Fermi level falls below the middle of the forbidden band. The second characteristic enables us to describe how the energy levels of the two types fit together at a *junction* between types.

If a crystal is part n-type and part p-type and if the crystal is isolated from any circuit, the various band levels shift so that the Fermi level is common to both types, Fig. 9–5(c). If the crystal is in a circuit, so that one type is held at a different potential than the other, then there is a discontinuity in their Fermi levels equal to this potential difference. These facts will help us explain semiconductor rectifiers and transistors.

9–12. SEMICONDUCTOR RECTIFIERS

Rectifiers are electrical components that have the property of passing an electric current in one direction much more easily than in the other. They are useful for the conversion of alternating currents into direct currents. We propose to discuss diffusion and then use the diffusion concept to show why the junction between n- and p-type semiconductors has rectifying properties.

The principal mechanism of conduction within a semiconductor is diffusion. An example of diffusion is the distribution of gas throughout a room even though there may be no convection or other gross motion of the air in the room. If a pungent substance like ammonia is spilled on the floor at one end of a room, the smell slowly diffuses across the room until the distribution of the gas is uniform. This motion is not due to any systematic force on the molecules. The ammonia NH_3 molecules are lighter than either the N_2 or O_2 molecules of the air, so that the gravitational force would tend to cause a layer of ammonia molecules at the ceiling. In spite of this tendency, the ammonia molecules will be found throughout the room because thermal agitation causes microscopic mixing of the gases. Because of diffusion, the molecules tend to move from a region of high concentration to a region of low concentration. Indeed, the diffusion current is proportional to the change in concentration per unit distance, or the *concentration gradient.* The diffusion current across any layer is equal to a constant times the concentration gradient, and this proportionality constant is called the *diffusion coefficient.*

Both the electrons and the holes in semiconductors move about because of thermal diffusion, but we shall fix our attention on the electrons, which are more mobile than the holes. If we refer to Fig. 9–5(c), we see that there are two electron currents between regions A and B. The one to the right, I_{AB}, results from the fact that region A is n-type whereas region B is p-type. The concentration of conduction electrons is far greater in A and they extend to higher energies in spite of the fact that the bottom of the conduction band in B is higher than that in A. The magnitude of this current is sensitive to the relative concentrations of electrons at the top of the conduction bands. The current to the left, I_{BA}, is due to electrons near the junction and at the bottom of the B conduction band "falling over" the potential "hill" into region A. This current is called a *saturation current* because its magnitude depends on the concentration of conduction electrons in region B and on *the slow rate at which they can diffuse toward the junction.* The saturation current is very small because the concentration of electrons in B, the p-type region, is small. *The saturation current does not depend on the height of the potential "hill."* In the situation pictured, these two opposing electron currents are both small and equal, since the saturation current *must* be small and since the two currents are in equilibrium. The exchange of positive holes has a similar explanation.

The equilibrium situation just described is upset if electrodes are connected to the right and left sides of the crystal and if these electrodes are maintained at different potentials. Figure 9–6(a) depicts the situation when the n-type end is made negative and the p-type end is made positive. This is called biasing in the forward direction. This bias increases the band height and consequently the energy of the electrons in A relative to those in B. The saturation electron current, I_{BA}, has the same magnitude as before, but now the more energetic electrons in A can surmount the junction barrier far more easily and the net electron current is overwhelmingly to the right. Of course, the transport of electrons to the right and holes to the left constitutes a conventional current to the left.

If the forward bias situation has been understood, there is hardly need to describe the reverse bias situation depicted in Fig. 9–6(b). The fact that the back current is small and

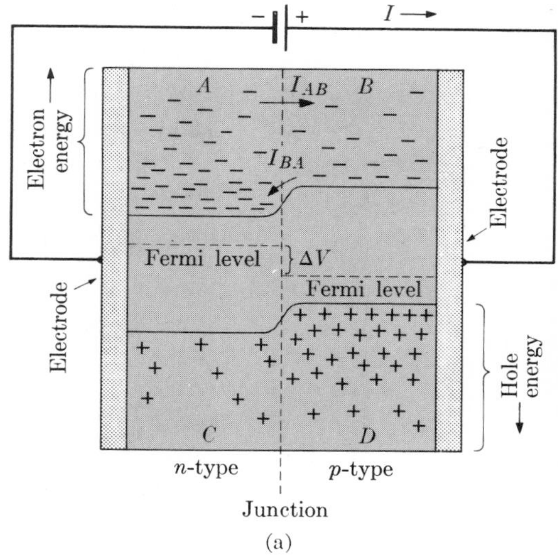

(a)

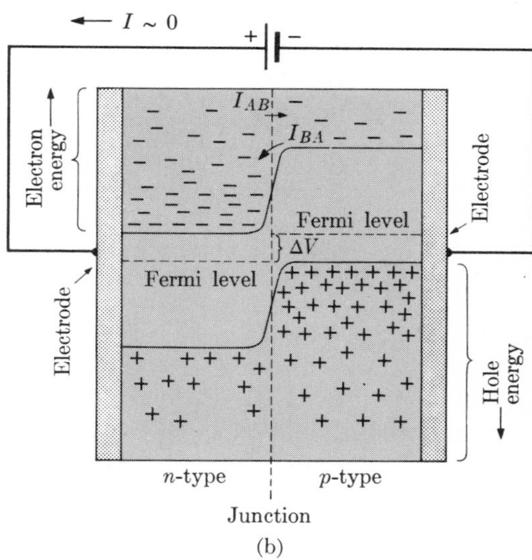

(b)

FIG. 9–6
Schematic energy-level diagrams of biased semiconductor junctions with (a) forward bias, ΔV, and (b) reverse bias, ΔV.

independent of the backward potential is clear, since it consists only of the saturation current. The characteristics of an *n-p* rectifier junction are shown in Fig. 9–7. Such a junction is often called a *crystal diode* because it performs the rectifying function of a two-electrode vacuum tube.

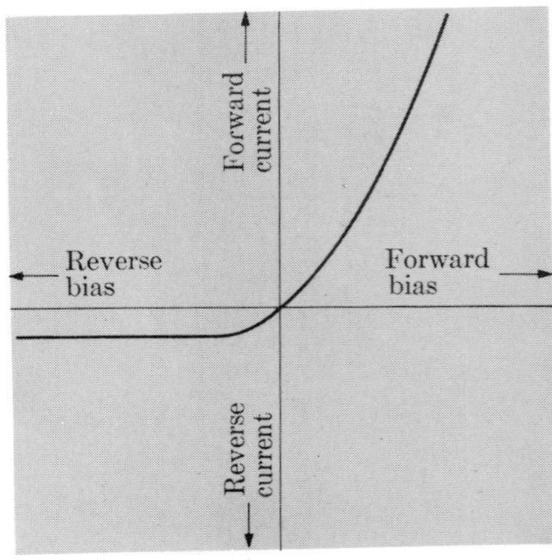

FIG. 9–7
Typical current-versus-bias characteristic curve of a rectifier semiconductor junction.

9–13. TRANSISTORS

We have discussed the diode rectifier rather fully because all of the ideas will be useful in describing a much more important device, the *transistor*. A transistor can transform a small electrical charge into a large electrical charge, that is, amplify. We shall consider the case of a junction transistor that is used so that a small amount of power can control a relatively large amount of power.

A schematic energy-level diagram of an isolated *n-p-n* junction transistor is shown in Fig. 9–8. A comparison with earlier figures shows that it may be regarded as two rectifiers back-to-back. An *n-p-n* junction transistor can be used as an amplifier when it is biased as shown in Fig. 9–9, where the left-hand *n*-type region, called the *emitter*, is heavily doped, so that it has a large electron conductivity. The central *p*-type region is called the *base*, and the junction between these regions, J_e, has a small forward bias of a few tenths of a volt. Considering this junction as a rectifier leads to the conclusion that the emitter-to-base resistance is small and the electron current is large.

The right-hand region is called the *collector*. It is made *n*-type and the junction between the base and the collector, J_c, has a strong reverse bias—up to fifty volts. Considered as a rectifier, we would conclude that the base-to-collector resistance is large and the electron current is small.

The next considerations lead us to alter what we can predict from rectifier behavior. The essential new fact is that the base region is *very*

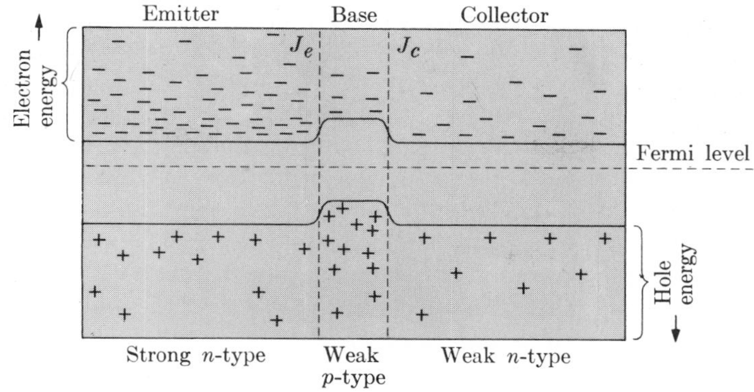

FIG. 9–8
Schematic energy-level diagram of an isolated n-p-n transistor.

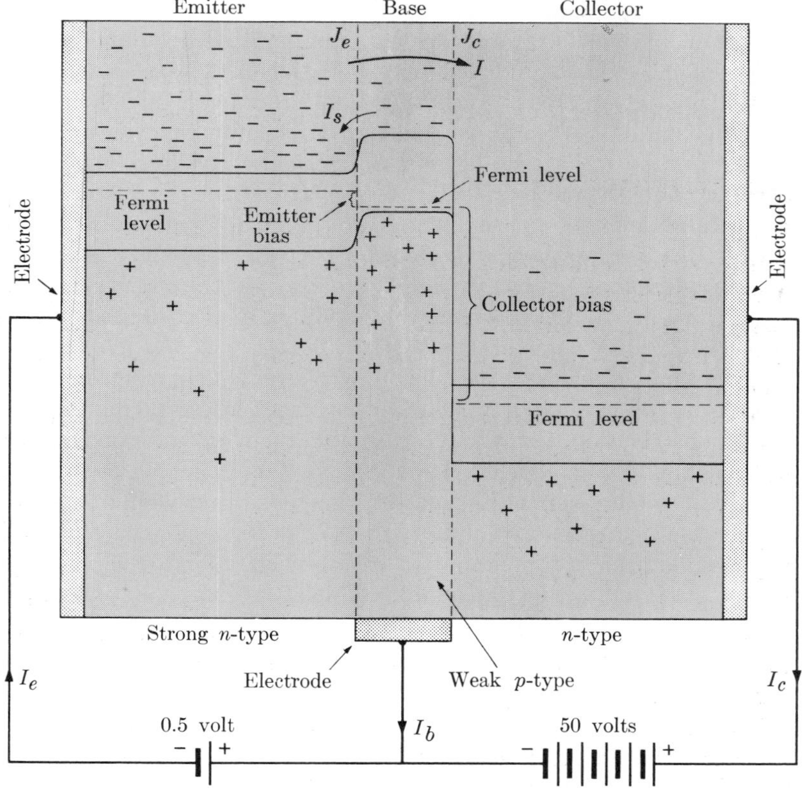

FIG. 9–9
Schematic energy-level diagram of a biased n-p-n transistor. (Arrows indicate the directions of electron flow.)

thin—about one-thousandth of an inch. This distance is small compared with the mean distance an electron moves before recombining with a hole. Almost all the electrons leaving the emitter and entering the base pass right through the base into the collector. The thinness of the base leads to an artificially increased "saturation current," I, to the collector. After the electrons penetrate the collector junction they are "over the wall" and cannot diffuse back into the base. Since the number of electrons which penetrate the base from emitter to collector includes almost all of the electrons in the large emitter current I_e, the current I to the collector is large also. In fact, the collector current is less than the emitter current by only the amount of the very small base current. Since this result is very different from what we would conclude from considering the transistor as two back-to-back rectifiers, we now review what happens in more detail.

The applied bias potentials appear mainly at the junctions, since each region of the transistor is a rather good conductor. Away from the junctions, the more important forces on electrons and holes are due to thermal agitation. When electrons pass into the emitter from the electrode an excess of electrons is produced at the left side of this region. This concentration of electrons has two effects. First, it upsets the equilibrium between electrons and holes, which are constantly recombining due to electrostatic attraction and re-forming because of thermal agitation. The excess electrons promote the recombination process, which tends to reduce the number of useful carriers in the emitter region. This effect is small, however, because this region is a heavily doped n-type material and there never were many positive holes anyway. The second effect of the added electrons is the establishment of an electron concentration gradient. This gradient causes electrons to diffuse toward the base.

When these electrons reach the emitter junction they are "emitted" into the base. We have already seen that electrons in n-type material have an energy distribution high enough to enable them to surmount an n-p junction barrier that is somewhat reduced, in this case by the small emitter bias. If the base were thick, most of these electrons would either migrate to the base electrode or combine with holes in the p-type base region. Such undesirable behavior is minimized in two ways. First, the base is a weak p-type material, so that the number of positive holes is not large. Second, the base is made very thin, so that the probability of an electron from the emitter encountering a hole in the base is very small. Once these electrons have been "collected" into the collector they suffer collisions that reduce the kinetic energy of most of them below the relatively high collector bias potential barrier, so that very few of them can diffuse back into the emitter. Thus the current to the electrode of the thin, weak, p-type base is kept small. This second discussion of transistor behavior leads to the same conclusion: almost all the current entering the transistor in the low-resistance emitter region finally emerges from the high-resistance collector region with only a small loss to the base electrode.

We must next show that these transistor characteristics can be used to accomplish amplification. Just as with vacuum tubes, there are many different circuits in which transistors are employed. A full analysis of any of these circuits is outside our present interest, but each analysis involves its own set of simplifying assumptions as different aspects of the transistor are emphasized. In what follows we will assume that the emitter-to-base resistance, R_e, is constant and low, that the base-to-collector resistance, R_c, is constant and large compared with R_e, and that the base current is zero, so that I_e equals I_c. If the emitter current is changed in some way by an amount ΔI_e, the power fed to the transistor is $2I_eR_e\,\Delta I_e$. This causes a change in the power developed in the collector circuit, equal to $2I_cR_c\,\Delta I_c$. Since we are assuming that I_e equals I_c, we see that the output collector power is greater than the input emitter power by the factor R_c/R_e. This is but one of several ways of explaining amplification by transistors. Of course, the output power really comes from the batteries; the small input power enables the transistor to liberate the greater power available from the batteries.

In the transistor we have the basis for amplification, oscillation, and electronic switching. Transistors can replace vacuum tubes in many applications. Compared with vacuum tubes, the advantages of transistors are small size, ruggedness, low power consumption, and reliability. All these advantages indicate the desirability of transistors in portable applications. Where portability is not important, reliability still gives transistors advantages over vacuum tubes. Electronic computers require thousands of electronic amplifiers, and if each tube has a probable life of one thousand hours, then there may be the costly nuisance of making replacements several times an hour. Transistors promise to greatly reduce this problem.

Transistors also have several disadvantages compared with vacuum tubes, of which we will discuss only one. We saw that the addition of an impurity into an intrinsic semiconductor shifts the Fermi level from the middle of the forbidden band. Both rectification and amplification depend on this fact. Raising the temperature of either an n- or p-type material greatly increases the number of conduction electrons and valence holes. Since the number of impurity conductors is largely independent of the temperature, the increased conductivity caused by a rise in temperature tends to swamp the impurity conductivity and the semiconductor loses its n- or p-type characteristics. As the temperature rises the conductivity becomes more nearly intrinsic, the Fermi levels tend toward the middle of the forbidden band, and the junctions between types lose their significance. The narrower temperature range of transistor operation compared with vacuum tubes is an important disadvantage.

9-14. OPTICAL PROPERTIES

It is interesting to note that the same quantum-mechanical explanation of the electrical properties of solids can be used to gain an understanding of their optical properties. In particular, when a beam of light containing photons of low energy strikes a metal in which there are many electrons in the conduction band with empty energy levels above them, the light will be absorbed. Thus a good conductor should be expected to have the opacity to visible light which experimentally it proves to have. On the other hand, since these low-energy photons cannot excite the electrons in the filled bands of an insulator to the next higher unfilled band, the light must pass through unabsorbed. Thus, in general, good insulators are also transparent to visible light. Experimentally, we know that as the wavelength of the electromagnetic radiation shortens toward the ultraviolet, these transparent solids become strongly absorbing. This is what one would expect for those insulators in which the energy gap in the forbidden region is just that corresponding to energies of ultraviolet photons. Because semiconductors have very narrow forbidden zones at room temperature, they are opaque to visible light but transparent in the far infrared.

9-15. DISLOCATIONS

While it is possible by the above methods to reach a good understanding of the thermal, electrical, and optical properties of solids, we must at the same time be aware of the fact that actual solids cannot truly be perfectly regular in atomic arrangement. Even the most carefully grown crystal has some imperfections in its structure. It has become increasingly evident over the past few years that imperfections play a vital role in the behavior of solids. In particular, it had long been known that the theoretical breaking strength of a solid is about 1000 times the actual maximum breaking strength attainable. Similarly, design engineers have been aware of the tendency of matter under continued stress to deform plastically, that is, to flow, even though the total stress was maintained well below the elastic limit. This *creep*, as it is called, can now be understood on the basis of a new theory of solids dealing with the imperfections alone. A complete comprehension of the mechanical properties of solids has by no means been reached, and such problems as the fatigue of metals under continued cyclic stressing have come increasingly to the fore in modern aircraft design.

9-16. CONCLUSION

It is remarkable that quantum theory, originally designed to explain the optical spectra of excited atoms, should, with suitable modifications, prove so successful in explaining such diverse problems as thermal, electrical, mechanical, and optical properties of solids. The application of the quantum theory of metals to their specific heat at low temperatures has brought order to what originally appeared to be an unrelated series of observations. And in this same field, too, a satisfactory theory has finally been found which accounts for the extraordinary phenomenon of superconductivity.

REFERENCES

BRATTAIN, W. H., "Developments of Concepts in Semiconductor Research," *Am. J. Physics* **24,** 421 (1956).

DEKKER, A. J., *Solid State Physics.* Englewood Cliffs, N. J.: Prentice-Hall, 1957.

GINSBURG, D. M., *Superconductivity,* Resource Letter Scy-1 and Selected Reprints. New York: American Institute of Physics, 1964.

GUY, A. G., *Elements of Physical Metallurgy.* 2nd ed. Reading, Mass.: Addison-Wesley, 1959.

KITTEL, CHARLES, *Introduction to Solid State Physics.* 2nd ed. New York: Wiley, 1956.

LE CROISSETTE, DENNIS, *Transistors.* Englewood Cliffs, N. J.: Prentice-Hall, 1963.

SEITZ, F., *The Physics of Metals.* New York: McGraw-Hill, 1943. A comprehensive book.

PROBLEMS

9–1. Verify the statement that at room temperature the order of magnitude of the electrical conductivity of copper is given by Eq. (9–12). The interatomic spacing for copper is 3.6 Å.

9–2. In a metal all of the electronic energy levels may be filled to an energy of the order of several eV. This means that the free electrons contain a great deal of energy. Can this energy be obtained from the metal by electron transition to a lower state? Why?

9–3. The photoelectric threshold for a certain metal is 2900 Å. (a) Assuming a Maxwellian distribution of speeds, what temperature would be assigned to the "electron gas" if the average kinetic energy of translation of the electrons in this metal is 70% of the work function? (b) What is the ratio of the root-mean-square speed of these electrons to that of light? Should relativistic equations be used for these electrons?

9–4. Consider a one-dimensional stretched spring of force constant k with a mass m attached at the end performing simple periodic motion. Show that the time average of the kinetic energy of the mass over a whole cycle is equal to the time average of its potential energy, and hence that the total energy, a constant, is twice the average kinetic energy.

9–5. Assuming that the interatomic distance in a metal crystal is 2 Å and that the conduction electrons have the root-mean-square speed of a Maxwellian distribution at 30,000°K, determine the ratio of the de Broglie wavelength of the electrons at this speed to the interatomic spacing.

9–6. (a) How could one determine experimentally whether electrical conduction is by holes or by electrons? [*Hint:* Consider the effect of a transverse magnetic field on an electric current.] (b) The ratio of the electric field per unit current density per unit magnetic induction, where all three are mutually perpendicular, is known as the Hall coefficient. Show that the Hall coefficient should be a sensitive measure of the number of conducting holes or electrons per unit volume. (c) Show that this coefficient has opposite signs for conduction by electrons and by holes.

9–7. The "mobility," μ, of an electric charge is defined as the velocity increment per unit accelerating electric field. If n is the number of conducting electrons or holes per unit volume of a solid, show that the electrical conductivity is given by

$$\sigma = ne\mu.$$

CHAPTER 10

NATURAL RADIOACTIVITY

10-1. DISCOVERY OF RADIOACTIVITY

We now go back to the year after Roentgen discovered x-rays when, in March 1896, Becquerel announced the discovery of radioactivity. Although Roentgen rays had been discovered less than four months earlier, it was already known that x-rays came from the fluorescent walls of the discharge tube, and thus it was thought that fluorescence and phosphorescence might be responsible for them. Becquerel knew that uranium salts became luminescent when exposed to bright sunlight, and he had heard that the phosphorescent radiations from these activated salts could penetrate opaque bodies. Upon studying these effects, he found that the radiations from light-activated uranium did cast shadows of metallic objects on photographic plates which were wrapped in black paper. But the particularly *new* thing Becquerel found was that the radiations came from uranium salts whether those salts had been excited by light or not. He found that uranium salts which had been protected from all known exciting radiations for months still emitted penetrating radiations without any noticeable weakening. He recognized the parallel between his discovery and the discovery of x-rays, and he found that the new radiations could discharge electrified bodies as x-rays do. He realized that these radiations were not due to fluorescence, but that the uranium metal was their source. This property of uranium, that it spontaneously emits radiation, is called *radioactivity.*

10-2. THE SEAT OF RADIOACTIVITY

In showing that radioactive radiations came from uranium metal, Becquerel worked with many uranium salts and the metal itself. He used these materials crystallized, cast, and in solution. In every case it appeared that the radiations were proportional to the concentration of the uranium. It has been found that this proportionality between radiation intensity and uranium concentration continues unchanged through variations of temperature, electric and magnetic fields, pressure, and chemical composition. Since the radioactive behavior of uranium is independent of the environment of the uranium atom or its electronic structure, which changes from compound to compound, the radioactive properties of uranium were attributed to its nucleus.

10-3. RADIUM

Becquerel's discovery of the radioactivity of uranium immediately raised the question whether other elements are radioactive. Pierre and Marie Curie investigated a uranium ore called pitchblende which contains uranium, bismuth, barium, and lead. Upon chemical separation, the uranium showed the expected activity, and the bismuth and barium fractions also showed activity. Since neither bismuth nor barium shows activity when pure, the Curies assumed that each fraction contained a new element, one chemically like bismuth and the other chemically like barium. They called these new elements polonium and radium and set out to isolate each. The skill, enthusiasm, and patience that went into this task has been beautifully told by her daughter in Madame Curie's biography.

Radium had to be separated from barium on the basis of its slightly different physical properties by the technique of fractional crystallization. The magnitude of the task may be seen from the fact that the Curies separated about one-fifth of a gram of a radium salt from a ton of pitchblende. They could trace their progress by noting the increased activity of the samples as they became more and more concentrated. Weight for weight, polonium is about 10 billion times more active than uranium, and radium is 20 million times more active than uranium.

10-4. THE RADIATIONS

Although the penetrating radiations from radioactive substances were immediately likened to x-rays, Rutherford found, in 1897, that the radiations were of more than one kind, some rays being more penetrating than others. He called the less penetrating rays alpha (α) rays and the more penetrating ones beta (β) rays. In 1899, several investigators found that the beta component of the radiation could be deflected by a magnetic field, and that it had about the same charge-to-mass ratio (e/m_e) as the cathode corpuscles which had been discovered by Thomson just two years earlier. We now know that beta rays* are electrons.

---

* During the 1930's it was discovered that positrons are emitted by some radioactive isotopes, and the term "beta ray" or "beta particle" has now come to mean an electron or a positron of nuclear origin. Their symbols, β^- and β^+, respectively, are used only in connection with nuclear reactions. Because of earlier custom, "beta ray" still usually means an electron when the sign of the charge is not stated.

Madame Curie deduced from their absorption properties that alpha rays were material particles and, in 1903, Rutherford succeeded in deflecting alpha particles with a magnetic field, where the deflection direction showed them to be positive. By causing alpha particles to discharge an electrometer and by simultaneously counting their scintillations, Rutherford determined that the alpha-particle charge was about twice the electronic charge. Since the charge was larger than that of electrons and the magnetic deflection much less, it was obvious that alpha particles are much more massive than electrons.

Conclusive proof that the alpha particles are helium nuclei was given by Rutherford and Royds in 1909. Their apparatus is shown in Fig. 10–1. The radioactive sample was placed inside a glass tube G which was so thin that alpha particles could pass through it. The tube G was inside an evacuated tube T which had a narrow end E with electrodes sealed into it. By raising the level of the mercury, any gas in T could be compressed into E. An electric discharge through this gas produced its spectrum and permitted the positive identification of the gas. With the tube G empty no helium was found, but with radioactive material in G, helium was found in the apparatus after two days. The alpha particles were trapped in the tube T, where they picked up electrons and became helium atoms.

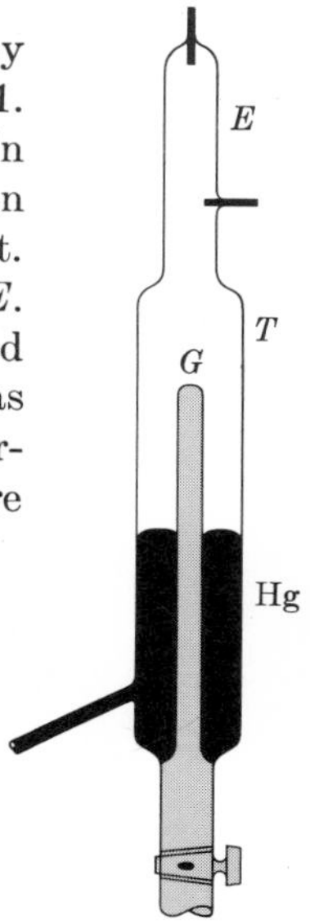

E

T

G

Hg

FIG. 10–1
Apparatus of Rutherford and Royds for identifying alpha particles.

The identification of alpha particles as helium nuclei makes it clear that radioactivity is a drastic process in which elements change in kind. If radium emits helium, it can no longer be radium. The *parent* or original substance becomes a new element called the *daughter* or *decay product* substance. Although the element lead cannot be changed into gold (the alchemists' dream of centuries before), the experiment by Rutherford and Royds leaves no doubt that one element may be transmuted into another element.

In 1900, Villard found a third kind of radiation from radioactive materials which is even more penetrating than either alpha or beta rays. These rays, called gamma (γ) rays, are not influenced by magnetic fields and therefore carry no charge. Their energy can be measured by measuring the energy of the photoelectrons they produce. They can be diffracted by crystals. Their wavelengths are found to range from about 0.5 to 0.005 Å. Crystal diffraction of these very short wavelengths is difficult and special techniques must be used, but we know that these rays are photons whose energy range overlaps that of x-rays and extends to several MeV. The term "gamma ray" is restricted to short-wavelength radiation from the nucleus, and "x-ray" is used only for similar radiation from the extranuclear part of the atom.

Most pure radioactive substances emit gamma rays accompanied by either alpha or beta rays, but since samples are seldom pure, one usually finds all three types of rays present. The three types of radiation can be characterized by the famous diagram of Fig. 10–2, which shows how the rays are deflected by a magnetic field.

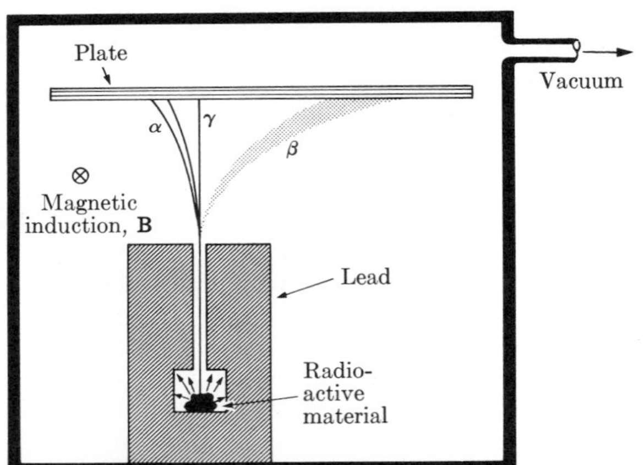

FIG. 10–2
Magnetic separation of the radiations from radioactive materials.

These radiations from the nucleus are absorbed as they pass through matter. Gamma rays are absorbed exponentially. Both the linear and mass absorption equations for x-radiation, Eqs. (6–10) and (6–14), respectively, are valid for gamma rays, although the absorption coefficients for gamma rays are usually much less than they are for x-rays. The absorption of beta rays is more complex. It is approximately exponential up to a certain thickness of the absorber and then absorption becomes complete. The exponential relation is entirely fortuitous in this case. As beta particles are stopped in an absorber some of their energy is converted into bremsstrahlung. The absorption coefficients for beta rays are less well defined than for x- and gamma rays, but in one respect the absorption of beta rays is simpler than photon absorption. Because the mass absorption coefficient is nearly the same for all absorbing materials, the mass per unit area, m_a, required to produce a given beta absorption is independent of the material. The value of m_a for an absorber of either alpha or beta rays is often called the "thickness" of the material. The absorption of alpha particles will be discussed in Chapter 11.

10-5. RADIATION DETECTORS

Before considering the laws of radioactive disintegration, we digress to summarize and extend our descriptions of the techniques used to observe the radiations from radioactive elements.

We have discussed fluorescence, which is sensitive to radiations with energies equal to or greater than those of visible photons. Fluorescence gives visible evidence of the radiation, but the radiation must be intense for this to occur.

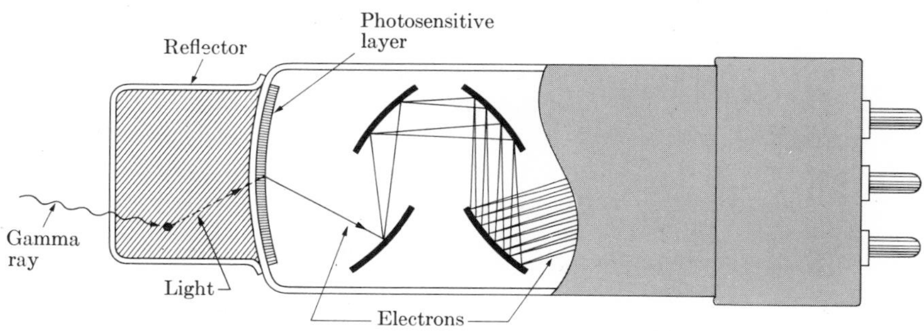

FIG. 10-3
Photomultiplier tube.

We have discussed ionization chambers, which are used to measure the ionization currents radiation produces. This technique is quite sensitive in the same energy range as fluorescence, and it is quantitative.

We have discussed the use of photographic materials, which are sensitive to visible radiations and to those of higher energies. Since the effect is cumulative, the method is very sensitive if long exposures can be employed. The method is quantitative if the density of the image is carefully measured.

We now turn to *scintillation*, which is really fluorescence examined closely. Because it can detect individual particles, this technique is extremely sensitive. If the scintillations are to be visually observed, the particles of radiation must have high energy, that is, they must be capable of producing a large number of visible photons when they strike the fluorescent material. However, very sensitive scintillation techniques developed during the past few years have eliminated the necessity for visual observation. Instead, the light is allowed to fall on a very sensitive photoelectric cell called a *photomultiplier tube* (Fig. 10-3), in which a single photoelectron is accelerated until it can knock "secondary" electrons from an anode. This anode is the cathode relative to another anode, where more secondary electrons are liberated. This process continues through several stages until the burst of electrons becomes great enough to constitute a disturbance that can be amplified by an electronic amplifier. Pulses from this amplifier are then electronically counted. By this technique, individual radiation particles of rather low energy can be counted.

A remarkable property of a scintillation detector and photomultiplier tube combination is that the electric pulses produced are usually proportional to the energy of the incident gamma rays. This permits not only gamma ray detection but a measurement of their energy. A sophisticated electronic device (really a specialized computer) called a *multi-channel analyzer* can accept the electrical pulses, sort them into size intervals, and keep track of the number of pulses of each size entering the instrument. The resulting data constitute an energy spectrum of the gamma rays entering the scintillation detector. (Energy spectra are discussed in Section 10–8.) Although this method of measuring gamma-ray energies is relative—the instrument must be calibrated with a source of gamma rays of known energy—and although the precision is low compared to diffraction from crystals, the method is quick and has many applications.

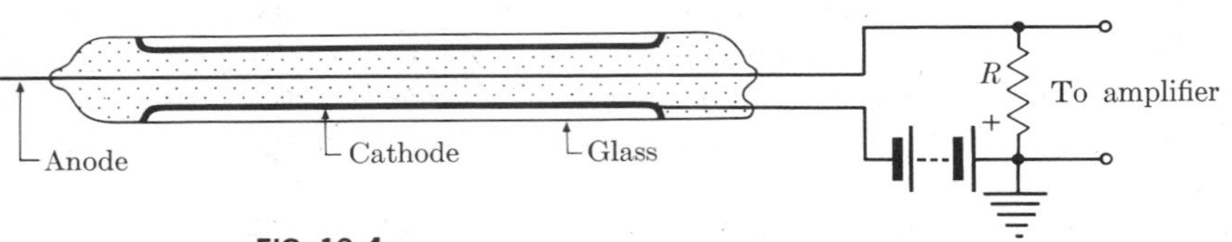

FIG. 10–4
Geiger-Mueller counter tube.

10–6. GEIGER-MUELLER COUNTERS

The Geiger-Mueller or G-M counter tube, Fig. 10–4, evolved from the ionization chamber. The tube, usually made of glass, is filled with a gas and contains two electrodes. One of these is a metal cylinder and the other a thin wire mounted coaxially with the cylinder. In a counter circuit, the d.c. potential applied between the wire and cylinder is often 1000 V or more.

In the Geiger tube, a measurable current is initiated by the ions formed by the radiation under study. The ions are swept out of the region between the cylinder and the wire by an electric field strong enough to prevent the recombination of the ions and weak enough so that the dielectric strength of the gas is not exceeded. If this electric field is made very intense, the tube can be brought to a critical condition, where the introduction of a single charge will initiate a discharge. Electrons from the original (primary) ion pair produce more ions by collision. In the intense electric field near the central electrode, the secondary ions produce still more ions, in a geometric progression. Thus a single electron may initiate a cascade of a million or more electrons. When this avalanche strikes the central electrode, the drop in potential is easily detected electronically. Meanwhile, the relatively massive positive ions migrate toward the outer cylinder, where they pick up electrons and become excited atoms or molecules. If these atoms emit photons which can liberate photoelectrons from the outer cylinder, these photoelectrons may initiate other avalanches which tend to keep the tube in continuous discharge. But if the gas of the tube is an organic gas which dissociates easily, the excitation energy may cause the molecule to come apart rather than to emit a photon. In the first case, the electric discharge must be stopped by reducing the potential across the tube to nearly zero. This available electrode potential difference does become very small each time a pulse of current resulting from ionization flows around the circuit containing the high resistance R shown in the figure. In the second case, the life of the tube is reduced as the chemical composition of the organic gas is changed. Thus in either type of Geiger counter the discharge must be "quenched" between counts. Since a finite time is required for quenching and for starting the voltage rise across the electrodes, there is always a dead time between possible counts. This is of the order of 10^{-4} s. Geiger counters readily respond to low-intensity radiations which have enough energy to enter their sensitive region. If, however, the intensity of the radiation is too high, a Geiger counter "freezes" or "jams" because the interval between successive primary ionizing events is less than the dead time.

Modern G-M counters can have no open window, since the gas is not air and the pressure is below atmospheric. High-energy particles can penetrate into the sensitive region within the metal cylinder from any direction, but lower energy particles enter more readily through the ends. Since particles of very low energy are stopped by the glass, the absorption of the glass is sometimes reduced by making an extremely thin window at one end. These windows are almost invisible and are very fragile. Their "thickness" is usually given in terms of their mass per unit area. Values of the order of 3 mg/cm^2 are common. Such windows are so thin that they easily transmit short-wavelength ultraviolet light. These photons have enough energy to "trigger" the tube by causing photoemission from the electrodes, or they even ionize the gas in a few cases. A match flame furnishes ample ultraviolet light to operate a thin-window, clear-glass tube.

10-7. CLOUD AND BUBBLE CHAMBERS

Certainly the most dramatic radiation detectors are cloud and bubble chambers. The principle of their operation is basically the same, although cloud chambers contain a supercooled vapor, while a bubble chamber contains a superheated liquid.

A supercooled vapor is most likely to condense on some discontinuity, and once condensation begins it continues readily by condensing onto droplets already formed. Supercooled water vapor may remain in the vapor phase and not condense, but in dusty and smoky regions there are discontinuities on which condensation can begin and therefore fog or smog form readily.

The cloud chamber illustrates the fact that charged particles are discontinuities on which condensation can be initiated. The schematic diagram of a Wilson cloud chamber is shown in Fig. 10–5. It consists of a gastight chamber having a large glass window, so that its volume can be illuminated and the events within seen or photographed. One wall of the chamber is a movable piston. The chamber contains a saturated vapor from an excess amount of some liquid, usually a mixture of alcohol and water. If the volume of the chamber is suddenly increased by moving the piston, the adiabatic expansion causes cooling which renders the vapor supersaturated, unstable, and likely to condense. If there are ions within the chamber, condensation occurs preferentially on them.

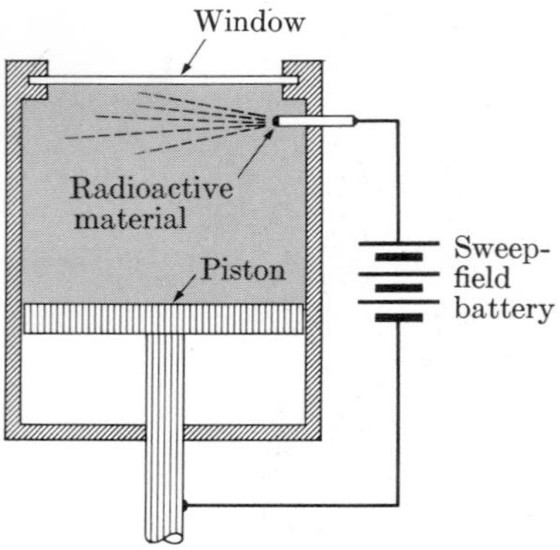

FIG. 10–5
A simple expansion cloud chamber.

Since we have no interest in stray ions that may happen to be about, the cloud chamber is provided with electrodes which permit maintaining a weak electric "sweep field."

If a high-energy particle enters the chamber just before the piston expansion makes the chamber sensitive, the ions will not have had time to be swept away (the sweep field is sometimes cut off just before the expansion), and the droplets that form make the location of the ions visible. The droplets tend to move because of gravity and gas turbulence, and since they also tend to evaporate as a new equilibrium condition is established, the "picture" soon spoils. If the piston is then returned to its original position for about a minute the chamber can again be expanded.

If a high-energy electron passes through this chamber the picture formed (Fig. 10–6) is a thin, beady line or "track" which shows very beautifully where the electron went. The droplet on each ion in its wake becomes visible. If a magnetic field is established in the space of the cloud chamber, the velocities of the electrons can be measured. Cloud-chamber photographs, such as Figs. 5–5 and 5–6 which accompanied the discussion of pair production, permit the measurement of the radius of curvature of the path of a charged particle, and since m_e, e, and B are known, the velocity can be computed from Eq. (2–5).

We first considered alpha particles in the Rutherford scattering experiment. Figure 10–7 shows that these doubly ionized helium nuclei produce

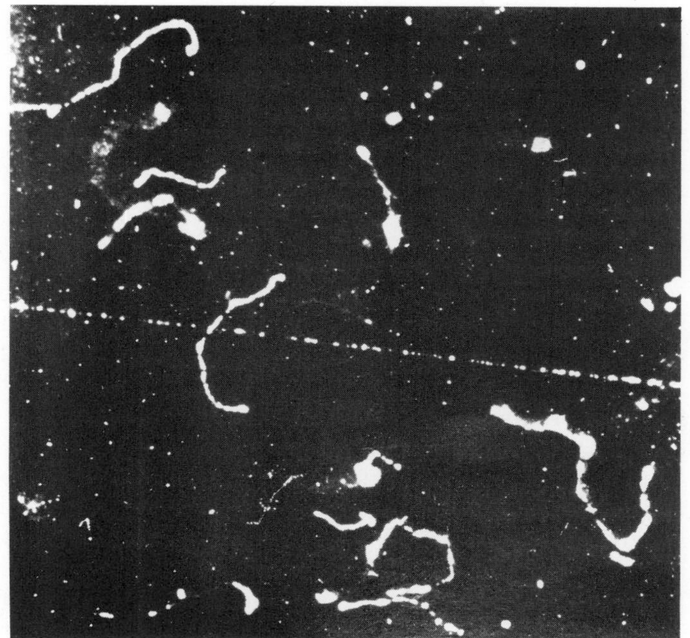

FIG. 10–6
Beady track produced by a high-energy electron. The broader tracks are due to low-energy photoelectrons. (Courtesy of C. T. R. Wilson and the Royal Society, London)

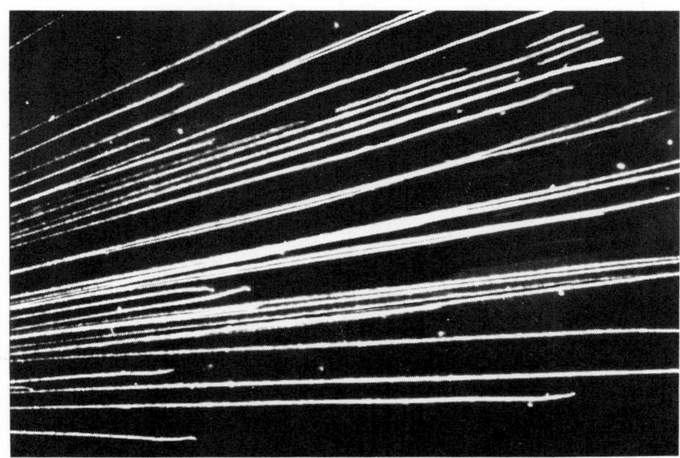

FIG. 10–7
Alpha-particle tracks. Note that there are two groups, which differ in range. (Courtesy of P. M. S. Blackett and D. S. Lees, Imperial College of Technology, London)

very dense cloud-chamber tracks that resemble the vapor trails of high-flying airplanes. Alpha trails are easily distinguished from those made by electrons. The former are more dense, primarily because alpha particles carry twice as much charge as do electrons. But since alpha particles are much more massive than electrons, they are very difficult to deviate in a magnetic field. It turns out, however, that alpha-particle energies are a function of the lengths of their tracks, or ranges. This energy-measuring stick has been calibrated in the following ingenious way. If the relative proportions of vapor and gas in the chamber are known, the average ionization energy of the chamber atoms can be computed. If there is a momentary lapse between the chamber expansion and the photographic exposure, the track becomes diffuse and it is then possible to count the individual drops. Determining the number of ions produced by the alpha particle permits calculating the energy loss along its path.

Alpha particles produce an average of about 50,000 ion pairs per centimeter in air at atmospheric pressure, while beta particles produce only about 50 ion pairs per centimeter. Since both kinds of particles lose energy by giving up about 34 eV per ion pair produced, alpha tracks are short and thick, and beta tracks are long and thin. If the two kinds of particles have the same initial energies, each produces about the same total number of ions. The slower the particle moves, the more ions it produces per unit length of path. (This will be discussed in Section 11–1.) Thus tracks are more dense near their ends, and the range-energy relationship is not linear. Only charged particles produce cloud-chamber tracks, although photoelectrons ejected by photons do leave short, feathery trails which give some idea of the photon path. To the practiced eye, cloud-chamber tracks due to different particles are as dissimilar as the tracks different animals leave in the snow.

Cloud-chamber studies have been exceedingly revealing. These chambers are much more than radiation detectors. They almost enable us to see atomic processes, and sometimes the analysis of a single picture has led to a basic discovery.

Extremely high-energy particles, such as are found in cosmic rays or are produced by the most modern nuclear accelerators, produce cloud-chamber tracks that are so long the chamber shows only a small fraction of the event. Such events can be studied better with bubble chambers, where the instrument is filled with a dense liquid instead of a gas.

FIG. 10–8
A 4-MeV electron slowing down in a liquid hydrogen bubble chamber traversed by a magnetic field. (Courtesy of Radiation Laboratory, University of California)

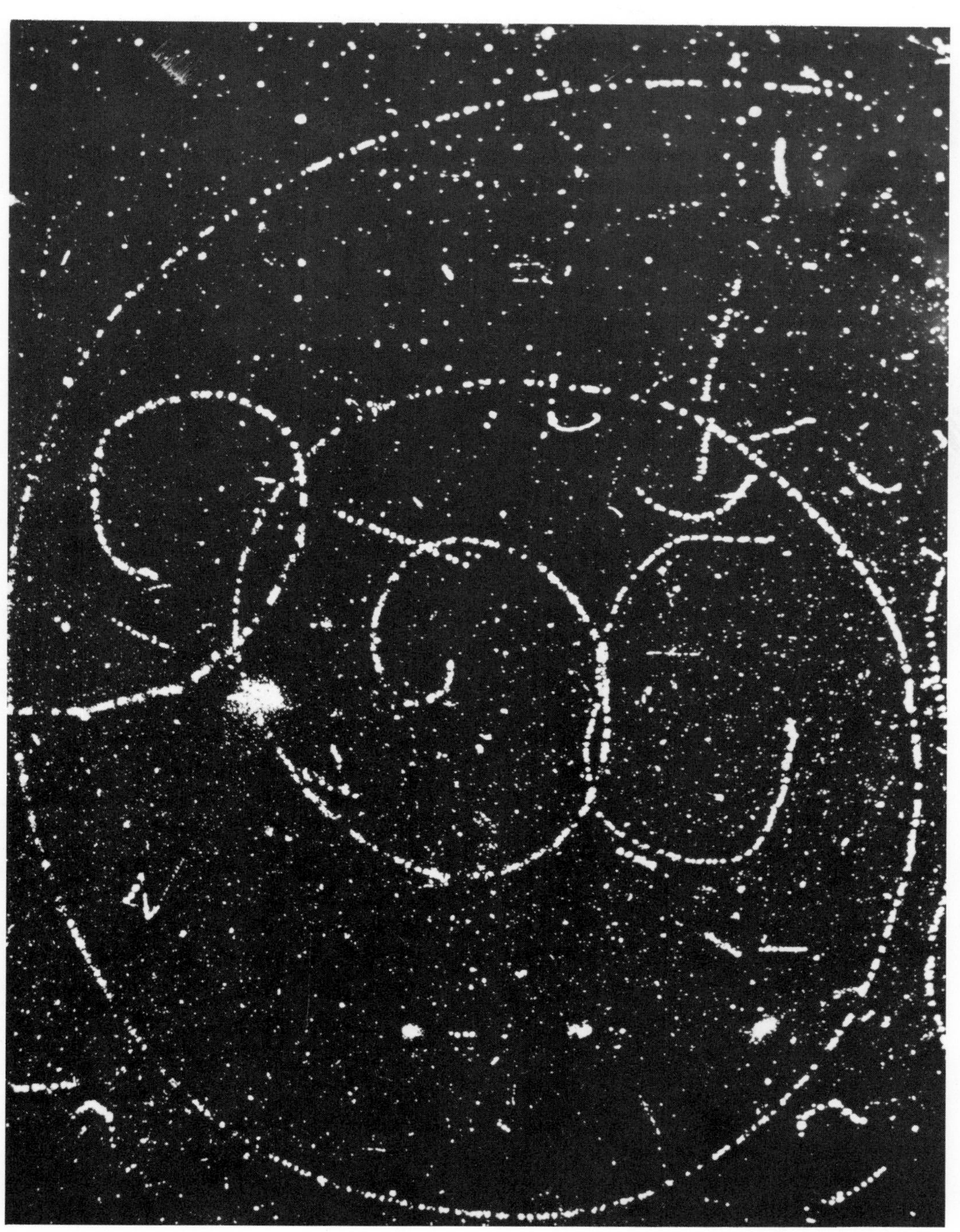

Consider for a moment a bottle of "pop." Before it is opened, the carbon dioxide over the drink and that dissolved in it are in equilibrium. When the bottle is opened, the pressure is reduced and the solution of gas in the liquid is supersaturated. Bubbling results from the unstable condition and we have effervescence. That the bubbles form most readily on discontinuities is best demonstrated by pouring the drink into a glass and noting that often there are streams of bubbles rising from some speck inside the glass. These bubbles are formed because the same discontinuity is a bubble nucleus over and over again. Adding sugar, salt, cracked ice, or ice cream provides many discontinuities, so that most of the carbon dioxide effervesces almost at once. Similarly, when ions are produced in a bubble chamber they act as nuclei for the formation of small bubbles.

In actual bubble chambers, a liquid such as isopentane or liquid hydrogen is maintained at a temperature above its normal boiling point but is prevented from boiling by the application of pressure. The chamber is made sensitive by suddenly reducing the pressure. Since boiling starts preferentially on the ions in the liquid, the bubbles formed along the path of an ionizing particle are visible, much as are the droplets in a cloud chamber. Such a trail of bubbles is shown in Fig. 10–8. A bubble track can be observed only during the brief period (a few milliseconds) before general boiling begins throughout the liquid.

10–8. ENERGIES OF THE RADIATIONS. NUCLEAR SPECTRA

One gram of radium gives off nearly two gram-calories of heat per second. This may not seem like much energy, since the oxidation of a gram of carbon to carbon dioxide liberates about 8000 gram-calories. But whereas the carbon might be oxidized in one second and be consumed, the gram of radium will continue to liberate two calories per second for years and the rate will only be reduced to one calorie per second in 1600 years. If we allow infinite time for the gram of radium to disintegrate, more than 10^{17} calories would be liberated, and this calculation does not take into account the energy liberated by the daughter products of the radium. If radium were cheap and plentiful, it would be a useful source of energy.

Since the energies of alpha, beta, and gamma rays from radioactive substances can be measured, we speak of the "spectra" of various radioactive substances. The nucleus is not nearly so well understood as the atom's electronic structure. There is no nice theory of the nucleus to compare with the Bohr analysis of the hydrogen atom. But the evidence from nuclear spectroscopy indicates that nuclei do have energy levels. Thus Pa^{231} (protactinium) emits alpha particles with nine distinct energies and also thirteen different gamma-ray wavelengths.

The existence of energy levels makes it clear that nuclear phenomena are quantum-mechanical. The spectra of beta particles, however, presented an

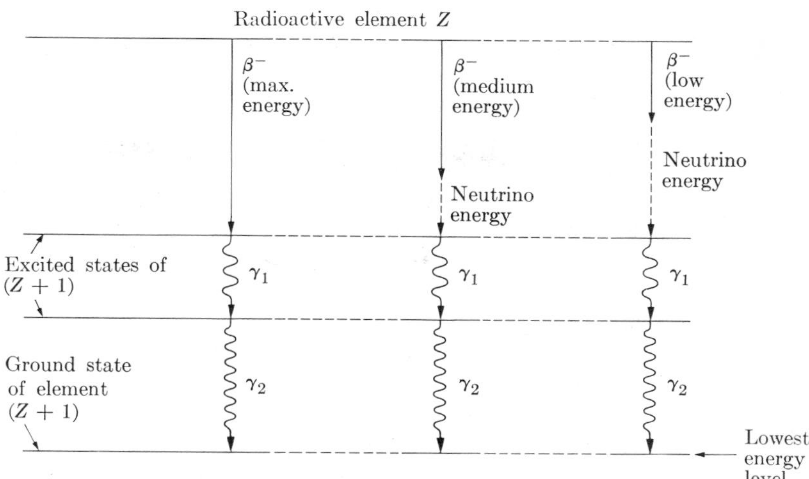

FIG. 10–9
Typical disintegration scheme of beta-particle emitter: (a) beta particle with maximum energy, and (b) two cases of beta-neutrino emission.

interesting problem. Although a given substance emits beta particles with a definite upper limit of energy, a continuous range of energies below that limit is found. The reason for an upper limit is evident from Fig. 10–9(a), which shows the energy-level diagram for a typical beta-ray emitter. In this case, a nucleus having an atomic number Z ejects an electron and thus changes to an excited state of a new nucleus, $Z + 1$. This then goes to the stable state by emitting two gamma-ray photons in cascade. The single energy transition of the electron accounts for the maximum beta-particle energy, but fails to account for the continuous range of smaller beta energies found when a large number of atoms of a substance are observed. It appeared, at first, that the law of conservation of energy was violated. Of all the generalizations of physics, conservation of energy is the most basic. We would like to think that this principle holds exactly and without exception.

The difficulty was resolved by supposing that there is a new particle, called a *neutrino* (ν), which carries away the energy difference (refer to Fig. 10–9b). Conservation of energy and of momentum required that this new particle have much less mass than the electron and that it be uncharged. This particle, proposed by Pauli in 1931, could carry away the energy difference undetected because of its small mass and zero charge. In addition to preserving the exactness of conservation of energy, the neutrino fitted into other theoretical considerations in the theory of beta decay developed by Fermi in 1934. The existence of this new particle was generally accepted long before it was experimentally observed in 1956 by Reines and Cowan.

Although nuclear energy states, unlike electronic energy states, cannot be computed theoretically, we apply the electronic language to the nucleus and speak of excited states, ground states, etc.

10-9. LAW OF RADIOACTIVE DISINTEGRATION

Every radioactive disintegration involves the emission of either an electron or a helium nucleus from the nucleus of the disintegrating atom. This leaves the original nucleus changed, so that the number of atoms of the original kind is reduced. Furthermore, scintillation observations show that these disintegrations occur in a random manner which can be discussed only statistically. The idea of treating the disintegrations statistically may have been originally a matter of convenience, but we have seen that wave mechanics makes the statistical analysis of such quantum phenomena a necessity.

The discovery of radioactivity antedated Planck's quantum idea, Einstein's quantum interpretation of the photoelectric effect, Bohr theory, and wave mechanics. The unpredictability of individual radioactive disintegrations was really the *first* experimental indication that the mechanistic determinism in the physical sciences of the 19th century was due for revision. Thus radioactivity had two shocking aspects: it showed that the elements were not inviolate but could change from one kind to another, and that radioactivity was not subject to causal analysis. This latter aspect was assimilated as an integral and useful part of physics, particularly after the advent of wave mechanics in 1925.

To present the quantitative description of radioactivity, let us begin at time $t = 0$ with a large number of radioactive atoms N_0. Experimentation shows that the number of disintegrations per unit time $(-dN/dt)$, or the *activity*, is proportional to the number of atoms N of the original kind present. The constant of proportionality λ is called the *disintegration* or *decay constant*. Therefore we have

ACTIVITY

$$\frac{-dN}{dt} = \lambda N. \tag{10-1}$$

The quantity dN must be an integer, but if it is a relatively small integer compared with N, then N will vary in an approximately continuous way and we may treat dN as a mathematical differential. Upon integrating and using the initial condition, we have for the *number of atoms of the original kind still present* at time t

$$N = N_0 e^{-\lambda t}. \tag{10-2}$$

$\frac{N}{N_0} = e^{-\lambda t}$

The statistical nature of the exponential law of radioactive decay, Eq. (10–2), is evident from the following derivation, in which the law is obtained without any special hypothesis about the structure of the radioactive atoms or about the mechanism of disintegration. Let us assume that the disintegration of an atom depends on the laws of chance, and that the probability P for an atom to disintegrate in a short time interval Δt is the same for all the atoms of the same kind and that it is independent of the past history or age of the atom. (Note that assuming that a quantity such as P is independent of age would not hold for life-experience tables.) Thus the probability of disintegration depends only on the time interval, and, for very short intervals, is proportional to the interval. Therefore $P = \lambda\,\Delta t$, where λ, the constant of proportionality, is the decay constant. The probability Q_1 that a given atom will *not* disintegrate during the interval Δt is $Q_1 = 1 - P = 1 - \lambda\,\Delta t$. The probability of the atom's not disintegrating in time $2\,\Delta t$, assuming that P is independent of the age of the atom, is

$$Q_2 = (1 - P)(1 - P) = (1 - \lambda\,\Delta t)^2.$$

Thus, in general, the probability of the atom's surviving n intervals is $Q_n = (1 - \lambda\,\Delta t)^n$. Considering a finite time t made up of a large number n of intervals Δt, we find that $t = n\,\Delta t$ or $\Delta t = t/n$. Therefore the probability of the atom's surviving or remaining unchanged for a time t is $Q_n = (1 - \lambda t/n)^n$. But the limit of this quantity as n becomes very large is N/N_0, which is the ratio of the number of atoms that remain unchanged at the end of time t to the total number of atoms originally present. To find this limit, we first expand the expression for Q_n by the binomial theorem. This gives

$$Q_n = \left(1 - \frac{\lambda t}{n}\right)^n = 1 - n\frac{\lambda t}{n} + \frac{n(n-1)}{2!}\frac{\lambda^2 t^2}{n^2} - \frac{n(n-1)(n-2)}{3!}\frac{\lambda^3 t^3}{n^3} + \cdots$$

$$= 1 - \lambda t + \left(1 - \frac{1}{n}\right)\frac{\lambda^2 t^2}{2!} - \left(1 - \frac{1}{n}\right)\left(1 - \frac{2}{n}\right)\frac{\lambda^3 t^3}{3!} + \cdots. \tag{10–3}$$

If n is very large, the preceding relation reduces to

$$Q_n = \frac{N}{N_0} = 1 - \lambda t + \frac{\lambda^2 t^2}{2!} - \frac{\lambda^3 t^3}{3!} + \cdots = e^{-\lambda t} \tag{10–4}$$

or

$$N = N_0 e^{-\lambda t}.$$

This decay law is thus the result of a large number of events subject to the laws of chance. Therefore the activity λN is an average value. The actual values during successive short time intervals fluctuate around this average. These variations must be taken into account when measurements are made with weak radioactive sources.

Each nuclear disintegration makes an equal contribution to the beam of radiation from a radioactive material. Thus the intensity I of the beam is directly proportional to $-dN/dt$. If we call I_0 the intensity when $t = 0$, then Eqs. (10–1) and (10–2) lead to

$$I = I_0 e^{-\lambda t}. \tag{10–5}$$

If we plot ln N or ln I against t, both Eqs. (10–2) and (10–5) become straight lines whose slopes are $-\lambda$. This sometimes is a convenient way of obtaining λ from experimental data. These equations are identical in form to that which describes the intensity of monochromatic x-rays as a function of absorber thickness. They indicate that the number of atoms and the intensity of radiation require infinite time to disappear completely. (It must be remembered that differential calculus breaks down and dN becomes a step function when N is small.)

Although we cannot get a meaningful answer to the question "How long will a given radioactive sample last?," we can compute how long it will be before the sample is reduced to some fraction of its initial amount. Just as we computed the thickness of absorber that will reduce the intensity of an x-ray beam to one-half its initial value, we can compute the *half-life*, $t_{1/2}$ or T, of a radioactive sample. If we let $N = N_0/2$ in Eq. (10–2), then $t = T$ and we have

$$\boxed{T = \frac{1}{\lambda} \ln 2 = \frac{0.693}{\lambda},} \tag{10–6}$$

which is similar to Eq. (6–12), the half-value layer for x-rays. Figure 10–10 is a typical graph of radioactive decay as a function of time in half-lives.

Before we can use any of these equations, the disintegration (decay) constant λ must be known. It depends on the radioactive material and varies widely from material to material, but this constant is completely independent of the environment, whether it is chemical combination, pressure, temperature, or whatever.

If λ is very small, the half-life is very large (1620 years for radium) and the reduction in N during an experiment is not significant. By measuring the disintegration rate dN/dt (with a scintillation counter, for example) from a sample of N atoms, λ can be computed from Eq. (10–1).

If λ is large enough so that the decay of the sample during an experiment is significant, then the reduction in intensity may be observed as a function of time and fitted to Eq. (10–5). A simple way to do this is to measure the time it takes for the intensity to be reduced to one-half (T) and solve for λ by means of Eq. (10–6).

If λ cannot be measured by either of the two methods described, then the decay constant must be determined from radioactive equilibrium, which will be discussed in Section 10–13.

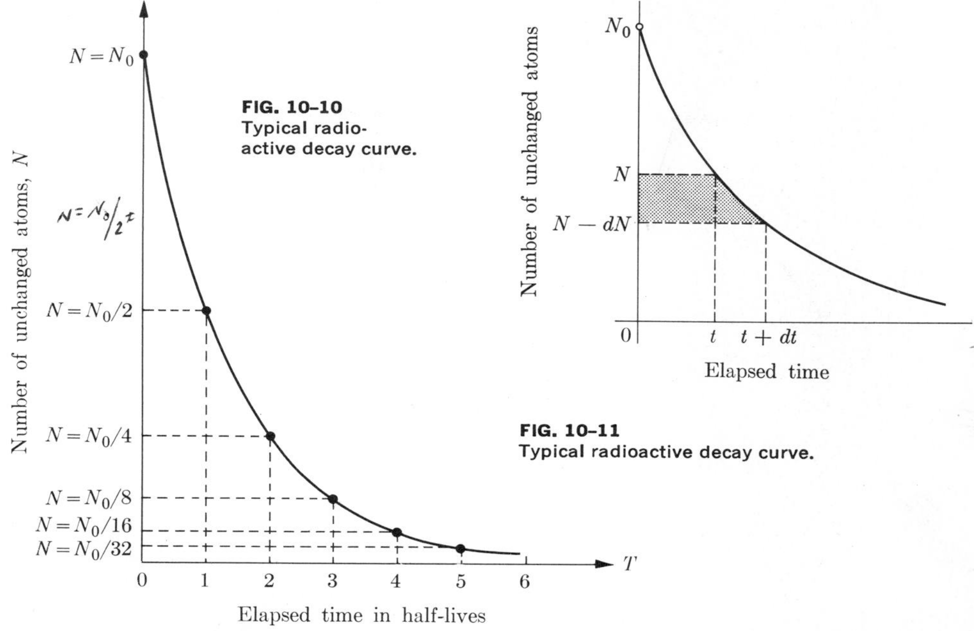

FIG. 10-10 Typical radioactive decay curve.

FIG. 10-11 Typical radioactive decay curve.

We can make another calculation which is conceptually important to the statistical view we are taking. Let us find the *mean* or *average life*, $\bar{t}$, of a radioactive atom. Some atoms survive much longer than others, so what we seek is the numerical average of the ages of the atoms as the number of atoms decreases from N_0 to 0. The expression for determining this can be found by considering the radioactive decay curve of Fig. 10–11. Let N be the number of atoms that have survived for a time t, and $N - dN$ those still existing at the time $t + dt$; then dN is the number of atoms that disintegrated during the time dt. Therefore the combined ages of the atoms in this group at the time of disintegration is $t\,dN$. The average life of an atom will be the sum of the combined ages of all the age groups of atoms from N_0 to 0 divided by the total number of atoms. Stated in mathematical terms, this is

$$\bar{t} = \frac{\int_{N_0}^{0} t\,dN}{\int_{N_0}^{0} dN}. \tag{10–7}$$

The integrals in this equation may be evaluated in the same way as in the derivation of Eq. (1–18). The result is

$$\boxed{\bar{t} = \frac{1}{\lambda}.} \tag{10–8}$$

Thus $\bar{t} = 1/\lambda$ is the mean life or life expectancy of an individual radioactive atom.

EXAMPLE. (a) If a radioactive material initially contains 3.00 mg of U^{234}, how much will remain unchanged after 62,000 y? (b) What will be its U^{234} activity at the end of that time? ($T = 2.48 \times 10^5$ y, $\lambda = 8.88 \times 10^{-14}\ s^{-1}$.)

Solution. (a) From Eq. (10–2), we have

$$N = N_0 e^{-\lambda t}.$$

The N's are computed from the Avogadro constant and the mass number (strictly isotopic mass) of the material. In this example, it will be convenient to express λ in terms of T. Let m be the mass of uranium remaining, then the quantities to be substituted for the terms in the equation are

$$N_0 = 3.00 \times 10^{-3}\ \text{g} \times \frac{6.02 \times 10^{23}\ \text{atoms}}{234\ \text{g}},$$

$$N = m \times \frac{6.02 \times 10^{23}\ \text{atoms}}{234\ \text{g}},$$

and

$$\lambda t = \frac{0.693}{T} t = \frac{0.693}{2.48 \times 10^5\ y} \times 6.2 \times 10^4\ y = 0.173.$$

Substituting these quantities in the radioactive decay equation, we get

$$m \times \frac{6.02 \times 10^{23}\ \text{atoms}}{234\ \text{g}} = 3.00 \times 10^{-3}\ \text{g} \times \frac{6.02 \times 10^{23}\ \text{atoms}}{234\ \text{g}} \times e^{-0.173},$$

or

$$m = 3.00 \times 10^{-3} \times e^{-0.173}\ \text{g}.$$

Therefore

$$\ln \frac{3.00 \times 10^{-3}}{m} = 0.173,$$

or

$$m = 2.52 \times 10^{-3}\ \text{g}.$$

(b) From Eq. (10–1) we have

$$\frac{dN}{dt} = \lambda N$$

$$= \frac{8.88 \times 10^{-14}}{1\ \text{s}} \times 2.52 \times 10^{-3}\ \text{g} \times \frac{6.02 \times 10^{23}\ \text{atoms}}{234\ \text{g}}$$

$$= 5.76 \times 10^5\ \text{dis/s}.$$

10–10. RADIOACTIVE SERIES

We have stated that when a radioactive disintegration occurs with the emission of an alpha or beta particle, the original atom, called the parent, changes into something else, called the daughter. In 1903 Rutherford and

Soddy proposed that the nature of the daughter could be inferred from the nature of the parent and the particle emitted. Since we have already established the concepts of the atomic nucleus, atomic number, and atomic mass numbers, we can state the Rutherford-Soddy rules for balancing nuclear reaction equations in modern terms. They are: (1) *the total electric charge (atomic number) or algebraic sum of the charges before the disintegration must equal the total electric charge after the disintegration*, and (2) *the sum of the mass numbers of the initial particles must equal the sum of the mass numbers of the final particles.* Thus if uranium with atomic number 92 emits an alpha particle with atomic number 2, the daughter must have an atomic number 90. This element is thorium. Similarly, since uranium has a mass number of 238 and the alpha particle a mass number of 4, the thorium must have a mass number of 234. Both rules are summarized in the equation

$$_{92}U^{238} \rightarrow {}_{90}Th^{234} + {}_{2}He^{4}.$$

If the parent is a beta emitter, the atomic number of the daughter must be one higher than that of the parent, since the beta particle is a negative electron. Furthermore, the electron is so light that the atomic mass numbers of the parent and daughter are the same. For example, Th^{234}, daughter of U^{238}, is radioactive and is a beta emitter. The daughter of Th^{234} must have the atomic number 91 (protactinium, Pa) and the mass number 234, as shown in the equation

$$_{90}Th^{234} \rightarrow {}_{91}Pa^{234} + {}_{-1}e^{0}.$$

The Rutherford-Soddy rules came before the establishment of the physical atomic weights or the discovery of isotopes by Thomson and Aston, described in Chapter 2. These rules predicted daughter atoms with atomic weights quite different from the then-accepted chemical atomic weights. Rutherford and Soddy had more faith in their rules (based on conservation of mass and charge) than they had in the chemical atomic weights. Their rules implied that the same chemical element could exist in forms having different masses and they predicted the discovery of isotopes, which we have already discussed.

We now know that there are three series of naturally radioactive elements which form a sequence of parent-daughter relationships: the uranium, actinium, and thorium series.* They are shown graphically in Figs. 10–12,

---

* The uranium series is also called the $4n + 2$ series because the mass numbers of the atoms in it are given by this expression. The quantity n is an integer which decreases by unity in going from any radioelement to the next one below it. The actinium series can be represented by $4n + 3$, and that of thorium by $4n$. The value of n is not the same for the first element in each series.

TABLE 10-1
THE URANIUM SERIES

Radioactive species Historic names	Nuclide	Type of disintegration	Half-life T	Disintegration constant λ, s^{-1}	Principal particle energy, MeV
Uranium I (UI)	$_{92}U^{238}$	α	4.51×10^9 y	4.87×10^{-18}	α, 4.18; γ, 0.05
Uranium X_1 (UX_1)	$_{90}Th^{234}$	β	24.1 d	3.33×10^{-7}	β, 0.19; γ, 0.03–0.09
Uranium X_2 (UX_2)	$_{91}Pa^{234}$	β	1.18 m	9.77×10^{-3}	β, 2.31; γ, 0.23–1.8
⋮					
Uranium Z (UZ)	$_{91}Pa^{234}$	β	6.7 h	2.88×10^{-5}	β, 0.5; γ, 0.04–1.7
Uranium II (UII)	$_{92}U^{234}$	α	2.48×10^5 y	8.88×10^{-14}	α, 4.76; γ, 0.5–0.12
Ionium (Io)	$_{90}Th^{230}$	α	7.6×10^4 y	2.89×10^{-13}	α, 4.69; γ, 0.07–0.25
Radium (Ra)	$_{88}Ra^{226}$	α	1620 y	1.36×10^{-11}	α, 4.78; γ, 0.19
Ra emanation (Radon)	$_{86}Rn^{222}$	α	3.82 d	2.10×10^{-6}	α, 5.49; γ, 0.51
*Radium A (RaA)	$_{84}Po^{218}$	β, α	3.05 m	3.78×10^{-3}	α, 6.00; β . . .
Astatine-218 (0.02)	$_{85}At^{218}$	α	1.3 s	0.53	α, 6.69
Radium B (RaB) (99^+)	$_{82}Pb^{214}$	β	26.8 m	4.31×10^{-4}	β, 0.65; γ, 0.05–0.35
*Radium C (RaC)	$_{83}Bi^{214}$	β, α	19.7 m	5.86×10^{-4}	α, 5.5; β, 0.40–3.2 γ, 0.61–2.43
Radium C′ (RaC′) (99^+)	$_{84}Po^{214}$	α	1.64×10^{-4} s	4.23×10^3	α, 7.68
Radium C″ (RaC″) (0.02)	$_{81}Tl^{210}$	β	1.30 m	8.75×10^{-3}	β, 1.99; γ, 0.10–2.43
Radium D (RaD)	$_{82}Pb^{210}$	β	22 y	1.00×10^{-9}	β, 0.02; γ, 0.05
*Radium E (RaE)	$_{83}Bi^{210}$	β, α	5.0 d	1.60×10^{-6}	α, 4.7; β, 1.16
Radium F (RaF) (99^+)	$_{84}Po^{210}$	α	138.4 d	5.80×10^{-8}	α, 5.30; γ, 0.80
Thallium-206 (2×10^{-4})	$_{81}Tl^{206}$	β	4.3 m	2.68×10^{-3}	β, 1.57
Radium G (RaG)	$_{82}Pb^{206}$	Stable			

10–13, and 10–14, respectively, and also are tabulated in Tables 10–1, 10–2, and 10–3. Each series ends with a stable isotope of lead. Note that some elements can emit either an alpha or a beta particle. Although no one atom can go both ways, some atoms can go either of two ways and cause *branches* in the series. No matter which way the parent goes, the daughter goes the other way, so that even though the series branches, they always come together again. The two types of disintegration of a given branch-point element always occur in the same proportion. This proportion is independent of the amount of the element.

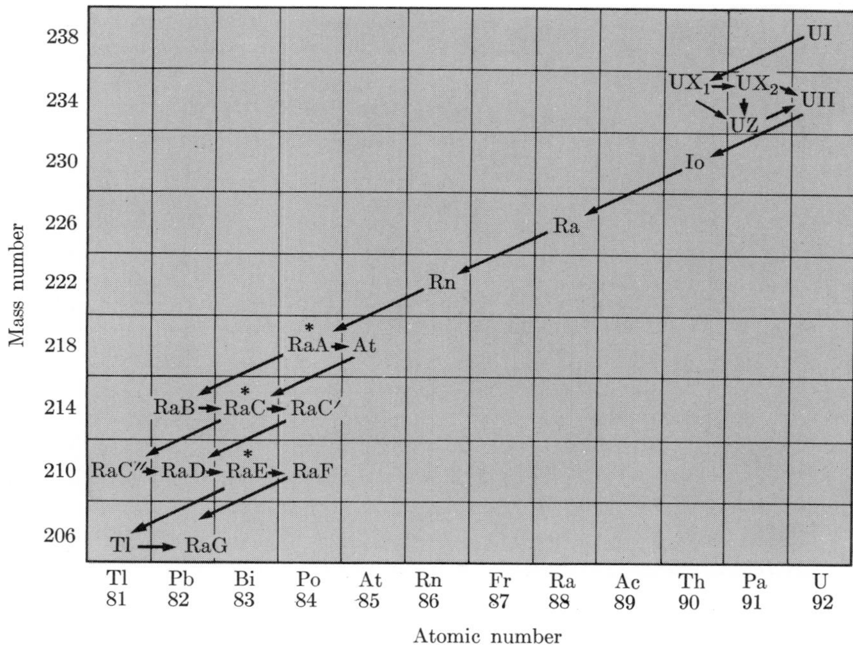

FIG. 10–12
The uranium ($4n + 2$) series.

Note: In Tables 10–1, 10–2, and 10–3 space limitations preclude listing more than the principal energy of the type of particle ejected and the energy range of the gamma-ray spectrum. A branch-point element is marked with an asterisk. The percentage proportion of each of the two daughter products of branch-point atoms is given in parentheses. Nuclear isomers are connected by a dotted line. The abbreviations under half-life are: s, second; m, minute; d, day; and y, year. (The data for these tables were obtained from the source listed in Appendix 5.)

TABLE 10–2

THE ACTINIUM SERIES

Radioactive species Historic names	Nuclide	Type of disintegration	Half-life T	Disintegration constant λ, s^{-1}	Principal particle energy, MeV
Actinouranium (AcU)	$_{92}U^{235}$	α	7.13×10^8 y	3.08×10^{-17}	α, 4.18; γ, 0.19
Uranium Y (UY)	$_{90}Th^{231}$	β	25.6 h	8.12×10^{-6}	β, 0.30; γ, 0.08–0.31
Protoactinium (Pa)	$_{91}Pa^{231}$	α	3.25×10^4 y	6.75×10^{-13}	α, 5.00, γ, 0.03–0.36
*Actinium (Ac)	$_{89}Ac^{227}$	β, α	21.2 y	1.08×10^{-9}	α, 4.94; β, 0.04
Radioactinium(RdAc)(1.2)	$_{90}Th^{227}$	α	18.2 d	4.40×10^{-7}	α, 5.97; γ, 0.05–0.34
Actinium K (AcK) (98.8)	$_{87}Fr^{223}$	β	22 m	5.25×10^{-4}	β, 1.15; γ, 0.05–0.08
Actinium X (AcX)	$_{88}Ra^{223}$	α	11.7 d	6.86×10^{-7}	α, 5.71; γ, 0.03–0.45
Ac emanation (Actinon)	$_{86}Rn^{219}$	α	4.0 s	0.174	α, 6.82; γ, 0.27–0.40
*Actinium A (AcA)	$_{84}Po^{215}$	β, α	1.8×10^{-3} s	3.86×10^2	α, 7.38
Astatine-215 (5×10^{-4})	$_{85}At^{215}$	α	10^{-4} s	7×10^3	α, 8.00
Actinium B (AcB) (99^+)	$_{82}Pb^{211}$	β	36.1 m	3.20×10^{-4}	β, 1.4; γ, 0.06–1.1
*Actinium C (AcC)	$_{83}Bi^{211}$	β, α	2.15 m	5.26×10^{-3}	α, 6.62; γ, 0.35
Actinium C′ (AcC′) (0.3)	$_{84}Po^{211}$	α	0.52 s	1.33	α, 7.45; γ, 0.89–1.06
Actinium C″ (AcC″) (99^+)	$_{81}Tl^{207}$	β	4.78 m	2.41×10^{-3}	β, 1.44; γ, 0.89
Actinium D (AcD)	$_{82}Pb^{207}$	Stable			

The time required for the nuclear transitions which produce gamma rays is usually of the order of 10^{-12} s. However, some *delayed* transitions with half-lives as long as several years have been observed. The delayed transitions occur in *nuclear isomers*, which are nuclides that have the same atomic number and same mass number, but exist for measurable times in different excited nuclear states with different energies and radioactive properties. In an *isomeric transition*, IT, an excited nucleus goes from a higher energy metastable state either to a lower energy metastable level or directly to the ground state with the emission of gamma rays. The isotope of Pa (protactinium) in the U^{238} series shows nuclear isomerism.

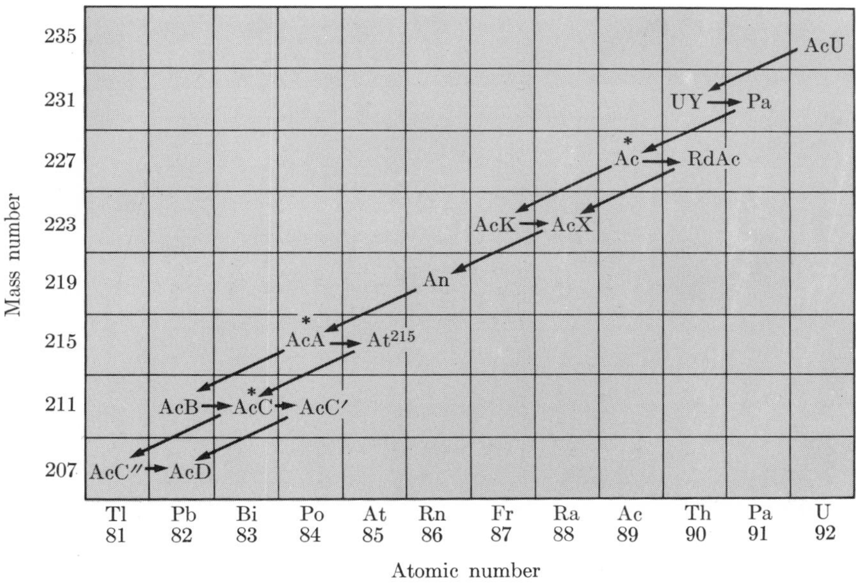

FIG. 10–13
The actinium ($4n + 3$) series.

Some of the nuclei of $_{91}Pa^{234}$ (UX_2) decay with a half-life of 1.18 m directly to the ground state of its isomer $_{91}Pa^{234}$ (UZ) with the emission of an energetic gamma ray. The remaining UX_2 nuclei disintegrate with the same half-life directly to $_{92}U^{234}$ (UII) by beta-particle and gamma-ray emission. On the other hand, all the nuclei of UZ decay with a half-life of 6.7 hours to $_{92}U^{234}$ accompanied by the emission of beta particles and rather low-energy gamma rays.

TABLE 10–3
THE THORIUM SERIES

Radioactive species Historic names	Nuclide	Type of disintegration	Half-life T	Disintegration constant λ, s^{-1}	Principal particle energy, MeV
Thorium (Th)	${}_{90}Th^{232}$	α	1.41×10^{10} y	1.56×10^{-18}	α, 4.01; γ, 0.06
Mesothorium 1 (MsTh 1)	${}_{88}Ra^{228}$	β	5.7 y	3.86×10^{-9}	β, 0.05
Mesothorium 2 (MsTh 2)	${}_{89}Ac^{228}$	β	6.13 h	3.14×10^{-5}	β, 1.11; γ, 0.06–1.64
Radiothorium (RdTh)	${}_{90}Th^{228}$	α	1.91 y	1.15×10^{-8}	α, 5.43; γ, 0.08–0.21
Thorium X (ThX)	${}_{88}Ra^{224}$	α	3.64 d	2.20×10^{-6}	α, 5.68; γ, 0.24
Th emanation (Thoron)	${}_{86}Rn^{220}$	α	56 s	1.23×10^{-2}	α, 6.29; γ, 0.54
Thorium A (ThA)	${}_{84}Po^{216}$	α	0.15 s	4.62	α, 6.78
Thorium B (ThB)	${}_{82}Pb^{212}$	β	10.6 h	1.82×10^{-5}	β, 0.35; γ, 0.11–0.41
*Thorium C (ThC)	${}_{83}Bi^{212}$	β, α	60.6 m	1.91×10^{-4}	α, 6.05; β, 2.25 γ, 0.04–2.2
Thorium C′ (ThC′) (66.3)	${}_{84}Po^{212}$	α	3.0×10^{-7} s	2.31×10^{6}	α, 8.78
Thorium C″ (ThC″) (33.7)	${}_{81}Tl^{208}$	β	3.1 m	2.52×10^{-3}	β, 1.80; γ, 0.23–2.61
Thorium D (ThD)	${}_{82}Pb^{208}$	Stable			

The tables of the radioactive series are redundant in the naming of the various isotopes. The names at the left antedate our full understanding of isotopes; they are names used in early discussion of the series. Even the more compact modern notation, like ${}_{92}U^{238}$, is redundant, since element 92 must be uranium.

The distribution of the natural radioactive materials is nonuniform over the surface of the earth. It has been estimated that, on the average over the earth's crust, each square mile has, within one foot of the surface, 8 tons of uranium and 12 tons of thorium.

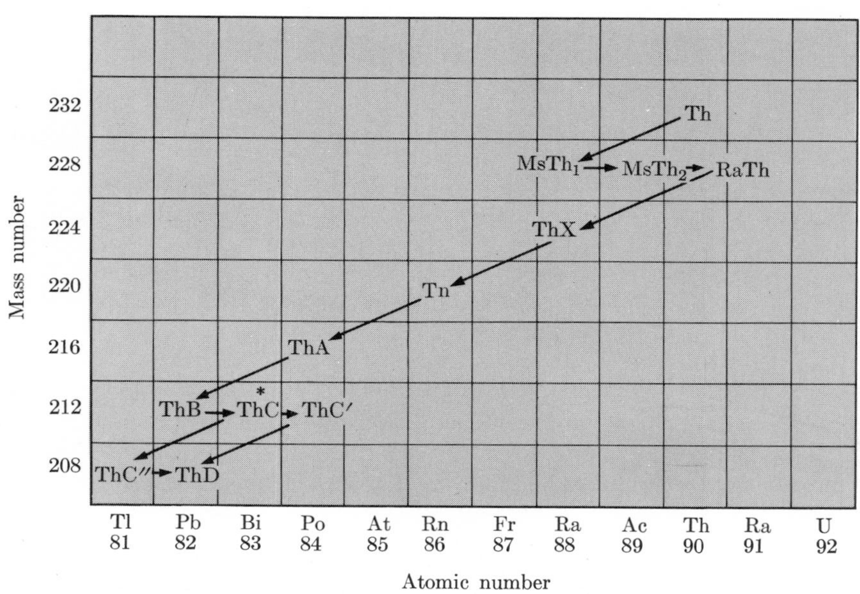

FIG. 10–14
The thorium (4*n*) series.

To complete the account of radioactive series, we now consider one that has been obtained only from an artificially produced radioactive material. The first element in this series is Pu^{241} and the stable end product is Tl^{205}. This series is called the neptunium or $4n + 1$ series because the longest-lived element in it is Np^{237}, which has a half-life of 2.14×10^6 years. If this element existed when the earth was formed about 4.4×10^9 years ago, the amount now remaining must be infinitesimal, because about 2000 half-lives of Np^{237} have elapsed since that time.

10-11. RADIOACTIVE GROWTH AND DECAY

These series of radioisotopes present interesting problems in relative abundances. Instead of considering a whole series, let us consider what happens to the abundance of substance B if A decays to B and B decays to C.

Let us call the number of A atoms at any instant N_1 and the initial number N_0. We can call the number of B atoms N_2 and assume that the initial number of B atoms is zero.

Every time an A atom disintegrates, it increases the number of B atoms. But every B disintegration reduces the number of B atoms. From Eq. (10–1) we have

$$\text{Number entering the } B \text{ category} = \frac{-dN}{dt} = \lambda_1 N_1 \tag{10–9}$$

and

$$\text{Number leaving the } B \text{ category} = \lambda_2 N_2, \tag{10–10}$$

where λ_1 is the decay constant for element A and λ_2 is that for element B. The net change in N_2 is the difference between Eqs. (10–9) and (10–10), or

$$\frac{dN_2}{dt}\,(\text{net}) = \lambda_1 N_1 - \lambda_2 N_2. \tag{10–11}$$

From Eq. (10–2) we can write

$$N_1 = N_0 e^{-\lambda_1 t}$$

and use this to eliminate N_1 from Eq. (10–11). This gives

$$\frac{dN_2}{dt} = \lambda_1 N_0 e^{-\lambda_1 t} - \lambda_2 N_2. \tag{10–12}$$

If we transpose $\lambda_2 N_2$ and multiply by the integrating factor $e^{\lambda_2 t} dt$, we have

$$e^{\lambda_2 t}\, dN_2 + \lambda_2 N_2 e^{\lambda_2 t}\, dt = \lambda_1 N_0 e^{(\lambda_2 - \lambda_1)t}\, dt. \tag{10–13}$$

This may be integrated at once to yield

$$N_2 e^{\lambda_2 t} = \frac{\lambda_1}{\lambda_2 - \lambda_1} N_0 e^{(\lambda_2 - \lambda_1)t} + C. \tag{10–14}$$

Since we are assuming that $N_2 = 0$ when $t = 0$, we have

$$0 = \frac{\lambda_1 N_0}{\lambda_2 - \lambda_1} + C. \tag{10–15}$$

Eliminating C from Eq. (10–14), we obtain

$$N_2 e^{\lambda_2 t} = \frac{N_0 \lambda_1}{\lambda_2 - \lambda_1} (e^{(\lambda_2 - \lambda_1)t} - 1), \tag{10–16}$$

and finally

$$N_2 = \frac{N_0 \lambda_1}{\lambda_2 - \lambda_1} (e^{-\lambda_1 t} - e^{-\lambda_2 t}). \tag{10–17}$$

The decay of A and the growth and decay of B are shown in Fig. 10–15.

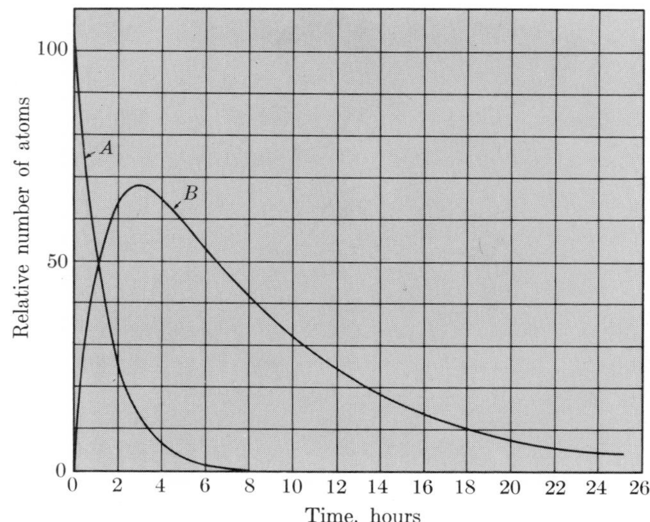

FIG. 10–15
Decay of element A ($T = 1$ h) and growth and decay of element B ($T = 5$ h).

We have considered the mathematics of but one link in the chain of events that constitute a radioactive series. The complete solution requires the simultaneous solution of as many differential equations as there are elements in the series. The general solution gives the number of daughter atoms as a function of time under the assumptions that at time zero there were no daughter atoms, and that there were a known number of atoms of the parent. These equations are called the Bateman equations, after the man who derived them.

It is possible to make a hydrodynamic model to demonstrate the growth and decay of elements in a radioactive series. Let us construct a number of tanks which can drain water from one to another in succession (Fig. 10–16). The tanks are shaped so that the velocity of efflux is proportional to the amount of water within the tank and the nozzles have openings proportional to λ. If the first tank is filled and the water is released at time zero, then the water in each tank is the same function of time as the concentration of radioactive atoms in a radioactive series. Since each radioactive series ends on a stable isotope, the bottom tank should have no outlet.

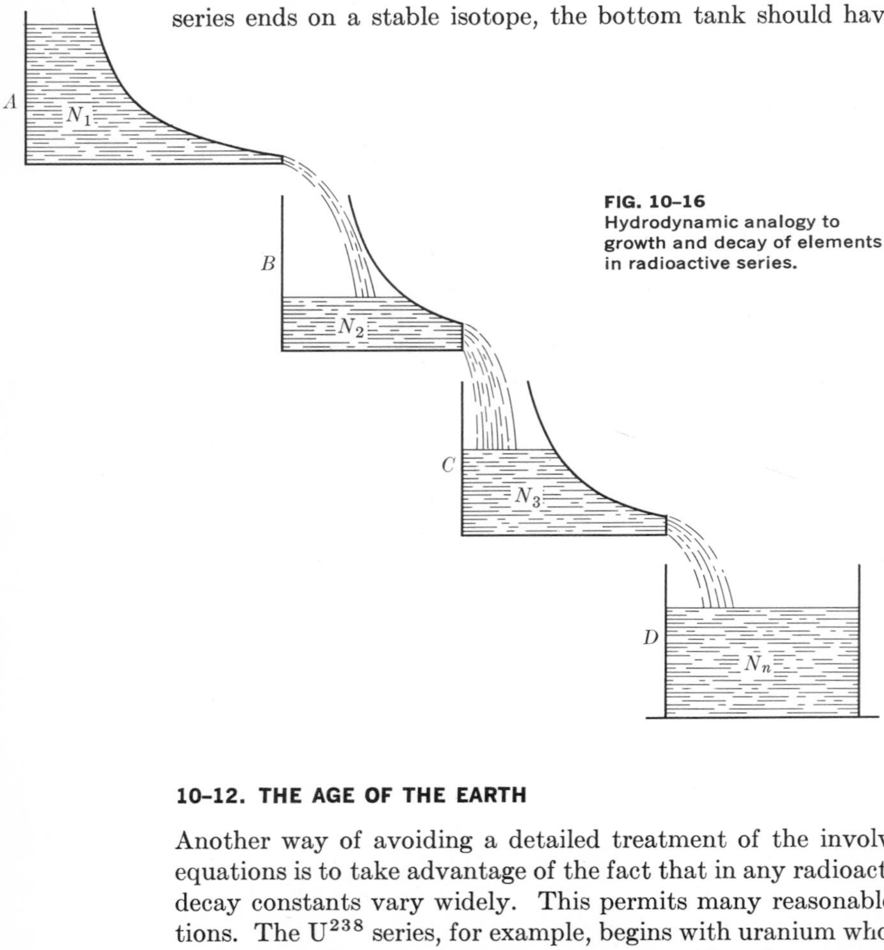

A N_1 B N_2 C N_3 D N_n

FIG. 10–16 Hydrodynamic analogy to growth and decay of elements in radioactive series.

10–12. THE AGE OF THE EARTH

Another way of avoiding a detailed treatment of the involved Bateman equations is to take advantage of the fact that in any radioactive series the decay constants vary widely. This permits many reasonable approximations. The U^{238} series, for example, begins with uranium whose half-life is 4.5 billion years and whose decay constant is less than 1/1000 that of any other element in the series, until we come to Pb^{206}, an isotope of lead that is uniquely associated with U^{238}. This lead is stable, has an infinite half-life, and has a decay constant of zero. This means that after a billion years or so the only elements present in any appreciable concentration will be uranium and lead. In terms of the water analogy, the intervening tanks are practically empty and serve only as a pipe through which the water in the first tank drains into the last. Thus it becomes reasonable to apply our Eq. (10–17) not only to the first and second elements of the series but also to the first and last. Since λ_2 becomes the decay constant of lead, which is zero, we have

$$N_{\text{Pb}} = N_{0\text{U}}(1 - e^{-\lambda_{\text{U}} t}). \tag{10–18}$$

U^{238} ore always contains Pb^{206} because this lead is the end element of the uranium radioactive series. It is reasonable to assume that this is the only reason lead is present, since there is no other compelling reason why uranium and lead should be found together. It then follows that the present number of lead atoms plus the present number of uranium atoms must equal the number of uranium atoms originally present, or

$$N_{\text{Pb}} + N_{\text{U}} = N_{0\text{U}}. \tag{10–19}$$

The present concentrations of Pb^{206} and U^{238} can be measured experimentally. With these data and the value of λ_{U}, we can find t after eliminating $N_{0\text{U}}$ between Eqs. (10–18) and (10–19). This t is the time that has passed since the earth solidified and the original pocket of uranium was sealed in the rock. There have been many assumptions in our argument, but they are justified by the fact that t has been computed for many ore samples from different parts of the earth and the results have consistently shown its age to be about four billion years. Since radioactive processes are independent of conditions of temperature and pressure, pockets of radioactive material constitute reliable clocks that have been running throughout geological history.

10–13. RADIOACTIVE EQUILIBRIUM

Another useful approximation in discussing the radioactive series is the concept of *radioactive* or *secular equilibrium.* We have seen that decay constants vary widely and it frequently happens that the parent substance has a much longer half-life than the daughter. In this case, if only the parent is present initially, the daughter product will grow and there will be a corresponding increase in the activity of this product. The more daughter atoms there are, the more daughters disintegrate, until the rates of production and disintegration of daughter atoms become equal. When these rates are equal, the decay product is in radioactive equilibrium with its parent.

The expression for this equality can be obtained from Eq. (10–17). This equation may be rewritten as

$$N_2 = \frac{N_0 \lambda_1 e^{-\lambda_1 t}}{\lambda_2 - \lambda_1}[1 - e^{-(\lambda_2 - \lambda_1)t}],$$

or, since $N_1 = N_0 e^{-\lambda_1 t}$, we have

$$N_2 = \frac{N_1 \lambda_1}{\lambda_2 - \lambda_1}[1 - e^{-(\lambda_2 - \lambda_1)t}]. \tag{10–20}$$

If $\lambda_2 \gg \lambda_1$, then to a good approximation

$$N_2 = \frac{N_1\lambda_1}{\lambda_2}(1 - e^{-\lambda_2 t}) = \frac{N_1\lambda_1}{\lambda_2}[1 - e^{-(0.693/T_2)t}]. \tag{10-21}$$

After several half-lives of the daughter product have elapsed, the value of the exponential term is negligible and Eq. (10–21) reduces to

$$N_1\lambda_1 = N_2\lambda_2. \tag{10-22}$$

If, instead of considering just one parent-daughter relationship, we treat a whole radioactive series in which the first element has a much longer half-life than any of the disintegration products, then the preceding analysis can be applied to each daughter product in the series. The resulting equilibrium relation is

$$\boxed{N_1\lambda_1 = N_2\lambda_2 = N_3\lambda_3 = \text{etc.}^*} \tag{10-23}$$

Each N in Eq. (10–23) decreases slowly and exponentially according to the small decay constant of the long-lived parent substance.

In terms of our water analog, if some tank leaks so slowly that the water level will not change noticeably while we watch it, then we can regard its leakage rate as constant. That tank will begin to fill the one below it. Presently the second tank will leak to a third, etc. Once equilibrium is established, the level in each tank will become constant at a level determined by its nozzle size. Tanks with large nozzles (λ's) will contain few water molecules (N's). The tank levels will not now be time-dependent.

Furthermore, if we leave the system running for a long time, so that the level in the original tank is noticeably decreased, then we will find the level in all the succeeding tanks decreased in the same proportion.

It must be appreciated, however, that the extent to which equilibrium is achieved depends on the relations among the half-lives and on the time, since all the atoms were of the original kind. The age of the earth is about one U^{238} half-life, which is much greater than any half-lives of uranium daughters except Pb^{206}. Thus for the past three billion years there has been very precise equilibrium between U^{238} and most of its daughters. The age of the earth can be computed only because Pb^{206} has an infinite half-life ($\lambda_{Pb} = 0$) and is *never* in equilibrium with uranium.

At the other extreme, equilibrium is very nearly established in a few seconds between such pairs as ${}_{83}Bi^{214}$ and ${}_{84}Po^{214}$. The bismuth half-life

---

* Note that the N's in Eq. (10–23) refer to different kinds of atoms and so their isotopic masses will *not* divide out of the equation as they did in the example at the end of Section 10–9.

is 19.7 m, while that of the polonium is 1.64×10^{-4} s. Equilibrium between these two permits the evaluation of the very short polonium half-life. The bismuth decay constant can be found by methods already described in Section 10–9. If the relative concentrations of bismuth and polonium can be measured after equilibrium has been established, then the very large polonium decay constant can be found from the equilibrium equation, Eq. (10–23).

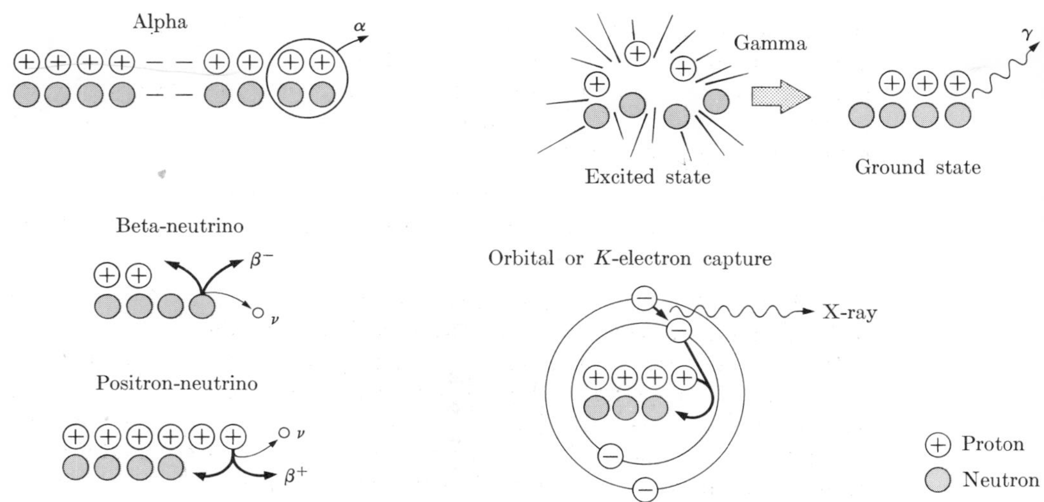

FIG. 10–17
Mechanisms of radiation in radioactivity.

10–14. SECONDARY RADIATIONS

Extensive study of the radiations from disintegrating nuclei have revealed two additional processes which result in the emission of electromagnetic radiation.

In some cases the nucleus will absorb one of the orbital electrons of the atom. Since this probability is greatest for the adjacent K-shell electron, the process is often called *K-electron capture*. When such electron capture occurs, the product atom will have an atomic number which is one less than that of the original nucleus. The remaining orbital electrons rearrange themselves to correspond to the structure of this new atom. X-rays characteristic of the product atom are emitted during the formation of the stable state of the new atomic system. Orbital electron capture is shown schematically in one part of Fig. 10–17.

The emission of a charged particle from a parent nucleus often leaves the daughter or product nucleus in an excited state. The latter usually goes to the ground state by emitting one or more gamma photons. In some elements, however, the transition energy is given to an orbital electron by direct interaction with the nucleus. This process is called *internal conversion,* IC. It causes a photoelectron to be ejected from a shell with a kinetic energy equal to the difference between the nuclear transition energy and the binding energy of the emitted electron. Thus internal conversion electrons give a line spectrum of beta particles. Note that these beta particles come from the extranuclear structure. Those from the nucleus always have a continuous energy distribution up to some maximum value. X-rays characteristic of the daughter element from which an electron was ejected are emitted when this element returns to the normal state.

The data obtained from the study of internal conversion show that the gamma rays in radioactive decay are often emitted *after* particle disintegration. Therefore they come from the product nucleus instead of the parent nucleus. This was represented in Fig. 10–9. Since the half-life of the parent element determines the rate of formation of the daughter element, and since gamma rays are emitted as the product "settles down," the gamma decay rate in this case follows the decay of the parent. This is the reason for a confusion in terminology. For example, when RaB, which has a half-life of 26.8 m, disintegrates to RaC, the associated gamma activity also decays with a half-life of 26.8 m. Thus it has become customary to call the radiation gamma rays from RaB even though they actually come from RaC.

10–15. RADIATION HAZARDS

All high-energy rays are hazardous to living tissue. They have enough energy to dissociate the complicated molecules of living tissue and to kill cells. We saw in Section 10–6 how the instability of organic molecules is utilized to quench Geiger counters. Thus high-energy radiation is able to convert the molecules of living structures into alien forms, or produce "burns."

Beta particles are not particularly hazardous because they are charged and therefore are not very penetrating. Clothing and skin provide adequate shielding for our vital organs.

Alpha particles are still less penetrating and on this basis are less dangerous than beta particles, although the specific ionization of the former greatly exceeds that of other radiations. There is a gaseous alpha-particle emitter in each one of the three natural radioactive series. Each of these gases is an isotope of the element having the atomic number 86. This element is usually called radon (sometimes emanation), and the gases are Rn^{219} (actinon), Rn^{220} (thoron), and Rn^{222} (radium emanation). Radon is chemically inert. These gases are liberated whenever radioactive ore is dug or crushed. Since they are inert, no gas mask can separate them from the air chemically. Those who work with radioactive ores must necessarily breathe these gases unless they are flushed away by very active ventilation.

In the lungs, these gases are very destructive, since there they are literally within a vital organ. The maximum permissible safe concentration of radon in the atmosphere is about $10^{-15}\%$ by volume. It is said that every miner who has worked in the Joachimstal uranium mines in Czechoslovakia for more than ten years has died of lung cancer. It may be that Pierre and Marie Curie were very fortunate that they had to perform their separation of radium from uranium ore in a dilapidated shed. The then-unwanted ventilation blew the radon away.

TABLE 10–4

Radiation	RBE
X-ray, gamma ray, beta ray	1
Thermal neutrons	2–5
Fast neutrons	10
Alpha particles and high-energy ions of O, N, etc.	10–20

Because beams of different radiations cause very different biological damage even when the body absorbs the same amount of energy from each type, it is necessary to define certain units in addition to the roentgen and the rad, which were discussed in Section 6–13. The absorbed dose of any ionizing radiation is the amount of energy imparted to matter by ionizing particles per unit mass of irradiated material at the place of interest. This leads to defining the *relative biological effectiveness*, RBE, of an ionizing radiation as the ratio of the absorbed dose in rads of x-rays of about 200 keV of energy which produces a specific biologic effect, to the absorbed dose in rads of the ionizing radiation which produces the same effect. A few RBE factors are listed in Table 10–4.

From the values listed we see that if the absorption in body tissue of 1 rad from a 200-keV x-ray beam produces a particular injury, then the absorption of only 0.1 rad from a beam of fast neutrons will produce the same injury. The unit of biological dose is the *roentgen equivalent man*, rem. The biological dose in rems is equal to the absorbed dose in rads times the RBE factor. The *millirem*, mrem, is a thousandth of a rem. The variations in body sensitivity to radiation are evident from the recommended maximum permissible doses. Some of these are: appendages of the body, 1500 mrem per week; and the lens of the eye, the gonads, and the blood-forming organs, each 100 mrem per week.

In general, the most hazardous radiations are the penetrating radiations. These include x-rays and gamma rays, together with neutral particles and the extremely high-energy particles produced by cosmic rays, which are penetrating despite their charge.

Not only are these penetrating radiations the ones that are destructively hazardous to vital organs, they are also the kind used purposefully when cancerous tissues are to be killed.

10-16. THE "RADIUM RADIATIONS" IN MEDICINE

The first radioactive material used in medical treatments was radium, Ra^{226}. But radium itself is primarily an alpha emitter. It was neither radium nor its daughter radon, Rn^{222}, that was effective in destroying cancerous tissue. Indeed, one must go to the fourth radioactive generation below radium before coming to a strong gamma emitter. Even when radium was administered to a patient, it was the Bi^{214} or RaC in equilibrium with the radium that was the important gamma-ray emitter.

Radium treatment required careful attention in the hospital. Radium has a half-life of 1620 years, which is certainly long compared with the duration of the treatment. In order to specify a given dose, the radiologist had to specify both the activity of the radium sample and the length of time it should be administered.

The radium daughter (radon or radium emanation, Rn^{222}) has a much shorter half-life, 3.82 days. In administering radon to a patient, the length of time of the application is often less critical or even inconsequential. A given amount of radon will give half its dose in 3.82 days and practically all its dose in 10 days or so. Thus the amount of radon to be administered can be specified with less concern for the removal of the radioactive material.

Hospitals usually keep radium as a solution of radium chloride. Periodically, the radon accumulated over the solution is chemically separated from the other gases over the solution and sealed into small glass or gold tubes called "radon seeds." These seeds grow in gamma activity as the radon descendants Po^{218}, Pb^{214}, and Bi^{214} come to equilibrium with the radon. Maximum gamma activity is reached in about 4 hr, since these products have half-lives of 3.05, 26.8, and 19.7 m, respectively. Once equilibrium is established, the number of Bi^{214} or RaC atoms is proportional to the number of the longer-lived radon atoms that are present. When the gamma activity begins to decline, the seed can be calibrated and its future activity can be calculated from the decay constant of radon.

10-17. UNITS OF RADIOACTIVITY AND EXPOSURE

The procedure we have just described led to the historical unit of radioactivity, the curie. Originally, it was the activity of 1 g of radium or the activity of the amount of radon in equilibrium with 1 g of radium. The original definition of the curie was replaced, in 1950, by a more general one which is closely equivalent to the old definition. By international agreement, *the curie*, Ci, is 3.7000×10^{10} disintegrations per second.* The *millicurie*, mCi, and *microcurie*, μCi, are also frequently encountered. Since the definition of the curie was peculiarly associated with radium, another unit of activity was proposed for all decaying elements. It is the *rutherford*, rd, which is 10^6 dis/s.

Medical doses are expressed in millicurie-hours. If a radon seed has reached its maximum activity and is found to have an activity of I_0 mCi at a time which we call zero, and if the seed is administered to a patient during the period from t_1 to t_2 (hours) later, the dose is

$$\text{dose in mCi-h} = I_0 \int_{t_1}^{t_2} e^{-\lambda t}\, dt, \tag{10–24}$$

where λ is the radon decay constant in reciprocal hours.

There is no one conversion factor between the activity of a radioactive material and the exposure rate in the surrounding medium. This is obvious from the fact that the ionization produced depends on the activity of the material and on the *kind* and the *energy* of the emitted radiation. It can be shown (Problem 10–27) that at a point P in air the *gamma-ray* exposure rate due to a *point* source of radioactive material is given approximately by

$$I = \frac{5.3 \times 10^3 CE}{d^2}. \tag{10–25}$$

In this expression, I is the exposure rate in roentgens per hour, C is the activity of the source in curies, E is the energy of the gamma photon in MeV, and d is the distance from the source to the point P, in centimeters. If more than one photon is emitted per disintegration, the total exposure rate will be the sum of the individual exposure rates. Since Eq. (10–25) is a point source law, it is essentially a differential expression which must be integrated subject to boundary conditions for an actual distribution of radioactive material.

10–18. CONCLUSION

In this chapter we have described some of the most important properties of natural radioactivity, together with some of the uses that have been found for it. Most of what was said also applies to artificial radioactivity, which is discussed in Chapter 11. In Chapter 11 we shall also discuss the reasons why some nuclei are radioactive and others are not.

---

* The value given for the number of disintegrations per second of one gram of radium had to be changed several times because of the increasing precision in the determinations of its half-life. To avoid further changes, the international curie was arbitrarily assigned a fixed value. According to current data, 1 g of Ra^{226} undergoes 3.61×10^{10} dis/s—about 2.5% less than the adopted curie.

REFERENCES

ANDERSON, CARL D., "The Positive Electron," in Beyer, R. T., *Foundations of Nuclear Physics.* New York: Dover, 1949, pp. 1–4.

BEERS, YARDLEY, *Introduction to the Theory of Error.* Reading, Mass.: Addison-Wesley, 1953. Chapter 7 is a concise account of the statistics employed in measurements of radioactivity.

BOHN, J. L. and F. H. NADIG, "Hydrodynamic Model for Demonstrations in Radioactivity," *The American Physics Teacher* (*Am. J. Phys.*) **6,** 320 (1938). Gives theory and directions for making a very instructive model of the decay and growth of the elements in a radioactive series.

CHOPPIN, GREGORY, *Nuclei and Radioactivity.* New York: Benjamin, 1964.

CONDON, E. U. and H. ODISHAW, *Handbook of Physics.* New York: McGraw-Hill, 1958. An extended account of cloud-chamber technique and the theory of drop formation begins on page **9**–167.

CURIE, EVE, *Madam Curie.* Vincent Sheean, translator. Garden City, N. Y.: Doubleday, 1937. An excellent biography.

FRIEDLANDER, G. and J. W. KENNEDY, *Nuclear and Radiochemistry.* New York: Wiley, 1955. Discussion and derivation of the statistical equations used in radioactivity measurements given in Chapter 10.

GENTNER, W., H. MAIER-LEIBNITZ, and H. BOTHE, *An Atlas of Typical Expansion Chamber Photographs.* London: Pergamon, 1954. An excellent collection of photographs of cloud chamber events with detailed description and interpretation of each. This pictorial history of nuclear physics should be examined by every student.

KAPLAN, IRVING, *Nuclear Physics.* 2nd ed. Reading, Mass.: Addison-Wesley, 1962. Wide coverage of nuclear phenomena.

MAGIE, W. F., *A Source Book in Physics.* Cambridge, Mass.: Harvard University Press, 1963. Contains the following papers: "The Radiation from Uranium," by Henri Becquerel, and "Polonium" and "Radium," by Marie S. Curie and Pierre Curie.

MANN, W. B. and S. B. GARFINKEL, *Radioactivity and Its Measurement,* Momentum Book No. 10. Princeton, N. J.: Van Nostrand, 1966.

NATIONAL BUREAU OF STANDARDS, Washington, D. C.: U. S. Government Printing Office:

Handbook 65, Safe Handling of Bodies Containing Radioactive Isotopes, 1958;
Handbook 69, Maximum Permissible Body Burdens and Maximum Permissible Concentrations of Radionuclides in Air and in Water for Occupational Exposure, 1963;
Handbook 86, Radioactivity, 1963.

REINES, F. and C. L. COWAN, JR., "The Neutrino," *Nature* **178,** 446 (1958). A good account of the work of these authors in verifying the existence of this elusive particle.

RUTHERFORD, E., J. CHADWICK, and C. D. ELLIS, *Radiations from Radioactive Substances*. New York: Macmillan, 1930. A complete account of the work in radioactivity during the first three decades after its discovery.

PROBLEMS

(The constants of various radioactive elements needed for solving problems can be found in Tables 10–1, 10–2, and 10–3, and in Appendix 6.)

10–1. A certain radioactive element has a half-life of 20 days. (a) How long will it take for $\frac{3}{4}$ of the atoms originally present to disintegrate? (b) How long will it take until only $\frac{1}{8}$ of the atoms originally present remain unchanged? (c) What are the disintegration constant and the average life of this element?

10–2. The original activity of a certain radioactive element is I_0. (a) Plot the subsequent activity, I, as ordinate against half-life as abscissa for a period of 6 half-lives. (b) Mark the average life, $\bar{t}$, on the abscissa and estimate the activity then in terms of I_0. (c) Assuming that the daughter product of the element is stable, plot the growth of this product on the graph for part (a).

10–3. (a) How many alpha particles are emitted per second by 1 microgram of Ra^{226}? (b) How many alpha particles will be emitted per second after the original microgram has disintegrated for 500 years? Neglect the contributions of daughter products. (c) What would be the answer to part (b) if the contributions of the daughter products were included? (Radioactive equilibrium will have been established in 500 years.)

10–4. What fraction of a given sample of radium emanation, Rn^{222}, will disintegrate in two days?

10–5. What is the lowest value and the highest value of the integer n for (a) the uranium $4n + 2$, (b) the actinium $4n + 3$, and (c) the thorium $4n$ series?

10–6. Radium C, Bi^{214}, is a branch point in the uranium series. Some of its atoms disintegrate by alpha emission and some by beta emission. Write the nuclear reaction equation (a) for each case, and (b) for the disintegration of each daughter product. (c) Identify the elements resulting from the decays in part (b).

10–7. A hypothetical radioactive series starts with element ${}_{50}A^{130}$. Element A^{130} is an alpha emitter, its "daughter" is a beta emitter, and its "grandaughter" is a branch point in the series. Find the atomic number and mass number of E, the "great-great grandaughter" of element A.

10–8. Discuss the changes in the amount of water in tank B after opening A in the hydrodynamic analog shown in Fig. 10–16. Correlate the growth and decay of a radioactive element with your discussion for the following sets of relative sizes of the nozzles: (a) A small, B large; (b) both equal; (c) A large, B small; (d) A any size, B zero (closed); and (e) both zero (closed).

10–9. How much radium B, Pb^{214}, will be present after 1 mg of radium A, Po^{218}, has been disintegrating for 20 m? Assume that all of RaA decays into RaB.

10–10. How much radium emanation, Rn^{222}, will be present after 1 g of radium, Ra^{226}, has disintegrated for one day?

10–11. Show from Eq. (10–17) that if there are initially N_0 radioactive atoms of the parent present, the time at which the number of radioactive daughter atoms is maximum is

$$t_{max} = \frac{\ln(\lambda_2/\lambda_1)}{\lambda_2 - \lambda_1}.$$

10–12. (a) How long after obtaining a pure sample of radium A, Po^{218}, will the amount of radium B, Pb^{214}, be maximum? (b) How long after obtaining a pure sample of radium will the amount of radium emanation be maximum?

10–13. If a radioactive element disintegrates for a period of time equal to its average life, (a) what fraction of the original amount remains, and (b) what fraction will have disintegrated?

10–14. Using particle accelerators, it is possible to produce radioactive nuclei. Assume that a particular type of nucleus whose decay constant is λ is produced at a steady rate of p nuclei per second. Show that the number of nuclei N present t s after the production starts is

$$N = \frac{p}{\lambda}(1 - e^{-\lambda t}).$$

10–15. How many alpha particles and how many beta particles are emitted in the actinium series, where ${}_{92}U^{235}$ decays to ${}_{82}Pb^{207}$?

10–16. (a) If a piece of uranium ore contains 2 g of U^{238} after disintegrating for a period equal to its average life, how much U^{238} did it contain originally? (b) If all the helium released during the disintegration of the original amount of uranium for one average life were collected, what would be its mass and its volume under standard conditions? (Note that substantially all the disintegrated uranium will have become stable lead in this period of time.)

10–17. Show that the slope of the curve of the logarithm to the base 10 of the activity of a radioactive substance plotted against time is equal to $(-\lambda/2.30)$. Start with Eq. (10–5).

10–18. The activity of a certain radioactive sample is measured every half-hour and the following activities in millicuries are measured: 78, 49, 32, 20, 13, 8, 5. Plot the activity against time and determine the half-life from the graph.

10–19. A certain radioactive element is a beta-particle emitter. When the beta activity of a freshly prepared sample of this element is measured as a function of time, the following data are obtained:

Time, m	Activity, units	Time, m	Activity, units
0	1080	100	173
20	730	130	106
40	504	160	63
70	303	190	35

(a) Plot the logarithm of the activity against time. (b) Determine the disintegration constant from the slope of the curve. (c) Find the half-life of the element.

10–20. When the activity of a mixture of radioactive elements is measured as a function of time, the following data are obtained:

Time, m	Activity, units	Time, m	Activity, units
10	252	110	11.5
30	105	130	9.1
50	46.8	160	6.8
70	25.0	190	5.2
90	15.5	220	4.0
		280	2.4

(a) Plot the logarithm of the activity against time. (b) How many radioactive elements are in the mixture? (c) Find the disintegration constant of each from the slopes of the corrected curve, and (d) the half-life of each. [*Hint:* To obtain the activity of each element from the combined curve, project the curve for the longest-lived element back to $t = 0$. The actual activity ordinates (not log activity) of this projected curve must be subtracted from those of the total activity curve to obtain the activity due solely to the other element present. The use of semilog paper will simplify plotting and solving.]

10–21. Lead-206 is found in a certain uranium ore. This indicates that the lead is of radioactive origin. What is the age of the uranium ore if it now contains 0.80 g of Pb^{206} for each g of U^{238}?

10–22. What is the mass of one curie of (a) U^{238}, (b) Pb^{210} (RaD), and (c) Po^{214} (RaC′)?

10–23. What is the mass and the volume under standard conditions of one curie of radium emanation, Rn^{222}?

10–24. A container holds 2 Ci of Po^{210}, which emits alpha particles with an energy of 5.30 MeV. If all of the alpha particles are stopped in the container, calculate the rate at which heat is evolved.

10–25. The activity of 20 g of Th^{232} is 2.18 μCi. Calculate the disintegration constant and the half-life of Th^{232}.

10–26. A hypothetical beta emitter X^{220} has a half-life of 6.93 days. Assume that a 20 mCi sample of this element is present at $t = 0$. (a) At what time will the activity be 5 mCi? (b) How many beta particles are emitted in the first second? (c) How many beta particles are emitted between $t = 10$ days and $t = 20$ days?

10–27. (a) Assuming that radioactive equilibrium has been established, how many grams of radium, Ra^{226}, and of radium emanation, Rn^{222}, are contained in a 2-lb piece of U^{238} ore which is 40% uranium oxide, U_3O_8? (b) What are the activities of these amounts of uranium, radium, and radium emanation in millicuries and in rutherfords?

10–28. Given two radioactive elements, a parent having a half-life T_1 and a daughter product having a half-life T_2. If $T_1 \gg T_2$ and if initially there is only the parent element, how long will it take, in terms of T_2, until these elements are (a) within 1 percent of their equilibrium value, and (b) within 0.1 percent of that value?

10–29. For health protection, the maximum permissible concentration of radium emanation in air for continuous exposure is 10^{-8} microcuries per milliliter of air. For this concentration, what is the Rn^{222} content of 1 ml of standard air (a) in percent by mass and (b) in percent by volume?

10–30. If the piece of uranium ore in Problem 10–27 is ground to a fine powder so that the entrapped radium emanation, Rn^{222}, is released, with how many cubic feet of air must this radon be mixed to reduce the concentration to the safe tolerance level of 10^{-8} microcuries per milliliter?

10–31. (a) Assuming that there are 8 tons of U^{238} and 12 tons of Th^{232} uniformly distributed in the first foot of depth under each square mile of the surface of the earth, find the combined activity of these two radioelements in microcuries in the cubic foot of soil under each square foot of surface. (b) Assuming radioactive equilibrium, find the mass of Ra^{226} in one cubic foot of soil. (c) What would be the kinds and energies of the radiations from this soil?

10–32. If either radium or radioactive strontium were swallowed, where would these elements concentrate in the body? To answer this, consider the chemical properties of the elements in the groups or columns in the periodic table.

10–33. (a) In the medical use of radon, Rn^{222}, it is actually the combined gamma radiation from Pb^{214} and especially from Bi^{214} which irradiates the part of the body treated. However, the half-life of radium emanation is used to calculate the exposure. Why? (b) After reaching its maximum gamma activity, calibration shows that a tube containing radium emanation, Rn^{222}, has an activity of 2 mCi. Assuming that it is placed in a tumor 24 h later, what is the dose in millicurie-hours if it remains there (1) 2 days, (2) 4 days, and (3) forever?

10–34. A point-source radioactive element has an activity of C curies and emits one gamma-ray photon having E MeV of energy per disintegration. (a) Neglecting the rather small absorption of the air, show that the gamma-ray energy flux at d cm from the source is $3.7 \times 10^{10}\, CE/4\pi d^2$ MeV/(cm^2/s). (b) The linear absorption coefficient of air for gamma rays in the 0.5- to 2-MeV range is about 3.4×10^{-5}/cm. Show, to two significant figures, that the energy absorbed per cm^3 of air in the region between d and $(d + 1)$ cm is $13 \times 10^5\, CE/4\pi d^2$ MeV/s, and (c) that the exposure rate in R/h at the distance d is $5.3 \times 10^3\, CE/d^2$ when d is in centimeters and $5.7\, CE/d^2$ when d is in feet.

10–35. What is the gamma-ray exposure in mR/hr in air at a distance of 10 ft from an unshielded 50-curie point source of Co^{60}? Each cobalt atom emits a 1.1- and a 1.3-MeV gamma-ray photon per disintegration.

CHAPTER 11

NUCLEAR REACTIONS AND ARTIFICIAL RADIOACTIVITY

11-1. PROTONS FROM NITROGEN

We have seen how Rutherford and his associates discovered the atomic nucleus in 1911 by analyzing the way in which alpha particles are scattered by thin metal foils. Their continued study of the interaction between alpha particles and matter led to another striking discovery in 1919. As before, Rutherford used alpha particles from a radioactive source and scintillation as the detection technique, but in this work attention was directed to the absorption of the alpha particles.

In the case of x-ray absorption, we found that the intensity of the rays penetrating an absorber depends exponentially on the thickness of the absorber, as in Eq. (6–10). This type of absorption equation applies whenever the absorption is due to a process or processes in which the incident rays are removed from a beam in one catastrophic event. We saw that for x-rays these processes could be the production of photoelectrons, Compton scattering, or pair production. As we mentioned in our discussion of the Wilson cloud chamber, the process of absorption of alpha particles is quite different. The energy of the alpha particle is usually lost by producing a very large number of low-energy ion pairs. Thus the alpha particle loses energy more or less continuously rather than in one catastrophic event. Instead of being absorbed according to an exponential law, the absorption is slight until the particles are stopped, as shown in Fig. 11–1. Alpha particles of a given energy are completely stopped by traversing a certain thickness of absorber and, for a particular kind of absorber, they have a rather definite range which is a function of their energy. Alpha particles lose more energy per unit length of path near the end of their range, since they are moving more slowly and have more time to interact with atoms near which they pass. This leads us to assume

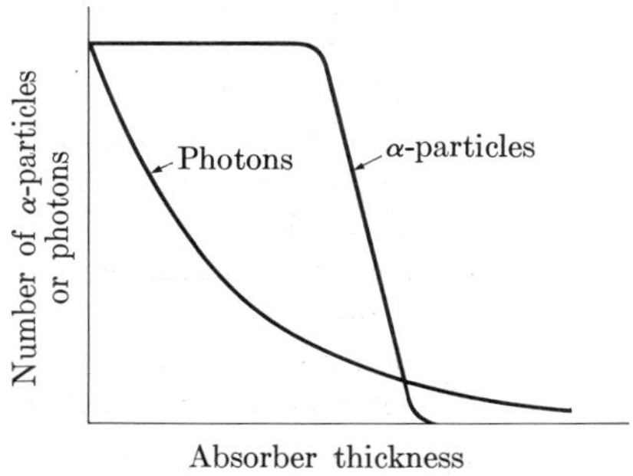

FIG. 11–1
Absorption of alpha particles and photons by matter.

that the energy loss per unit length of path is inversely proportional to the velocity, and therefore we have

$$-\frac{dE_k}{dx} = \frac{k}{v}, \tag{11-1}$$

where k is a constant that depends on the absorbing material. Since the velocity of alpha particles is considerably less than the velocity of light, we can eliminate dE_k by substituting the derivative of the classical expression for kinetic energy, $E_k = \frac{1}{2}mv^2$. We obtain

$$-mv\frac{dv}{dx} = \frac{k}{v}. \tag{11-2}$$

Since the particle is stopped in a distance R, its range, we have $v = V_0$ for $x = 0$ and $v = 0$ for $x = R$, and we can separate the variables in Eq. (11–2) and write

$$\int_{V_0}^{0} -mv^2\,dv = \int_0^R k\,dx. \tag{11-3}$$

Upon integration, this becomes

$$\frac{mV_0^3}{3} = kR$$

or

$$\boxed{V_0^3 = k'R,} \tag{11-4}$$

where k' is another constant depending on the absorbing material. This is *Geiger's rule*, which is approximately correct for any charged particles in any medium if the velocities of the particles are small compared with the velocity of light.

If we use alpha particles from the same source, V_0 is then a constant, and variations in k' can be studied by observing the range in different materials. The interaction between alpha particles and the absorbing medium is electrical, and it is not surprising that k' depends both on the atomic number of the absorber and its concentration. Indeed, if the absorber is a gas, it is convenient to study alpha-particle velocities by having a chamber in which the source and detector are separated a fixed distance and the "stopping power" of the absorber is varied by changing the pressure. Since k' is proportional to the pressure P and since R here is constant, Eq. (11–4) may be written

$$V_0^3 = k''P, \tag{11-5}$$

where k'' depends on the kind of gas and the fixed distance, and where P is the gas pressure.

Rutherford found that his variable-pressure gas chamber worked very nicely with dried oxygen or carbon dioxide as the gas. But with dried air in the chamber, he found *more scintillations than when the chamber was evacuated.* He was using alpha particles from radium-C′, whose range in air at atmospheric pressure is about 7 cm. When Rutherford increased the pressure in the chamber until the absorption of the gas was equivalent to 19 cm of atmospheric air, the number of scintillations was about twice that observed when the chamber was evacuated. The brightness of the scintillations suggested that the long-range scintillations might be due to hydrogen nuclei (protons) rather than to alpha particles.

Here was an effect obviously related to the particular gas in the chamber. Since the scintillations appeared to be hydrogen nuclei, Rutherford first suspected that the alpha particles were knocking protons off water-vapor molecules in the air. But since the effect was greatest in nitrogen even when the nitrogen was prepared so as to be scrupulously free of any hydrogen, Rutherford decided the new particles came from the nitrogen itself.

It remained to be shown whether the scintillations that appeared to be due to protons were actually protons or were nitrogen atoms. Rutherford attempted to make this distinction by deflecting the particles in a magnetic field. The geometry of his beam was poorly defined because narrowing the beam further reduced the already small number of scintillations. He felt reasonably sure, however, that the mass of the particles was two or less—certainly not 14.

Thus Rutherford became sure that either the alpha particle was breaking a proton fragment off the nitrogen nucleus or the nitrogen was breaking a fragment off the alpha particle. He sensed that the alpha particle is particularly stable, since he regarded the nitrogen nucleus as consisting of three alpha particles and two hydrogen nuclei. He felt that the alpha particles preserved their identity and that the fragment he was observing was one or both of the hydrogens which he called "outriders of the main system of mass 12."

Although Rutherford was uncertain about the details of what he observed, he was aware that he had produced the first man-controlled nuclear rearrangement and that by similar techniques "we might expect to break down the nuclear structure of many of the lighter elements."

Rutherford could speculate about the kind of nucleus remaining after the alpha particle knocked a proton from nitrogen. The alpha particle might simply knock the proton off and continue with reduced energy according to the equation

$$ {}_{2}\mathrm{He}^{4} + {}_{7}\mathrm{N}^{14} \rightarrow {}_{1}\mathrm{H}^{1} + {}_{2}\mathrm{He}^{4} + {}_{6}\mathrm{C}^{13}, \qquad (11\text{–}6) $$

or the alpha particle could penetrate the nucleus and remain with it, as in

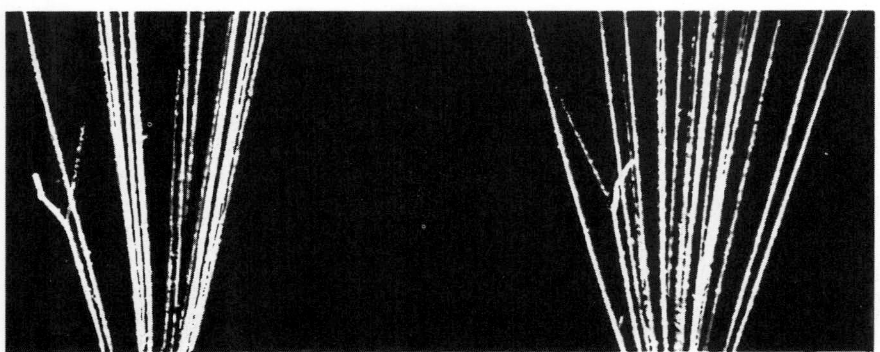

FIG. 11–2
Cloud-chamber stereograph of the ejection of a proton from a nitrogen nucleus. The slight curvature of the tracks is due to mass motion of the gas caused by a small leak in the expansion chamber. (Courtesy of P. M. S. Blackett)

equation

$$_2He^4 + {}_7N^{14} \rightarrow {}_9F^{18} \rightarrow {}_1H^1 + {}_8O^{17}. \qquad (11\text{–}7)$$

A choice between these possibilities could have been made if carbon, fluorine, or oxygen could be found in the gas by chemical or spectrographic means. But the amount of any of these elements formed was far too small to identify.

In 1925 Blackett reported that out of more than 20,000 cloud chamber photographs, including 400,000 alpha-particle tracks in nitrogen, eight were like Fig. 11–2, which shows the nuclear disintegration taking place. The two parts of the figure are two pictures of the same event taken from different angles—a stereograph. As in all Blackett's pictures, most of the tracks are straight alpha-particle tracks. But in Fig. 11–2 one alpha track ends abruptly. At the end of the alpha track may be seen two tracks, a thin one caused by a proton and a thick one due to a heavy nucleus. There appear to be two rather than three products of the nuclear reaction.

Furthermore, the stereographic picture permits a complete analysis of the three-dimensional situation. The fact that the two product-particle tracks determine a plane which includes the original alpha-particle track makes it highly unlikely that a third particle escaped undetected. Thus we can be sure that Eq. (11–7) is the correct one.

Blackett's photographs also show that the probability of having an alpha particle collide with another atom while traveling several centimeters in a gas is only about 1 in 50,000. This verifies the concept that most of the volume occupied by a gas is free space and that the atomic nucleus is actually a very minute target.

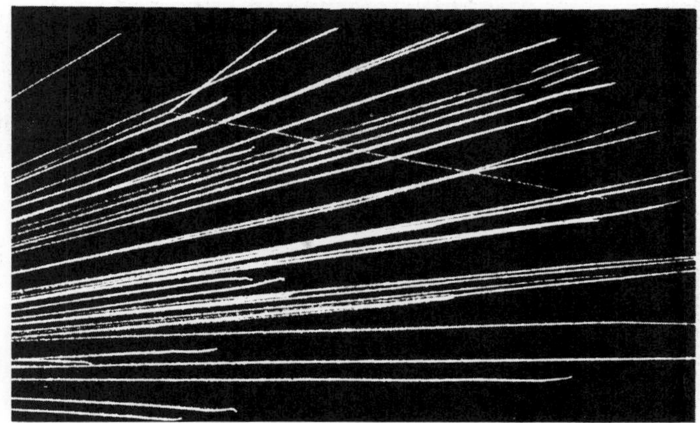

FIG. 11–3
Elastic collision of an alpha particle with hydrogen—the long, thin track. (Courtesy of P. M. S. Blackett and D. S. Lees, Imperial College of Science and Technology, and the Royal Society, London)

Rutherford and Chadwick bombarded all the light elements with alpha particles. They found about ten cases where protons were produced. By studying the range of these protons, they determined that in some cases the proton has more energy than the original alpha particle. This further refutes the idea that the proton is knocked off the bombarded nucleus and supports the idea that the alpha particle is first absorbed by the nucleus, which then ejects a proton whose energy is largely determined by the instability of the intermediate nucleus. These experiments also indicated that the proton is a fundamental nuclear particle.

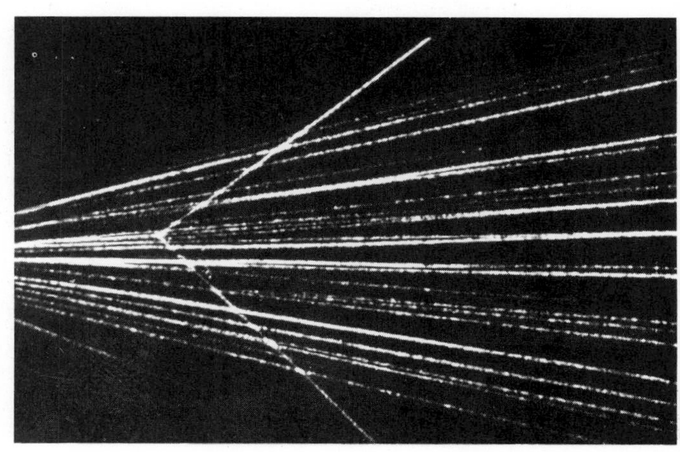

FIG. 11–4
Elastic collision of an alpha particle with helium. (Courtesy of P. M. S. Blackett)

11-2. PENETRATING RADIATION PUZZLE

One of the light elements which failed to give protons under alpha-particle bombardment was beryllium. In 1930, Bothe and Becker in Germany reported that beryllium bombarded with alpha particles from polonium produced a very penetrating radiation which they assumed to be gamma rays. Others showed that boron also had this property, and the energy of the photons was estimated to be about 10 MeV—greater than that of any gamma rays from radioactive sources.

In 1932, the daughter and son-in-law of Madame Curie, Mme. Curie-Joliot and M. Joliot, announced that these radiations could eject rather high-energy protons from matter containing hydrogen. When paraffin wax was placed in front of their ionization chamber, the ionization was nearly doubled. They *assumed* (incorrectly as it turned out) that this process was analogous to the Compton effect. Instead of an x-ray knocking out an electron, they thought a gamma ray was striking a proton. They could compute the energy the gamma-ray photon must have in order to give the proton its observed energy. Assuming a direct hit in which the gamma ray scatters back on itself, they computed the gamma-ray energy to be about 50 MeV, a tremendous amount.

It is interesting to see how this type of calculation is made. Repeating the Compton equation, Eq. (6–30), which gives the change in wavelength of a photon upon being deflected through an angle α by a particle of mass m_0, we have

$$\Delta\lambda = \frac{h}{m_0 c}(1 - \cos\alpha). \tag{11–8}$$

The photon suffers the largest change of wavelength and transfers the greatest amount of energy when the angle of scattering is 180°, in which case Eq. (11–8) becomes

$$\Delta\lambda = \frac{2h}{m_0 c}\cdot \tag{11–9}$$

Since we shall show that only a small fraction of the photon energy E_f is transferred to the proton, the changes in the wavelength, frequency, and energy of the photon are slight, and these quantities may be treated as differentials. Thus the maximum change in the energy of the photon is

$$dE_f = h\,df = h\,d\left(\frac{c}{\lambda}\right) = -\frac{hc\,d\lambda}{\lambda^2}\cdot \tag{11–10}$$

Using $\lambda = c/f$ and Eq. (11–9) to eliminate λ and $d\lambda$, we have

$$dE_f = -\frac{2(hf)^2}{m_p c^2} = -\frac{2E_f^2}{m_p c^2}\cdot \tag{11–11}$$

Now, the energy lost by the photon is just that gained by the proton,

dE_p. Furthermore, m_0c^2 is the rest mass-energy of the proton being hit, which is about 1 u. Thus we have

$$dE_p = 2E_f^2 \text{ in u.} \tag{11–12}$$

The observed proton ranges were up to about 26 cm in air, which corresponds to an initial proton velocity of about 3×10^7 m/s. This velocity permits a nonrelativistic calculation of the kinetic energy, which is 7.5×10^{-13} joule or 4.7 MeV. Since Eq. (11–12) is in u, we must convert the observed proton energy of 4.7 MeV to 0.005 u. This is the change in the proton energy, dE_p; therefore we can compute the photon energy, E_f:

$$E_f = \sqrt{dE_p/2} = \sqrt{0.005/2} = 0.05 \text{ u} = 47 \text{ MeV}. \tag{11–13}$$

Thus it would take about a 50-MeV gamma-ray quantum to give a proton having about 5 MeV of kinetic energy in an ideal Compton-type collision.

It was found that the penetrating radiations could also expel nitrogen nuclei from nitrogenous compounds. The kinetic energy of the nitrogen was 1.2 MeV, and a calculation such as we have just made indicated that the necessary photon energy was about 95 MeV. It appears that these gamma-ray energies are not only very great, but are not unique.

By 1932 the isotopic masses of the elements had been measured fairly accurately. All these masses were very nearly integers, and the minor differences were attributed to the relativistic equivalence of mass and energy, $E = mc^2$. Thus if the mass of an atom was found to be less than the sum of the masses of its parts (then assumed to be alpha particles, protons, and electrons), the difference was assigned to the mass-energy that would be released if the nucleus were assembled from those parts.

We have seen that the equivalence of mass and energy permits the use of mass units for the measurement of energy and that one atomic mass unit is equivalent to 931 MeV. Using the available data for the energies of the particles, one might, in the early thirties, have made the following calculation. The alpha particle from polonium-210 has a kinetic energy of 5.26 MeV, which may be expressed as 0.00565 u. The rest energy or isotopic mass of an alpha particle was then thought to be 4.00106 u. Let this alpha particle strike a boron atom assumed to be at rest and to have a mass-energy of 11.00825 u.* There would then be a total mass-energy of 15.01496 u going into the reaction. If the products of the reaction are assumed to be a gamma ray and a new nucleus, the new nucleus would have to be ${}_7N^{15}$, whose rest mass-energy was thought to exceed 14.999. Assuming that the nitrogen nucleus is left with no kinetic energy and that the gamma ray carries away all the energy difference, the gamma ray would

---

* Since it was known that alpha particles produced protons from ${}_5B^{10}$, it was logical to assume that the penetrating radiation came from ${}_5B^{11}$.

have an energy of 15.015 — 14.999 = 0.016 u. Thus the maximum energy the gamma ray could have would be 0.016 × 932 = 14.9 MeV, far less than 50 or 90 MeV.

Similar results were obtained from the alpha-particle bombardment of beryllium.

11-3. DISCOVERY OF THE NEUTRON

Ever since 1924, Rutherford and Chadwick had believed that there was an electrically neutral particle and Chadwick had been looking for experimental proof of its existence. He found that proof in the data we are discussing. Chadwick assumed that when boron is bombarded with alpha particles, the products are not ${}_7N^{15}$ and a gamma ray, but ${}_7N^{14}$ and ${}_0n^1$, a *neutron*. Such a particle would have tremendous penetrating ability, since it has no charge. Whereas charged particles interact with the electric fields of the nuclei in the matter they penetrate, an uncharged particle would interact with nuclei only when within the influence of the very short-range nuclear forces. Most important, however, Chadwick assumed that the mass of the neutron was nearly equal to the mass of the proton, so that if it hit a proton, the transfer of its energy to the proton could be complete. To find this energy transfer let us consider in detail an *elastic* collision between two bodies. We assume a direct (head-on) hit so that both bodies move on along the extension of the line of incidence. Let the first body have a mass m_1 and a velocity v_1. Then its kinetic energy, E_{k_1}, is $\frac{1}{2}m_1v_1^2$. Its momentum p_1 is m_1v_1, which may be conveniently expressed as $\sqrt{2m_1E_{k_1}}$. After the collision, v_1, E_{k_1}, and p_1 change to v_1', E_{k_1}', and p_1'. Let the target particle have a mass m_2 which is initially at rest, so that

$$v_2 = E_{k_2} = p_2 = 0.$$

After the collision, these become v_2', E_{k_2}', and p_2', respectively.

Since the collision is elastic, kinetic energy is conserved, and

$$E_{k_1} = E_{k_1}' + E_{k_2}' \qquad \text{or} \qquad E_{k_1}' = E_{k_1} - E_{k_2}'. \tag{11–14}$$

Because the collision is linear, the momenta add algebraically and we have

$$p_1 = p_1' + p_2'$$

or

$$\sqrt{2m_1E_{k_1}} = \sqrt{2m_1E_{k_1}'} + \sqrt{2m_2E_{k_2}'}. \tag{11–15}$$

We can eliminate one quantity between these two equations, and we choose to eliminate E_{k_1}' since it is difficult to measure. Then, substituting E_{k1}' from Eq. (11–14) into Eq. (11–15), we have

$$\sqrt{2m_1E_{k1}} = \sqrt{2m_1(E_{k1} - E_{k2}')} + \sqrt{2m_2E_{k2}'}. \tag{11–16}$$

Squaring and dividing by 2, we obtain

$$m_1E_{k1} = m_1E_{k1} - m_1E'_{k2} + m_2E'_{k2} + 2\sqrt{m_1m_2(E_{k1} - E'_{k2})E'_{k2}}. \tag{11–17}$$

Canceling, transposing, squaring again, and dividing by E'_{k2}, we get

$$(m_1^2 - 2m_1m_2 + m_2^2)E'_{k2} = 4m_1m_2E_{k1} - 4m_1m_2E'_{k2}. \tag{11–18}$$

Upon rearranging, we obtain

$$(m_1 + m_2)^2E'_{k2} = 4m_1m_2E_{k1},$$

or

$$E'_{k2} = \frac{4m_1m_2E_{k1}}{(m_1 + m_2)^2}. \tag{11–19}$$

By putting E'_{k2} from Eq. (11–14) into Eq. (11–19), we can get E'_{k1} in terms of E_{k1} and thus

$$E'_{k1} = E_{k1} - \frac{4m_1m_2E_{k1}}{(m_1 + m_2)^2} = \left(\frac{m_1 - m_2}{m_1 + m_2}\right)^2 E_{k1}, \tag{11–20}$$

which shows that E'_{k1} is zero for $m_1 = m_2$ so that in this circumstance $E_{k1} = E'_{k2}$, and all the incident energy is transferred.

We now follow Chadwick's procedure and apply these equations to collisions between particles of unknown mass and energy and particles of known mass and energy, namely hydrogen and nitrogen nuclei. (These may be considered at rest because the thermal kinetic energies are relatively insignificant.) If m_2 is a proton, $m_2 = 1$ u, $E'_{k2} = 4.7$ MeV and Eq. (11–19) becomes

$$4.7 = \frac{4m_1E_{k1}}{(m_1 + 1)^2}, \tag{11–21}$$

whereas if m_2 is a nitrogen nucleus, $m_2 = 14$ u, $E'_{k2} = 1.2$ MeV, and we have

$$1.2 = \frac{4m_1 \times 14E_{k1}}{(m_1 + 14)^2}. \tag{11–22}$$

Solving Eqs. (11–21) and (11–22) simultaneously, we find that $m_1 = 1.03$ u and $E_{k1} = 4.7$ MeV.

Thus Chadwick proposed that the penetrating radiation was not gamma radiation but neutral particles, each having a mass of about 1 u. Such a particle would transfer energy to other nuclei more efficiently than photons, and both recoil protons and nitrogen nuclei can be attributed to a particle having a single reasonable energy, instead of the two large contradictory values required by the puzzling gamma-photon proposal.

Our equations have been applied to the collision between neutrons and other observable nuclei. Chadwick went back and analyzed the nuclear reaction itself to obtain a further check of his hypothesis and a better estimate of the neutron's mass.

For the case of boron, the reaction equation assumed was

$$_2\mathrm{He}^4 + {}_5\mathrm{B}^{11} \rightarrow {}_7\mathrm{N}^{15} \rightarrow {}_7\mathrm{N}^{14} + {}_0\mathrm{n}^1. \qquad (11\text{–}23)$$

To apply the law of conservation of mass-energy, he had to know the various particle energies. The alpha particle was known to have a kinetic energy equivalent to 0.00565 u. The boron was assumed to be at rest. The neutrons were detected by letting them collide with protons. The proton range indicated their energy to be 0.0035 u, and since neutrons and protons have nearly the same mass, Chadwick could take the measured proton energy to be the neutron energy. The nucleus of ${}_7\mathrm{N}^{15}$ recoils as it emits the neutron and becomes ${}_7\mathrm{N}^{14}$. From the law of conservation of momentum we have

$$\sqrt{2m_\alpha E_{k\alpha}} = \sqrt{2m_N E_{kN}} + \sqrt{2m_n E_{kn}}. \qquad (11\text{–}24)$$

Here everything is known except E_{kN}, which turns out to be 0.00061 u. Aston had measured the isotopic masses of ${}_2\mathrm{He}^4$, ${}_5\mathrm{B}^{11}$, and ${}_7\mathrm{N}^{14}$ to be 4.00106, 11.00825, and 14.0042, respectively. Chadwick wrote the complete mass-energy equation:

$$_2\mathrm{He}^4 + E_{k\alpha} + {}_5\mathrm{B}^{11} = {}_7\mathrm{N}^{14} + E_{kN} + {}_0\mathrm{n}^1 + E_{kn}, \qquad (11\text{–}25)$$

where every quantity was known except the mass of ${}_0\mathrm{n}^1$. Substituting, we have

$$4.00106 + 0.00565 + 11.00825 = 14.0042 + 0.00061 + {}_0\mathrm{n}^1 + 0.0035.$$

Solving for the mass of the neutron, Chadwick obtained 1.0067 u. This is surprisingly near the best modern value, which is 1.008665 u on the C^{12} scale.

We have treated the discovery of the neutron in great detail for several reasons. The neutron is a new fundamental particle so important that we shall devote Chapter 12 to its effects. The discovery is an excellent example of discovery by inference. [Neutrons do not cause fluorescence, do not cause photographic images, do not make cloud-chamber tracks, and do not trip Geiger counters. Neutrons cause weak ionization by colliding with electrons (about one ion per meter of path) and neutrons interact with nuclei.] Furthermore, the methods used in this discovery, the mechanics of impact and conservation of mass-energy, will clarify material that still lies before us.

In 1950, it was found that a free neutron, one outside the nucleus, is radioactive. It has a half-life of 12 min and disintegrates into a proton, an electron, and a neutrino.

11-4. ACCELERATORS

Although important discoveries came from nuclear reactions produced by alpha-particle bombardment, the only elements disintegrated by alpha particles were the light elements. In our discussion of the Rutherford alpha-particle scattering, we obtained an equation for the distance of closest approach when an alpha particle is directly aimed at a nucleus, Eq. (4–3). In words, the equation states that the minimum distance is proportional to the charge of the alpha particle times the charge of the target nucleus divided by twice the kinetic energy of the alpha particle. The reason it is easy to disintegrate light elements is that these elements have small nuclear charges. If the heavier elements repel alpha particles and therefore resist disintegration, the remedy is to use bullets with more energy, less charge, or both.

Protons are not ejected by radioactive atoms, but they are easily formed by ionizing hydrogen. They have half the charge of alpha particles and should be good bullet particles if given enough energy. Because of their charge, they can be accelerated by causing them to fall through a potential difference.

It is not our purpose to discuss here the inspired engineering that has gone into the design of accelerators; we shall discuss only the two generic types of instrument.

The first particle accelerators or "atom smashers" consisted of an ion source such as Thomson used to produce canal rays, an evacuated region in which the ions could be accelerated, and a source of high potential to do the accelerating. With such machines the main problem is to develop the necessary high voltage. One scheme of historic interest involved charging capacitors in parallel and then connecting them to the accelerator in series. Another scheme was to use transformers such as had been developed for producing x-rays, either singly or in cascade. The only high-voltage system in wide use today is the Van de Graaff electrostatic generator, in which charges are transported to bodies by means of moving belts, to supply the required large potentials. Of these schemes, the Van de Graaff generator is unique in that it provides a constant potential which accelerates the ions to a well-defined and measurable energy. Accelerators of this type can supply ions with energies of a few million electron volts at most.

The second generic type of accelerator is the cyclotron, developed by Lawrence and Livingston in 1932. Rather than obtain higher energies by

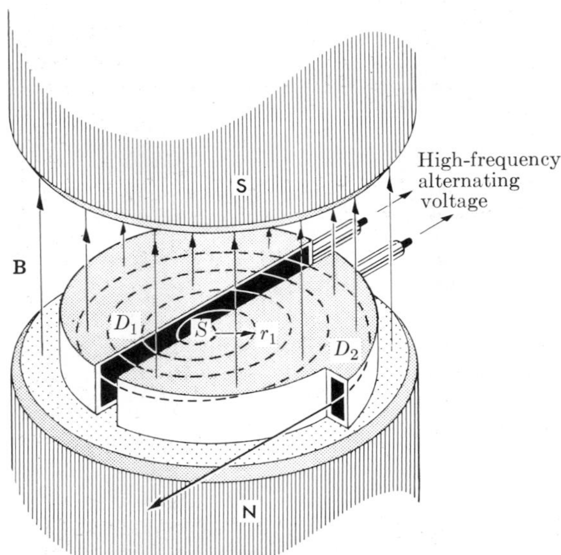

FIG. 11–5
Schematic diagram of the D's of a cyclotron, and an ion path within them.

going to higher potentials, these investigators devised an ingenious method for using a comparatively low voltage over and over.

We have seen (Eq. 2–5) that a charged particle moving with a velocity perpendicular to a magnetic field moves in a circular path. The time it takes for the particle to move around the circle once is the length of the path divided by the particle's tangential velocity, or $T = 2\pi r/v$. Using Eq. (2–5), $mv = qBr$, to eliminate the velocity, we have

$$T = \frac{2\pi m}{qB}.$$

Since ω, the angular velocity of the particle, is equal to $2\pi/T$, then

$$\boxed{\omega = \frac{qB}{m}.} \tag{11–26}$$

Thus the angular velocity of the charged particles is independent of their translational velocity. Lawrence utilized this property of moving charges in a magnetic field with the device shown schematically in Fig. 11–5. The heart of the instrument consists of two chambers called "D's" (because of their shape) which have roughly the size and shape of a circular metal cake box sawed in two. The ions to be accelerated are introduced near the center of these D's at S. If D_2 is negative relative to D_1, the positive ion is attracted into D_2 where, because it is within a conducting chamber, it

(a)

(b)

FIG. 11-6
(a) The M.I.T. cyclotron. (b) The D's removed from the gap between the poles of the electromagnet.

experiences no electric field. A magnetic field perpendicular to the plane of the D's forces the ions into a path with the angular velocity given by Eq. (11–17). In a time $T/2$, the ions return to the gap between the D's, and if by this time the potentials of the D's have been reversed; the ions are again accelerated in going from D_2 into D_1. This process is repeated over and over again. It might appear difficult to switch the potentials back and forth in time with the ions, but this is not so. If the D's are connected to a source of alternating voltage of frequency $f = qB/2\pi\text{m}$, the potential and the ions remain in phase. The tangential velocity v of the ion is equal to $r\omega$. Therefore the kinetic energy of an ion emerging from the cyclotron along a path of radius r is

$$E_k = \tfrac{1}{2}mv^2 = \tfrac{1}{2}mr^2\omega^2 = \frac{q^2B^2r^2}{2m}. \tag{11–27}$$

The magnitude of the potential used does not appear in the equation. The main problem in building a cyclotron is to produce a very intense magnetic field which is uniform over a large area. Large cyclotrons (Fig. 11–6) can produce 10-MeV protons, 20-MeV deuterons, and 40-MeV alpha particles.

The real upper limit of cyclotrons is set by the relativistic mass variation of the ions accelerated. As the velocity of the ions approaches the velocity of light, the mass increases and the angular velocity is no longer constant. Some instrument designs correct for this effect by making the magnetic induction nonuniform or by frequency modulating the exciting voltage, but the greatest advances have been made with radically new designs. The largest recent machines can produce energies measured in GeV.

11–5. THE COCKCROFT-WALTON EXPERIMENT

The first nuclear disintegrations induced without the help of alpha particles from radioactive substances were made by Cockcroft and Walton in the same year that Chadwick discovered the neutron, 1932. They used protons as bombarding particles. The protons were accelerated by a high voltage which was achieved by charging capacitors in parallel and discharging them in series, a technique in use at that time in x-ray technology. Proton energies up to 700,000 eV were obtained in this way.

Cockcroft and Walton found reactions which were often the converse of reactions already known. It was known that alpha bombardment often produced protons, but they also found that proton bombardment often produced alpha particles. Although they bombarded 16 different elements, the quantitative data they reported for lithium and fluorine are of particular interest.

The target region of their apparatus is shown in Fig. 11–7. The fast protons struck the target material at A. A zinc-sulphide fluorescent screen

FIG. 11-7
Schematic diagram of the Cockcroft-Walton apparatus.

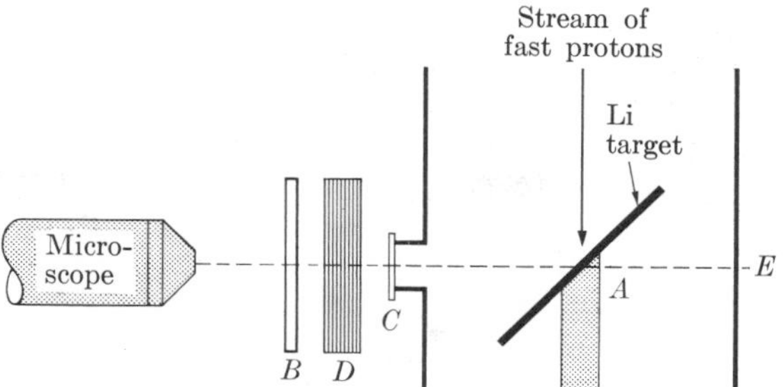

was placed at B, and any scintillations produced could be observed with the microscope at the left. The mica sheet at C held the vacuum in the accelerating tube and it had sufficient stopping power to prevent scintillations due to scattered protons. With lithium as the target, scintillations were observed which appeared to be alpha particles. When additional mica sheets, D, of known stopping power were introduced, the rays were found to have a range equivalent to about 8 cm of air. An ionization chamber and a cloud chamber were also used to positively identify the alpha particles and to determine that their range was 8.4 cm (Fig. 11–8). This range indicated an energy of 8.5 MeV for each particle. Evidently

FIG. 11–8
Cloud-chamber photograph of pairs of alpha particles from proton disintegration of lithium. The lithium target is close to a thin mica window at the center of the chamber. (Courtesy of P. I. Dee, University of Glasgow)

the reaction taking place was

$$_1H^1 + {_3Li^7} \rightarrow {_4Be^8} \rightarrow {_2He^4} + {_2He^4}. \qquad (11\text{–}28)$$

Since the lithium atom had only thermal motion, it could be regarded as being at rest, and since the proton velocity was small compared with that of the observed alpha particle, Cockcroft and Walton assumed that the two product alpha particles must acquire equal and opposite velocities (conservation of energy and of momentum require this if the initial energy and momentum may be neglected). Cockcroft and Walton tested this assumption in the following way. They made the target very thin and added another scintillation detector at E, opposite the original one, as shown in Fig. 11–7. They could then observe *simultaneous* scintillations, which indicated that both alpha particles came from a single disintegration.

They also checked their assumption by setting up the mass-energy equation of the reaction, as Chadwick did:

Proton mass + lithium mass + input energy in mass units
= 2(alpha-particle mass + alpha-particle energy in mass units).

When they substituted numerical values in this equation, they disregarded the kinetic mass-energy of the proton because this was less than the uncertainty in the isotopic mass of lithium known at that time. Using values from *their* data, we have, in u,

$$1.0072 + 7.0134 = 2(4.0011 + E_{k\alpha})$$

or

$$E_{k\alpha} = 0.0092 \text{ u} = 8.6 \text{ MeV per particle.} \qquad (11\text{–}29)$$

This is in good agreement with the energy determined from the range, 8.5 MeV.

In case of fluorine, the reaction was

$$_1H^1 + {_9F^{19}} \rightarrow \text{(excited) } {_{10}Ne^{20}} \rightarrow {_8O^{16}} + {_2He^4}. \qquad (11\text{–}30)$$

They again wrote the mass-energy equation to find the energy liberated. By dividing the energy between the oxygen and the alpha particle according to the laws of conservation of energy and of momentum, they again obtained good agreement between the measured and computed alpha-particle energies.

The significance of the Cockcroft-Walton experiments can hardly be overemphasized. They were the first to use an ion accelerator to produce nuclear disintegrations, and they proved, beyond any doubt, the equivalence of mass and energy. Together with Chadwick, their effective use of the mass-energy equation to account for the mass-energy balance of nuclear reactions provided a powerful tool for nuclear physics.

It is appropriate that we pause here and reflect on the power of abstract reasoning. When the search for the ether drift failed, Einstein decided to make physics valid without an ether medium. His theory of relativity gave a unique significance to the speed of light. He proposed that mass and energy are different manifestations of the same thing, with the speed of light squared as the constant of proportionality. This consequence of relativity was in violent disagreement with the highly successful Newtonian mechanics, and Einstein had not a shred of experimental evidence to support him. In the Cockcroft-Walton experiment we find that Einstein is not only justified, but also that his concept of mass-energy equivalence provides the basic key to the application of conservation of energy to nuclear processes. Einstein, who helped Perrin establish the atomic view of matter, had never heard of the nucleus when he proposed the theory of relativity.

11-6. NUCLEAR MASS-ENERGY EQUATIONS. Q-VALUE

We have already given examples of the mass-energy equation; we now consider its use in detail. This equation is the relativistic combination of the two classical principles of conservation of mass and of energy. The combination is effected through the use of the relation $E = mc^2$, which expresses the equivalence of mass and energy. Mass-energy may be measured in either mass or energy units. The conversion factor is c^2. Using units convenient to nuclear physics, we may say that 1 u has an energy of 931 MeV or that an energy of 931 MeV has a mass of 1 u. Nuclear mass-energy equations are usually written in atomic mass units, although any consistent units of mass or energy can be used. The fundamental equation is simply that the total mass-energy before a reaction equals the total mass-energy after the reaction.

Let us consider a general kind of reaction involving moving particles and photons. Let m_1 and m_2 be the rest masses of the initial particles and E_{k1} and E_{k2} their respective kinetic energies; let m_3 and m_4 be the rest masses of the final particles and E_{k3} and E_{k4}, their respective kinetic energies; finally, let f_0 be the frequency of an initial photon and f be the frequency of a final photon. Upon equating the total mass-energy before the reaction to that afterwards, we obtain

$$m_1c^2 + E_{k1} + m_2c^2 + E_{k2} + hf_0 = m_3c^2 + E_{k3} + m_4c^2 + E_{k4} + hf \tag{11–31}$$

or, transposing and collecting similar terms,

$$[(m_1 + m_2) - (m_3 + m_4)]c^2 = [(E_{k3} + E_{k4}) - (E_{k1} + E_{k2})] + h(f - f_0). \tag{11–32}$$

The right-hand side of this equation is the change in energy resulting from the reaction. The energy that is released or absorbed in a nuclear reaction is called the *Q-value* or *disintegration energy* of the reaction. Thus from Eq. (11–32) we have

$$\begin{aligned} Q &= [(m_1 + m_2) - (m_3 + m_4)]c^2 \\ &= [(E_{k3} + E_{k4}) - (E_{k1} + E_{k2})] + h(f - f_0). \end{aligned} \qquad (11\text{–}33)$$

If the rest masses are in u, then the Q-value in u is

$$\boxed{Q = (m_1 + m_2) - (m_3 + m_4).} \qquad (11\text{–}34)$$

The Q-value may be either positive or negative. If it is positive, then rest mass-energy is converted to kinetic mass-energy, radiation mass-energy, or both and the reaction is said to be exoergic or exothermic. If it is negative, the reaction is endoergic or endothermic and then either kinetic mass-energy or radiation mass-energy must be supplied if the reaction is to take place.

We are particularly interested in applying the equations we have just obtained to nuclear reactions. When we calculate the Q-value of a nuclear reaction, the masses used in either Eq. (11–33) or Eq. (11–34) must be the rest masses of the nuclei. However, the isotopic masses given in tables such as that in Appendix 5 are the rest masses of the neutral atoms. To obtain the rest mass of the nucleus of an atom, we must subtract the rest masses of all the orbital electrons from the isotopic mass. Are there any circumstances in which the rest masses of the neutral atoms may be used instead of those of the bare nuclei to calculate the disintegration energy?

Consider the reaction

$${}_5\mathrm{B}^{11} + {}_1\mathrm{H}^1 \rightarrow {}_4\mathrm{Be}^8 + {}_2\mathrm{He}^4. \qquad (11\text{–}35)$$

To obtain the Q-value of this reaction, we will substitute the nuclear masses into Eq. (11–34). Let the chemical symbols represent the isotopic masses of the atoms in u and let e represent the rest mass of an electron in the same units; then we have

$$\begin{aligned} Q &= [(\mathrm{B} - 5\mathrm{e}) + (\mathrm{H} - \mathrm{e})] - [(\mathrm{Be} - 4\mathrm{e}) + (\mathrm{He} - 2\mathrm{e})] \\ &= (\mathrm{B} + \mathrm{H} - 6\mathrm{e}) - (\mathrm{Be} + \mathrm{He} - 6\mathrm{e}) \\ &= (\mathrm{B} + \mathrm{H}) - (\mathrm{Be} + \mathrm{He}). \end{aligned}$$

This result shows that the *isotopic masses of the atoms can be used* to compute the Q-value in this case.

Next we consider a radioactive substance undergoing beta decay, that is, emitting an electron. An example is

$${}_{47}\mathrm{Ag}^{106} \rightarrow {}_{48}\mathrm{Cd}^{106} + {}_{-1}\mathrm{e}^0 \quad (\text{or } \beta^-). \qquad (11\text{–}36)$$

Following the same procedure as before, we find

$$\begin{aligned} Q &= [(\text{Ag} - 47\text{e})] - [(\text{Cd} - 48\text{e}) + \text{e}] \\ &= (\text{Ag} - 47\text{e}) - (\text{Cd} - 47\text{e}) \\ &= \text{Ag} - \text{Cd}. \end{aligned}$$

Therefore, *to find the disintegration energy in reactions involving electron emission, the isotopic masses may be used if the mass of the emitted electron is disregarded.* It should be remembered that the kinetic energy of the beta particle calculated from this and similar reactions is the maximum value it can have under the circumstances. In general, this amount of energy is actually shared between the emitted electron and the neutrino. Since the latter has negligible rest mass, it does not enter into the calculation of the Q-value.

In some cases positrons are emitted from nuclei. A reaction of this type is

$$_7\text{N}^{13} \rightarrow {}_6\text{C}^{13} + {}_{+1}\text{e}^0 \quad (\text{or } \beta^+). \tag{11–37}$$

In this case, we have

$$\begin{aligned} Q &= [(\text{N} - 7\text{e})] - [(\text{C} - 6\text{e}) + \text{e}] \\ &= (\text{N} - 7\text{e}) - (\text{C} - 7\text{e} + 2\text{e}) \\ &= \text{N} - (\text{C} + 2\text{e}). \end{aligned}$$

Therefore, to compute the energy change *in reactions involving positron emission, the isotopic masses may be used if two electron masses are included with those of the product particles.* In this case too, in general, the positron shares kinetic energy with the neutrino.

In following these rules for reactions involving alpha particles, α, or protons, p, we use the isotopic masses of $_2\text{He}^4$ and $_1\text{H}^1$.

In obtaining data to determine isotopic masses, we find that the results of mass spectroscopy provide only a starting point. We saw how Chadwick used the mass-energy equation to compute the mass of the neutron. In Eq. (11–25) the rest-mass terms are much larger than the kinetic energy-mass terms; therefore Chadwick could obtain the mass of the neutron to five significant figures. The repeated application of the mass-energy equation to a large number of nuclear reactions has enabled the computation of isotopic masses with great precision. Many are now known to either seven or eight significant figures.

In nuclear collisions having a Q-value the treatment is like that given in Eqs. (11–14) through (11–20), except that the sum of the kinetic energies of the final particles is equal to $E_{k1} + Q$ instead of E_{k1}. Also, if the collision is not head-on, the law of conservation of momentum must be written in component form.

11-7. CENTER-OF-MASS COORDINATE SYSTEM. THRESHOLD ENERGY

When the law of conservation of energy and the law of conservation of momentum were written to solve the nuclear collision discussed in Section 11–3, it was *implicit* that the velocities in the various expressions were the values in the *laboratory coordinate system*. In this *laboratory*, or *L-system* it was assumed that the target nucleus was at rest before the collision and that only the bombarding particle was in motion. It turns out that the treatment of nuclear collisions is easier if the center of mass of the particles is taken as the reference system. In the *center-of-mass* or *C-system*, the center of mass of the incident particle and the target nucleus is considered to be at rest and both particles are approaching it. The equations obtained by an observer moving with the center of mass are relatively simple and may be readily solved. The results can easily be transformed back to the L-system.

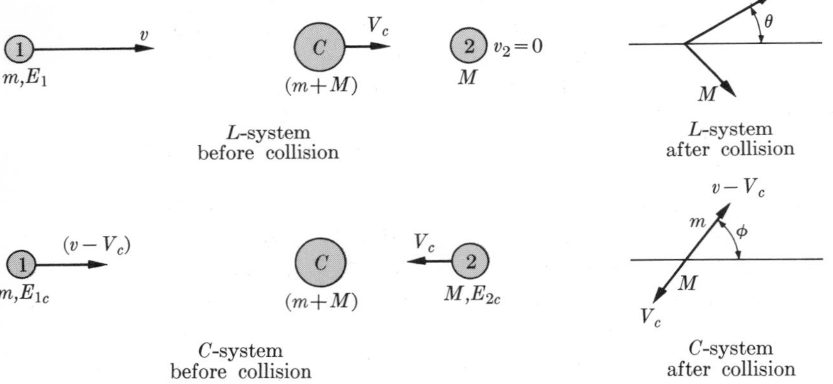

FIG. 11-9
An elastic collision between two particles as described in the laboratory system and the center-of-mass system.

The relationship between the L- and C-systems is shown in Fig. 11–9. According to the law of conservation of momentum, the momentum of the center of mass in the L-system must equal the sum of the momenta of the particles in the same system. Therefore, assuming low velocities so that classical mechanics may be used, we have

$$mv = (m + M)V_c$$

or

$$V_c = \frac{m}{m + M} v. \qquad (11\text{–}38)$$

In the C-system, before collision, the incident particle has a velocity of $v - V_c$ toward the right and the target has a velocity of V_c toward the left. Therefore the total momentum p_C in this reference frame is

$$p_c = m(v - V_c) - MV_c.$$

Substituting V_C from Eq. (11–38) into the last equation, we get

$$\begin{aligned} p_c &= m\left(v - \frac{mv}{m+M}\right) - M\frac{mv}{m+M} \\ &= m\frac{Mv}{m+M} - M\frac{mv}{m+M} = 0. \end{aligned}$$

Because momentum is conserved, it follows that after collision the magnitude of the momentum of each particle will be the same but oppositely directed, as seen from the C-system. If the collision is *elastic,* kinetic energy is also conserved and therefore the speeds of the particles in the C-system will be unchanged by the collision. Thus an observer in the C-system will describe an elastic collision as one in which the magnitude of each velocity is unchanged but in which the direction of each changes by the same amount, ϕ in Fig. 11–9. The velocity of each particle in the L-system is obtained by adding V_c vectorially to the velocity of each in the C-system.

Let us now consider an *inelastic* collision, particularly an endoergic nuclear reaction. As before, the total momentum in the C-system before collision is zero and so is the momentum afterwards. In the L-system, the initial momentum of the system is not zero and therefore the momentum of the final particles in it cannot be zero. Thus these final particles must have kinetic energy. Because of this, not all of the energy of the bombarding particle in a nuclear reaction is available to provide the Q-value required to cause the reaction. The minimum amount of energy that a bombarding particle must have in the L-system in order to initiate an endoergic reaction is called the *threshold energy,* E_{th}. The expression for this energy can be found from the Q-value relation, Eq. (11–33), which is valid in all inertial reference frames. Writing this for the C-system, we have

$$Q = (E_{3c} + E_{4c}) - (E_{1c} + E_{2c}), \tag{11–39}$$

where E_{3c} and E_{4c} are the kinetic energies of the product particles. At threshold, these particles are formed with no velocity in the C-system. Therefore, $(E_{3c} + E_{4c}) = 0$, and Eq. (11–39) becomes

$$Q = -(E_{1c} + E_{2c})$$

or

$$Q = \tfrac{1}{2}m(v - V_c)^2 + \tfrac{1}{2}MV_c^2. \tag{11–40}$$

When V_c in this equation is replaced by its equivalent from Eq. (11–38) and the resulting expression is simplified, we find that

$$-Q = \frac{1}{2} mv^2 \left(\frac{M}{m + M}\right). \tag{11–41}$$

The quantity $\frac{1}{2}mv^2$ is the kinetic energy of the bombarding particle in the L-system, and in this particular case it equals E_{th}. When E_{th} is substituted in Eq. (11–41), the equation becomes

$$-Q = E_{\text{th}} \left(\frac{M}{m + M}\right)$$

or

$$\boxed{E_{\text{th}} = (-Q)\left(1 + \frac{m}{M}\right).} \tag{11–42}$$

Thus the kinetic energy of the incident particle in the laboratory system must exceed the Q-value to cause an endoergic reaction.

11-8. ARTIFICIAL (INDUCED) RADIOACTIVITY

Although the work of Cockcroft and Walton stimulated the investigation of many nuclear reactions with better and better ion accelerators, we have not come to the end of important discoveries made with alpha particles from radioactive sources. In 1934, while I. Curie-Joliot and F. Joliot were bombarding aluminum with alpha particles from polonium, they observed neutrons, protons, and positrons coming off the aluminum. We have seen that alpha bombardment of light elements often produces protons and sometimes neutrons. Even the presence of positrons was not too surprising, since protons could be thought of as a neutron and positron combined. Their new discovery was made when they noted that the emission of positrons continued *after* the alpha bombardment was stopped. The positron activity decreased with time according to an exponential law and the phenomenon was clearly just like natural radioactivity. They assumed that the artificially radioactive element was ${}_{15}\text{P}^{30}$, formed according to the reaction

$${}_{13}\text{Al}^{27} + {}_{2}\text{He}^{4} \rightarrow {}_{15}\text{P}^{30} + {}_{0}\text{n}^{1}. \tag{11–43}$$

The phosphorus decays with a half-life of 2.55 m into silicon and a positron:

$${}_{15}\text{P}^{30} \rightarrow {}_{14}\text{Si}^{30} + {}_{+1}\text{e}^{0}. \tag{11–44}$$

To justify their assumption, the Joliots irradiated aluminum for a long period and then separated the phosphorus chemically. Since the radioactivity went with the phosphorus fraction, their assumption was verified.

In their original paper, the Joliots also reported the formation of radioactive nitrogen and silicon isotopes by bombardment of boron and magnesium, respectively. These new isotopes have half-lives of 10.1 m and 4.9 s. We can call these isotopes "new" since, if they were ever abundant in nature, they are now present in undetectable amounts.

In the years since the original discovery of artificial radioactivity, it has been found possible to form radioactive isotopes of all the elements, and for most elements there are many radioactive isotopes. Thus many more than half the known isotopes are radioactive.

When Madame Curie separated radium, she could follow the progress of the separation by observing the increased activity of the radium fraction. The radium was radioactively "tagged." In the discovery of artificial radioactivity, a crucial point was that in the separation of the aluminum and phosphorus the radioactivity went with the phosphorus. The actual amount of phosphorus was exceedingly small, but radioactivity is easy to detect and small amounts can be traced through successive chemical processes. This property of radioisotopes has been developed into a fine art called "tracer technique." We give a few examples of its applicability.

When iodine is taken into the body, it tends to collect in the thyroid gland. If a patient has cancer of the thyroid, he may be given radioactive iodine. The radioactivity goes just where it should to fight the malignancy.

One pipeline may be used to transport, successively, several kinds of petroleum products with surprisingly little mixing at the boundaries. At the source end of the line, a radioactive material is introduced between products. At the receiving end of the line, Geiger counters outside the pipe announce the arrival of the new product.

If a small amount of a radioelement is mixed with a large amount of the same stable element, the relative proportions of these remain fixed except for the predictable decay of the radioelement. Thus the concentration of that element can be followed through all kinds of chemical and physical processing by measuring the intensity of the activity. It may well be that the new knowledge of phenomena in the basic sciences, obtained by the use of radioactive tracers, will be more important than power from nuclear reactors.

11-9. CARBON DATING

One of the radioactive isotopes that is midway between natural radioactivity and artificial radioactivity is C^{14}, which disintegrates by beta decay into stable N^{14}. It is natural radioactivity in the sense that it occurs in nature, and it is artificial in the sense that it would not occur in nature if it were not constantly being re-formed. Carbon-14 is formed in the atmosphere by high-energy particles from outer space called cosmic rays (refer to Problem 11–31). Carbon-14 has a half-life of 5730 years.

In the atmosphere there is a kind of "radioactive equilibrium" between the production of C^{14} by cosmic rays and its diminution by radioactive disintegration. Fortunately for us, the concentration of C^{14} in the air we breathe and the food we eat is very small. In the body the concentration is only about $10^{-6}\%$ of the C^{12} in living tissue.

But we, the plants, and every living thing contain carbon, and all living things have an amount of C^{14} which is in equilibrium with the C^{14} in the atmosphere. When death comes, living things stop breathing and stop eating food. The intake of C^{14} stops. From the time of death, the C^{14} disintegrates without further replacement. Thus, at death, equilibrium is ended and exponential radioactive decay is the only process remaining.

Old wood contains less C^{14} than new wood. (Problem 11–32). Old bones have less C^{14} than new bones. By measuring the concentration of C^{14}, the time since death occurred can be computed. Thus C^{14} provides a radioactive clock for anthropologists just as uranium provides a radioactive clock for geologists. The C^{14} half-life is suitable for the dating of cultural history, just as the half-life of uranium is suitable for the dating of the history of the earth.

11-10. NUCLEAR BINDING ENERGY

In Chapter 10 on radioactivity we said nothing about the *cause* of radioactivity. Why are some isotopes stable while others are not? Why do those that are unstable postpone their disintegration and have probable lives ranging from microseconds to billions of years? We can say something definite about the first question and something plausible about the second.

We shall limit our discussion to radioactive alpha emitters, although a similar argument can be made for the other cases. Let us inquire whether ordinary aluminum, ${}_{13}Al^{27}$, is alpha radioactive on the basis of mass-energy. The assumed reaction would be

$$ {}_{13}\mathrm{Al}^{27} \rightarrow {}_{11}\mathrm{Na}^{23} + {}_{2}\mathrm{He}^{4} + Q. \tag{11–45}$$

From the table of isotopic masses, we find for this case that

$$26.98153 = 22.98977 + 4.00260 + Q$$

or

$$Q = -0.01084\ \mathrm{u} = -10.05\ \mathrm{MeV}. \tag{11–46}$$

The negative Q-value indicates that instead of taking place spontaneously, this reaction cannot proceed unless we supply 10.05 MeV of energy per disintegration. This isotope of aluminum is very stable against alpha decay. Now consider radium,

$$ {}_{88}\mathrm{Ra}^{226} \rightarrow {}_{86}\mathrm{Rn}^{222} + {}_{2}\mathrm{He}^{4} + Q. \tag{11–47}$$

This gives

$$226.0254 = 222.0175 + 4.0026 + Q$$

or

$$Q = +0.0053 \text{ u} \qquad \text{or} \qquad +4.93 \text{ MeV}. \tag{11–48}$$

The positive Q-value indicates that radium is unstable and can emit an alpha particle, giving off 4.93 MeV of energy in the process. This result is the combined kinetic energy of the alpha particle and the radon nucleus. (The combined kinetic energy actually is not quite as high as 4.93 MeV because this Q-value includes the energy of the gamma-ray photon emitted during the disintegration.)

Rather than carry out in detail energy calculations like those of Eqs. (11–46) and (11–48) for every reaction in which we might be interested, it is very helpful to view the matter graphically. Every energy calculation involves an energy difference and requires that the energy reference level be specified. The total *binding energy* of a nucleus is defined as the energy that would be required to separate the nucleus into isolated particles. Before the discovery of the neutron, nuclei were thought to be composed of protons and electrons. This view presented certain theoretical difficulties which were resolved by regarding the nucleus as composed of protons and neutrons—collectively called *nucleons.* When an atom is discussed with particular emphasis on its nuclear composition, it is called a *nuclide.*

The total binding energy is equal to the difference between the isotopic mass of the atom and the sum of the masses of its neutrons, protons, and extranuclear electrons. (The mass of a proton and an extranuclear electron is equal to that of hydrogen.) It is obvious that the total binding energy of heavy nuclei is greater than that of light nuclei because they have more nucleons to separate. But one might expect the binding energy per nucleon to be about constant, and indeed it is about constant for heavy elements. However, the variations are significant, as is shown in Fig. 11–10, which is based on data for stable and nearly stable nuclides. This figure is *usually* presented with the curve concave downward, and is then called the binding-energy curve. Our figure is consistent with the scheme we used in discussing the Bohr energy levels in hydrogen. The distance from the curve up to the zero energy level is proportional to the energy per nucleon that would be required to take the nucleus apart, just as the distance from a Bohr energy level to the top is proportional to the energy required to free an electron from an atom. The important characteristic of this curve is that it is concave upward with a minimum around mass number 56. From this curve we can see, qualitatively, the results we obtained for the alpha emission of aluminum and radium.

Alpha-particle emission always produces a new nucleus with a mass number that is four less than that of the original nucleus. If we find aluminum with mass number 27 on the curve and move to the left to sodium with mass number 23, the ordinate becomes less negative. Therefore, in going from aluminum to sodium, we must have done some of the work necessary to take aluminum apart. Since we must do work to make the reaction proceed, the reaction is endothermic and cannot take place spontaneously. In going from radium to radon, however, we go to a new element that is harder to take apart than the element with which we started. Since the new element is more stable than the original element, the reaction is exothermic and can take place spontaneously. This graph makes it evident that those reactions which produce new nuclei with mass numbers nearer the minimum of the curve are the ones that are likely to be exothermic.

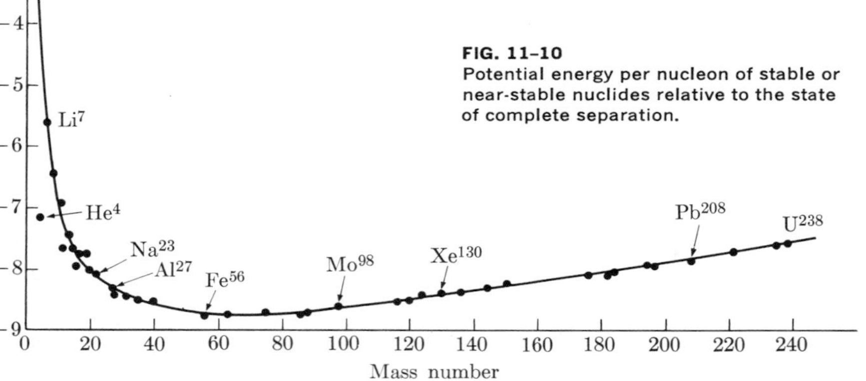

FIG. 11-10
Potential energy per nucleon of stable or near-stable nuclides relative to the state of complete separation.

In general, to release energy a nuclide having a mass number greater than about 56 has to disintegrate, whereas one below about 56 will have to combine with a particle. The fission of uranium is an outstanding example of the release of energy by disintegration; the fusion of hydrogen is a case of releasing energy by nuclear combination. From this curve we can understand too why there is a limit to the number of naturally occurring elements. Since the naturally radioactive series of elements are all at the right where the curve bends upward, it is logical to assume that if there were any heavier elements at the time of creation they were very unstable and have long since ceased to exist. We shall see in Chapter 12 that several of these elements have been man-made and have been found to be unstable, with short half-lives compared with the age of the earth.

In addition to the mass-energy equation, there are subtler factors that influence the stability of a nuclide. For example, a nuclide of atomic number Z and mass number A has A nucleons, of which Z are protons and $A - Z$ are neutrons. Nuclides having the same value of Z are *isotopes;* those having the same value of A are called *isobars;* those having the same number of neutrons are called *isotones;* and those having the same excess of neutrons over protons, $A - 2Z$, are *isodiapheres.* Thus $_{17}Cl^{37}$ is an isotope of $_{17}Cl^{35}$, an isobar of $_{16}S^{37}$, an isotone of $_{19}K^{39}$, and an isodiaphere of $_{18}Ar^{39}$.

A survey of 281 stable nuclides shows that 165 of them have an even number of protons and an even number of neutrons; 53 have an odd number of protons and an even number of neutrons; 57 have the number of neutrons odd and the number of protons even; and only 6 have odd numbers of both kinds of nucleon. All of these odd-odd nuclides except $_7N^{14}$ are rare. The even-even type are more stable and are, therefore, the least likely to be radioactive. Six nuclides of this type together comprise about 80% of the earth's crust: $_8O^{16}$, $_{12}Mg^{24}$, $_{14}Si^{28}$, $_{20}Ca^{40}$, $_{22}Ti^{48}$, and $_{26}Fe^{56}$. Further light on the question of nuclear stability comes from wave mechanics.

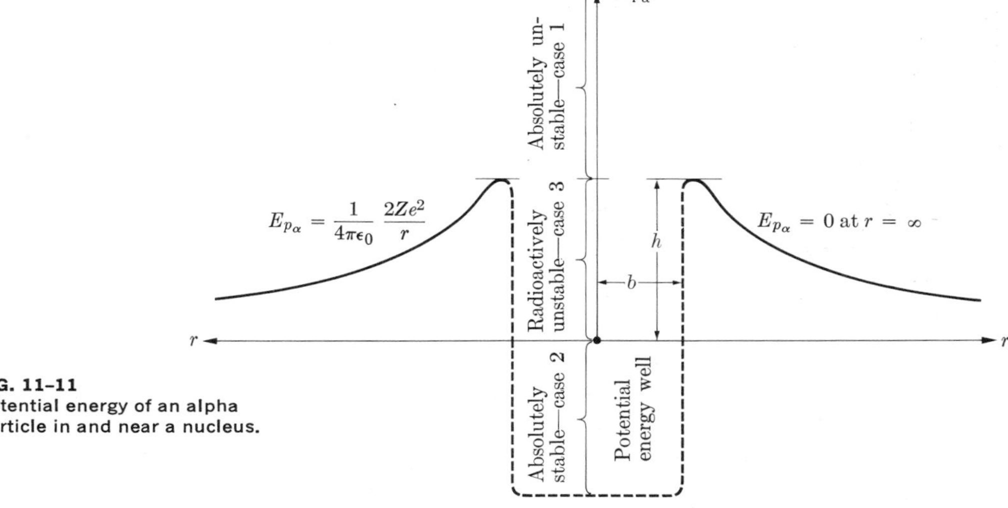

FIG. 11-11
Potential energy of an alpha particle in and near a nucleus.

11-11. RADIOACTIVITY AND WAVE MECHANICS

No elements except hydrogen could be stable were it not for very strong, short-range attractive forces between nucleons. Without such forces, the protons would fly apart by coulomb repulsion. Figure 11-11 is an approximate potential energy diagram for an alpha particle in or near a nucleus having an effective radius b. When the alpha particle is far from the nucleus $r > b$, it is repelled according to Coulomb's law, and its electric potential energy is then given by $E_{p\alpha} = (1/4\pi\epsilon_0)(2Ze^2/r)$. In the figure, the solid curve sloping away from the nucleus represents $E_{p\alpha}$ as a function of r. The Coulomb force on the alpha particle is equal to the slope of the potential energy curve, $F = -dE_{p\alpha}/dr$. The central part of the curve is shown dashed because we do not know its shape, but we do know that the curve must dip as the strong attractive short-range nuclear forces overcome the

Coulomb repulsion. The distance b is equal to the nuclear radius introduced in our discussion of Rutherford's alpha-particle scattering. This potential energy wall around the nucleus tends to prevent the entrance or exit of alpha particles whose energies are less than the height h of the potential barrier. The barrier is higher for nuclei of larger atomic number, and would be half as high for protons as it is for alpha particles. Thus this barrier illustrates the difficulty of disintegrating heavy elements with alpha particles, and shows why protons with a given energy can disintegrate more nuclides than can alpha particles. The situation is much like the problem of rolling a ball up a volcanic cone into the crater.

Although the alpha particle emitted by a radioactive substance like radium probably forms from two protons and two neutrons at the moment of ejection, we can think of the alpha particle as an entity within the radium nucleus. There are three cases we must consider. If this alpha particle had more energy than the height of the potential barrier, it could easily overcome the attractive force and thus nuclear disintegration would occur at once. If the alpha particle in the nucleus has a negative energy, the barrier is insurmountable and the particle can never get out. If the alpha particle has positive energy but it is less than h, the nucleus is radioactive.

An example of the first case is the intermediate nuclide, (excited) Ne^{20}, Eq. (11–30). The neon nucleus immediately emits an alpha particle. An example of the second case is the fluorine, F^{19}, of this same reaction, which is completely stable. Samples of the third case include all the radioactive nuclides, both natural and artificial. Radium C′, Po^{214}, is very unstable, with a half-life of 1.64×10^{-4} s, and U^{238} is nearly stable, with a half-life of 4.5×10^{9} yr.

Cases one and two are perfectly understandable from the classical energy viewpoint. Particles with more energy than the barrier are not hindered by it. Particles with energy far below the barrier are completely stopped by it. But, according to the classical energy viewpoint, the nuclides of case three should also be stable. If the nuclear alpha particle has less energy than the height of the barrier, it should be confined just as certainly as if its energy were negative. If this classical view were correct, there would be no radioactivity.

In our discussion of wave mechanics, we compared the classical and wave-mechanical treatments of a harmonic oscillator such as a mass oscillating on a spring. The mass has a fixed energy which is alternately potential and kinetic, and it moves subject to an attractive force determined by the spring. This force may be described as producing a potential barrier which prevents the mass from getting away. According to the classical solution of the problem, the mass never gets farther from its central position than its amplitude, which is the distance to the potential barrier. But in the wave-mechanical solution, we found that there is a finite probability that the mass can have a displacement from the center greater than the amplitude. Some authors attempt to salvage some of the classical view by saying that there are a few holes in the potential barrier around the nucleus and that a nuclear particle can "tunnel through" it. We prefer to dispense with classical models and simply say that classical physics does not apply. Because of the wave nature of matter, there is a finite probability that the particle will escape. Every radioactive disintegration demonstrates this.

The energy of the emitted alpha particle (and the recoil nucleus) does not depend on the height of the potential barrier. The kinetic energy after the disintegration is the same as the energy of the nucleus and its alpha particle before the disintegration. But this energy does have a bearing on the decay process. If the energy is nearly equal to the energy height of the potential barrier, then disintegration is likely, the decay constant is large, and the half-life is small. If the energy is small compared with the barrier height, then the radioactive nucleus is relatively stable and has a long half-life. This point is not covered by our analogy between an alpha particle in a nucleus and a mass on a spring. The reason is that whereas the spring exerts an attractive force on the mass everywhere the mass moves, the alpha particle experiences a repulsive force after r exceeds b. If the potential energy well for the oscillating mass had had a shape like that in which the alpha particle moves, the probability of finding the mass beyond its amplitude would increase as its energy increased.

One might try to explain radioactivity classically by imagining that the nucleons in a nuclide are like gas molecules having a Maxwellian energy distribution. Every once in a while some nucleons "gang up" on one or more of their fellows and give them enough energy to eject them from the nuclide. If this were so, the ejected particle would have energy at least as great as the potential barrier. This is not observed experimentally.

We may use wave-mechanical ideas to resolve another dilemma. If radium disintegrates into radon and an alpha particle with the release of 4.78 MeV of energy, why should not radium be formed when radon is bombarded with alpha particles of about this energy? No doubt this reverse reaction can and does occur, but consider how unlikely it is. If you could place a radium atom on the table and watch it disintegrate, you might be rewarded in a few seconds. On the other hand, you might have to watch for millions of years. If you want a 50-50 chance of seeing the disintegration, you must be prepared to sit with rapt attention for about 2340 years, the average life. The probability of finding the radium alpha particle outside the nucleus in a short time is exceedingly small. Conversely, suppose you were to try to introduce a 4.78-MeV alpha particle into radon to form radium. To have a 50-50 chance of succeeding, you would have to keep the alpha particle near the radon nucleus for about 2340 years. This would be even more difficult than waiting for radium to disintegrate. It would be necessary to bombard the radon nucleus untold number of times. If one must penetrate the radon nucleus with an alpha particle, it is far more promising to bombard it with particles whose energy exceeds the radon potential barrier. Recall that alpha particles from radioactive materials were used successfully in disintegrating light nuclei whose potential barriers are relatively low.

11-12. MÖSSBAUER EFFECT

The Mössbauer effect provides a means for producing and studying gamma rays whose energies are extremely well defined. We will begin this discussion by reviewing the factors which cause a gamma ray to be emitted with a range of energies.

Cobalt-57 decays to an excited state of iron-57 by K electron capture. Nine percent of the excited iron-57 nuclei go directly to the ground state with the emission of a 137-keV gamma-ray photon, but 91 percent of the transitions involve two steps. After emitting a 123-keV gamma ray, the nucleus is in an intermediate excited state which has a half-life of 10^{-7} s. In the transition from this state to the ground state a gamma ray of low energy, 14.4 keV, is emitted. What factors limit the precision of the energy value of this transition?

Knowing the half-life of the excited state, we can use Heisenberg's uncertainty principle to estimate the "inherent uncertainty" of the energy. This turns out to be about 4×10^{-8} eV, and comparing this to the energy of the gamma ray, we find an uncertainty of about three parts in 10^{12}, a very small fraction.

Another consideration in discussing the energy of the emitted gamma ray is the energy imparted to the recoiling nucleus. This always reduces the energy of the gamma ray. The data for computing the recoil energy can be obtained from momentum considerations. If a gamma-ray photon has a nominal energy E (14.4 keV for iron-57), then its momentum is given by the relativistic expression $p = h/\lambda = E/c$. Because momentum is conserved, the emitting system must also have E/c units of momentum imparted to it in a direction opposite to that of the emitted quantum of radiation. *If* this recoil momentum is given only to the emitting nucleus, we can calculate its recoil energy. For iron-57, which has a mass of about 10^{-25} kg, we find the recoil velocity to be about 80 m/s. This is low enough to justify a classical calculation. Thus the kinetic energy of the recoiling nucleus is about 2×10^{-3} eV, and the energy of the emitted gamma ray is *less* than the transition energy by this amount. It is to be noted that this recoil energy is about 5×10^{4} times greater than the uncertainty in the gamma-ray energy calculated from Heisenberg's principle.

A third influence upon the gamma-ray energy would seem to be the Doppler effect of the source nucleus. This effect depends upon the velocity of the source. The thermal velocity of a *free* atom may be computed by the classical methods discussed in Chapter 1. The root-mean-square speed of an atom which has a mass of 57 u and which is in the gaseous state at room temperature is about 350 m/s. But an atom which is *in a crystal* cannot be treated classically. Since it is bound in a crystal, the atom has only quantized vibrational energy states. These quantized states are variously populated and widely separated compared to the low recoil energy of our example. At reduced temperature the occupancy of the lowest energy state is greatly increased. In the discussion which follows we will assume that the temperature is so low that the effects of thermal motion and crystal lattice vibrations are negligible.

Returning to gamma-ray emission, if the energy of the photon is low enough, as in iron-57, then the recoil energy is insufficient to separate the atom from the crystal lattice or to change the vibrational state of the emitting atom. In this case the recoil energy is not transferred to *one* emitting atom but to a *group* of atoms. This means that the recoiling mass is relatively tremendous, the recoil velocity is practically zero, and the gamma ray carries away substantially all the energy of the transition. As a consequence, the remaining uncertainty of the energy of the gamma ray is almost completely determined by the Heisenberg principle alone.

This is the *Mössbauer effect,* which is also called *recoilless emission.* Mössbauer emitters (there are others in addition to iron-57) must emit gamma rays of low enough energy so that the emitting atom is neither freed from the crystal lattice nor excited to a high vibrational energy state.

The Mössbauer effect is important in realizing the necessary conditions for the absorption of gamma rays by nuclei which are identical to those of the emitter. A nucleus can absorb only that gamma-ray photon whose energy is equal, within the limits set by the uncertainty principle, to the difference in energy between an excited state and the ground state of the nucleus. This is called *resonance absorption.* It will occur only under Mössbauer conditions because, unless the gamma-ray emission is recoilless, the gamma ray has its energy reduced so much that it cannot resonate with a nucleus of the same kind. A Mössbauer emitter, however, produces gamma rays with an extremely narrow energy spread and these gamma rays can be detected with equal energy precision by means of resonance absorption.

To emphasize the degree of precision obtainable and the importance of the Mössbauer effect we cite two of its many applications.

In discussing radioactivity we stated that magnetic fields and other environmental influences had no detectable effect on nuclear events. However, infinitesimal effects have actually been observed by using a Mössbauer emitter whose radiation is being detected by a Mössbauer absorber. Placing either the emitter or the absorber in a magnetic field will put the system out of resonance because of slight changes of the nuclear energy levels. Indeed if one is in a different chemical state from the other, the difference can be detected. The change is usually measured by means of the Doppler effect. Resonance may be restored by moving the emitter relative to the absorber with almost unbelievable slowness, of the order of a few centimeters per minute. By varying this motion in a carefully controlled way, the system can be "tuned" over a very narrow energy spectrum.

One of the phenomena predicted by the general theory of relativity is the gravitational red shift which may be considered as the change in energy of a photon as it travels from one place to another of different gravitational potential. For a photon of energy E, traveling between two points which are separated by a height h near the earth where the acceleration due to gravity is g, the change in energy of the photon is given by $(E/c^2)gh$. Obviously, the corresponding fractional change of energy, gh/c^2, is minute. Nevertheless this change has been verified by placing an absorber about 75 ft above a Mössbauer emitter, and then determining the energy shift from the rate at which one or the other must be moved to achieve resonance absorption. This experiment provides a terrestrial test of the general theory of relativity.

11-13. THE BOMBARDING PARTICLES

Thus far we have discussed the use of alpha particles and protons to bombard nuclei. With the advent of particle accelerators the study of nuclear physics became very popular and a great number of reactions were studied. Many of these transitions were excited by high-energy electrons and high-energy photons or gamma rays. Many new isotopes were discovered. One of these, which deserves special mention, is deuterium.

As early as 1920, Harkins and Rutherford predicted that there should be a heavy isotope of hydrogen having mass number 2. It was not found in mass spectroscopy because it was masked by the hydrogen molecule H_2. In 1931 Urey, Brickwedde, and Murphy separated heavy hydrogen from ordinary hydrogen by evaporating liquid hydrogen. Since light hydrogen is more volatile, it evaporated a little more readily than the heavy hydrogen. This isotope was positively identified spectroscopically. The atomic spectrum of heavy hydrogen is shifted from that of ordinary hydrogen because its greater mass causes a small but measurable difference in the Rydberg constant. Soon after its discovery, heavy hydrogen was separated in considerable quantities by the electrolysis of water, where again the molecules formed with light hydrogen were more mobile and electrolyzed more readily. Since heavy hydrogen has twice the mass of its common isotope, it has many striking properties. Water formed from heavy hydrogen, called "heavy water," has a specific gravity of 1.108, which is quite different from that of ordinary water. Because of its special importance, heavy hydrogen has been given a special name, *deuterium*, and the symbol

D. Thus $_1H^2$ and $_1D^2$ have identical meaning. (Ordinary hydrogen, $_1H^1$, also has a special name, *protium*, but it is seldom used.) Just as ionized hydrogen is called a proton, p, ionized deuterium is called a *deuteron*, d. A third hydrogen isotope, $_1H^3$, is called *tritium* and is sometimes represented by the symbol T. The tritium nucleus is called *triton* and is indicated by t. Tritium is radioactive, with a half-life of 12.26 years.

One of the important uses of deuterium was as a bombarding particle. It has unit charge like the proton, but it has twice the mass of a proton. Deuterium bombardment led to still more nuclear reactions, such as

$$_1D^2 + {}_1D^2 \rightarrow {}_2He^3 + {}_0n^1 + Q. \qquad (11\text{–}49)$$

We choose this particular reaction as an example because it is one of many reactions which produce neutrons.

It is easy to see why neutrons were also used as bombarding particles. Because they have no charge, they are not subject to coulomb repulsion. With no potential barrier to penetrate, neutrons can enter almost any nucleus with ease. The number of nuclear transitions that can be produced with neutrons is greater than the number induced by all other particles combined.

11-14. NEUTRON REACTIONS. MODES OF NUCLIDE DECAY

Nuclear reactions are often represented in a shorthand notation in which the participating particles are identified in this order: target particle, bombarding particle, ejected particle, and product particle. The symbols for the bombarding and ejected particles are enclosed in parentheses. Thus Eq. (11–49) may be written $D^2(d, n)He^3$. Equation (11–43) becomes $Al^{27}(\alpha, n)P^{30}$. Any reaction in which a deuteron produces a neutron is called a (d, n) reaction, and the second example is called an (α, n) reaction.

The first type of neutron reaction observed was of the (n, α) type, where the target was, successively, nitrogen, oxygen, fluorine, and neon. A second type is called simple *radiative capture*, (n, γ). The first observed reaction of this type was the formation of deuterium from hydrogen and a neutron with the emission of a gamma photon. Nearly all elements undergo radiative capture of neutrons and it is probably the most common nuclear process. It is particularly important that the probability of this type of reaction taking place is greater for slow neutrons than for fast ones.

A third type of neutron reaction is the (n, p) type. Except for the lighter elements, the potential barrier inhibiting the escape of the proton causes these reactions to have negative Q-values. Therefore the incident neutron must be fast, with an energy of 1 MeV or more. Although they are less probable than (n, p) reactions, (n, d) and (n, t) reactions are also found.

The second type of reaction, (n, γ), is especially interesting. It adds a neutron to the bombarded nucleus and frequently makes it radioactive.

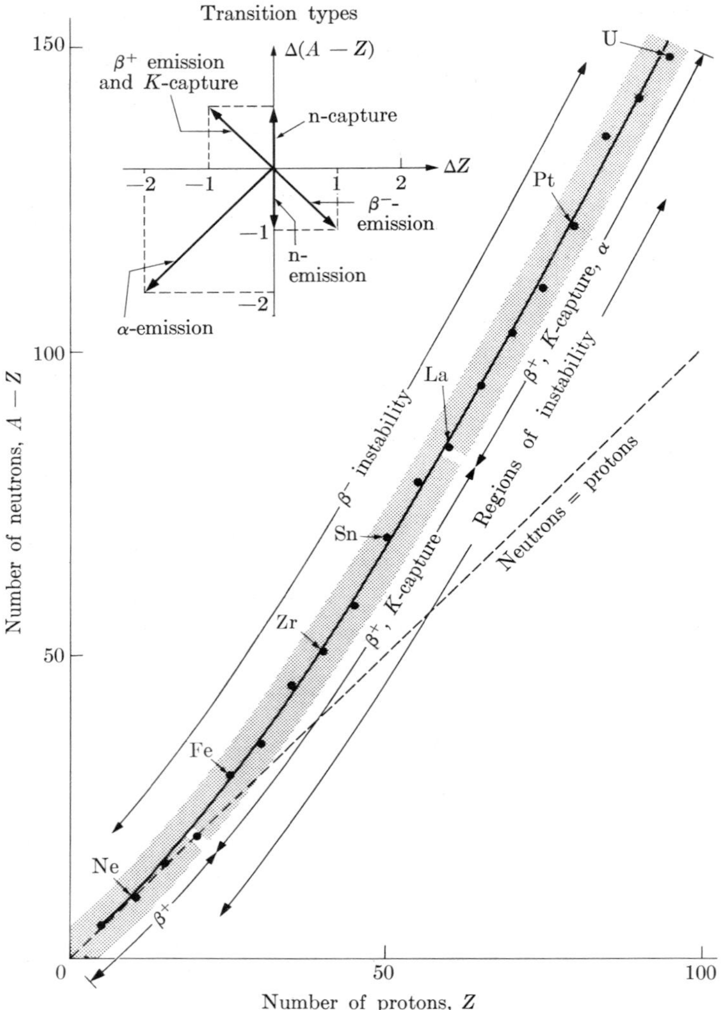

FIG. 11–12
Modes of decay of nuclides.

An examination of the periodic table shows that atomic weight is not proportional to atomic number. The higher the atomic number of a nuclide, the more the number of neutrons exceeds the number of protons. If we plot the number of neutrons, $A - Z$, against the number of protons, Z, for the stable nuclides, we obtain the result shown in Fig. 11–12.

In the inset of this figure, the abscissa is the change ΔZ in atomic number due to a nuclear reaction, and the ordinate is the corresponding change

$\Delta(A - Z)$ in the neutron content of the nucleus. Five arrows are shown in this inset. One, directed upward, represents the change in a nuclide for simple neutron radiative capture, (n, γ). Another, downward and to the right, represents beta decay in which the emission of a negative electron is accompanied by the conversion of a neutron into a proton. The one pointing directly downward represents neutron emission. The fourth arrow represents the emission of an alpha particle in which two neutrons and two protons are removed. The remaining arrow, upward and to the left, represents both positron emission and orbital electron or K-electron capture, where a proton is converted to a neutron. Nuclides off the full-line curve tend to be unstable, and the particle each emits is likely to be one which brings the new nuclide nearer the curve. Thus the region above the curve is the region of likely beta decay. A nuclide in the region below the curve disintegrates in such a way that its atomic number decreases. In general, elements of low mass number disintegrate by positron emission; those in the middle range, either by positron emission or by orbital electron capture; and the heavy elements, by positron emission or by orbital electron capture or by ejecting an alpha particle. The natural radioactive series involve alpha and beta decay in such a way that the series of nuclides tend to follow the curve of stability. Since neutron capture makes a new nuclide above the original one on the chart of Fig. 11–12, the new nuclide tends to be beta-unstable. As a typical example, we have

$$_{45}\mathrm{Rh}^{103} + {}_0\mathrm{n}^1 \rightarrow {}_{45}\mathrm{Rh}^{104} + \gamma. \tag{11–50}$$

The Rh^{104} is beta-unstable, with a half-life of 42 s, and disintegrates according to the equation

$$_{45}\mathrm{Rh}^{104} \rightarrow {}_{46}\mathrm{Pd}^{104} + {}_{-1}\mathrm{e}^0. \tag{11–51}$$

11–15. ACTIVATION ANALYSIS

Since the addition of a neutron to a nucleus converts nearly every element to a radioactive form and since radioactivity can be readily detected, a method which supplements chemical analysis is possible.

Suppose that one has a sample of material which might contain a trace of, say, indium. The problem is to determine whether indium is present and, if it is present, its concentration. We assume that the amount present is too small to detect by normal chemical methods. Indeed, there is a possibility that reagents used in the chemical analysis might introduce more indium than exists in the given sample.

Now let this sample be bombarded with neutrons. Then some of the atoms of each of the elements present will be converted to a radioactive form. In particular if any indium is present, some of it will become radioactive. Next a solution containing a high concentration of stable indium is added to the unknown. We now know that the sample contains a lot of indium of which most is not radioactive. But if indium was present originally, then some of the indium in the sample is radioactive and this portion of it will, of course, emit the gamma-ray spectrum characteristic of that element. Through a chemical procedure all of the indium is next separated by standard techniques. This is comparatively easy because now indium is abundantly present. If the separated indium is found to be radioactive, then the original sample must have contained indium. Since the intensities of the lines in the gamma-ray spectrum are directly proportional to the number of radioactive atoms present and since this spectrum can readily be obtained with a multi-channel analyzer, the amount of indium in the original sample is easily determined.

This is but one example of neutron activation analysis, which is a whole new technique that can be used to determine, both qualitatively and quantitatively, the chemical elements present in very minute samples of a substance.

11-16. THE DISCOVERY OF FISSION

Since neutron capture followed by beta decay produces a new nuclide of higher atomic number, the process suggested to Fermi an intriguing possibility. Why not cause neutron capture in uranium, the last (then known) element of the periodic system? If uranium formed a beta-unstable isotope, the disintegration product would be element number 93, a "transuranic" element. In 1934, Fermi and his associates attempted the experiment. Apparently the uranium did undergo neutron capture. Apparently the product was beta-unstable. But the beta activity included no less than four different half-lives. As in the discoveries of natural and artificial radioactivity, an attempt was made to identify the product nuclide by chemical separation. Early in 1938, Hahn and Strassmann showed that the beta-active product was chemically like radium and concluded that the new nuclides were new isotopes of radium. In order to get from uranium to radium, whose atomic number is four less, the uranium would have to emit two alpha particles. These alpha particles were not found.

The chemical analysis of such tiny amounts of material is very difficult, but in September 1938 Mme. Joliot-Curie and P. Savitch published a report of their work that identified one activity as due, apparently, to a rare-earth element *lanthanum* which has a much smaller mass number than uranium. This led Hahn and Strassmann to repeat their own earlier work very carefully. In December 1938 they concluded, "As chemists we should replace the symbols Ra, Ac, and Th . . . in [our] scheme . . . by Ba, La, and Ce . . . As nuclear chemists, closely associated with physics, we cannot decide to take this step in contradiction to all previous experience in nuclear physics."

The correct interpretation of the puzzling data came in January 1939 when Meitner and Frisch wrote, "It seems possible that the uranium nucleus has only small stability of form, and may, after neutron capture, divide itself into two nuclei of roughly equal size." Because of its resemblance to the splitting of one living cell into two of equal size, this nuclear process was named after the biological process, *fission*.

All previous nuclear reactions had had as their products no particles more massive than alpha particles and no nuclei farther removed from the original nucleus than two atomic numbers. Even in this decade of rapid advance, it took five years to break away from the limitation of the old conceptions and break through to the conception of this strikingly new process.

Once the fission process was regarded as plausible, the chemical evidence for its correctness was almost immediately reinforced by physical evidence. There were literally scores of laboratories around the world that were already equipped to verify fission. If one uranium atom splits into two atoms near the middle of the periodic series, a glance at Fig. 11–10 or a few moments of calculation will show that the Q-value of the reaction is tremendous. It is 200 MeV, which is about ten times that of the most energetic reaction previously known. Within a short time after hearing the news, many laboratories confirmed this energetic reaction.

REFERENCES

BEYER, ROBERT T., *Foundations of Nuclear Physics*. New York: Dover, 1949. Everyone interested in the scientific method and how it revealed properties of the nucleus should read this book. It contains facsimiles of the fundamental studies as originally reported in the scientific journals of Europe and the United States.

Some of the articles are: "The Positive Electron," by Carl D. Anderson, pp. 1–4; "Experiments with High Velocity Positive Ions," by J. D. Cockcroft and E. T. S. Walton, pp. 23–38; "Un Nouveau Type de Radioactivité," by Mme. Irene Curie and M. F. Joliot, pp. 39–41; "Possible Production of Elements of Atomic Number Higher than 92," by Enrico Fermi, pp. 43–44; "Uber den Nachweis und das Verhalten der bei der Bestrahlung des Urans mittels Neutronen entstehenden Erdalkalimetalle," by O. Hahn and F. Strassman, pp. 87–91; "The Production of High Speed Light Ions without the Use of High Voltages," by E. O. Lawrence and M. S. Livingston, pp. 93–109; "Collision of α Particles with Light Atoms. An Anomalous Effect in Nitrogen," by E. Rutherford, pp. 111–137.

BLEWETT, J. P., "Resource Letter PA-1 on Particle Accelerators," *Am. J. Phys.* **34,** 742 (1966).

CURTISS, LEON F., *Introduction to Neutron Physics*. Princeton, N. J.: Van Nostrand, 1959. Broad coverage.

LIBBY, WILLARD F., *Radiocarbon Dating*. 2nd ed. Chicago: University of Chicago Press, 1955. A very interesting book.

LIBBY, WILLARD F., "Chemistry and the Atomic Nucleus," *Am. J. Phys.* **26,** 524 (1958). An informal account of radiochemistry and carbon dating.

LIVINGSTON, M. STANLEY, *High-energy Accelerators*. New York: Interscience, 1954. A good discussion of these important devices.

MURPHY, GEORGE M., editor, *Production of Heavy Water*. New York: McGraw-Hill, 1955.

TAYLOR, DENIS, *Neutron Irradiation and Activation Analysis*. Princeton, N. J.: Van Nostrand, 1964.

WERTHEIM, G. K., *Mössbauer Effect,* Resource Letter ME-1 and Selected Reprints. New York: American Institute of Physics, 1963.

WERTHEIM, G. K., *Mössbauer Effect: Principles and Applications*. New York: Academic Press, 1964.

PROBLEMS

11–1. Show that the range of a charged particle in a material is proportional to the 3/2 power of its kinetic energy at the beginning of its range.

11–2. Radium-C′, Po^{214}, emits some of the most energetic alpha particles observed from naturally radioactive elements. The energy of these particles is 7.68 MeV and their range is 6.90 cm in air at 15°C and 1 atm pressure. (a) Show that the moving mass and the relativistic velocity of an alpha particle just ejected from RaC′ do not differ from the rest mass and the classical velocity, respectively, by more than 1%. (b) Repeat part (a) for a 7.68-MeV proton. (c) Do you think that the differences between the classical and relativistic results are so significant that relativistic mechanics must be used in calculations involving high-energy particles such as those in this problem?

11–3. Using the data of Problem 11–2, calculate (a) the classical velocity of an alpha particle from RaC′, Po^{214}, and (b) the total number of ion pairs it produces over its range in air. (It requires 34 eV to produce an ion pair in air.) (c) If all the ion pairs produced by the alpha particles from a microcurie of RaC′ constitute the current in an ionization chamber, what would be the current in amperes?

11–4. An alpha particle just ejected from RaC′, Po^{214}, undergoes an elastic collision with one of the nuclei of molecular hydrogen in a region filled with this gas. If both particles continue in the original direction of motion of the alpha particle, find classically, (a) the ratio of the velocity of the proton after collision to that of the incident particle, and (b) the kinetic energy of the proton. (It requires about 4 eV of energy to dissociate the hydrogen molecule. The data for Po^{214} are given in Problem 11–2.)

11–5. An alpha particle going through a cloud chamber undergoes an elastic collision with a nucleus of unknown mass number initially at rest. A photograph of the event shows a forked track. It is seen from measurements of the track that the collision deviated the alpha particle 60°, and the struck nucleus went off at an angle of 30° with the direction of motion of the incident particle. What is the mass number of the unknown nucleus? (Since this is not a head-on collision and since momentum is a vector quantity, the law of conservation of momentum must be applied in component form.)

11–6. A 7.68-MeV alpha particle has an elastic collision with a hydrogen nucleus. If the path of the alpha particle is deflected upward 10° from its original direction and this particle then has 2.57 MeV of energy, what is the velocity and direction of motion of the proton?

11–7. (s) Show from Eq. (11–20) that when a particle of mass m_1 has a head-on elastic collision with a particle of mass m_2 which is initially at rest, then the fraction of the energy lost by m_1 is $4m_1m_2/(m_1 + m_2)^2$. (b) Considering m_1 constant and m_2 variable, show that this loss is maximum for $m_1 = m_2$. (c) How much is this maximum loss in percent?

11–8. How many head-on, elastic collisions must a neutron have with other particles to reduce its energy from 1 MeV to 0.025 eV if the particles are atoms of (a) deuterium, (b) carbon, and (c) lead?

11–9. A nucleus of Ni^{60} in an excited state decays to its ground state by emitting a 1.33-MeV photon. What is the recoil energy and recoil speed of the Ni^{60} nucleus?

11–10. Before the discovery of the neutron, it was thought that the nucleus consisted of protons and electrons. The wave nature of the electron requires that a large energy is necessary to confine an electron in a region as small as the nucleus. If an electron has a wavelength of 10^{-13} cm, calculate its kinetic energy. (Relativistic expressions must be used.)

11–11. The beam in a cyclotron has a maximum diameter of 1.6 m. Given that the magnetic induction is 0.75 T, calculate the kinetic energies of the particles if they are (a) protons and (b) deuterons.

11–12. Deuterons in a cyclotron describe a circle of radius 32.0 cm just before emerging from the D's. The frequency of the applied alternating voltage is 10 MHz. Neglecting relativistic effects, find (a) the flux density of the magnetic field, and (b) the energy and speed of the deuterons upon emergence.

11–13. Sulfur-32 is bombarded with neutrons. Write the following reactions: (n, p), (n, α), (n, d), (n, γ).

11–14. Complete each of the following by writing the nuclear reaction equation:

$$B^{10}(n, \alpha)?,\quad N^{14}(\alpha, n)?,\quad P^{31}(\gamma, n)?,\quad Li^{7}(p, n)?,$$
$$Al^{29} \rightarrow Si^{29} + ?,\quad Al^{27}(n, \gamma)?,\quad P^{31}(d, p)?,\quad C^{12}(\gamma, \alpha)?,$$
$$N^{14}(?, p)C^{14},\quad Ca^{39} \rightarrow K^{39} + ?,$$
$$I^{120} \rightarrow ? + {}_{+1}e^{0}\ (\text{or } \beta^{+}),\quad Co^{59}(n, ?)Co^{60}$$

11–15. A certain photonuclear reaction is $Mg^{24}(\gamma, n)Mg^{23}$. What is the least energy the photon must have to produce this reaction, *assuming* that the products have negligible kinetic energy? The isotopic mass of the product nuclide is 23.001453. (To conserve momentum, the product particles must actually have velocities and therefore some kinetic energy.)

11–16. The Q-value of the disintegration of Ra^{226} by alpha-particle emission to form Rn^{222} was found, from Eq. (11–48), to be 4.93 MeV. Using this result and the laws of conservation of energy and of momentum, find (a) the recoil energy of the radon nucleus, and (b) the kinetic energy of the alpha particle. Assume that all of the disintegration energy appears as kinetic energy of the particles.

11–17. (a) Calculate the Q-value of the reaction in the Cockcroft-Walton experiment in which Li^7, when bombarded with 0.7-MeV protons, produced two alpha particles having equal kinetic energies. (b) What is the kinetic energy of each alpha particle? (c) What is the range of each of these alpha particles in air at 15°C and 1 atm pressure? [Use either Eq. (11–1) or the result of Problem 11–1. Comparative data for alpha particles from Po^{214} are given in Problem 11–2.]

11–18. (a) Calculate the Q-value for the decay of a free neutron into a proton and an electron. (b) Calculate the Q-value for the decay of a proton into a neutron and a positron.

11–19. The maximum beta-particle energy from the decay of Be^{10} to B^{10} is 0.56 MeV. Calculate the isotopic mass of the Be^{10} atom.

11–20. Suppose that a thermal neutron with an energy of 0.025 eV is absorbed by a S^{32} nucleus to produce S^{33}. What is the energy of the photon that is radiated if the S^{33} nucleus goes to the ground state? The isotopic mass of S^{33} is 32.97146.

11–21. One student claims that he has observed N^{16} to decay by alpha-particle emission. Another student claims that N^{16} decays by beta-particle emission. Which student is correct?

11–22. The Q-value of the reaction $Li^6(p, \alpha)He^3$ is found experimentally to be 3.945 MeV. Calculate the isotopic mass of He^3 from this information.

11–23. Alpha particles from radium have an energy of 4.78 MeV and can transmute ${}_{16}S^{32}$ into ${}_{17}Cl^{35}$. (a) Write the equation for this reaction. (b) What is the total kinetic energy of the products?

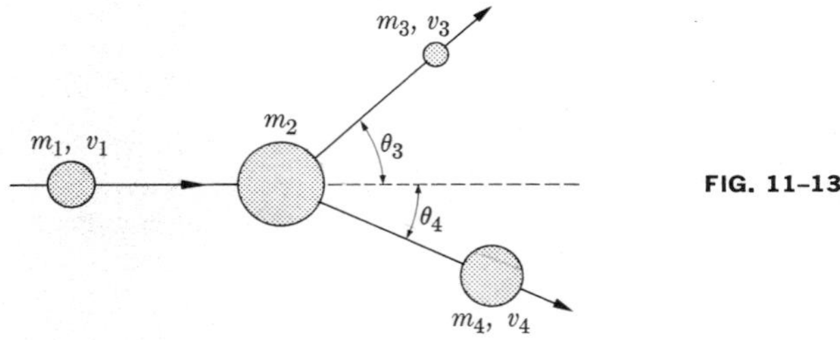

FIG. 11–13

11–24. A bombarding particle of mass m_1 and kinetic energy E_{k1} strikes a nucleus at rest having a mass m_2. The resulting nuclear reaction produces two particles having masses m_3 and m_4 and kinetic energies E_{k3} and E_{k4}, respectively. The directions of their motions are shown in Fig. 11–13.

Using classical mechanics, show that

$$Q = \left(1 + \frac{m_3}{m_4}\right) E_{k3} - \left(1 - \frac{m_1}{m_4}\right) E_{k1} - \frac{2\sqrt{m_1 m_3 E_{k1} E_{k3}}}{m_4} \cos \theta_3.$$

11–25. A neutron beam is incident on a stationary target of F^{19} atoms. The reaction $F^{19}(n, p)O^{19}$ has a Q-value of -3.9 MeV. What is the lowest neutron energy which will cause this reaction to take place?

11–26. (a) Compute the Q-value of the reaction $N^{14}(\alpha, p)O^{17}$, which occurred in Rutherford's alpha-range-in-nitrogen experiment. (b) If a 7.68-MeV alpha particle causes the reaction in part (a) in a nitrogen atom which is initially at rest, what is the kinetic energy in the laboratory system of the proton produced, assuming that it and the product nucleus continue in the direction of the motion of the incident alpha particle? (c) What is the threshold energy of the alpha particle for the reaction in part (a), assuming that the nitrogen target atom is at rest? (d) If the nitrogen atom were moving with the most probable speed in a Maxwellian distribution at 20°C, what would be its kinetic energy? Would this be significant with respect to the threshold energy of the alpha particle?

11–27. Under the proper conditions two alpha particles can combine to produce a proton and a nucleus of Li^7. (a) Write the equation for this reaction. (b) What minimum kinetic energy must one of the alpha particles have in the laboratory system in order to make the reaction proceed if the other alpha particle is at rest? (c) Now assume that both alpha particles are moving, each directly toward the other, with identical speeds. What is the minimum energy of the alpha particles which will allow the reaction to proceed? (Note that laboratory system and the center-of-mass system are the same in this part.)

11–28. A common means of obtaining neutrons in the laboratory is to bombard tritium with deuterons. Assume that the incident deuterons have an energy of 1 MeV. (a) Write the equation for this reaction. (b) Compute the Q-value. (c) Calculate the energy of the neutrons which are emitted at an angle of 90° with respect to the beam of incident deuterons. (Mass of H^3 = 3.01605 u.)

17.6 MeV

11–29. A neutron of zero kinetic energy interacts with lithium at rest and the reaction ${}_3Li^6(n, \alpha){}_1H^3$ occurs. (a) Write the equation for this reaction. (b) Calculate the Q-value. (c) Calculate the kinetic energy of the ejected alpha particle. (Isotopic mass of H^3 = 3.01605 u.)

11–30. A sample of P^{31} weighing 25 g is bombarded with neutrons until the activity of the P^{32} produced is 3mCi. What is the ratio of the number of radioactive phosphorus atoms to the number of stable atoms?

11–31. Two reactions which occur in the upper atmosphere due to the effects of cosmic "rays" are $N^{14}(n, p)?$; and $N^{14}(n, ?)C^{12}$. (a) Complete the nuclear reaction equation for each and calculate the Q-value of each. (b) Is the nuclide formed in each case stable or radioactive? (See Appendixes 5 and 6. Isotopic mass of C^{14} = 14.003233.)

11–32. Due to C^{14}, the charcoal from the fire pit in an ancient Indian camp site has an average beta activity of 12.9 counts or disintegrations per minute per gram of carbon in the sample (cpm/g). The absolute specific activity of the C^{14} in the wood from living trees is independent of the species of the tree and it averages 15.3 cpm/g. What is the age of the charcoal sample?

11–33. If each of the average counts in Problem 11–32 has a numerical uncertainty of ± 0.1 cpm/g and if the half-life of C^{14} is 5730 $\pm$ 30 years, what is the numerical uncertainty in the determination of the age of the charcoal sample?

11–34. The tritium content of the water from certain deep wells is only 33% of that in the water from recent rains. (a) How long has it been since the water in the well came down as rain? (b) What assumption did you make about tritium in solving this problem?

11–35. What must be the activity of a radiosodium (Na^{24}) compound when it is shipped from Oak Ridge National Laboratory so that upon arrival at a hospital 24 h later its activity will be 100 millicuries?

11–36. The steel compression ring for the piston of an automobile engine has a mass of 30 gm. The ring is irradiated with neutrons until it has an activity of 10 microcuries due to the formation of Fe^{59} (T = 45.0 days). Nine days later the ring is installed in an engine. After being used for 30 days, the crankcase oil has an average activity due to Fe^{59} of 12.6 disintegrations per minute per 100 cm^3. What was the mass of iron worn off this piston ring if the total volume of the crankcase oil is 6 qt?

11–37. (a) What energy must be supplied to overcome the electrostatic repulsion in order to bring two widely separated protons together until they are separated by 2×10^{-13} cm, which is of the order of nuclear dimensions? (b) What electrostatic force acts on two protons separated by 2×10^{-13} cm?

11–38. Determine, by referring to Fig. 11–10, whether energy must be supplied or will be released if we assume that the following transmutations can be accomplished: Fe $\rightarrow$ Xe, Li $\rightarrow$ Na, and Pb $\rightarrow$ Al.

11–39. (a) Find the total binding energy and the binding energy per nucleon of H^2, He^4, and Fe^{56}. (b) How much energy would be involved in removing the nucleons in He to infinity? Would energy be released or have to be supplied for this dispersion?

11–40. Assuming that the lifetime of the excited state involved in the 14.4-keV transition in the nucleus of Fe^{57} is 10^{-7} s, calculate (a) the associated uncertainty in the energy using Heisenberg's principle, Eq. (7–32), (b) the corresponding fractional uncertainty in the energy of the emitted gamma ray, and (c) the fractional uncertainty in its frequency. (d) What is the absolute value of the uncertainty in its frequency and in its wavelength?

11–41. What is the recoil velocity and the recoil energy of a free Fe^{57} nucleus when it emits a 14.4 keV gamma-ray photon?

11–42. When the nucleus of a Mössbauer emitter is placed in a magnetic field, there is fractional increase in the energy of the emitted gamma-ray photon of 10^{-12}. What must be the magnitude and the direction of the velocity, with respect to the emitter, of an absorber of the same material so that resonance absorption will occur? (The fractional change in observed frequency in the Doppler effect when a receiver moves toward a wave source is v/c, where v is the velocity of the receiver with respect to the source and c is the speed of propagation of the waves.)

CHAPTER 12

NUCLEAR ENERGY

12-1. NUCLEAR ENERGY

Although one wink of your eye requires the expenditure of many billions of electron volts of chemical energy, nuclear processes are potentially far more energetic than chemical processes. The Q-values we have been discussing are energies per single disintegration, whereas the energy to wink an eye comes from the chemical conversion of many billions of molecules. The liberation of chemical energy is usually expressed on a per-gram, per-pound, or per-kmole basis, and the conversion factor between per kmole and per molecule basis is the Avogadro constant, about 6×10^{26}. In Section 10–8 we compared the energy from oxidizing carbon and from radioactive disintegration on a per-gram basis. Table 12–1 lists some energies on a per-particle basis. The first and third processes of Table 12–1 account for most of the energy utilized in industrial and life processes.

TABLE 12–1
ENERGIES PER PARTICLE

Kinetic energy of one water molecule, 450-ft waterfall	0.00025	eV
Average kinetic energy of a gas molecule at room temperature	0.025	eV
Carbon atom oxidized to CO_2	4.0	eV
Visible photon	2.0	eV
Ultraviolet photon	3.0 to 100.0	eV
Hard x-ray photon	0.1 to 1.0	MeV
Gamma-ray photon	1.0 to 3.0	MeV
Radium disintegration	4.8	MeV
Fission disintegration, uranium	200.0	MeV
Cosmic-ray particle	1.0 to 10.0	GeV

We have seen that radioactive disintegrations are nuclear processes in which an appreciable amount of rest mass-energy is converted to kinetic mass-energy. The mass-energy equation which we have repeatedly applied to nuclear reactions is perfectly general and can be applied to chemical reactions as well. We can write

$$C + 2(O) = CO_2 + Q, \tag{12-1}$$

but nothing useful can be learned from applying the equivalence of mass-energy to a chemical equation. The Q-value in u is $4.0/(931 \times 10^6)$ or 4.3×10^{-9} u. In the chemical reaction the carbon and oxygen would be mixtures of isotopes, but even if pure isotopes were used, the uncertainty of the isotopic masses far exceeds the Q-value. Thus in chemical reactions it is proper to assume that mass and energy are conserved separately. We write Eq. (12–1) in order to emphasize the contrast between chemical and nuclear reactions.

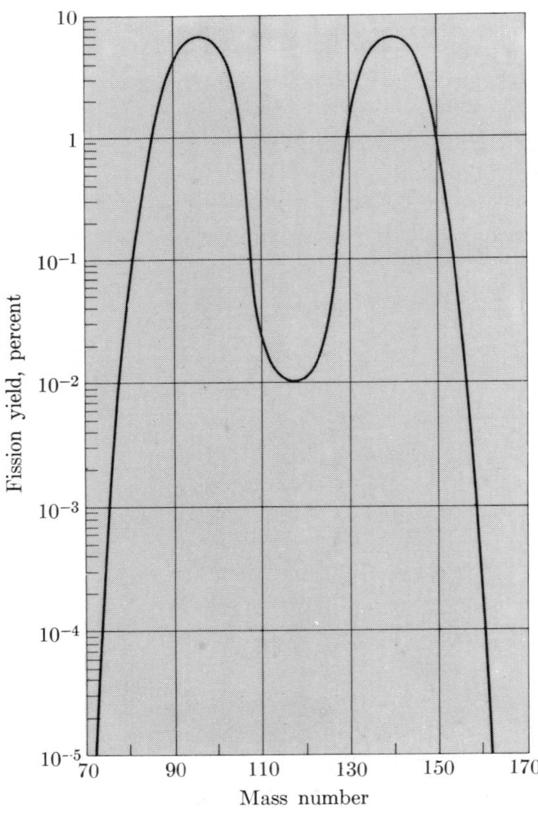

FIG. 12–1
Fission yield from U^{235}.

12–2. CHAIN REACTION

Highly energetic radioactive disintegrations are not very promising as practical sources of energy. Although man-made radioactive nuclides are available in far greater quantities than the naturally radioactive nuclides, there is none whose abundance would justify fleeting consideration as a prime source of energy in a class with coal or oil. Nevertheless, the United States Atomic Energy Commission has announced that radioactivity may be utilized to make electric batteries of exceptionally long life. Thus energy from radioactivity will find limited use in special applications.

Some nuclear transmutations provide more energy per event than the natural radioactive disintegrations. The $Li^6(d, \alpha)He^4$ process has a Q-value of 22.4 MeV. But this reaction requires accelerated deuterons, and the energy required to operate the apparatus is millions of times greater than the energy derived from the process.

Although fission reactions release more energy per event than any type of reaction known earlier, it is another property of fission that makes it a practical source of energy.

The principal isotope of uranium that undergoes fission is U^{235}. This nuclide, which has about a 0.7% concentration in natural uranium, fissions in many ways. The *fission yield* of a fission product is the percentage of the fissions that lead to the formation of a particular nuclide or a group of isobars. Since there are two nuclides produced per fission, the total of all the fission yields for a given process is 200%. The fission yield from U^{235} of nuclides of different mass numbers is shown in Fig. 12–1 (note that the ordinate scale is logarithmic). Obviously, we can write no unique reaction, but a typical one is

$$_{92}U^{235} + {}_0n^1 \rightarrow {}_{92}U^{236} \rightarrow {}_{54}Xe^{140} + {}_{38}Sr^{94} + 2\ {}_0n^1 + \gamma + 200 \text{ MeV}. \tag{12–2}$$

The fission energies and the fission-yield curves for U^{238} and Pu^{239} do not differ appreciably from U^{235}. Some heavy nuclei undergo *spontaneous fission* in which the nucleus divides in the ground state without being bombarded by neutrons or other particles. The half-life of U^{238} and certain other heavy nuclei for this process is of the order of 10^{16} years. Therefore, only about 20 nuclei per gram of these elements undergo fission every hour. Their contribution to the operation of a reactor is insignificant and will be neglected in further discussion.

Figure 11–12 is a graph of the number of nuclide neutrons as a function of nuclide protons for the stable elements. This figure shows that the heavier the element, the greater its neutron excess. When a fission occurs, the product nuclides are lighter than the parent element and, to be stable, they must have a smaller excess of neutrons. The nuclides first formed have neutron excesses which make them either absolutely or radioactively unstable against neutron emission (see the first and third cases of Section 11–11). Both Xe^{140} and Sr^{94} are radioactive beta emitters; therefore Eq. (12–2) does not take us to the end of the disintegration. Before these two nuclides were formed, they had a parent nuclide which was a neutron emitter. Its two neutrons appear on the right-hand side of the equation along with the large Q-value.

The *average* number of neutrons from U^{235} is 2.5 per fission. They provided the real potentiality of fission as a power source because they made a chain reaction appear feasible. Here was a reaction that had among its products the same kind of particle that initiated it. If, on the average, at least one neutron from the first fission could produce another fission, and at least one neutron from the second could produce a third, etc., then a self-sustaining chain of events would take place with no necessity for accelerators or other devices to keep the process going. The first fission reactions were obviously not self-sustaining chain reactions, since the fission activity stopped when the neutron source was removed. Although some of the neutrons from the first fission reactions undoubtedly caused other fission reactions, it was evident that too many of the neutrons produced were being absorbed in some nonfission-producing manner. The problem that faced nuclear physicists can be stated simply: Under what experimental conditions—if any—could the fission reaction be made self-sustaining?

Soon after the fission of U^{235} was accomplished, two more reactions were found. These produced the transuranic nuclides Fermi set out to produce in his original experiments in 1934. Just as he expected, simple neutron capture is followed by beta activity and the formation of the transuranic element neptunium, Np:

$$_{92}U^{238} + {_0}n^1 \rightarrow {_{92}}U^{239} \rightarrow {_{93}}Np^{239} + {_{-1}}e^0. \qquad (12\text{–}3)$$

The neptunium, which is radioactive with a half-life of 2.35 days, becomes plutonium, Pu:

$$_{93}Np^{239} \rightarrow {_{94}}Pu^{239} + {_{-1}}e^0. \qquad (12\text{–}4)$$

The plutonium is also radioactive, but its half-life is 24,360 yr. It, like U^{235}, is fissionable, and we shall see that it is a very important nuclide.

Enough has now been said to account for the fact that a chain reaction did not occur when the first uranium sample was bombarded with neutrons. In the first place, we know that neutrons are very penetrating. Thus the most likely fate of neutrons from the first fission was escape from the uranium sample. The original uranium samples were mixtures of U^{235} and U^{238}. Since U^{238} is much more abundant than U^{235}, the neutrons that did not escape from the first fission reactions had an excellent chance of nonfission capture by U^{238}. It was anticipated that if U^{235} could be separated from U^{238} and other neutron-capturing nuclides, and if enough of this pure U^{235} could be prepared so that neutrons had a good chance of causing fission before they could escape, then a chain reaction would probably occur. If the chain reaction grew within a large body of U^{235}, energy should be liberated in explosive amounts.

12-3. NEUTRON CROSS SECTIONS

In our previous discussion of nuclear reactions we have been concerned with "what happens" rather than with "how much happens." Although the reaction products have been so minute that identification was difficult, the minuteness did not detract from interest in "what happened." In discussing a chain reaction where one event must successfully cause another, "how much" is just as important as "what."

In Chapter 1 we introduced the concept of mutual collision cross section as a way of describing the probability of one particle hitting another. We showed that this cross section was a property of both the bullet and the target particles, since it is defined as

$$\sigma = \pi(r_{\text{bullet}} + r_{\text{target}})^2. \tag{12-5}$$

By assigning this area to the target particle, we could regard the bullet particle as a point and still have a measure of the likelihood of a collision. When we introduced this concept we were discussing the kinetic theory of gases. The particles were whole, uncharged atoms, and we were thinking of the collisions as being pure mechanical hits, like those between billiard balls. If we considered collisions between positive ions, the cross section for deflection would be bigger. The coulomb force of repulsion would extend beyond the physical limits of the particles, so that the bullet particle would be deflected without coming so near the target particle. Although the ion-deflection cross section would be bigger than if the same atoms were not ionized, the cross section for chemical interaction would be less. The mutual repulsion that increases the probability of deflection diminishes the probability of near approach. Thus the concept of cross section becomes a general and useful measure of the probability of many classes of "collisions." If two particles can interact in more than one way, the probabilities of the various interactions can be measured in terms of the cross section of each.

TABLE 12–2

Process name	Reaction type	Cross-section symbol
Radiative capture	(n, γ)	σ_r
Fission	(n, f)	σ_f
Elastic scattering	(n, n)	σ_s
Proton capture, etc. (absorption)	(n, p), (n, d), etc.	σ_a
Total	$\sum\sigma$'s	σ_t

If, instead of thinking of a gas molecule moving through a gas, we think of a neutron moving through matter, we find that a variety of kinds of collision interactions are possible. Each has its own probability, which is directly proportional to its cross section. Some of the possibilities are listed in Table 12–2.

In Chapter 1 we showed that the probability of a particle making a simple collision in going through a gas for a distance dx was $n\sigma\,dx$, where

σ is the mutual collision cross section, and n is the number of particles per unit volume (Eq. 1–20). In complete analogy, we may now say that the probability of a neutron causing fission in moving a distance dx through matter composed of an element whose atoms are at rest is

$$P_f = \sigma_f N\, dx, \tag{12–6}$$

where the nuclear concentration, N, is the number of nuclei per unit volume. The *microscopic cross sections*, σ's, become the *macroscopic cross sections*, $N\sigma$'s, when multiplied by N. The total probability that some type of interaction will occur is

$$P_t = (\sigma_f + \sigma_s + \sigma_a + \text{etc.})N\, dx. \tag{12–7}$$

If the material is composed of more than one element, then the neutron may collide with any nuclide present, depending on the concentration of the nuclide and its cross sections. Thus if there are two nuclides present in concentration N and N', the *total probability* that the neutron will undergo some interaction in going a distance dx is

$$P_t = (N\sigma_t + N'\sigma_t')\, dx. \tag{12–8}$$

Before these equations can be used quantitatively, the values of the σ's must be known. Although many σ's were known before the period of secrecy about the fission process, one of the first tasks in calculating the feasibility of a chain reaction was the determination of many more cross sections. Mathematically, this process is very much like the determination of the probability of radioactive decay, the decay constants.

If I_0 incident neutrons per second produce I_f fissions per second, then the probability of fission is

$$P_f = \frac{I_f}{I_0}\cdot \tag{12–9}$$

But P_f is also given by Eq. (12–6), so we may write

$$\frac{I_f}{I_0} = N\sigma_f\, dx$$

or

$$\sigma_f = \frac{I_f}{I_0 N\, dx}\cdot \tag{12–10}$$

Thus if we know the number of fissions produced per unit time by a known number of neutrons per unit time in a material of known concentration and thickness, the fission cross section can be determined. By measuring the number of absorptions per unit time or the number of scatters per unit time, etc., these cross sections can be determined.

In this differential method it is assumed that dx is so small that the intensity of the neutron beam can be considered constant throughout the thickness of material, and that the nuclei in the material cannot hide behind one another.

If the slab of material has finite thickness, then the neutron intensity is a variable, I, and we must integrate to determine its value. In penetrating a slab of the material of thickness dx, the intensity of the neutron beam is changed an amount $(-dI)$ by any interactions that remove neutrons from the beam. The total probability of interaction is then

$$P_t = -\frac{dI}{I} = N\sigma_t\, dx. \tag{12–11}$$

Upon integrating, and noting that for $x = 0$ we have $I = I_0$, we obtain

$$\boxed{I = I_0 e^{-N\sigma_t x},} \tag{12–12}$$

where I is the intensity of the *transmitted* beam. This equation permits the evaluation of the total cross section from data such as those given in Fig. 12–2 and provides a check on the more difficult differential method through the equation

$$\boxed{\sigma_t = \sigma_f + \sigma_s + \sigma_a + \text{etc.}} \tag{12–13}$$

As in Eq. (1–24), we can now speak of the neutron mean free path:

$$\boxed{L_t = \frac{1}{N\sigma_t}.} \tag{12–14}$$

Or for the mean neutron path per fission we may write

$$L_f = \frac{1}{N\sigma_f}. \tag{12–15}$$

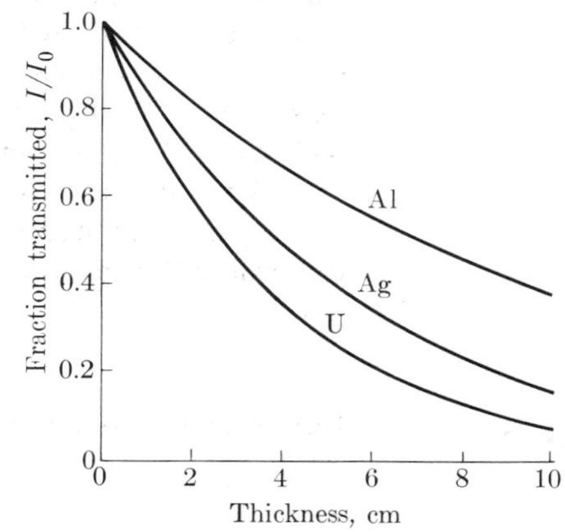

FIG. 12–2
Neutron absorption curves.

If the material is a mixture of two fissionable materials having different fission cross sections and different nuclear concentrations, the mean neutron path per fission becomes

$$L_f = \frac{1}{N\sigma_f + N'\sigma'_f}. \tag{12–16}$$

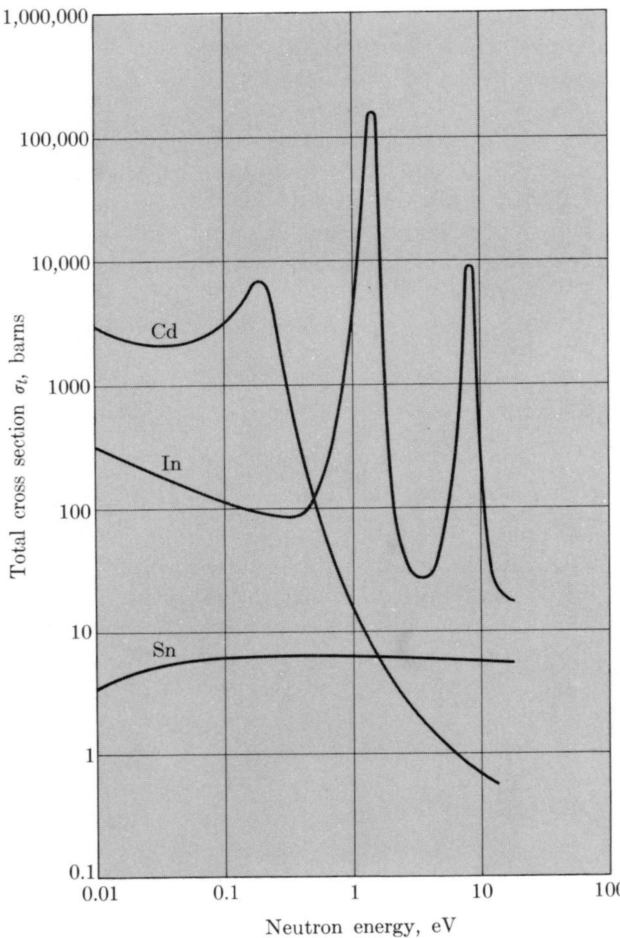

FIG. 12–3
Total cross section of several nuclei for neutrons as a function of energy.

The measurement of cross sections is complicated by the fact that not only does each nuclide have several cross sections but also cross sections are often complicated functions of the neutron energy. The total cross sections of three nuclides are shown in Fig. 12–3. Since the microscopic cross sections are usually very small quantities, it is convenient to have a special unit for them, the *barn*. One barn is an area equal to 10^{-24} cm^2. Cross sections are proportional to the probabilities of neutron reactions. Although cross sections have area units, they are not the physical cross sections of the nuclei. Nevertheless, it is interesting to note that most nuclides have "diameters" of about 10^{-12} cm and consequently their physical cross sections are about 1 barn.

Although every cross section should be expressed as a function of energy, as in Fig. 12–3, many elements have cross sections which can be repre-

TABLE 12–3

NEUTRON CROSS SECTIONS IN BARNS

Values in parentheses are less certain than the others. Data taken or deduced from *Neutron Cross Sections*, Brookhaven National Laboratory (325).

	σ_t (slow)	σ_t (fast)	σ_s (slow)	σ_a (slow)	σ_f (slow)	σ_f (fast)
H^1	(38)	(4.3)	38	0.33	0	0
H_2O	(110)					
D^2	(7)	(3.4)	7	0.0005	0	0
D_2O	(14.5)				0	0
T^3	5400	(1.9)	1	5400	0	0
Li	(72)	(1.7)	1.4	71	0	0
Li^6	(951)	(0.26)	(6)	945	0	0
Li^7	(1.4)		(1.4)	(0)	0	0
B	(760)		4	755	0	0
C	(4.8)		4.8	0.0032	0	0
O	(4.2)		4.2	0.0002	0	0
Zr	(8)		8	0.18	0	0
Cd	(2560)		7	2550	0	0
Hf	(113)		8	105	0	0
U (natural)		(7)	4.7 (fast)	(2) (fast)		0.5
U^{235}	697	6.5	10	107	580	1.3
U^{238}	2.8		0	2.8	0.0005	0.5
Pu^{239}	1075		9.6	315	750	

sented by a simple equation. These nuclides have cross sections which vary inversely as the velocity of the neutron. Such materials are called *1/v-absorbers*. The most important neutron energy distinction is that between fast neutrons and thermal neutrons. Neutrons from a fission reaction are fast and have an average energy of about 2 MeV. Neutrons which have traversed enough matter to be in thermal equilibrium with the molecular motions of the material are called *slow* or *thermal* neutrons, and these have energies distributed about the value of 0.025 eV. Table 12–3 gives values of the cross sections of several important nuclides for these two energies.

Obtaining neutron cross sections as a function of velocity requires methods for measuring neutron velocities. We cannot discuss these methods in detail, but it is easy to see that for uncharged particles many familiar methods must be excluded. We can diffract neutrons by crystals and thus obtain the de Broglie wavelength, from which the velocity can be computed. Also there are mechanical devices which can "chop" a slow neutron beam so as to eliminate all neutrons outside a narrow energy range.

12-4. REACTOR CRITICALITY

We are now in a position to make simple calculations that give some idea of what we may expect in various circumstances. Consider first that we have a very large body of pure U^{238}. It is known that each fission produces about 2.5 neutrons having energies of about 2 MeV. Since the body of metal is very large, we can neglect the loss of neutrons by escape and consider the probable fate of neutrons within the metal. We consider three possibilities.

(1) Neutrons may scatter elastically within the metal. These collisions have a cross section of 4.7 barns for fast neutrons, so that scattering collisions are rather common. That scattering collisions deflect the neutrons is of no consequence. So long as the neutrons remain within the metal, their direction of motion makes no difference. Scattering collisions slow the neutrons but, since the uranium nuclei are 238 times as massive as the neutrons, it takes many collisions to slow the neutrons appreciably. We therefore ignore the influence of scattering.

(2) The neutrons may be absorbed in some nonfissioning manner, such as that described in Eq. (12–3). These absorptions have a cross section of 2 barns for high-energy (fast) neutrons and represent neutron loss so far as the possibility of a chain reaction is concerned.

(3) The neutrons may produce fission reactions for which the cross section is 0.5 barn. These neutrons are productively absorbed and tend to maintain the chain reaction. Under our assumptions, the probability that one neutron will produce another fission is the microscopic fission cross section divided by the total cross section, or

$$P_f = \frac{\sigma_f}{\sigma_a + \sigma_f} = \frac{0.5}{2 + 0.5} = 0.2.$$

Thus about one out of five neutrons produces a fission. Since each fission produces 2.5 neutrons on the average, each neutron probably produces $0.2 \times 2.5 = 0.5$ neutron. We can summarize this result in the following way. If there were ten neutrons in one neutron generation, eight of these would be unproductively absorbed and two would produce fission. Each fission would produce 2.5 neutrons, so that in the next generation there would be five neutrons. Thus each generation has half as many neutrons as the preceding one and no chain reaction can occur. The ratio of the number of neutrons in one generation to the number in the preceding generation is called the production factor or *multiplication constant*, k. When the body of material is so large that leakage may be neglected, the multiplication constant is represented by the symbol k_∞, called *k-infinity* or *infinite* (k). Since we have found $k_\infty = 0.5$ for a body of U^{238}, we know that any amount of U^{238} will be *subcritical*. This means that k is less than unity and that any fission process that may be initiated will quickly die out.

If we repeat these calculations for U^{235} under the same assumptions, we find a very different result. Every U^{235} nucleus that absorbs a high-energy neutron produces a fission. Since every absorbed neutron produces another fission, the multiplication constant is simply the number of neutrons per fission, or $k_\infty = 2.5$, and it is evident that each neutron generation is much larger than the preceding one. A large mass of U^{235} would explode. The only way to prevent a body of pure U^{235} from exploding would be to break it into pieces so small that 60% of the neutrons would escape without producing fission. These pieces would each be under the critical size for U^{235}.

The concept of critical size results from the fact that fissions occur throughout the volume of a reacting body and neutrons escape through the surface of that body. As the size of a body is changed, the volume changes according to the cube of its dimensions, while its area changes as the square of its dimensions. Thus the ratio of neutron production to neutron leakage varies as the first power of the dimensions. If the configuration of the reactor is such that k_∞ is greater than unity, there is always a smaller size for which the material will be just critical, with k equal to unity. Below this size k is less than unity. In summary: above the critical size, k is greater than unity and the neutron population increases exponentially with time; *at the critical size, k is equal to one* and the neutron population is constant; below the critical size, k is less than unity and the neutron population falls off exponentially with time.

The calculations we have made are sufficient to show that natural uranium, which is about 99.3% U^{238} and 0.7% U^{235}, cannot chain react and that the concentration of the U^{235} must be increased. In order to estimate the degree of enrichment required, we make a third calculation under the same assumptions as before except that we now have a metal composed of a mixture of U^{235} and U^{238} in which the number of nuclei per unit volume is designated by N_5 and N_8, respectively. The calculation is made just as before, but now we must use the fast-neutron cross sections of both nuclides weighted according to their concentrations. Since $\sigma_a = 0$ for fast neutrons in U^{235}, we have

$$k_\infty = 2.5\,\frac{N_5\sigma_{f5} + N_8\sigma_{f8}}{N_5(\sigma_{a5} + \sigma_{f5}) + N_8(\sigma_{a8} + \sigma_{f8})} \tag{12-17}$$

$$= 2.5\,\frac{1.3N_5 + 0.5N_8}{(0 + 1.3)N_5 + (2 + 0.5)N_8}\,. \tag{12-18}$$

Letting $N_5 = 0$ gives the result of our first calculation and letting $N_8 = 0$ gives our second result. Letting $k_\infty = 1$, we can find the concentration ratio that will make a large mass of uranium critical. Solving for N_8/N_5, we obtain 1.5, which tells us that for a large body of uranium metal to be critical it must be about 40% U^{235}. For a body of finite size, the enrichment of U^{235} would have to be greater than 40%. Calculations such as these led to the decision to attempt the enrichment of U^{235} in uranium metal.

The separation of isotopes is always difficult, but since uranium is the heaviest natural element, the fractional mass difference between U^{235} and U^{238} is especially small.

The most effective method of separation found was the gaseous diffusion of uranium hexafluoride, UF_6. Although the enrichment of U^{235} to about 99% purity requires about 4000 diffusion stages, a tremendous plant based on this principle was built at Oak Ridge, Tennessee, which has been in production since 1945. A measure of the difficulty of this process is contained in the fact that in 1955 the AEC evaluated natural uranium at $40 per kg and 95% U^{235} at $16,258 per kg. With fuel enriched with U^{235}, the construction of a fast-neutron bomb reactor is, in principle, simple. The uranium parts must each be less than critical size lest they explode spontaneously from stray neutrons or an occasional spontaneous fission. Detonation is accomplished by bringing the subcritical parts together into a *supercritical* whole. The critical size of the U^{235} enriched uranium is about the size of a grapefruit.

12-5. MODERATORS

A glance at Table 12–3 shows that the uranium cross sections for slow neutrons are very different from those for fast neutrons. Whereas most nuclear reactions are best induced with high-energy particles, neutron-induced reactions usually have cross sections which increase as the neutron energy is decreased. This may be phrased in a somewhat naive way by saying that since there is no coulomb repulsion, slow neutrons spend more time near the nuclei they pass and therefore the short-range nuclear attractive forces have a better chance to take effect. This is why many cross sections vary as $1/v$.

The nonfission capture cross section of U^{238} increases as the neutron energy is reduced, but the fission cross section of U^{235} increases even more. Thus the relative probability of fission in natural uranium is much greater for low-energy neutrons than for high-energy neutrons. To show how marked this difference is, we next compute the production factor that a large body of natural uranium *would* have *if* each fission produced 2.5 *thermal* neutrons. We again use Eq. (12–17), except that we substitute the thermal cross sections and take $N_8/N_5 = 1/0.007 = 143$, which yields

$$k_\infty = 2.5 \frac{1 \times 580 + 143 \times 0.0005}{1 \times (107 + 580) + 143(2.8 + 0.0005)} = 1.33. \quad (12\text{–}19)$$

Since k_∞ is considerably greater than unity, it appears that even a finite size of natural uranium would chain react if the fission neutrons were slow. Since fission neutrons are fast rather than slow, the calculation we have just made is purely academic unless neutrons can be slowed. Even before the period of secrecy, it was proposed that if the fast-fission neutrons could be slowed (*moderated*), critical conditions might be achieved in natural uranium.

Another important consideration is brought out qualitatively in Fig. 12–4(a). For neutrons with an energy of about 7 eV, U^{238} has a neutron resonance capture in which the U^{238} nucleus has a very large probability of going into an excited state. Thus, in a natural mixture of U^{235} and U^{238}, there is very little chance that a neutron will be slowed below 7 eV, to the energy for which the probability of U^{235} fission capture is very great.

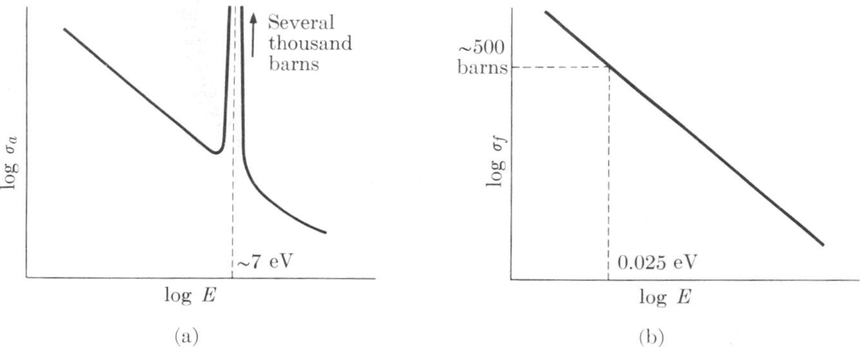

FIG. 12–4
(a) Capture cross section of U^{238} for slow neutrons, and (b) fission cross section of U^{235} for slow neutrons, assuming a $1/v$ relationship.

The idea of using a moderator to increase the probability of a chain reaction was based on these variations of cross section with neutron velocity. If the fast-fission neutrons could escape from a lump of uranium and be slowed to very low energies, and were then allowed to fall on natural uranium, the relative probability of fission might be increased enough so that a chain reaction would result.

There are reactions in which a neutron is absorbed and then one, two, or even three neutrons of lower energy are emitted. Such reactions have the effect of slowing neutrons and even increasing their number, but these reactions are most easily produced in heavy nuclei with high-energy neutrons. For the slowing of neutrons with the average energy of fission neutrons, about 2 MeV, elastic scattering is much more suitable.

In our discussion of the discovery of the neutron, we worked out the mechanics of head-on collisions between moving particles and particles at rest (Section 11–3). We found that the energy transfer depends on the relative masses of the bodies involved in the collision and, in particular, that if a body strikes another body having the same mass, all the energy of the first body is transferred to the second one. If the collision is not head-on, the fraction of the energy transferred is certainly less. It is not too difficult to compute the *average fractional energy loss* between two bodies whose relative masses are known when the effect of *glancing collisions* is taken into account. This fraction is greatest when the two bodies have the same mass.

If a neutron enters a material, it loses a certain average fraction of its energy with each collision so long as the material nuclei can be considered at rest. If the neutron survives capture until its energy is reduced nearly to that due to thermal motion of the material molecules, then its energy loss per collision is much less than before. Indeed, at these energies there is a chance that the neutron may acquire energy from the molecules. Thus thermal agitation sets a lower limit to the energy a neutron acquires by the collision process. In Chapter 1 we showed that the average kinetic energy of a gas molecule depends only on its temperature, and we can consider neutrons in equilibrium with a moderator to be a gas at the same temperature as the moderator. Thus no moderator can reduce the energies of neutrons below the energies of the moderator molecules. The energies of the molecules of a gas at room temperature, 20°C, are distributed about a value of 0.025 eV. Neutrons in thermal equilibrium with matter at 20°C will have the same energy distribution as such gas molecules. These neutrons are called thermal neutrons.

Assuming that fission neutrons are emitted with an average energy of 2 MeV and have an energy of 0.025 eV when they have been "thermalized," we find the total fractional energy remaining to be $0.025/(2 \times 10^6) = 1.3 \times 10^{-8}$. The fractional energy remaining after a single collision can be calculated from Eq. (11–20). This value, raised to the nth power, where n is the number of collisions required to produce thermalization, is equal to the total fractional energy remaining. Since the absorption cross sections are functions of the neutron energy, it is awkward to compute the chance a neutron has to survive the thermalization process, but if criticality is to be achieved this chance times the number of neutrons per fission must be at least one.

Table 12–4 shows some of the characteristics of light elements as neutron moderators. Mechanically, hydrogen is the best moderating nuclide, since it requires, on the average, only 18 collisions to achieve thermalization. But since hydrogen can absorb neutrons to form deuterium, the chance of a neutron surviving the 18 collisions is not too good. We shall see, however, that hydrogen in ordinary water can be used as a moderator.

Deuterium requires 25 collisions on the average, but it has more promise as a moderator because its capture cross section is small. Oxygen also has a small capture cross section. Since a gas moderator has a very small

TABLE 12-4

Element	σ (absorption), barns	σ (scatter), barns	Average fractional energy loss per collision	Average number of collisions to thermalize, n
H	0.33	38	0.63	18
D	0.0005	7	0.52	25
He	0.000	1	0.35	42
Li	71	1.4	0.27	62
Be	0.01	7	0.18	90
B	755	4	0.17	98
C	0.0032	4.8	0.14	114
N	1.7	10	0.12	132
O	0.0002	4.2	0.11	150

number of nuclei per unit volume, both hydrogen and deuterium are used as moderators in the form of water.

Helium would be a superior moderator except that its inertness prevents its concentration in a chemical compound.

Both lithium and boron have absorption cross sections that are too high. Thus beryllium and carbon are the lowest atomic weight elements that are solid at room temperature and whose neutron absorption cross sections are low enough for consideration as moderators. Of these, carbon is by far the more abundant and thus it is used more frequently.

After neutrons have undergone a number of random glancing collisions in a material, the motions of some of them will have been deviated so much that they return to the region from which they came. This "back scattering" is equivalent to diffuse reflection. Any material that has a low neutron absorption cross section can be used as a *neutron reflector*. If the "reflected" beam is to be thermalized, then the scatter cross section of the atoms of the reflector must be large.

12-6. THE FIRST REACTOR

In 1942 many physicists believed that chain reaction would occur in pure U^{235} and possibly in ordinary unenriched uranium in a properly moderated reactor. The first alternative required the very difficult separation of a considerable amount of U^{235} from its isotope, U^{238}. The second alternative appeared to be the faster way of testing the theoretical calculations.

Graphite for the moderator had to be prepared in an especially pure form, since any impurities such as boron and cadmium (neutron absorbers) would reduce the number of neutrons that survived the slowing-down process.

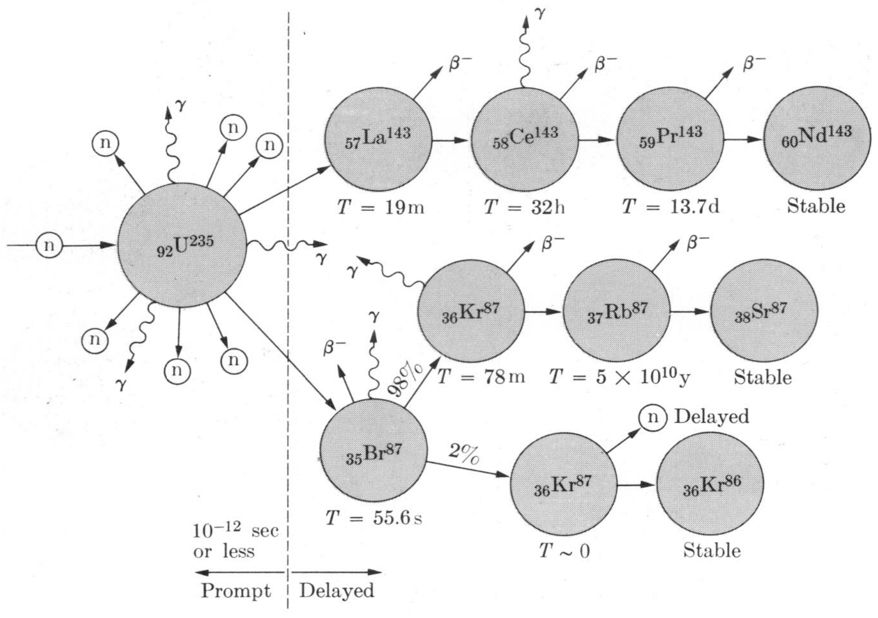

FIG. 12–5
Uranium fission with delayed neutron emission in the bromine fission fragment decay chain. The six fission neutrons shown in this example are many more than the 2.5 average.

Uranium had never been used commercially, and in 1941 there were only a few grams of the pure metal in the United States. By the fall of 1942 about six tons of uranium metal had been prepared. It, too, had to be free of neutron-absorbing impurities.

Before the reactor (pile) could be built, there had to be instrumentation for its control. In the first pile and in all controlled reactors that have followed it, the basis for control is the presence of materials which are good neutron absorbers. We have seen that boron has a large absorption cross section and cadmium is another good neutron absorber. These elements, alloyed with steel in the form of rods, can be pushed into the pile to make it subcritical and to adjust the multiplication constant k.

Before the pile could be built, it was important to know how rapidly it would respond to control-rod manipulation. Fortunately, not all the neutrons emitted by fission appear at once. Although most neutrons are produced without delay, i.e., *prompt neutrons*, some of the fission products are radioactive neutron emitters that produce *delayed neutrons* as shown in Fig. 12–5. Measurements showed that 0.4% of the neutrons are delayed

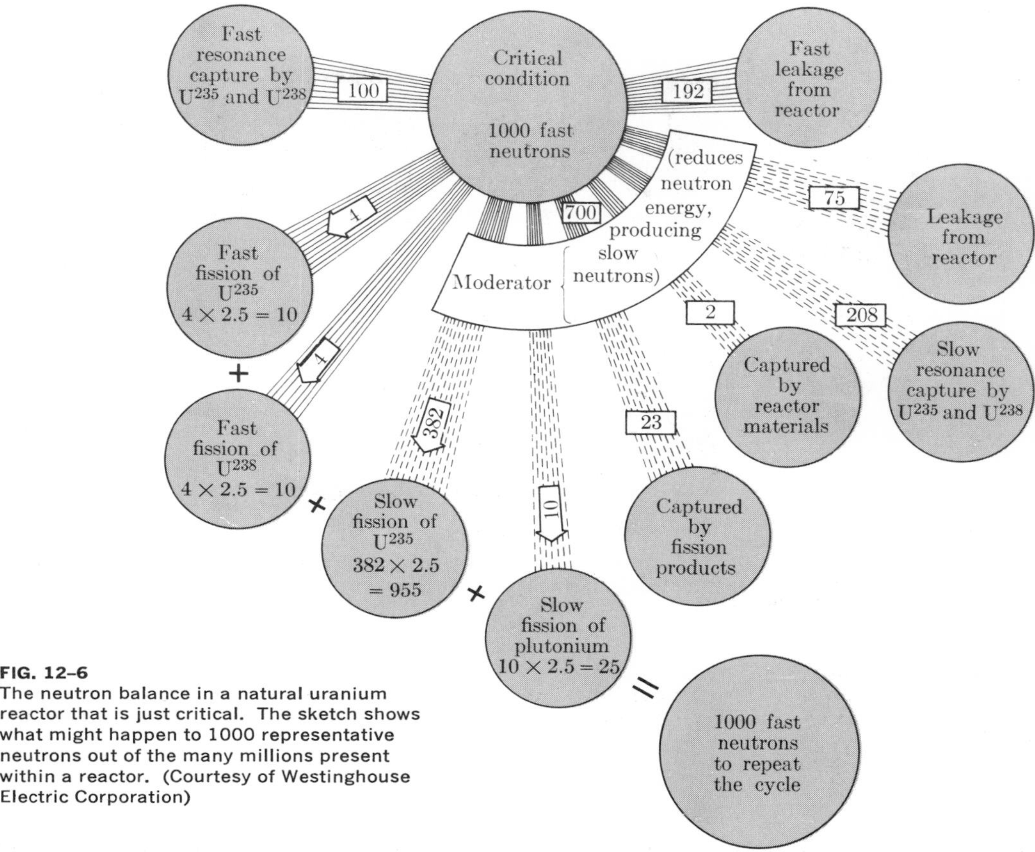

FIG. 12–6
The neutron balance in a natural uranium reactor that is just critical. The sketch shows what might happen to 1000 representative neutrons out of the many millions present within a reactor. (Courtesy of Westinghouse Electric Corporation)

at least 0.1 s and that 0.01% are delayed about a minute. Thus a reaction which is subcritical for prompt neutrons can be critical for prompt and delayed neutrons together. (This latter state is sometimes called *delayed critical.*) The idea was that by putting the reactor together with the control rods in place, the rods could be carefully withdrawn at frequent intervals during construction. By seeing to it that the reactor never went critical for prompt neutrons, that is, *prompt critical,* it was assumed that the reactor would be sufficiently sluggish so that human responses could prevent the reactor from "running away" destructively. (See Fig. 12–6.)

This first pile (Fig. 12–7) was assembled by Fermi and his associates in a squash court under the west stand of Stagg Field, the stadium of the University of Chicago. It was a cubic lattice of lumps of uranium in a *pile* of graphite blocks. The lump size was a compromise. They were made large, so as to include as many U^{235} nuclei as possible, and yet small

FIG. 12–7
Sketch of the first nuclear reactor. (Courtesy General Electric Company) An inscription on a plaque on the wall of the building (now torn down) read: "On December 2, 1942 man achieved here the first self-sustaining chain reaction and thereby initiated the controlled release of nuclear energy."

enough so that most of the fast-fission neutrons escaped without capture by U^{238}. They were about 10 cm in diameter. The graphite space between the lumps was chosen so that most of the neutrons from one lump would be thermalized by the time they reached the next lump about 30 cm away. On the average, then, fission neutrons left the lumps at high energy, were in the graphite when their energy was 7 eV, and reentered the uranium at 0.025 eV. Since the neutrons were not in the uranium when their energy was at the resonance absorption of U^{238}, the loss of neutrons to U^{238} was small.

The reactor was built around a neutron source using alpha particles from a radioactive source to excite an (α, n) reaction, as in beryllium. As the blocks of graphite and the lumps of uranium were assembled, some fissions occurred and the number of neutrons increased. The neutrons that were not absorbed in the pile material escaped. As the size of the structure grew, the chance that a neutron from one lump would find another lump became greater and the neutron intensity increased. By observing the neutron intensity as the pile grew, Fermi and his associates could anticipate at what stage of the construction the pile would become self-sustaining. The pile first became critical on the afternoon of December 2, 1942, and operated at a power level of $\frac{1}{2}$ watt. It was later increased to 200 watts—the maximum considered safe in that location.

Fermi was concerned that nitrogen of the air in the gaps of the pile would absorb enough neutrons to prevent criticality. He was prepared to encase the entire pile in balloon cloth and flush out the nitrogen with helium or some other gas that has a low absorption cross section. This cloth can be seen in the sketch reproduced in Fig. 12–7.

This first reactor was of tremendous significance. It proved that a chain reaction could be produced and controlled. It is the prototype of all the controlled power-production reactors that have followed it. But in 1942 the motivation was to produce a destructive reactor, and the Chicago pile facilitated this end by making the conversion process practical.

12–7. THE CONVERSION PROCESS

We have noted that the U^{238}, which absorbs neutrons and hinders the operation of a thermal pile, is ultimately converted to plutonium according to Eqs. (12–3) and (12–4). Plutonium is fissionable and, since it is a different chemical element, it can be separated from the uranium by chemical methods. Once the Chicago reactor proved that a chain reaction could be maintained, reactors specifically designed for the manufacture of plutonium were built at Oak Ridge, Tennessee and at Hanford, Washington. In these cases, the reactors served only as prolific neutron sources for the bombardment of U^{238}. Although these reactors were designed to permit easy insertion and removal of fuel elements and the removal of heat energy, they used natural uranium and a graphite moderator, just as in the first reactor. A fuel element is left in the reactor until a point of diminishing returns is reached in the conversion of U^{238}. This point comes well before all the U^{235} is used up because the fission products have large neutron capture cross sections and begin *poisoning* the reactor. The impurities absorb an appreciable fraction of the neutrons that are needed to maintain the reaction and to convert U^{238} into plutonium. Once a fuel rod is spent, it is dissolved and the plutonium is separated by chemical processes. This seemingly straightforward operation is complicated by the fact that tremendous quantities of radioactively dangerous material must be handled. This radioactivity is due principally to the fission products which poisoned the fuel element. The mere disposal of these radioactive fission products is a problem of mammoth proportions.

A large amount of this transuranic, man-made element was prepared by this method during World War II. The first atomic bomb was of uranium, but the second contained plutonium.

12–8. CONVERTER REACTORS

Although there is enough fissionable material in the earth's crust to operate reactors for a long time to come, there is an exciting possibility which may extend that source many fold. Experiments indicate that it is possible to operate a reactor on U^{235} or on Pu^{239}, which will not only produce power, but will also convert U^{238} into Pu^{239}. An element like U^{238} which can be transformed into a fissionable substance is called a *fertile material*. The new fuel that is formed in a layer or *blanket* of fertile material placed within a *converter* reactor* can supply a significant portion of the total energy output of the system. Eventually, however, the new fissionable material can no longer be utilized effectively because of poisoning in the blanket. This must then be removed and reprocessed. Nevertheless, since U^{238} is otherwise useless, this conversion operation may greatly reduce the net cost of fissionable material.

Thorium is another important fertile material. Nonfissioning thorium is more plentiful than uranium, and it can be converted to a fissionable nuclide by neutron bombardment. The reaction steps are

$$_{90}\text{Th}^{232} + {}_0\text{n}^1 \rightarrow {}_{91}\text{Pa}^{233} + {}_{-1}\text{e}^0, \tag{12-20}$$

$$_{91}\text{Pa}^{233} \xrightarrow{27.4\text{d}} {}_{92}\text{U}^{233} + {}_{-1}\text{e}^0. \tag{12-21}$$

Uranium-233 is fissionable and radioactive. It has a half-life of 1.62×10^5 years.

12-9. RESEARCH REACTORS

Chain reactors may be used for explosions, research, and for power. The primary value of reactors for research is that they are prolific sources of neutrons. Neutrons are produced from reactions excited by particles from radioactive elements and by those produced with accelerators, but the neutrons from chain reactions are many orders of magnitude more numerous than from these other sources. Of all the bombarding particles, neutrons are the best able to produce nuclear rearrangements, and an intense source of neutrons is a powerful tool for the study of nuclear reactions.

In some reactors a window in the protective shielding permits the escape of an intense beam of neutrons that can be used in many ways to learn more about nuclear physics. Another technique is to insert a slug of mate-

---

* At present, the word "converter," as applied to reactors, has two meanings that are somewhat different. The broad, general meaning is any reactor that converts fertile atoms into fuel by neutron capture. We have used converter in that sense. But *converter* can also refer to a reactor that uses one kind of fuel and produces another (for example, U^{235} produces Pu^{239} from U^{238}), and then *breeder* is used to designate a reactor in which the primary and the produced fuels are the same (for example, Pu^{239} produces more Pu^{239} from U^{238}). Finally, a *breeder* can also mean a converter that produces more fissionable atoms than it comsumes. It usually requires careful reading to determine which of these meanings is being used in articles on reactors.

rial into the reactor, where it is subjected to intense neutron irradiation. Practically every element thus irradiated forms radioactive isotopes in significant amounts. Such man-made radioisotopes are used in medical and tracer applications, where they have virtually replaced the naturally radioactive elements. These artificial radionuclides are available to qualified users at very reasonable prices.

The availability of intense radioactive sources permits the study of the effects of their radiations. Cobalt-60 is an example of a man-made isotope which is an intense gamma-ray source. One effect of gamma-ray irradiation under active study is the sterilization of food. Beta or electron irradiation produces promising structural changes in molecules, particularly in high polymers. By this means rubber can be vulcanized, and the melting temperature of polyethylene can be raised.

Since Co^{60} is an intense source of gamma radiation, it is dangerous to be near this material. In laboratories equipped with "gamma facilities" the Co^{60} is immersed to a depth of about ten feet in a deep tank filled with water, to shield the personnel from its radiation. Materials to be irradiated are either lowered into the water near the gamma source or, more commonly, placed near the top of the tank, and the Co^{60} is raised out of the water by remote control. Upon disintegrating, Co^{60} nuclei emit beta particles whose maximum energy is 0.31 MeV and gamma rays having energies of 1.17 MeV and 1.33 MeV.

In this section we have seen that a research reactor provides a series of research possibilities. One may study the characteristics of the reactor itself, the nuclear effects produced by excess neutrons, and the properties and behavior of materials made by the reactor.

12-10. CHERENKOV RADIATION

When an intense radioactive source ejects high-energy charged particles into a transparent material such as water, plastic, or glass, a ghostly bluish glow extending some distance into the medium can be seen. This phenomenon is easily observed when a room containing a swimming pool reactor or a gamma facility is darkened. This *Cherenkov radiation* has an interesting explanation and use. Let us consider a particular case, Co^{60} immersed in water. The most energetic beta particles emitted by this radioelement have a relativistic velocity of about $0.8c$. But these electrons move through water, where the velocity of light is about $0.75c$ (computed from c/n, where n is the index of refraction of water). Thus the electrons in the water travel faster than the phase velocity of light *in that medium*. The situation is much like that of a boat going through water faster than the velocity of water waves, or of a jet plane going through air faster than the speed of sound in air. The electron produces a "bow-wave" of light. Let a projectile (boat, plane, or electron) move with a velocity v from A to B in time t, as

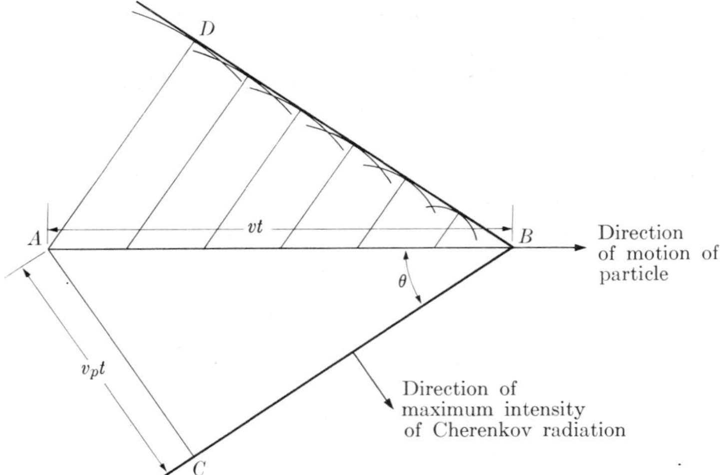

FIG. 12–8
Bow wave in Cherenkov radiation.

shown in Fig. 12–8. This projectile causes a sequence of disturbances in the medium which combine constructively to form the wavefronts represented by the lines from B to C and from B to D. These wavefronts move with a phase velocity v_p, and while the projectile goes from A to B the Huygens' wavelet (disturbance) from A goes from A to C, a distance $v_p t$. Thus the angle θ between the direction of motion of the projectile and the wavefront is given by

$$\sin \theta = \frac{v_p t}{vt} = \frac{c}{nv} \cdot \qquad (12\text{–}22)$$

It can be shown that the number of quanta emitted as Cherenkov radiation in a wavelength interval $\Delta\lambda$ is proportional to $[1 - (c^2/n^2v^2)]/\lambda^2$. It is evident from this expression that the short wavelengths will be predominant, and therefore that the color will be bluish white.

Cherenkov radiation can be used to measure electron velocities. The radiation is most intense in the direction of advance of the wavefront, as shown in Fig. 12–8. If the electron beam is collimated so that its direction is known and if the direction of maximum radiation intensity is measured, then θ is determined. Since c and n are known, v can be calculated. The value of n depends on the wavelength of the light. For precision measurements, the direction of Cherenkov radiation should be measured for a particular color.

12–11. POWER REACTORS

Any chain reactor can be adjusted to produce tremendous amounts of energy. It is only necessary to let the chain reaction become supercritical and wait for it to grow. The Chicago reactor was operated first at one-half watt and then at 200 watts, but the power level could have been run much higher. The main reason that the first reactor was operated at so low a level was that it was inadequately shielded and the neutron leakage might have endangered the personnel. The second reason was that there was no provision for cooling the reactor, and if it became too hot it would destroy itself. (There was little chance that the reactor would explode violently. The entrapped gases could have become hot enough to push the pile apart, but this would have made the reactor subcritical and stopped the reaction.) Thus the main physical limitation on the power a reactor can develop is the provision for the removal of heat.

Research reactors can be aircooled because no attempt is made to utilize the heat energy developed, but for power reactors a more dense material is indicated. We naturally think of water as a cooling material, especially since we shall see that water can also serve as a moderator. Boiling-water reactors will be discussed in Section 12–12. An efficient coolant should have a large specific heat capacity (specific heat times density). Since it must flow through the reactor, the use of dense liquid is indicated; and since it should have good thermal conductivity, metals appear attractive. Mercury, a dense liquid metal, has been used for the transfer of thermal energy and has replaced water in some boiler-turbine power systems. But mercury has a slow-neutron absorption cross section of 380 barns for the natural mixture of its isotopes, and this is too great. Not only would mercury absorb neutrons from the chain reactor, but it would also form radioactive isotopes which would render all the apparatus near which it passed radioactive. It turns out that despite its chemical activity, especially when molten, liquid sodium is a good reactor coolant. It is a rather dense material with good thermal conductivity and with a thermal neutron absorption cross section of only 0.5 barn. The sodium is usually alloyed with potassium to lower the melting temperature. The essentials of a sodium-cooled reactor power system are sketched in Fig. 12–9. The reactor itself is cooled by the molten metal and the energy of the molten metal is

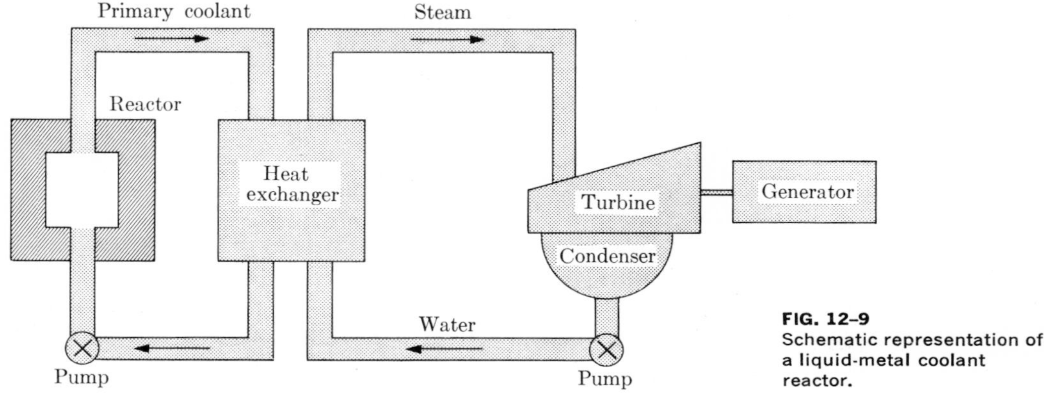

FIG. 12–9
Schematic representation of a liquid-metal coolant reactor.

transferred to water in a heat exchanger. The steam that is generated operates a conventional turbine and electric generator.

The way in which the molten metal is pumped is an ingenious application of basic electrical theory. It is a well-known fact that good thermal conductors are good electrical conductors, and molten sodium is a good electrical conductor. Electric charges moving perpendicular to a magnetic field experience a force perpendicular to both the current and the field, as expressed in Eq. (2–3). Figure 12–10 is a sketch of a liquid-metal pump based on this law. The electric current from M to N flows partly through the pipe but largely through the conducting liquid metal. Since there is a magnetic field directed toward the bottom of the page, the moving charges experience a force parallel to the pipe. By varying the current or the field or both, the force and pumping action can be controlled. There is obviously no need for sealed bearings, since there are no moving parts.

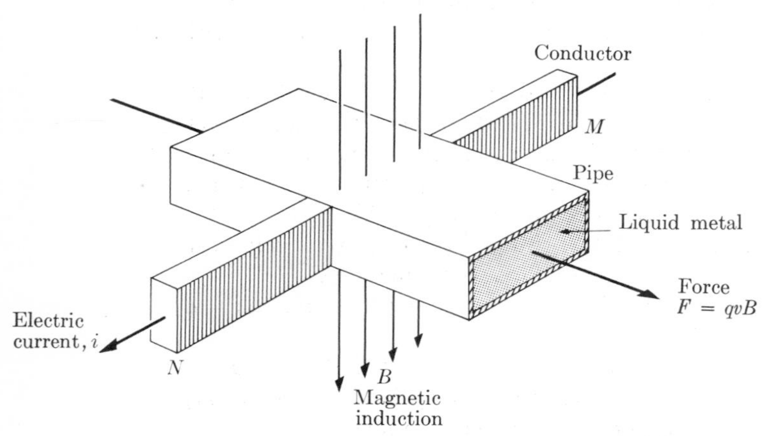

FIG. 12–10 Schematic diagram of an electromagnetic pump for liquid metals.

A heat-transfer "fluid" consisting of a high concentration of finely divided graphite in a stream of compressed air has been developed recently. Because it is inert and has good thermal properties, it may replace corrosive liquid metals as well as coolants that are made radioactive by neutrons.

On the average, the fission energy of either uranium or plutonium is 200 MeV. This is distributed approximately as follows: kinetic energy of the fission fragments, 167 MeV; kinetic energy of the fission neutrons, 5 MeV; prompt gamma rays and delayed gamma rays from the fission fragments, each 6 MeV; and the beta particles and neutrinos from the fission products, 5 MeV and 11 MeV, respectively. All of the neutrino energy and some of the gamma-ray and neutron energy escape through the walls of the reactor. This loss is just about compensated by the heat produced in moderating the neutrons plus the energy released in various parts of the reactor by the radioactive decay of atoms that were activated by gamma rays and neutrons. Because the fission fragments and the beta particles are easily absorbed, their energies are converted into heat close to their points of origin. Therefore, about 75% of the heat appears in the fuel rods and their cladding.

12–12. THE BOILING-WATER REACTOR

We have reserved the boiling-water reactor for more detailed discussion because the first large all-nuclear-powered generating station in the United States is of this type.

Although it is impossible to achieve a chain reaction with unenriched uranium and ordinary water as the moderator, ordinary water can moderate a reactor with enriched fissionable fuel. The boiling-water reactor is particularly attractive for two reasons. The moderating water can be converted to steam within the reactor and fed to a turbine without an intermediate heat exchanger, and a boiling-water reactor has an inherent stability which contributes to safety.

If the control rods in a reactor having a solid moderator are set so that the multiplication factor is greater than one, the chain reaction grows without practical limit until the reactor destroys itself. It is not too difficult to prevent this from happening by having safety control rods which can be thrust into the reactor quickly in the event anything goes amiss. (Such a shutdown is called a *scram.*) But a reactor having a solid moderator is basically unstable and always presents the possibility of a minor explosion that might break the shielding walls and spread dangerous radioactive materials in its vicinity.

In a water-moderated reactor even this unlikely hazard is eliminated. If the reactor gets out of hand, the water moderator will boil excessively, and since the bubbles of vapor are not dense enough to moderate the neutrons effectively, the reactor automatically goes *subcritical.* To test this point, the USAEC built a boiling-water reactor and tried to blow it up. They were able to produce a moderate explosion only when all the control rods were jerked out of the reactor simultaneously.

Indeed, a simple boiling-water reactor is too stable. If it is made supercritical for a short time in order to increase the level of power production, the water vaporizes and the reactor goes subcritical. Thus a simple boiling-water reactor cannot meet a sudden demand for increased power.

The Dresden reactor is designed to meet this problem in an ingenious way that utilizes the thermodynamic properties of water. It is a dual-cycle system, as shown in Fig. 12–11. The reactor proper, at the left, operates normally at a pressure of 600 psi (lb/in^2). Half the thermal energy leaves the reactor as steam, which is applied to the high-pressure blades of the turbine. The other half of the thermal energy leaves the reactor as

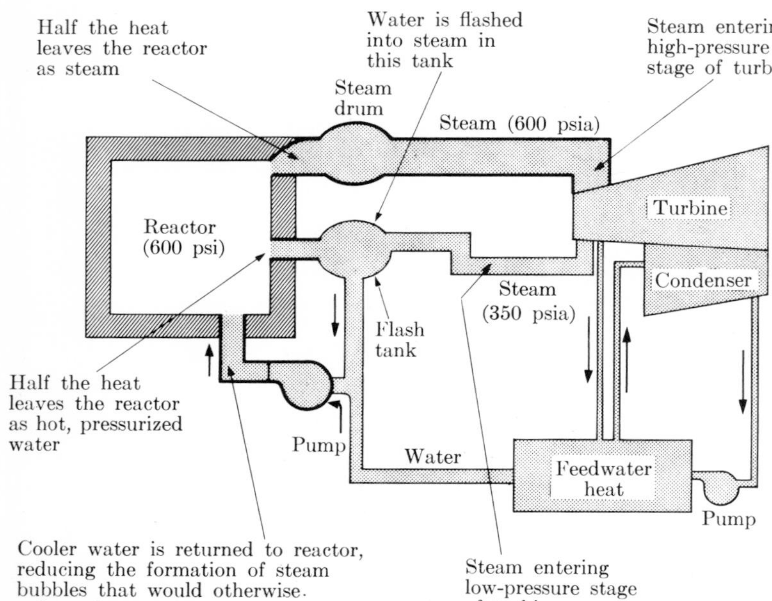

FIG. 12–11
Schematic diagram of a dual-cycle, boiling-water reactor. (Courtesy of General Electric Company)

water, which is removed from below the water level of the reactor. This water is "throttled" down to a pressure of 350 psi in what is called a "flash tank." At the lower pressure, much of the water vaporizes into steam, which is applied to the low-pressure blades of the turbine. But in converting water to steam, the heat of vaporization of the steam is provided at the expense of the temperature of the water not converted to steam. This low-temperature water is then mixed with the condensate from the turbine and pumped back into the reactor. The cool water injected into the reactor prevents boiling and keeps the moderator liquid.

The entire system is really two systems in parallel. The high-pressure system is inherently stable, as we have described. The low-pressure system is inherently unstable, since it tends to maintain the moderator as a liquid. The combination is still stable, but enough "positive feedback" is provided to give the reactor a reasonable "response time."

The nuclear reactor supplies only thermal energy, and this must be converted to mechanical energy in the turbine and electrical energy in the generator. As with conventional power plants, the overall utilization of the prime thermal energy is low, about 25%. If the plant is to deliver 180,000 kW to the electrical lines, the nuclear reactor must supply 720,000 kW of thermal power, of which 540,000 kW is extracted from the condenser by water from a nearby river.

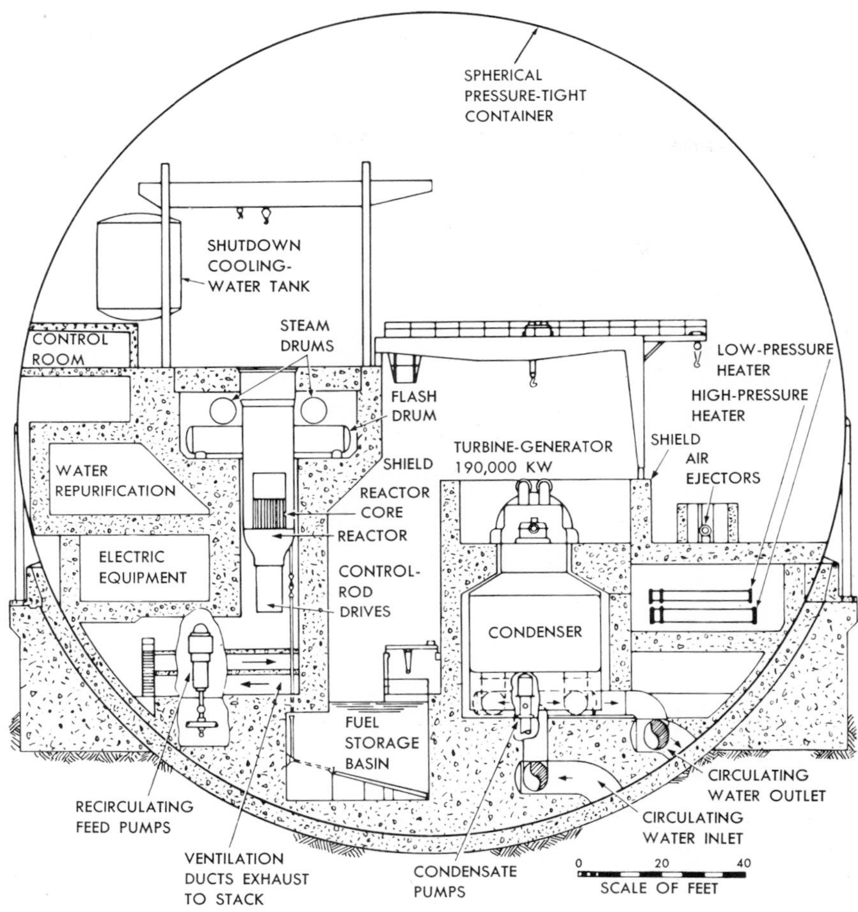

FIG. 12–12
Cross section of preliminary arrangement of equipment in the Dresden plant. (Courtesy of General Electric Company)

The following quotation* describes the nuclear-powered plant at Dresden, Illinois:

"The principal station components of Commonwealth Edison's nuclear power plant are housed in an airtight sphere 200 feet in diameter [Fig. 12–12]. This assures safety to the surrounding area in case of an

* Quoted by permission from the *General Electric Review*, November, 1955. (The design of the Dresden power station has been changed somewhat since 1955.)

incident because the sphere is designed to contain the internal pressure that would result if all the water in the reactor were to escape in the form of steam. The sphere is ventilated via a stack that can be blocked off during an emergency.

"The reactor and associated equipment are surrounded by a thick concrete shield. The control room—the plant's nerve center—is located on the upper level to give a good top view of the reactor during maintenance and reloading. (Throughout reloading, enough water is maintained above the reactor core to provide biological shielding.)

"During a power failure, heat from the reactor is removed by condensing steam and returning the condensate by gravity to the reactor vessel. Evaporating the water at atmospheric pressure cools the condenser. An overhead tank stores water for cooling the shut-down condenser.

"During operation, isotopes of oxygen that are formed when neutrons strike oxygen atoms are the major source of radioactivity in the steam-and-water system. When the plant is shut down, the value of these isotopes in the water becomes insignificant in about five minutes.

"Corrosion products and other impurities in the water that have become radioactive in passing through the reactor core are an annoying source of radioactivity. Tests have shown that the only solid impurities are those that are entrained by minute droplets of water. But with efficient moisture separators, the steam will contain only about one ten-thousandth of the concentration of radioactive solids present in the water.

"To reduce this residual radioactivity, the concentration of impurities in the reactor water is maintained at a low level by continuously bypassing a portion of the water through a cleanup filter and demineralizer.

"Only a small fraction of the impurities in the steam will adhere to the turbine. Thus any difficulty appears unlikely in maintaining the turbine and associated equipment under normal conditions."

12-13. NATURAL FUSION

In 1928, Professor A. A. Knowlton wrote a physics text in which a unifying theme was the question: Where does the sun's energy come from? He considers the possibility that the sun is leaking energy stored within it during some stage of astronomical evolution. He treats the sun as a coal pile deriving its energy from chemical reaction. He estimates the energy brought to the sun by falling meteors. He assumes that possibly the sun is contracting, with the potential energy of its parts becoming kinetic and thermal. He concludes as follows:*

"The English astronomer Eddington has calculated that if $\frac{1}{10}$ of the hydrogen now present in the sun were to be converted into helium, the

---

* Quoted by permission from *Physics for College Students*, 1st ed., by A. A. Knowlton, McGraw-Hill Book Co., Inc., 1928.

reaction would liberate a supply of heat great enough to keep the sun radiating at its present rate for a billion years. If such a process is going on anywhere, it must be at some place where atomic nuclei are forced to approach one another either as in the impact of swiftly moving atoms or by enormous external pressure. A few years ago, we had not the slightest ground for believing that either atomic disintegration or atom building was possible. Perhaps there are other things happening in the great solar crucible of which we have as yet no suspicion. At any rate we have here a plausible and fascinating hypothesis as to the source of solar energy which, unlike any of those previously considered, is quantitatively adequate."

By 1939, enough was known about nuclear reactions for Bethe to propose a set of reactions by which hydrogen might be converted to helium in the sun. It is called the carbon cycle, since carbon serves as a sort of nuclear catalyst. The steps are:

Reaction	Product half-life
${}_6\mathrm{C}^{12} + {}_1\mathrm{H}^1 \rightarrow {}_7\mathrm{N}^{13} + Q$	9.96 m
${}_7\mathrm{N}^{13} \rightarrow {}_6\mathrm{C}^{13} + {}_{+1}\mathrm{e}^0$	
${}_6\mathrm{C}^{13} + {}_1\mathrm{H}^1 \rightarrow {}_7\mathrm{N}^{14} + Q$	
${}_7\mathrm{N}^{14} + {}_1\mathrm{H}^1 \rightarrow {}_8\mathrm{O}^{15} + Q$	124 s
${}_8\mathrm{O}^{15} \rightarrow {}_7\mathrm{N}^{15} + {}_{+1}\mathrm{e}^0$	
${}_7\mathrm{N}^{15} + {}_1\mathrm{H}^1 \rightarrow {}_6\mathrm{C}^{12} + {}_2\mathrm{He}^4$	

(12–23)

If we add these equations, cancel out nuclides that appear on both sides, and calculate the total Q-value, we have

$$4({}_1\mathrm{H}^1) \rightarrow {}_2\mathrm{He}^4 + 2({}_{+1}\mathrm{e}^0) + 24.7\ \mathrm{MeV}. \tag{12–24}$$

The annihilation of the positrons supplies an additional 2 MeV of energy, so that the total released is actually 26.7 MeV.

To produce these reactions on earth, the carbon and nitrogen nuclides must be bombarded with accelerated protons. But the temperature at the center of the sun is so great that some thermal protons at the high-energy end of their Maxwellian velocity distribution are able to react. The fraction of high-energy thermal protons is so small that it may take a million years to go through this cycle once with particular atoms, but the total number of protons able to produce the reactions is so great that it accounts for the enormous solar energy. Measurements indicate that there is enough hydrogen in the sun to maintain this carbon cycle for about 30 billion years.

The sun is in equilibrium because of two opposing effects. If the temperature of the sun were to rise, the fraction of the protons able to cause the cycle would increase. This would increase the rate of enery production and further increase the solar temperature. On the other hand, if the temperature increases, the sun tends to expand and the concentration of the nuclides tends to decrease, making reactions less likely. These two effects balance each other.

There is every reason to believe that the sun is not unique in this respect, and that all stars are nuclear furnaces of one sort or another. Sometimes weak or unknown stars grow to great brilliance quite suddenly. These *new stars* (*nova*) are unquestionably unstable stars whose nuclear reactions have gone wild.

Since all animals derive their energy ultimately from plants, which get their energy from photosynthesis of the sun's light, since coal and oil were formed from plant life, and since water power is derived from solar heat, we can trace almost all the energy we use to the sun. Thus we live and breathe and have our being because solar hydrogen is converted to helium.

Radioactivity, atom smashers, and nuclear fission cause processes in which matter is taken apart. The solar carbon cycle is a "putting-together process" in which matter is fused, and the process is called *fusion*.

12-14. MAN-MADE FUSION

If fusion can take place in the stars, is it possible to cause fusion here on earth? Before Bethe proposed the carbon cycle for the sun, each of the reactions of that cycle had been produced on the earth. Thus by 1939 there had been man-made fusion in a limited sense. But like many other exothermic nuclear reactions, there was no possibility of making practical use of the energy liberated so long as it took extensive apparatus using much *more* energy to make the reaction proceed. The key to practical nuclear power is a chain reaction which runs itself. The links of the fission chain are neutrons, while the links of a fusion chain are protons.* Fission proceeds best with thermal *neutrons*, where thermal means at or near room temperature. Fusion proceeds best with thermal *particles*, where thermal means temperatures measured in millions of degrees. If the solar fusion of hydrogen into helium is a slow process in the sun, where Eddington estimated the temperature to be 15,000,000°K, the prospect of fusion on earth looks dim.

Our best hopes lie in some of the possible fusion reactions listed below, together with their Q-values in MeV:

$$\begin{array}{llll} \mathrm{D(d, p)T} & 4.02 & \mathrm{Li^6(d, \alpha)He^4} & 22.4 \\ \mathrm{T(p, \gamma)He^4} & 19.6 & \mathrm{Li^7(p, \alpha)He^4} & 17.3 \\ \mathrm{T(d, n)He^4} & 17.6 & \mathrm{Li^7(d, n)Be^8} & 14.9 \\ \mathrm{Li^6(n, \alpha)T} & 4.96 & & \end{array} \qquad (12\text{–}25)$$

* There are other stellar cycles besides the Bethe carbon cycle we have described and there are fusion reactions in which deuterons, alpha particles, etc. provide the "links" of the chain.

A fusion reactor might utilize several of the reactions in this list. For example, the fourth reaction could be initiated with a neutron. If the tritium product struck deuterium, the third reaction would provide a neutron. This neutron might excite the fourth reaction again, etc. Hydrogen isotopes have been ignited to fusion in a fission bomb. This so-called hydrogen bomb (fusion bomb) is a hideous thing because its energy release is "open-ended." Once the fusion is started, the fusion itself maintains the temperature to keep the process going. The energy liberated is a function of how much fusible material is present, and there is no theoretical limit. Recall that in a fission bomb, the parts, before detonation, must each be smaller than the critical size lest the parts explode spontaneously. A fusion bomb cannot explode until it is "ignited" and any amount of fusible material is safe until ignited. Thus the amount of fusible material in a fusion bomb is not limited.

The constructive utilization of the energy released from fusion is now a major field of research endeavor, and there is still some doubt that a way can be found. Fusion on the sun is "contained" by the tremendous gravitational field of the sun, and therefore a continuous reaction takes place. Fusion in an H-bomb is a destructive explosion, since it is not contained. At first glance it appears entirely hopeless, not only to achieve the temperatures necessary for fusion, but at the same time to contain the reaction at a temperature many orders of magnitude above that at which earth materials vaporize.

One answer to the achievement of suitable temperature is already implied by our discussion. We do not need temperature as such. What we need is high-velocity particles. Although both the sun and the H-bomb produce these velocities with thermal energy, all the fusion reactions have been produced on a small scale by using particle accelerators. We have seen that the kinetic energy of a particle moving with the most probable speed in a Maxwellian distribution for a group of particles at room temperature, 293°K, is 0.025 eV. *By stretching the concept of temperature from a statistical property associated with the random motion of all the atoms or molecules in a piece of gross matter to a description of individual particle energy* we get a new concept, called *kinetic temperature.* Thus any particle with 0.025 eV of energy has a kinetic temperature of 293°K. Multiplying each of these quantities by 4×10^4, we find any particle having 1 keV of energy has a kinetic temperature of 11.6 million °K. A well-thrown baseball may have a kinetic temperature of 10^{20} electron volts or about 10^{24} degrees Kelvin. Calculations show that it may be possible to maintain fusion continuously at a kinetic temperature as low as 45 million degrees. Charged particles may be given this kinetic temperature by accelerating them through a mere 4000-V potential difference.

If an electric arc is maintained by a potential of this order of magnitude, any gas of low atomic number in the arc becomes completely ionized.

The space between the electrodes is then filled with a mixture of two gases, one consisting of negative electrons and the other consisting of positive nuclei. This mixture of charged-particle gases is called a *plasma.*

Even though the number of particles per unit volume in a plasma may be much smaller than in the atmosphere, a kinetic temperature of 100 keV may produce a pressure as high as 1000 atm.

If we assume that a plasma can be produced in which fusion can take place, how might a body of gas with a pressure of 1000 atm and a kinetic temperature of about a billion degrees be contained? Obviously, contact with any material container will result in the absorption of energy from the plasma, thereby reducing its kinetic temperature. One important mechanism that causes the plasma to lose energy rapidly is radiative, not thermal, in character. When a high-energy particle from the plasma strikes the wall of the container, some of the material (usually stainless steel) is vaporized. The atoms of high atomic number that are released diffuse into the plasma stream and cause a marked increase in the rate at which the moving, colliding, charged particles lose energy by bremsstrahlung. Even minute amounts of impurities cause a large increase because the power radiated by bremsstrahlung varies as the square of the atomic number of the atom causing deceleration. Since the presence of elements of high atomic number lowers the kinetic temperature of the plasma, a container must be fantastically clean before deuterium or some other gas is introduced. The best hope of preventing contact between a stream of charged particles and the walls of the container lies in the restraining effect that can be produced by a magnetic field. If the plasma arc is along the axis of a solenoid, any charges moving parallel to this axis experience no magnetic force. If charges are knocked out of their parallel paths, however, their paths become helices, since their velocities now have components perpendicular to the field (see Problem 2–17). Thus a strong enough magnetic field turns escaping charged particles back into the body of the plasma and constitutes a "bottle."

The high kinetic temperature necessary to establish fusion conditions is reached by shrinking the plasma with what is known as the *pinch effect.* We know that two parallel wires carrying currents in the same direction attract each other. In a plasma carrying a current we may think of the current as taking place in a bundle of wires each attracting all the others. If a large bank of highly charged capacitors is discharged through the plasma, there may be a transient current of many million amperes. Such a current, properly directed through the plasma parallel to the axis of the existing magnetic field produces an intense pinch effect. Charges at the outside edges of the plasma are thrown toward its axis, where they collide, with high kinetic temperature. Fusion conditions have been produced momentarily in this way for deuterium nuclei. The energy released, has been small compared with the energy required to operate the apparatus.

If a pinch-initiated fusion reaction were to sustain itself continuously, or long enough for the energy liberated to exceed the energy input to the apparatus, then electromagnetic radiation and uncharged particles (principally neutrons) could carry away energy to heat a boiler. There is also hope that because of the electrical nature of the fusion plasma it may be possible to generate electricity by pulsing the plasma and thereby inducing currents in a secondary coil.

It is easy to see why research on fusion continues in spite of its many discouraging aspects. Although fission presents a far greater energy source than our remaining fossil fuels, this source could be used up in a few centuries of modern civilization. Furthermore, fission produces radioactive wastes which are certain to be an increasingly serious problem, possibly requiring that these "poisons" be projected into space. On the other hand, the product materials of fusion are almost harmless, and the known fuel supply is sufficient for man's extravagant use for billions of years. If hydrogen nuclides are what we need, every water molecule in the oceans has two of them.

REFERENCES

BEKEFI, G. and S. C. BROWN, "Resource Letters PP-2 on Plasma Physics: Waves and Radiation Processes in Plasmas," *Am. J. Phys.* **34,** 1001 (1966).

BISHOP, AMASA S., *Project Sherwood—The U. S. Program in Controlled Fusion.* Reading, Mass.: Addison-Wesley, 1958. An excellent account of the work in this field.

BOLEY, F. I., *Plasmas-Laboratory and Cosmic,* Momentum Book No. 11, Princeton, N. J.: Van Nostrand, 1966.

BONILLA, CHARLES F., editor, *Nuclear Engineering.* McGraw-Hill, 1957. A comprehensive book.

BROWN, S. C., *Plasma Physics,* Resource Letter PP-1 and Selected Reprints. New York: American Institute of Physics, 1963.

CLAUSER, F. H., editor, *Plasma Dynamics.* Reading, Mass.: Addison-Wesley, 1960. An inclusive and authoritative book.

COMPTON, ARTHUR H., *Atomic Quest, A Personal Narrative.* New York: Oxford University Press, 1956.

FERMI, E., et al., *Neutronic Reactor,* U. S. Patent No. 2,708,656; dated May 17, 1955; filed December 19, 1944 (57 pages). Reveals the state of the art of reactors at the time of filing. Well worth reading.

FERMI, LAURA, *Atoms in the Family.* Chicago: University of Chicago Press, 1954. A delightful biography of ENrico Fermi.

GLASSTONE, SAMUEL, *Principles of Nuclear Reactor Engineering.* Princeton, N. J.: Van Nostrand, 1955. Good coverage.

GLASSTONE, S. and R. H. LOVBERG, *Controlled Thermonuclear Reactions.* Princeton, N. J.: Van Nostrand, 1960.

HOAG, J. BARTON, editor, *Nuclear Reactor Experiments.* Princeton, N. J.: Van Nostrand, 1958.

HUGHES, D. J. and R. B. SCHWARTZ, *Neutron Cross Sections, BNL 325.* 2nd ed. Washington, D. C.: U. S. Government Printing Office, 1958. The most extensive data published.

MURRAY, RAYMOND L., *Reactor Physics.* Englewood Cliffs, N. J.: Prentice-Hall, 1957.

POST, RICHARD F., "Controlled Fusion Research—An Application of the Physics of High Temperature Plasmas," *Rev. Mod. Phys.* **28,** 338 (1956). An excellent discussion of this field.

ROCKWELL, THEODORE, III, editor, *Reactor Shielding Design Manual.* New York: McGraw-Hill, 1956.

SCHULTZ, M. A., *Control of Nuclear Reactors and Power Plants.* New York: McGraw-Hill, 1956.

SMYTH, HENRY D., *Atomic Energy for Military Purposes.* Princeton, N. J.: Princeton University Press, 1945. The official report of the development of the nuclear bomb. A very good historical account.

SPITZER, LYMAN, *Physics of Fully Ionized Gases.* 2nd ed. New York: Wiley, 1962.

United States Atomic Energy Commission, *The Effects of Nuclear Weapons.* Washington, D. C.: U. S. Government Printing Office, 1957. An extensive account.

PROBLEMS

12–1. (a) If Sr^{90} constitutes 5% of the fission-fragment yield of uranium (see Fig. 12–1), how many grams of this strontium are formed when a uranium bomb having the explosive equivalent of one megaton of TNT is exploded? (All the atoms in 50 kg of uranium must fission to produce the equivalent blast effect.) (b) If all this strontium is distributed uniformly throughout the atmosphere over the whole earth within a few months, what is the approximate radioactivity in microcuries above each square mile of the surface of the earth due to this one kind of fission fragment? (c) Comment on the likelihood of the uniform distribution assumed. (d) What happens to this strontium?

12–2. Show that the graph of the logarithm of the microscopic cross section of a "$1/v$-absorber" against the logarithm of the kinetic energy of the bombarding neutron is a straight line with a negative slope.

12–3. Assume a neutron causes fission in ${}_{92}U^{235}$ producing ${}_{40}Zr^{97}$, Te^{134}, and some neutrons. (a) Determine the atomic number of Te from the data. (b) How many neutrons are released?

12–4. A beam of 0.025-eV neutrons passes through an apparatus 20 cm long at the rate of 10^9/s. From the neutron half-life of 12 m, calculate the rate at which neutron decays are expected in the apparatus.

12–5. A nuclear reactor containing U^{235} is operating at a power level of 2 W. Calculate the fission rate assuming that 200 MeV of useful energy is released in each fission.

12–6. A beryllium target with a thickness of 6.5×10^{-5} cm is bombarded with a 20 μÅ beam of 3-MeV alpha particles. Neutrons are produced from the reaction $Be^9(\alpha, n)C^{12}$, which has a cross-section of 9.00×10^{-2} barns. What is the rate at which neutrons are produced?

12–7. A very thin sheet of gold foil (thickness $= 4 \times 10^{-3}$ cm) is irradiated for 2000 s in a neutron beam of flux 2×10^{10} neutrons/s. The gold is transmuted to Au^{198} by an (n, γ) reaction. After the irradiation has ended, the measured activity of the Au^{198} is 10^{10} disintegrations/h. Calculate the neutron capture cross section of gold for the neutrons used in this experiment.

12–8. A piece of Co^{59} (relative density, 8.9) whose dimensions are 3 cm $\times$ 4 cm $\times$ 0.5 mm is to be irradiated by neutrons in a graphite pile until its total activity is 2 curies, due to the formation of Co^{60} by the reaction $Co^{59}(n, \gamma)Co^{60}$. (a) How many atoms of Co^{60} does the activated piece of metal contain? (b) How many neutrons must interact with the Co^{59}, according to the reaction given, to produce the number of atoms in part (a)? (c) The microscopic activation cross section of Co^{59} for the given reaction is 34 barns. If the irradiating beam is 2×10^{12} neutrons/cm^2/s normal to the 3×4 cm face, how long will it take to achieve the total activity of 2 curies? Assume that the radioactive disintegration of the Co^{60} being formed during the time of bombardment is negligible. (d) Under what conditions is the assumption in part (c) reasonable?

12–9. (a) What fraction of the total number of atoms in the sheet of activated cobalt in Problem 12–8 changes into the daughter product in a day? (b) In 1965 the population of the United States was 194 million and the annual birthrate was 3.8 million. What fraction of the total population became "daughter products" each day? (c) How much greater is this fractional birthrate than the fractional rate of radioactive disintegration in the cobalt sheet in part (a)? (d) Is radioactive "blowing up" a commonplace event from the viewpoint of the atoms in this piece of cobalt? (The cobalt sheet in this problem actually has a high activity for a radioactive material.)

12–10. A borated-steel sheet (relative density 7.81) which is used as a control "rod" in a reactor is 1/16 inch thick and contains 2% boron by weight. The microscopic absorption cross section of iron for neutrons is 2.5 barns and for boron, 755 barns. (a) Find the macroscopic absorption cross sections of each of these elements in the borated-steel sheet. (b) What fraction of a neutron beam is absorbed in passing through this sheet?

12–11. 100 ml of a solution contains 2 gm of boric acid, H_3BO_3, dissolved in 100 g of water. (a) What is the macroscopic neutron absorption coefficient of each element in the solution, and (b) what is the total absorption cross section of the solution? (The microscopic absorption cross sections for these elements are given in Table 12–3.)

12–12. Using the same method as that used to obtain the average life of a radioactive atom, Eq. (10–8), show from Eq. (12–12) that the neutron mean free path in a material is $1/N\sigma$.

12–13. What is the absorption mean free path of (a) neutrons in iron (σ_a = 2.5 barns), (b) neutrons in borated-steel (Problem 12–10)? (c) What fraction of a beam of neutrons is transmitted by a sheet of material having a thickness equal to their mean free path?

12–14. (a) It is said that the effective neutron shielding property of concrete depends primarily on its hydrogen content but that its gamma-ray absorbing property depends mostly on its aluminum and silicon content. Do Eq. (6–15) and the data in Table 12–4 confirm this? (b) What would be the effect on the scattering property for neutrons and the absorption for gamma radiations if the concrete were made denser by using steel punchings in the mix instead of gravel?

12–15. The index of refraction of water at 20°C for red light is 1.33 and for blue light is 1.34. (a) What is the half-angle, θ, of the bow wave for red light and for blue light in the Cherenkov radiation produced by a beta particle emitted from a sheet of Co^{60} with a speed of $0.8c$? (b) At what particle speed will (1) the red light and (2) the blue light in Cherenkov radiation in water disappear?

12–16. Each fission of a certain hypothetical nucleus produces two identical fragments, each with a kinetic energy of 50 MeV. What is the kinetic temperature of the "gas" of fission fragments?

12–17. The energy received from the sun by the earth and its surrounding atmosphere is 2.00 cal/(cm^2 · m) on a surface normal to the rays of the sun. (a) What is the total energy received in joules per minute by the earth and its atmosphere? (The diameter of earth-atmosphere system is 1.27×10^4 km.) (b) What is the total energy radiated in joules per minute by the sun to the universe? (Distance from sun to earth is 1.49×10^8 km.) (c) At what rate, in tons/m, must hydrogen be consumed in the fusion reaction, Eq. (12–24), to provide the sun with the energy it radiates? (d) At what rate, in tons/m, must the mass of the sun decrease to provide the radiated energy calculated in part (b)?

12–18. A deuterium reaction that occurs in experimental fusion reactors is D(d, p)T followed by a reaction of the tritium product with another deuterium atom. This second reaction is T(d, n)He^4. (a) Compute the combined Q-value of these successive reactions. (b) What is the energy released in the combined reactions per unit mass of the three initial deuterium particles? (c) What percent of the total rest mass-energy of these three initial particles is released in the combined reactions? (d) Compare the answers for parts (b) and (c) with the corresponding calculations for uranium fission in Problem 5–38. (e) Compare U^{235} fission with deuterium fusion as a source of energy on the basis of the availability of the isotopes and of the energy released per unit mass.

12–19. It is believed that the microscopic absorption cross section of iron for neutrinos is of the order of 10^{-20} barn. In kilometers, what is the half-value layer of iron for neutrinos?

CHAPTER 13

HIGH-ENERGY PHYSICS

13-1. INTRODUCTION

In this concluding chapter we return again to the early part of this century to describe the discovery and some of the properties of cosmic rays. We shall see how studies of these natural phenomena, together with studies of events produced by the modern superaccelerators, serve to clarify some matters already discussed. But this chapter will not tie up a neat package. In March, 1966 Nobel prizewinner Lee stated in *Physics Today*, "The more we learn about symmetry operations—space inversion, time reversal, and particle-antiparticle conjugation—the less we seem to understand them. At present, although still very little is known about the true nature of these discrete symmetries, we have, unfortunately, already reached the unhappy state of having lost most of our previous understanding." Our hope is that the reader will gain some comprehension of the problems that are most challenging to the physicists of today.

13-2. COSMIC RAYS

Cosmic rays were discovered during the search for a cause of ionization chamber leakage. Of course no insulator is perfect, and some ionization chamber leakage was expected. When it was found that the leakage rate was greater than could be accounted for by imperfect insulation, it was assumed that the difference was due to ionization from traces of radioactivity in the earth. To test this point, ionization chambers were carried above the earth in balloons. During the first part of the ascent, the ionization did indeed decrease slightly. But in 1909, Gockel found that at greater heights the ionization began to increase, and at 2.5 miles the ionization was greater than at the surface of the earth. In 1910, Hess and Kohlhoerster carried a chamber even higher and, after extensive study, they proposed that there must be a very penetrating radiation coming from outside the earth—*cosmic rays*. It was estimated that these rays were at least ten times as penetrating as the most energetic gamma rays known.

Just as cosmic-ray research was begun by moving ionization chambers up into the air, so it was continued by moving them into the earth and over its surface. The presence of cosmic rays was traced deep inside mines, where the effect of any radioactive material was excluded with lead shields. They were found deep in lakes. Glacier-fed lakes were chosen because their water is "distilled" and free of radioactive materials. Recording ionization chambers were sent all over the world on ships.

As a result of these geographic explorations, it was concluded that cosmic rays are extremely penetrating, and that although they come to the earth from all directions, a few more reach the earth in the latitudes of the magnetic poles than at the equator.

The fact that the intensity is greater toward the polar regions is called the *latitude effect* and it suggests that at least some of the cosmic rays are charged. Charged particles approaching the earth at the equator must move perpendicular to the magnetic field of the earth. Depending on the sign of the charge, these particles experience a force either toward the east or toward the west which may cause some rays to miss the earth that would otherwise have hit it. Charged particles approaching the high-latitude portions of the earth move nearly parallel to the magnetic field and experience a negligible deflecting force.

Although this analysis indicated that at least some of the rays were charged, it did not indicate the sign of the charge. It also follows from the analysis that at the magnetic equator charged particles should strike the earth from a preferred direction, either east or west of the vertical, depending on the sign of the charge. Ionization chambers have no directional sensitivity, therefore clever arrangements of Geiger-Mueller tubes were used to investigate the east-west effect. A single G-M tube has no directional sensitivity but several of them, connected electronically so that no count is recorded unless they all count in coincidence, may be physically arranged like the rungs of a ladder. With such an apparatus, called a *cosmic-ray telescope*, the direction of a ray capable of firing all the tubes could be determined.

The east-west experiment was performed by observing the counting rate of a cosmic-ray telescope set at a variety of angles east and west of the vertical. A symmetrical distribution was found at 35° south latitude (Buenos Aires). At the magnetic equator, however, although the distribution had the same general shape it skewed a little to the west. The slightly greater intensity from the west indicated a preponderance of positive charges, but it was impossible to conclude from this that there were no negative particles or uncharged particles present.

13-3. COSMIC-RAY SHOWERS

In 1928 Skobeltsyn photographed cosmic-ray tracks in a Wilson cloud chamber. He found that the tracks appeared in groups and that all the particles in a group seemed to originate from some point near and above

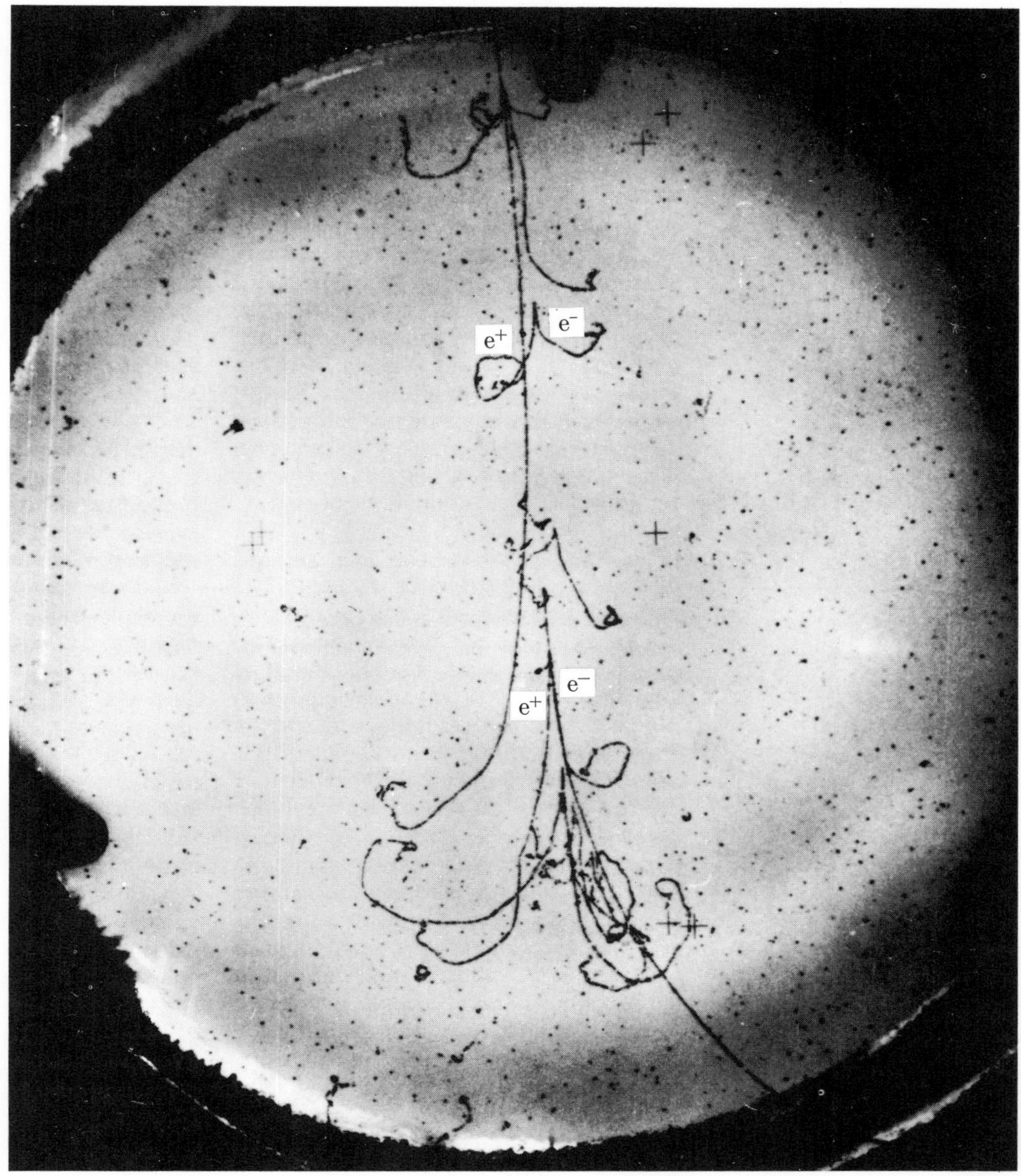

Fig. 13–1
Electron-gamma ray shower formation. An electron with about 1 GeV of energy entered a methyl iodide propane bubble chamber, initiating a shower of gamma rays and electron-positron pairs. The entire development can be seen because of the high atomic number and high density of the medium. (Courtesy of the "Cambridge Bubble Chamber Group," Brown University, Harvard, M.I.T., and Brandeis)

the apparatus, such as the ceiling. It was later found that G-M tubes that were widely separated horizontally had more coincidence counts than could be accounted for by chance. These observations led to the discovery of what are called *cosmic-ray showers.* The mechanism of these bursts of rays is well understood. Somewhere in the atmosphere a cosmic ray may produce a photon with several GeV of energy. This photon may produce an electron-positron pair. (Recall that positrons were discovered by Anderson while using a cloud chamber in cosmic-ray research.) These very high-energy electrons can next produce high-energy photons by the mechanisms of x-ray production and also by the eventual production of annihilation radiation. Consequently, one initiating particle (either photon or electron) may be multiplied into many shower particles, as shown in Fig. 13–1.

The discovery of showers points up the difficulty of establishing the nature of the real primary cosmic radiation that comes to the earth from outer space. The nature of the rays changes as they come down through the atmosphere, and sea-level cosmic rays may bear little resemblance to their initial state.

Furthermore, the known absorption characteristics of photons and electrons precluded the possibility that they were the only components of cosmic rays. These particles simply do not possess the observed penetrating ability. The heavier particles like protons and neutrons were considered, until help came from another direction—theoretical physicists.

13–4. DISCOVERY OF MU-MESONS

In trying to explain the binding force between neutrons and protons in the nucleus, Fermi had attributed it to the "exchange" of electrons between the two types of particles. It developed that the magnitude of a force from this cause was too small to fit the experimental evidence. In 1935, Yukawa followed the same general idea but postulated that the exchange particle was not an electron but a new particle having a mass intermediate between those of an electron and a proton. Since this postulated particle, if it existed, would have the penetrating properties consistent with cosmic-ray data, cosmic-ray researchers set out to find it. In 1936, Anderson and Neddermeyer observed a cloud-chamber track of such a particle, now called a *mu-meson,* but it was shown some years later that it was not identical with the one assumed by Yukawa. In 1940, Leprince-Ringuet obtained the first good cloud-chamber picture that showed a collision between a mu-meson and an electron. Since both tracks were curved by a magnetic field, the particle energies could be calculated. The mass of the mu-meson was calculated from the mechanics of the collision; the modern accepted value is 207 electron rest masses.

The discovery of the mu-meson, also called a *muon*, accounted for the penetrating ability of cosmic rays, but it was soon evident that the mu-meson could not be the primary radiation from outer space. It developed that the mu-meson is itself radioactive, with a very short half-life. Measurement of this half-life is complicated by the fact that mu-mesons formed in the upper atmosphere by cosmic-ray bombardment are projected toward the earth with a speed close to that of light. In our discussion of relativity we saw that there is a time transformation equation which takes into account the relative velocity of two observers. An observer moving with a mu-meson would conclude that its half-life was a few microseconds. To an observer on earth, the half-life appears considerably longer—long enough, in fact, to let a mu-meson traverse the atmosphere but not long enough for it to have come from outer space. Thus mu-mesons cannot be primary radiation but must be created by something else at the top of the atmosphere. Present data indicate that most of the primary cosmic rays are protons, but there are also some positive nuclei or stripped atoms of elements up to $Z = 26$. These very high-energy primary particles produce a large variety of other particles by *spallation* (nuclear smashing) of the atoms in the outer atmosphere.

Although much more is known about the primary radiation than we have discussed here, there is still speculation going on. Earth satellites are now sending us information from beyond the atmosphere. We will soon know more of the nature of primary cosmic rays and the hazards they may present to space travel.

13-5. NUCLEAR EMULSION TECHNIQUE

It is difficult to study penetrating radiations with Wilson cloud chambers because the absorption of the gas within the chamber is so small. We have already pointed out that bubble chambers containing dense material are better in this respect, but both these devices are sensitive only intermittently and neither is portable. These considerations led to the development of special photographic emulsions. Ordinary photographic materials are unsuitable because they contain too small a fraction of sensitive silver halides and because they are too thin. C. F. Powell and others developed satisfactory emulsions that contain about 80% silver bromide by dry weight and are about 1 mm thick. These can be stacked to make a sensitive region of practically any size. Such blocks of emulsion are continuously sensitive and are portable enough to be carried by balloons and other high-altitude craft. It requires elaborate techniques to develop, piece together, and read nuclear emulsions, but the resultant "tracks" closely resemble those formed in cloud and in bubble chambers. The important characteristics of a track are its density, range, direction, and deviation in direction, and these characteristics often permit the positive identification of the particle causing the track. The emulsion record of a nuclear disintegration is shown in Fig. 13.2.

Fig. 13–2
Photomicrograph, enlarged 100 times, of photographic emulsion film exposed to 2Bev nuclear particles in the cosmotron. The incoming particle, presumably a neutron, which is invisible, hit the nucleus of an atom of the emulsion and exploded it into 17 visible particles which formed tracks in a star-shaped pattern. In general, the broad tracks are made by slow particles (protons) and the narrow tracks by fast particles. The spots in the background are grains in the emulsion. (Courtesy of Brookhaven National Laboratory)

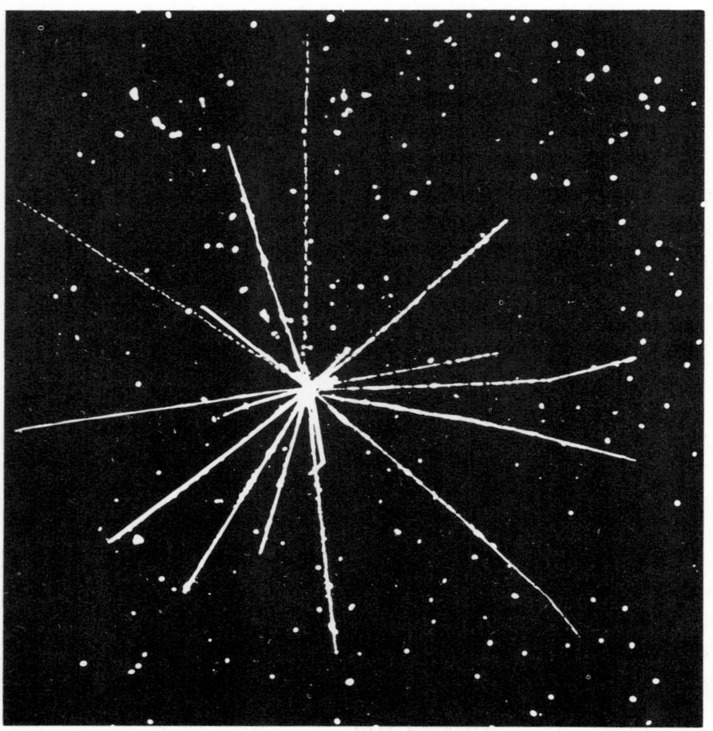

13–6. DISCOVERY OF PI-MESONS

One of the first discoveries made with nuclear emulsions was that of the pi-meson. A block of emulsions was exposed for several weeks on a mountain. One mosaic of track pieces showed a meson slowing down and then emitting a new high-speed meson in a new direction. The second meson could be identified as the familiar mu-meson. Since the first could create a mu-meson and give it kinetic energy as well, this first meson had to be more massive. It is called the *pi-meson* (primary meson) and its mass is now known to be 270 electron masses. Three types of pi-mesons, or *pions*, have been discovered. One has a positive charge, another is negative, and the third is neutral. All are radioactive. The charged ones have mean lives* of the order of 10^{-8} s, but the life of a neutral pion is only about 10^{-16} s.

In flight, the positive and negative pi-mesons decay into the positive and negative mu-mesons, respectively, and also into neutrinos. When negative pions are slowed down in matter, they are captured by atoms, make x-ray transitions down to the K level, and then are absorbed by the nucleus, causing violent explosions called "stars." Positive pi-mesons stopped by matter decay into mu-mesons, all of which have about the same energy. Experiments show that neither the positive nor the negative mu-meson has an affinity for atomic nuclei. Because of this, it is difficult to assign to muons the role of providing the nuclear exchange forces of the Yukawa theory. According to present nuclear theory, all three types of pi-mesons are found in the nucleus, and it is believed that both protons and neutrons are continuously emitting or absorbing positive, negative, and neutral pi-mesons. The particular process depends upon which nucleons are paired in experiencing exchange forces at the moment, that is, proton-neutron, proton-proton, or neutron-neutron. Thus the general view is that each nucleon has an associated meson field through which it interacts with other nucleons. This is analogous to the action through an electromagnetic field of one electrically charged body on another. The pions play the same role in the meson field as photons do in an electromagnetic field.

13-7. SUPERACCELERATORS

Both pi- and mu-mesons were discovered in cosmic-rays, but some of their properties that we have already discussed were discovered with the aid of superaccelerators. The chief limitation of the cyclotron already described is the relativistic mass variation of the accelerated particle. Superaccelerators like the Berkeley synchrocyclotron or Bevatron and the Brookhaven Cosmotron take the relativistic mass variation into account and take advantage of another basic relativistic fact—the simplification that results from all high-energy particles having approximately the same velocity, which is very close to the free space velocity of light. These machines can accelerate particles into the cosmic-ray range of giga-electron volts.

When protons or alpha particles are given such extreme energies, they can do remarkable things to nuclei. Instead of "nudging" nuclei into new and possibly unstable forms, these particles can knock nuclear particles out of a nucleus as easily as though the nuclear particles were "free."

---

* In describing the fundamental particles, mean lives rather than half-lives are usually stated. The definition of mean life is given in Section 10-9.

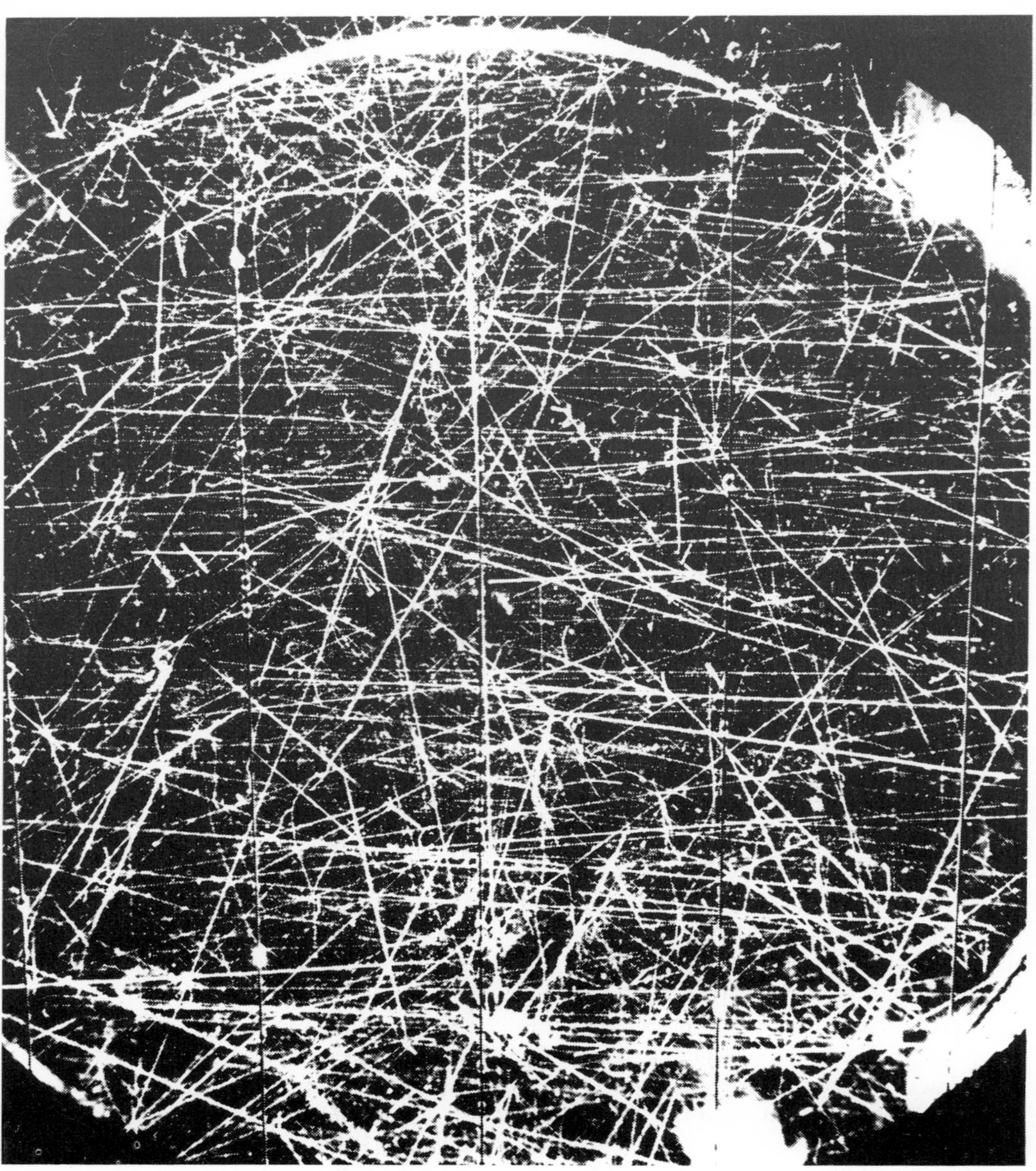

Fig. 13–3
Cloud-chamber picture showing the high-energy particles resulting from a single pulse of 1.2 GeV protons striking a one-eighth-inch brass target inside the Cosmotron. Particles reaching the chamber, located 195 ft from the Cosmotron, traversed an inch of stainless steel in the outside wall of the vacuum chamber of this superaccelerator, two heavy doors, and the quarter-inch stainless steel wall of the cloud chamber. (Courtesy of Brookhaven National Laboratory)

Often, unexpected particles are knocked out or formed during the collision process. It is evident from the nuclear "wreckage" shown in Fig. 13–3 that high-energy particles are destructive even though they have traveled far from their source.

13-8. THE STRANGE PARTICLES

Electrons, protons, and neutrons are familiar in our scheme of things. In earlier chapters we introduced positrons and neutrinos. The mu- and pi-mesons facilitate our understanding of the penetrating ability of cosmic rays, and pi-mesons fit into the theory of nuclear binding forces. But the total number of elementary particles now known is about 100. Those we have discussed in this book thus far are "useful" in the sense that they fit into a rational scheme for the structure of matter, and they "explain" events which are more or less common and natural. The plethora of new particles present an enigma which has led them to be called the *strange* particles. One can make an analogy between these particles and the transuranic elements. Elements with atomic numbers above 92 did not exist in nature in measurable amounts before they were "made" by man. They are all unstable with half-lives short compared to the age of the earth, so that if they ever existed, they have long since disappeared. Thus these elements are strange elements. Having made them, however, man seeks to measure their chemical and physical properties, understand them, and find uses for them. Similarly, the strange elementary particles are man-made. They are created (or liberated) by collisions of the extremely energetic particles from the superaccelerators. They are all unstable with mean lives from about 10^{-8} to 10^{-16} s. They are not elementary in the sense that they are simple and basic like elementary algebra. They are elementary in the sense that they leave their tracks in bubble chambers and nuclear emulsions the way elementary electrons and protons do. Indeed, it is only from these tracks that we know of them at all. An analysis of these tracks—their density, their relative directions, their curvatures in magnetic fields, and their branchings—serves to specify their unique identities and their physical properties. See Fig. 13–4 which we will discuss in some detail shortly. As in the case of the transuranic elements, having made them we seek to understand them. Because we believe in the *oneness-of-knowledge*, until we understand these particles and can incorporate them into our theoretical scheme, the strange particles constitute a threat to much we think we now understand about the physical world. Note again the statement of Professor Lee quoted at the beginning of this chapter.

Figure 13–5 represents 32 of the elementary particles in a kind of "periodic table." Masses increase upward from the bottom of the figure.

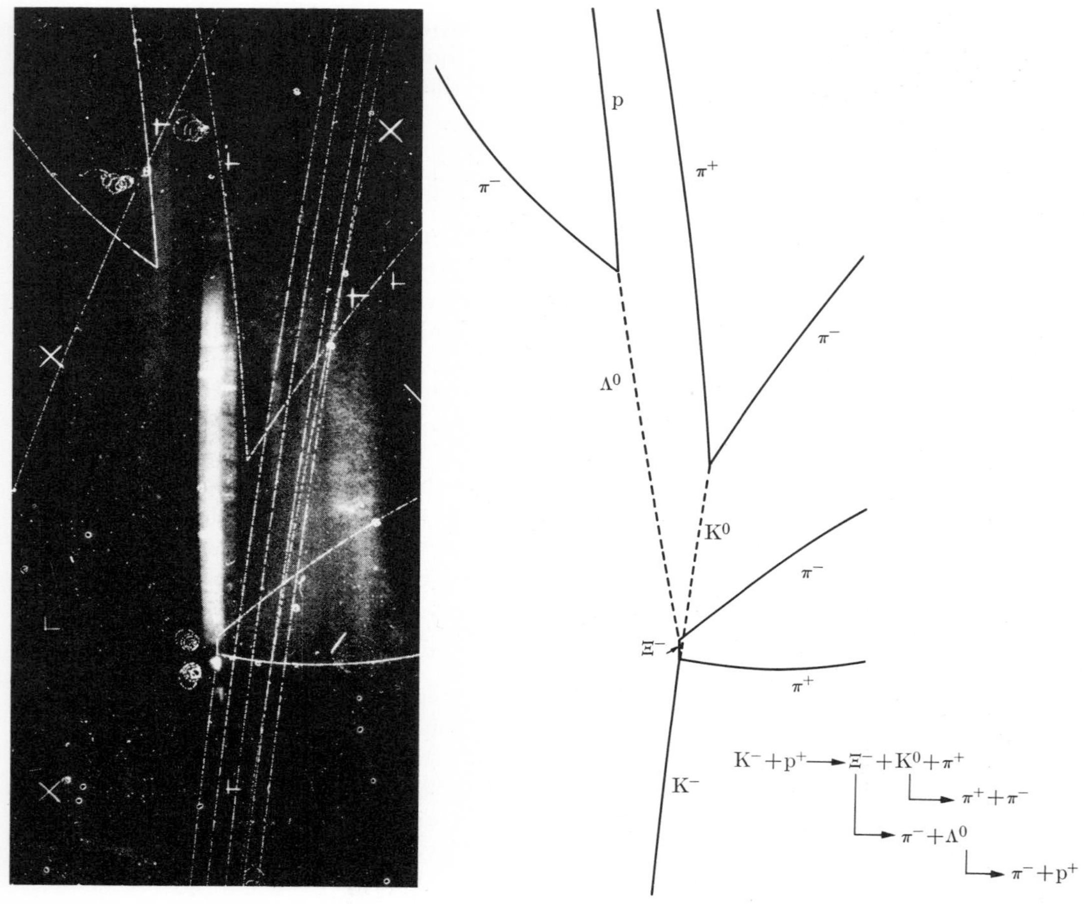

FIG. 13–4
Charged particles produced in K° decay. (Courtesy of Brookhaven National Laboratory)

FIG. 13–5
The mirror symmetry of elementary particles. Only seven of the thirty-two particles shown are intrinsically stable: photons, protons, electrons, neutrinos, and antineutrinos, antiprotons, and positrons. The last two, however, share the usual fate of antiparticles that wander into this world: they are annihilated when they meet their counterparts. Photons and neutral pi-mesons commute freely between both worlds, since each is its own antiparticle—although the neutral pi-meson has little time for traveling because it is the second shortest-lived of all the known particles. A neutron is stable if bound in an atomic nucleus; otherwise it has a half-life of 12 minutes, decaying into a proton, an electron, and a neutrino. Similarly, all unstable particles decay, directly or indirectly, into two or more of the seven stable ions. Their mean lifetimes range from about 10–16 seconds to less than 10^{-16} seconds. (Adapted by permission from Fortune Magazine, copyright 1957 by Time, Inc.)

Group	Subgroup	Mass (electron units)	Strangeness no.	Particles: Positive	Particles: Neutral	Particles: Negative	Antiparticles: Positive	Antiparticles: Neutral	Antiparticles: Negative	Strangeness no.
Baryons	Hyperons	3300	−3			OMEGA Ω^-	OMEGA $\bar{\Omega}^+$			+3
Baryons	Hyperons	2575	−2		XI Ξ^0	XI Ξ^-	XI $\bar{\Xi}^+$	XI $\bar{\Xi}^0$		+2
Baryons	Hyperons	2336	−1	SIGMA Σ^+	SIGMA Σ^0	SIGMA Σ^-	SIGMA $\bar{\Sigma}^+$	SIGMA $\bar{\Sigma}^0$	SIGMA $\bar{\Sigma}^-$	+1
Baryons	Hyperons	2182	−1		LAMBDA Λ^0			LAMBDA $\bar{\Lambda}^0$		+1
Baryons	Nucleons	1836	0	PROTON p	NEUTRON n			ANTI-NEUTRON $\bar{n}$	ANTI-PROTON $\bar{p}$	0
Mesons		970	+1	K^+	K^0			$\bar{K}^0$	$\bar{K}^-$	−1
Mesons		270	0	PI π^+	PI π^0				PI π^-	0
Leptons		207	0			MU μ^-	MU μ^+			0
Leptons		1	0			ELECTRON e^-	POSITRON e^+			0
Leptons		0	0		NEUTRINO ν			ANTI-NEUTRINO $\bar{\nu}$		0
		0	0		PHOTON γ					0

Particles on the left are considered "ordinary" particles whereas those on the right are antiparticles. In Section 5–17 we related the discovery of the positron which is the antiparticle of the electron. As the chart shows, nearly every particle has its antiparticle. There is no significance as to which is which except that the rarer particle is chosen to be the antiparticle. We have discussed the annihilation of positive and negative electrons, and whenever any particle and its antiparticle collide, they annihilate each other. This is the "particle-antiparticle conjugation" referred to by Professor Lee in the passage we quoted.

Although these new particles constitute the major puzzle of nuclear physics, we see that much information has been gathered and a few generalizations have been made which are appropriate to a book of this level. The masses, spins, and charges are known. The mean lifetimes of the unstable particles are known as are many of the decay schemes. Thus, for example, there is the "radioactive" decay

$$\Sigma^0 \rightarrow \Lambda^0 + \gamma.$$

Similarly, a wide variety of elementary particle reactions are known which are produced by bombarding one particle with another. Thus we have

$$\pi^+ + \mathrm{p} \rightarrow \Sigma^+ + \mathrm{K}^+.$$

In this reaction mass-energy and charge are conserved as in any nuclear reaction. Translational momentum and spin are conserved. Another property of the particles, discovered by Murray Gell-Mann and called *strangeness*, is also conserved. Figure 12–5 includes columns headed "Strangeness numbers." Gell-Mann found it possible to assign these numbers in such a way that for all known reactions (as distinguished from mere disintegrations) the total strangeness before and after the reaction are the same. Thus for the reaction given

$$0 + 0 = -1 + 1.$$

Having introduced the symbols of the particles, let us discuss the events depicted in Fig. 13–4 more fully. A high energy heavy meson K^- entered the hydrogen filled bubble chamber from below. Unlike its companions which traversed the chamber without encountering a nuclear collision, this one struck a proton hydrogen nucleus. A reaction took place and the products were Ξ^-, K^0, and π^+, according to the reaction

$$\mathrm{K}^- + \mathrm{p} \rightarrow \Xi^- + \mathrm{K}^0 + \pi^+. \tag{13–1}$$

Since the product particles have about 1000 more electron masses than the initial particles, we know the incident K^- must have had at least 500 MeV of energy. It is obvious from the length of the tracks that the product particles had great kinetic energy; therefore the incident K^- particle must have had far more than 500 MeV of energy.

The π^+ left the region photographed at the right. The charge free K^0 left no track, but it evidently continued in the direction of the original K^- until it produced a π pair at the upper right. The Ξ^- went only a short distance before it disintegrated into a neutral Λ^0 and a π^-. The neutral Λ^0 proceeded upward and to the left leaving no track, but it disintegrated into a proton and a π^-. By sighting along the tracks, it is evident that the magnetic field curved the paths of positive particles one way and the paths of negative particles the other. Now it may seem that a great deal has been read into this photograph, but it must be realized that the photograph contains a vast amount of data. Using conservation of charge, conservation of translational momentum, conservation of angular momentum (spin), conservation of energy, and conservation of strangeness, no other interpretation of this diagram is reasonable. It is through such an analysis of nuclear emulsions and bubble chamber photographs that most of the new particles have been identified and their properties determined.

13-9. CONCLUSION

One way of summarizing what we have said in this book is to point out that we have considered events of higher and higher energy. We started with molecules, whose thermal energies are about 0.025 eV. We discussed the electronic structure of atoms, where energies have magnitudes of a few electron volts. The x-rays that interact with inner atomic electrons have energies of thousands of electron volts, and the energies of nuclear transformations fall in the MeV range. Finally, with our discussion of cosmic rays and events artificially produced with the superaccelerators, we introduced super-high-energy phenomena in the GeV range.

The story we have told has, in general, followed a familiar pattern: a large body of information had been gathered before some genius saw that the data led to new concepts and new generalizations. These interpretations of data have lifted the human intellect so that the past could be seen in perspective and the future could present exciting challenge. Avogadro, Planck, Einstein, Rutherford, Bohr—all stood on the strong shoulders of experimentalists and breathed the heady air of speculation. De Broglie, on the other hand, exemplifies the genius who speculated before the fact.

Just as Rutherford discovered the nucleus of the atom by probing with high-energy alpha particles, today the interior of the nucleus is being probed with particles of still higher energy. A great deal of information has been gathered, and much of it is reasonably well understood. But there are far more questions than answers and the time is ripe for some genius to put the house in order.

Except for mention of the mesons and the neutrino, we have discussed the nucleus as though it involved only protons and neutrons, whereas there have now been identified no less than thirty nuclear particles. We spoke of short-range attractive nuclear forces, potential barriers, and binding energies, but we said almost nothing about the nature of the forces within the nucleus. There are partial theories about these nuclear forces and the thirty or so particles certainly have a lot to do with the ways in which these forces act. But the impressive thing about nuclear forces (if indeed the term force is to be retained) is not what we know about them but what we do not know about them.

With the exception of continued work on the process of fusion, physicists have turned away from devising nuclear reactors, and are back at the task of unraveling the basic problems of the structure of matter. The newest particles found in high-energy processes are completely foreign to our present scheme of things. Let us hope that out of this chaotic riddle will come a profound and simplifying answer. We may be likened to those who knew only Ptolemy's complex description of the solar system. What we need is a Copernicus to assimilate and interpret the data with a generalization which will not only solve the riddle, but lift our sights to levels we cannot now foresee.

REFERENCES

Cook, C. S., *Structure of Atomic Nuclei*, Momentum Book No. 8. Princeton, N. J.: Van Nostrand, 1964.

Frisch, D. H. and A. M. Thorndike, *Elementary Particles*, Momentum Book No. 1. Princeton, N. J.: Van Nostrand, 1964.

Gell-Mann, M. and E. P. Rosenbaum, "Elementary Particles," *Scientific American*, **197,** 72–88 (1957). A very good article.

Hofstadter, R. and L. I. Schiff, *Nucleon Structure*. Stanford, Cal.: Stanford University Press, 1964.

Preston, M. A., *Nuclear Structure*, Resource Letter NS-1 and Selected Reprints. New York: American Institute of Physics, 1965.

Rochester, G. D. and J. D. Wilson, *Cloud Chamber Photographs of the Cosmic Radiation*. London: Pergamon Press, 1952. (New York: Academic Press.) An excellent collection of photographs.

Swartz, C. E., "Resource Letter SAP-1 on Subatomic Particles," *Am. J. Phys.* **34**, 1079 (1966).

Swartz, C. E., *The Fundamental Particles*. Reading, Mass.: Addison-Wesley, 1965.

Winckler, J. R., and D. J. Hofmann, "Resource Letter CR-1 on Cosmic Rays," *Am. J. Phys.* **35** (January, 1967).

PROBLEMS

13–1. A π^0-meson (mass $= 264\ m_e$) at rest decays into two gamma-ray photons. (a) What is the energy of each photon? (b) Why must each photon have the same energy? (c) What is the direction of emission of one photon with respect to the other?

13–2. Calculate the kinetic energy of the μ^--meson (mass $= 207\ m_e$) emitted in the decay of a stationary π^--meson (mass $= 270\ m_e$) according to the equation $\pi^- \rightarrow \mu^- + \bar{\nu}$.

13–3. A proton-antiproton pair is created from a photon. If the kinetic energy of each particle is 15 MeV, what was the energy and wavelength of the photon?

13–4. What is the sum of the strangeness numbers on each side of Eq. (13–1)?

13–5. Determine the change in strangeness number for the following decays: (a) $\Sigma^0 \rightarrow \Lambda + \gamma$, and (b) $\Lambda \rightarrow p^+ + \pi^-$.

13–6. When a stationary target containing protons is bombarded by a proton beam, the reaction

$$p + p \rightarrow n + p + \pi^+$$

takes place. What is the threshold energy for the incident proton? (Mass of the π^+-meson $= 273\ m_e$.

APPENDIXES

APPENDIX 1

A CHRONOLOGY OF THE ATOMIC VIEW OF NATURE

c. 550 B.C. THALES of Miletus (Greece, c. 640–546 B.C.) recorded the attractive properties of rubbed amber and of lodestone.

c. 450 B.C. LEUCIPPUS (Greece) proposed an atomic concept of matter.

c. 400 B.C. DEMOCRITUS of Abdera (Greece, c. 460–357 B.C.), pupil of Leucippus, was the most famous of the atomists in ancient times. He taught: "The only existing things are the atoms and empty space; all else is mere opinion."

c. 335 B.C. ARISTOTLE (Greece, 384–322 B.C.) held that all matter was basically composed of the same continuous primordial stuff.

c. 300 B.C. EPICURUS of Samos (Greece, c. 342–270 B.C.) founded a philosophical system based on the atomism of Democritus.

c. 300 B.C. ZENO of Cition (Greece, c. 336–264 B.C.) founded the Stoic school of philosophy which held that matter, space, etc. were continuous.

c. 60 B.C. TITUS LUCRETIUS CARUS (Rome, c. 96–55 B.C.) attempted to formulate a rational explanation of natural phenomena by extending the beliefs of Democritus and Epicurus. His poem, *De Rerum Natura,* is the most complete record of Greek atomism extant.

The atomism of antiquity was primarily a system of metaphysics. The atomic view of matter in the modern sense was barely introduced in its most elementary form by the beginning of the 19th century.

c. 400 SAINT AUGUSTINE (Aurelius Augustinus) (North Africa, 354–430) was the first to report that the forces exerted by rubbed amber and by lodestone are different properties.

c. 1600 WILLIAM GILBERT (England, 1540–1603) made the first detailed study of magnetism and also showed that, in addition to amber, many other materials can be electrified.

1638 GALILEO GALILEI (Italy, 1564–1642) published *Discorsi e Dimostrazioni Matematiche intorno a due nuove Scienze attenti alla Mecanica e Movimenti locali* (*Discourses and Mathematical Demonstrations concerning Two New Sciences pertaining to Mechanics and Local Motions,* usually contracted to *Two New Sciences*). This account of Galileo's contributions to science establishes him as the founder of dynamics. He was the first to make extensive use of the experimental method to study natural phenomena. From his time on, induction from experiment replaced the teleology of the scholastics as a guiding principle in the organization of the natural sciences.

1650–1700 ROBERT BOYLE (England, 1627–1691), ROBERT HOOKE (England, 1635–1703), and ISAAC NEWTON (England, 1642–1727) gave qualitative explanations of Boyle's law by assuming a kinetic theory of gases.

1675 JEAN PICARD (France, 1620–1682) observed the luminous glow in the Torricellian vacuum of a barometer produced by motion of the mercury when the instrument was carried from place to place.

1675 ISAAC NEWTON (England) developed a corpuscular theory of light.

1676 OLE CHRISTENSEN ROEMER (Denmark, 1644–1710) was the first to show that the velocity of light is finite. His conclusion was based on the variations of the time intervals between consecutive eclipses of one of the moons of Jupiter during the course of the revolution of the earth around the sun.

1678 CHRISTIAN HUYGENS (CHRISTIAAN HUYGHENS) (Netherlands, 1629–1695) developed a wave theory of light in which light was regarded as composed of longitudinal "pulses" consisting of compressions and rarefactions, similar to sound, in an extremely thin, all-pervading medium which he called the *aether*. The concept that light is a periodic wave motion was introduced in about 1750 by Leonard (Leonhard) Euler (Switzerland, Germany, Russia, 1707–1783).

Not only did Huygens correctly account for the refraction of light by transparent bodies by means of spherical emanations (wavelets), but also, by using both spherical and spheroidal wavelets, he became the first one to explain double refraction, a phenomenon that was discovered in 1669 by Erasmus Bartholinus (Denmark, 1625–1692).

1687 ISAAC NEWTON (England, 1642–1727) published *Philosophiae Naturalis Principia Mathematica* (*Mathematical Principles of Natural Philosophy*) which contains the fundamental laws of classical dynamics and the law of gravitation. The synthesis involved in obtaining these laws is one of the greatest achievements of the human mind.

1705 FRANCIS HAUKSBEE (England, d. 1713) made a "powerful" electrostatic generator and discovered the conditions for producing luminous electric discharges in gases.

1728 JAMES BRADLEY (England, 1693–1762) explained the aberration of light from stars by taking the vector sum of the orbital velocity of the earth v and the free-space velocity of light c, and showed that the angle of aberration was a function of the ratio of these velocities, v/c. On this basis, he also showed that the revolution of the earth around the sun correctly accounted for the observed cyclic change in the aberration of starlight. (The Anti-Copernicans, still numerous in the first half of the 18th century, were unable to refute this explanation of the change.) Bradley's work is the first of many instances that *seemed* to show that the value of the velocity of light depends on the motion of the observer.

1731 STEPHEN GRAY (England, 1666/7–1736) discovered the conduction of electricity.

1734 CHARLES FRANCOIS DE CISTERNAY DUFAY (France, 1698–1739) showed that there are two kinds of electrification, resinous and vitreous, and then proposed a two-fluid theory of electric discharge. He also found that the air in the vicinity of a hot body is conducting.

1738 DANIEL BERNOULLI (Switzerland, 1700–1782) was the first to devise a quantitative kinetic theory of gases.

1745 EWALD JURGEN VON KLEIST (Germany, d. 1748) and PEITER VAN MUSSCHENBROEK (Netherlands, 1692–1761) independently made the first capacitors, called Leyden jars.

1752 BENJAMIN FRANKLIN (USA, 1707–1790) experimentally verified the electrical nature of lightning and introduced the one-fluid theory of flow of electricity—from surplus or positive to deficiency or negative. His theory contained the first clear statement of the law of conservation of electric charge.

1753 JOHN CANTON (England, 1718–1772) discovered the facts of electrostatic induction.

1766 HENRY CAVENDISH (England, 1731–1810) discovered hydrogen. Within the next score of years he found the inverse square law of force action between electric charges and other important laws of electricity but, because of excessive shyness, he withheld announcement of his experiments. The great extent of his work was not known until James Clerk Maxwell published Cavendish's papers in 1879.

1785 CHARLES AUGUSTIN COULOMB (France, 1736–1806) determined the law of force action between electric charges.

1789 ANTOINE LAURENT LAVOISIER (France, 1734–1794) published a book containing a well-founded concept of chemical elements and the verification of the law of conservation of matter in chemical reactions.

1791 BRYAN HIGGINS (Ireland, 1737–1820) and WILLIAM HIGGINS (Ireland, c. 1769–1825) reported the first of a series of experiments leading to the laws of chemical combination.

1799 JOSEPH LOUIS PROUST (France, Spain, 1754–1826) established the law of definite proportions for chemical compounds.

1800 ALESSANDRO GUISEPPE ANTONIO ANASTASIO VOLTA (Italy, 1745–1827) made the first voltaic pile (battery), based on his discovery of the fundamental conditions necessary to produce the "animal electricity" that had first been observed in 1780 by Aloisio (or Luigi) Galvani (Italy, 1737–1798).

1801 THOMAS YOUNG (England, 1773–1829) showed that his interference experiments verified the wave theory of light.

1803 JOHN DALTON (England, 1766–1844) published the first of a series of papers introducing atomic weights, establishing the law of multiple proportions, and founding the atomic theory of matter.

1808 JOSEPH LOUIS GAY-LUSSAC (France, 1778–1850) discovered the law of combining volumes of gases.

1810– ETIENNE LOUIS MALUS (France, 1775–1812), DOMINIQUE FRANCOIS JEAN
1875 ARAGO (France, 1786–1853), AUGUSTIN JEAN FRESNEL (France, 1788–1827), JEAN BERNARD LEON FOUCAULT (France, 1819–1868), HIPPOLYTE LOUIS FIZEAU (France, 1819–1896), and MARIE ALFRED CORNU (France, 1841–1902) established conclusively, through many experiments, especially in physical optics, that light is a transverse wave. Several of these men made precise measurements of the velocity of light in various media. In 1818 Arago found that the refraction of a prism for starlight was the same for light incident in the direction of the earth's orbital veloclty v as for that coming in the opposite direction. This unexpected null result was explained that same year by Fresnel's ether-drag theory, which assumed partial ether entrainment in transparent media by an amount depending upon the *first* power of v. This theory appeared fully verified by the measurements of the speed of light in moving water by Fizeau in 1851 and, in 1871, by the observations of the aberration of starlight with a water-filled telescope by GEORGE BIDDELL AIRY (England, 1801–1892).

1811 LORENZO ROMANO AMADEO AVOGADRO (Italy, 1776–1856) introduced Avogadro's hypothesis and differentiated between atoms and molecules.

1813 JONS JACOB BERZELIUS (Sweden, 1779–1848) introduced the present symbols for the chemical elements.

1815 WILLIAM PROUT (England, 1875–1850) proposed that all elements are composed of an integral number of hydrogen atoms.

1815– JOSEPH FRAUNHOFER (Germany, 1787–1826) noted the spectral lines of
1820 several elements, obtained the first grating spectra, and observed the Fraunhofer (absorption) lines in solar spectra.

1819 PIERRE LOUIS DULONG (France, 1785–1838) and ALEXIS THERÈSE PETIT (France, 1791–1821) found the law of constancy of molar specific heat capacities of elements.

1820 HANS CHRISTIAN OERSTED (Denmark, 1777–1851) discovered that an electric current produces a magnetic field. This initiated the study of electromagnetism.

1821 THOMAS JOHANN SEEBECK (Russia, Germany, 1770–1831) discovered thermoelectricity.

1823 ANDRÈ MARIE AMPÈRE (France, 1775–1836) published his mathematical theory of electromagnetism and the laws of magnetic field produced by currents. Some of these laws were also discovered independently by JEAN BAPTISTE BIOT (France, 1774–1862) and FELIX SAVART (France, 1791–1841).

1826 Georg Simon Ohm (Germany, 1787–1854) discovered Ohm's law.

1827 Robert Brown (England, 1773–1858) discovered Brownian movement.

1831 Michael Faraday (England, 1791–1867) and Joseph Henry (USA, 1797–1878) independently discovered electromagnetic induction.

1833 Michael Faraday (England) discovered the laws of electrolysis and introduced the terms "anode" and "cathode."

1835 Joseph Henry (USA) discovered self-induction and, in 1842, oscillatory electric discharge.

1842 Johann Christian Doppler (Austria, 1803–1853) deduced a relation that showed that the observed frequency of waves depends upon the relative motion of the source and the observer.

1842 Julius Robert Mayer (Germany, 1814–1878) calculated the mechanical equivalent of heat theoretically from the specific heats of gases and vaguely proposed a law of conservation of energy based on "*Ex nihilo, nihil fit.*" His work was not published for several years.

1843 James Prescott Joule (England, 1814–1889) published the first of a series of reliable experimental results that showed the constancy of the relation between mechanical energy and heat—a basic step toward the law of conservation of energy.

1847 Hermann Ludwig Ferdinand von Helmoltz (Germany, 1821–1894) proposed the law of conservation of "force" (energy).

1848 William Thomson (Lord Kelvin, 1st Baron) (Ireland, Scotland, 1824–1907) introduced absolute temperature.

1850 Rudolph Julius Emanuel Clausius (Germany, 1822–1888) announced the second law of thermodynamics. Lord Kelvin independently found the same law in 1852.

1850–1900 August Karl Kroenig (Germany, 1822–1879), Rudolph Julius Emanuel Clausius (Germany), James Clerk Maxwell (Scotland, England, 1831–1879), Ludwig Boltzmann (Austria, 1844–1906), and Josiah Willard Gibbs (USA, 1839–1903) developed the kinetic theory of gases and founded statistical mechanics. Maxwell derived his speed distribution law in 1860, Clausius introduced the concept of entropy in 1865, and Boltzmann related entropy to thermodynamic probability in 1877.

1858 Stanislao Cannizzaro (Italy, 1826–1910) resolved the conflicting values of atomic weights by clarifying the terms "atomic," "molecular," and "equivalent" weights.

1859 Gustav Robert Kirchhoff (Germany, 1824–1887) showed that the ratio of the emittance to the absorptance for a given wavelength of radiation is the same for all surfaces at the same temperature, and introduced the concept of cavity (Hohlraum) or blackbody radiation.

1859 HEINRICH GEISSLER (Germany, 1814–1879) and JULIUS PLUECKER (Germany, 1801–1868) discovered the "rays" (now called cathode rays) from the negative electrode in gaseous discharge tubes.

1863 JAMES ALEXANDER REINA NEWLANDS (England, 1837–1898) stated the law of octaves, a limited and elementary form of the periodic table of the elements.

1864 JAMES CLERK MAXWELL (Scotland, England, 1831–1879) wrote *A Dynamical Theory of the Electromagnetic Field*, a paper synthesizing electricity, magnetism, and light. This was probably the greatest work since Newton's *Principia*.

1865 JOSEPH LOSCHMIDT (Germany, 1821–1895) used the equations of the kinetic theory of gases to make the first determination of Avogadro's number and of molecular diameters.

1869 DMITRI IVANOVICH MENDELEEV (Russia, 1834–1907) and JULIUS LOTHAR MEYER (Germany, 1830–1895) independently introduced the periodic table of the elements, a concise summary of years of experimental and theoretical chemistry. The table is both mnemonic and heuristic.

1869 JOHANN WILHELM HITTORF (Germany, 1824–1914) observed the deflection of rays from the cathode in a discharge tube, by means of a magnetic field.

1871 CROMWELL FLEETWOOD VARLEY (England, 1828–1883) found that the rays from the cathode are negatively charged.

1876 EUGEN GOLDSTEIN (Germany, 1850–1930) introduced the name "cathode rays" and began experiments leading eventually to the discovery of the positive counterpart, Kanalstrahlen (channel or canal rays). In 1886 he suggested that the aurora is due to cathode rays from the sun.

1877 WILLIAM RAMSAY (England, 1852–1916) and, independently, JOSEPH DELSAULX (France, 1828–1891) and IGNACE J. J. CARBONELLE (France, 1829–1889) advanced the first rather complete qualitative explanation of Brownian movement by attributing it to molecular impact. Some years later Ramsay discovered several of the noble gases, and made important contributions to the study of radioactivity. He was awarded the Nobel prize for chemistry in 1904.

1879 EDWIN HERBERT HALL (USA, 1855–1938) discovered the existence of a potential difference between the opposite edges of a metal strip carrying a longitudinal electric current, when the plane of the strip is set normal to a magnetic field. This is called the Hall effect.

1879 WILLIAM CROOKES (England, 1832–1919) began a long series of brilliant experiments on the discharge of electricity through gases.

1879 JOSEF STEFAN (Austria, 1835–1893) announced Stefan's law, which gives the total energy radiated by a blackbody. This was the first successful attempt to connect absolute temperature and radiation.

1881 JULIUS ELSTER (Germany, 1854–1920) and HANS GEITEL (1855–1923) started a long, systematic investigation of electrical effects produced by incandescent solids.

1883 THOMAS ALVA EDISON (USA, 1847–1931) discovered the Edison effect, the emission of negative electricity from incandescent filaments in a vacuum.

1884 JOHANN JAKOB BALMER (Switzerland, 1825–1898) found an empirical wavelength relation for a spectral series of hydrogen. This was the first series equation found for any spectrum.

1887 SVANTE AUGUST ARRHENIUS (Sweden, 1859–1927) conclusively established the ion dissociation theory of electrolytes which grew from suggestions made by Clausius in 1857. Arrhenius was awarded the Nobel prize for chemistry in 1903.

1887 ALBERT ABRAHAM MICHELSON (Germany, USA, 1852–1931) and EDWARD WILLIAMS MORLEY (USA, 1838–1923) performed the first precision experiment that showed that the earth has no ether drift. In a letter to *Nature* in 1879 Maxwell pointed out that evidence of ether drift had to be sought in second-order effects—those depending on v^2/c^2. These are involved in interference methods. The first trial by Michelson in 1881 gave inconclusive results. Michelson was awarded the Nobel prize for physics in 1907.

1887 HEINRICH RUDOLPH HERTZ (Germany, 1857–1894) discovered the photoelectric effect while verifying the existence of the electromagnetic waves predicted by Maxwell.

1888 WILHELM HALLWACHS (Germany, 1859–1922) showed that only negative charges are emitted in the photoelectric effect.

1890 JOHANNES ROBERT RYDBERG (Sweden, 1854–1919) found an empirical wavelength relation for complex series of spectral lines.

1891 JOHNSTONE STONEY (England, 1826–1911) introduced the name "electron" for an elementary unit of negative charge in electrolysis.

1892 GEORGE FRANCIS FITZGERALD (Ireland, 1851–1901) and HENDRIK ANTOON LORENTZ (Netherlands, 1853–1929) independently made the *ad hoc* assumption of contraction of length to account for the null result of the Michelson-Morley experiment. As Lorentz successively refined his electric theory of matter to conform with the results of new experiments, he obtained the space and time transformations later derived by Einstein. For later work Lorentz was awarded the Nobel prize for physics jointly with P. Zeeman in 1902.

1893 WILHELM WIEN (Germany, 1864–1928) derived his blackbody radiation displacement law. His blackbody radiation law was announced in 1896. He was awarded the Nobel prize for physics in 1911.

1893 Philipp Eduard Anton von Lenard (Hungary, Germany, 1862–1947) investigated cathode rays by passing them through a Lenard window (thin-window) tube into air. For this and later work he was awarded the Nobel prize for physics in 1905.

1895 Jean Baptiste Perrin (France, 1870–1942) demonstrated conclusively that cathode rays are negatively charged. For this and later work he was awarded the Nobel prize for physics in 1926.

1895 Wilhelm Conrad Roentgen (Germany, 1845–1923) discovered x-rays. He was awarded the Nobel prize for physics in 1901.

1896 Antoine Henri Becquerel (France, 1852–1908) discovered the radioactivity of uranium. He was awarded the Nobel prize for physics jointly with the Curies in 1903.

1896 Pieter Zeeman (Netherlands, 1865–1943) observed the splitting of spectral lines radiated by excited atoms in an intense magnetic field. The early theory of this effect was derived by H. A. Lorentz (Netherlands). They were jointly awarded the Nobel prize for physics in 1902.

1896 Oliver Lodge (England, 1851–1940) reported that, contrary to expectations, there was no detectable ether drag on light passing between two large closely spaced disks of steel rotating at enormous speeds even when the disks were strongly magnetized or electrified.

1897 Joseph John Thomson (England, 1856–1940) determined q/m for cathode rays. He was awarded the Nobel prize for physics in 1906.

1897 Ernest Rutherford (Lord Rutherford of Nelson, 1st Baron) (New Zealand, Canada, England, 1871–1937) showed that the radiation from uranium was complex, consisting of "soft" (alpha) and "hard" (beta) rays. He was awarded the Nobel prize for chemistry in 1908.

1898 Pierre Curie (France, 1859–1906) and Marie Sklodowska Curie (Poland, France, 1867–1934) isolated radium and polonium. They were awarded the Nobel prize for physics jointly with H. Becquerel in 1903. Marie Curie was also awarded the Nobel prize for chemistry in 1911.

1899 Henri Becquerel (France), Stefan Meyer (Austria, 1872–1949) and Egon von Schweidler (Austria, 1873–1948), and Frederick Otto Giesel (Germany, 1852–1927) independently observed the magnetic deflection of alpha and beta rays.

1899 Julius Elster (Germany) and Hans Geitel (Germany) determined the law of radioactive decay experimentally.

1899 Philipp Lenard (Germany) showed that photoelectric emission is due to electrons.

1899 J. J. Thomson (England) showed that the Edison effect is due to electrons.

1899 Otto Lummer (Russia, Germany, 1860–1925) and Ernst Georg Pringsheim (Germany, 1881–1917), and also Ferdinand Kurlbaum (Germany, 1857–1927) and Heinrich Rubens (Germany, 1865–1922) made precise measurements of the intensity-wavelength distribution of blackbody radiation.

1900 John William Strutt (Lord Rayleigh, 3rd Baron), (England, 1842–1919) announced a blackbody radiation law. The derivation of this law was re-examined in collaboration with James Hopwood Jeans (England, 1877–1946) and, after publication in 1905, became known as the Rayleigh-Jeans law. Rayleigh was awarded the Nobel prize for physics in 1904.

1900 Max Karl Ernst Ludwig Planck (Germany, 1858–1947) introduced the quantum theory of radiation—a revolutionary concept. He was awarded the Nobel prize for physics in 1918.

1900 Henri Becquerel (France) showed that beta rays are identical with cathode-ray corpuscles.

1900 Paul Villard (France, 1860–1934) discovered gamma rays.

1902 Philipp Lenard (Germany) discovered photoelectric threshold frequency and also that the kinetic energy of photoelectrons is independent of the intensity of the incident light.

1903 Frederic Thomas Trouton (Ireland, England, 1863–1922) and H. R. Noble (England) were unable to observe any orienting torque on a suspended, charged capacitor as predicted on the basis of an ether drift (a second-order effect).

1903 Ernest Rutherford (England) and Frederic Soddy (England, 1877–1956) showed that every radioactive process is a transmutation of elements. Soddy was awarded the Nobel prize for chemistry in 1921.

1903 William Crookes (England), and, independently, Julius Elster (Germany) and Hans Geitel (Germany) found that the luminescence produced when alpha particles strike zinc sulfide consists of discrete flashes of light or scintillations. This led to a method of counting individual alpha particles.

1904 William Ramsay (England) and Frederic Soddy (England) discovered the remarkable occurrence of helium in all radium compounds.

1904 Dewitt Bristol Brace (USA, 1859–1905) found no trace of the double refraction predicted for an isotropic transparent body when it is rotated from parallel to the ether drift to normal to it (a second-order effect). This type of experiment had been suggested by the elder Lord Rayleigh.

1904 John Ambrose Fleming (England, 1849–1945) applied the Edison effect to make the first thermionic valve ("radio" tube).

1904 Maryan von Smoluchowski (Austria, 1872–1919) proposed a statistical theory of Brownian movement.

1905 ALBERT EINSTEIN (Germany, Switzerland, USA, 1879–1955) completed the statistical theory of Brownian movement, introduced the quantum explanation of the photoelectric effect, and announced the *special* theory of relativity. He was awarded the Nobel prize for physics in 1921. The citation stated that the award was "for his contributions to mathematical physics, and especially for his discovery of the law of the photoelectric effect."

1905 EGON VON SCHWEIDLER (Austria) derived the law of radioactive decay from probability theory—not obtainable from causality.

1906 OWEN WILLANS RICHARDSON (England, 1879–1959) began a long series of important investigations on the emission of electricity from hot bodies (thermionic emission). He was awarded the Nobel prize for physics in 1928.

1906 LEE DE FOREST (USA, 1873–1961) made the first audion (triode) by introducing a grid into a Fleming valve.

1907–1912 J. J. THOMSON (England) devised methods of positive-ray analysis. This was the beginning of mass spectroscopy.

1908 WALTER RITZ (Switzerland, 1878–1909) announced the combination principle for computing the frequencies of spectral lines.

1908 LOUIS CARL HEINRICH FRIEDRICH PASCHEN (Germany, 1865–1947) experimentally verified the existence of a spectral series of hydrogen in the near infrared predicted by the Rydberg-Ritz relation.

1908 CHARLES GLOVER BARKLA (England, 1877–1944) discovered from absorption experiments that the secondary x-rays of various elements are composed of groups of characteristic x-rays which he called the K, L, and M radiations, and demonstrated the polarization of x-rays. He was awarded the Nobel prize for physics in 1917.

1908 JEAN PERRIN (France) verified experimentally the several equations for Brownian movement, obtained good values of Avogadro's number, and showed that equipartition of energy held for small particles suspended in a stationary liquid.

1908 HERMANN MINKOWSKI (Lithuania, Germany, 1864–1909) developed a geometrical interpretation of the special theory of relativity in which time and the three space coordinates all had the same validity in a four-dimensional continuum.

1908–1910 ALFRED HEINRICH BUCHERER (Germany, 1863–1927), E. HUPKA (Germany), and CHARLES EUGENE GUYE (France, 1866–1942) and SIMON RATNOWSKY (Russia, Switzerland, 1884–1945) independently made precision measurements of the mass of an electron as a function of its velocity. The results verified the Lorentz-Einstein mass variation relation.

1909 GUGLIELMO MARCONI (Italy, 1874–1937) and CARL FERDINAND BRAUN (Germany, 1850–1918) were jointly awarded the Nobel prize for physics—

the former for combining the basic knowledge about Hertzian waves to produce wireless telegraphy, and the latter for the study, production, and use of electrical oscillators. Braun also developed the Braun tube, called the "cathode-ray" tube in the USA.

1909 ERNEST RUTHERFORD (England) and THOMAS ROYDS (England, 1884–1955) showed that alpha particles are doubly ionized helium atoms.

1909–1910 T. WULF (France) observed the rate of leak of charge from a highly insulated electroscope placed at the top of the Eiffel Tower, and ALBERT GOCKEL (Switzerland, 1860–1927) studied the same effect in balloon ascents up to 4500 meters. Both found the leakage rate greater than at the surface of the earth. Their results were unexpected because the effect at ground level had been ascribed to local radioactivity of the soil.

1909–1911 ROBERT ANDREWS MILLIKAN (USA, 1868–1953) established the law of multiple proportions for electric charges and made the first precise determination of the electronic charge. He was awarded the Nobel prize for physics in 1923.

1910–1912 VICTOR FRANZ HESS (Austria, USA, 1883–1964) and WERNER KOLHOERSTER (Austria, 1887–1945) discovered cosmic rays. Hess was awarded the Nobel prize for physics jointly with C. D. Anderson in 1936.

1911 PETER JOSEPH WILHELM DEBYE (Netherlands, Switzerland, Germany, USA, 1884–1966) used the quantum theory to obtain a rather complete theory of specific heats, and later applied the quantum concept to many problems in physical chemistry. He was awarded the Nobel prize for chemistry in 1936.

1911–1913 ERNEST RUTHERFORD (England), HANS GEIGER (Germany), and ERNEST MARSDEN (England, b. 1889) showed that a nuclear model of the atom was required to explain their experiments on alpha-particle scattering by thin metal foils.

1911 CHARLES THOMSON REES WILSON (Scotland, England, 1869–1959) made the first expansion cloud chamber. This is the most important device in nuclear physics. He was awarded the Nobel prize for physics jointly with A. H. COMPTON in 1927.

1912 MAX FELIX THEODOR VON LAUE (Germany, 1879–1960) with WALTER FRIEDRICH (Germany, b. 1883) and PAUL C. M. KNIPPING (Germany, 1883–1935) established the wave nature of x-rays by crystal diffraction. Laue was awarded the Nobel prize for physics in 1914.

1912 HANS GEIGER (Germany, 1882–1945) and JOHN MITCHELL NUTTALL (England, 1890–1958) obtained an empirical law relating the energy of an emitted alpha particle to the disintegration constant of the parent nucleus.

1913 HANS GEIGER (Germany) published a detailed description of the point-discharge counter tube which was developed from a simpler form first made in 1908. This instrument was greatly improved in 1928.

1913 Frederick Soddy (England) and Kasimir Fajans (Poland, Germany, b. 1887) announced the laws of displacement in the periodic table for elements undergoing radioactive decay. Soddy introduced the term "isotopes."

1913 George Charles de Hevesy (Hungary, Germany, Sweden, 1885–1966) and Fritz Adolf Paneth (Austria, 1887–1959) used radium-D, an isotope of lead, to study the solubility and the chemistry of lead compounds. This was the first use of an isotope as a tracer element. Hevesy was awarded the Nobel prize for chemistry in 1943.

1913 Niels Henrik David Bohr (Denmark, 1885–1962) developed the first successful theory of atomic structure. He was awarded the Nobel prize for physics in 1922.

1913 Johannes Stark (Germany, 1874–1957) observed the splitting of spectral lines radiated by excited atoms in an intense electric field. He was awarded the Nobel prize for physics in 1919.

1913 James Franck (Germany, USA, 1882–1964) and Gustav Hertz (Germany b. 1887) supported the Bohr atomic theory with their measurements of ionization and resonance potentials. They were awarded the Nobel prize for physics in 1925.

1913 William Henry Bragg (England, 1862–1942) and son, William Lawrence Bragg (Australia, England, b. 1890), studied x-ray "reflection" from crystals and devised an x-ray spectrometer. They were awarded the Nobel prize for physics in 1915.

1914 Henry Gwyn Jeffrey Moseley (England, 1884–1915) made x-ray spectrograms of the elements and established the identity of the ordinal number of an element in the periodic table with its nuclear charge (atomic number).

1914 Karl Manne Georg Siegbahn (Sweden, b. 1886) began a long series of pioneer researches in the theory and application of precision x-ray spectroscopy. He was awarded the Nobel prize for physics in 1924.

1915 Arnold Johannes Wilhelm Sommerfeld (Germany, 1868–1951) improved the Bohr atomic model by introducing elliptical orbits and relativistic effects.

1915 William Duane (USA, 1872–1935) and Franklin Livingston Hunt (USA, b. 1883) showed that the short-wavelength limit of emitted x-radiation is determined by the quantum theory.

1915 Albert Einstein (Germany, USA) announced the *general* theory of relativity. It considers the observations of phenomena on accelerated reference frames.

1916 R. A. Millikan (USA) experimentally verified Einstein's photoelectric equation.

1916 P. J. W. DEBYE (Netherlands, Switzerland, USA) PAUL SCHERRER (Switzerland, b. 1890), and, independently, ALBERT WALLACE HULL (USA, 1880–1966) obtained the first x-ray powder diffraction patterns.

1916 THEODORE LYMAN (USA, 1874–1954) found the Lyman series lines predicted by Bohr's theory of the hydrogen atom. Lyman had observed at least one of these lines as early as 1906.

1919 ERNEST RUTHERFORD (England) produced hydrogen and oxygen by alpha-particle bombardment of nitrogen, the first "man-made" transmutation of an element.

1919 FRANCIS WILLIAM ASTON (England, 1877–1945) made the first high-precision determinations of isotopic masses. He was awarded the Nobel prize for chemistry in 1922.

1919 The observations made during a total solar eclipse in this year by an expedition from the Royal Astronomical Society and the Royal Society of London confirmed the deviation of starlight in the gravitational field of the sun as predicted by the general theory of relativity. The strongest support for this theory came later from the agreement between the calculated and observed values of the precession of the perihelion of Mercury.

1921 OTTO STERN (Germany, USA, b. 1888) and WALTER GERLACH (Germany, b. 1889) verified the space quantization of silver atoms in a magnetic field and measured their magnetic moment. Stern was awarded the Nobel prize for physics in 1943.

1923 ARTHUR HOLLY COMPTON (USA, 1892–1962) discovered the Compton effect, which showed that a photon has momentum. He was awarded the Nobel prize for physics jointly with C. T. R. WILSON in 1927.

1924 EDWARD VICTOR APPLETON (England, 1892–1965) began a series of experiments that established the existence and properties of ionized layers in the high atmosphere. Such layers had been postulated in 1902 by ARTHUR EDWIN KENNELLY (India, USA, 1861–1939) and, independently, by OLIVER HEAVISIDE (England, 1850–1925) to account for long-distance wireless telegraphy. Appleton was awarded the Nobel prize for physics in 1947.

1924 SATYENDRANATH BOSE (India, b. 1894) and A. EINSTEIN independently developed the statistics "obeyed" by bosons, a collective name for photons, nuclei of even mass number, and certain other particles.

1924 LOUIS VICTOR, DUC DE BROGLIE (France, b. 1892) introduced the concept of de Broglie waves, the beginning of the wave theory of matter. He was awarded the Nobel prize for physics in 1929.

1925 WALTER M. ELSASSER (Germany, USA, b. 1904) predicted from de Broglie theory that electrons could be diffracted by crystals.

1925 Charles Drummond Ellis (England, b. 1895) and W. A. Wooster (England) established that in a number of elements the emission of either an alpha or a beta particle precedes the radiation of gamma rays, and thus the latter should be associated with the daughter product, not with the parent.

1925 Pierre Victor Auger (France, b. 1899) discovered a type of energy transition in which an atom goes from a higher to a lower state by ejecting one of its own electrons, without the emission of electromagnetic radiation.

1925 George Eugene Uhlenbeck (Java, Netherlands, USA, b. 1900) and Samuel Abraham Goudsmit (Netherlands, USA, b. 1902) introduced spin and magnetic moment of the electron into atomic theory.

1925 Wolfgang Pauli (Austria, Switzerland, 1900–1958) announced the exclusion principle. He was awarded the Nobel prize for physics in 1945.

1925 Patrick Maynard Stuart Blackett (England, b. 1897) obtained the first cloud-chamber tracks of the induced transmutation of nitrogen and of other elements, and later made many cosmic-ray studies. He was awarded the Nobel prize for physics in 1948.

1925 Max Born (Bermany, b. 1882), Werner Karl Heisenberg (Germany, b. 1901), and Pascual Jordan (Germany, b. 1902) developed quantum mechanics. Later, Born originated the statistical interpretation of wave mechanics, and he was awarded the Nobel prize for physics jointly with W. Bothe in 1954.

1926 Erwin Schroedinger (Austria, Ireland, 1887–1961) proposed the wave-mechanical theory of the hydrogen atom. He was awarded the Nobel prize for physics jointly with P. A. M. Dirac in 1933.

1926 Enrico Fermi (Italy, USA, 1901–1954) and Paul Adrien Maurice Dirac (England, b. 1902) independently developed the statistics "obeyed" by fermions, a collective name for nuclei of odd mass number, some particles, and electrons, particularly the electron gas in a conductor. Each was awarded the Nobel prize for work listed later in this chronology.

1926 Eugene Paul Wigner (Hungary, USA, b. 1902) published the first of a long series of important papers on the application of group theory in quantum mechanics. He was awarded the Nobel prize for physics jointly with Maria Mayer and J. H. D. Jensen in 1963.

1927 Werner Heisenberg (Germany) announced the "Unbestimmtheit Prinzip" (indeterminacy or uncertainty principle). He was awarded the Nobel prize for physics in 1932.

1927 Clinton Joseph Davisson (USA, 1881–1958), and Lester Halbert Germer (USA, b. 1896) obtained electron diffraction from single crystals, and George Paget Thomson (England, b. 1892) obtained powder diffraction patterns using electrons. Their work verified the existence of de Broglie waves. Davisson and Thomson were awarded the Nobel prize for physics in 1937.

1928 Edward Uhler Condon (USA, b. 1902) and Ronald Wilfrid Gurney (England, USA, 1898–1953) and, independently, George Gamow (Russia, USA, b. 1904) solved the nuclear problem of alpha-particle emission by means of wave mechanics and derived the Geiger-Nuttall law.

1928 P. A. M. Dirac (England) developed relativistic quantum mechanics and predicted the existence of the positron. He was awarded the Nobel prize for physics jointly with E. Schroedinger in 1933.

1928 Chandrasekhara Venkata Raman (India, b. 1888) discovered the Raman effect. This is the presence, in light scattered from molecules, of frequencies differing from that of the incident light by amounts characteristic of the scattering substance and independent of the incident frequency. He was awarded the Nobel prize for physics in 1930.

1928 Dmitri Vladimirovich Skobeltsyn (Russia, b. 1892) obtained the first cloud-chamber photographs of cosmic rays. These showed that the rays either were, or produced, many charged, high-energy particles.

1928 Hans Geiger (Germany) and W. Mueller (Germany) developed the Geiger point counter (1913) into a greatly improved form, called the Geiger-Mueller counter.

1928 Walther Wilhelm Georg Franz Bothe (Germany, 1891–1957) and W. Kolhoerster (Austria) applied G-M tubes to make coincidence counters and other ingenious devices for cosmic-ray study. Bothe was awarded the Nobel prize for physics jointly with M. Born in 1954.

1929 Otto Stern (Germany) obtained crystal diffraction of a beam of helium atoms.

1930 W. Bothe (Germany) and H. Becker (Germany) observed a puzzling penetrating "radiation" from beryllium bombarded with alpha particles.

1930 Jacob Clay (Netherlands, b. 1882) discovered that cosmic-ray intensity decreased in going toward the geomagnetic equator. This latitude effect was investigated exhaustively by A. H. Compton, R. A. Millikan, and others.

1930–1934 Fredrik Carl Muelertz Stoermer (Norway, 1874–1957) applied his theory of the motion of charged particles in the magnetic field of the earth, originally developed to account for the aurora borealis, to cosmic rays. This theory of the cause of geomagnetic effects was greatly expanded in 1933 by Georges Lemaitre (Belgium, 1894–1966) and Manuel Sandoval Vallarta (Mexico, b. 1899). Further theoretical work was done by William Francis Gray Swann (England, USA, 1884–1962) and others.

1930–1933 Ira Forry Zartman (USA, b. 1899) and C. C. Ko (China, USA) experimentally verified the Maxwell distribution law of molecular speeds.

1931 Thomas Hope Johnson (USA, b. 1899) obtained crystal diffraction of a beam of hydrogen atoms.

1931 Robert Jemison Van de Graaff (USA, 1901–1967) constructed the first reliable, high-voltage, electrostatic generator for nuclear research.

1931 Wolfgang Pauli (Austria, Switzerland) proposed a hypothesis of beta-decay processes postulating that a "new," small, neutral particle was emitted simultaneously with the electron.

1931 Harold Clayton Urey (USA, b. 1893), Ferdinand Graft Brickwedde (USA, b. 1903), and George Moseley Murphy (USA, b. 1903) discovered deuterium and made the first heavy water. Urey was awarded the Nobel prize for chemistry in 1934.

1932 Roy James Kennedy (USA, b. 1897) and Edward Moulton Thorndike (USA, b. 1905) sought to detect the ether drift with a very stable and refined form of interferometer having arms of unequal length. Both the length and the time transformations of the special theory of relativity had to be used to account for the null result.

1932 Ernest Orlando Lawrence (USA, 1901–1958) and Milton Stanley Livingston (USA, b. 1905) made the first cyclotron. Lawrence was awarded the Nobel prize for physics in 1939.

1932 John Douglas Cockcroft (England, b. 1897) and Ernest Thomas Sinton Walton (Ireland, b. 1903) accomplished the transmutation of lithium by bombarding it with high-energy protons, and so obtained the first direct verification of Einstein's law of mass-energy equivalence. This was also the first time a high-voltage accelerator was used successfully to produce a nuclear reaction. They were awarded the Nobel prize for physics in 1951.

1932 James Chadwick (England, b. 1891) discovered the neutron. This particle accounted for Bothe and Becker's penetrating "radiation." He was awarded the Nobel prize for physics in 1935.

1932 Bruno Benedetto Rossi (Italy, USA, b. 1905) found an initial increase with thickness in the cosmic-ray intensity "transmitted" by an absorber and explained this by cosmic-ray showers. A transition to decreasing intensities was observed beyond a certain thickness.

1932 Carl David Anderson (USA, b. 1905) discovered the positron during cosmic-ray research. He was awarded the Nobel prize for physics jointly with V. F. Hess in 1936.

1933 P. M. S. Blackett (England) and G. P. S. Occhialini (England) obtained the first cloud-chamber photographs of electron-positron pair production.

1933 Jean Valentin Thibaud (France, 1901–1960) and Frederic Joliot (France, 1900–1958) observed the radiation produced by electron-positron annihilation. They also showed that the mass of the positron is equal to that of the electron.

1933 T. H. JOHNSON (USA) and JABEZ CURRY STREET (USA, b. 1906) observed that the cosmic-ray intensity from the west exceeded that from the east. This east-west asymmetry shows that there is an excess of positively charged particles in the primary cosmic-ray beam.

1934 PAVEL ALEKSEJEVIC CHERENKOV (Russia, b. 1904) observed the weak, bluish glow in transparent substances when irradiated with high-energy beta particles. The theory of this Cherenkov radiation was given by IGOR JEVGENEVIC TAMM (Russia, b. 1895) and ILYA MICHAJLOVIC FRANK (Russia, b. 1908) three years later. These three scientists were jointly awarded the Nobel prize for physics in 1958.

1934 S. MOHOROVICIC (Yugoslavia) predicted the existence of a transitory non-nuclear "element" (later called positronium) preceding electron-positron annihilation.

1934 IRENE JOLIOT-CURIE (France, 1897–1956) and FREDERIC JOLIOT (France) discovered artificial (induced) radioactivity. They were awarded the Nobel prize for chemistry in 1935.

1934 ENRICO FERMI (Italy, USA) developed Pauli's theory of beta decay and named the "new" particle the neutrino (little neutron). It is postulated in Fermi's theory that the neutron is radioactive, disintegrating into a proton with the formation of an electron and a neutrino just before beta emission. He also began a series of experiments in collaboration with EDUARDO AMALDI (Italy, b. 1908), OSCAR D'AGOSTINO (Italy), FRANCO RASETTI (Italy, USA, b. 1901) and EMILIO GINO SEGRÈ (Italy, USA, b. 1905) to produce transuranic elements by irradiating uranium with neutrons. They were granted a patent on the graphite moderator in 1955. Fermi was awarded the Nobel prize for physics in 1938.

1935–1939 ISIDOR ISAAC RABI (Austria, USA, b. 1898) made precise determinations of nuclear magnetic moments in beams of atoms by his radiofrequency resonance method. He was awarded the Nobel prize for physics in 1944.

1935 HIDEKI YUKAWA (Japan, b. 1907) announced his theory of nuclear binding forces involving the postulate of a particle having a mass intermediate between that of the electron and the proton. He was awarded the Nobel prize for physics in 1949.

1936 C. D. ANDERSON (USA) and SETH HENRY NEDDERMEYER (USA, b. 1907) discovered, during cosmic-ray research, a particle of the type postulated by Yukawa. They called it the "mesotron" (later changed to "meson").

1936 MARIETTA BLAU (Austria, b. 1894) was the first to use nuclear track plates.

1937 NIELS BOHR (Denmark) introduced the liquid-drop model of the nucleus.

1938 IRENE JOLIOT-CURIE (France) and PAVLE SAVITCH (Yugoslavia, b. 1908) found indications of the existence of lanthanum in uranium after it was irradiated with neutrons.

1938 Otto Hahn (Germany, b. 1879) and Fritz Strassmann (Germany, b. 1902) discovered that bombarding uranium with neutrons produces alkali earth elements. Hahn was awarded the Nobel prize for chemistry in 1944.

1939 Lise Meitner (Austria, Germany, Sweden, b. 1878) and Otto Richard Frisch (Austria, Germany, England, b. 1904) proposed nuclear splitting to explain Hahn's results on the disintegration of uranium by neutrons and predicted the release of an enormous amount of energy per fission.

1939 Niels Bohr (Denmark) and John Archibald Wheeler (USA, b. 1911) developed the theory of nuclear fission.

1939 Hans Albrecht Bethe (Germany, USA, b. 1906) and Carl Friedrich von Weizsaecker (Germany, b. 1912) independently proposed two sets of nuclear reactions to account for stellar energies: the carbon cycle and the proton-proton chain.

1940 John Ray Dunning (USA, b. 1907), Eugene Theodore Booth (USA, b. 1912), and Aristid V. Grosse (Russia, USA, b. 1905) showed that it is U^{235}, the less abundant isotope of uranium, that is fissioned by slow neutrons.

1940 Louis Leprince-Ringuet (France, b. 1901) obtained the first cloud-chamber photograph of a meson-electron collision, from which the mass of the meson could be deduced.

1940 Donald William Kerst (USA, b. 1911) made the first betatron, an induction type accelerator.

1940 Edwin Mattison McMillan (USA, b. 1907) and Philip Hague Abelson (USA, b. 1913) produced the first transuranic element, neptunium; and Glenn Theodore Seaborg (USA, b. 1912), Edwin Mattison McMillan (USA), Joseph William Kennedy (USA, b. 1917), and Arthur Charles Wahl (USA, b. 1917) prepared the second transuranic element, plutonium. McMillan and Seaborg were awarded the Nobel prize for chemistry in 1951.

1942 Enrico Fermi (Italy, USA), Leo Szilard (Hungary, Germany, USA, 1898–1964), and associates built the first successful self-sustaining fission reactor. It was first put into operation on December 2 and operated at a power level of one-half watt. It was located in Chicago, Illinois.

1945 J. Robert Oppenheimer (USA, b. 1904), then Director of the Los Alamos Scientific Laboratory, and the many scientists engaged there in a "crash" program of basic nuclear research and development saw the culmination of their work in the detonation of the first nuclear bomb at Almagordo, New Mexico, on July 16.

1945 E. M. McMillan (USA) and Vladimir Iosifovich Veksler (Russia, 1907–1966) independently proposed the principle of the synchrotron, the type of accelerator which produces very high-energy particles, as in the Cosmotron, Bevatron, etc.

1946 FELIX BLOCH (Switzerland, USA, b. 1905) devised the magnetic induction method and, independently, EDWARD MILLS PURCELL (USA, b. 1912) originated the magnetic resonance absorption method for determining nuclear magnetic moments, using liquids or solids in bulk (not beams). This led to the nuclear resonance spectrometer. They were awarded the Nobel prize for physics in 1952.

1947 POLYKARP KUSCH (Germany, USA, b. 1911) made high-precision determinations of the magnetic moment of the electron and found a small but theoretically significant difference between the predicted value and the experimental results. Kusch was awarded the Nobel prize for physics jointly with W. R. LAMB in 1955.

1947 WILLIS EUGENE LAMB, JR., (USA, b. 1913) and ROBERT E. RETHERFORD (USA) observed, during the course of spectral measurements of the fine structure of hydrogen in the microwave region, a small displacement (the "Lamb shift") of an energy level from its theoretical position as predicted by Dirac's quantum theory of the electron. Lamb was awarded the Nobel prize for physics jointly with P. KUSCH in 1955.

1947 H. A. BETHE (Germany, USA) and, independently, JULIAN SEYMOUR SCHWINGER (USA, b. 1918) explained the discrepancies found by KUSCH and by LAMB as resulting from an interaction of electrons with the radiation field. Schwinger was awarded the Nobel prize for physics jointly with R. P. FEYNMAN and S. TOMONAGA in 1965.

1947 HARTMUT PAUL KALLMANN (Germany, USA, b. 1896) and, independently, JOHN WESLEY COLTMAN (USA, b. 1915) and FITZ-HUGH BALL MARSHALL (USA, b. 1912) developed scintillation counters.

1947 CECIL FRANK POWELL (England, b. 1903), G. P. S. OCCHIALINI (England), and CESARE MANSUETO GIULIO LATTES (Brazil, England, b. 1924) discovered the pi-meson. Powell was awarded the Nobel prize for physics in 1950.

1947 GEORGE DIXON ROCHESTER (England, b. 1908) and CLIFFORD CHARLES BUTLER (England, b. 1922) discovered V-particles and hyperons.

1948 EUGENE GARDNER (USA, 1913–1950) and C. M. G. LATTES (Brazil, England, USA) were the first to produce mesons in the laboratory.

1948–1950 WILLARD FRANK LIBBY (USA, b. 1908) and collaborators developed the techniques of radiocarbon dating. He was awarded the Nobel prize for chemistry in 1960.

1949 MARIA GOEPPERT MAYER (Germany, USA, b. 1906) and, independently, OTTO HAXEL (Germany, b. 1909), JOHANNES HANS DANIEL JENSEN (Germany, b. 1907), and HANS EDUARD SUESS (Austria, Germany, USA, b. 1909) developed the shell theory of the nucleus, which assumes a spherical distribution of nucleons. Mayer and Jensen were awarded the Nobel prize for physics jointly with E. P. WIGNER in 1963.

1950 ARTHUR HAWLEY SNELL (Canada, USA, b. 1911) and associates at the Oak Ridge National Laboratory and JOHN MICHAEL ROBSON (England, Canada, b. 1920) and his associates at the Chalk River Laboratory experimentally verified that the free neutron is radioactive.

1950 Scientists began intensified research on light-element fusion reactions.

1951 MARTIN DEUTSCH (Austria, USA, b. 1917) experimentally confirmed the prediction of the existence of positronium.

1952 Brookhaven National Laboratory was the first to achieve the acceleration of particles to the giga-electron-volt energy range: 2.3-Gev protons.

1952 AAGE BOHR (Denmark, b. 1922) and BEN ROY MOTTELSON (Denmark b. 1926) developed the unified (collective) shell model of the nucleus, which assumes a nonspherical nuclear core. The possibility of a distorted core had been suggested in 1950 by LEO JAMES RAINWATER (USA, b. 1917).

1952 DONALD ARTHUR GLASER (USA, b. 1926) made the first bubble chamber. He was awarded the Nobel prize for physics in 1960.

1952 The first large-scale, terrestrial thermonuclear reaction was produced when a "hydrogen fusion device" was tested at Einewetok atoll on November 1.

1953 MURRAY GELL-MANN (USA, b. 1929) introduced the strangeness numbers for nucleons, mesons, and hyperons, and found that strangeness is conserved in strong interactions.

1953 ROBERT HOFSTADTER (USA, b. 1915) and collaborators started a series of experiments on the scattering of high-energy electrons by atoms. The results led to the determination of the charge distribution and structure of nuclei and nucleons. Hofstadter was awarded the Nobel prize for physics jointly with R. L. Mössbauer in 1961.

1954 JAMES POWER GORDON (USA, b. 1928), H. J. ZEIGER (USA, b. 1925) and CHARLES HARD TOWNES (USA, b. 1915) made the first *maser* [molecular (formerly, microwave) amplification by stimulated emission of radiation]. In this device, many molecules which have been put into high energy states are induced to emit their energy as radiation by a weak incoming signal of the same frequency. Townes was awarded the Nobel prize for physics jointly with N. BASOV and A. PROKHORV in 1964.

1955 OWEN CHAMBERLAIN (USA, b. 1920), EMILIO GINO SEGRÈ (Italy, USA, b. 1905), CLYDE EDWARD WIEGAND (USA, b. 1915), and THOMAS JOHN YPSILANTIS (USA, b. 1928) created proton-antiproton pairs. Chamberlain and Segrè were awarded the Nobel prize for physics in 1959.

1956 LUIS WALTER ALVAREZ (USA, b. 1911) and collaborators accomplished cold fusion of deuterium with the negative mu-meson as a catalyst.

1956 JOHN BARDEEN (USA, b. 1908), WALTER HOUSER BRATTAIN (China, USA, b. 1902), and WILLIAM SHOCKLEY (England, USA, b. 1910) were awarded the Nobel prize for physics in recognition of their work in theory of the solid state, particularly semiconductors.

1956 FREDERIC REINES (USA, b. 1918) and CLYDE LORRAIN COWAN, JR. (USA, b. 1919) and collaborators experimentally confirmed the existence of the neutrino.

1956 The world's first full-scale nuclear power plant was put into operation on October 17 at Calder Hall, England. The gas-cooled reactors develop 360 megawatts of thermal power to deliver 78 megawatts of electrical power.

1956 TSUNG DAO LEE (CHINA, USA, b. 1926) and CHEN NING YANG (China, USA, b. 1922) deduced theoretically that the law of conservation of parity (the invariance of spatial inversion) is invalid for weak interactions. They were awarded the Nobel prize for physics in 1957.

1956 CHIEN-SHIUNG WU (China, USA, b. 1915) and collaborators performed the first experiment that demonstrated the violation of conservation of parity. They observed the beta emission from Co^{60} at very low temperatures.

1957 JOHN BARDEEN (USA), LEON N. COOPER (USA, b. 1930) and JOHN ROBERT SCHRIEFFER (USA, b. 1931) announced the first comprehensive theory of superconductivity.

1958 C. H. TOWNES (USA), JOHN PERRY CEDARHOLM (USA, b. 1927), GEORGE FRANCIS BLAND (USA, b. 1927), and BYRON LUTHER HAVENS (USA, b. 1914) employed maser beams in the most precise ether-drift experiment yet performed. The results showed that if the effect exists, it is less than one-thousandth of the earth's orbital speed or less than one ten-millionth of the speed of light. The precision in the comparison of the frequencies of the masers was about one part in a million million.

1958 RUDOLPH L. MÖSSBAUER (Germany, b. 1929) predicted and found an extremely small frequency spread in the emission of low-energy gamma rays from nuclei bound in a crystal lattice. This effect results from giving the gamma-ray recoil momentum to the whole lattice instead of to an individual nucleus. The effect provides a very high-precision frequency standard suitable for testing several predictions of the special and general theories of relativity. He was awarded the Nobel prize for physics jointly with R. HOFSTADTER in 1961.

1959 JAMES ALFRED VAN ALLEN (USA, b. 1914) showed from the data obtained from instruments carried by artificial satellites that the earth is encircled by two zones, called VAN ALLEN radiation belts, of high-energy charged particles which are trapped by the earth's magnetic field.

1960 THEODORE HAROLD MAIMAN (USA, b. 1927) made the first ruby laser.

1960 ALI JAVAN (Iran, USA, b. 1926) made the first helium-neon laser.

1960 Vernon Willard Hughes (USA, b. 1921), D. W. McColm, Klaus Otto Ziock (Germany, USA, b. 1925) and R. Prepost made and studied muonium, a short-lived atom having a positive mu-meson nucleus and an orbiting electron.

1962 B. D. Josephson (England) discovered and theoretically analyzed a number of unexpected phenomena occurring at a "Josephson junction," an arrangement consisting of two superconductors separated by a very thin layer of insulating material.

1965 Jerome V. V. Kasper and George Claude Pimentel (USA, b. 1922) made the first chemical laser, a device in which pumping energy is supplied by chemical reactions instead of by an external source of power.

APPENDIX 2

NOBEL PRIZE WINNERS

PHYSICS		CHEMISTRY
	1901	
W. C. Roentgen (Germany)		J. H. van't Hoff (Netherlands)
	1902	
H. A. Lorentz (Netherlands) Pieter Zeeman (Netherlands)		Emil Fischer (Germany)
	1903	
Henri Becquerel (France) Pierre Curie (France) Marie S. Curie (France)		S. A. Arrhenius (Sweden)
	1904	
J. W. Strutt (Lord Rayleigh) (England)		William Ramsay (England)
	1905	
Philipp Lenard (Germany)		Adolf von Baeyer (Germany)
	1906	
J. J. Thomson (England)		Henri Moissan (France)
	1907	
A. A. Michelson (USA)		Eduard Buchner (Germany)
	1908	
Gabriel Lippman (France)		Ernest Rutherford (England)
	1909	
Guglielmo Marconi (Italy) K. F. Braun (Germany)		Wilhelm Ostwald (Germany)
	1910	
J. D. van der Waals (Netherlands)		Otto Wallach (Germany)
	1911	
Wilhelm Wien (Germany)		Marie S. Curie (France)
	1912	
N. G. Dalen (Sweden)		Victor Grignard (France) Paul Sabatier (France)

(*cont.*)

PHYSICS		CHEMISTRY
	1913	
Kamerlingh Onnes (Netherlands)		Alfred Werner (Switzerland)
	1914	
Max von Laue (Germany)		T. W. Richards (USA)
	1915	
W. H. Bragg (England) W. L. Bragg (England)		Richard Willstaetter (Germany)
	1916	
No Award		No Award
	1917	
C. G. Barkla (England)		No Award
	1918	
Max Planck (Germany)		Fritz Haber (Germany)
	1919	
Johannes Stark (Germany)		No Award
	1920	
C. E. Guillaume (Switzerland)		Walther Nernst (Germany)
	1921	
Albert Einstein (Germany)		Frederick Soddy (England)
	1922	
Niels Bohr (Denmark)		F. W. Aston (England)
	1923	
R. A. Millikan (USA)		Fritz Pregl (Austria)
	1924	
K. M. G. Siegbahn (Sweden)		No Award
	1925	
James Franck (Germany) Gustav Hertz (Germany)		Richard Zsigmondy (Germany)
	1926	
Jean Perrin (France)		Theodor Svedberg (Sweden)
	1927	
A. H. Compton (USA) C. T. R. Wilson (England)		Heinrich Wieland (Germany)

PHYSICS		CHEMISTRY
	1928	
O. W. Richardson (England)		Adolf Windaus (Germany)
	1929	
Louis de Broglie (France)		Arthur Harden (England) H. von Euler-Chelpin (Sweden)
	1930	
C. V. Raman (India)		Hans Fischer (Germany)
	1931	
No Award		Carl Bosch (Germany) Friedrich Bergius (Germany)
	1932	
Werner Heisenberg (Germany)		Irving Langmuir (USA)
	1933	
Erwin Schroedinger (Austria) P. A. M. Dirac (England)		No Award
	1934	
No Award		H. C. Urey (USA)
	1935	
James Chadwick (England)		Frederic Joliot (France) Irene Joliot-Curie (France)
	1936	
V. F. Hess (Austria) C. D. Anderson (USA)		Peter J. W. Debye (Germany)
	1937	
C. J. Davisson (USA) G. P. Thomson (England)		W. N. Haworth (England) Paul Karrer (Switzerland)
	1938	
Enrico Fermi (Italy)		Richard Kuhn (Germany)
	1939	
E. O. Lawrence (USA)		Leopold Ruzicka (Switzerland) Adolf Butenandt (Germany)
	1940	
No Award		No Award
	1941	
No Award		No Award

(cont.)

PHYSICS		CHEMISTRY
	1942	
No Award		No Award
	1943	
Otto Stern (Germany)		George de Hevesy (Hungary)
	1944	
I. I. Rabi (USA)		Otto Hahn (Germany)
	1945	
Wolfgang Pauli (Austria)		A. I. Virtanen (Finland)
	1946	
P. W. Bridgman (USA)		J. B. Sumner (USA) J. H. Northrop (USA) W. M. Stanley (USA)
	1947	
E. V. Appleton (England)		Robert Robinson (England)
	1948	
P. M. S. Blackett (England)		Arne Tiselius (Sweden)
	1949	
Hideki Yukawa (Japan)		W. F. Giauque (USA)
	1950	
C. F. Powell (England)		Otto Diels (Germany) Kurt Alder (Germany)
	1951	
J. D. Cockcroft (England) E. T. S. Walton (Ireland)		E. M. McMillan (USA) G. T. Seaborg (USA)
	1952	
Felix Bloch (USA) E. M. Purcell (USA)		A. J. P. Martin (Canada) E. L. M. Synge (England)
	1953	
Fritz Zernike (Netherlands)		Herman Staudinger (Germany)
	1954	
Max Born (Germany) Walter Bothe (Germany)		Linus Pauling (USA)
	1955	
Polykarp Kusch (USA) W. E. Lamb (USA)		Vincent du Vigneaud (USA)

PHYSICS		CHEMISTRY
	1956	
John Bardeen (USA) W. H. Brattain (USA) William Shockley (USA)		C. N. Hinshelwood (England) N. N. Semenov (USSR)
	1957	
C. N. Yang (China, USA) T. D. Lee (China, USA)		Alexander Todd (England)
	1958	
P. A. Cherenkov (USSR) I. Y. Tamm (USSR) I. M. Frank (USSR)		Frederic Sanger (England)
	1959	
Owen Chamberlain (USA) E. G. Segrè (USA)		Jaroslav Heyrovsky (Czechoslovakia)
	1960	
D. A. Glaser (USA)		W. F. Libby (USA)
	1961	
Robert Hofstadter (USA) R. L. Mössbauer (Germany)		Melvin Calvin (USA)
	1962	
L. D. Landau (USSR)		J. C. Kendrew (England) M. F. Perutz (England)
	1963	
J. H. D. Jensen (Germany) Maria G. Mayer (USA) E. B. Wigner (USA)		Karl Ziegler (Germany) Giulio Natta (Italy)
	1964	
Nikolai Basov (USSR) Alexander Prokhorv (USSR) C. H. Townes (USA)		Dorothy C. Hodgkin (England)
	1965	
R. P. Feynman (USA) J. S. Schwinger (USA) Sin-itiro Tomonaga (Japan)		R. B. Woodward (USA)
	1966	
Alfred Kastler (France)		R. S. Mullikan (USA)

APPENDIX 3

PERIODIC TABLE OF THE ELEMENTS

(Numbers in parentheses indicate the mass number of the longest-lived isotope of radioactive elements)

Outer electrons are in the	I	II	III	IV	V	VI	VII	VIII	0	Electrons per shell
First or K-shell	1 H 1.00797								2 He 4.0026	2
Second or L-shell	3 Li 6.939	4 Be 9.0122	5 B 10.811	6 C 12.01115	7 N 14.0067	8 O 15.9994	9 F 18.9984		10 Ne 20.183	2,8
Third or M-shell	11 Na 22.9898	12 Mg 24.312	13 Al 26.9815	14 Si 28.086	15 P 30.9738	16 S 32.064	17 Cl 35.453		18 Ar 39.948	2,8,8
Fourth or N-shell	19 K 39.102	20 Ca 40.08	21 Sc 44.956	22 Ti 47.90	23 V 50.942	24 Cr 51.996	25 Mn 54.9380	26 Fe 55.847 27 Co 58.9332 28 Ni 58.71		
	29 Cu 63.54	30 Zn 65.37	31 Ga 69.72	32 Ge 72.59	33 As 74.9216	34 Se 78.96	35 Br 79.909		36 Kr 83.80	2,8,18,8
Fifth or O-shell	37 Rb 85.47	38 Sr 87.62	39 Y 88.905	40 Zr 91.22	41 Nb 92.906	42 Mo 95.94	43 Tc (99)	44 Ru 101.07 45 Rh 102.905 46 Pd 106.4		
	47 Ag 107.870	48 Cd 112.40	49 In 114.82	50 Sn 118.69	51 Sb 121.75	52 Te 127.60	53 I 126.9044		54 Xe 131.30	2,8,18, 18,8
Sixth or P-shell	55 Cs 132.905	56 Ba 137.34	57–71 La series*	72 Hf 178.49	73 Ta 180.948	74 W 183.85	75 Re 186.2	76 Os 190.2 77 Ir 192.2 78 Pt 195.09		
	79 Au 196.967	80 Hg 200.59	81 Tl 204.37	82 Pb 207.19	83 Bi 208.980	84 Po (210)	85 At (210)		86 Rn (222)	2,8,18, 32,18,8
Seventh or Q-shell	87 Fr (223)	88 Ra (226)	89–103 Ac series**	104 (260)						

*Lanthanide series:	57 La 138.91	58 Ce 140.12	59 Pr 140.907	60 Nd 144.24	61 Pm (145)	62 Sm 150.35	63 Eu 151.96	64 Gd 157.25	65 Tb 158.924	66 Dy 162.50	67 Ho 164.930	68 Er 167.26	69 Tm 168.934	70 Yb 173.04	71 Lu 174.97	2,8,18, 32,9,2
**Actinide series:	89 Ac (227)	90 Th 232.038	91 Pa (231)	92 U 238.03	93 Np (237)	94 Pu (244)	95 Am (243)	96 Cm (247)	97 Bk (247)	98 Cf (251)	99 Es (254)	100 Fm (257)	101 Md (256)	102 No (255)	103 Lw (257)	2,8,18, 32,32,9,2

APPENDIX 4

PROPERTIES OF ATOMS IN BULK

The values in parentheses in the column of atomic weights are the mass numbers of the longest-lived isotopes of those elements which are radioactive. Melting points and boiling points in parentheses are uncertain.

All the physical properties are given for a pressure of one atmosphere except where otherwise specified.

The data for gases are valid only when these are in their usual molecular state, such as H_2, He, O_2, Ne, etc. The specific heats of the gases are the values at constant pressure.

The data for this table were obtained from Key to Periodic Chart of the Atoms, 1963 edition, by William F. Meggers. (Courtesy of the Welch Scientific Company, Skokie, Illinois.)

Element	Symbol	Atomic number, Z	Atomic weight	Density, g/cm^3 at 20°C	Melting point, °C	Boiling point, °C	Specific heat, cal/(g · C°) at 20°C
Actinium	Ac	89	(227)	...	1050	...	...
Aluminum	Al	13	26.9815	2.699	660	2450	0.215
Americium	Am	95	(243)	11.7	...	...	...
Antimony	Sb	51	121.75	6.62	630.5	1380	0.049
Argon	Ar	18	39.948	1.6626×10^{-3}	−189.4	−185.8	0.125
Arsenic	As	33	74.9216	5.72	817 (28 at.)	613	0.082
Astatine	At	85	(210)	...	(302)	...	...
Barium	Ba	56	137.34	3.5	714	1640	0.068
Berkelium	Bk	97	(247)	...	...	...	...
Beryllium	Be	4	9.0122	1.848	1277	2770	0.45
Bismuth	Bi	83	208.980	9.80	271.3	1560	0.0294
Boron	B	5	10.811	2.34	2030	...	0.309
Bromine	Br	35	79.909	3.12 (liquid)	−7.2	58	0.070
Cadmium	Cd	48	112.40	8.65	320.9	765	0.055
Calcium	Ca	20	40.08	1.55	838	1440	0.149
Californium	Cf	98	(251)	...	...	...	...
Carbon	C	6	12.01115	2.25	3727	4830	0.165
Cerium	Ce	58	140.12	6.768	804	3470	0.045
Cesium	Cs	55	132.905	1.9	28.7	690	0.04817
Chlorine	Cl	17	35.453	3.214×10^{-3} (0°C)	−101	−34.7	0.116
Chromium	Cr	24	51.996	7.19	1875	2665	0.11
Cobalt	Co	27	58.9332	8.85	1495	2900	0.099
Copper	Cu	29	63.54	8.96	1083	2595	0.092
Curium	Cm	96	(247)	...	...	...	...
Dysprosium	Dy	66	162.60	8.55	1407	2330	0.041
Einsteinium	Es	99	(254)	...	...	...	...
Erbium	Er	68	167.26	9.15	1497	2630	0.040
Europium	Eu	63	151.96	5.245	826	1490	0.039
Fermium	Fm	100	(257)	...	...	...	...
Fluorine	F	9	18.9984	1.696×10^{-3} (0°C)	−219.6	−188.2	0.18
Francium	Fr	87	(223)	...	(27)	...	...

(cont.)

PROPERTIES OF ATOMS IN BULK *(cont.)*

Element	Symbol	Atomic number, *Z*	Atomic weight	Density, g/cm³ at 20°C	Melting point, °C	Boiling point, °C	Specific heat, cal/(g · C°) at 20°C
Gadolinium	Gd	64	157.25	7.86	1312	2730	0.071
Gallium	Ga	31	69.62	5.907	29.8	2237	0.079
Germanium	Ge	32	72.59	5.323	937.4	2830	0.073
Gold	Au	79	196.967	19.32	1063.0	2970	0.0312
Hafnium	Hf	72	178.49	13.09	2222	5400	0.0351
Helium	He	2	4.0026	0.1664×10^{-3}	−269.7	−268.9	1.25
Holmium	Ho	67	164.930	8.79	1461	2330	0.039
Hydrogen	H	1	1.00797	0.08375×10^{-3}	−259.19	−252.7	3.45
Indium	In	49	114.82	7.31	156.2	2000	0.057
Iodine	I	53	126.9044	4.94	113.7	183	0.052
Iridium	Ir	77	192.2	22.5	2454	(5300)	0.0307
Iron	Fe	26	55.847	7.87	1536.5	3000	0.110
Krypton	Kr	36	83.80	3.488×10^{-3}	−157.3	−152	...
Lanthanum	La	57	138.91	6.189	920	3470	0.048
Lawrencium	Lw	103	(257)	...	...	...	...
Lead	Pb	82	207.19	11.36	327.4	1725	0.0309
Lithium	Li	3	6.939	0.534	180.5	1300	0.79
Lutetium	Lu	71	174.97	9.849	1652	1930	0.037
Magnesium	Mg	12	24.312	1.74	650	1107	0.245
Manganese	Mn	25	54.9380	7.43	1245	2150	0.115
Mendelevium	Md	101	(256)	...	...	...	...
Mercury	Hg	80	200.59	13.55	−38.36	357	0.033
Molybdenum	Mo	42	95.94	10.22	2610	5560	0.066
Neodymium	Nd	60	144.24	7.00	1019	3180	0.045
Neon	Ne	10	20.183	0.8387×10^{-3}	−248.6	−246.0	...
Neptunium	Np	93	(237)	19.5	637	...	...
Nickel	Ni	28	58.71	8.902	1453	2730	0.105
Niobium or (Columbium, Cb)	Nb	41	92.906	8.57	2468	4927	0.065
Nitrogen	N	7	14.0067	1.1649×10^{-3}	−210	−195.8	0.247
Nobelium	No	102	(255)	...	...	...	...
Osmium	Os	76	190.2	22.57	(2700)	5500	0.031
Oxygen	O	8	15.9994	1.3318×10^{-3}	−218.83	−183.0	0.218
Palladium	Pd	46	106.4	12.02	1552	3980	0.058
Phosphorus	P	15	30.9738	1.83	44.25	280	0.177
Platinum	Pt	78	195.09	21.45	1769	4530	0.0314
Plutonium	Pu	94	(244)	...	640	3235	0.033
Polonium	Po	84	(210)	9.24	254	...	...
Potassium	K	19	39.102	0.86	63.7	760	0.177
Praseodymium	Pr	59	140.907	6.769	919	3020	0.045

Promethium	Pm	61	(145)	...	(1027)	...	...
Protactinium	Pa	91	(231)	...	(1230)	...	...
Radium	Ra	88	(226)	5.0	700	...	...
Radon (or Emanation, Em)	Rn	86	(222)	9.96×10^{-3} (0°C)	(−71)	−61.8	...
Rhenium	Re	75	186.2	21.04	3180	5900	0.033
Rhodium	Rh	45	102.905	12.44	1966	4500	0.059
Rubidium	Rb	37	85.47	1.53	39	688	0.080
Ruthenium	Ru	44	101.107	12.2	2500	4900	0.057
Samarium	Sm	62	150.35	7.49	1072	1630	0.042
Scandium	Sc	21	44.956	2.99	1539	2730	0.134
Selenium	Se	34	78.96	4.79	217	685	0.084
Silicon	Si	14	28.86	2.33	1410	2680	0.162
Silver	Ag	47	107.870	10.49	960.8	2210	0.056
Sodium	Na	11	22.9898	0.9712	97.8	892	0.295
Strontium	Sr	38	87.62	2.60	768	1380	0.176
Sulfur	S	16	32.064	2.07	119.0	444.6	0.175
Tantalum	Ta	73	180.948	16.6	2996	5425	0.034
Technetium	Tc	43	(99)	11.46	(2130)	...	...
Tellurium	Te	52	127.60	6.24	450	990	0.047
Terbium	Tb	65	158.924	8.25	1356	2530	0.044
Thallium	Tl	81	204.37	11.85	303	1457	0.031
Thorium	Th	90	(232)	11.66	1750	(3850)	0.034
Thulium	Tm	69	168.934	9.31	1545	1720	0.038
Tin	Sn	50	118.69	7.2984	231.9	2270	0.054
Titanium	Ti	22	47.90	4.507	1668	3260	0.124
Tungsten (or Wolfram)	W	74	183.85	19.3	3410	5930	0.033
Uranium	U	92	(238)	19.07	1132	3818	0.028
Vanadium	V	23	50.942	6.1	1900	3400	0.119
Xenon	Xe	54	131.30	5.495×10^{-3}	−111.9	−108	...
Ytterbium	Yb	70	173.04	6.959	824	1530	0.035
Yttrium	Y	39	88.905	4.472	1509	3030	0.071
Zinc	Zn	30	65.37	7.133	419.5	906	0.0915
Zirconium	Zr	40	91.22	6.489	1852	3580	0.067

APPENDIX 5

PARTIAL LIST OF ISOTOPES

The values of the isotopic masses are based on carbon 12. Naturally occurring radioactive isotopes are indicated by (R). The mass numbers given for the radioactive elements are those of the longest-lived isotopes.

The data for this table were obtained from the Chart of the Nuclides, 8th edition, revised to March, 1965 by David T. Goldman. (Courtesy of Knolls Atomic Power Laboratory, Schenectady, N.Y., operated by the General Electric Company for the United States Atomic Energy Commission.)

At. no. Z	Element	Symbol	Mass no., A	Isotopic mass, u	Relative abundance, %	No. of Isotopes: Stable	No. of Isotopes: Radio-active
0	Neutron	n				0	1
			1 (R)	1.008665			
1	Hydrogen	H				2	1
			1	1.007825	99.985		
		D					
			2	2.01410	0.015		
		T					
			3 (NR)				
2	Helium	He				2	3
			3	3.01603	0.00013		
			4	4.00260	100		
3	Lithium	Li				2	3
			6	6.01513	7.42		
			7	7.01601	92.58		
4	Beryllium	Be				1	6
			9	9.01219	100		
5	Boron	B				2	4
			10	10.01294	19.78		
			11	11.00931	80.22		
6	Carbon	C				2	6
			12	12.00000	98.89		
			13	13.00335	1.11		
7	Nitrogen	N				2	5
			14	14.00307	99.63		
			15	15.00011	0.37		
8	Oxygen	O				3	5
			16	15.99491	99.759		
			17	16.99914	0.037		
			18	17.99916	0.204		
9	Fluorine	F				1	5
			19	18.99840	100		
10	Neon	Ne				3	5
			20	19.99244	90.92		
			21	20.99395	0.257		
			22	21.99138	8.82		
11	Sodium	Na				1	6
			23	22.98977	100		
12	Magnesium	Mg				3	5
			24	23.98504	78.70		
			25	24.98584	10.13		
			26	25.98259	11.17		
13	Aluminum	Al				1	7
			27	26.98153	100		
14	Silicon	Si				3	5
			28	27.97693	92.21		
15	Phosphorus	P				1	6
			31	30.97376	100		
16	Sulfur	S				4	6
			32	31.97207	95.0		

At. no. Z	Element	Symbol	Mass no., A	Isotopic mass, u	Relative abundance, %	No. of isotopes	
						Stable	Radio-active
17	Chlorine	Cl				2	7
			35	34.96885	75.53		
			37	36.96590	24.47		
18	Argon	Ar				3	6
			40	39.96238	99.60		
19	Potassium	K				2	9
			39	38.96371	93.10		
			40 (NR)		0.0118		
			41	40.96184	6.88		
20	Calcium	Ca				6	8
			40	39.96259	96.97		
			44	43.95594	2.06		
21	Scandium	Sc				1	11
			45	44.95592	100		
22	Titanium	Ti				5	5
			48	47.94795	73.94		
23	Vanadium	V				1	9
			50 (NR)	49.9472	0.24		
			51	50.9440	99.76		
24	Chromium	Cr				4	7
			52	51.9405	83.76		
			53	52.9407	9.55		
25	Manganese	Mn				1	8
			55	54.9381	100		
26	Iron	Fe				4	6
			56	55.9349	91.66		
			57	56.9354	2.19		
27	Cobalt	Co				1	10
			59	58.9332	100		
28	Nickel	Ni				5	7
			58	57.9353	67.88		
			60	59.9303	26.23		
29	Copper	Cu				2	9
			63	62.9298	69.09		
			65	64.9278	30.91		
30	Zinc	Zn				5	8
			64	63.9291	48.89		
			66	65.9260	27.81		
			68	67.9249	18.57		
31	Gallium	Ga				2	12
			69	68.9256	60.4		
			71	70.9247	39.6		
32	Germanium	Ge				4	10
			70	69.9242	20.52		
			72	71.9221	27.43		
			74	73.9212	36.54		
33	Arsenic	As				1	14
			75	74.9216	100		
34	Selenium	Se				6	11
			78	77.9173	23.52		
			80	79.9165	49.82		
35	Bromine	Br				2	16
			79	78.9183	50.54		
			81	80.9163	49.46		

(cont.)

PARTIAL LIST OF ISOTOPES (*cont.*)

At. no. Z	Element	Symbol	Mass no., A	Isotopic mass, u	Relative abundance, %	No. of isotopes Stable	Radio-active
36	Krypton	Kr				6	16
			82	81.9135	11.56		
			83	82.9141	11.55		
			84	83.9115	56.90		
			86	85.9106	17.37		
37	Rubidium	Rb				1	16
			85	84.9117	72.15		
			87		27.85		
38	Strontium	Sr				4	12
			88	87.9056	82.56		
39	Yttrium	Y				1	14
			89	88.9056	100		
40	Zirconium	Zr				5	9
			90	89.9047	51.46		
			92	91.9050	17.11		
			94	93.9063	17.40		
41	Niobium (or Columbium, Cb)	Nb				1	13
			93	92.9064	100		
42	Molybdenum	Mo				7	10
			92	91.9068	15.84		
			95	94.9058	15.72		
			96	95.9047	16.53		
			98	97.9054	23.78		
43	Technetium	Tc				0	14
			99 (R)				
44	Ruthenium	Ru				7	9
			102	101.9043	31.61		
			104	103.9054	18.58		
45	Rhodium	Rh				1	14
			103	102.9055	100		
46	Palladium	Pd				6	12
			105	104.9051	22.23		
			106	105.9035	27.33		
			108	107.9039	26.71		
47	Silver	Ag				2	14
			107	107.9051	51.82		
			109	108.9047	48.18		
48	Cadmium	Cd				8	11
			110	109.9030	12.39		
			111	110.9042	12.75		
			112	111.9028	24.07		
			113	112.9046	12.26		
			114	113.9034	28.86		
49	Indium	In				1	18
			113	112.9043	4.28		
			115 (NR)	114.9039	95.72		
50	Tin	Sn				10	15
			116	115.9017	14.30		
			118	117.9016	24.03		
			119	118.9033	8.58		
			120	119.9022	32.85		
51	Antimony	Sb				2	22
			121	120.9038	57.25		
			123	122.9042	42.75		

At. no. Z	Element	Symbol	Mass no., A	Isotopic mass, u	Relative abundance, %	No. of isotopes: Stable	No. of isotopes: Radio-active
52	Tellurium	Te				8	16
			123 (NR)	122.9043	0.87		
			126	125.9033	18.71		
			128	127.9045	31.79		
			130	129.9062	34.48		
53	Iodine	I				1	22
			127	126.9045	100		
54	Xenon	Xe				9	16
			129	128.9048	26.44		
			131	130.9051	21.18		
			132	131.9042	26.89		
55	Cesium	Cs				1	20
			133	132.9051	100		
56	Barium	Ba				7	14
			137	136.9056	11.32		
			138	137.9050	71.66		
57	Lanthanum	La				1	20
			138 (NR)	137.9068	0.089		
			139	138.9061	99.911		
58	Cerium	Ce				3	16
			140	139.9053	88.48		
			142 (NR)	141.9090	11.07		
59	Praseodymium	Pr				1	14
			141	140.9074	100		
60	Neodymium	Nd				6	8
			142	141.9075	27.11		
			144 (NR)	143.9099	23.85		
			146	145.9127	17.22		
61	Promethium	Pm				0	14
			145 (R)				
62	Samarium	Sm				4	14
			147 (NR)	146.9146	14.97		
			148 (NR)	147.9146	11.24		
			149 (NR)	148.9169	13.83		
			152	151.9195	26.72		
			154	153.9209	22.71		
63	Europium	Eu				2	16
			151	150.9196	47.82		
			153	152.9209	52.18		
64	Gadolinium	Gd				6	12
			152 (NR)	151.9195	0.20		
			156	155.9221	20.47		
			158	157.9241	24.87		
			160	159.9271	21.90		
65	Terbium	Tb				1	17
			159	158.9250	100		
66	Dysprosium	Dy				6	13
			156 (NR)	155.9238	0.052		
			162	161.9265	25.53		
			163	162.9284	24.97		
			164	163.9288	28.18		
67	Holmium	Ho				1	18
			165	164.9303	100		
68	Erbium	Er				6	12
			166	165.9304	33.41		
			167	166.9320	22.94		
			168	167.9324	27.07		

(*cont.*)

PARTIAL LIST OF ISOTOPES (*cont.*)

At. no. *Z*	Element	Symbol	Mass no., *A*	Isotopic mass, u	Relative abundance, %	No. of Isotopes: Stable	No. of Isotopes: Radio-active
69	Thulium	Tm				1	17
			169	168.9344	100		
70	Ytterbium	Yb				7	10
			172	171.9366	21.82		
			174	173.9390	31.84		
71	Lutetium	Lu				1	15
			175	174.9409	97.41		
			176 (NR)		2.59		
72	Hafnium	Hf				5	13
			174 (NR)	173.9403	0.18		
			178	177.9439	27.14		
			180	179.9468	35.24		
73	Tantalum	Ta				2	13
			181	180.9480	99.988		
74	Tungsten (Wolfram)	W				5	10
			182	181.9483	26.41		
			184	183.9510	30.64		
			186	185.9543	28.14		
75	Rhenium	Re				1	14
			185	184.9530	37.07		
			187 (NR)	186.9560	62.93		
76	Osmium	Os				7	8
			190	189.9586	26.4		
			192	191.9612	41.0		
77	Iridium	Ir				2	15
			191	190.9609	37.3		
			193	192.9633	62.7		
78	Platinum	Pt				5	16
			190 (NR)	189.9600	0.0127		
			194	193.9628	32.9		
			195	194.9648	33.8		
			196	195.9650	25.3		
79	Gold	Au				1	18
			197	196.9666	100		
80	Mercury	Hg				7	14
			199	198.9683	16.84		
			200	199.9683	23.13		
			202	201.9706	29.80		
81	Thallium	Tl				2	18
			203	202.9723	29.50		
			205	204.9745	70.50		
			207 (NR)				
82	Lead	Pb				3	18
			204 (NR)	203.9731	1.48		
			206	205.9745	23.6		
			207	206.9759	22.6		
			208	207.9766	52.3		
83	Bismuth	Bi				1	18
			209	208.9804	100		
			210 (NR)				
84	Polonium	Po				0	27
			210 (NR)	209.9829			
85	Astatine	At				0	20
			210 (NR)				
			211 (NR)	210.9875			
86	Radon	Rn				0	18
			222 (NR)	222.0175			

At. no. Z	Element	Symbol	Mass no., A	Isotopic mass, u	Relative abundance, %	No. of Isotopes	
						Stable	Radio-active
87	Francium	Fr	223 (NR)	223.0198		0	18
88	Radium	Ra	226 (NR)	226.0254		0	13
89	Actinium	Ac	227 (NR)	227.0278		0	11
90	Thorium	Th	232 (NR)	232.0382		0	13
91	Protactinium	Pa	231 (NR)	231.0359		0	12
92	Uranium	U	234 (NR)	234.0409	0.0057	0	14
			235 (NR)	235.0439	0.72		
			238 (NR)	238.0508	99.27		
93	Neptunium	Np	237 (R)	237.0480		0	11
94	Plutonium	Pu	239 (R)	239.0522		0	15
			242 (R)	242.0587			
			244 (R)				
95	Americium	Am	241 (R)	241.0567		0	10
			243 (R)	243.0614			
96	Curium	Cm	243 (R)	243.0614		0	13
			247 (R)				
97	Berkelium	Bk	247 (R)	247.0702		0	8
98	Californium	Cf	251 (R)			0	11
99	Einsteinium	Es	254 (R)	254.0881		0	11
100	Fermium	Fm	257 (R)			0	11
101	Mendelevium	Md	256 (R)			0	2
102	Nobelium	No	255 (R)			0	3
103	Lawrencium	Lw	257 (R)			0	1
104			260 (R)			0	1

APPENDIX 6

PARTIAL LIST OF RADIOISOTOPES

Element	Nuclide	Half life, T	Decay constant λ, s^{-1}	Principal particle energy, MeV
Antimony	$_{51}Sb^{122}$	2.80 d	2.87×10^{-6}	β^- 1.40; γ, 0.56, 0.70
Argon	$_{18}Ar^{37}$	35.1 d	2.29×10^{-7}	*K*-capture
Arsenic	$_{33}As^{76}$	26.5 h	7.26×10^{-6}	β^- 2.97; γ, 0.56, 1.21
Barium	$_{56}Ba^{140}$	12.8 d	6.26×10^{-7}	β^- 1.02; γ, 0.03, 0.54
Bismuth	$_{83}Bi^{210}$ (NR)	5.0 d	1.60×10^{-6}	β^- 1.16
Cadmium	$_{48}Cd^{115}$	2.3 d	3.49×10^{-6}	β^- 1.11; γ, 0.52
Calcium	$_{20}Ca^{45}$	165 d	4.87×10^{-8}	β^- 0.25
Carbon	$_{6}C^{14}$	5730 y	3.83×10^{-12}	β^- 0.156
Cerium	$_{58}Ce^{141}$	32.5 d	2.47×10^{-7}	β^- 0.44; γ, 0.15
Cesium	$_{55}Cs^{134}$	2.1 y	1.05×10^{-8}	β^- 0.65; γ, 0.60
	$_{55}Cs^{137}$	30 y	7.32×10^{-10}	β^- 0.51; γ, 0.66
Chlorine	$_{17}Cl^{36}$	3.0×10^{5} y	7.32×10^{-14}	β^- 0.71
Chromium	$_{24}Cr^{51}$	27.8 d	2.89×10^{-7}	γ, 0.32; *K*-capture
Cobalt	$_{27}Co^{58}$	71 d	1.13×10^{-7}	β^+ 0.48; γ, 0.81, 1.65
	$_{27}Co^{60}$	5.26 y	4.17×10^{-9}	β^- 0.31; γ, 1.17, 1.33
Gold	$_{79}Au^{198}$	64.8 h	2.97×10^{-6}	β^- 0.96; γ, 0.41, 0.67
Hafnium	$_{72}Hf^{181}$	45 d	1.78×10^{-7}	β^- 0.41; γ, 0.48
Hydrogen	$_{1}H^{3}$	12.26 y	1.79×10^{-9}	β^- 0.018
Iodine	$_{53}I^{131}$	8.05 d	9.96×10^{-7}	β^- 0.61; γ, 0.36, 0.72
Iron	$_{26}Fe^{59}$	45 d	1.78×10^{-7}	β^- 0.46; γ, 1.10, 1.29
Krypton	$_{36}Kr^{85}$	10.76 y	2.04×10^{-7}	β^- 0.67; γ, 0.52
Lanthanum	$_{57}La^{140}$	40.2 h	4.79×10^{-6}	β^- 1.34; γ, 1.60, 0.49
Mercury	$_{80}Hg^{203}$	47 d	1.71×10^{-7}	β^- 0.21; γ, 0.28
Molybdenum	$_{42}Mo^{99}$	66 h	2.92×10^{-6}	β^- 1.23; γ, 0.04, 0.14
Neptunium	$_{93}Np^{237}$	2.14×10^{6} y	1.03×10^{-14}	α, 4.50; γ, 0.03, 0.09
	$_{93}Np^{239}$	2.35 d	3.42×10^{-6}	β^- 0.72; γ, 0.05, 0.33
Nickel	$_{28}Ni^{63}$	92 y	2.39×10^{-10}	β^- 0.07
Phosphorus	$_{15}P^{32}$	14.3 d	5.61×10^{-7}	β^- 1.71
Plutonium	$_{94}Pu^{239}$	2.44×10^{-4} y	9.01×10^{-13}	α, 5.15; γ, 0.013, 0.038
Potassium	$_{19}K^{40}$ (NR)	1.3×10^{9} y	1.69×10^{-17}	β^- 1.32; γ, 1.46
	$_{19}K^{42}$	12.4 h	1.55×10^{-5}	β^- 3.53; γ, 1.52
Selenium	$_{34}Se^{75}$	120 d	6.70×10^{-8}	γ, 0.27; *K*-capture
Silver	$_{47}Ag^{111}$	7.5 d	1.07×10^{-6}	β^- 1.05; γ, 0.34
Sodium	$_{11}Na^{24}$	15.0 h	1.28×10^{-5}	β^- 1.39; γ, 1.37, 2.75
Strontium	$_{38}Sr^{89}$	50.4 d	1.59×10^{-7}	β^- 1.46
	$_{38}Sr^{90}$	28 y	7.85×10^{-10}	β^- 0.54
Sulfur	$_{16}S^{35}$	86.7 d	9.25×10^{-8}	β^- 0.17
Tantalum	$_{73}Ta^{182}$	115 d	6.98×10^{-8}	β^- 0.51; γ, 0.10, 1.12
Xenon	$_{54}Xe^{135}$	9.2 h	2.09×10^{-5}	β^- 0.91; γ, 0.25, 0.61
Zinc	$_{30}Zn^{65}$	245 d	3.28×10^{-8}	β^- 0.33; γ, 1.12

APPENDIX 7

THE MKSA SYSTEM

Conversion Factors

A conversion factor is a dimensionless ratio used to make a change in units. Thus the conversion factor

$$\frac{12 \text{ in.}}{1 \text{ ft}}$$

(read, "There are 12 inches in 1 foot") may be used to convert 10 feet to inches by direct multiplication and cancellation of the units, feet. To convert 15 inches to feet we evidently must invert the conversion factor to produce the desired result and multiply by

$$\frac{1 \text{ ft}}{12 \text{ in.}}$$

(read, "In 1 foot there are 12 inches"). When the conversion factor is used incorrectly we notice immediately that the desired cancellation of units is not obtained and this is the signal to invert the factor.

To convert speed in $\text{mi} \cdot \text{h}^{-1}$ to $\text{m} \cdot \text{s}^{-1}$ we proceed as follows:

$$1 \frac{\text{mi}}{\text{h}} \times \frac{1 \text{ h}}{3600 \text{ s}} \times \frac{5280 \text{ ft}}{1 \text{ mi}} \times \frac{12 \text{ in.}}{1 \text{ ft}} \times \frac{2.549 \text{ cm}}{1 \text{ in.}} \times \frac{1 \text{ m}}{100 \text{ cm}} = 0.4470 \text{ m} \cdot \text{s}^{-1}$$

The conversion factor then is

$$\frac{0.4470 \text{ m} \cdot \text{s}^{-1}}{1 \text{ mi} \cdot \text{h}^{-1}}.$$

The conversion factors in the following table can be considered exact except where they are given to five significant figures.

	in one unit	there are	
Length s	cm	10^{-2}	meter, m
	in.	2.54×10^{-2}	
	ft	0.3048	
	mi (U.S. statute)	1609.3	
	μ (micron)	10^{-6}	
	nm (nanometer)	10^{-9}	
	Å (angstrom)	10^{-10}	
	f (fermi)	10^{-15}	

(cont.)

	in one unit	there are	
Wave number $\frac{1}{\lambda}$, $\bar{f}$, $\bar{\nu}$	kayser	10^{-2}	m^{-1}
Time t	d (day, mean solar)	8.64×10^4	second, s
	y (year, calendar)	3.1536×10^7	
	shake	10^{-8}	
Frequency f, ν	cycle · s^{-1}	1	hertz, Hz
Velocity u, v	ft · s^{-1}	0.3048	m · s^{-1}
	mi · h^{-1}	0.44704	
Acceleration a	g (free fall, standard)	9.8067	m · s^{-2}
	gal	10^{-2}	
Mass m	g (gram)	10^{-3}	kilogram, kg
	slug	14.594	
Force F and weight w, F	dyne	10^{-5}	newton, N
	poundal	0.13826	
	lb (avoirdupois)	4.4482	
Energy W, E	erg	10^{-7}	joule, J
	eV	1.6021×10^{-19}	
	kWh	3.6×10^6	
	cal (thermochemical)	4.184	
	kcal	4.184×10^3	
	ft · lb	1.3558	
	Btu (thermochemical)	1054.4	
Power P	erg · s^{-1}	10^{-7}	watt, W
	cal · s^{-1}	4.184	
	Btu · h^{-1}	0.29288	
	ft · lb · s^{-1}	1.3558	
	hp (electric)	746.00	
Pressure p	dyne · cm^{-2}	10^{-1}	N · m^{-2} or pascal
	lb · in^{-2} (psi)	6.8948×10^3	
	bar	10^5	
	atm (normal)	1.0133×10^5	
	cm-mercury (0°C)	1.3332×10^3	
	torr (0°C)	1.3332×10^2	

	in one unit	there are	
Density ρ	$g \cdot cm^{-3}$ $lb \cdot ft^{-3}$	10^3 16.018	$kg \cdot m^{-3}$
Specific heat c	$cal \cdot g^{-1} \cdot (C°)^{-1}$ $Btu \cdot lb^{-1} \cdot (F°)^{-1}$	4.184×10^3 4.184×10^3	$J \cdot kg^{-1} \cdot (C°)^{-1}$
Charge q, Q	statcoulomb(esu) abcoulomb(emu)	$1/(3 \times 10^9)$ 10	coulomb, C
Potential difference V	statvolt (esu) abvolt (emu)	300 10^{-8}	volt, V
Capacitance C	statfarad (esu) abfarad (emu)	$1/(9 \times 10^{11})$ 10^9	farad, F
Permittivity ϵ_0	esu	$1/(36\pi \times 10^9)$	$F \cdot m^{-1}$
Electric field intensity E	$V \cdot cm^{-1}$ dyne · stat-coulomb^{-1} (esu)	10^2 3×10^4	$V \cdot m^{-1}$
Electric flux density or electric displacement D	esu emu	$1/(12\pi \times 10^5)$ $10^5/4\pi$	$C \cdot m^{-2}$
Current i, I	statampere (esu) abampere (emu)	$1/(3 \times 10^9)$ 10	ampere, A
Resistance r, R	statohm (esu) abohm (emu)	9×10^{11} 10^{-9}	ohm, Ω
Resistivity ρ	ohm · cm ohm · (mil-ft)$^{-1}$	10^{-2} 1.6624×10^{-9}	ohm · m, Ω · m
Magnetic flux ϕ	maxwell (emu) esu	10^{-8} 3×10^2	weber, Wb
Magnetic flux density or magnetic induction B	gauss (emu) gamma line · in^{-2} esu	10^{-4} 10^{-9} 1.5500×10^{-5} 3×10^6	$Wb \cdot m^{-2}$ or tesla, T
Inductance L	abhenry (emu) stathenry (esu)	10^{-9} 9×10^{11}	henry, H
Permeability μ_0	emu	$4\pi \times 10^{-7}$	$H \cdot m^{-1}$
Magnetic field intensity H	ampere (turn) · cm^{-1} oersted (emu)	10^2 $10^3/4\pi$	$A \cdot m^{-1}$

Multiples and Submultiples

The names of multiples and submultiples of the units are formed with the following prefixes:

Factor by which unit is multiplied	Prefix and pronunciation	Symbol	Factor by which unit is multiplied	Prefix and pronunciation	Symbol
10^{12}	tera (tĕr′ȧ)	T	10^{-2}	centi (sĕn′tĭ)	c
10^{9}	giga (jĭ′gȧ)	G	10^{-3}	milli (mĭl′ĭ)	m
10^{6}	mega (mĕg′ȧ)	M	10^{-6}	micro (mī′krō)	μ
10^{3}	kilo (kĭl′ō)	k	10^{-9}	nano (năn′ō)	n
10^{2}	hecto (hĕk′tō)	h	10^{-12}	pico (pē′kō)	p
10	deka (dĕk′ȧ)	da	10^{-15}	femto (fĕm′tō)	f
10^{-1}	deci (dĕs′ĭ)	d	10^{-18}	atto (ăt′tō)	a

APPENDIX 8

PHYSICAL CONSTANTS AND CONVERSION FACTORS

(The values are within ±0.1 percent of the best ones presently known. For the more precise values consult the references given in the footnote in Section 1–3, Chapter 1.)

Constant	Symbol	Système International (MKSA)	Centimeter-gram-second (cgs)
Speed of light in vacuum	c	3.00×10^{8} m · s^{-1}	3.00×10^{10} cm · s^{-1}
Avogadro constant	N_A	6.02×10^{26} kmole^{-1}	6.02×10^{23} mole^{-1}
Faraday constant	F	9.65×10^{4} C · mole^{-1}	9.65×10^{3} cm$^{1/2}$ · g$^{1/2}$ · mole^{-1} (emu)
Planck constant	h	6.63×10^{-34} J · s	6.63×10^{-27} erg · s
Elementary charge	e	1.602×10^{-19} C	1.602×10^{-20} cm$^{1/2}$ · g$^{1/2}$ (emu) 4.803×10^{-10} cm$^{3/2}$ · g$^{1/2}$ · s^{-1} (esu)
Unified atomic mass unit, 1/12 mass $C^{12} = 1/N_A$	u	1.6604×10^{-27} kg 931 MeV	1.6604×10^{-24} g 931 MeV
Electron rest mass	m_e	9.11×10^{-31} kg 5.49×10^{-4} u 0.511 MeV	9.11×10^{-28} g 5.49×10^{-4} u 0.511 MeV
Proton rest mass	m_p	1.673×10^{-27} kg 1.0073 u 938.3 MeV	1.673×10^{-24} g 1.0073 u 938.3 MeV

(*cont.*)

PHYSICAL CONSTANTS AND CONVERSION FACTORS *(cont.)*

Constant	Symbol	Système International (MKSA)	Centimeter-gram-second (cgs)
H-atom rest mass	m_H	1.673×10^{-27} kg 1.0078 u 938.8 MeV	1.673×10^{-24} g 1.0078 u 938.8 MeV
Neutron rest mass	m_n	1.675×10^{-27} kg 1.0087 u 939.6 MeV	1.675×10^{-24} g 1.0087 u 939.6 MeV
Charge to mass ratio for electron	e/m_e	1.760×10^{11} C · kg^{-1}	1.760×10^{7} cm$^{1/2}$ · g$^{1/2}$ (emu)
Quantum-charge ratio	h/e	4.14×10^{-15} J · s · C^{-1}	1.38×10^{-17} cm$^{3/2}$ · g$^{1/2}$ · s^{-1} (esu) 4.14×10^{-7} cm$^{3/2}$ · g$^{1/2}$ · s^{-1} (emu)
Compton wavelength of electron (h/m_ec)	λ_C	2.426×10^{-12} m	2.426×10^{-10} cm
Compton wavelength of proton (h/m_pc)	$\lambda_{C,P}$	1.321×10^{-15} m	1.321×10^{-13} cm
Rydberg constant	R_∞	1.097×10^{7} m^{-1}	1.097×10^{5} cm^{-1}
Bohr radius	a_0	5.29×10^{-11} m	5.29×10^{-9} cm
Electron radius	r_e	2.82×10^{-15} m	2.82×10^{-13} cm
Wien displacement constant ($\lambda_{max}T$)	b	2.90×10^{-3} m · °K	2.90×10^{-1} cm · °K
Stefan-Boltzmann constant	σ	5.67×10^{-8} W · m^{-2} · °K^{-4}	5.67×10^{-5} erg · cm^{-2} · s^{-1} · °K^{-4}
Boltzmann constant	k	1.38×10^{-23} J · °K^{-1}	1.38×10^{-16} erg · °K^{-1}
Normal volume, perfect gas	V_0	22.4 m^3 · kmole^{-1}	2.24×10^{4} cm^3 · mole^{-1}
Gas constant	R	8.31×10^{3} J · °K^{-1}, kmole^{-1}	8.31×10^{7} erg · °K^{-1} · mole^{-1}
Gravitational constant	G	6.67×10^{-11} N · m^2 · kg^{-2}	6.67×10^{-8} dyn · cm^2 · g^{-2}

ANSWERS TO SELECTED ODD-NUMBERED PROBLEMS

CHAPTER 1

1–5. $p = mvr/A$
1–9. (a) 10 m/s, 10 m/s; 6.25 m/s, 6.72 m/s; 7.62 m/s, 8.04 m/s; 5 m/s, 5 m/s
(b) No
1–11. (b) $2N/V^2$
(c) $2V/3$, $V/\sqrt{2}$, V
(d) 56%, 50%
1–13. (a) H_2
(b) Cl_2
1–17. (a) 927°C
(b) —10.5°C
1–19. (a) and (b) 1.91×10^3 m/s, 1.76×10^3 m/s, 1.56×10^3 m/s
1–21. 0.66 cm
1–23. (a) 8.47×10^6m/s
(b) 5.8×10^9 °K
1–25. (a) 7.479×10^6J
(b) 7.503×10^6J
(c) 2.5×10^4J
1–27. (b) $2.69 \times 10^{25}/\text{m}^3$
(c) $4.59 \times 10^{25}/\text{m}^3$
1–31. (a) 2.24×10^{-7} m
(b) 2.99×10^{-7} m
(c) 2.24×10^{-9} m
(d) 7.84×10^9/s, 6.82×10^9/s, 7.84×10^{11}/s
(e) 0.407
1–35. (a) 8.24×10^{-20} J
(b) 9.95×10^3 m/s
1–37. (d) 3

CHAPTER 2

2–1. (a) 3.58×10^{13}
(b) 1.94×10^{10}
(c) Yes
2–3. 3.67×10^6 V/m downward
2–5. (a) 0.56 mm, $\phi_x = 4°2'$
(b) 1.11 cm
2–7. 5.34×10^{-6} N
2–9. (a) 1.67 cm
(b) 7.95 ns
(c) No
2–11. (b) 2.00×10^7 m/s
(c) 5.69 mm
2–13. 4.8×10^7 C/kg
2–15. $x = 0$, $y = 3.65$ mm
2–19. 2.24×10^7 m/s
2–21. (b) No
2–25. (a) 1.61×10^{-6} m
(b) 1.41×10^{-14} kg
(c) 0.23
2–27. (a) 2.50×10^{-4} m/s
(b) 1.11×10^{-4} m/s
3.48×10^{-4} m/s,
2.30×10^{-4} m/s,
5.30×10^{-4} m/s,
1.67×10^{-4} m/s,
(c) 3.61×10^{-4} m/s,
5.98×10^{-4} m/s,
4.80×10^{-4} m/s,
7.80×10^{-4} m/s,
4.17×10^{-4} m/s
(d) 6, 10, 8, 13, 7
(e) 1.63×10^{-19} C
2–29. (a) 9.00×10^{21}
(b) 4.50×10^{21}
(c) 1.05×10^{-22} g
(d) 6.04×10^{23}
(e) 6.04×10^{23}
(f) 9.67×10^4 C
(g) 1.65×10^{-24} g
2–31. (a) Inner segment
(b) Outer segment
2–33. (b) Case 2
2–35. 14.6 cm
2–37. 70%, 30%
2–39. 1.625 m/s, 1.632 m/s

CHAPTER 3

3–1. (a) 4.02×10^{17}
(b) 5.90×10^{-15} J

3–3. 5660 °K

3–5. (a) 1500%
(b) 46%
(c) 4%
(d) 0.4%

3–7. 9.67×10^4 Å, 2.90×10^4 Å, 0.483×10^4 Å

3–11. 1.62×10^3/s

3–19. (a) $n_3/n_1 = 0.135$, $n_2/n_1 = 0.368$
(b) 0.036 eV

3–21. 7000 atoms

3–23. (a) 1.05 g
(b) 1265 tons
(c) 110 lb

3–25. (a) 1.04 eV
(b) 11,900 Å
(c) 2.52×10^{14} Hz
(d) 4.15×10^{-7} eV

3–27. (a) 1.44 V
(b) 1.44 eV
(c) 7.10×10^5 m/s

3–29. (a) 1.2 V
(b) 4.3 V
(c) No photoemission, 2.4 V

3–31. 2260 Å

CHAPTER 4

4–3. (a) 0.532 Å, 2.13 Å, 4.79 Å
(b) 1.06 Å

4–5. (a) $r^2 = n\hbar/\sqrt{km}$
(b) $E = n\hbar\sqrt{k/m}$

4–7. 21.8×10^{-19} J, 13.6 eV; 5.44×10^{-19} J, 3.40 eV; 2.42×10^{-19} J, 1.51 eV; 0, 0

4–9. 1.10×10^{-14} m

4–13. 4866.4 Å; 4865.1 Å; 4864.7 Å

4–15. (a) −122 eV, −30.5 eV, −13.6 eV
(b) 122 V
(c) 91.5 V

4–17. Lyman: 912 Å, 13.6 eV;
Balmer: 3640 Å, 3.40 eV;
Paschen: 8200 Å, 1.51 eV;
Brackett: 14550 Å, 0.85 eV;
Pfund: 22,800 Å, 0.54 eV

4–19. (a) 6.20 eV, 8.27 eV
(b) −8.27 eV, −2.07 eV, −0.93 eV
(c) 8.27 V

4–21. 973 Å, 1026 Å, 1216 Å, 4860 Å, 6560 Å, 18800 Å

4–23. (a) 4630 Å, 3910 Å, 3700 Å
(b) 3470 Å
(c) −3.57 eV, −0.89 eV, −0.38 eV, −0.23 eV
(d) 0.38 eV
(e) (1) excite $n = 2$, (2) none

4–25. (a) 24.6 eV, 3.94×10^{-18} J
(b) 2.85×10^5 °K

4–29. 8.22×10^6 rev

4–31. (a) 8560 °K

CHAPTER 5

5–5. (a) 6.28×10^5 mi/s
(b) 1.86×10^5 mi/s
(c) 0
(d) No

5–11. (a) 72 m
(b) 4.0×10^{-7} s
(c) 1.8×10^8 m/s

5–13. (a) 5 μs
(b) 1200 m
(c) 720 m

5–15. (a) 4.00 m/s
(b) 2.77×10^8 m/s

5–17. $L_0(\sin^2\theta + (\cos^2\theta)/k)^{1/2}$, $\tan\theta = k\tan\theta'$

5–21. (a) 4.65×10^{-12} kg
(b) 4.65×10^{-12}
(d) 3.72×10^{-12} kg

5–27. 0.5 m

5–29. 7.81×10^{-16}%

5–31. 47.8 cm, 58.2 cm

5–33. (a) 235 MeV, 0.128 MeV
(b) 1.80×10^8 m/s

5–35. (a) 6.84 kV
(b) 12.6 MV

5–37. 9.0×10^{16} J, 2.5×10^{10} kWh

5–39. (a) 8.03 u
(b) 7.47×10^3 MeV
(c) 17.3 MeV, 1.86×10^{-2} u

5–41. (a) 1.73 MeV

5–43. (a) 25.8×10^{-31} kg
(b) 2.81×10^8 m/s
(c) 0.038 T normal to the trajectories

5–45. (b) $\frac{1}{2}$

CHAPTER 6

6–1. (a) 3.53 Å
(b) 7.06 Å

6–3. (a) 1.24 Å
(b) 0.413 Å
(c) x-ray

6–5. (a) 355 cal/s
(b) 33.8 C°/s

6–7. (a) 24, 29
(b) Cr, Cu

6–9. (a) 10.4 keV, 69.4 keV, 77.7 keV
(b) 79.7 to 80.0 keV

6–17. (a) 70.7%, 0.098%
(b) 10 cm

6–19. (a) 0.664
(b) 0.091
(c) 0.464

6–21. (a) 0.023 ft²/lb, 0.022 ft²/lb, 0.021 ft²/lb, 0.022 ft²/lb
(b) 154 lb/ft², 156 lb/ft², 163 lb/ft², 156 lb/ft²

6–25. (a) 2.28
(b) 3.41×10^{-2}
(c) 4.85×10^{-6}
(d) No significant change for (a) and (b). Impossible for (c), but for whole atom, 2.19×10^{-10}.

6–27. (a) 570 eV
(b) 0
(c) 44°2′; no recoil

6–29. (a) 5.44×10^{-27} kg · m/s
(b) 5.44×10^{-7}%

6–33. 1.56×10^{-22} kg · m/s

6–37. 2.66×10^{-13} A

CHAPTER 7

7–3. (a) 1.23 Å
(b) 0.14 Å

7–5. (a) 1.47 Å, (b) 1.30 Å
(c) 20.5°, 18.0°

7–7. (a) 5.93×10^{-5} Å
(b) Smaller

7–9. (a) Less
(b) Phase; equal
(c) No

7–15. 2.21×10^{-18} Å, 145 Å

7–17. 4.12×10^{-7} eV. No

CHAPTER 8

8–1. (a) 2.5, 7.5
(b) 2.5, 6.83

8–3. 9.45 eV, 37.8 eV, 84.9 eV

8–7. (a) $1/\pi$
(b) 0, $a^2/2$

8–9. $\Delta x = a/\sqrt{2}$, $\Delta p = \hbar/\sqrt{2}\,a$

8–11. (a) 1.8×10^{-3} J
(b) 1.7×10^{30}
(c) 3.5×10^{-30} cm
(d) No

CHAPTER 9

9–3. (a) 2.31×10^4 °K
(b) 3.4×10^{-3}, no

9–5. 3.1

CHAPTER 10

10–1. (a) 40 d
(b) 60 d
(c) 0.0346/d, 28.9 d

10–3. (a) 3.62×10^4/s
(b) 2.92×10^4/s
(c) 1.46×10^5/s

10–5. (a) 51, 59
(b) 51, 58
(c) 52, 58

10–7. $_{48}E^{122}$

10–9. 6.48×10^{-4} g

10–13. (a) 0.368
(b) 0.632

10–15. 7α, 4β

10–19. (b) 2.94×10^{-4} s
(c) 39.2 m

10–21. 4.25×10^9 y

10–23. 6.50×10^{-6} g, 0.655 mm³

10–25. 1.55×10^{-18}/s, 1.42×10^{10} y

10–27. (a) 1.05×10^{-4} g, 6.65×10^{-10} g
(b) 0.103 mCi, 3.81 Rd
10–29. (a) $5.0 \times 10^{-15}\%$
(b) $6.5 \times 10^{-16}\%$
10–31. (a) 0.13 μCi
(b) 8.9×10^{-8} g
(c) All in U^{238} and Th^{232} series
10–33. (b) 67.4 mCi · h, 114 mCi · h, 220 mCi · h
10–35. 6840 mR/h

CHAPTER 11

11–3. (a) 1.92×10^{7} m/s
(b) 2.26×10^{5} ip
(c) 1.34×10^{-9} A
11–5. 4
11–7. (c) 100%
11–9. 1.59×10^{-5} MeV, 7.14×10^{3} m/s
11–11. (a) 17.3 MeV
(b) 8.65 MeV
11–13. P^{32}, Si^{29}, P^{31}, S^{33}
11–15. 23.3 MeV
11–17. (a) 17.3 MeV
(b) 9.0 MeV
(c) 8.76 cm
11–19. 10.01354 u
11–21. β-decay
11–23. (b) 2.91 MeV
11–25. 4.1 MeV
11–27. (b) 34.6 MeV
(c) 8.65 MeV for each
11–29. (b) 4.8 MeV
(c) 2.06 MeV
11–31. (a) 0.63 MeV, −3.99 MeV
(b) Radioactive, stable
11–33. ±125 y
11–35. 303 mCi
11–37. (a) 0.720 MeV
(b) 57.6 N
11–39. (a) 2.22 MeV, 1.11 MeV; 28.2 MeV, 7.07 MeV; 492 MeV, 8.79 MeV
(b) Supply 28.2 MeV
11–41. 81.5 m/s, 1.97×10^{-3} eV

CHAPTER 12

12–1. (a) 945 g
(b) 663 μCi/mi^2
12–3. (a) 52
(b) 5
12–5. 6.24×10^{10}/s
12–7. 99.5 barns
12–9. (a) 1.17×10^{-7}
(b) 5.37×10^{-5}
(c) 458
12–11. (a) H, 0.022/cm; B, 0.147/cm; 0, 6.82×10^{-6}/cm
(b) 0.169/cm
12–13. (a) 4.75 cm
(b) 0.15 cm
(c) 0.368
12–15. (a) Red, 69°38′; blue, 68°53′
(b) Red, 0.75 c; blue, 0.746 c
12–17. (a) 1.07×10^{19} J/m
(b) 2.36×10^{28} J/m
(c) 4.10×10^{10} tons/m
(d) 2.88×10^{8} tons/m
12–19. 1.18×10^{16} km

CHAPTER 13

13–1. (a) 67.5 MeV
(c) 180°
13–3. 1907 MeV, 6.53×10^{-6} Å
13–5. (a) 0
(b) +1

INDEX

 BCDE698